Phosphors, Up Conversion Nano Particles, Quantum Dots and Their Applications

Ru-Shi Liu
Editor

Phosphors, Up Conversion Nano Particles, Quantum Dots and Their Applications

Volume 1

Springer

Editor
Ru-Shi Liu
Department of Chemistry
National Taiwan University
Taipei
Taiwan

and

Department of Mechanical Engineering
and Graduate Institute of Manufacturing
Technology
National Taipei University of Technology
Taipei
Taiwan

ISBN 978-3-662-52769-6 ISBN 978-3-662-52771-9 (eBook)
DOI 10.1007/978-3-662-52771-9

Library of Congress Control Number: 2016943379

Printed on acid-free paper

This Springer imprint is published by Springer Nature
The registered company is Springer-Verlag GmbH Berlin Heidelberg

Preface

This book is divided into two volumes. In Volume 1, we start with an introduction to the basic properties of luminescent materials (phosphors) before addressing the principle of energy transfer and the pressure effect of phosphor. Moreover, we present the theoretical first-principles calculation of luminescent materials. After having established a basic understanding of phosphors, we then discuss a variety of phosphors of oxides, nitrides, (oxy)nitrides, fluorides, etc. In Volume 2, we shift focus to the applications of phosphors in light-emitting diodes, field-emission displays, agriculture, solar spectral convertors, and persistent luminescent materials. We then demonstrate through the basic upconversion of nanoparticles as wel as their applications in biomedical contexts. Last, we introduce readers to the basics and applications of quantum dots.

Taken together, the two volumes offer essential insights on the basics and applications of phosphor at the bulk and nano scales.

Taipei, Taiwan — Ru-Shi Liu

Contents

Chapter 1
Introduction to the Basic Properties of Luminescent Materials

Chun Che Lin and Ru-Shi Liu

Abstract Climate change affects people's lives by altering temperature, water supply, and rainfall and by influencing the frequency of natural disasters. Researchers are actively looking for alternative energy sources that do not emit carbon dioxide. White light-emitting diodes (WLEDs) have been considered one of the most promising next-generation lighting technologies because they significantly reduce global power requirements and the use of fossil fuels. WLEDs have attracted considerable attention because of their significant luminous efficiency, low power consumption, reliability, and environmental friendliness. This chapter discusses the fundamental principles and optical properties of phosphors and quantum dots (QDs) for light-emitting diodes (LEDs). The discussions are mainly focused on the luminescent mechanisms, phosphor and QD components, and the corresponding effects on their optical properties and prospect. We also tackle a number of concepts involved in the nephelauxetic effect, crystal field splitting, energy transfer, thermal effect, and quantum confinement effect, which leads to luminescence. Illustrative examples from luminescent materials applied in lighting are used. Phosphors for ultraviolet (UV) LEDs and blue LEDs are treated separately because the processes leading to excitation, emission, color, bandwidth, and thermal stability are comparable with each other in fluorescent lamps. The production of semiconducting

C.C. Lin
Condensed Matter and Interfaces, Debye Institute for Nanomaterials Science,
Utrecht University, Princetonplein 5, 3584 CC Utrecht, The Netherlands

C.C. Lin · R.S. Liu (✉)
Department of Chemistry, National Taiwan University,
No. 1, Sec. 4, Roosevelt Road, Taipei 10617, Taiwan
e-mail: rsliu@ntu.edu.tw

R.S. Liu
Department of Mechanical Engineering and Graduate Institute of Manufacturing Technology,
National Taipei University of Technology, No. 1, Sec. 3, Zhongxiao East Road,
Taipei 10608, Taiwan

C.C. Lin
Institute of Organic and Polymeric Materials, National Taipei University of Technology,
No. 1, Sec. 3, Zhongxiao East Road, Taipei 10608, Taiwan

R.S. Liu (ed.), *Phosphors, Up Conversion Nano Particles, Quantum Dots and Their Applications*, DOI 10.1007/978-3-662-52771-9_1

QDs will be discussed with the concept of “the separation of nucleation and growth.” The exploration of QD-emission colors emphasizes the core material and nanocrystal size for tuning from near-UV to near-infrared spectra.

1.1 History and Classification of LEDs

An important challenging application, namely, the replacement of fluorescent and incandescent lamps, has recently attracted considerable attention because of their energy savings and positive environmental issues. These conventional lighting devices rely on either discharge or incandescent in gases, which usually accompanies large energy loss because of the simultaneous occurrence of large Stokes shift and high temperature. Therefore, researchers have focused on the use of light-emitting diodes (LEDs), which offer an alternative approach for illumination, as a novel approach for the generation of solid-state lighting sources with high efficiency to significantly reduce worldwide energy consumption [1–6]. The low electrical power consumption of LEDs is ascribed to the recombination of electrons and holes produced from current injection with small energy losses [3].

In 1907, Round [7] developed the first LED, which is an SiC junction diode. Subsequently, Losev [8] independently discovered an emission phenomenon from an SiC and ZnO diode in 1927. However, during these two decades, the potential of LED applications was not maximized, and inventions remained unclear. In 1961, the first inorganic semiconductor LED was fabricated by Baird and Pitman at Texas Instruments. This device emitted near-infrared (NIR) radiation because the material is InGaP with a band gap of 1.37 eV. After 1 year, Holonyak, who worked as a consulting scientist at General Electric Company in New York, invented the first practically useful visible LED device on the basis of GaAsP material in red light; he is called “the father of LEDs.” The main yellow phosphor, cerium-doped $Y_3Al_5O_{12}$ (YAG:Ce) [9], which is now used in phosphor-converted LEDs (pc-LEDs), was reported in 1967 [10]; YAG:Ce was primarily used in cathode ray tubes during that

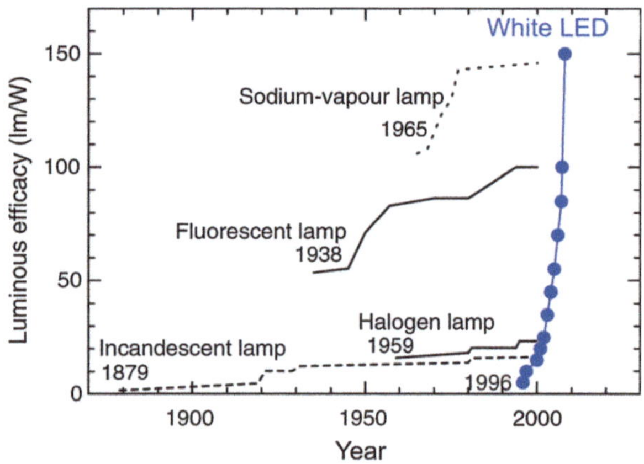

Fig. 1.1 History of luminous efficacy for incandescent, halogen, fluorescent, sodium-vapor lamps, and white LED. Adapted from Ref. [12] © IOP Publishing. Reproduced with permission. All rights reserved

time. From 1993 to 1996, Dr. Nakamura, an employee of Nichia Chemical Industries, developed the world's first high-brightness InGaN blue and green LEDs and then developed WLEDs by covering an InGaN chip with YAG:Ce as a pc-LED device to generate a pseudo-white light [9] (i.e., mixing the blue light and down-converted yellow light to create white light). Therefore, this development from Nichia Company enabled the use of LEDs as an everyday application for general lighting.

Given the high growth rate in the LED-related industry, current LEDs possess sophisticated technology and can compete with fluorescent and incandescent lamps because they have high efficiency, low cost, superior lifetime, portable size, and fast response properties. Figure 1.1 shows the history of the luminous efficiency of WLEDs compared with traditional forms of lighting [11, 12]. Currently, LEDs are utilized not only as indoor and outdoor illumination but also as backlights of liquid crystal displays (LCDs), indicators, and other related applications because of their advantages of high color rendition and brightness.

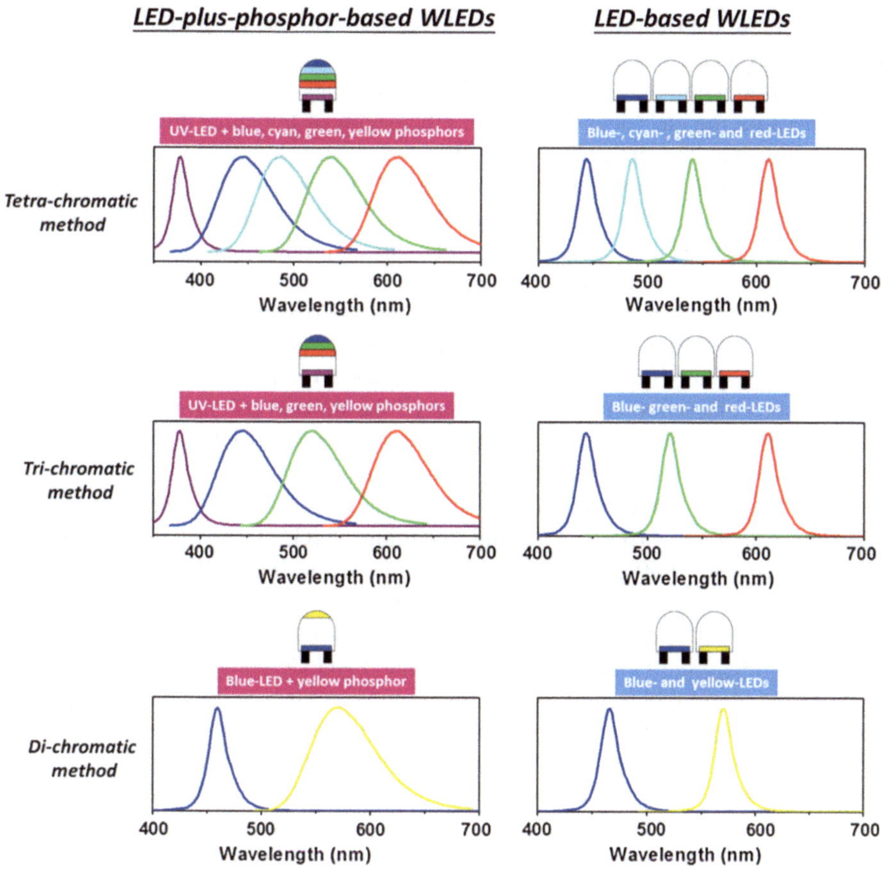

Fig. 1.2 LED-based and LED-plus-phosphor-based approaches for white light performed as di-, tri-, and tetra-chromatic sources [5]

Figure 1.2 shows the three methods for generating white light (di-, tri-, and tetra-chromatic methods) by using semiconductor-based LEDs and phosphors [5]. These three methods have different luminous efficiencies (i.e., radiation quantum output per electrical power input) and color renditions (i.e., the ability of illumination to present the true colors of objects). Two of the above properties are compared with each other because of their fair trade-off. Therefore, the lighting source with high luminous efficiency must have poor color rendition and vice versa. Two types of fabricating strategies for creating white light, namely, the LED-based strategy and LED-plus-phosphor-based strategy, are also shown in this section. The LED-based strategy uses light sources of different wavelengths from semiconductor-based LED chips, whereas the LED-plus-phosphor-based strategy involves the use of phosphors to partially convert the original light sources from semiconductor-based LED chips into complementary lights with long wavelengths. Moreover, LED-plus-phosphor-based white light has a well-known drawback: Phosphor materials suffer from unavoidable Stokes shift energy losses because of the conversion process for photons from short to long wavelengths. These energy losses decrease the overall efficiency of the systems by 10–30 % on the basis of the phosphors excited by semiconductor-based LEDs.

LED-based white light has high luminous efficiency without large Stokes energy losses because the electrical conversion efficiencies of the chips are nearly close to 100 %. However, the circuit design of LED-based white light is complicated and the lifetime span of the chips varies. Therefore, LED-plus-phosphor-based white light sources are more popular than LED-based white light. Di-, tri-, and tetra-chromatic methods are adopted to optimize the wavelength for producing pseudo-white light in the visible spectrum. The di-chromatic LED-plus-phosphor-based white light source also has high luminous efficiency. However, the color-rendering index (CRI) is low (<80), thus rendering the colors of the objects poorly. The tetra-chromatic LED-plus-phosphor-based white light source has excellent CRI (>90); however, the luminous efficiency is lower than that of the di- and tri-chromatic white light sources. Thus, the tri-chromatic LED-plus-phosphor-based white source can have both good color rendition and high luminous efficiency.

These methods for combining phosphors and semiconductor-based LEDs are generally called "pc-LEDs." After the invention of (Ga, In)N-based UV and blue LEDs in the late 1990s, near-UV wavelengths from 360 to 410 nm, as well as blue-light wavelengths from 420 to 480 nm, were adopted as the primary light sources to convert single, double, or triple phosphors into white light. The most common strategy is to combine blue LED and YAG:Ce phosphor, which partially converts the original blue-light source into complementary yellow emission to yield a cool white light. This type of device has several advantages such as simple fabrication, high brightness, and low cost, but it also has disadvantages such as low CRI value because of the low contribution in the red-spectrum region [13]. To avoid the low-color rendition and patent infringement from the Nichia Company, three colors of phosphor can be embedded in UV LEDs to achieve a high CRI value.

1.2 Fundamentals of Phosphors

Phosphors are composed of the host lattice and the activator, which are discussed in detail in the following section. The optical properties of phosphors are mainly affected by several effects including the nephelauxetic effect, crystal field splitting, energy transfer, and thermal effect. These factors must be considered in practical and actual applications. We also list the suitable phosphors combined with UV- or blue-chip sources that have been developed and extensively used in lighting according to the excitation spectra. Other important factors, such as intensity (quantum efficiency), durability, thermal stability, and bandwidth, are discussed for the representative applications of phosphors.

1.2.1 Host Lattice

The general compositions of a luminescent material are host lattice plus activator, which has different optical properties with the variations in crystal structure. Some factors are considered in this subsection. A reduction in the interaction among electrons increases covalency because the electrons spread out over wide orbitals [14]. Accordingly, the energy of electronic transition decreases with increasing covalency. This phenomenon is known as the "nephelauxetic effect" and will be explained later. For example, the charge-transfer absorption band of Eu^{3+} in YF_3 has higher energy than that in Y_2O_3 because oxide has a high covalent structure [15]. Another factor is the crystal field, which denotes the influence of the host lattice on the optical properties of a dopant. For example, Cr_2O_3 and Al_2O_3:Cr^{3+} are green and red, respectively. Cr^{3+} ions occupy smaller Al^{3+} sites, thus resulting in a stronger crystal field than that in Cr_2O_3. The mechanism of the interactions is expressed as follows: different host lattices → different crystal fields → different splittings [16]. The host consists of one or several cations and anions as shown in Fig. 1.3 [17]. Both of the ions are required to have rare-gas configuration to be optically inactive. In the $Y_3Al_5O_{12}$ lattice, the amount of positive oxidation state is as follows: Y $(+3 \times 3)$ + Al $(+3 \times 5) = +24$. The amount of negative oxidation state is: O $(-2 \times 12) = -24$; thus, it is charge balanced for the $Y_3Al_5O_{12}$ lattice. Replacing the cation with other cations that have the same charge is possible: the Ce^{3+}-doped $Lu_3Al_5O_{12}$ is isostructure with YAG. When the activators are doped into the host lattice, the activators replace some of the cations. The activators should have the same charges, but an exception emerges with the advancement of phosphor technology. In commercialized green β-SiAlON:Eu^{2+}, the Eu^{2+} ions are located at the interstitial sites [18]; however, the phosphor still has high efficiency.

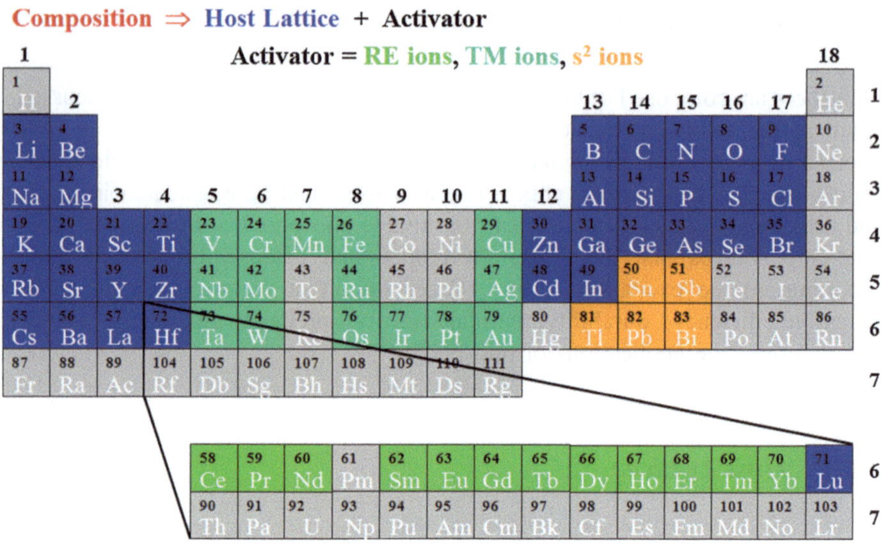

Fig. 1.3 Phosphor composition in the periodic table [17]

1.2.2 *Activator*

The phrase "activator" means that an object can emit light after excitation. Examples of an activator include donor–acceptor pairs, self-active anion groups, defects, s^2 ions, rare-earth ions, and transition-metal ions. Sometimes, even the host itself, with a band gap in the visible region, can emit light. Transition metals and rare-earth ions will be discussed in this section.

1.2.2.1 Transition-Metal Ions

The electron configuration of transition metal ions is d^n ($0 < n < 10$). Their energy levels can be predicted from the Tanabe–Sugano diagram [19]. The free-ion terms for different electron configurations can be expressed as ^{2S+1}L where S represents the total spin quantum number, and L is the total orbital angular momentum. L can be 0 (marked as S), 1 (P), 2 (D), 3 (F), 4 (G), and so on. In an octahedral ligand field, the free-ion terms will split into states corresponding to the irreducible representation. Figure 1.4 shows the Tanabe–Sugano diagram for the d^5 Mn^{2+} ion where E is the energy of excited states above the ground state; B is the Racah parameter; and Δ_0 is the octahedral ligand field splitting [20]. Although the transitions between each state in Mn^{2+} are forbidden according to the Laporte selection and spin selection rules, the rules are relaxed by coupling the electronic transition with vibrations of suitable symmetry. In $BaMgAl_{10}O_{17}$:Mn^{2+}, five excitation bands

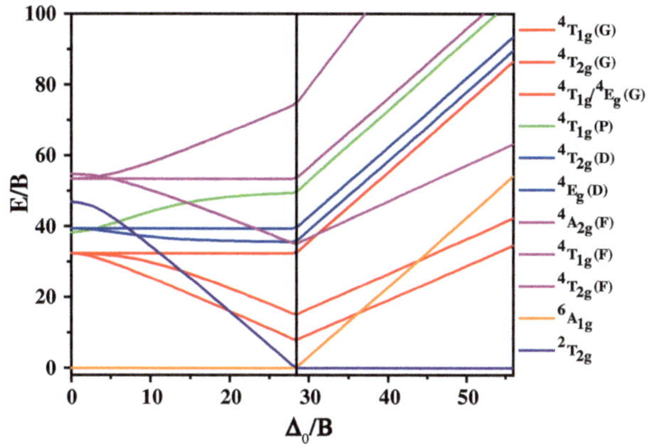

Fig. 1.4 Tanabe–Sugano diagram for the d^5 configuration [19]

at 279, 360, 386, 427, and 452 nm correspond to the transitions of ${}^6A_1 \rightarrow {}^4T_1$, ${}^6A_1 \rightarrow {}^4T_2$, ${}^6A \rightarrow {}^4E$, ${}^4A_1$ (${}^4G$), ${}^6A_1 \rightarrow {}^4T_2$, and ${}^6A_1 \rightarrow 4E$ (${}^4D$), respectively [21].

A total of 10 transition-metal ions are available, and most of the ions have different charges; however, only a few of them—including Cr^{3+} (d^3), Mn^{4+} (d^3), Ti^{3+} (d^1), and Ni^{2+} (d^8)—have been applied in phosphors.

1.2.2.2 Rare-Earth Ions ($4f \rightarrow 4f$ Transition)

Trivalent rare-earth ions RE^{3+} have the electron configuration of $4f^n5\,s^25p^6$ where $n = 1$–13. The $4f^n$ electrons are shielded by the $5s^2$ and $5p^6$ orbitals; therefore, the influence of the host lattice on the transition with $4f^n$ configuration is small. The energy levels can be predicted precisely according to the "Dieke diagram" shown in Fig. 1.5 [22]. The surrounding crystal field induces Stark splitting of the energy levels in the order of 10^2 cm^{-1}. In the configuration coordinate diagram, the ground and excited states appear as parallel parabola ($\Delta R = 0$); thus, the $4f \rightarrow 4f$ transitions have sharp emission lines. The intensities of intraconfigurational $4f$ transitions are usually weak because of the prohibited-selection rule. Trivalent rare-earth ions can emit strong light if the excitation source is strong (electron beams) or is imposed on the $5d$ bands.

The $5d$ bands of the trivalent rare-earth ions can be excited if the location of the $5d$ bands is reachable. The location of the $5d$ bands can be predicted by the following [23]:

$$E(Ln, A) = 49340\ \text{cm}^{-1} - D(A) + \Delta E^{Ln,Ce} \tag{1.1}$$

where D(A) denotes the crystal field depression when the lanthanide ion Ln^{3+} is doped in compound A. $\Delta E^{Ln,Ce}$ means the energy between the spin allowed for the

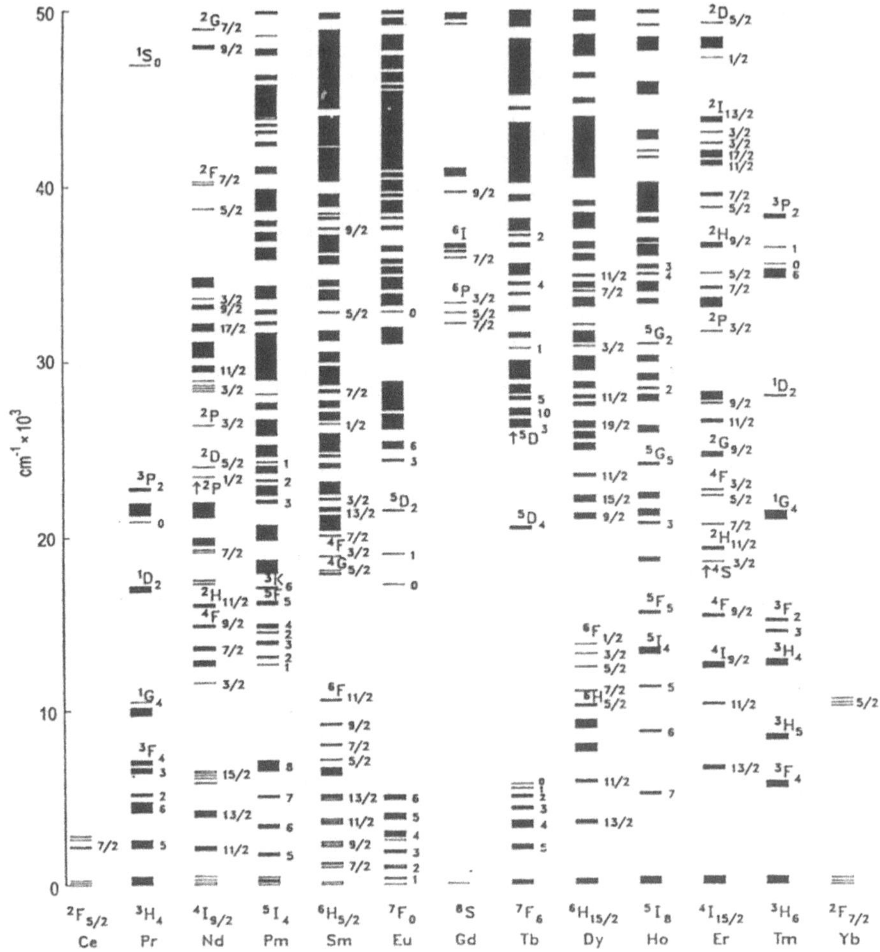

Fig. 1.5 Dieke's diagram: free-ion energy levels of trivalent rare earth ions. Reproduced from Ref. [22] by permission of John Wiley & Sons Ltd

f-*d* transition with that of Ce^{3+}. Table 1.1 lists the $\Delta E^{Ln,Ce}$ for all trivalent lanthanide ions.

The value of D(A) is the same for the same compound, that is, once it is calculated for a certain lanthanide ion, the 5*d* bands of all other lanthanide ions can also be predicted. By simple unit conversion, observing the $4f \rightarrow 5d$ bands at vacuum UV (VUV) region for Pr^{3+} and Tb^{3+} ions is relatively easy. The value of D (A) will be affected by the type of anions/cations, type of ionic complexes, and amount of condensation between complexes. With the sufficiently large number of compounds, the rough position of the lowest 5*d* bands of all the trivalent lanthanide ions can be predicted.

Table 1.1 $\Delta E^{Ln,Ce}$ for all trivalent lanthanide ions

Ln	$\Delta E^{Ln,Ce}$ (cm^{-1})	[a]*N*
Ce	–	–
Pr	12,240 ± 750	64
Nd	22,700 ± 650	18
Pm	–	–
Sm	26,500 ± 460	2
Eu	35,900 ± 380	4
Gd	45,800	1
Tb	13,200 ± 920	30
Dy	25,100 ± 610	4
Ho	31,800 ± 1,400	5
Er	30,000 ± 1,300	8
Tm	29,300 ± 1,100	9
Yb	38,000 ± 570	3
Lu	49,170	1

[a]The number of different compounds

1.2.2.3 Rare-Earth Ions (5*d* → 4*f* Transition)

The Ce^{3+} ion has a $4f^1$ electron configuration; thus, the only possible excited state configuration is $5d^1$. This result indicates that the emission always allows transition. The Ce^{3+} ground state yields two levels by approximately 2000 cm^{-1} because of spin-orbital coupling. When an ion is placed in an octahedral environment, the electron in the orbitals will be repelled by the field. The five levels of 5*d* bands will split into two sets where the d_{xy}, d_{xz}, and d_{yz} orbitals have lower energy than d_z^2 and d_{x-y}^{22} orbitals. The three lower-energy orbitals and two higher-energy orbitals are referred as "t_{2g}" and "e_g," respectively. The resulting energy difference is identified as Δ_o. In some cases, the fitting of the excitation and emission spectrum can obtain five peaks and two peaks, respectively [24].

Another well-known example of 5*d* → 4*f* transitions is Eu^{2+} ($4f^7$) ion. The influence of host lattice to the energy of the 5*d* band is the same as Ce^{3+} ion, but the Eu^{2+} ion has 4*f* levels. When the Eu^{2+} ion is located in a weak crystal field, the energy of its 5*d* bands shifts to high energy, and the $^6P_{7/2}$ level of the 4*f* configuration lies below it. Thus, sharp-line emission from the $^6P_{7/2} \rightarrow {}^8S_{7/2}$ transition occurs. This phenomenon also occurs when the temperature is cooled down. The Eu^{2+} ion also has a slow decay time because the excited state is a sextet, whereas the ground state is an octet; thus, the spin-selection rule slows down the optical transition rate.

Other 5*d* → 4*f* transitions of lanthanide are available for Sm^{2+} ($4f^6$) and Tm^{2+} ($4f^{13}$) ions. They are rare to observe and difficult to synthesize even under a

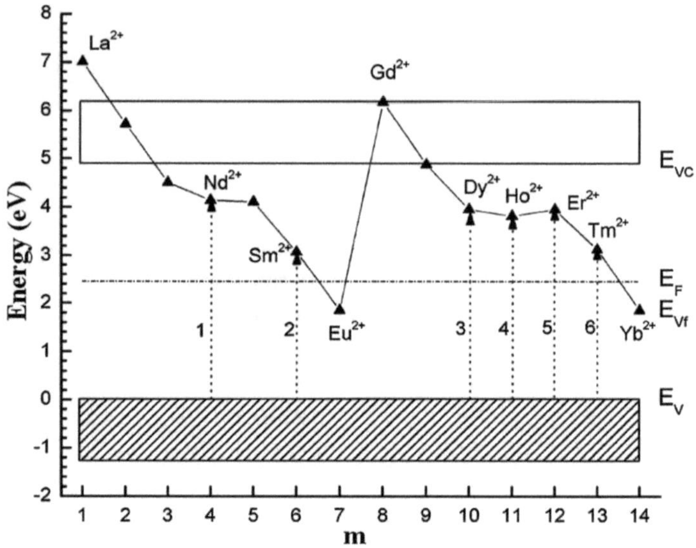

Fig. 1.6 Position of the 4*f* state of divalent lanthanide ions in $CaGa_2S_4$. *m* is the number of $4f^m$ electron, and E_f is the Fermi energy level. Reprinted with the permission from Ref. [25]. Copyright 2005 American Chemical Society

reducing environment owing to the valence stability issue. These transitions can be evaluated from the "energy difference" as follows [25]:

$$E_{Ff} = E^{CT} - 0.54E^{ex} \quad (1.2)$$

where E^{CT} indicates the energy needed to transfer an electron from the valence band of an inorganic compound to a trivalent lanthanide impurity, thereby providing information on the location of the ground state of the corresponding divalent lanthanide ion. E^{ex} indicates the energy needed to excite the host lattice, that is, the energy to create a bound electron–hole pair. Lanthanide ions tend to prefer the divalent state when $E_{Ff} < 0$ and the trivalent state when $E_{Ff} > 0$. Figure 1.6 shows the positions of the lowest 4*f* state of divalent lanthanide ions in $CaGa_2S_4$. Although different hosts will have larger or smaller band gaps, the trend of the 14 rare-earth ions is the same. Intuitively, the values of E_{Ff} of Sm^{2+} and Tm^{2+} ions will be larger than that of Eu^{2+} ion, thus explaining why Sm^{2+} and Tm^{2+} ions are rarely observed.

1.2.3 Effect-Dependent Luminescence

Figure 1.7 illustrates the overall factors of the host lattice affecting the emission properties of $5d \rightarrow 4f$ transitions [3]. The nephelauxetic effect is the result of the covalency of the host lattice and ligand polarization. The polarization of nitrogen

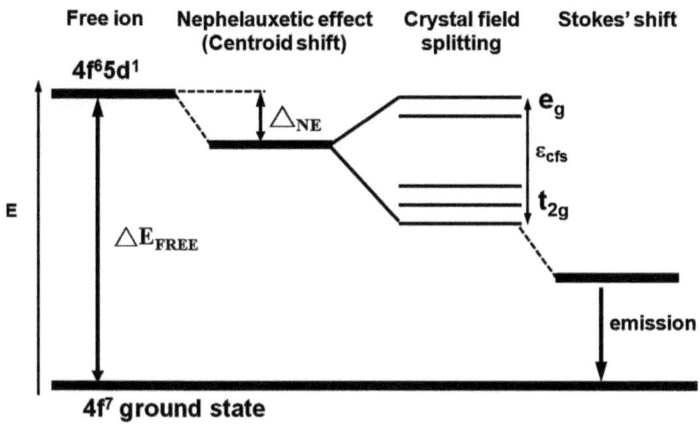

Fig. 1.7 Schematic of the energy level shifts for $5d \rightarrow 4f$ transitions in octahedral coordination [3]

ions has a larger effect than that of the oxygen ions; thus, the emission of nitride compounds tends to be a red shift. The covalency concept of the host lattice is the same as the "D(A)," that is, the crystal field depression in a certain host. By collecting thousands of data, the general sequence of the host lattice effect is as follows: oxides > aluminates > silicates > borates > phosphates > carbonates > sulfates > flourides. In each class of the host lattice, the amount of condensation also affects the value: oxy-ortho-borates > ortho-borates > pyro-borates > condensed borates > meta-borates. Crystal field splitting is mostly controlled by the anion polyhedral shape, coordination number, and type of coordinated anion. The shape of the excitation spectrum denotes the splitting of 5*d* orbitals. In most cases, the octahedral, dodecahedral, tricapped trigonal prism, and cuboctahedral types of coordination are studied in more detail [26]. The emission spectrum shifts toward a long-wavelength region because of the centroid shifting from the nephelauxetic effect and the lowest 5*d* orbital position from the crystal field splitting.

1.2.4 Energy Transfer

The exciting radiation is not only absorbed by the activator because defects and other ions exist in many luminescent materials. For example, the sensitizer can absorb the exciting radiation and subsequently transfer it to the activator. In Fig. 1.8, the $S \rightarrow S^*$ transition is the excitation or absorption, and the $A_2^* \rightarrow A$ transition is the emission. The $A_1^* \rightarrow A_2^*$ is a nonradiative process and prevents back transfer. In particular, UV radiation is absorbed by Sb^{3+} but not by Mn^{2+} in the lamp phosphor $Ca_5(PO_4)_3F{:}Sb^{3+}$, Mn^{2+}. Hence, the Mn^{2+} ion is not excited directly; rather, the excitation energy is transferred from Sb^{3+} to Mn^{2+}. For the

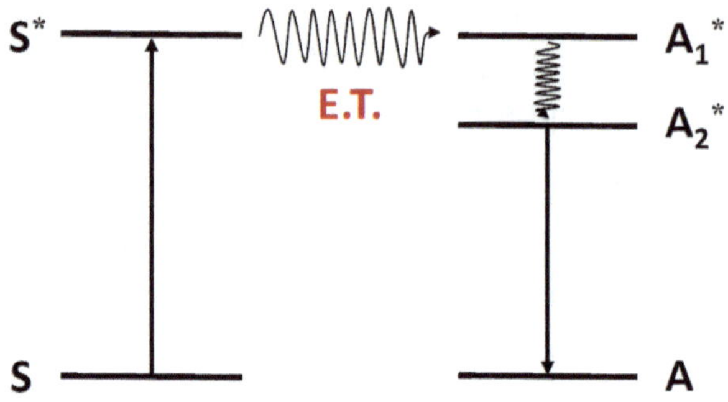

Fig. 1.8 Energy transfer from sensitizer (S) to activator (A) [16]

$Ca_5(PO_4)_3F{:}Sb^{3+}$, Mn^{2+} compound, the emission consists partly of blue Sb^{3+} emission and yellow Mn^{2+} emission under UV irradiation. The luminescent processes can be written as follows [16]:

$$Sb^{3+} + h\nu \rightarrow (Sb^{3+})^*$$
$$(Sb^{3+})^* + Mn^{2+} \rightarrow Sb^{3+} + (Mn^{2+})^*$$
$$(Mn^{2+})^* \rightarrow Mn^{2+} + h\nu$$

where hν is the radiation with frequency ν, and the asterisk denotes an excited state. These reactions indicate absorption, energy transfer, and emission, respectively. Their own blue emission occurs when the Sb^{3+} ion has no Mn^{2+} ions in its vicinity.

1.2.5 *Thermal Effect*

In high-power semiconductor-based LEDs, approximately 60 % of the electrical input power is converted into heat, thus increasing the temperature of the entire LED device, including the phosphors. Local heat can reach up to 400–450 K; thus, the thermal quenching (TQ) property of phosphors is important for practical WLED devices. The emission peak of activator-doped phosphors does not only decrease the intensity as temperature increases but also broadens the spectral bandwidth because excited electrons rise to higher vibrational microstates. In this subsection, four mechanisms of the thermal effects on phosphors are introduced. The perfect luminescence of dopants proceeds from the lowest position of the excited state to the ground state without thermal effects as exhibited by the emission spectrum shown in Fig. 1.9a (*solid green line*) [27]. However, a number of lanthanide activators display emission spectra with spectral intensities and positions that are easily affected by environmental temperature. Heat is generally detrimental to phosphors, and phosphor efficiency decreases through nonradiative relaxation as device temperature increases. This phenomenon indicates TQ, and, consequently, phosphors shift to emission peak wavelengths, which decrease luminescent intensity. Excited electrons can relax through radiative (Fig. 1.9b [*dashed green line*]) and nonradiative (Fig. 1.9b [*gray line*]) processes such as photon emission and collisional quenching, respectively. The TQ property can also be elucidated by using the configurational coordinate diagram, in which thermal energy assists excited electrons across the intersection point between the excited and ground state parabola curves, thereby resulting in nonradiative decay. Figure 1.9c shows the relative positions between the localized 5*d* electron states of the activators and the delocalized conduction band states of the hosts. First, auto-ionization spontaneously occurs, and no 5*d*–4*f* emission is observed when the lowest 5*d* state is above the bottom of the conduction band. Such cases include $Ba_{10}(PO_4)_4(SiO_4)_2:Eu^{2+}$ [28], $Ln_2O_3:Ce^{3+}$ [29], $LaAlO_3:Ce^{3+}$ [30], and the Eu^{2+} on trivalent RE sites in oxide compounds [31]. Second, the 5*d* states of the activators are below the conduction band of the hosts in most 5*d*–4*f* emission situations. The 5*d* electrons are ionized to the conduction band through thermal ionization, which depends on the energy E_{dC} between the 5*d* state (d) of the activator and the bottom of the conduction band (C) [32, 33]. The activator Eu^{2+}, located in the fluffy structure, is easily oxidized to the trivalent species at high temperatures. Therefore, the existence of Eu^{3+} can be observed in the PL and X-ray absorption spectra. This phenomenon is called the "thermal-degradation effect" (Fig. 1.9d).

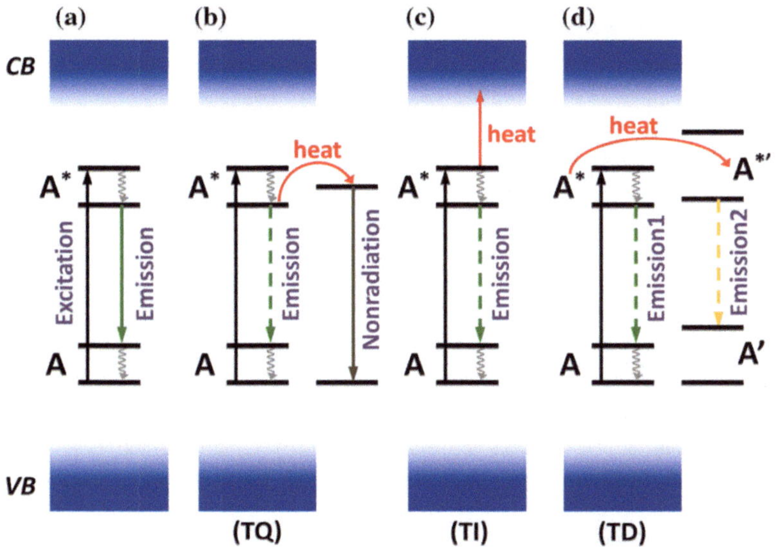

Fig. 1.9 Luminescent mechanisms of thermal effects on inorganic solids. **a** Emission from a luminescent activator on excitation. **b** TQ results in a nonradiative pathway associated with heat. **c** TI excites electrons toward the conduction band through heat. **d** TD can lead to other emissions as a result of heat. A and A* represent the ground and excited states of the activator, respectively. A′ and A*′ represent the ground and excited states of the activator with different charges, respectively. VB and CB represent the valence and conduction bands of the host, respectively. Reprinted from Ref. [27], Open Access

1.2.6 Classification of Phosphors for Pc-WLEDs

To date, phosphors—such as $Y_3Al_5O_{12}$:Ce (yellow), $Lu_3Al_5O_{12}$:Ce (greenish-yellow), (Sr, Ba)$_2$SiO$_4$:Eu (yellowish-green), β-SiAlON:Eu (green), $Sr_2Si_5N_8$:Eu (orange-red), and $CaAlSiN_3$:Eu (red)—have been commercialized and employed in

Table 1.2 UV LED-excited phosphors for WLEDs (◎ = best, ○ = good, △ = poor)

LED	Color	Composition	Emission properties				Excitation bandwidth (nm)
			Intensity[a]	Durability[b]	Thermal stability[c]	Band-width[d]	
UV LED	Blue	(Sr, Ba)$_3$MgSi$_2$O$_8$: Eu [34]	○	○	○	Middle	250–450
		BaMgAl$_{10}$O$_{17}$: Eu [35]	◎	△	○	Middle	200–400
	Green	β-SiAlON:Eu [36]	○	◎	○	Middle	250–500
		(Sr, Ba)$_2$SiO$_4$: Eu [37]	◎	△	△	Middle	250–500
		(Sr, Ba) Si$_2$O$_2$N$_2$:Eu [38]	○	○	○	Middle	250–525
		Ba$_3$Si$_6$O$_{12}$N$_2$: Eu [39]	○	○	○	Middle	200–500
		SrAl$_2$O$_4$:Eu [40]	○	○	○	Broad	200–450
		BaMgAl$_{10}$O$_{17}$: Eu, Mn [41]	◎	◎	◎	Narrow	225–400
		SrGa$_2$S$_4$:Eu [42]	◎	△	△	Middle	300–500
	Red	Sr$_2$Si$_5$N$_8$:Eu [43]	○	○	○	Middle	250–600
		CaAlSiN$_3$:Eu [44]	◎	◎	◎	Middle	200–600
		(Sr, Ca)S:Eu [45]	◎	△	△	Broad	250–620
		La$_2$O$_2$S:Eu [46]	○	○	○	Narrow	250–400
		Ba$_3$MgSi$_2$O$_8$: Eu, Mn [47]	◎	○	○	Broad	250–450
		CaSc$_2$O$_4$:Eu [48]	◎	◎	◎	Narrow	350–410

[a]The emission intensity of phosphor is better than (◎), equal to (○), or poorer (△) than other phosphors of the same color

[b]The emission intensity of phosphor after the durability task (at 80 °C and 80 % relative humidity for >12 h) is equal to (◎), 80 % (○), and 50 % (△) compared with the original intensity

[c]The decay percentage of phosphor-emission intensity compared with the original intensity after the 300 °C task is <10 % (◎), approximately 10–30 % (○), and >30 % (△)

[d]The bandwidth of the emission peak is narrow (<80 nm), middle (approximately 80–120 nm), or broad (>120 nm)

either 1-pc-WLEDs or 3-pc-LEDs for illumination or for backlighting LCDs. Tables 1.2 and 1.3 summarize and classify the popular phosphors for UV LEDs and blue LEDs, respectively. We measured and compared these data ourselves. The other details of these phosphors are described in [34–58].

Table 1.3 Blue LED-excited phosphors for WLEDs (◎ = best, ○ = good, △ = poor)

LED	Color	Composition	Emission Properties				Excitation bandwidth (nm)
			Intensity[a]	Durability[b]	Thermal stability[c]	Band-width[d]	
Blue LED	Green	β-SiAlON:Eu [36]	○	◎	○	Middle	250–500
		$Lu_3Al_5O_{12}$:Ce [49]	△	○	◎	Broad	300–500
		$(Sr, Ba)_2SiO_4$:Eu [37]	◎	△	△	Middle	250–500
		(Sr, Ba)$Si_2O_2N_2$:Eu [38]	◎	○	◎	Middle	250–525
		$Y_3(Al, Ga)_5O_{12}$:Ce [50]	○	◎	○	Broad	300–500
		$SrGa_2S_4$:Eu [51]	◎	△	△	Middle	300–500
		$Ca_3Sr_2Si_3O_{12}$:Eu [52]	◎	◎	◎	Broad	300–500
	Yellow	Ca-α-SiAlON:Eu [53]	△	○	○	Middle	250–550
		$(Y, Gd)_3Al_5O_{12}$:Ce [54]	◎	○	○	Broad	400–500
		$Tb_3Al_5O_{12}$:Ce [55]	○	○	○	Broad	400–520
		$(Ca, Sr, Ba)_2SiO_4$:Eu [37]	○	△	△	Middle	250–500
		$La_3Si_6N_{11}$:Ce [56]	○	○	○	Broad	300–500
		$CaGa_2S_4$:Eu [57]	◎	△	△	Middle	350–550
	Red	$Sr_2Si_5N_8$:Eu [43]	○	○	○	Middle	250–600
		$CaAlSiN_3$:Eu [44]	◎	◎	◎	Middle	200–600
		(Sr, Ca)S:Eu [45]	◎	△	△	Broad	250–620
		$(Sr, Ba)_3SiO_5$:Eu [58]	◎	△	◎	Broad	300–550

[a]The emission intensity of phosphor is better than (◎), equal to (○), or poorer than (△) other phosphors of the same color

[b]The emission intensity of phosphor after the durability task (at 80 °C and 80 % relative humidity for over 12 h) is equal to (◎), 80 % (○), and 50 % (△) compared with the original intensity

[c]The decay percentage of phosphor emission intensity compared with the original intensity after the 300 °C task is <10 % (◎), approximately 10–30 % (○), and >30 % (△)

[d]The bandwidth of the emission peak is narrow (<80 nm), middle (approximately 80–120 nm), or broad (>120 nm)

1.3 Fundamentals of Nanomaterials

Nanomaterials have been widely investigated because of their remarkable potential in different applications such as LEDs, solar cell, biolabeling, and lasers. Moreover, the basic physical and chemical properties of nanomaterials have been increasingly understood in the last two decades [59–62]. Bulk materials and nanomaterials are classified according to their particle sizes. Bulk materials, such as gold, silver, and copper, are widely used in jewelry, electronics, and other industries. Nanomaterials possess particle sizes that are generally <100 nm; these materials exhibit different properties from their bulk form, particularly in terms of their distinct optical [63, 64], electrical [65, 66], thermodynamic [67], and magnetic properties [68].

Nanomaterials (from the Greek word "nanos," meaning dwarf) possess at least one dimension measuring in the range of 1–100 nm. With various nanoscale dimensions, a nanomaterial is categorized as 0D nanoparticles or dots, 1D nanowires or rods, and 2D thin films or nanowells. Nanomaterials are composed of a group of atoms and molecules and can bridge the gap between bulk and molecular levels, thereby resulting in an entirely new chemical and physical properties. Figure 1.10 shows the merged routes between the "top-down" (physical methods) and "bottom-up" (chemical methods) processes in fabricating nanomaterials. Top-down processes often include relevant mechanical facilities, such as etching, cutting, or grinding, to split bulk precursors into nanoparticles. By contrast, the bottom-up approach includes building atomic and molecular components by self-assembly, which can be time-consuming [69]. To control the particle size and uniformity of nanomaterials, fine manufacturing is indispensable. Thus, the bottom-up process is suitable in improving particle miniaturization.

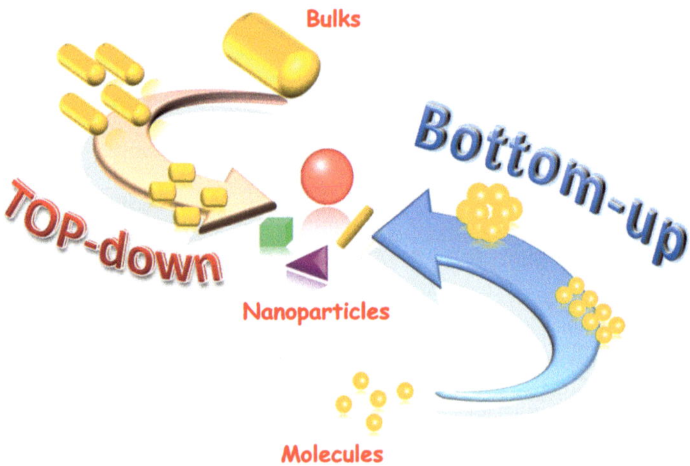

Fig. 1.10 Top-down and bottom-up approaches for the synthesis of nanoparticles

1.3.1 Quantum-Confinement Effect

A distinct property of QDs is quantum confinement, which modifies the density of states (DOS) near the band edges. Figure 1.11 shows the schematic of the number of states per unit energy (DOS as a function of energy) and per unit volume for different dimensionalities: bulk (3D), quantum well (2D), quantum wire (1D), and QD (0D). QDs between the atomic and molecular level limits have distinct electrical and optical properties depending on their size. An electron–hole pair is regularly bound within a characteristic distance in a semiconductor, which is called the "exciton *Bohr* radius" (R_B). At this state, a particle behaves at a certain limit where its dimension is smaller than the wavelength of the particle. The change in energy state into a discrete state results in size-dependent band gap. This physical phenomenon is a form of the quantum–confinement effect, and this conclusive result obtains the blue shift of optical illumination with the decrease in size of the particles. On the contrary, bulk materials are not confined by their dimension, which can be larger than the wavelength of the particle with a band gap under a continuous-energy state.

The band gap of QDs can be expressed by the *Schrödinger* wave equation where m_e and m_h, ε, $\hbar$, and e are the effective masses of electrons and holes, optical dielectric constant, reduced *Planck's* constant, and charge of an electron, respectively.

$$r_B = \frac{\hbar^2 \varepsilon}{e^2}\left(\frac{1}{m_e} + \frac{1}{m_h}\right) \tag{1.3}$$

When the radius (R) of a QD approaches R_B, that is, $R \approx R_B$ or $R < R_B$, the motion of the electrons and holes are confined spatially to the QD dimension, thereby increasing the excitonic transition energy and the observed blue shift in the QD band gap and luminescence.

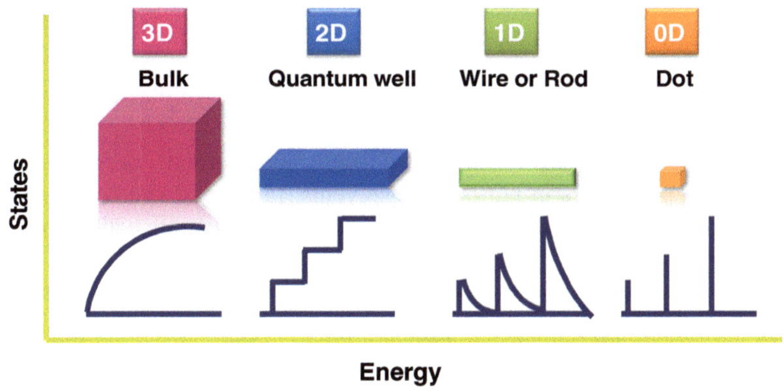

Fig. 1.11 Schematic of DOS for one band of 3D, 2D, 1D, and 0D semiconductors

Quantum confinement can be predicted by the particle-in-box model. Efros et al. [70] were the first to propose this approach in 1982, which was later modified by Brus [71]. This model assumes a particle in a potential well with an infinite potential barrier at the particle boundary. For a particle that is free to assume any position in the box, the relationship between its energy (*E*) and wave vector (*k*) is given by the following:

$$E = \frac{\hbar^2 k^2}{2m^*} \tag{1.4}$$

This relationship is assumed to hold for an electron or hole in a semiconductor. Thus, the energy band is parabolic near the band edge. The shift of band gap energy (ΔE_g) caused by confinement of the exciton in QDs with diameter R can be expressed as Eq. 1.5 where μ is the reduced mass of an electron–hole pair, and E^*_{Ry} is the *Rydberg* energy.

$$\Delta E_g = \frac{\hbar^2 \pi^2}{2\mu R^2} - \frac{1.8e^2}{\varepsilon R} = \frac{\hbar^2 \pi^2}{2R^2}\left(\frac{1}{m_e} + \frac{1}{m_h}\right) - \frac{1.78e^2}{\varepsilon R} - 0.248E^*_{R_y} \tag{1.5}$$

The first term in Eq. 1.5 represents a relation between the “particle-in-a-box” quantum localization energy or confinement energy and the radius of the QDs (R), whereas the second term shows the Coulombic interaction energy with $R - 1$ dependence. The *Rydberg* energy term is size-independent and generally negligible except for semiconductors with small dielectric constant [72]. On the basis of Eq. 1.5, the band gap of excitonic transition increases with decreasing R {quantum localization term shifts to higher energy with lower R value ($R - 2$), and the Coulombic terms shift the excited electronic state to lower value ($R - 1$)}. However, the EMA model breaks down in the small QD regime [72, 73] because the *E–k* relationship can no longer be approximated as parabolic. Figure 1.12 shows such a deviation of theoretically predicted band gaps for CdS QDs from the experimental values [72].

1.3.2 Nucleation and Growth

Semiconducting QDs are synthesized by the thermal decomposition of organometallic precursors in hot organic solvents. In general synthetic procedures, anionic precursors are rapidly injected into a solution containing cationic precursors at high temperature; this process is called the “hot-injection method” [74]. The growth of NCs is relatively slow at high temperatures, thus resulting in well-passivated QDs with reduced defects [75–80]. Nucleation is the extremely localized budding of a distinct thermodynamic phase. LaMer et al. [81] proposed the concept of “the separation of nucleation and growth,” and this process is schematically illustrated in Fig. 1.13. The concentration of monomer precursors

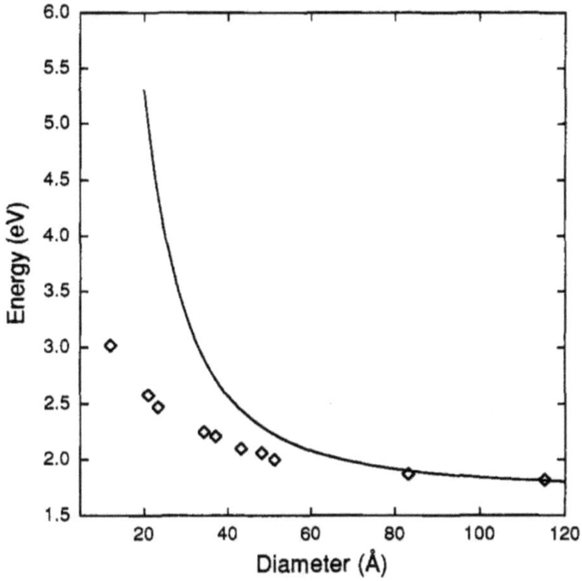

Fig. 1.12 Band gap of CdS QDs as a function of size in the experimental data (*diamonds*) and theoretical prediction (*solid line*). Adapted with the permission from refernce [72]. Copyright 1991 American Chemical Society

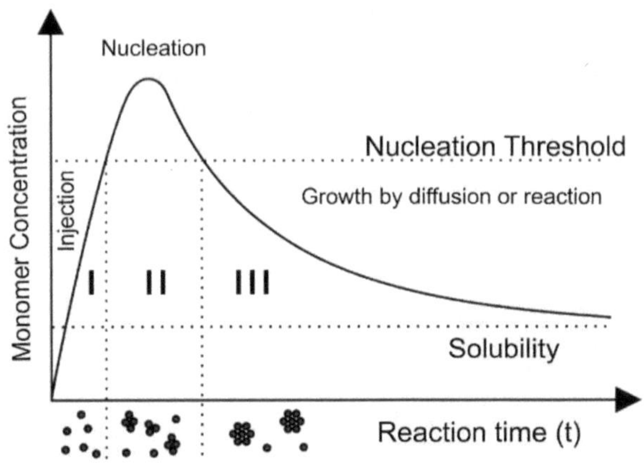

Fig. 1.13 Schematic processes of nucleation and subsequent growth. Reprinted with the permission from Ref. [81]. Copyright 1950 American Chemical Society

increases depending on various times such that no precipitation can occur in stage I even in a supersaturated solution ($S > 1$) because of the extremely high energy barrier for spontaneous homogeneous nucleation. In stage II, nucleation processes occur, and the level of supersaturation achieves a certain value and conquers the energy barrier for nucleation. The monomer concentration decreases below zero nucleation rates as the rate of monomer consumption induced by the nucleation and growth process exceeds the monomer supply rate. In stage III, the particle grows continuously when the system is supersaturated. When the monomer concentration plunges below the critical supersaturation state for nucleation, nucleation stops, and nuclei growth starts.

1.3.3 II–VI, III–V, and I–III–VI Semiconducting QDs

Figure 1.14 shows the different types of QDs such as II–VI (e.g., CdTe), III–V (e.g., InP), and I–III–VI (e.g., $CuInS_2$). The core material determines the range of QD-emission colors that can be tuned by NC size and with the different materials emitting across portions of the near-UV, visible, and NIR spectra [82–84]. Therefore, the results describe the development in the areas of fundamental concepts and properties, which are then applied to LEDs and bioimaging.

1.3.3.1 Binary II–VI QDs

The II–VI semiconductors are composed of Zn^{2+}, Cd^{2+}, and Hg^{2+} with O^{2-}, S^{2-}, Se^{2-}, and Te^{2-} such as CdSe [85], CdTe [86], HgTe [87], and ZnSe [88]. These QDs typically crystallize in either face-centered cubic (zinc blende) or hexagonal (wurtzite) crystal structures. The II–VI QDs can exhibit very good luminescence because they possess a direct band gap. The PL bands of these QDs are situated typically close to the absorption thresholds and sufficiently narrow full-width at half-maximum (FWHM) from 35 to 60 nm with increasing particle size. Grabolle et al. [89] improved the PL-quantum yield (QY) of CdTe from 40 to 60 % with the use of thiol ligand.

Band gaps can be engineered by alloying the core [90–95]. The optoelectronic properties of QDs can be controlled by manipulating the composition of the core materials and the ratio of the alloying materials. The emission colors of alloyed $Zn_{1-x}Cd_xS_{1-y}Se_y$ across portions of the near-UV, visible, and NIR spectra can be

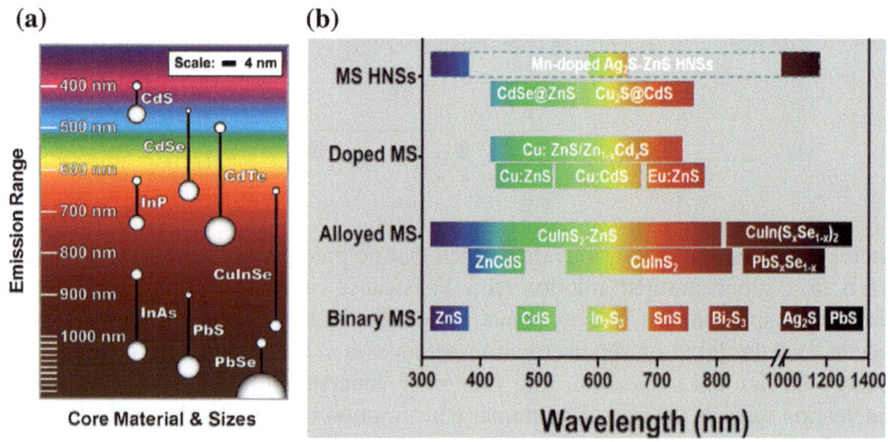

Fig. 1.14 a PL ranges of several semiconductors with various sizes. Reprinted with the permission from Ref. [82]. Copyright 2011 American Chemical Society; **b** PL ranges of composition control and doping metal sulfide QDs. Reprinted with the permission from Ref. [84]. Copyright 2013 American Chemical Society

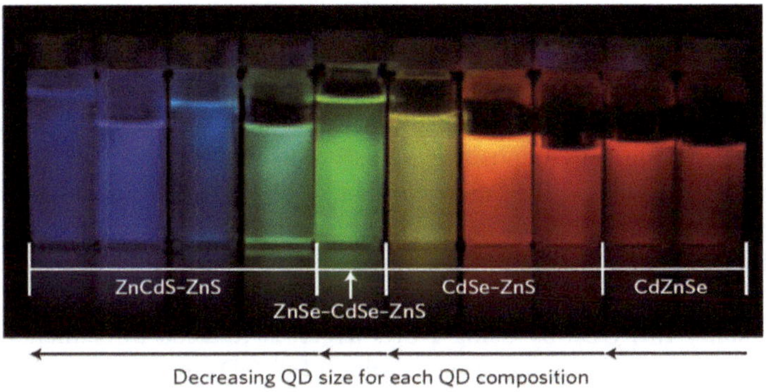

Fig. 1.15 Photograph of $Zn_{1-x}Cd_xS_{1-y}Se_y$ QDs with varying sizes and compositions under UV light. Adapted with the permission from Ref. [96]. Copyright 2009 American Chemical Society

tuned by various compositions (Fig. 1.15) [96]. These QDs are fabricated in QD-LEDs with high efficiency. The alloys of CdHgTe QDs exhibit NIR emission (600–1350 nm) by varying the stoichiometric ratio of two binary semiconductors [97–101]. These QDs are also used in many applications including biological imaging [102–104].

1.3.3.2 Binary III–V QDs

Studies on semiconducting NCs have focused on the fabrication of versatile groups II–VI QDs because of their potential application in lasers [105], LEDs [106], and biological studies [107, 108]. However, the presence of highly intrinsic toxic Cd limits their biological application. Therefore, overcoating the Cd-based core with a less toxic shell, such as ZnS, has been introduced to overcome the toxicity problem. The shell can prevent the release of the toxic Cd cations. However, this shell cannot ensure the complete nontoxicity of II–VI QDs under UV light or oxidation. Cd cations can be released under such conditions by surface oxidation [109].

Semiconducting III–V QDs have been investigated in the past decade, particularly InP QDs, which possess a band gap of 1.35 eV. The principal attraction to these semiconductors focuses on the robustness of the covalent bond in III–V semiconductor groups compared with the ionic bond in II–VI semiconductor groups. The formation of covalent bond enhances the optical stability of the QD systems. Thus, the reduction of the toxicity derived from the noncorrosive composition elements enables the use of QD systems in a biological field [110, 111]. The resultant defects, also called "surface sites," act as traps for nonradiative decay of the QDs under the excited state [112]. Some of the excited electrons can cross to the surface states located in the intraband gap and subsequently recombine nonradiatively with the holes in the valence band, thereby decreasing the PL-QY. Several syntheses have been developed to enhance the PL-QY. To passivate surface defects, an inorganic shell made from large-band-gap material is epitaxially grown around each core QDs, thus resulting in an improved PL-QY and enhanced photostability [113–115]. Lim et al. [113] reported that the InP/ZnSeS core/shell exhibits the highest PL-QY, hence demonstrating that surface traps are effectively depressed after epitaxially depositing ZnSeS onto the InP bare core. The type and thickness of the inorganic shell are important in tailoring the optical and electronic properties of QDs [116, 117]. Moreover, chemically eliminating the particle surface of the InP QD core entails etching them with dilute ethanolic or butanolic solutions of HF [118, 119] or NH_4F [120]. Liu et al. [121] demonstrated a colloidal synthesis of good-quality InP QDs by using PCl_3, which is reduced to elemental P with a superhydride solution. The QY of the resulting InP QDs reaches only 0.25 %, but it is substantially increased to approximately 20 % after an appropriate HF-based photo-etching treatment. The FWHM of PL emission for typical InP QDs is broader (50–80 nm) than that of CdSe (15–40 nm), which induced a worse color saturation for InP compared with CdSe QDs. Yang et al. [117] synthesized InP/ZnS NCs with tunable emission wavelength by changing the InP:ZnS without multiple injections, thus achieving an impressive FWHM of PL (38 nm) emission (Fig. 1.16). Nevertheless, obtaining InP QDs with high optical performance and pure color as good as CdSe QDs remains a remarkable challenge.

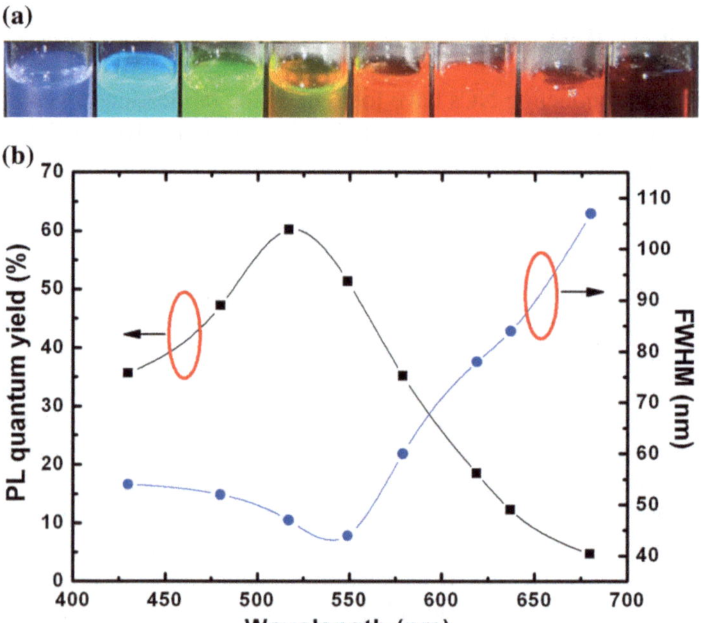

Fig. 1.16 **a** Photograph of samples with different InP:ZnS ratios under UV light. **b** PL-QY and FWHM of InP/ZnS NCs with various ratios of InP:ZnS. Reproduced from Ref. [117] by permission of John Wiley & Sons Ltd

1.3.3.3 Ternary I–III–VI QDs

Ternary I–III–VI QDs have attracted less research interest compared with binary III–V compounds (InP). The I–III–VI semiconductors composed of group I (Cu, Ag), III (Al, Ga, In, Tl), and VI (S, Se, Te) elements can be conceptually derived from II–VI binary compounds by replacing two divalent cations with one monovalent and one trivalent cation; these elements are potentially less toxic. These compounds have a wide range of optical and electronic properties and depend on one another for their chalcopyrite structure (Fig. 1.17) [85].

The development of synthesis methods for I–III–VI NCs with controlled size, shape, composition, and surface chemistry has attracted significant attention in the past 5 years (2008–2013), thus showing an exponential increase in publications. Particularly, CIS, $CuInSe_2$, and $AgInS_2$ have been intensively investigated [122–124].

The I–III–VI QDs exhibit tunable spectroscopic properties from the NIR region through visible region and up to the UV region [85]; they also demonstrate other features including large Stokes shifts (that is, the energy difference between the absorbance band gap and emission peak energy), long PL lifetime, and low toxicity [125–128]. Elucidating the luminescence mechanism, understanding the factors affecting luminescence properties, and optimizing the luminescence efficiency of I–III–VI QDs are necessary. The basic quantum confinement effects in I–III–VI

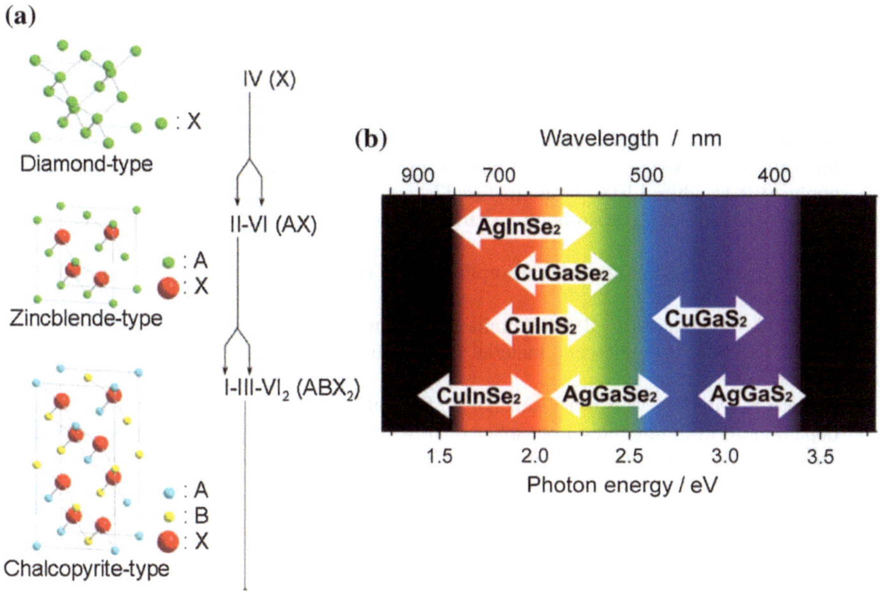

Fig. 1.17 **a** Schematic of the structural relationship among groups IV, II–VI, and I–III–VI_2 semiconducting materials and **b** optical band gap of different chalcopyrite-type I–III–VI_2 semiconducting QDs. Reproduced from Ref. [85] with the permission of AIP Publishing

NCs provide size-tunable luminescent properties [85, 127]. However, the fluorescence feature of the I–III–VI QDs is entirely different from those of II–VI or III–V QDs [129].

References

1. Nakamura S, Fasol G (1996) The blue laser: GaN based light emitters and lasers. Springer, Berlin
2. Jüstel T, Nikol H, Ronda C (1998) New developments in the field of luminescent materials for lighting and displays. Angew Chem Int Ed 37:3084
3. Shionoya S, Yen WM (1998) Phosphor handbook. CRC Press, Florida
4. Xu R, Su M (2004) Luminescence and luminescent materials. Chemical Industry Press, Beijing
5. Schubert EF, Kim JK (2005) Solid-state light sources getting smart. Science 308:1274
6. Phillips JM, Coltrin ME, Crawford MH, Fischer AJ, Krames MR, Regina MM, Mueller GO, Ohno Y, Rohwer LES, Simmons JA, Tsao JY (2007) Research challenges to ultra-efficient inorganic solid-state lighting. Laser Photon Rev 1:307
7. Round HJ (1907) A note on carborundum. Electrical World 49:308
8. Zheludev N (2007) The life and times of the LED—a 100-year history. Nat Photonics 1:189
9. Shimizu Y, Sakano K, Noguchi Y, Moriguchi T (1999) Light emitting device having a nitride compound semiconductor and a phosphor coating a garnet fluorescent material. US Patent 5,998,925
10. Blasse G, Bril A (1967) A new phosphor for flying-spot cathode-ray tubes for color television—yellow-emitting $Y_3Al_5O_{12}$: Ce^{3+}. Appl Phys Lett 11:53
11. Smet PF, Parmentier AB, Poelman D (2011) Selecting conversion phosphors for white light-emitting diodes. J Electrochem Soc 158:R37
12. Narukawa Y, Ichikawa M, Sanga D, Sano M, Mukai T (2010) White light emitting diodes with super-high luminous efficacy. J Phys D Appl Phys 43:354002. doi:10.1088/0022-3727/43/35/354002
13. Jang HS, Im WB, Lee DC, Jeon DY, Kim SS (2007) Enhancement of red spectral emission intensity of $Y_3Al_5O_{12}$: Ce^{3+} phosphor via Pr co-doping and Tb substitution for the application to white LEDs. J Lumin 126:371
14. JØrgensen CK (1962) Absorption spectra and chemical bonding in complexes. Pergamon, Oxford
15. Blasse G (1972) The ultraviolet absorption bands of Bi^{3+} and Eu^{3+} in oxides. J Solid State Chem 4:52
16. Blasse G, Grabmaier BC (1994) Luminescent materials. Springer
17. Jüstel T (2005) Luminescent materials for high brightness LEDs. FH Munster-Philips Research Aachen
18. Kimoto K, Xie RJ, Matsui Y, Ishizuka K, Hirosaki N (2009) Direct observation of single dopant atom in light-emitting phosphor of β-SiAlON: Eu^{2+}. Appl Phys Lett 94:041908
19. Tanabe Y, Sugano S (1954) On the absorption spectra of complex ions. I. J Phys Soc Japan 9:766
20. Kamimura A, Sugano S, Tanabe Y (1969) Ligand field theory and its applications. Shokabo
21. Yang P, Yao GQ, Lin JH (2004) Energy transfer and photoluminescence of $BaMgAl_{10}O_{17}$ co-doped with Eu^{2+} and Mn^{2+}. Opt Mater 26:327
22. Dieke GH (1968) Spectra and energy levels of rare earth ions in crystals. Interscience, New York
23. Dorenbos P (2000) The 5d level positions of the trivalent lanthanides in inorganic compounds. J Lumin 91:155

24. Im WB, Fellows NN, DenBaars SP, Seshadri R, Kim YI (2009) $LaSr_2AlO_5$, a versatile host compound for Ce^{3+}-based yellow phosphors: structural tuning of optical properties and use in solid-state white lighting. Chem Mater 21:2957
25. Dorenbos P (2005) Valence stability of lanthanide ions in inorganic compounds. Chem Mater 17:6452
26. Dorenbos P (2001) 5d-level energies of Ce^{3+} and the crystalline environment. III. Oxides containing ionic complexes. Phys Rev B 64:125117
27. Lin CC, Liu RS (2014) Thermal effects in (oxy)nitride phosphors. J Solid State Light 1:16
28. Yu JJ, Gong WT, Xiao ZG, Ning GL (2012) Spectral structure of barium–phosphate–silicate phosphor $Ba_{10}(PO_4)_4(SiO_4)_2$: Eu^{M+}. J Lumin 132:2957
29. Hirosaki N, Ogata S, Kocer C (2003) Ab initio calculation of the crystal structure of the lanthanide Ln_2O_3 sesquioxides. J Alloys Comp 351:31
30. van der Kolk E, de Haas JTM, Bos AJJ, van Eijk CWE, Dorenbos P (2007) Luminescence quenching by photoionization and electron transport in a $LaAlO_3$: Ce^{3+} crystal. J Appl Phys 101:083703
31. Dorenbos P (2003) Energy of the first $4f^7$-$4f^65d$ transition of Eu^{2+} in inorganic compounds. J Lumin 104:239
32. Lyu LJ, Hamilton DS (1991) Radiative and nonradiative relaxation measurements in Ce^{3+} doped crystals. J Lumin 48&49:251
33. Dorenbos P (2005) Thermal quenching of Eu^{2+} $5d$–$4f$ luminescence in inorganic compounds. J Phys: Condens Matter 17:8103
34. Hana JK, Piqutte A, Hannah ME, Hirata GA, Talbot JB, Mishra KC, McKittrick J (2014) Analysis of $(Ba,Ca,Sr)_3MgSi_2O_8$:Eu^{2+}, Mn^{2+} phosphors for application in solid state lighting. J Lumin 148:1
35. Yadav RS, Pandey SK, Pandey AC (2010) Blue-shift and enhanced photoluminescence in $BaMgAl_{10}O_{17}$: Eu^{2+} nanophosphor under VUV excitation for PDPs application. J Rare Earths Mater Sci Appl 1:25
36. Xie RJ, Hirosaki N, Li HL, Li YQ, Mitomo M (2007) Synthesis and photoluminescence properties of beta-sialon: Eu^{2+} ($Si_{6-z}Al_zO_zN_{8-z}$: Eu^{2+}). J Electrochem Soc 154:J314
37. Kim JS, Park YH, Kim SM, Choi JC, Park HL (2005) Temperature-dependent emission spectra of M_2SiO_4: Eu^{2+} (M = Ca, Sr, Ba) phosphors for green and greenish white LEDs. Solid State Comm 133:445
38. Lei BF, Machida KI, Horikawa T, Hanzawa H (2011) Preparation of $(Sr_{0.5}Ba_{0.5})Si_2N_2O_2$: Eu^{2+} phosphor and its luminescence properties. Chem Lett 40:140141
39. Tang JY, Chen JH, Hao LY, Xu X, Xie WJ, Li QX (2011) Green Eu^{2+}-doped $Ba_3Si_6O_{12}N_2$ phosphor for white light-emitting diodes: synthesis, characterization and theoretical simulation. J Lumin 131:1101
40. Peng TY, Liu HJ, Yang HP, Yan CH (2004) Synthesis of $SrAl_2O_4$:Eu, Dy phosphor nanometer powders by sol–gel processes and its optical properties. Mater Chem Phys 85:68
41. Zhou J, Wang YH, Liu BT, Li F (2010) Energy transfer between Eu-Mn and photoluminescence properties of $Ba_{0.75}Al_{11}O_{17.25}$-$BaMgAl_{10}O_{17}$: Eu^{2+}, Mn^{2+} solid solution. J Appl Phys 108:033106
42. Do YR, Bae JW, Kim Y, Yang HG (2000) Preparation and optical properties of $SrGa_2S_4$:Eu phosphor. Bull Kor Chem Soc 21:295
43. Li HL, Xie RJ, Hirosaki N, Yajima Y (2008) Synthesis and photoluminescence properties of $Sr_2Si_5N_8$: Eu^{2+} red phosphor by a gas-reduction and nitridation method. J Electrochem Soc 155:J378
44. Piao XQ, Machida KI, Horikawa T, Hanzawa H (2007) Preparation of $CaAlSiN_3$: Eu^{2+} phosphors by the self-propagating. Chem Mater 19:4592
45. Van Haecke JE, Smet PF, De Keyser K, Poelman D (2007) Single crystal CaS: Eu and SrS: Eu luminescent particles obtained by solvothermal synthesis. J Electrochem Soc 154:J278
46. Yap SV, Ranson RM, Cranton WM, Koutsogeorgis DC, Hix GB (2009) Temperature dependent characteristics of La_2O_2S: Ln [Ln = Eu, Tb] with various Ln concentrations over 5–60 °C. J Lumin 129:416

47. Kim JS, Lim KT, Jeong YS, Jeon PE, Choi JC, Park HL (2005) Full-color $Ba_3MgSi_2O_8$: Eu^{2+}, Mn^{2+} phosphors for white-light-emitting diodes. Solid State Commun 135:21
48. Hao ZD, Zhang JH, Zhang X, Wang XJ (2011) $CaSc_2O_4$: Eu^{3+}: a tunable full-color emitting phosphor for white light emitting diodes. Opt Mater 33:355
49. Li HL, Liu XJ, Huang LP (2007) Luminescent properties of LuAG: Ce phosphors with different Ce contents prepared by a sol-gel combustion method. Opt Mater 29:1138
50. Hansel RA, Allison SW, Walker DG (2009) Temperature-dependent luminescence of Ce^{3+} in gallium-substituted garnets. Appl Phys Lett 95:114102
51. Arai Y, Kominami H, Nakanishi Y, Hatanaka Y (2005) Luminescent properties of $SrGa_2S_4$: Eu thin film phosphors deposited by two electron beam evaporation. Appl Surf Sci 244:473
52. Nagai H (2011) Light-emitting device. US patent 08288790
53. Li HL, Xie RJ, Hirosaki N, Suehiro T, Yajima Y (2008) Phase purity and luminescence properties of fine Ca-alpha-SiAlON: Eu phosphors synthesized by Gas reduction nitridation method. J Electrochem Soc 155:J175
54. Park JY, Jung HC, Raju GSR, Moon BK, Jeong JH, Son SM, Kim JH (2009) Sintering temperature effect on structural and luminescence properties of 10 mol% Y substituted $Gd_3Al_5O_{12}$:Ce phosphors. Opt Mater 32:293
55. Zorenko Y, Gorbenko V, Voznyak T, Zorenko T, Kuklinski B, Turos-Matysyak R, Grinberg M (2009) Luminescence properties of phosphors based on $Tb_3Al_5O_{12}$ (TbAG) terbium-aluminum garnet. Opt Spectrosc 106:365
56. Seto T, Kijima N, Hirosaki N (2009) A new yellow phosphor $La_3Si_6N_{11}$: Ce^{3+} for white LEDs. ESC Trans 25:247
57. Najafov H, Kato A, Toyota H, Iwai K, Bayramov A, Iida S (2002) Effect of Ce co-doping on $CaGa_2S_4$:Eu phosphor: I. energy transfer from Ce to Eu ions. Jpn J Appl Phys 41:1424
58. Jang HS, Won YH, Vaidyanathan S, Kim DH, Jeon DY (2009) Emission band change of $(Sr_{1-x}M_x)_3SiO_5$: Eu^{2+} (M = Ca, Ba) phosphor for white light sources using blue/near-ultraviolet LEDs. J Electrochem Soc 156:J138
59. Bruchez M, Moronne M, Gin P, Weiss S, Alivisatos AP (1998) Semiconductor nanocrystals as fluorescent biological labels. Science 281:2013
60. Klimov VI, Ivanov SA, Nanda J, Achermann M, Bezel I, McGuire JA, Piryatinski A (2007) Single-exciton optical gain in semiconductor nanocrystals. Nature 447:441
61. Sun Q, Wang YA, Li LS, Wang DY, Zhu T, Xu J, Yang CH, Li YF (2007) Bright, multicoloured light-emitting diodes based on quantum dots. Nat Photon 1:717
62. Huynh WU, Dittmer JJ, Alivisatos AP (2002) Hybrid nanorod-polymer solar cells. Science 295:2425
63. Kelly KL, Coronado E, Zhao LL, Schatz GC (2003) The optical properties of metal nanoparticles: the influence of size, shape, and dielectric environment. J Phys Chem B 107:668
64. Ni WH, Ambjornsson T, Apell SP, Chen HJ, Wang JF (2010) Observing plasmonic-molecular resonance coupling on single gold nanorods. Nano Lett 10:77
65. Kim MG, Hennek JW, Kim HS, Kanatzidis MG, Facchetti A, Marks TJ (2012) Delayed ignition of autocatalytic combustion precursors: low-temperature nanomaterial binder approach to electronically functional oxide films. J Am Chem Soc 134:11583
66. Liang YY, Wang HL, Zhou JG, Li YG, Wang J, Regier T, Dai HJ (2012) Covalent hybrid of spinel manganese-cobalt oxide and graphene as advanced oxygen reduction electrocatalysts. J Am Chem Soc 134:3517
67. Schmid G, Corain B (2003) Nanoparticulated gold: syntheses, structures, electronics, and reactivities. Eur J Inorg Chem 2003:3081
68. Dutta P, Pai S, Seehra MS, Shah N, Huffman GP (2009) Size dependence of magnetic parameters and surface disorder in magnetite nanoparticles. J Appl Phys 105:07B501
69. Chen HM, Liu RS (2011) Architecture of metallic nanostructures: synthesis strategy and specific applications. J Phys Chem C 115:3513
70. Efros AL (1982) Interband absorption of light in a semiconductor sphere. Sov Phys Semicond 16:772

71. Brus LE (1983) A simple-model for the ionization-potential, electron-affinity, and aqueous redox potentials of small semiconductor crystallites. J Chem Phys 79:5566
72. Wang Y, Herron N (1991) Nanometer-sized semiconductor clusters–materials synthesis, quantum size effects, and photophysical properties. J Phys Chem 95:525
73. Murray CB, Norris DJ, Bawendi MG (1993) Synthesis and characterization of nearly monodisperse CdE (E = S, Se, Te) semiconductor nanocrystallites. J Am Chem Soc 115:8706
74. Donegá CM, Liljeroth P, Vanmaekelbergh D (2005) Physicochemical evaluation of the hot-injection method, a synthesis route for monodisperse nanocrystals. Small 1:1152
75. Qu L, Yu WW, Peng X (2004) In situ observation of the nucleation and growth of CdSe nanocrystals. Nano Lett 4:465
76. Talapin DV, Rogach AL, Haase M, Weller H (2001) Evolution of an ensemble of nanoparticles in a colloidal solution: theoretical Study. J Phys Chem B 105:12278
77. Puzder A, Williamson AJ, Zaitseva N, Galli G, Manna L, Alivisatos AP (2004) The effect of organic ligand binding on the growth of CdSe nanoparticles probed by Ab initio calculations. Nano Lett 4:2361
78. Liu H, Owen JS, Alivisatos AP (2007) Mechanistic study of precursor evolution in colloidal group II–VI semiconductor nanocrystal synthesis. J Am Chem Soc 129:305
79. Xie R, Li Z, Peng X (2009) Nucleation kinetics vs chemical kinetics in the initial formation of semiconductor nanocrystals. J Am Chem Soc 131:15457
80. Owen JS, Chan EM, Liu H, Alivisatos AP (2010) Precursor conversion kinetics and the nucleation of cadmium selenide nanocrystals. J Am Chem Soc 132:18206
81. LaMer VK, Dinegar RH (1950) Theory, production and mechanism of formation of monodispersed hydrosols. J Am Chem Soc 72:4847
82. Algar WR, Susumu K, Delehanty JB, Medintz IL (2011) Semiconductor quantum dots in bioanalysis: crossing the valley of death. Anal Chem 83:8826
83. Gur I, Fromer NA, Geier ML, Alivisatos AP (2005) Air-stable all-inorganic nanocrystal solar cells processed from solution. Science 310:462
84. Shen S, Wang Q (2013) Rational tuning the optical properties of metal sulfide nanocrystals and their applications. Chem Mater 25:1166
85. Omata T, Nose K, Otsuka-Yao-Matsuo S (2009) Size dependent optical band gap of ternary I–III–VI_2 semiconductor nanocrystals. J Appl Phys 105:073106
86. Rogach AL, Kornowski A, Gao MY, Eychmuller A, Weller H (1999) Synthesis and characterization of a size series of extremely small thiol-stabilized CdSe nanocrystals. J Phys Chem B 103:3065
87. Gao MY, Kirstein S, Mohwald H, Rogach AL, Kornowski A, Eychmuller A, Weller H (1998) Strongly photoluminescent CdTe nanocrystals by proper surface modification. J Phys Chem B 102:8360
88. Rogach A, Kershaw S, Burt M, Harrison M, Kornowski A, Eychmuller A, Weller H (1999) Colloidally prepared HgTe nanocrystals with strong room-temperature infrared luminescence. Adv Mater 11:552
89. Fang Z, Li Y, Zhang H, Zhong X, Zhu L (2009) Facile synthesis of highly luminescent UV-blue-emitting ZnSe/ZnS core/shell nanocrystals in aqueous media. J Phys Chem C 113:14145
90. Grabolle M, Spieles M, Lesnyak V, Gaponik N, Eychmueller A, Resch-Genger U (2009) Determination of the fluorescence quantum yield of quantum dots: suitable procedures and achievable uncertainties. Anal Chem 81:6285
91. Kulkarni SK, Winkler U, Deshmukh N, Borse PH, Fink R, Umbach E (2001) Investigations on chemically capped CdS, ZnS and ZnCdS nanoparticles. Appl Surf Sci 169:438
92. Wang WZ, Germanenko I, El-Shall MS (2002) Room-temperature synthesis and characterization of nanocrystalline CdS, ZnS, and $CdxZn_{1-x}S$. Chem Mater 14:3028
93. Petrov DV, Santos BS, Pereira GAL, Donega CD (2002) Size and band-gap dependences of the first hyperpolarizability of $CdxZn_{1-x}S$ nanocrystals. J Phys Chem B 106:5325

94. Bailey RE, Nie SM (2003) Alloyed semiconductor quantum dots: Tuning the optical properties without changing the particle size. J Am Chem Soc 125:7100
95. Zhong XH, Han MY, Dong ZL, White TJ, Knoll W (2003) Composition-tunable $Zn_xCd_{1-x}Se$ nanocrystals with high luminescence and stability. J Am Chem Soc 125:8589
96. Anikeeva PO, Halpert JE, Bawendi MG, Bulović V (2009) Quantum dot light-emitting devices with electroluminescence tunable over the entire visible spectrum. Nano Lett 9:2532
97. Gurusinghe NP, Hewa-Kasakarage NN, Zamkov M (2008) Composition-tunable properties of CdS_xTe_{1-x} alloy nanocrystals. J Phys Chem C 112:12795
98. Harrison MT, Kershaw SV, Burt MG, Eychmuller A, Weller H, Rogach AL (2000) Wet chemical synthesis and spectroscopic study of CdHgTe nanocrystals with strong near-infrared luminescence. Mater Sci Eng, B 69:355
99. Sargent EH (2005) Infrared quantum dots. Adv Mater 17:515
100. Rogach AL, Eychmueller A, Hickey SG, Kershaw SV (2007) Infrared-emitting colloidal nanocrystals: synthesis, assembly, spectroscopy, and applications. Small 3:536
101. Gaponik N, Radtchenko IL, Gerstenberger MR, Fedutik YA, Sukhorukov GB, Rogach AL (2003) Labeling of biocompatible polymer microcapsules with near-infrared emitting nanocrystals. Nano Lett 3:369
102. Qian H, Dong C, Peng J, Qiu X, Xu Y, Ren J (2007) High-quality and water-soluble near-infrared photoluminescent CdHgTe/CdS quantum dots prepared by adjusting size and composition. J Phys Chem C 111:16852
103. Bailey RE, Nie SM (2003) Alloyed semiconductor quantum dots: tuning the optical properties without changing the particle size. J Am Chem Soc 125:7100
104. Smith AM, Gao XH, Nie SM (2004) Quantum dot nanocrystals for *in vivo* molecular and cellular imaging. Photochem Photobiol 80:377
105. Bailey RE, Smith AM, Nie SM (2004) Quantum dots in biology and medicine. Physica E 25:1
106. Artemyev MV, Woggon U, Wannemacher R, Jaschinski H, Langbein W (2001) Light trapped in a photonic dot: microspheres act as a cavity for quantum dot emission. Nano Lett 1:309
107. Gaponik NP, Talapin DV, Rogach AL, Eychmuller A (2000) Electrochemical synthesis of CdTe nanocrystal/polypyrrole composites for optoelectronic applications. J Mater Chem 10:2163
108. Chan WCW, Nie SM (1998) Quantum dot bioconjugates for ultrasensitive nonisotopic detection. Science 281:2016
109. Michalet X, Pinaud FF, Bentolila LA (2005) Quantum dots for live cells, in vivo imaging, and diagnostics. Science 307:538
110. Bharali DJ, Lucey DW, Jayakumar H, Pudavar HE, Prasad PN (2005) Folate-receptor-mediated delivery of InP quantum dots for bioimaging using confocal and two-photon microscopy. J Am Chem Soc 127:11364
111. Yong KT, Ding H, Roy I, Law WC, Bergey EJ, Maitra A, Prasad PN (2009) Imaging pancreatic cancer using bioconjugated InP quantum dots. ACS Nano 3:502
112. Mohapatra P, Dung MX, Choi JK, Jeong S, Jeong HD (2011) Effects of curing temperature on the optical and charge trap properties of InP quantum dot thin films. Bull Korean Chem Soc 32:263
113. Lim J, Bae WK, Lee D, Nam MK, Jung J, Lee C, Char K, Lee S (2011) InP@ZnSeS, Core@composition gradient shell quantum dots with enhanced stability. Chem Mater 23:4459
114. Xie R, Battaglia D, Peng X (2007) Colloidal InP nanocrystals as efficient emitters covering blue to near-infrared. J Am Chem Soc 129:15432
115. Xu S, Ziegler J, Nann T (2008) Rapid synthesis of highly luminescent InP and InP/ZnS nanocrystals. J Mater Chem 18:2653
116. Cao YW, Banin U (2000) Growth and properties of semiconductor core/shell nanocrystals with InAs cores. J Am Chem Soc 122:9692

117. Yang X, Zhao D, Leck KS, Tan ST, Tang YX, Zhao J, Demir HV, Sun XW (2012) Full visible range covering InP/ZnS nanocrystals with high photometric performance and their application to white quantum dot light-emitting diodes. Adv Mater 24:4180
118. Blackburn JL, Ellingson RJ, Micic OI, Nozik AJ (2003) Electron relaxation in colloidal InP quantum dots with photogenerated excitons or chemically injected electrons. J Phys Chem B 107:102
119. Talapin DV, Gaponik N, Borchert H, Rogach AL, Haase M, Weller H (2002) Etching of colloidal InP nanocrystals with fluorides: photochemical nature of the process resulting in high photoluminescence efficiency. J Phys Chem C 106:12659
120. Micic OI, Sprague J, Lu ZH, Nozik AJ (1996) Highly efficient band-edge emission from InP quantum dots. Appl Phys Lett 68:3150
121. Liu Z, Kumbhar A, Xu D, Zhang J, Sun Z, Fang J (2008) Coreduction colloidal synthesis of III-V nanocrystals: the case of InP. Angew Chem Int Ed 47:3540
122. Li L, Pandey A, Werder DJ, Khanal BP, Pietryga JM, Klimov VI (2011) Efficient synthesis of highly luminescent copper indium sulfide-based Core/Shell nanocrystals with surprisingly long-lived emission. J Am Chem Soc 133:1176
123. Zhong HZ, Wang ZB, Bovero E, Lu ZH, Van Veggel FCJM, Scholes GD (2011) Colloidal $CuInSe_2$ nanocrystals in the quantum confinement regime: synthesis, optical properties, and electroluminescence. J Phys Chem C 115:12396
124. Mao BD, Chuang CH, Wang JW, Burda C (2011) Synthesis and photophysical properties of ternary I–III–VI $AgInS_2$ nanocrystals: intrinsic versus surface states. J Phys Chem C 115:8945
125. Allen PM, Bawendi MG (2008) Ternary I–III–VI quantum dots luminescent in the red to near-infrared. J Am Chem Soc 130:9240
126. Zhong HZ, Zhou Y, Ye MF, He YJ, Ye JP, He C, Yang CH, Li YF (2008) Controlled synthesis and optical properties of colloidal ternary chalcogenide $CuInS_2$ nanocrystals. Chem Mater 20:6434
127. Li L, Daou TJ, Texier I, Chi TTK, Liem NQ, Reiss P (2009) Highly luminescent $CuInS_2$/ZnS core/shell nanocrystals: cadmium-free quantum dots for in vivo imaging. Chem Mater 21:2422
128. Chen BK, Zhong HZ, Zhang WQ, Tan ZA, Li YF, Yu CR, Zhai TY, Bando Y, Yang SY, Zou BS (2012) Highly emissive and color-tunable $CuInS_2$-based colloidal semiconductor nanocrystals: off-stoichiometry effects and improved electroluminescence performance. Adv Funct Mater 22:2081
129. van Sark W, Frederix P, Bol A, Gerritsen H, Meijerink A (2002) Blueing, bleaching, and blinking of single CdSe/ZnS quantum dots. ChemPhysChem 3:871

Chapter 2
Phosphors for White-Light LEDs Through the Principle of Energy Transfer

Teng-Ming Chen and Yoan-Jen Yang

Abstract This chapter is an attempt to investigate and develop single-phased white light-emitting phosphors essentially for potential applications in white light-emitting diodes (WLEDs) based on the principle of resonant-type energy transfer. We first review the fundamentals of energy-transfer processes and propose a set of semi-empirical protocols for the design of white-emitting phosphors on the basis of bonding nature of hosts and the resonant-type energy transfer. The luminescent properties and energy-transfer processes for four categories of single-phased white light-emitting phosphors and a multi-composition phosphor are used as examples and discussed in detail. In summary, white-light emission can be realized by adopting single-composition two-complementary systems such as blue/yellow-emitting ($CaAl_2Si_2O_8$:Eu^{2+}, Mn^{2+}), cyan/red-emitting ($BaGa_4S_7$:Eu^{2+}, Mn^{2+}), or green-yellow/red-emitting phosphor (Ba_2ZnS_3:Ce^{3+}, Eu^{2+}) under blue or ultraviolet (UV) excitation. On the other hand, white-light emission can also be realized in three-primary phosphors such as blue/green/red-emitting $SrZn_2(PO_4)_2$: Eu^{2+}, Mn^{2+} under blue excitation. Furthermore, the most intriguing examples of white-light generation with a high color-rendering index through resonant energy transfer are trichromatic inorganic phosphors such as $Ca_3Y(GaO)_3(BO_3)_4$:Ce^{3+}, Mn^{2+}, Tb^{3+}, $Mg_2Y_8(SiO_4)_6O_2$:Ce^{3+}, Mn^{2+}, Tb^{3+}, $NaCaBO_3$:Ce^{3+}, Mn^{2+}, Tb^{3+}, and $BaMg_2Al_6Si_9O_{30}$:Eu^{2+}, Tb^{3+}, Mn^{2+}, which can effectively convert UV, near-UV, or blue light into a combination of RGB. To support and confirm the occurrence of energy transfer from a sensitizer (energy donor) to an activator (energy acceptor), one must provide two evidences, namely, spectral overlapping between a sensitizer and activator from steady-state spectra and shortening of luminescence decay rate of a sensitizer with increasing content of an activator. With transient spectra of the white-emitting phosphors, the mechanism of electrical multipolar resonance-type energy transfer can also be deduced.

T.-M. Chen (✉) · Y.-J. Yang
Phosphors Research Laboratory, Department of Applied Chemistry,
National Chiao Tung University, Hsinchu 30010, Taiwan
e-mail: tmchen@mail.nctu.edu.tw

R.S. Liu (ed.), *Phosphors, Up Conversion Nano Particles, Quantum Dots and Their Applications*, DOI 10.1007/978-3-662-52771-9_2

2.1 Introduction

White light-emitting diodes (LEDs) have been recognized as green lightings for the next generation to replace conventional lamps and backlights due to the advantages of their low power consumption and being free of mercury. Although white light could be generated by two or three complementary LED chips at least, the cost of production is higher. Therefore, the economical method to produce white light is one-wavelength LED-converted phosphor, such as blue LED-pumped yellow $Y_3Al_5O_{12}:Ce^{3+}$ (YAG:Ce) [1]. However, the color-rendering index (R_a) of a white-light source made by the complementary blue and yellow emission is deficient because of the lack of red-light contribution resulting in nonrealization of the true color rendition. Hence, several red phosphors were developed to add into the above-mentioned system to improve the R_a value [2–5]. Unfortunately, the extreme difference in degradation between different host phosphors will produce color aberration. Accordingly, it is important to investigate a single-host phosphor with green-to-red emission bands for blue LEDs. The best choices of activators in phosphors are the ions with *f–d* or *d–d* electron configurations because they could emit visible and broad-band light under the influence of crystal field and nephelauxectic effect [6, 7]. Furthermore, a phosphor could emit a couple of radiation by co-doping these ions in a single host such as Eu^{2+}/Mn^{2+} [8–18], Ce^{3+}/Eu^{2+} [19–21], and Ce^{3+}/Mn^{2+} [22], and the energy transfer would exist between activators in a phosphor by effective resonant-type by way of a multipolar interaction [8, 9]. Nevertheless, in the past few years, coactivated single-composition phosphors with blue absorption or for blue LEDs have rarely been investigated. In this chapter, we will review a single-phased phosphor, Ce^{3+}/Eu^{2+} co-doped Ba_2ZnS_3, which shows ultraviolet-to-blue absorption and green-to-red emission, thus exhibiting great potential for application in white LEDs while a blue chip is coupled.

2.2 Theory of Electronic Transition and Luminescence

An inorganic phosphor is composed of a host lattice and an activator, and sometimes a sensitizer is also required. The host usually comprises a combination of optically inert cations and anions; the activator generally comprises optically active cations.

The energy levels of an activator can be defined by the spectroscopic term, $^{2S+1}L_J$, where S is the spin-angular momentum; L is orbital angular momentum; and J is the spin-orbit angular momentum. The Hamiltonian (***H***) of an activator could be written as

$$\boldsymbol{H}_{\text{activator}} = H_o + H_c + H_{so} + H_{cf} \tag{2.1}$$

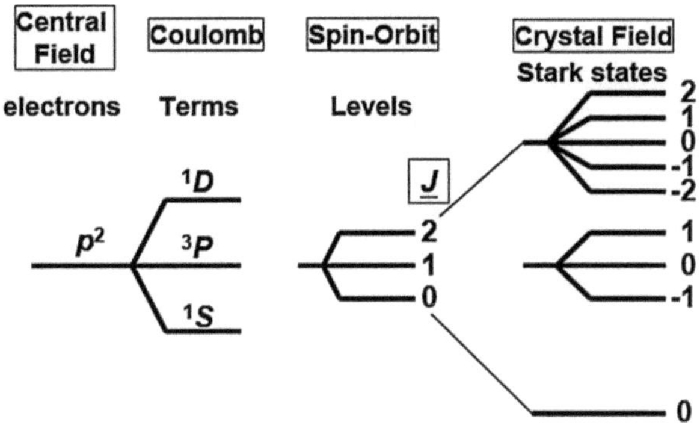

Fig. 2.1 Hamiltonians for a carbon atom with $1s^2 2s^2 2p^2$ configuration [4]. Reprinted from Ref. [4]. Copyright 2004, with permission from Elsevier

where H_o is the central field Hamiltonian; H_c is the coulomb Hamiltonian; H_{so} is the spin-orbit Hamiltonian; and H_{cf} is crystal field Hamiltonian.

As shown in Fig. 2.1, the diagram gives the relation between various Hamiltonians. The allowed transitions are governed by the following selection rules:

$$L = \pm 1; \quad S = 0; \quad J = \pm 1, 0$$

but not $J = 0$ to $J = 0$. In fact, these selection rules are only suitable for dipole transitions, i.e., transitions between electric dipole states. In addition, other possible multipolar states can exist as magnetic dipoles, electric quadrupoles, magnetic quadrupoles, and higher multipole orders. The easiest way to visualize the types of multipoles is to imagine a p-electron as shown in Fig. 2.2.

The electric field vector $\overrightarrow{F}$, directs at the greatest electron field density. On the contrary, the magnetic field vector $\overrightarrow{M}$, is at right angles to that density. Therefore, the transition intensities between the electric and magnetic multipoles are lower than those involving electric-to-electric multipoles. Moreover, one pole can influence another at fair distances in a lattice, and this is called an "electric multipolar interaction" [4]. The transition probability can be written as [7]

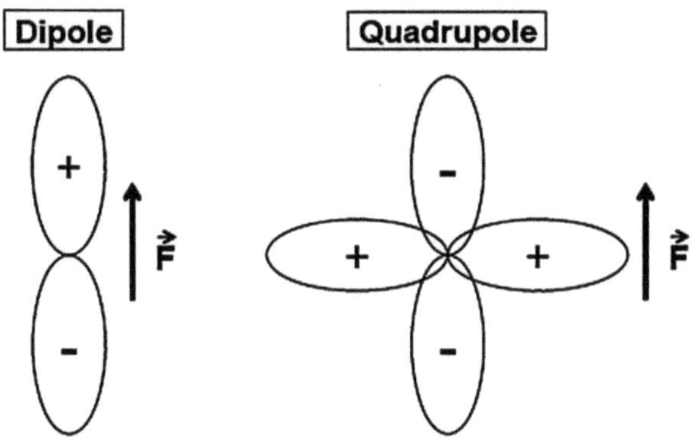

Fig. 2.2 Diagrams for electropole radiators. Reprinted from Ref. [4]. Copyright 2004, with permission from Elsevier

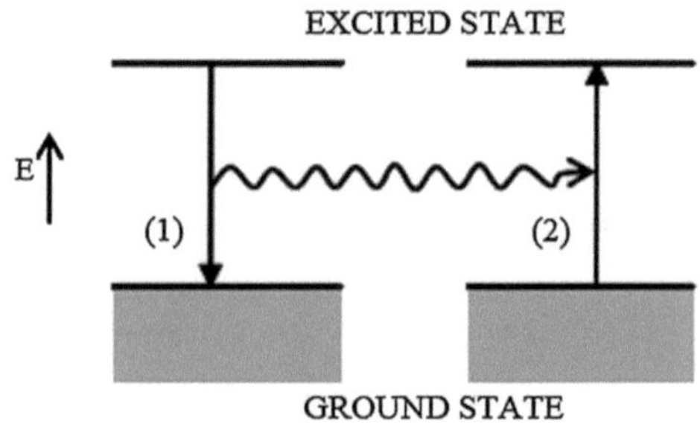

Fig. 2.3 Diagram of virtual photon exchange between activators. Reprinted from Ref. [4]. Copyright 2004, with permission from Elsevier

$$(P_k)_{fi} = \left\langle \Psi_f^* | \mathrm{M} | \Psi_i \mathrm{d}\Omega \right\rangle$$

where M is the operator form of a multipole. Accordingly, the types of multipoles are related to the orientations of electric or magnetic field.

The energy-transfer mechanisms between activators could be classified into radiative and nonradiative types chiefly. Radiative energy transfer occurs between the two activators, A_1 and A_2, separated by a few unit-cell distances. A_1 absorbs enough energy to be excited, and then it will emit a photon, which may be captured by an adjacent activator, A_2, which then also becomes excited. Therefore, the radiative energy transfer could be labeled as virtual photon exchange as shown in Fig. 2.3. The nonradiative energy transfer is also classified into resonance-energy transfer (RET), spin-coupling energy exchange (SCEE), and non-resonance-energy transfer (NRET) [4].

Resonance energy transfer occurs between a sensitizer (S) and an activator (A) with the same radiative frequency, and the mechanism could be of exchange interaction or electric multipolar interaction.

In fact, the phenomenon of energy transfer could be demonstrated by the overlap of emission spectrum of S and excitation spectrum of A. Dexter's energy transfer formula is described as [8]:

$$P_{\mathrm{SA}} = 2\pi/\hbar \,|\, \langle \mathrm{S},\, \mathrm{A}^* \,|\, H_{\mathrm{SA}} \,|\, \mathrm{S}^*,\, \mathrm{A} \rangle \,|^2 \int \mathrm{F_S}(E)\mathrm{F_A}(E)\mathrm{d}E \tag{2.2}$$

where the integral describes the spectral overlap; $\mathrm{F_S}(E)$ and $\mathrm{F_A}(E)$ represent the normalized shapes of the sensitizer and activator, respectively; the matrix element represents the interaction between initial state $|\mathrm{S}*, \mathrm{A}\rangle$ and final state $\langle \mathrm{S}, \mathrm{A}*|$; H_{SA} is the Hamiltonian of the interaction; P_{SA} is the probability of energy transfer and, when the spectral overlap is absent, the phenomenon of resonance energy-transfer vanishes, as shown in Figs. 2.4 and 2.5, respectively. Furthermore, the energy-transfer probability is related to the type of interaction. For electric multipolar interaction, the distance dependence is given by $R^{-\alpha}$ where α is 6 and 8 corresponding to electric dipole–dipole and dipole–quadrupole interactions,

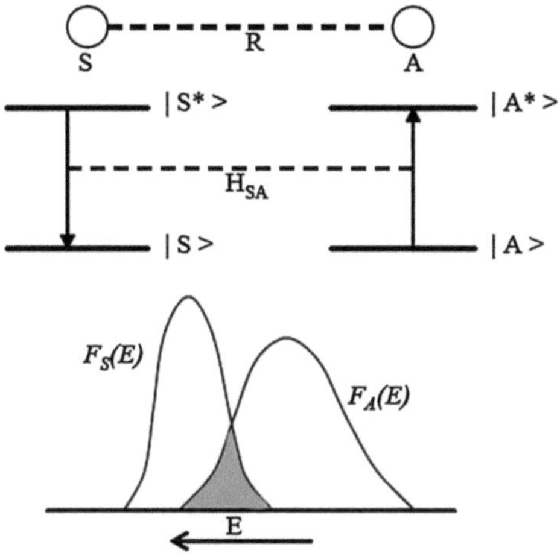

Fig. 2.4 Diagram of the energy transfer between a sensitizer and an activator. Reprinted from Ref. [5], with kind permission from Springer Science+Business Media

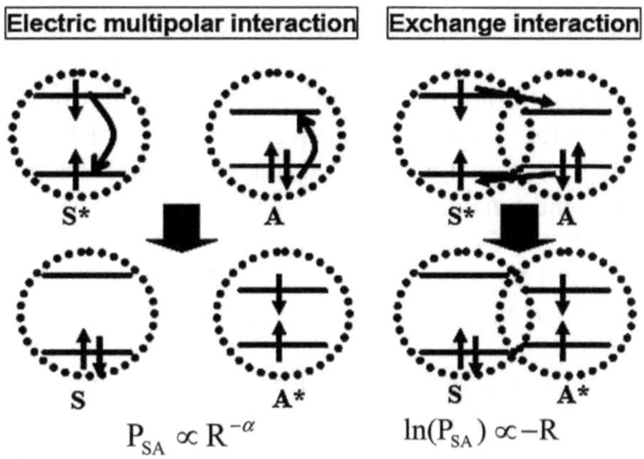

Fig. 2.5 Diagrams of the energy-transfer mechanisms of electric multipolar and exchange interactions

respectively. For exchange interaction, the distance dependence is exponential, and the wavefunction overlap between S and A is required [8, 9].

On the other hand, spin-coupling energy exchange occurs between two adjacent activators, A_1 and A_2, which are no further than approximately two to three lattice sites apart [4]. As shown in Fig. 2.6, the initial state of A_1 and A_2 are both in the excited state, and they exchange energy by way of spin-coupling. Finally, A_1 returns to ground state, and A_2 transits to a higher level doubling that of initial state. The process is usually observed in the activators, which can absorb infrared radiation, which is then converted to visible light (also called an “anti-Stokes process”).

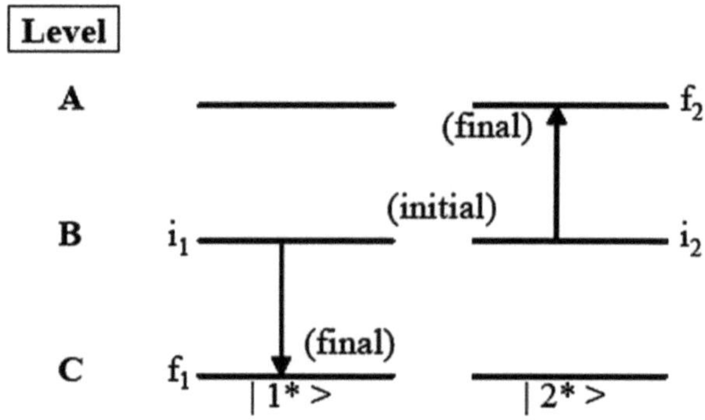

Fig. 2.6 Diagram of spin-coupling energy exchange. Reprinted from Ref. [4]. Copyright 2004, with permission from Elsevier

2.2.1 Literature Review

It is well-known that energy transfer mostly occurs between a sensitizer and an activator in a phosphor, e.g., Eu^{2+}/Mn^{2+}, Ce^{3+}/Eu^{2+}, Ce^{3+}/Mn^{2+}, $Ce^{3+}/Tb^{3+}/Mn^{2+}$, $Eu^{2+}/Tb^{3+}/Mn^{2+}$, and so on. The activator acts as an efficient sensitizer to transfer energy to the co-doped activator. In 1970, Barry observed that an effective energy transfer from Eu^{2+} to Mn^{2+} occurred in $BaMg_2Si_2O_7:Eu^{2+}$, Mn^{2+} phosphor, in which Eu^{2+} and Mn^{2+} occupied the Ba^{2+} and Mg^{2+} sites, respectively [10]. In the 1980s, Rubio et al. reported that energy transfer from Eu^{2+} to Mn^{2+} took place in the sodium halide lattices, in which Eu^{2+} and Mn^{2+} formed close pairs [11–14]. In the 1990s, Caldino et al. described that the $Eu^{2+} \rightarrow Mn^{2+}$ energy transfer process occurred by way of an electric dipole–quadrupole or exchange interaction in the small Eu–Mn clusters formed in the calcium halide and KBr crystals [15–17]. In 1996, Lin et al. discussed the energy-transfer process between the luminescent centers in $Ca_8Zn(SiO_4)_4Cl_2:Eu^{2+}$, Mn^{2+} phosphor and demonstrated the substitution of Eu^{2+} and Mn^{2+} for Ca^{2+} and Zn^{2+} sites, respectively [18]. In 1999, Tan et al. determined that the $Ce^{3+} \rightarrow Eu^{2+}$ energy transfer in $BaLiF_3$ occurs by way of a dipole–dipole interaction [19]. In 2002, Najafov et al. observed that the energy-transfer process from Ce^{3+} to Eu^{2+} in $CaGa_2S_4$ is a dipole–dipole interaction [20]; Lin et al. found that the energy-transfer mechanism from Ce^{3+} to Eu^{2+} in $Ca_8Mg(SiO_4)_4Cl_2$ is electric dipole–dipole interaction [21]. In 2003, Caldino discussed that the energy-transfer mechanisms between Ce^{3+}/Eu^{2+} and Ce^{3+}/Mn^{2+} in CaF_2 are of the electric dipole–dipole type and electric dipole–quadrupole type, respectively [22]. Until recently, single-composition phosphors under UV and blue radiations have rarely been investigated for the fabrication of white LEDs. In 2004, Setlur et al. declared that $Sr_2P_2O_7:Eu^{2+}$, Mn^{2+} and $Ca_5(PO_4)_3F:Eu^{2+}$, Mn^{2+} phosphors could be excited by UV/blue LEDs and produce white-light emission [23]. Kim et al. reported that $Ba_3MgSi_2O_8:Eu^{2+}$, Mn^{2+} [24] and $Sr_3MgSi_2O_8:Eu^{2+}$, Mn^{2+} [25] phosphors could be pumped by 375-nm InGaN LED chips to generate white light. In 2011, Huang and Chen reported a novel single-phased trichromatic white-emitting $Ca_3Y(GaO)_3(BO_3)_4:Ce^{3+}$, Mn^{2+}, Tb^{3+} for UV-LEDs and studied the energy transfer from $Ce^{3+} \rightarrow Mn^{2+}$ as well as the green emission from Tb^{3+} [26].

Zhang and Gong reported a series of single-phased white-emitting $NaCaBO_3$:Ce^{3+}, Tb^{3+}, Mn^{2+} phosphors and proposed them for LED applications [27]. Lü et al. reported a tunable full-color trichromatic-emitting $BaMg_2Al_6Si_9O_{30}$:Eu^{2+}, Tb^{3+}, Mn^{2+} phosphor, in which white light can be generated from RGB emissions with blue at 450 nm, green at 542 nm, and red at 610 nm, respectively, through $Eu^{2+}\rightarrow Tb^{3+}$ and $Eu^{2+}\rightarrow Mn^{2+}$ energy-transfer processes [28].

Consequently, this has triggered many active research efforts devoted to the quest for new UV/blue LED-converting single-phased and white-emitting phosphors. In the following sections we will describe and discuss the principle of white-light generation for four categories of single-phased white-emitting phosphors.

2.3 Design Principles and Preparation Protocol of White-Emitting Phosphors

Regarding the design of single-phased white-emitting phosphors for UV/blue LEDs, we will describe two groups of commonly encountered materials as examples, namely, oxides and sulfides. The activators generally selected are Eu^{2+}, Ce^{3+} and Mn^{2+} ions because their emission wavelengths are tunable by changing coordination environment and site symmetry as well as the emission spectra exhibit broadband. According to the effective ionic radii of cations with different coordination number reported by Shannon [29], we propose that Eu^{2+} and Ce^{3+} ions tend to occupy the Ca^{2+}, Sr^{2+} and Ba^{2+} sites, and Mn^{2+} ions reasonably substitute for the Ca^{2+}, Sr^{2+}, Ba^{2+}, Mg^{2+}, and Zn^{2+} site and emit green and red light when occupying the four- and six-coordinated lattice sites in a dual-element oxide matrix, respectively [5]. However, in an oxide matrix with more than two elements, the Mn^{2+} ions are affected by not only substitution sites but also adjacent cations, generally resulting in emission in the green-to-red spectral range. Therefore, based on semiempirical results, we propos flowcharts for the design and preparation of white-emitting oxide and sulfide phosphors for blue LEDs; these flowcharts are summarized in Fig. 2.7, respectively.

For sulfide phosphors, the host lattices are mostly tri-element compounds in nature, and they can be represented by $M'_aM''_bS_m$, where M′ and M″ can be selected from the following:

$$M' : Ca^{2+}, Sr^{2+} \text{ and } Ba^{2+}$$
$$M'' : Al^{3+}, Ga^{3+}, In^{3+}, Mg^{2+} \text{ and } Zn^{2+}$$

In the sulfide matrix, due to the influence of the nephelauxetic effect resulting from the lower electronegativity of sulfur atom, the energy level of center of gravity of the *d* level of an activator is lower than that in oxide matrix. Accordingly, Eu^{2+}

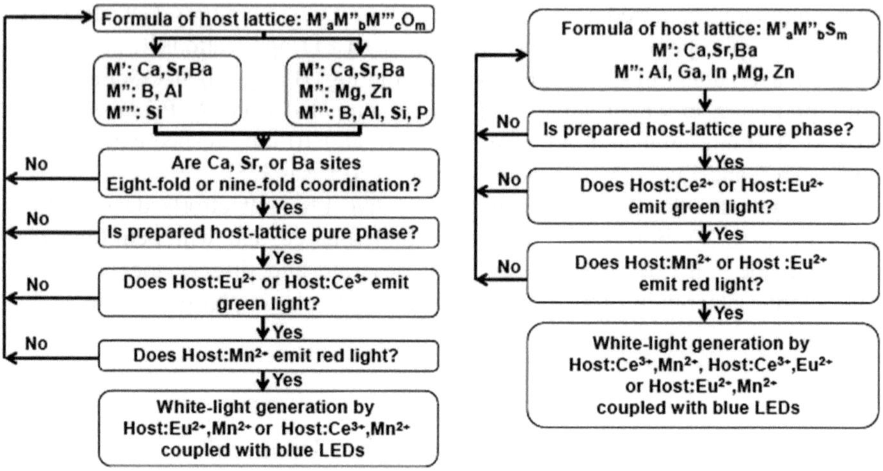

Fig. 2.7 Flow diagram describing the design of white-emitting oxide and sulfide phosphors for blue LEDs

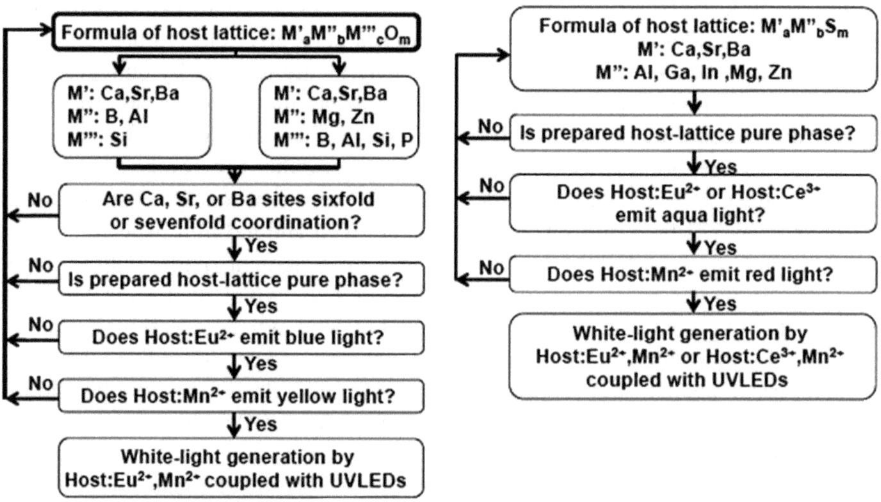

Fig. 2.8 Flow diagram describing the design of white-emitting oxide and sulfide phosphors for UVLED

could emit aqua to red light; Ce^{3+} ions could emit blue to green light; and Mn^{2+} ions could emit mostly red light.

Furthermore, the flowcharts of design and preparation of white-emitting oxide and sulfide phosphors for UV LEDs are also shown, respectively in Fig. 2.8.

According to the flowcharts showing the design and preparation protocol, we have demonstrated that the white-emitting phosphors for UV and blue LEDs can be obtained and used to generate white light. In addition, a series of trichromatic white-emitting phosphor for UVLEDs has also been developed and will be discussed in the next section. The luminescence properties and energy-transfer mechanisms between two or among three luminescent centers have been widely investigated and is described in following sections.

2.4 White-Emitting Phosphors with Predesigned Energy-Transfer Mechanisms

To illustrate the principle of white light emission through the energy-transfer principle and to verify the proposed protocol of designing white-emitting phosphors, we will describe how white light is generated and give representative examples to demonstrate that white-emitting phosphors can be predesigned and realized using the energy-transfer principle.

2.4.1 White-Emitting Phosphors with Energy Transfer from Eu^{2+} to Mn^{2+}

It has been investigated that Eu^{2+} may serve as an efficient sensitizer that absorbs and transfers energy to Mn^{2+} in different host lattices. As reported in the literature, Caldino et al. reported the $Eu^{2+}\rightarrow Mn^{2+}$ energy-transfer process in the single crystals of $CaCl_2:Eu^{2+}$, Mn^{2+} under photoexcitation. The investigators suggested that the Eu^{2+}-to-Mn^{2+} energy-transfer process observed in $CaCl_2$:Eu, Mn can be rationalized by the formation of small complexes of Eu–Mn in the lattice [15, 16]. Similar energy transfer was also observed by Barry in the $BaMg_2Si_2O_7:Eu^{2+}$, Mn^{2+} phosphor [10]. Yao et al. investigated the luminescence and decay lifetimes of $BaMg_2Si_2O_7:Eu^{2+}$, Mn^{2+} as a function of Mn^{2+} concentrations and confirmed the occurrence of energy transfer from Eu^{2+} to Mn^{2+} [30]. Rubio et al. used an ionic radius criterion to predict paring between two impurity dopant ions in alkali halide host matrice, which may provide a reasonable basis for selecting appropriate impurity dopant ions for developing efficient phosphor materials [12]. Furthermore, the $Eu^{2+}\rightarrow Mn^{2+}$ energy-transfer mechanism in KBr:Eu^{2+}, Mn^{2+} phosphor was described by Mendez et al. [17] who proposed the possible formation of small Eu^{2+}–Mn^{2+} clusters in the KBr lattice. Mendez et al. also proposed that the $Eu^{2+}\rightarrow Mn^{2+}$ energy transfer may be rationalized by assuming that a possible dipole–quadrupole or exchange (superexchange) interaction mechanism is active in the Eu^{2+}–Mn^{2+} cluster formation [17]. Recently, Kim et al. reported that $Ba_3MgSi_2O_8$:Eu, Mn can serve as a phosphor for fabrication of a warm-white LED [24]. They observed that with optimal excitation wavelength at 375 nm, $Ba_3MgSi_2O_8$:Eu, Mn shows three emission bands centered at 442 nm from the first Eu^{2+}, 505 nm from the second Eu^{2+}, and 620 nm from Mn^{2+} [24], respectively. Table 2.1 summarizes a variety of representative examples of white-emitting

Table 2.1 Single-phased white-emitting phosphors with $Eu^{2+} \rightarrow Mn^{2+}$ energy transfer

Phosphor	$CaAl_2Si_2O_8$: Eu^{2+}, Mn^{2+}	$SrZn_2(PO_4)_2$: Eu^{2+}, Mn^{2+}	$BaGa_4S_7$: Eu^{2+}, Mn^{2+}	$Ca_2MgSi_2O_7$: Eu^{2+}, Mn^{2+}	$BaMgSiO_4$:Eu^{2+}, Mn^{2+}
Excitation range (nm)	280–395 (UV)	310–395 (UV)	330–390 (UV)	300–450 (B)	350–410 (UV-VIS)
Emission wavelength (nm)	425 (B), ~560 (Y)	416 (B), 538 (G), 613 (R)	475 (A), 653 (R)	~450 (B)-650 (R)	477 (B), 626 (O-R)
Energy transfer mechanism	Dipole–quadrupole	Dipole–quadrupole	Exchange	Dipole–quadrupole	Dipole–quadrupole and exchange
Energy transfer critical distance (Å)	10.8	11.4	5.9	11.90	9.46, 4.32
Reference	[31]	[32]	[33]	[34]	[35]
Phosphor	$Ca_{10}K(PO_4)_7$: Eu^{2+}, Mn^{2+}	$Ca_9La(PO_4)_7$: Eu^{2+}, Mn^{2+}	$Ca_9Y(PO_4)_7$: Eu^{2+}, Mn^{2+}	$Ca_9Gd(PO_4)_7$: Eu^{2+}, Mn^{2+}	$Ca_4Si_2O_7F_2$:Eu^{2+}, Mn^{2+}
Excitation range (nm)	250–420 (n-UV)	280–395 (UV)	310–395 (UV)	380 (n-UV)	240–500 (UV-Blue)
Emission wavelength (nm)	460 (B), 635 (O-R)	425 (B), ~560 (Y)	486 (B), 638 (R)	494 (B), 652 (O-R)	460 (B), 576 (Y-O)
Energy transfer mechanism	Dipole–quadrupole	Dipole–quadrupole	Dipole–quadrupole	Dipole–quadrupole	Dipole-quadrupole
Energy transfer critical distance (Å)	12.2	11.36	11.04	11.10	11.66
Reference	[36]	[37]	[38]	[39]	[40]

phosphors with energy transfer $Eu^{2+} \rightarrow Mn^{2+}$ reported in various hosts such as silicates, phosphates, oxyfluoride, and sulfide. In general, the white light consists of blue and yellow or blue/green/red radiations resulting from blue, near-UV, or ultraviolet light source pumping. Furthermore, the energy-transfer mechanism found in almost all the phosphors is of dipole–quadrupole type, which is supported by the critical distance of energy transfer indicated in Table 2.1, with a few cases being exchange type due to an intrinsic short interatomic distance between the sensitizer and the activator.

Table 2.2 Single-phased white-emitting phosphors with Ce^{3+}→Eu^{2+} energy transfer

Phosphor	Ba_2ZnS_3: Ce^{3+}, Eu^{2+}	$Sr_3Al_2O_5Cl_2$: Ce^{3+}, Eu^{2+}	Ca_2BO_3Cl: Ce^{3+}, Eu^{2+}	$Sr_3B_2O_6$: Ce^{3+}, Eu^{2+}
Excitation range (nm)	400–450 (B)	320–370 (UV-nUV)	325–400 (UV-nUV)	365 (UV)
Emission wavelength (nm)	498–540 (GY), 655 (R)	438 (B), 583 (Y-O)	422 (B), 573 (Y)	434 (B), 574 (O-Y)
Energy transfer mechanism	Dipole–dipole	Dipole–dipole	Dipole–dipole	Dipole–dipole
Energy transfer critical distance (Å)	32.7	28.6	32.2	30
Reference	[36]	[37]	[38]	[39]

2.4.2 *White-Emitting Phosphors with Energy Transfer from Ce^{3+} to Eu^{2+}*

Table 2.2 summarizes several representative examples of single-phased white-emitting phosphors with energy transfer Ce^{3+}→Eu^{2+} reported in the literature. The host composition ranges in a variety from highly covalent sulfide, oxychloride, and more-ionic borate. The principle to generate white light generally employs phosphors that emit a couple of radiation by co-doping activators with *f–d* or *d–d* electron configurations such as Ce^{3+}/Eu^{2+} [19–22] the energy transfer would essentially occur between activator/coactivator couples by effective resonant type by way of a multipolar interaction. As indicated in Table 2.2, we have observed that the white light consists of G/Y/R, B/Y/O, or B/Y radiations resulted from blue, near-UV, or ultraviolet light-source pumping, respectively. In addition, the energy-transfer mechanism determining the four phosphors is essentially of resonant dipole–dipole type with very long interatomic distances (i.e., ≥ 30 Å) between sensitizer Ce^{3+} and activator Eu^{2+} regardless of the type of host compound.

Yang and Chen [41] reported the Ba_2ZnS_3:Ce^{3+}, Eu^{2+} sulfide phosphor, which shows intense blue absorption and tunable green-to-red emission. The Ce^{3+}→Eu^{2+} energy transfer has been demonstrated to be resonant type by way of an electric dipole–dipole mechanism. Because Ba_2ZnS_3:Ce^{3+}, Eu^{2+} is able to generate white light under excitation at 420 nm, the CIE chromaticity coordinates were found to be (0.34, 0.49). The sulfide phosphor would be a great potential application as a blue radiation–converting phosphor for white LEDs. Chang also investigated a series of cold-and-warm white LEDs using the $Sr_3Al_2O_5Cl_2$:Ce^{3+},Eu^{2+} phosphour [42]. In the Ca_2BO_3Cl:0.06Ce^{3+}, 0.01Eu^{2+} phosphor [43], the mechanism of resonance-type Ce^{3+}→Eu^{2+} energy transfer was established to be of an electric dipole–dipole nature, and the critical distance was estimated to be 31 Å based on the spectral overlap and concentration quenching model. A white light was obtained from Ca_2BO_3Cl:0.06Ce^{3+}, 0.01Eu^{2+} phosphor with chromaticity coordinates (0.31, 0.29) and relative color temperature of 7,330 K on excitation with a wavelength of 360 nm, and therefore it is potentially a good candidate as an UV-convertible phosphor for white LEDs. The $Sr_3B_2O_6:Ce^{3+}$, Eu^{2+} borate phosphors reported by Chang and Chen [44] exhibit varied hues from blue through white and eventually to yellow-orange via resonance-type $Ce^{3+} \rightarrow Eu^{2+}$ energy transfer by properly tuning the relative proportion of Ce^{3+}/Eu^{2+}. The investigators claimed that electric dipole–dipole interaction dominates the energy-transfer mechanism in $Sr_3B_2O_6:Ce^{3+}$, Eu^{2+} phosphor, and the critical distance of energy transfer has been estimated to be approximately 30 Å by both spectral-overlap and concentration-quenching methods. Under the excitation of UV radiation, white light is generated by coupling 434- and 574-nm emission bands attributed to Ce^{3+} and Eu^{2+} radiations, respectively.

2.4.3 White-Emitting Phosphors with Energy Transfer from Ce^{3+} to Mn^{2+} [45–48]

Table 2.3 summarizes four of the representative single-phased white-emitting phosphors with energy transfer from Ce^{3+} to Mn^{2+}. Hsu et al. described that in the PL/PLE spectra of $MgY_4Si_3O_{13}:Ce^{3+}$, Mn^{2+}, the increment of Mn^{2+} concentration would lead to a systematic decrease of Ce^{3+} emission due to the efficient $Ce^{3+} \rightarrow Mn^{2+}$ energy transfer [45], which occurs because of the spectral overlap between the emission band of Ce^{3+} and the excitation band of Mn^{2+}. The resonant $Ce^{3+} \rightarrow Mn^{2+}$ energy transfer is of the dipole–quadrupole type, which was supported by decay-lifetime data, and the critical distance of the energy transfer was calculated to be 14.13 Å. The color-tunable emission in Ce^{3+}/Mn^{2+} co-doped system can be realized by continuous shifting of the emission colors from blue to white and eventually to the orange-red region. Müller and Jüstel investigated the photoluminescence and measured the transient spectral from 100 to 500 K to understand the thermal behavior of $Ca_3Y_2(Si_3O_9)_2:Ce^{3+}$, Mn^{2+} [46]. It turned out that the mechanism of $Ce^{3+} \rightarrow Mn^{2+}$ energy transfer is due to dipole–quadrupole interaction and $Ca_3Y_2(Si_3O_9)_2:Ce^{3+}$, Mn^{2+} also shows white emission and exhibits excellent thermal stability. Liu et al. reported a new tunable full color-emitting $Ca_3Sc_2Si_3O_{12}:Ce^{3+}$,

Table 2.3 Single-phased white-emitting phosphors with $Ce^{3+} \rightarrow Mn^{2+}$ energy transfer

Phosphor	$MgY_4Si_3O_{13}$: Ce^{3+}, Mn^{2+}	$Ca_3Y_2(Si_3O_9)_2$: Ce^{3+}, Mn^{2+}	$Ca_3Sc_2Si_3O_{12}$: Ce^{3+}, Mn^{2+}	$Ca_2Gd_8(SiO_4)_6O_2$: Ce^{3+}, Mn^{2+}
Excitation range (nm)	328 (UV)	300 (UV)	450 (B)	352 (UV)
Emission Wavelength (nm)	455 (B), 587 (O-R)	395 (V-B), 545 (Y-O)	574 (Y), 680 (R)	428 (B), 588 (R-O)
Energy transfer mechanism	Dipole–quadrupole	Dipole–quadrupole	Dipole–quadrupole	Dipole–quadrupole
Energy transfer critical distance (Å)	14.13	8.1, 8.8	NA	9.4
Reference	[45]	[46]	[47]	[48]

Mn^{2+} phosphor [47] by controlling the distribution of Mn^{2+} in the Ca^{2+} and Sc^{3+} lattice sites, respectively, through the addition of doped La^{3+}, Gd^{3+}, Lu^{3+}, and Y^{3+} lanthanides ions as charge compensators with the aid of $Ce^{3+}{\rightarrow}Mn^{2+}$ energy transfer. By using this single-phased $Ca_3Sc_2Si_3O_{12}:Ce^{3+}$, Mn^{2+} phosphor and blue (450 nm) InGaN LED chips, pcWLEDs with a high R_a of 91–92 and a color temperature of 5379–6954 K could be achieved [47]. Li et al. investigated the $Ce^{3+}{\rightarrow}Mn^{2+}$ energy transfer in $Ca_2Gd_8(SiO_4)_6O_2:Ce^{3+}$, Mn^{2+} phosphors [48] and demonstrated that the $Ce^{3+}{\rightarrow}Mn^{2+}$ transfer is of a resonant type by way of a dipole–quadrupole mechanism, and the critical distance calculated by the quenching-concentration and spectral-overlap methods are 9.4 and 9.2 Å, respectively. A color-tunable emission in $Ca_2Gd_8(SiO_4)_6O_2:Ce^{3+}$, Mn^{2+} phosphors can be realized by the modulation of excitation wavelengths, namely, the change of Ce^{3+} emission at different lattice sites and the relative PL intensity of Ce^{3+} and Mn^{2+}. Wide-ranging white light with varied hues was obtained in $Ca_2Gd_8(SiO_4)_6O_2:Ce^{3+}$, Mn^{2+} by utilizing the principle of energy transfer, properly designed activator contents, and the selection of excitation wavelength in the range of 287–352 nm [48].

2.4.4 Trichromatic White-Emitting Phosphors with Dual-Energy Transfer

Tremendous efforts have been made to generate white-light emission with a high color-rendering index (R_a) using inorganic phosphors to convert ultraviolet (UV), near-UV, or blue light into a combination of RGB. However, in the three-component (RGB) converting system, the manufacturing cost is high and the blue-emission efficiency poor because of the strong reabsorption of blue light by the red- and green-emitting phosphors. In an attempt to circumvent these disadvantages, there have been many investigations on efficient, durable, and single-phased white-emitting phosphors with RGB components (i.e., green from $Ce^{3+}{\rightarrow}Tb^{3+}$ and $Eu^{2+}{\rightarrow}Tb^{3+}$, red from $Ce^{3+}{\rightarrow}Mn^{2+}$ or $Eu^{2+}{\rightarrow}Mn^{2+}$) through energy-transfer couples, and this is drawing investigators' attention. Caldino et al. [22] first reported the $Ce^{3+}{\rightarrow}Mn^{2+}$ energy transfer and mechanism involved with $CaF_2:Ce^{3+}$, Eu^{2+}, Mn^{2+} phosphor. Such energy transfer occurs by way of a short-range interaction mechanism into Ce^{3+}–Eu^{2+} and Ce^{3+}–Mn^{2+} clusters formed in the crystalline matrix. In contrast to the $Eu^{2+}{\rightarrow}Mn^{2+}$ energy transfer usually observed in $CaF_2:Eu^{2+}:Mn^{2+}$, manganese ions do not appear to be sensitized by europium ions, which might suggest that Eu^{2+} ions prefer to cluster with tetragonal Ce^{3+} ions when these ions are also present in the lattice [22]. Table 2.4 summarizes four examples of the representative single-phased white-emitting phosphors with multiple energy transfer.

Regarding the mechanism, we shall describe two types of triply-doped white-emitting phosphor systems to illustrate the generation of white light through energy-transfer couples. First, the energy-level scheme illustrating the energy-transfer models of $Ce^{3+}{\rightarrow}Tb^{3+}$ and $Ce^{3+}{\rightarrow}Mn^{2+}$ in the $Ce^{3+}/Mn^{2+}/Tb^{3+}$ system is summarized in Fig. 2.9 [27].

Table 2.4 Single-phased white-emitting phosphors with multiple energy transfer

Phosphor	$Ca_3Y(GaO)_3(BO_3)_4$: Ce^{3+}, Tb^{3+}, Mn^{2+}	$Mg_2Y_8(SiO_4)_6O_2$: Ce^{3+}, Tb^{3+}, Mn^{2+}	$NaCaBO_3$: Ce^{3+}, Tb^{3+}, Mn^{2+}	$BaMg_2Al_6Si_9O_{30}$: Eu^{2+}, Tb^{3+}, Mn^{2+}
Excitation wavelength (nm)	365 (UV)	324 (UV)	347 (UV)	330 (UV)
Emission Wavelength (nm)	409 (V-B), 544 (G), 589 (O)	456 (B), 544 (G), 613 (R)	427 (B), 545 (G), 600 (R)	450 (B), 542 (G), 610 (R)
Energy transfer mechanism	Dipole–quadrupole (Ce^{3+}–Mn^{2+}) Dipole–quadrupole (Ce^{3+}–Tb^{3+})	Dipole–quadrupole (Ce^{3+}–Mn^{2+}) Dipole–quadrupole (Ce^{3+}–Tb^{3+})	Dipole–quadrupole (Ce^{3+}–Mn^{2+}) Dipole–quadrupole (Ce^{3+}–Tb^{3+})	Dipole–quadrupole (Eu^{2+}–Mn^{2+}) Dipole–quadrupole (Eu^{2+}–Tb^{3+})
Energy transfer critical distance (Å)	10.3 (Ce^{3+}–Mn^{2+})	10.5 (Ce^{3+}–Mn^{2+})	NA	NA
Reference	[26]	[49]	[27]	[28]

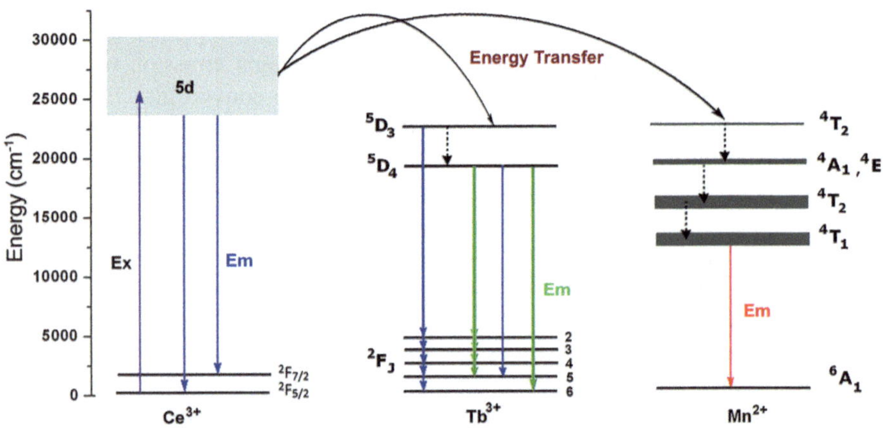

Fig. 2.9 Energy-level scheme illustrating the energy-transfer models of $Ce^{3+} \rightarrow Tb^{3+}$ and $Ce^{3+} \rightarrow Mn^{2+}$. Reproduced from Ref. [27] by permission of The Royal Society of Chemistry

In the ET model for $Ce^{3+}/Tb^{3+}/Mn^{2+}$ in various hosts, Ce^{3+} ions can be pumped from the $^2F_{5/2}$ ground state to the excited states by UV radiation and then transfer the energy to the Tb^{3+} $^5D_3$ level; subsequently, the $^5D_3$ level gives its characteristic transitions or continues to transfer the energy to the $^5D_4$ level by way of cross-relaxation. Then the green emission occurs from a set of characteristic optical transitions $^5D_4 \rightarrow ^7F_J$. The situation of the $Ce^{3+} \rightarrow Mn^{2+}$ energy transfer, which

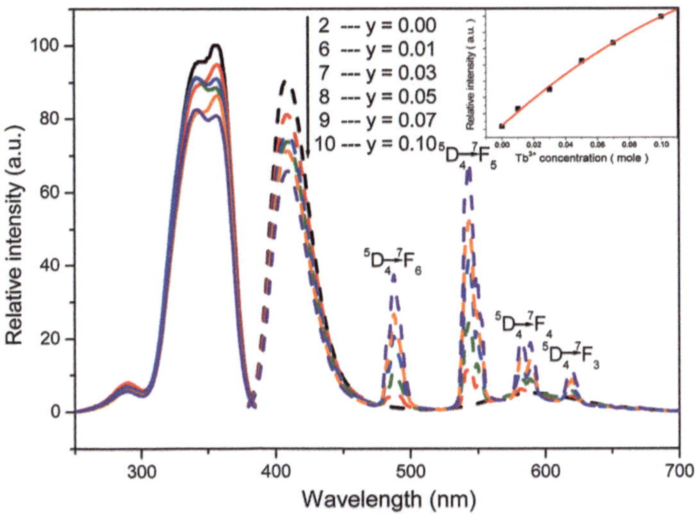

Fig. 2.10 Dependence of PL and PLE spectra of CYGB:0.01Ce^{3+}, 0.03Mn^{2+}, $y$$Tb^{3+}$ on y; the inset shows the relative PL intensity as a function of y at $\lambda_{ex} = 365$ nm. Reprinted with the permission from Ref. [26]. Copyright 2011 American Chemical Society

generates red emission, is similar: Mn^{2+} receives the energy transferred from excited Ce^{3+}. The Mn^{2+} ion can be pumped from the ground state into its higher-energy levels. The excited state then relaxes to the excited state of $^4T_1$ (^4G) through ^4E (^4D), $^4T^2$ (^4D), (^4E, $^4A_1$) (^4G) and $^4T_2$ (^4G) intermediate energy levels in a nonradiative process followed by a radiative transition from the excited state of $^4T_1$ (^4G) to the ground state of $^6A_1$ (^6S), thus giving rise to the typical emission of Mn^{2+} attributed to $^4T_1 \rightarrow {}^6A_1$ in the host lattice. Three examples of white light-emitting phosphors are illustrated below.

(a) Single-phased white-emitting $Ca_3Y(GaO)_3(BO_3)_4$:Ce^{3+}, Mn^{2+}, Tb^{3+} phosphor. The first novel single-composition white-emitting phosphor is a galloborate. $Ca_3Y(GaO)_3(BO_3)_4$:Ce^{3+}, Mn^{2+}, Tb^{3+}, which has been reported by Huang et al. [26]. Figure 2.10 reports the dependence of PL and PLE spectra of $Ca_3Y(GaO)_3(BO_3)_4$:0.01Ce^{3+}, 0.03Mn^{2+}, $y$$Tb^{3+}$ on y, and the inset shows the relative PL intensity as a function of y at $\lambda_{ex} = 365$ nm. The spectral overlap between the Ce^{3+} emission band and the Mn^{2+} excitation band supports the occurrence of energy transfer from Ce^{3+} to Mn^{2+}. This phosphor has been investigated and found to be a resonant type by way of a dipole–quadrupole mechanism. Because there is no spectral overlap between the PL spectrum of Ce^{3+} and PLE spectrum of Tb^{3+} in Huang's work, no energy transfer from Ce^{3+} to Tb^{3+} was observed, thus indicating that Ce^{3+} and Tb^{3+} were co-excited simultaneously.

Through effective resonance-type energy transfer and co-excitation, the chromaticity coordinates of $Ca_3Y(GaO)_3(BO_3)_4$:Ce^{3+}, Mn^{2+}, Tb^{3+} phosphors

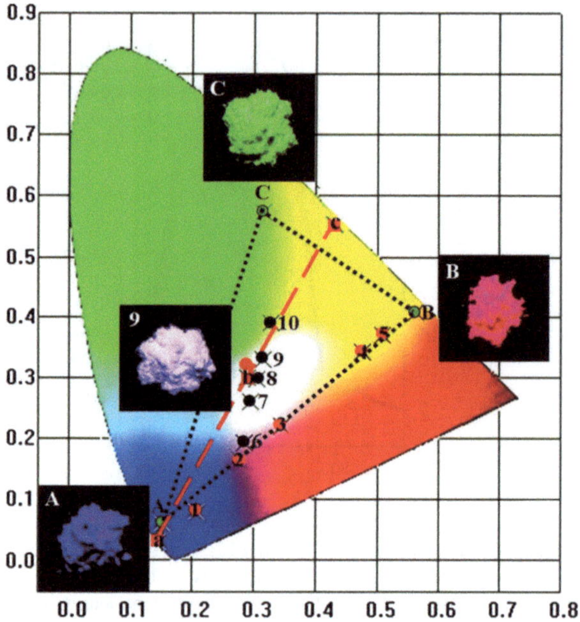

Fig. 2.11 Representation of the CIE chromaticity coordinates of CYGB:0.01Ce^{3+}, $x$$Mn^{2+}$ (points 1 through 5), CYGB:0.01Ce^{3+} and CYGB:0.03Mn^{2+}, $y$$Tb^{3+}$ (points 6 through 10). Reprinted with the permission from Ref. [26]. Copyright 2011 American Chemical Society

can be effectively tuned from (0.152, 0.061) for $Ca_3Y(GaO)_3(BO_3)_4:Ce^{3+}$ to (0.562, 0.408) for $Ca_3Y(GaO)_3(BO_3)_4:Mn^{2+}$, eventually reaching (0.314, 0.573) for $Ca_3Y(GaO)_3(BO_3)_4:Tb^{3+}$. A WLED can be fabricated by using the white-emitting single-composition $(Ca_{0.97}Mn_{0.03})_3(Y_{0.92}Ce_{0.01}Tb_{0.07})(GaO)_3(BO_3)_4$ pumped by a 365-nm U-chip. These results revealed that the CIE chromaticity coordinates and correlated color temperature (CCT) for white UV-LEDs were (0.31, 0.33) and 6524 K, respectively. Therefore, the white-emitting $Ca_3Y(GaO)_3(BO_3)_4:Ce^{3+}$, Mn^{2+}, Tb^{3+} may serve as a promising material for phosphor-converted white-light UV-LEDs (Fig. 2.11).

(b) White-emitting triply-doped $Mg_2Y_8(SiO_4)_6O_2:Ce^{3+}$, Mn^{2+}, Tb^{3+} phosphor [49]. The second triply-doped example deals with oxyapatite-type oxosilicate, $Mg_2Y_8(SiO_4)_6O_2$ (MYSO):Ce^{3+}, Mn^{2+}, Tb^{3+}, which was reported by Li et al. [49] In this research, the dual-energy transfer-induced $Ce^{3+}/Mn^{2+}/Tb^{3+}$-triactivated MYSO phosphors were investigated. The investigators demonstrated that $Ce^{3+} \rightarrow Mn^{2+}$ energy transfer in MYSO:Ce^{3+}, Mn^{2+} phosphors is resonant type by way of a dipole–quadrupole mechanism, and the critical distances were found to be 10.5 and 9.7 Å, respectively, based on the calculation using both quenching-concentration and spectral-overlap methods. By the dual-energy transfer of $Ce^{3+} \rightarrow Mn^{2+}$ and $Ce^{3+} \rightarrow Tb^{3+}$, the emission colors of the investigated phosphors can be adjusted from blue to orange-red and from blue to green, respectively. Moreover, a wide-range tunable white light emission from cool to warm white with high quantum yields were obtained in $Mg_2Y_8(SiO_4)_6O_2:Ce^{3+}$, Mn^{2+}, Tb^{3+} samples by controlling the dopant

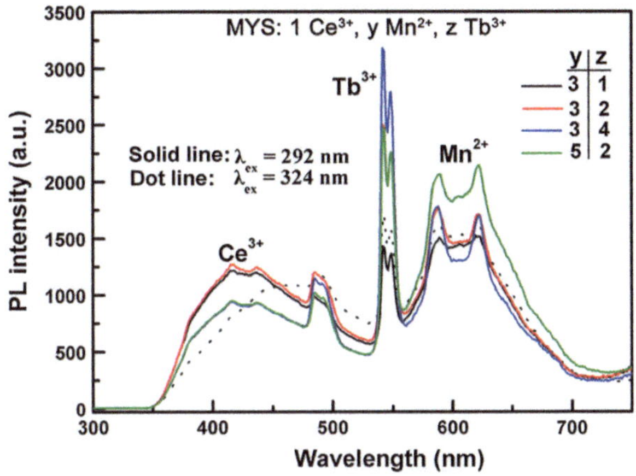

Fig. 2.12 PL spectra of MYSO: $1Ce^{3+}$, yMn^{2+}, zTb^{3+} samples ($y = 3$, 5 mol%; $z = 1$, 2, 4 mol%) under UV excitation: **a** 292 nm and **b** 324 nm. Reprinted with permission from Ref. [49]. Copyright 2011 American Chemical Society

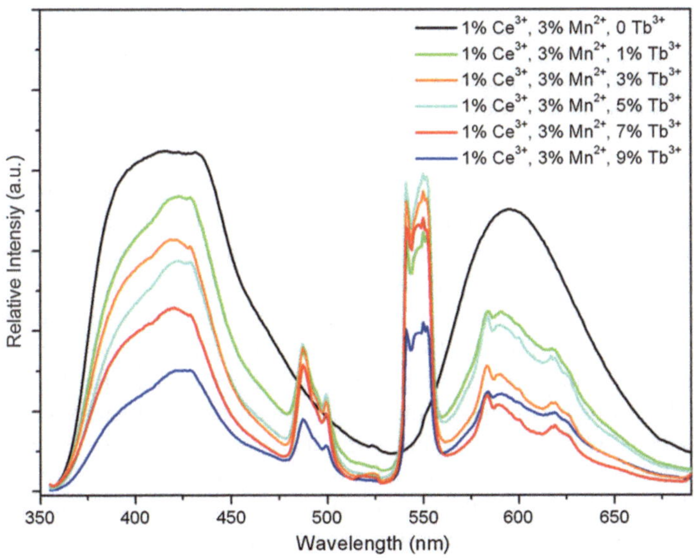

Fig. 2.13 Dependence of PL on Tb^{3+} content in $NaCaBO_3$:1%Ce^{3+}, 3%Mn^{2+}, z%Tb^{3+} ($z = 0$, 1, 3, 5, 7, 9). Reproduced from Ref. [27] by permission of The Royal Society of Chemistry

contents of Ce^{3+}, Mn^{2+}, and Tb^{3+} ions. In addition, a color-tunable emission from MYSO:Ce^{3+}, Mn^{2+}, Tb^{3+} phosphors can be obtained by the modulation of excitation wavelength from 292 to 324 nm [49] (Fig. 2.12).

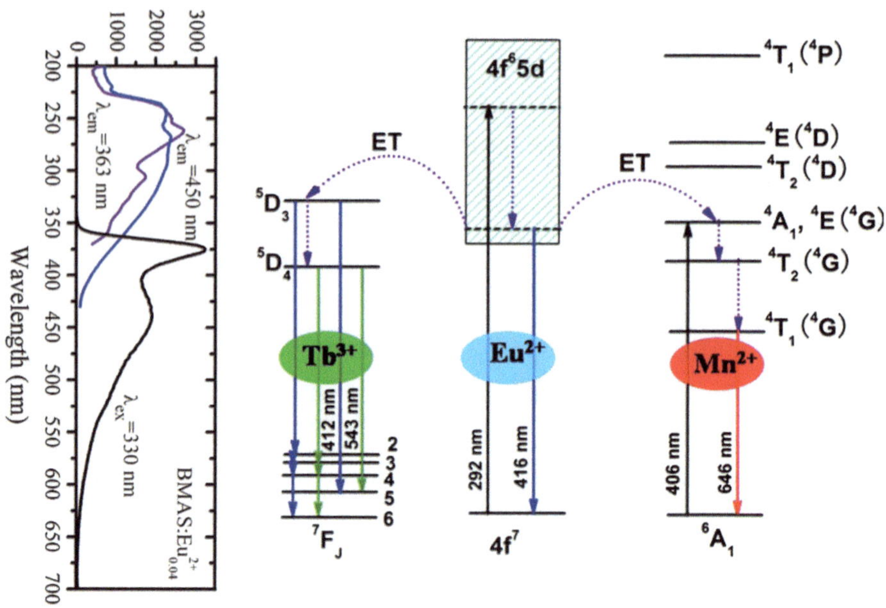

Fig. 2.14 Schematic-level diagram for the energy-transfer process in the BMASO:Eu^{2+}, Tb^{3+}, Mn^{2+} phosphor. The PLE (λ_{em} = 450 nm) and PL (λ_{ex} = 330 nm) spectra of BMASO:0.3%Eu^{2+} phosphor are also shown as a reference. Reprinted with permission from Ref. [50] by permission of The Royal Society of Chemistry

The investigators also demonstrated that a wide-range tunable white light emission from cool to warm white with high quantum yields (37–47 %) could also be obtained by properly controlling the contents of the three dopants [49].

(c) White-emitting triply-doped $NaCaBO_3$:Ce^{3+}, Mn^{2+}, Tb^{3+} phosphor. The single-phased tri-chromatic white light–emitting phosphor $NaCaBO_3$:Ce^{3+}, Mn^{2+}, Tb^{3+} was reported by Zhang and Gong [27]. A wide-range-tunable trichromatic emission was obtained by precisely controlling the Ce^{3+}, Mn^{2+}, and Tb^{3+} dopant contents, which is attributed to the efficient resonance-type $Ce^{3+} \rightarrow Tb^{3+}$/$Ce^{3+} \rightarrow Mn^{2+}$ energy-transfer processes. Figure 2.13 shows the dependence of PL on Tb^{3+} content in $NaCaBO_3$:1%Ce^{3+}, 3%Mn^{2+}, z%Tb^{3+} (z = 0, 1, 3, 5, 7, 9) and the optimization of dopant compositions. The optimal composition of the phosphor was found to be $NaCaBO_3$:1%Ce^{3+}, 3%Mn^{2+}, 5%Tb^{3+}, which emits white light on a 347-nm near-UV excitation. This white emission has a lower color temperature of 4898 K, and its CIE coordinates (0.344, 0.313) are closer to those of warm white light.

(d) White-emitting triply-doped $BaMg_2Al_6Si_9O_{30}$:Eu^{2+}, Tb^{3+}, Mn^{2+} phosphor [28]. The second type of triply-doped white-emitting phosphor contains mainly a Eu^{2+}/Tb^{3+}/Mn^{2+} or $Eu^{2+} \rightarrow Tb^{3+}$/$Eu^{2+} \rightarrow Mn^{2+}$ system to generate white light through the energy-transfer process. Figure 2.14 illustrates the relative energy level schemes of Eu^{2+}/Tb^{3+}/Mn^{2+} and the possible

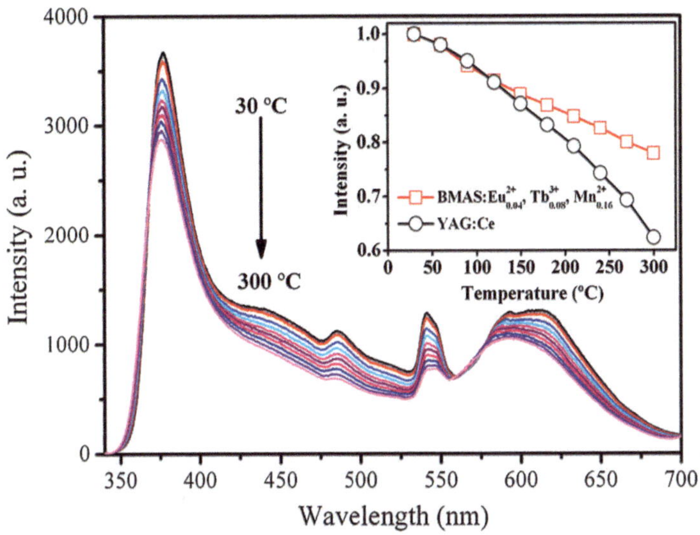

Fig. 2.15 Temperature-dependent emission spectra of the optimized BMAS:4%Eu^{2+}, 8%Tb^{3+}, 16%Mn^{2+} sample. Reprinted with the permission from Ref. [28]. Copyright 2011 American Chemical Society

energy-transfer models for Eu^{2+}→Tb^{3+} and Eu^{2+}→Mn^{2+} in a variety of hosts [50]. The PLE ($ë_{em}$ = 450 nm) and PL ($ë_{ex}$ = 330 nm) spectra of $BaMg_2Al_6Si_9O_{30}$:Eu^{2+}, Tb^{3+}, Mn^{2+} is also shown on the left in Fig. 2.14 as a reference.

As shown in Fig. 2.14, the 5*d* band of Eu^{2+} overlaps partially the excited-state energy levels of Mn^{2+} ($^4A_1$, ^4E (^4G)) and Tb^{3+} ($^5D_3$), thus indicating that the two energy-transfer processes, Eu^{2+}→Mn^{2+} and Eu^{2+}→Tb^{3+}, may occur by way of a nonradiative transition. When the Eu^{2+} ion is excited to higher component of 5*d* level by UV light, it relaxes to the lowest 5*d* crystal field-splitting level non-radiatively, and consequently it returns to the ground state ($4f^7$) with a 450-nm emission. Because they have similar values of energy levels, an energy transfer is expected to take place from the excited 5*d* state of Eu^{2+} to $^4A_1$, ^4E (^4G) level of Mn^{2+} and ($^5D_3$) level of Tb^{3+} in the $BaMg_2Al_6Si_9O_{30}$ host. The Mn^{2+} ion receives the energy transferred from the excited Eu^{2+}, and the Mn^{2+} ion is then excited into the $^4A_1$ and ^4E (^4G) energy levels from the ground state. Furthermore, the excited Mn^{2+} relaxes to the excited $^4T_1$ (^4G) state through $^4T_2$ (^4G) intermediate energy level in a nonradiative process followed by a radiative transition from excited state of $^4T_1$ (^4G) to the ground state of $^6A_1$ (^6S) with a typical emission of Mn^{2+} located at 610 nm [28]. The Eu^{2+}→Tb^{3+} energy transfer is similar to the Eu^{2+}→Mn^{2+} energy transfer. The excited Eu^{2+} ions transfer energy to the $^5D_3$ level of Tb^{3+} ions; subsequently, the $^5D_3$ level gives its characteristic transitions or continues to transfer the energy to the $^5D_4$ level by way of cross-relaxation. Then a set of characteristic transitions of $^5D_4$→$^7F_{3–6}$ attributed to Tb^{3+} are observed. By properly tuning the relative composition of Tb^{3+}/Mn^{2+} in $BaMg_2Al_6Si_9O_{30}:Eu^{2+}$, Tb^{3+}, Mn^{2+}, the chromaticity coordinates of (0.31, 0.30), high color-rendering index $R_a = 90$, and correlated color temperature (CCT) = 5,374 K can be achieved on the excitation of UV light. Figure 2.15 shows the temperature-dependent PL spectra of to investigate the thermal-quenching behavior of $BaMg_2Al_6Si_9O_{30}$:4%Eu^{2+}, 8% Tb^{3+}, 16%Mn^{2+}. This investigation reveals excellent quenching characteristics of the current phosphors better than that of YAG:Ce.

2.5 Summary and Perspectives

In summary, this chapter has reviewed four categories of single-phased white light-emitting phosphors with potential applications for WLEDs. We have also proposed the design protocols on the basis of bonding nature of hosts and the resonant-type energy-transfer principle described previously. The luminescence properties and energy-transfer processes for the four types of singled-composition phosphors and a multi-composition phosphor have been discussed. As a result, light emission can be realized by adopting single-composition two-complementary or three-primary phosphors under ultraviolet excitation such as blue/yellow-emitting ($CaAl_2Si_2O_8:Eu^{2+}$, Mn^{2+}), aqua/red-emitting ($BaGa_4S_7:Eu^{2+}$, Mn^{2+}), or blue/green/red-emitting ($SrZn_2(PO_4)_2:Eu^{2+}$, Mn^{2+}) phosphors. On the other hand, white-light generation can also be realized by blue radiation coupled with a green-yellow/red-emitting phosphor ($Ba_2ZnS_3:Ce^{3+}$, Eu^{2+}), which will generate warm white light under ultraviolet excitation. Based on the observed significant overlap between the emission spectrum of the sensitizer and the excitation spectrum of the activator, the energy-transfer mechanism is undoubtedly an electric dipole–dipole interaction when the sensitizer and activator are both allowed transitions, e.g., the Ce^{3+}/Eu^{2+} couple, whereas the electric dipole–quadrupole or exchange interaction occurs when the sensitizer and activator are allowed and forbidden transitions, respectively, e.g., the Eu^{2+}/Mn^{2+} couple. In general, the magnitude of critical distance for energy transfer can be compared and listed as electric dipole–dipole > electric dipole–quadrupole > exchange interaction. The most intriguing examples of white-light generation with high color-rendering index (R_a) through resonant energy transfer are trichromatic inorganic phosphors to convert ultraviolet (UV), near-UV, or blue light into a combination of RGB.

In the investigation of energy transfer in inorganic phosphors, the mismatch in decay lifetimes among the sensitizers and the activators has been noticed, and the effect of decay mismatching on the efficiency of energy transfer from the sensitizer to the activator is beyond the scope of this discussion and requires detailed studies in the future.

Acknowledgments We acknowledge the generous financial support to this research from the Ministry of Science and Technology of Taiwan (ROC) during the past years. Part of the content here is adapted from the Ph.D. dissertation of Dr. Yoan-Jen Yang of Department of Applied Chemistry, National Chiao Tung University, Taiwan.

References

1. Shimizu Y, Sakano K, Noguchi Y, Moriguchi T (1998) Japan Patent 10-56208
2. Bogner G, Botty IG, Braune B, Hintzen HT, van Krevel JWH, Waitl G (2003) Light source using a yellow-to-red-emitting phosphor. United States Patent 6,649,946
3. Hideki M, Koji K, Akihiko S (1999) Japan Patent 19990132924
4. Ropp RC (2004) Luminescence and the solid state, 2nd edn. Elsevier Publishers B. V., The Netherlands
5. Blasse G, Grabmaier BC (1994) Luminescent materials. Springer, Berlin, p 91
6. Douglas BE, McDaniel DH, Alexander JJ (1994) Concepts and models of inorganic chemistry, 3rd edn. Wiley, New York
7. Bartolo BD (1968) Optical interactions in solids. Wiley, New York
8. Dexter DL (1953) A theory of sensitized luminescence in solids. J Chem Phys 21:836
9. Blass G (1969) Energy transfer in oxidative phosphors. Philips Res Repts 24:131
10. Barry TL (1970) Luminescent properties of Eu^{2+} and Eu^{2+} +Mn^{2+} activated $BaMg_2Si_2O_7$. J Electrochem Soc 117:381
11. Rubio JO, Murrieta HS, Powell RC, Sibley WA (1985) $Eu^{2+}\rightarrow Mn^{2+}$ energy transfer in NaCl. Phys Rev B 31:59
12. Rubio JO, Munoz FA (1987) Energy transfer between europium and manganese close pairs in monocrystalline sodium bromide. Phys Rev B 36:8115
13. Rubio JO (1989) $Eu^{2+}\rightarrow Mn^{2+}$ energy transfer in monocrystalline NaCl and NaBr. Phys Rev B 39:1962
14. Camarillo E, Rubio JO (1989) A spectroscopic study of NaI containing europium and manganese. J Phys: Condens Matter 1:4873
15. Caldino UG, Munoz AF, Rubio JO (1990) Luminescence and energy transfer in CaF_2 slightly doped with europium and manganese. J Phys: Condens Matter 2:6071
16. Caldino UG, Munoz AF, Rubio JO (1993) Energy transfer in $CaCl_2$: Eu: Mn crystals. J Phys: Condens Matter 5:2195
17. Mendez A, Ramos FL, Riveros H, Camarillo E, Caldino UG (1999) Energy transfer mechanisms in the KBr: Eu^{2+}: Mn^{2+} phosphor. J Mater Sci Lett 18:399
18. Lin H, Liu X (1996) Chinese. J Lumin 17:121
19. Tan Y, Shi C (1999) $Ce^{3+}\rightarrow Eu^{2+}$ energy transfer in $BaLiF_3$ phosphor. J Phys Chem Solids 60:1805
20. Najafov H, Kato A, Toyota H, Iwai K, Bayramov A, Iida S (2002) Effect of Ce co-doping on $CaGa_2S_4$: Eu phosphor: I. energy transfer from Ce to Eu ions. Jpn J Appl Phys 41:1424
21. Lin H, Liu XR, Pun EYB (2002) Sensitized luminescence and energy transfer in Ce^{3+} and Eu^{2+} codoped calcium magnesium chlorosilicate. Opt Mater 18:397
22. Caldino UG (2003) Energy transfer in CaF_2 doped with Ce^{3+}, Eu^{2+} and Mn^{2+} ions. J Phys: Condens Matter 15:7127
23. Setlur AA, Srivastava AM, Comanzo HA, Doxsee DD (2004) Phosphor blends for generating white light from near-UV/blue light-emitting devices. The United States Patent 6,685,852
24. Kim JS, Jeon PE, Choi JC, Park HL, Mho SI, Kim GC (2004) Warm-white-light emitting diode utilizing a single-phase full-color $Ba_3MgSi_2O_8$: Eu^{2+}, Mn^{2+} phosphor. Appl Phys Lett 84:2931
25. Kim JS, Jeon PE, Park YH, Choi JC, Kim HL, Kim GC, Kim TW (2004) White-light generation through ultraviolet-emitting diode and white emitting phosphor. Appl Phys Lett 85:3696

26. Huang CH, Chen TM (2011) A novel single-Composition trichromatic white-light $Ca_3Y(GaO)_3(BO_3)_4$: Ce^{3+}, Mn^{2+}, Tb^{3+} phosphor for UV-LED. J Phys Chem C 115(5):2349
27. Zhang X, Gong M (2014) Single-phased white-light-emitting NaCaBO3: Ce^{3+}, Tb^{3+}, Mn^{2+} phosphors for LED applications. Dalton Trans 43:2465
28. Lü W, Zhang Z, Zhang X, Luo Y, Wang C, Zhang J (2011) Tunable full-color emitting BaMg2Al6Si9O30: Eu^{2+}, Tb^{3+}, Mn^{2+} phosphors based on energy transfer. Inorg Chem 50:7846
29. Shannon RD, Prewitt CT (1969) Effective ionic radii in oxides and fluorides. Acta Crystallogr B 25:925
30. Yao GQ, Lin JH, Zhang L, Lu GX, Gong ML, Su MZ (1998) Luminescent properties of $BaMg_2Si_2O_7$: Eu^{2+}, Mn^{2+}. J Mater Chem 8:585
31. Yang WJ, Chen TM (2005) Luminescence of Eu- and Mn-coactivated $CaAl_2Si_2O_8$ as a potential white-light phosphor for UVLED. Chem Mater 17:3883
32. Yang WJ, Chen TM (2006) Luminescence and energy transfer of Eu- and Mn-coactivated $SrZn_2(PO_4)_2$ as a single-composition white-emitting phosphor for UVLEDs. Appl Phys Lett 88(10):101903
33. Yang WJ, Chen TM (2006) Unpublished results
34. Chang CK, Chen TM (2007) White-light generation under violet-blue excitation from tunable green-to-red emitting $Ca_2MgSi_2O_7$: Eu, Mn through energy transfer. Appl Phys Lett 90 (16):161901
35. Chang K, Chen TM (2007) Unpublished results
36. Liu WR, Chiu YC, Yeh YT, Jang SM, Chen TM (2009) Luminescence and energy transfer mechanism in Ca10K(PO4)7: Eu^{2+}, Mn^{2+} phosphor. J Electrochem Soc 156(7):J165
37. Huang CH, Chen TM (2010) Ca9La(PO4)7: Eu^{2+}, Mn^{2+}: an emission-tunable phosphor through efficient energy transfer for white light-emitting diodes. Opt Express 18(5):5089
38. Huang CH, Chen TM, Liu WR, Chiu YC, Yeh YT, Jang SM (2010) A single-phased emission-tunable phosphor $Ca_9Y(PO_4)_7$: Eu^{2+}, Mn^{2+} with efficient energy transfer for white light-emitting diodes. ACS Appl Mater Interf 2(1):259
39. Huang CH, Liu WR, Chen TM (2010) Single-phased white-light phosphors $Ca_9Gd(PO_4)_7$: Eu^{2+}, Mn^{2+} under near-ultraviolet excitation. J Phys Chem C 114:18698
40. Huang CH, Chan TS, Liu WR, Wang DY, Chiu YC, Yeh YT, Chen TM (2012) Crystal structure and blue-white-yellow color-tunable $Ca_4Si_2O_7F_2$: Eu^{2+}, Mn^{2+} phosphor through energy transfer for single-phased white-light near-ultraviolet LEDs. J Mater Chem 22:20210
41. Yang WJ, Chen TM (2007) Ce^{3+}/Eu^{2+} codoped Ba_2ZnS_3: a blue radiation-converting phosphor for white light-emitting diodes. 90(17):171908
42. Chang CK, Chen TM (2007) Unpublished results
43. Xiao F, Xue YN, Zhang QY (2009) Ca_2BO_3Cl: Ce^{3+}, Eu^{2+}: a potential tunable yellow-white-blue-emitting phosphors for white light-emitting diodes. Phys B 404:3743
44. Chang CK, Chen TM (2007) $Sr_3B_2O_6$: Ce^{3+}, Eu^{2+}: a potential single-phased white-emitting borate phosphor for ultraviolet light-emitting diodes. Appl Phys Lett 91(8):081902
45. Hsu CH, Das S, Lu CH (2012) Color-tunable, single-phased $MgY_4Si_3O_{13}$: Ce^{3+}, Mn^{2+} phosphors with efficient energy transfer for white-light-emitting diodes. J Electrochem Soc 159(5):J193
46. Müller M, Jüstel T (2014) On the luminescence and energy transfer of white emitting $Ca_3Y_2(Si_3O_9)_2$: Ce^{3+}, Mn^{2+} phosphor. J Lumin 155:398
47. Liu Y, Xia Zhang X, Hao Z, Wang X, Zhang J (2011) Tunable full-color-emitting $Ca_3Sc_2Si_3O_{12}$: Ce^{3+}, Mn^{2+} phosphor via charge compensation and energy transfer. Chem Commun 47:10677
48. Li G, Geng D, Sheng M, Peng C, Cheng Z, Lin J (2011) Tunable luminescence of Ce^{3+}/Mn^{2+}-coactivated $Ca_2Gd_8(SiO_4)_6O_2$ through energy transfer and modulation of excitation: potential single-phase white/yellow-emitting phosphors. J Mater Chem 21:13334

49. Li G, Geng D, Shang M, Zhang Y, Peng C, Cheng Z, Lin J (2011) Color tuning luminescence of Ce^{3+}/Mn^{2+}/Tb^{3+}-triactivated $Mg_2Y_8(SiO_4)_6O_2$ via energy transfer: potential single-phase white-light-emitting phosphors. J Phys Chem C 115:21882
50. Jiang L, Pang R, Li D, Sun W, Jia Y, Li H, Fu J, Li C, Zhang S (2015) Tri-chromatic white-light emission from a single-phase $Ca_9Sc(PO_4)_7$: Eu^{2+}, Tb^{3+}, Mn^{2+} phosphor for LED applications. Dalton Trans 44:17241

Chapter 3
Energy Transfer Between Luminescent Centers

Jiahua Zhang

Abstract This chapter presents an elementary analytical method of energy transfer between luminescent centers based on spectroscopy data. It includes observation of energy transfer as well as determination of efficiency, mechanism, critical distance, and coefficient of energy transfer.

3.1 Introduction

Energy transfer from one luminescent center to another is a common process in phosphors. The quantitative theories of energy transfer have been given by Förster [1] and Dexter [2]. The luminescent centers include rare-earth ions, transition-metal ions, and color centers, etc. The luminescent centers doped in a phosphor may either be identical or different. Energy transfer between the identical centers result in the well-known phenomena of concentration quenching. Energy transfer from one type of center to another type of center then produces sensitized luminescence. Sensitized luminescence refers to energy transfer that enables a luminescent center (activator) having no appreciable absorption band covering a given excitation wavelength to emit radiation when another type of center (sensitizer) is effectively excited at the given wavelength. This is the typical result of energy transfer from the sensitizer to the activator. The sensitizer is also called a "*donor*," which is denoted hereafter by D. The activator is also called an "*acceptor*" and is denoted hereafter by A.

As the luminescent centers D and A are codoped in a phosphor, energy transfer may allow both D and A to emit, thus generating two color emissions on a single color excitation. Such a doubly doped system, or even triply doped system, exhibits attractive application for full-color white-light generation on blue and/or UV LED (light-emitting diode) excitation.

J. Zhang (✉)
Changchun Institute of Optics, Fine Mechanics and Physics,
Chinese Academy of Sciences, Changchun, China
e-mail: zhangjh@ciomp.ac.cn

R.S. Liu (ed.), *Phosphors, Up Conversion Nano Particles, Quantum Dots and Their Applications*, DOI 10.1007/978-3-662-52771-9_3

3.2 Spectroscopic Evidence for Energy Transfer

In D–A energy-transfer systems, pronounced photoluminescence (PL) of A can be observed as only D is excited. As a result, the photoluminescence excitation (PLE) spectrum of A contains the characteristic excitation spectrum of D. This is strong spectroscopic evidence for the occurrence of D–A energy transfer. The other spectroscopic effect of energy transfer is an intensity decrease of D luminescence followed by an intensity increase of A luminescence.

Figure 3.1 [3] shows PL and PLE spectra of α-$Ca_2P_2O_7$:Eu^{2+} (a), α-$Ca_2P_2O_7$:Mn^{2+} (b), and α-$Ca_2P_2O_7$:Eu^{2+}, Mn^{2+} (c). One finds the Eu^{2+} singly doped sample exhibits a strong blue emission band centered at 416 nm with its PLE bands within the UV and near UV regions originating from the $4f \rightarrow 5d$ transition of Eu^{2+}. The Mn^{2+} singly doped sample exhibits a weak orange emission band peaking at 600 nm due to the $^4T_1 \rightarrow {}^6A_1$ transition of Mn^{2+} with its PLE peaks at 355 and 406 nm corresponding to the forbidden transitions from the ground state $^6A_1(^6S)$ to $^4T_2(^4D)$ and $[^4A_1(^4G), {}^4E(^4G)]$ levels of Mn^{2+}, respectively. As expected, the emission spectrum of Eu^{2+} and Mn^{2+} codoped α-$Ca_2P_2O_7$ generates not only the blue band of Eu^{2+} but also a strong orange band of Mn^{2+}. It is clearly shown that the

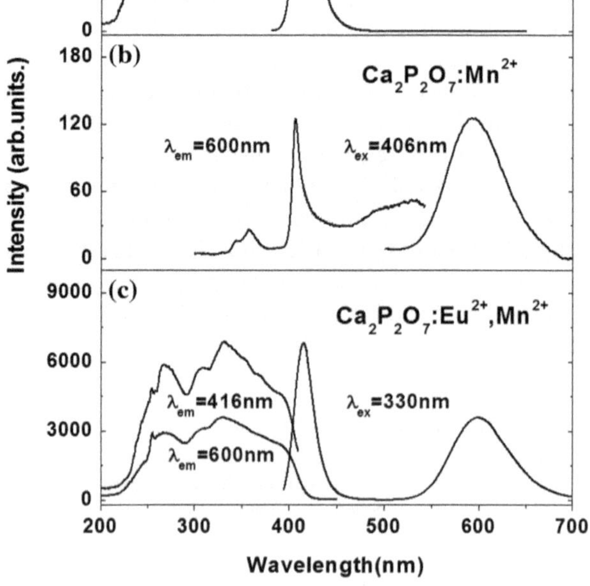

Fig. 3.1 PL and PLE spectra of α-$Ca_2P_2O_7$:Eu^{2+} (**a**), α-$Ca_2P_2O_7$:Mn^{2+} (**b**), and α-$Ca_2P_2O_7$:Eu^{2+}, Mn^{2+} (**c**). Reprinted with permission from Ref. [3]. Copyright 2007, American Institute of Physics

PLE spectrum monitoring the orange band of Mn^{2+} reflects the characteristics of Eu^{2+} in the codoped material, thus demonstrating effective energy transfer from Eu^{2+} (D) to Mn^{2+} (A). The PLE spectrum covers the spectral region of 225–425 nm, thus implying that the doubly doped phosphor is suitable to near-UV LED excitation.

One can also see that the PL intensity of donor Eu^{2+} is decreased followed by considerable enhancement of acceptor Mn^{2+} emission in the codoped material with respect to their singly doped materials.

3.3 Efficiencies of Donor Luminescence and Energy Transfer

If D–A energy transfer occurs, the transfer obviously can decrease the intensity of D luminescence because a part of excitation energy of D is transferred to A. Accordingly, the efficiency of energy transfer (η_{ET}) can be determined by comparing the intensity of D luminescence in the presence of A with that in the absence of A. It is simple and convenient to use the steady-state PL measurement for determination of η_{ET}. Note that only D is excited in the measurement. If the intensity of the D steady-state luminescence is S_0 in the absence of A and S in the presence of A, the D luminescence yield (η) is simply given by

$$\eta/\eta_0 = S/S_0 \tag{3.1}$$

where η_0 is the D luminescence yield in the absence of A. Thus, the efficiency of energy transfer can be calculated by

$$\eta_{ET} = 1 - S/S_0 \tag{3.2}$$

The D–A transfer can speed up the decay of D in its excited state and thus shorten the decay time of D. As a result, η_{ET} can also be obtained from the decay curve of the D luminescence after flash excitation of D. The decay curve represents the decay of transient luminescence intensity, which can be regarded as the number of photons emitted per time at time t. Denote the D-luminescence decay function by $I_0(t)$ for the absence of A and by $I(t)$ for the presence of A. The decay function can be obtained by normalizing the initial intensity of the experimentally measured luminescence decay curve, meaning $I_0(0) = I(0) = 1$. The integrals of the intensity-decay function over time are proportional to the number of total photons emitted after excitation, i.e., it is just the measure of the steady-state luminescence intensity. From Eqs. (3.1) and (3.2), one thus has [4]

$$\eta/\eta_0 = \frac{\int_0^\infty I(t)\mathrm{d}t}{\int_0^\infty I_0(t)\mathrm{d}t} \tag{3.3}$$

$$\eta_{\rm ET} = 1 - \frac{\int_0^\infty I(t)\mathrm{d}t}{\int_0^\infty I_0(t)\mathrm{d}t} \tag{3.4}$$

Here we introduce τ_0 and τ, defined by

$$\tau_0 = \int_0^\infty I_0(t)\mathrm{d}t \tag{3.5}$$

$$\tau = \int_0^\infty I(t)\mathrm{d}t \tag{3.6}$$

Note that if $I_0(t)$ is a single exponential decay function; and τ_0 is just the decay time. Here, Eqs. (3.3) and (3.4) can be rewritten as

$$\eta/\eta_0 = \frac{\tau}{\tau_0} \tag{3.7}$$

$$\eta_{\rm ET} = 1 - \frac{\tau}{\tau_0} \tag{3.8}$$

Figure 3.2 [3] shows the luminescence decay curves of Eu^{2+} in α-$Ca_2P_2O_7$:0.04Eu^{2+}, xMn^{2+}. The inset presents the dependence of the energy-transfer efficiency, which is denoted by η_T, on Mn^{2+} concentration x. The transfer efficiency was calculated using Eqs. (3.5), (3.6), and (3.8). With increasing Mn^{2+} concentrations, the energy-transfer efficiencies increase gradually and reach to as high as 70 % for Mn^{2+} concentrations of 14 mol%.

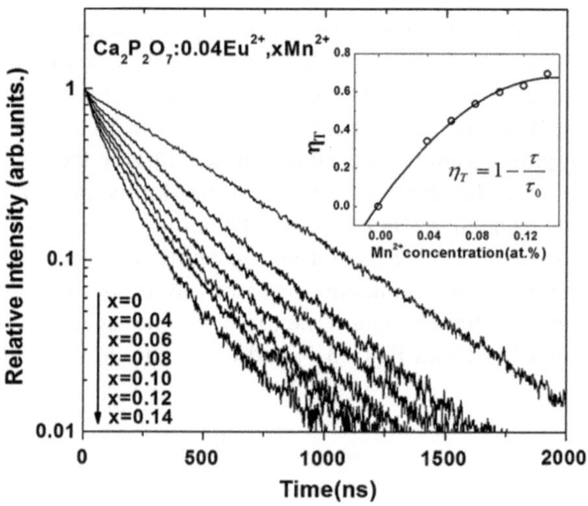

Fig. 3.2 Luminescence decay of Eu^{2+} in α-$Ca_2P_2O_7$:0.04Eu^{2+}, xMn^{2+}. The *inset* shows the dependence of energy-transfer efficiency in α-$Ca_2P_2O_7$:0.04Eu^{2+}, xMn^{2+} on Mn^{2+} content x. Reprinted with permission from Ref. [3]. Copyright 2007, American Institute of Physics

3.4 Lifetimes

3.4.1 Excited-State Lifetime

The excited-state lifetime refers to the average time the molecule stays in its excited state after the removal of excitation. Suppose D is excited with flash light at $t = 0$. When no A is present, the excited-state population of D decays exponentially after the excitation.

$$N_0(t) = N_0(0)\exp(-t/\tau_0) \tag{3.9}$$

where $N_0(t)$ and τ_0 are the excited-state population of D at time t and the excited state lifetime in the absence of A, respectively. The inverse of the lifetime is the total decay rate of excited D.

When A is present, the decay rate of excited D is increased by energy transfer, which is strongly dependent on the D–A distance. If D and A ions distribute randomly in space, there is a transfer-rate distribution. This means a lifetime distribution resulting in multiexponential or nonexponential decay of the D excited-state population $N(t)$ in the presence of A. In this case, the average excited-state lifetime ($\langle\tau\rangle$) can be calculated by

$$\langle\tau\rangle = \frac{\int_0^{\infty} N(t)\mathrm{d}t}{N(0)} \tag{3.10}$$

In case of exponential decay for the absence of A as described in Eq. (3.9), one obtains

$$\langle\tau\rangle = \tau_0 \tag{3.11}$$

In isolated luminescent centers, the radiative transition is a single molecular process, i.e., the radiative rate is a constant. If all of the centers have the same radiative rate, for instance, in the case of identical centers, the intensity is proportional to the excited-state population. This means that the intensity-decay function is identical to the decay function of the excited-state population. In this case, the right side of Eq. (3.10) is equal to the integral of the intensity decay function expressed as.

$$\langle\tau\rangle = \int_0^{\infty} I(t)\mathrm{d}t = \tau \tag{3.12}$$

In an energy-transfer system, the multiexponential or nonexponential decay of D luminescence generally results from different transfer rates. The radiative rate is the same for all of the donors, thus indicating that D-luminescence intensity is proportional to its excited-state population. Therefore, Eq. (3.12) is suitable for calculating the average lifetime of the excited donors in an energy-transfer system.

If a nonexponential decay of luminescence results from different radiative rates, then $I(t)$ no longer follows the same function as $N(t)$, achieving the average excited-state lifetime using Eq. (3.12) is invalid. Luminescent materials with different radiative rates may be caused by codoping different types of luminescent centers or the same center on different lattice sites or mixing different phosphors, etc.

3.4.2 Fluorescence Lifetime

In the case of exponential decay of fluorescence intensity, determination of the fluorescence lifetime is simple. In the case of complex decay, one may find an average fluorescence lifetime ($\bar{\tau}$) defined by [4]

$$\bar{\tau} = \frac{\int_0^\infty tI(\mathrm{t})\mathrm{d}t}{\int_0^\infty I(t)\mathrm{d}t} \tag{3.13}$$

The denominator is the number of total photons emitted; and the numerator is the sum of the delay times for each emitted photon after flash excitation. Hence, $\bar{\tau}$ means the average delay time for emitting a photon after excitation. The physical meaning of $\langle\tau\rangle$ as defined by Eq. (3.10) is different from that of $\bar{\tau}$ as defined by Eq. (3.13). Only in case of exponential decay are the values of both identical.

$$\bar{\tau} = \langle\tau\rangle = \tau_0 \tag{3.14}$$

One finds the transfer efficiency calculated by Eq. (3.8) using $\bar{\tau}$ instead of τ, which is not correct.

3.5 Theory of Energy Transfer

D–A energy transfer takes place generally by multipolar or exchange interaction. Both mechanisms rely on the separation between D and A. The former is a long-range interaction through the Coulomb field, and the later is a short-range interaction through the overlapping of electronic wave functions of D and A. Quantitative theories of resonant energy transfer by electric multipole–multipole interaction or exchange interaction have been given by Förster and Dexter.

3.5.1 Electric Multipolar Interaction

In electric multipolar interaction, the D–A transfer rate is proportional to an inverse power of distance r, which is written as

$$W_{\mathrm{ET}} = \frac{\alpha}{r^m} \tag{3.15}$$

where α is a rate constant for energy transfer and m = 6, 8, and 10 corresponds to electric dipole–dipole interaction, electric dipole–quadrupole interaction, and electric quadrupole–quadrupole interaction, respectively.

In case of no diffusion among donors, or the average D–D transfer rate being much less than the average D–A transfer rate, the decay function of D luminescence is given by Inokuti and Hirayama [4]

$$I(t) = \exp\left[-\frac{t}{\tau_0} - \frac{4\pi}{3}\Gamma\left(1 - \frac{3}{m}\right)x\alpha^{3/m}t^{3/m}\right] \tag{3.16}$$

where x is the concentration of A. From Eq. (3.16), $\ln\{\ln[I_0(t)/I(t)]\}$ is proportional to $\ln(t)$ with a slope of $3/m$, and $\ln[I(t)/I_0(t)]$ is proportional to $t^{3/m}$ with a slope of $-4\pi\Gamma(1-3/m)x\alpha^{3/m}/3)$ where $I_0(t) = \exp(-t/\tau_0)$. The parameters m and α can be obtained from the slops by fitting the measured decay curve. It is important to note that the fitting should be focused on the tail of the decay because Eq. (3.16) is obtained by assuming the nearest distance between a donor and an acceptor to be 0, thus leading to an infinite initial transfer rate.

Transfer critical distance (R_c) is an important parameter. This distance is defined as the distance between a donor and an acceptor for which the transfer rate is equal to the decay rate of the excited D in the absence of energy transfer; then

$$\frac{\alpha}{R_c^m} = \tau_0^{-1} \tag{3.17}$$

If m and α are obtained by fitting optical data using Eq. (3.16), then R_c can be calculated by Eq. (3.17). Here we present an example [5] of energy transfer from Ce^{3+} to Pr^{3+} in YAG ($Y_3Al_5O_{12}$). The luminescence decay of Ce^{3+} is speeded up with increasing Pr^{3+} concentration and becomes nonexponential as shown in Fig. 3.3. This is the typical effect of energy transfer.

The analysis of the measured decay curves based on Eq. (3.16) is performed as shown in Fig. 3.4. In Fig. 3.5, the parameters m, α, and R_c are determined to be 6, $\alpha = 4.5 \times 10^{-36}$ cm^6 s^{-1}, and 0.81 nm, respectively.

The energy transfer between Pr^{3+} ions is also observed. A pronounced decay-time shortening of the red $^1D_2 \rightarrow {}^3H_4$ emission of Pr^{3+} with the increase of Pr^{3+} concentration is shown in Fig. 3.6 [5]. Using the same method as mentioned previously, the fitting of decay curves are performed as shown in Figs. 3.7 and 3.8. The parameters m, α, and R_c for Pr^{3+}–Pr^{3+} energy transfer (generally called "energy

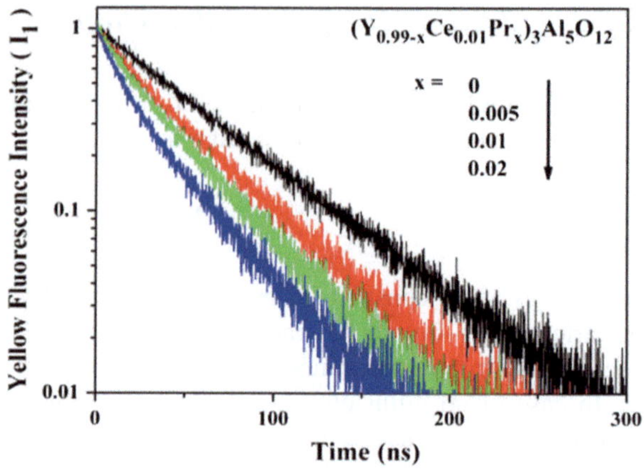

Fig. 3.3 Decay curves of Ce^{3+} *yellow* emission in $(Y_{0.99-x}Ce_{0.01}Pr_x)_3Al_5O_{12}$ for $x = 0$, 0.005, 0.01, and 0.02. Reprinted with permission from Ref. [5]. Copyright 2010, American Institute of Physics

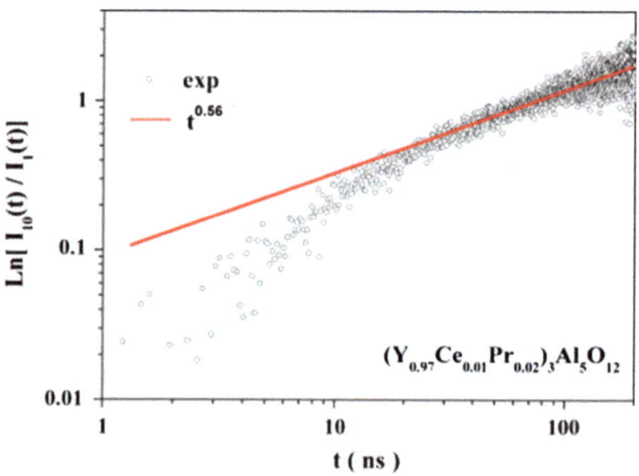

Fig. 3.4 ln-ln plot of ln $[I_{10}(t)/I_1(t)]$ versus t for Ce^{3+} *yellow* emission in $(Y_{0.99-x}Ce_{0.01}Pr_x)_3Al_5O_{12}$ with $x = 0.02$. Reprinted with permission from Ref. [5]. Copyright 2010, American Institute of Physics

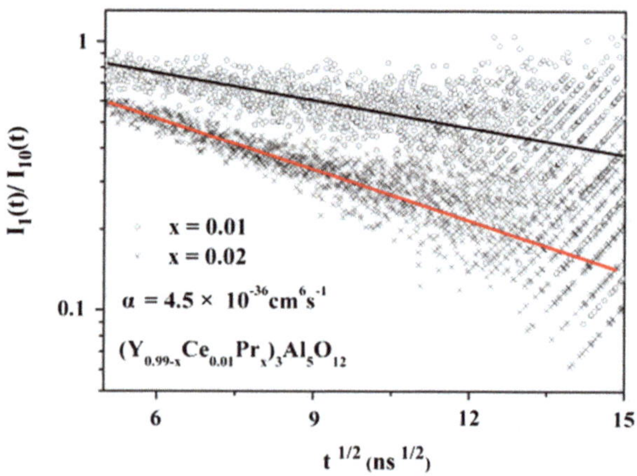

Fig. 3.5 Plotted $\ln[I_1(t)/I_{10}(t)]$ versus $t^{1/2}$ for Ce^{3+} *yellow* emission in $(Y_{0.99-x}Ce_{0.01}Pr_x)_3Al_5O_{12}$ with $x = 0.01$ and 0.02. Reprinted with permission from Ref. [5]. Copyright 2010, American Institute of Physics

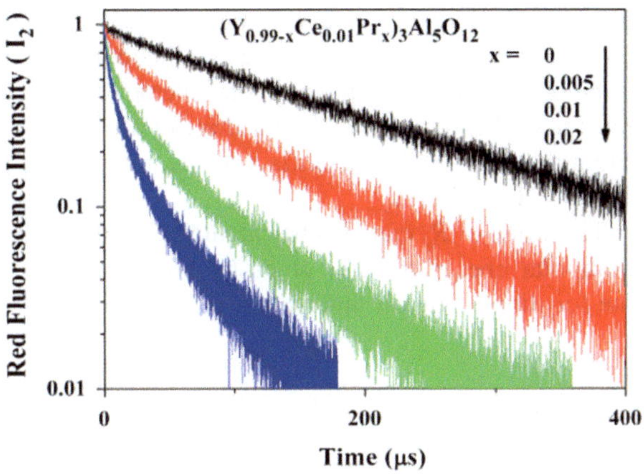

Fig. 3.6 Decay curves of Pr^{3+} *red* emission in $(Y_{0.99-x}Ce_{0.01}Pr_x)_3Al_5O_{12}$ for $x = 0$, 0.005, 0.01, and 0.02. Reprinted with permission from Ref. [5]. Copyright 2010, American Institute of Physics

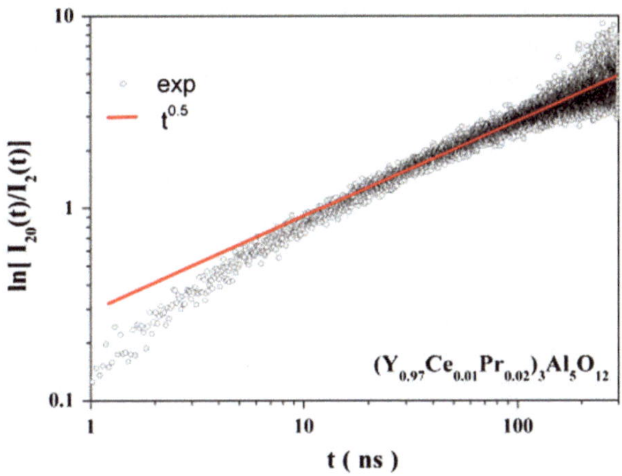

Fig. 3.7 ln-ln plot of ln $[I_{20}(t)/I_2(\mathrm{t})]$ versus t for Pr^{3+} emission in $(Y_{0.99-x}Ce_{0.01}Pr_x)_3Al_5O_{12}$ with $x = 0.02$. Reprinted with permission from Ref. [5]. Copyright 2010, American Institute of Physics

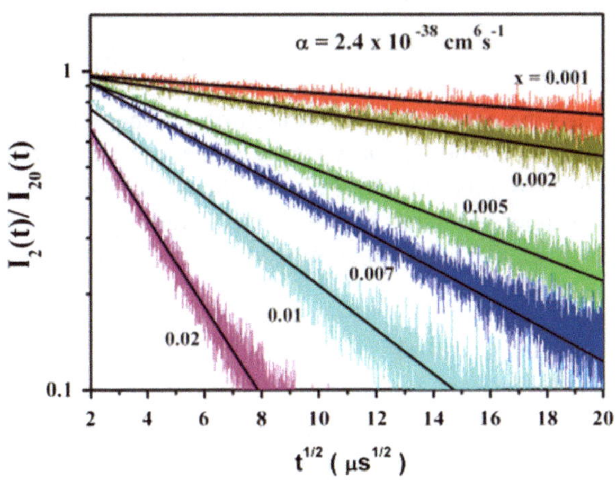

Fig. 3.8 Plotted $\ln[I_2(t)/I_{20}(\mathrm{t})]$ versus $t^{1/2}$ for Pr^{3+} emission in $(Y_{0.99-x}Ce_{0.01}Pr_x)_3Al_5O_{12}$ with various x ($x = 0$–0.02). Reprinted with permission from Ref. [5]. Copyright 2010, American Institute of Physics

migration for identical centers") are obtained. They are 6, 2.4×10^{-38} cm^6 s^{-1}, and 1.3 nm, respectively. One can see that the transfers for Ce^{3+}–Pr^{3+}, and for Pr^{3+}–Pr^{3+} all belong to electric dipole–dipole interaction. The critical distance for Pr^{3+}–Pr^{3+} transfer (1.3 nm) is longer than that for Ce^{3+}–Pr^{3+} transfer (0.81 nm), thus indicating that Pr^{3+}–Pr^{3+} transfer is a strong competitive process with respective to Pr^{3+} $^1D_2 \rightarrow {}^3H_4$ emission.

The value of R_c can also be calculated by using the quantitative theories for D–A multipolar energy transfer given by Förster and Dexter. For instance, the transfer rate for dipole –dipole interaction is given by the Dexter equation [2]:

$$W_{ET} = \frac{3\hbar^4 c^4 \sigma_A}{4\pi n^4 \tau_D r^6} \int \frac{f_D(E) F_A(E)}{E^4} dE \quad (3.18)$$

where n is the refractive index of the material; σ_A is the absorption cross-section of A; and τ_D is the radiative lifetime of D, which can be regarded as τ_0 if the decay of excited D in the absence of energy transfer is governed by radiative process rather than multiphonon relaxation. $f_D(E)$ and $F_A(E)$ represent the shape of D emission and A absorption spectra, respectively, which are normalized in spectral area. The integrals are the energy overlap of the two spectra. One can see from Fig. 3.1 that the PL spectrum of α-$Ca_2P_2O_7$:Eu^{2+} along with the PLE spectrum of α-$Ca_2P_2O_7$:Mn^{2+} exhibit a significant spectral overlap of Eu^{2+} to Mn^{2+}. The overlap means that the transfer is a resonant-energy transfer, it is necessary that the transition energies of D and A be equal. If $W_{ET}\tau_D = 1$ is inserted into Eq. (3.18), this equation can be solved for R_c.

There is another simple method for determining R_c from the critical concentration (x_c) of acceptors for the D–A system. The critical concentration is defined as the concentration at which the efficiency of energy transfer is equal to the efficiency of D luminescence, i.e.. the efficiency of D luminescence reaches one-half of its efficiency in the absence of A. The value of x_c can be obtained with steady-state luminescence measurement as the intensity of D luminescence is decreased to one-half of its intensity in the absence of A, which is expressed as

$$S(x_c) = S_0/2 \quad (3.19)$$

At the acceptor concentration of x_c, one may think there is, on average, one A ion per volume $V/x_c n$, where V is the volume of the crystallographic unit cell, and n is the number of lattice sites in the unit cell that can be occupied by A. The radius of a sphere centered at D with this volume can be calculated. Then the D–A critical distance, R_c, is proximately equal to the radius derived from

$$R_c = \left(\frac{3V}{4\pi x_c n}\right)^{1/3} \quad (3.20)$$

Let us consider energy transfer between identical centers, each of which can act as D and A. Energy transfer between two identical centers may result in concentration quenching. If the critical concentration of concentration quenching of the luminescence can be determined experimentally, the radius of a sphere with a volume occupied on average by one center is achievable using Eq. (3.20). When a space is fully filled by spheres, the critical distance between the two centers is proximately twice the radius of the sphere [6]:

$$R_c = 2\left(\frac{3V}{4\pi x_c n}\right)^{1/3} \tag{3.21}$$

3.5.2 Exchange Interaction

Exchange interaction is a short-range interaction because overlapping of the electron clouds of D and A is required. The rate of exchange interaction is given by [2]:

$$W_{ET} = \left(\frac{2\pi}{\hbar}\right) K^2 \exp\left(-\frac{2r}{L}\right) \int f_D(E) F_A(E) dE \tag{3.22}$$

where K is a constant with dimension of energy; L is a constant called "the effective average Bohr radius"; and the integrals are the spectral overlap of D emission and A absorption. It is convenient to rewrite Eq. (3.22) by introducing R_c in the form:

$$W_{ET} = \tau_0^{-1} \exp[2(R_c - r)/L] \tag{3.23}$$

The decay of D luminescence is given by [4]

$$I(t) = \exp\left[-\frac{t}{\tau_0} - \gamma^{-3} \frac{4\pi}{3} R_c^3 g\left(e^{\gamma} \frac{t}{\tau_0}\right) x\right] \tag{3.24}$$

where $\gamma = 2R_c/L$, g() is a function.

The exponential decay of transfer rate indicates that the exchange interaction is very sensitive to distance. The exchange transfer prefers the nearest or the second nearest D–A pairs. Such a behavior of exchange interaction is similar to the Perrin model, in which the transfer rate is assumed to be constant if A exists within a critical distance and is zero outside the range. This model is thought to be the most applicable to the exchange interaction. In the Perrin model, the steady-state luminescence of D as a function of concentration of A is simply written as [4]

$$S/S_0 = \exp(-x/x_c) \tag{3.25}$$

We have discussed resonant energy transfer. If the spectral overlap of D and A is small, D–A energy transfer by both electronic multipolar or exchange mechanisms requires phonon emission or absorption for compensating the energy difference between D and A transitions; this is called "phonon-assisted energy transfer." In this case, the transfer rate is strongly influenced by the energy separation, phonon energy, and temperature, and the reader is referred to [7] for details.

3.5.3 *Diffusion-Limited Energy Transfer*

In the previous discussion, the energy transfer between donors is not taken into account. The energy transfer between identical centers is called "energy migration." This migration speeds up the D–A transfer rate. Yokota and Tanimoto [8] obtained a general solution for the donor decay function including both diffusion within the D system and the D–A energy transfer by way of dipole–dipole coupling. Their expression is as follows:

$$I(t) = \exp\left[-\frac{t}{\tau_0} - \frac{4}{3}\pi^{\frac{3}{2}}x(\alpha t)^{\frac{1}{2}}\left(\frac{1 + 10.87y + 15.50y^2}{1 + 8.743y}\right)^{\frac{3}{4}}\right] \quad (3.26)$$

where $y = D\alpha^{-1/3}t^{2/3}$, D is the energy diffusion constant between donors depending on donor concentration. If the migration rate is much faster than the D–A transfer rate, for instance, in case of high D concentration, the D–A transfer rate can maintain its initial rate over time, thus resulting in a single exponential decay of D luminescence [9]. This is called "rapid diffusion-limited energy transfer."

References

1. Forster Th (1948) Zwischenmolekulare energiewanderung und fluoreszenz. Ann Physik 2:55
2. Dexter DL (1953) A theory of sensitized luminescence in solids. J Chem Phys 21:836
3. Hao Z, Zhang J, Zhang X, Sun X, Luo Y, Lu S (2007) White light emitting diode by using α-$Ca_2P_2O_7$:Eu^{2+}, Mn^{2+} phosphor. Appl Phys Lett 90:261113
4. Inokuti M, Hirayama F (1965) Influence of energy transfer by the exchange mechanism on donor luminescence. J Chem Phys 43:1978
5. Wang L, Zhang X, Hao Z, Luo Y, Zhang J (2010) Interionic energy transfer in $Y_3Al_5O_{12}$:Ce^{3+}, Pr^{3+} phosphor. J Appl Phys 108(1):093515
6. Blasse G (1968) Energy transfer in oxidic phosphors. Phys Lett 28A:444
7. Miyakawa T, Dexter DL (1970) Phonon sidebands, multiphonon relaxation of excited states, and phonon-assisted energy transfer between ions in solids. Phys Rev B 1:2961
8. Yokota M, Tanimoto O (1967) Effects of diffusion on energy transfer by resonance. J Phys Soc Jpn 22:779
9. Huber DL (1979) Fluorescence in the presence of traps. Phys Rev B 20(6):2307

Chapter 4
Principles of Energetic Structure and Excitation-Energy Transfer Based on High-Pressure Measurements

Marek Grinberg

Abstract In this chapter, we summarize recent accomplishments in the area of high-pressure luminescence spectroscopy of phosphor materials. The effect of pressure on the luminescence related to f-f, d-d, and d-f transitions is discussed. Several recent examples from the literature are presented to illustrate the influence of pressure on luminescence energy, intensity, lineshape, luminescence kinetics, and luminescence efficiency. Especially, the unique ability of pressure to investigate the influence of impurity-trapped exciton states, which are created after ionization and charge-transfer transitions, on the luminescence of TM and RE ions in solids and energy-transfer processes are presented.

4.1 Introduction

Luminescent solid-state materials comprised of insulating host lattices (normally oxides, fluorides, nitrides, and oxy-nitrides) activated by rare-earth and transition-metal ions continue to be an active area of research due to their application as phosphors, scintillators, and functional materials.

The optical properties of metal ion–doped insulators are controlled by the interaction between the metal ion and the host lattice and by interactions between metal-ion dopants. The dopant–lattice interaction establishes a local dopant-bonding environment as well as the strength of coupling of the dopant ion to lattice vibrations. These factors determine the energy-level positions, crystal-field splitting, coupling of electronic states, and nonradiative-decay pathways that are responsible for the wavelength, lineshape, intensity, and dynamics of absorption and emission. The most important interactions occur in the nearest neighbor coordination shell where factors—such as bond length, bond angles, covalence, and coordination number—determine the energy, splitting, and mixing of the electronic

M. Grinberg (✉)
Faculty of Mathematics, Physics and Informatics, Institute of Experimental Physics,
University of Gdańsk, Wita Swosza 57, 80-952 Gdańsk, Poland
e-mail: fizmgr@ug.edu.pl

R.S. Liu (ed.), *Phosphors, Up Conversion Nano Particles, Quantum Dots and Their Applications*, DOI 10.1007/978-3-662-52771-9_4

states involved in the luminescence. Local interactions are also responsible for the location of the transition-metal and lanthanide states in the band gap of the host lattice. The design of luminescence materials for new applications requires a systematic understanding of the relationship between the bonding environment of the luminescence centers and their optical properties such as absorption and emission, excitation-energy transfer, and efficiency of luminescence. The customary approach for investigating these relationships is to vary the chemical composition of the host. Examples of this approach include the spectroscopic studies of a given transition metal dopant in a series of host lattices.

The alternative approach to gain information on structure and local environment on luminescent transition-metal and lanthanide ions is based on using high hydrostatic pressure to systematically influence the surrounding luminescence center.

The primary effect of pressure is to decrease the volume and increase the crystal-field strength through bond-length compression. As a consequence, the pressure provides an ability to tune the crystal-field strength of dopants and systematically examine the variation of optical properties with crystal-field strength. High pressure also influences the energy of metal ion–defect states in the band gap of a host lattice relative to the band edges. This effect alters the activation energy as well as the efficiency of energy transfer from the host lattice to the dopant and provides an opportunity to identify and resolve competing excitation and decay pathways in many phosphors and scintillators.

A series of chemically inequivalent, isostructural host lattices, representing a set of discrete points along the crystal-field strength and band-structure continua, is obtained when the crystal-field strength and band structure is altered by the chemical composition of the host lattice. An important feature of high pressure is its ability to continuously vary the crystal-field strength. As a result, high-pressure studies offer a meaningful complement to host-lattice studies because of the finer resolution of crystal-field strength and band structure that they provide. This finer resolution is especially beneficial for studying the properties of materials near boundaries corresponding to abrupt changes in properties (e.g., at phase transitions, crossover points of electronic energy levels, or resonance of dopant states with host-lattice band edges). High-pressure studies also avoid potential complications associated with differences in dopant distribution, multisite behavior, and chemical impurities that may exist over a series of host lattices.

Chemical-composition and high-pressure studies share the goal to identify the physical and chemical factors that control the optical properties of the materials. The final objective is to progress toward the rational design of novel phosphors with required properties. High-pressure spectroscopic study also provides fundamentally unique information on quantum mechanical mixing of the localized states related to transition-metal and lanthanide ions with the band-delocalized states of the host lattice, which are responsible for the creation of intermediate states of the impurity-trapped exciton.

In the following review, we restrict our attention to the luminescence properties of transition-metal and rare-earth ions in insulating inorganic solids. At the beginning, the high-pressure phenomenological effects on the properties of solid state and the equipment for high-pressure generation and control are described. Then, the influence of pressure on basic luminescence properties is discussed. The main part of the review concerns the recent literature, which presents a range of phenomena observable at high pressure where special attention is paid to inorganic phosphors materials doped with transition-metal and rare-earth ions. We conclude with several remarks about the future.

4.2 High-Pressure Generation and Equipment

4.2.1 Hydrostatic Pressure as Experimental Variable

High hydrostatic pressure, such as temperature, electric, and magnetic fields, is one of the external parameters that can alter the system and, such as temperature and fields, can be used for tuning their physical properties. The basic effect of pressure on the system is a consequence of thermodynamic stability described as the second law expressed as [1, 2]

$$\left(\frac{\partial V}{\mathrm{d}p}\right)_T < 0 \tag{4.2.1}$$

which states that the volume of material decreases in isothermal process. The effect of pressure on a system depends on the magnitude of pressure and on isothermal compressibility K (i.e., reciprocal bulk modulus B_o)

$$K = \frac{1}{B_0} = -\frac{1}{V}\left(\frac{\partial V}{\mathrm{d}p}\right)_T \tag{4.2.2}$$

Gasses are very compressible; therefore, pressure of a few bars can strongly influence gas properties. In contrast to gasses, condensed phases (liquids and solids) experience strong intermolecular interactions and are much less compressible. As the result, pressure on the order of kbar is necessary to noticeably alter the properties of condensed matter.

The precise manner in which material reduces its volume in response to applied pressure is described by the equation of state, which relates volume reduction with the bulk modulus of the material B_o and its first derivative with respect to pressure B_o'. The most frequently used equation is universal the Birch–Murnaghan equation of state (EOS) [3–5] which can be expressed as follows:

$$p = \frac{3}{2} B_o x^{\frac{5}{3}} (x^{\frac{2}{3}} - 1)\left[1 + \frac{3}{4}\left(B_o^{'} - 4\right)\left(x^{2/3} - 1\right)\right] \tag{4.2.3}$$

where

$$x = \left(\frac{V}{V_o}\right)^{1/3} \tag{4.2.4}$$

and V, and V_o are the volume of the material at a given pressure and at ambient pressure, respectively. Equation (4.2.3) relates the pressure generated in the material by reduction of the volume to the volume. The simpler equation of state valid for uniform pressure, which describes the volume reduction as the uniform response of the material to an applied pressure, which was delivered by Murnaghan [6] is often used [7]:

$$\frac{V}{V_o} = \left(\frac{pB_o'}{B_o} + 1\right)^{-\frac{1}{B_o'}} \tag{4.2.5}$$

The relations (4.2.3) through (4.2.5) describe the macroscopic quantities such as volume and density. The manner in which pressure influences other properties of matter, such as chemical, electrical, optical, and magnetic properties, requires consideration on the microscopic molecular level.

Actually, microscopic molecular attributes—such as intermolecular forces, molecular angles and lengths, and types of chemical bonds—are responsible for energetic structure and spectroscopic properties of the system. Reducing the volume pressure continuously varies all microscopic attributes of the solid. Alternatively, pressure induces rapid changes that are characteristics for phase transitions. In both cases, continuous and rapid changes in high-pressure experiments allow correlation of the spectroscopic and other properties of the material to bond length and crystal structure. Quantitative analysis of influence of pressure on the crystal structure and energetic-band structure usually involves extended quantum mechanical calculations such as ab initio calculations of band structure [8, 9] and total energy calculations [3, 5] where pressure effect is simulated according to the EOS.

In the case of localized states related to the dopants—including transition-metal (TM) and rare-earth (RE) ions—the effect of hydrostatic pressure on the local energetic structure and electronic transitions is simulated by the reduction of the size of the cluster, which includes the dopant ion and the ligands. The influence of pressure on the energetic structure of the TM and RE ions can be simulated using crystal-field phenomenological model calculations [10–12]. However, a more advanced approach, i.e., ab initio model potential embedded-cluster method, also has been used [13].

4.2.2 High-Pressure Cells

Methods and instruments that have been used to generate high pressure are probably more diverse than those in other fields of instrumentation. Depending on the particular purpose, different devices are used for the generation of static hydrostatic pressure and also different for generating uniaxial stress. Different types of pumps and compressors are used for the generation of hydrostatic pressure <6 kbar, whereas higher pressure is generated in piston-cylinder systems. The extended reviews on high pressure–generation instruments and methods can be found in monographs [14, 15].

The most commonly used equipment for the generation of pressure >10 kbar is a diamond anvil cell (DAC). Diamond is known as the hardest substance and also it is quite transparent to x-ray and visible light. This is a reason why DACs have been used since the 1950s for performing high-pressure x-ray diffraction [16, 17]. Since then, DACs of different types have been successfully used for IR spectroscopic measurements, visible spectroscopy including photoluminescence, and Raman as well as Brillouin scattering. DACs were also used to carry out resistivity and Mössbauer high-pressure measurements. There are several review articles devoted to application of DACs [18–20].

The basic principle of the modern DAC for optical experiments is simple [14, 21]. The construction consists of two brilliant-cut stones beveled at the tip (culet) with their larger faces (table) resting on the hard sits. See Fig. 4.1. A metal gasket with a drilled hole placed between culets contains a sample and a pressure sensor embedded in a pressure-transmitting medium.

A sample is subjected to pressure when a force pushes two opposite anvils together. Variation in the DAC arises from the different designation of the anvil alignment and the force-generating mechanisms [19, 22]. The principal idea is based on a piston-cylinder assembly with different mechanisms of force generation. One of the simplest and therefore infallible systems is the Merrill–Bassett (MB) DAC [23], presented in Fig. 4.1a, b, in which the force is generated by screws that pull two plates together when tightened. The alignment is assured by the same screws supported by additional walls situated perpendicular to the plates. The sample and pressure sensor is excited by UV from one side, and photoluminescence can be collected in backscattering and and/or transmitting mode.

The MB DAC is relatively small; therefore, it can be easy placed in a standard cryostat or in a magnetic field. In spectroscopic experiments with DAC of an MB system, the sample is excited by light and information on the physical properties of the sample, as well as the value of pressure, is available from the analysis of optical signals.

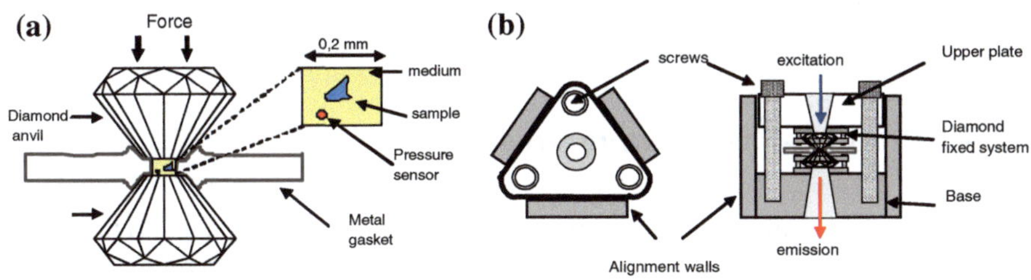

Fig. 4.1 **a** The principle of DAC. **b** The Merrill–Bassett optical DAC design and performance

4.2.3 Anvils and Gaskets

Diamond is the hardest material known to man. The hardness of diamond on the Mohs hardness scale is 10. The second hardest material is corundum (sapphire [Al_2O_3]), which has a hardness of 9. Corundum, as well as diamond, is transparent to visible radiation. The real difference in hardness is quite large because the absolute hardness of diamond is 7,000 Knoop and of the corundum is 2,000 Knoop [24]. Bulk modulus of diamond is 4420 kbar [25] but is only 2400 kbar for synthetic sapphire. However, because the cost of sapphire is much lower than that of the diamond and also sapphire is more transparent to UV light, sapphire anvil cells (SAC) are often used for an optical cell for pressure <100 kbar. Sometimes cubic zirconia anvils are also used for pressure <30 kbar [26].

Diamond-polishing is a specialized technique. Two principal designs of diamond cutting are available on the market: (i) high pressure end–surface flat and (ii) high pressure end–faces beveled (see Ref. [15]). Typically diamonds vary from 1/8 to 1/2 carat depending on the cell type. The culet diameter *d* ranges from 0.2 to 0.7 mm. The design of sapphire cutting is usually simpler [15], and synthetic sapphire also usually has a larger culet. The important for sapphire anvil cell (SAC) performance is that the both culet and the table should be perpendicular to the c-axis of the sapphire crystal.

The maximum pressure that can be generated by DAC is 5000 kbar, whereas only 100 kbar can be generated with SAC [27, 28]. Sapphire as well as diamond anvils should be selected with respect to transparency and lack of luminescence and fluorescence.

The choice for gasket material should be related to the anvil material as well as the pressure range. The hard steels—e.g., Inconel X750, tempered T30, stainless steel (RC45), and Waspalloy—are used for gasket materials for diamond anvils [15]. In the case of sapphire anvils, copper gaskets are more suitable. The initial thickness of the metal sheet is 100–250 μm. It is important to indent the gasket blank by compression in the cell and then drill the hole in the center of the indentation. The hole diameter should not be larger than 0.6 indentations.

The maximum pressure generated by DAC is determined by the gasket material and the diameter of the culet. The limit is not precisely defined. Different, but similar, phenomenolgical rules for the inverse and inverse-square relationships P_{max} (kbar) = 100/d (mm) [14] and P_{max} (kbar) = $125/d^2$ (mm^2) [21], respectively, have been proposed.

4.2.4 Pressure-Transmitting Media

If the sample is a liquid or a gas at ambient conditions, it can be loaded to the gasket hole directly. If the sample is a solid, the pressure-transmitting medium should be used to ensure homogeneous pressure distribution in the sample chamber. In general, the pressure-transmitting media can be soft solids, liquids, or gasses at ambient pressure. The extended review on the properties of different transmitting media can be found in [15].

The required properties to be looked for in an ideal medium include the following: (1) zero shear strength, (2) chemically inert, (3) zero penetration into the sample and the construction of high-pressure apparatus, (4) zero compressibility, (5) easy to handle (including nontoxic and nonflammable), (6) easy to seal within the high-pressure enclosure, and (7) inexpensive and readily available. In addition, for spectroscopic experiments, only optically neutral materials without intrinsic absorption and emissions are acceptable.

Of course, no materials meet all of these requirements; for each experiment the choice should be made in terms of adequate performance depending on pressure range and temperature. It is clear that no material satisfies the requirement (4), but it is enough to use the medium for which compressibility is lower or comparable with the compressibility of the gasket. The last requirement excludes spectroscopic measurements for most solids, such as metals, which absorb light and many of the alkali and silver halides, the optical properties of which are related to lattice defects and can depend on pressure.

Due to above-mentioned requirements, the most frequently used pressure-transmitting media in spectroscopy are He, Ne, Ar, Xe, He_2, D_2, N_2, and O_2 [19] which are gasses under ambient conditions. All of them need cryogenic filling at ambient pressure. Liquids (e.g., methanol–ethanol mixture [29], methanol, ethanol, water mixture [30], oils [31–33], and other organic materials (e.g., propanol, glycerin, daphne 7474 n-pentane:isopentane [1:1 mixture], and petroleum ether) [34]) are the most convenient for application. Important information on the selected materials can be found in Table 4.1.

The idea of high-pressure experiments is to alter the system by uniform hydrostatic pressure. Therefore, the ideal medium would distribute the pressure homogenously in the sample chamber in response to the force exerted by anvils. Any nonhydrostatic effects, such as pressure gradients and shear stress, should be eliminated or reduced as much as possible because they alter the system in an uncontrolled way. According to the Pascal rule, hydrostaticity is achieved because the medium maintains fluidity. When pressure increases, the liquid medium solidifies to a glassy or crystalline phase. After solidification, the mechanical strength of the medium can create pressure gradients and stresses. The crucial parameter is the yield strength, and only liquids that have low-yield strengths after solidification are suitable for pressure transmission.

The process of selecting pressure-transmitting media is largely empirical. Determination of the hydrostatic limit, which is related to the pressure of solidification, is crucial. Several methods exist that can detect the phase transition of a liquid medium. The most important include (1) gigahertz ultrasonic interferometry [35], (2) x-ray diffraction line-broadening using a single quartz crystal embedded in the investigated medium [36], and (3) the oldest one, which is based on deviations of ruby fluorescence spectra [29]. The gigahertz experiments provide the information on the solidification pressure, which is not equivalent to the lack of

Table 4.1 Useful data on selected pressure media

Medium	Freezing pressure at RT (kbar)	Melting temperature at ambient pressure (K)	Pressure range of nearly hydrostatic behavior (kbar)	Reference
He	118	0.95	>600	[19]
Ne	47	24.56	160	[19]
Ar	12	83.81	300	[33]
Xe		161.40	200	[39]
Kr		115.78		[15]
CO_2	6	194.7[a]		[15]
H_2	57	19.95	>600	[19]
D_2	53		>600	[19]
N_2	24	63.15	130	[40]
O_2	59	54.36		[41]
Methanol:ethanol (4:1 ratio)	104[b]		200	[29]
Methanol:ethanol: water (16:3:1 ratio)	145[b]		200	[30]
Silicate oil Dow Corning 200			150	[31]
Paraffin oil	50–70[b]		150	[32]
Polydimethylsiloxanes $[(CH_3)_3\ SiO[SiO(CH_3)_2]_nSi(CH_3)_3$	100[b] [33] 9[c] [37]	<188	640	[33]

[a]Sublimation temperature, [b] glass transition pressure, [c]hydrostatic limit pressure from x-ray diffraction

hydrostaticity. The two other methods yield very different results: X-ray diffraction method provides hydrostatic limits for much lower pressures than the spectroscopic methods [37]. The extreme example is polydimethylsiloxane, where the hydrostatic limit obtained from ruby spectroscopy is 100 kbar [33]; however, for x-ray experiments it is only 9 kbar. This ambiguity in determination of solidification pressure is discussed in [34] and explained as being related to the difference in the value of the bulk modulus, which is equal to 2530 kbar for ruby [38] and only 371 kbar for quartz [37]. As a result, quartz is more sensitive than ruby in terms of pressure gradients and shear-strength. Actually, according to the result of one paper [33], polydimethylsiloxane is applicable for spectroscopic high-pressure experiments ≤ 650 kbar. This also concerns other liquids. According to the data in Table 4.1, many substances, especially silicate oils even after solidification, are suitable as pressure-transmitting media for spectroscopic investigations.

4.2.5 High-Pressure Sensors

Determination of the value of high hydrostatic pressure inside the DAC is performed indirectly by the measurements of pressure-induced changes of the magnitudes of physical parameters of a known system (pressure sensor) encapsulated in the DAC together with the investigated sample.

The main developments in experimental techniques for measuring high pressure to obtain reliable pressure sensors are extensively discussed by Decker et al. [42]. These include (1) the establishment of a primary pressure scale using a free piston gauge; (2) the selection and precise measurement of identifiable phase transitions as fixed pressure points; and (3) the use of interpolation and extrapolation techniques for continuous-pressure calibration based on changes in resistance, volume, or optical spectra (based on an equation of state). An alternative method of estimating absolute pressure in isotropically compressed materials is based on measurements of ultrasonic velocity [43, 44].

The sensor material and the detection method are selected depending on previously performed experiments. In x-ray experiments, pressure-induced changes in the unit-cell volume of standard materials are used. The most frequently used standard is NaCl [45–47], but other compounds (e.g., KCl and CsCl [46]) as well as metals (Cu, Mo, Ag, Pa, and Al [15, 48]) have been proposed.

The spectroscopic method is based on pressure-induced changes in absorption or/and emission spectra. The idea is to relate the pressure-induced shift of the fluorescence lines of the specific material to the value of the pressure. The material selected for the luminescence pressure sensor should be characterized by strong intensity of the emission line(s), which should be stable at a broad range of pressures and temperatures and the energy of which is possibly related linearly to pressure. It is also important that the emission of the sensor does not overlap the emission of the sample. Considering the above-mentioned requirements, the Raman fluorescence and photoluminescence of transition-metal and rare-earth ions were used. Raman modes of nitrogen [49], which is the pressure-transmitting medium, and Raman frequencies of diamond chips [51] have been used. Recently, a pressure-induced shift of the Raman line 1332 cm^{-1} of the face of the DAC culet was proposed to estimate pressure $\leq 1{,}000$ kbar [50, 52].

Several high-pressure luminescence gauges based on pressure shifts of sharp emission lines of RE and TM ions—such as Sm^{2+}, Eu^{3+}, Tm^{3+}, Nd^{3+}, and Cr^{3+}—in solids have been developed (see the Ref. in [2]). It was found that Sm^{2+} in $Y_3Al_5O_{12}$ (YAG) can be applied ≤ 1200 kbar [53, 54], whereas YAG:Nd^{3+} can be applied ≤ 120 kbar and work with a near-infrared pressure gauge [55, 56].

However, the most commonly used standard is the luminescence of ruby (Al_2O_3:Cr^{3+}) crystal. Ruby yields the sharp-lines emission that consists of the R_1 and R_2 lines (694.2 and 692.81 nm, respectively) related to the $^2E \rightarrow {}^4T_2$ transitions in Cr^{3+} ions, the energies of which decrease when pressure increases. Pressure-induced shifts of the R_1 and R_2 lines is almost linear with the pressure [57, 58]. The more precise dependence of R-line luminescence on pressure was investigated in the early 1970s [59], and the following numerical formula has been developed for dependence of the R_1-line wavelength on pressure [57, 58].

$$\frac{\Delta\lambda}{\lambda_0} = \left\{ [\frac{B}{A}p + 1]^{\frac{1}{B}} - 1 \right\} \tag{4.2.6}$$

where p is pressure; λ_0 is emission wavelength at ambient pressure; and A and B are constants and $\Delta\lambda = \lambda(p) - \lambda_0$, respectively. It should be noted that wavelengths of the R_1 and R_2 lines depend on temperature [60], and a weak dependence (the blue shift) of energies of the R_1 and R_2 lines on Cr^{3+} concentration was observed [61, 62]. Therefore, the values of A and B vary (A from 1820 to 1904 and B from 1 to 14) depending on the experimental conditions, and other calibration formulas have been proposed [38, 63]. For Cr^{3+} concentration >04 wt% apart of the R_1 and R_2 lines, additional lines related to Cr^{3+} pairs and clusters appear, the energies of which also change with pressure [63–65]. Using ruby crystal as the pressure sensor has an additional advantage. The gradient of pressure and shear stress in the DAC result in broadening of the R_1 and R_2 lines as well as an increase of R_1 R_2 splitting [29]. Thus luminescence of ruby can be also used as an indicator of the nonhydrostaticity of pressure.

In the last decades, a new type of pressure gauge, based on InAsP semiconductor quantum wells luminescence, was proposed [66]. Because InAsP quantum wells emit sharp-line luminescence peaking at 1200 nm, which is characterized by a pressure blue shift equal to approximately 9 nm/kbar, an InAsP quantum well system is very sensitive pressure indicator for medium pressure <100 kbar.

4.3 Fundamentals of High-Pressure Luminescence Phenomena

The energetic structure of a system of solid matrices with a luminescent center, usually comprising transition-metal (TM) or rare-earth ions (RE), consists of delocalized band states represented by a conduction band (c.b.) and a valence band (v.b.) as well as localized states of the luminescence center, i.e., the ground state $X^{\alpha+}$ and excited states $X^{\alpha+}$, and $X^{(\alpha-1)+}$ (X = TM, RE, and α is a valence of the ion). The example of such a structure is presented in the diagram shown in Fig. 4.2a. In the ground state of the system, the valence band is fully occupied, and the conduction band is empty. The highest occupied state can be the ground state of the luminescence center—either $X^{\alpha+}$ or $X^{(\alpha-1)+}$—which determines the position the Fermi level in the band gap. The typical situation is when the highest occupied state is $X^{\alpha+}$, and the Fermi level is located between the $X^{\alpha+}$ and $X^{(\alpha-1)+}$ states.

Possible electronic transitions that excite the system are represented by the arrows in Fig. 4.2a. One can distinguish the transitions where the electron does not change the localization and the transitions where the electron does change its

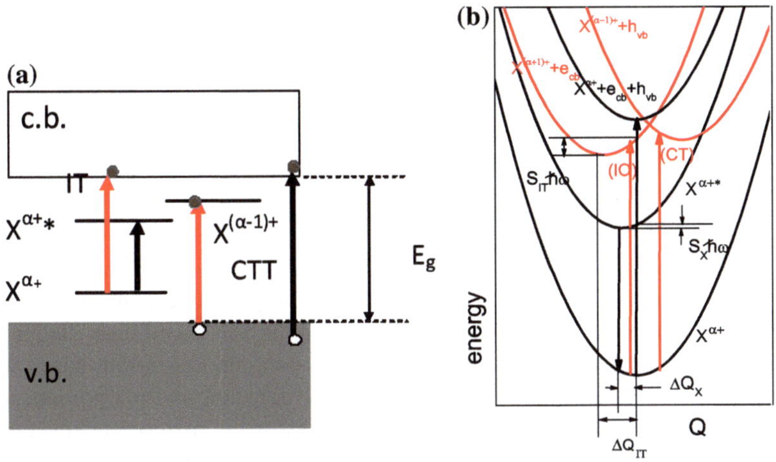

Fig. 4.2 **a** Energetic structure of the luminescence center in the lattice and possible electronic transitions. The CTT and the IT are indicated by *red arrows*; band-to-band and internal transitions in the impurity are indicated by *black arrows*. **b** Configurational coordinate diagram representing the ground and excited states of the system from Fig. 4.2a

localization. The first type is the valence band–conduction band transition, and internal transitions in the luminescence center are presented by black arrows. The second type of transition (red arrows) belongs to the ionization transition (IT), in which an electron is excited from the luminescence center to the conduction band and the charge-transfer transition (CTT) in which an electron is excited from the valence band to the luminescence center, particularly to the $X^{(\alpha-1)+}$ state.

A consequence of changes in electron localization is lattice relaxation. After the IT, there is one electron less in the luminescence center, which causes shrinkage of the lattice-surrounding center, whereas after CTT, one extra electron in the center causes the to expand. Lattice relaxation is presented in the configurational-coordinated diagram in Fig. 4.2b. The model is based on adiabatic approximation and harmonic potential wells. In such an approach, the electronic energies are represented by parabolas [10]. Configuration coordinate *Q* represents the average distance between the luminescence center and the negative ligands. The electronic manifold, labeled "$X^{(\alpha+1)+} + e_{cb}$," represents the ionized luminescence center and electron in the conduction band, whereas the state labeled "$X^{(\alpha-1)+} + h_{vb}$" represents the system after CTT with one extra electron at the luminescence center and a hole in the valence band. Shrinkage of the lattice results in diminishment of the minimum energy of the $X^{(\alpha+1)+} + e_{cb}$ state and the shift toward the smaller value of *Q*. In the case of lattice expansion, the reduction of the minimal energy of the $X^{(\alpha-1)+} + h_{vb}$ state is accompanied by the shift toward a larger value of *Q* [67, 68].

The electron-lattice relaxation is described by the quantity $S_i\hbar\omega$ where $\hbar\omega$ is a phonon energy, and S_i is a Huang–Rhys parameter, $i =$ X, IT, CTT. In the case of the IT and CTT states, *S* is much larger than 1 (usually >10), whereas for internal transitions S varies from 0 for transitions inside the $4f^n$ electronic configuration of RE to 4–5 for transitions between $4f^n$ and $4f^{n-1}5d$ electronic configurations of RE and for d-d transitions in TM. The shift of the electronic energy minimum in the configurational space can be calculated as follows:

$$\Delta Q_i = \sqrt{\frac{2S_i\hbar\omega}{K}} \tag{4.3.1}$$

where K is the elastic constant. Using the square root of electron–lattice interaction energy to describe the shift of the excited-state parabola with respect to the ground-state parabola, one can obtain the configuration coordinate diagram considering only energies of the maxima of the emission and absorption bands without the necessity of knowing the value of the elastic constant and phonon energy.

Pressure influences the spectroscopic properties of the material in several ways. First, it compresses the lattices and changes their volume. This bulk effect, described by the equation of state (see Eq. 4.2.5), also alters the energetic structure of the system, especially the energies of the conduction and valence bands. Such an effect can change the energy of the band-to-band transition. Compression also alters phonon frequencies, which is described by the Grüneisen relation [69]:

$$\gamma = -\frac{\partial \ln \omega}{\partial \ln V} \tag{4.3.2}$$

where γ is the Grüneisen parameter, and V is the volume of the matrix. Usually the energy of phonons increases with increasing pressure. This effect can be responsible for increasing the probability of nonradiative processes as well as changes in electron–lattice coupling described by $S\hbar\omega$.

The pressure-induced changes of the energies of localized states of dopant, i.e., the ground $X^{\alpha+}$ and the excited $X^{\alpha+*}$ states and the energies of the internal transitions, are important. In the case of RE ions, the localized states belong to the $4f^n$ and $4f^{n-1}5d$ electronic configurations. Because the electrons occupying the f orbitals are very effectively shielded from the influence of the crystal lattice by $5s^25p^6$ electrons, the small effect of pressure on $4f^n$–$4f^n$ transitions is anticipated regardless of orbital occupancy n. In contrast, the $4f^n \rightarrow 4f^{n-1}5d$ transitions include one configuration that interacts strongly with the lattice ($4f^{n-1}5d$) and one that interacts weakly with the lattice. As a result, the energies of these transitions are expected to vary with increasing pressure.

In the case of TM ions, the localized electrons occupy the 3d electronic orbitals. The 3d orbitals of TM are more spatially extended than the f orbital states of RE; in addition, 3d electrons are not shielded from the lattice. In consequence, their energies strongly depend on coordination number, local symmetry, and crystal-field strength and can be influenced to a great degree by pressure.

Pressure also can shift the energies of the localized states of the luminescence centers in the band gap and change the energies of the IT and the CTT. Obviously, because it is a localized electron–delocalized electron transition during the IT and

CTT process, the interaction with the lattices in the initial and the final state is different. However, because the pressure shifts of the energies of the band edges can compensate the shifts related to different electronic localization, it is not necessarily that strong dependence of the energies of the IT and the CTT on pressure appear.

4.3.1 *Pressure-Induced Shifts of the Band States*

The most frequently studied were semiconductors (IV-IV, III-VI, and II-VI) in which pressure causes changes the band-gap magnitude as well as induces the transition from the direct to indirect gap semiconductor [70]. Much less works was performed on the pressure dependence of the band structure of dielectrics. Lacomba-Perales et al. [9] studied the evolution of the band gap with pressure for AWO_4 (A = Ca, Sr, Ba, Pb) and found that in the low-pressure (<60 kbar) scheelite phase, the direct ($\Gamma \rightarrow \Gamma$) band gap changes slightly with pressure with the rates –0.21, 0.37, and 0.89 meV/kbar for $CaWO_4$, $SrWO_4$, and $BaWO_4$, respectively, whereas in the high-pressure fergusonite phase >100 kbar, the indirect band gap decreases more rapidly with the rates –7.3, –8.1, and –1.1 meV/kbar, respectively. In the case of $PbWO_4$ in both cases—the scheelite phase and the high-pressure $PbWO_4$-III phase—both the direct and indirect band gap diminish with pressure by –6.2 and –3 meV/kbar, respectively.

In the AVO_4 (A = Yb, Y, Lu and Nd), the band gaps increase in the low-pressure zircon phase (by 1–2 meV/kbar) and decrease with pressure in the high-pressure scheelite phase (by 0.7–2.2 meV/kbar) [71]. In wurtzite and rocksalt InN [72], the absorption threshold energy was found to increase by 3.0–3.2 meV/kbar, which corresponds to the increase of the band gap with increasing pressure. A similar dependence of 2.1–2.7 meV/kbar was obtained from the luminescence measurements [73].

4.3.2 *Pressure Dependence of Transition-Metal Ion Luminescence*

TM are the elements that do not have a completely filled 3d electronic manifold. When TM appears as an atom, the core electronic configuration is [Ar], and the valence electrons are $3d^n4s^m$ where n varies from 1 for Sc {$[Ar]3d^14\ s^2$} to 10 for Cu {$[Ar]3d^{10}4\ s^1$} and m is equal 1 or 2. Actually, because Sc exists only as Sc^{3+} in solid matrices' its 3d orbital is always empty, and because Cu is a typical metal, it exists as Cu^+ with a completely filled 3d orbital. The real TM ions that do not have completely filled 3d orbitals are $Ti^{2+}(3d^2)$, $Ti^{3+}(3d^1)$, $V^{2+}(3d^1)$, $V^{3+}(3d^2)$, $V^{4+}(3d^1)$, $Cr^{2+}(3d^4)$, $Cr^{3+}(3d^3)$, $Cr^{4+}(3d^2)$, $Cr^{5+}(3d^1)$, $Mn^{2+}(3d^5)$, $Mn^{3+}(3d^4)$, $Mn^{4+}(3d^3)$, $Mn^{5+}(3d^2)$, $Mn^{6+}(3d^1)$, $Fe^{2+}(3d^6)$, $Fe^{3+}(3d^5)$, $Fe^{4+}(3d^4)$, $Fe^{6+}(3d^2)$, $Co^{2+}(3d^7)$, $Co^{3+}(3d^6)$, Co^{4+} $(3d^5)$, $Ni^{2+}(3d^8)$, $Ni^{3+}(3d^7)$, and $Ni^{4+}(3d^6)$.

Hamiltonian that yields the energetic structure of localized states of TM ions is a sum of the Coulomb, exchange, and spin-orbit interactions of $3d^n$ electrons in the ion core and their interaction with the crystal surrounding. The free-ion energetic terms can be calculated using three Racah parameters—*A*, *B*, and C—as well as the spin-orbit coupling strength parameter, ξ. The states obtained are labeled by atomic terms ${}^{2S+1}L_J$, where $L = \sum_i l_i$ is the resultant angular momentum quantum number (l_i is the angular momentum of the *i*-th electron); $S = \sum_i s_i$ is the resultant spin (s_i is the spin of the *i*-th electron); and $J = L + S$ is the total angular momentum of the electrons [74].

In the crystal-field model, the interaction with the lattice is represented by the interaction with the neighboring charged ions, called "ligands," considered as point charges and the averaged potential of remaining ions. The crystal-field potential is given by:

$$V_{cr}(r) = \sum_{i=1}^{N} \frac{e^2 Z_i}{|r - R_i|} - \sum_{j=N+1}^{\infty} \frac{e^2 Z_j}{|r - R_j|} \tag{4.3.3}$$

where R_i and Z_i are the *i*-th ligand position and charge, respectively; and *N* is the number of ligands (coordination number).

The first term can be presented as a sum of Legendre polynomials $P_k(\cos\omega_i)$, and ω_i is an angle between *r* and R_i [75]

$$\frac{e^2 Z_i}{|r - R_i|} = e^2 Z_i \left\{ \frac{1}{R_i} + \sum_{k=1}^{\infty} \frac{r^k}{R_i^{k+1}} P_k(\cos\omega_i) \right\} \tag{4.3.4}$$

For negative ligands, the average spherical negative potential would destabilize the system by increasing the energy of each electron by the same quantity. Obviously this destabilization is compensated by an attraction term coming from the second sum that includes all remaining ions in the crystal. Thus, the effective crystal potential that influences the energy of the d electron is given by:

$$V_{cr}(r) = \sum_{k=1}^{\infty} \sum_{i=1}^{N} \frac{r^k}{R_i^{k+1}} P_k(\cos\omega_i) \tag{4.3.5}$$

To consider the splitting of the d state, it is convenient to present five orthogonal d orbitals in symmetrized form as d_{xy}, d_{yz}, d_{zx}, $d_{x^2-y^2}$, and d_{z^2}. Under interaction with an octahedral field of six ligands, the fivefold degenerated d state splits into a threefold degenerated t_{2g} state build with d_{xy}, d_{yz}, d_{zx} electronic orbitals and a twofold degenerated e_g state build with $d_{x^2-y^2}$, d_{z^2} electronic orbitals. In spectroscopy, the splitting energy is called "crystal-field strength" and is equal to 10 *Dq*, where

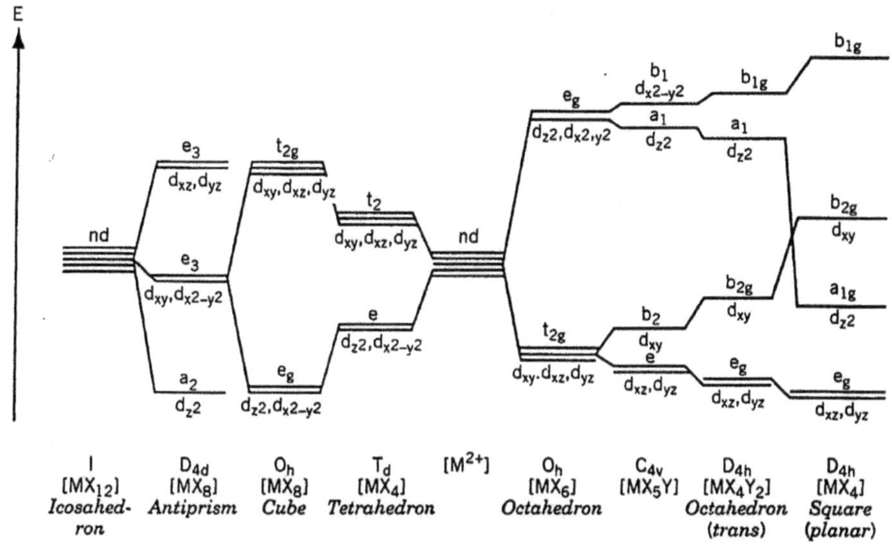

Fig. 4.3 Splitting of d-orbital energy levels in a crystal field of different symmetries. Reproduced from Ref. [74] by permission of John Wiley & Sons Ltd

$$D = \frac{35Ze^2}{4R^5} \text{ and } q = \frac{2}{105}\int r^4 \varphi_{3d}^2(r)\,\mathrm{d}r \qquad (4.3.6)$$

In Eq. (4.3.6), $\varphi_{3d}(r)$ is the 3d orbital radial wave function; and R is the average distance between the central ion and the ligands. It should be emphasized that the additional splitting appears when symmetry diminishes. Figure 4.3 shows the spitting of d-orbital energy levels in the crystal field of different symmetries.

It should be emphasized that in an octahedral field (octahedron with point-group symmetry O_h), the splitting is equal to 10 Dq, the lower state is the triply degenerated t_{2g} state, and the upper state is doubly degenerated e_g, whereas in a cubic field (cube with O_h point-group symmetry) and tetragonal field (T_d point-group symmetry), the lower state is the doubly degenerated e_g and e state, respectively. The spitting energy is equal to 80/9 Dq and 40/9 Dq for a cubic and a tetragonal field, respectively.

If more than one electron is captured at the d orbitals, the total energy is a sum of the energy of the Coulomb and exchange interaction and the crystal field. As the result, the degenerated atomic terms with the quantum number $J > 1$ will split. In the field described by point-group symmetry T_d and O_h, the energetic structure of the TM ion can be presented as a function of the parameter Dq/B in the form of Tanabe–Sugano diagrams [75].

High hydrostatic pressure compresses the lattice and diminishes the central ion–ligand distance R_i. In the first approximation, the pressure is isotropic; therefore, the reduction of R_i does not change the symmetry of the system. As the result, pressure causes an increase of the value of the parameters D and Dq. Relation (4.3.6) allows obtaining [76] the following:

$$\frac{\mathrm{d}Dq}{\mathrm{d}p} = -\frac{nDq}{R}\frac{\mathrm{d}R}{\mathrm{d}p} = \frac{nDq}{3B_0} \tag{4.3.7}$$

Diminishing of the electron–ligand distance R under pressure, as well as the pressure-induced changes in the phonon energies, also influences the electron–lattice interaction $S\hbar\omega$. When the interaction with a totally symmetric breathing mode is considered, the pressure dependence of the electron–lattice interaction is given by [76, 77]:

$$\frac{\mathrm{d}S\hbar\omega}{\mathrm{d}p} = -[2(n+1) - 6\gamma]\frac{S\hbar\omega}{R}\frac{\mathrm{d}R}{\mathrm{d}p} = [2(n+1) - 6\gamma]\frac{S\hbar\omega}{3B_0} \tag{4.3.8}$$

In Eqs. (4.3.7) and (4.3.8), n is the exponent that describes the dependence of crystal-field strength on the central ion–ligand distance R, and it should be equal 5. The pressure-induced changes of Racah parameters, spin–orbit coupling are much smaller than the changes of the crystal-field strength; therefore, significant changes of the electronic-transition energies are expected between the states belonging to the different electronic configurations (e.g., t_2 and e in a cubic field). In other cases, weak dependence of the energies on pressure are expected.

4.3.2.1 Pressure Dependence of Ti^{3+} ($3d^1$) Luminescence

Transition-metal ions with a $3d^1$ electronic configuration, such as Ti^{3+}, V^{4+}, Cr^{5+}, and Mn^{6+}, have the simplest energetic scheme. In an octahedral environment, fivefold degenerated d orbital splits into threefold degenerated ground state $^2T_{2g}$ and twofold degenerated excited state ^2E_g. The energetic distance between them is approximately represented by the crystal-field splitting 10 Dq. Both states are usually influenced by the Jahn–Teller effect related to the coupling to t he two-dimensional lattice mode, ε, which causes additional splitting of the ^2E_g state into $^2A_{1g}$ and $^2B_{1g}$ states and the $^2T_{2g}$ state into $^2B_{2g}$ and ^2E_g states as well as additional homogeneous broadening of the emission and absorption bands. A detailed description of the Jahn–Teller effect in the case of $Al_2O_3:Ti^{3+}$ and related systems are presented in [78, 79].

High hydrostatic pressure was applied to the Ti sapphire $Al_2O_3:Ti^{3+}$ [77, 80] and to $YAlO_3:Ti^{3+}$ [80]. In both materials, blue shifts of the emission and absorption bands were observed, which is in accordance with Eq. (4.3.7). To analyze the pressure-induced shifts of all bands, the interaction with totally symmetric lattice mode a_1 and two-dimensional lattice mode $\varepsilon(Q_u, Q_v)$ have been considered. In the paper [80], the contributions from the linear and quadratic Jahn–Teller effect have been considered.

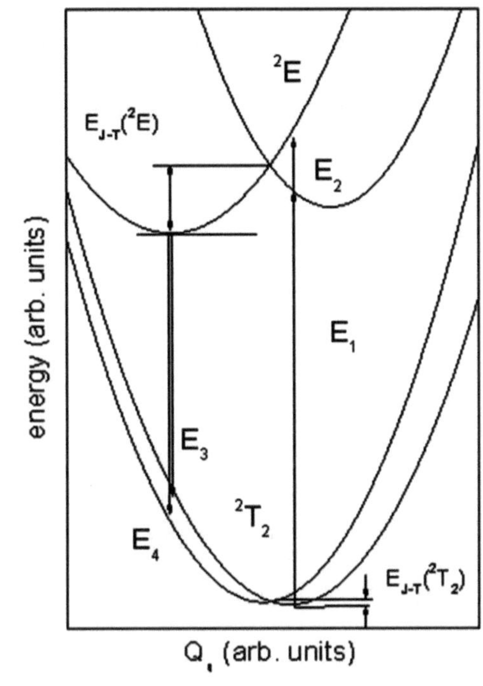

Fig. 4.4 Configurational coordinate diagram for the TiO_6 complex with the linear and quadratic Jahn–Teller effect. In the diagram, the cross-section along the Q_u variable is presented. Adapted from Ref. [80] by permission of Taylor & Francis Ltd

A configurational-coordinate diagram representing of the Jahn–Teller effect in $Al_2O_3:Ti^{3+}$ is presented in Fig. 4.4. As the result of Jahn–Teller splitting, both, the absorption and emission spectra consist of two overlapping broad bands, which correspond to transition to the upper and lower electronic manifolds. In Fig. 4.4, these transitions are indicated by arrows, the lengths of which are proportional to the energies of the maximums of the bands. Arrows labeled E_1 and E_2 correspond to energies of the absorption bands, and arrows labeled E_3 and E_4 corresponds to the energies of the emission bands.

The results obtained by García-Revilla et al. [77] for $Al_2O_3:Ti^{3+}$ showed that the pressure-induced shifts for the E_1 and E_2 transitions are different each other; the same concerns the E_3 and E_4 transitions. That means that on increasing pressure, the ground and excited electronic manifold splitting diminishes. Such an effect is expected when the interaction with the symmetrical lattice vibration mode increases with pressure and reduction of the Jahn–Teller coupling on increasing pressure takes place [77]. Detailed analysis of the temperature dependence of the $YAlO_3$: Ti^{3+} luminescence obtained for pressure ≤ 181 kbar led to the conclusion that high pressure induces the nonlinear Jahn–Teller distortion [80]. Energies of the absorption and emission bands in Ti^{3+} ion in $Al_2O_3:Ti^{3+}$ are presented in Table 4.2.

Table 4.2 Pressure dependence of the energy of the $^4T_{2g} \rightarrow {}^4A_{2g}$ transitions in $3d^3$ ions in dielectric materials and calculated effective dependence of crystal-field strength on 10 *Dq* on *R*

Material	E_{em} (cm^{-1})	dE_{em}/dp (cm^{-1}/kbar)	E_{abs} (cm^{-1})	dE_{abs}/dp (cm^{-1}/kbar)	B_0 (kbar)	n
LLGG:Cr^{3+} [84]	11,700	13	15,033		1330 [110]	3.45
Al_2O_3:Cr^{3+} [101]	–	–	17,950	8.4[a]	2420 [102]	3.40
Al_2O_3:Ti^{3+} [77]	13,670 (E_4)	5.40	20,590 (E_2)	6.96	2420 [102]	2.73
	12,450 (E_3)	5.93	17,690 (E_1)	8.52		
Al_2O_3:Ti^{3+} [80]	13,350	7.1	20,290			2.54
			17.800			
$YAlO_3$:Ti^{3+} [100]	16,330	11.4 [100]	22,880		2004 [104]	3.18
			20,400			
$LiNbO_3$:Cr^{3+} [99]	10,800	13.5	15,000	–	1058 [103]	2.85
$LiTaO_3$:Cr^{3+} [99]	11,020	13.5	15,820	–	1057 [103]	2.7
YAG:Cr^{3+} [90]	–	–	16,500	9.0	1870 [106]	3.06
$KZnF_3$:Cr^{3+} [105]	12,869	16.7	15,730			
K_2NaGaF_6:Cr^{3+} [107]	13,180	15.7	16,120		513 [108]	1.5
GGG:Mn^{4+} [96]						
Site A			19,300	13[a]	1710 [109]	3.46
Site B			19,700	18[a]		4.69
$LiSc(WO_4)_2$: Cr^{3+} [83]	9000	15 for $p < 70$ kbar				
		19 for $p > 80$ kbar				

[a]Pressure shift calculated from analysis of the dependence of the *R*-line lifetime on pressure

4.3.2.2 High-Pressure Spectroscopy of the $3d^3$ and $3d^2$ Systems

Spectacular effects can be observed when the luminescence of Cr^{3+}, Mn^{4+} ($3d^3$ system), and V^{3+} ($3d^2$ system) is measured as a function of high hydrostatic pressure, especially for sixfold coordination, when local symmetry of the ion can be approximated by octahedron (O_h point group). According to Fig. 4.3, the d orbital splits into the threefold degenerated t_{2g} electronic manifold (the lower one) and the twofold degenerated e_g electronic manifold (the upper one). In the octahedrally

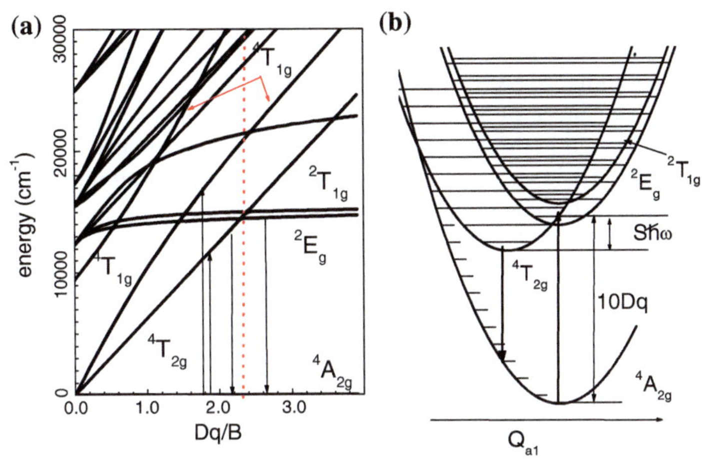

Fig. 4.5 a Tanabe–Sugano diagram for $3d^3$ system in an octahedral crystal field, **b** one-dimensional configurational coordinate diagram representing $3d^3$ system of the low field. Parabolas represent the electronic states. The vibronic states are indicated by *horizontal lines*. **b** Adapted from Ref. [81], copyright 2006, with permission from Elsevier

coordinated Cr^{3+} ($3d^3$), the ground state is $^4A_{2g}$, in which all three electrons occupy the lower electronic manifold t_{2g} with parallel spins. The Cr^{3+} system can be excited either by excitation of one electron to the e_g electronic manifold without a change of spin, thus resulting in the excited state $^4T_{2g}$, or by inversion of the spin of one electron without changing the electronic manifold. The latter case corresponds to excitation of the ion to the ^2E_g and $^2T_{1g}$ excited states. The Tanabe–Sugano diagram, which represents the lowest optically active states of octahedrally coordinated Cr^{3+}, is presented in Fig. 4.5a.

Energy of the $^4T_{2g}$ state with respect to the ground state is equal to the crystal-field strength 10 Dq, whereas the energy of the lowest doublet ^2E_g with respect to the ground state weakly depends on the crystal field and for the reasonable values of Dq/B between 1.5 and 3.5 is equal to [10]:

$$E(^2E_g) \cong 7.90B + 3.05C - 1.80\left(\frac{B^2}{Dq}\right) \quad (4.3.9)$$

Depending on the crystal-field strength, represented in the diagram by ratio Dq/B, the first excited state is either $^4T_{2g}$ (the low-field case for $Dq/B < 2.4$) or ^2E_g (the high-field case for $Dq/B > 2.4$). The configurational coordinate diagram describing the structure of the ground state and the first three excited states of the low-field $3d^3$ system is presented in Fig. 4.5b.

Because the excited $^4T_{2g}$ and the ground $^4A_{2g}$ states are split by the energy proportional to the crystal-field strength, which depends on the ion–ligand distance as R^{-5} (Eq. 4.3.6), the blue pressure shift of the $^4T_{2g} \rightarrow {}^4A_{2g}$ emission band, of the order of tenths cm^{-1}/kbar, was observed in low-field material [81]. A small red shift of the *R* lines emission, corresponding to the $^2E_g \rightarrow {}^4A_{2g}$ transition, in the Cr^{3+} ions in high-field systems [82–85] is also well known.

The combined effect of a large increase in the energy of the $^4T_{2g}$ state and essentially no change in the energy of the ^2E_g state with pressure is responsible for the observed electronic crossover. The change from a spin-allowed $^4T_{2g} \rightarrow {}^4A_{2g}$ emission to a spin-forbidden $^2E_g \rightarrow {}^4A_{2g}$ emission leads to the observed transformation from a smooth, broadband emission spectrum to a sharp, structured emission spectrum. The $^4T_{2g}$–^2E_g crossover was observed in many materials. The latest reviews on high-pressure spectroscopy of the Cr^{3+}-doped materials can be found in [81, 87]. A typical example is presented in Fig. 4.6a–e where the luminescence spectra and luminescence decays of the $LiSc(WO_4)_2{:}Cr^{3+}$ system, obtained at different pressures between ambient and 260 kbar, are presented. As pressure was applied and increased from ambient pressure to 88 kbar, Cr^{3+} retained its broadband emission related to the $^4T_{2g} \rightarrow {}^4A_{2g}$ transition, which exhibited a strong blue shift with pressure. For pressure >88 kbar, $^4T_{2g} \rightarrow {}^4A_{2g}$ emission continued to show the blue shift, and an additional narrowband emission related to the $^2E_g \rightarrow {}^4A_{2g}$ transition began to appear approximately 13400 cm^{-1} at 116 kbar. With further increase of pressure from 116 to 260 kbar, the narrowband emission showed a red shift and gradually grew in intensity at the expense of the broadband emission. A pressure-induced spectral transformation in the Cr^{3+} emission from the broadband to the narrowband was completed at 227 kbar. As seen in Fig. 4.6b through e, the R line luminescence decays are multiexponential and become slower when pressure increases, whereas for fixed pressure they become faster with

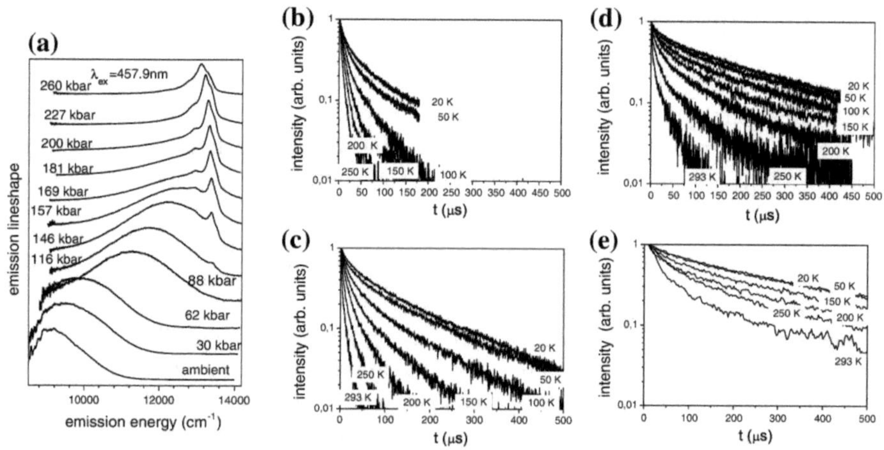

Fig. 4.6 **a** RT emission spectra of $LiSc(WO_4)_2{:}Cr^{3+}$ obtained for different pressures. $LiSc(WO_4)_2{:}Cr^{3+}$ luminescence decays of the R-line emission related to the $^2E \rightarrow {}^4A_2$ transition at different temperatures at 146 (**b**), 157 (**c**), 181 (**d**), and 227 kbar (**e**). **a** adapted from Ref. [83], copyright 2006, with permission from Elsevier. **b**, **c** Adapted from Ref. [86], copyright 2006, with permission from Elsevier

increasing temperature. Because the $^2E_g \rightarrow {}^4A_{2g}$ transition is spin forbidden, the R lines luminescence takes place mainly due to the spin-orbit coupling between the ^2E_g and $^4T_{2g}$ states. Finally, the R-line luminescence lifetime is dependent on the energy difference between the ^2E_g and $^4T_{2g}$ states, and the nonexponential decay of the R line luminescence is the result of the multisite effect (different energy of the $^4T_{2g}$ state in different sites). By analysis of the pressure dependence of the R line to the broadband ratio and R-line luminescence kinetics, the multisite effect resulted in inhomogeneous broadening of the emission as described by distribution of the crystal-field strength $10Dq$, which is equal to energy of the $^4T_2$ state and distribution of the electron–attice coupling strength $S\hbar\omega$ in $LiSc(WO_4)_2{:}Cr^{3+}$ [83, 86].

To recover the distribution of crystal-field strength in Cr^{3+} ions in the lattice, the most complete model describing the spectral properties of octahedrally coordinated Cr^{3+} ions developed for analysis of pressure dependence of the emission and emission kinetics of Cr^{3+} in $LiTaO_3{:}Cr^{3+}$ [99] and $LiNbO_3{:}Cr^{3+}$ [11, 84] was used. This model considers the $\Gamma_8(^4A_2)$ ground electronic manifold and the $\Gamma_8(^4T_2)$, $\Gamma_8'(^4T_2)$, $\Gamma_7(^4T_2)$, Γ_6 $(^4T_2)$, $\Gamma_8(^2E)$, $\Gamma_8(^2T_1)$, and $\Gamma_6(^2T_1)$ excited electronic manifolds, which mix with each other through the spin-orbit interaction. The simplified structure is presented in Fig. 4.5b. The recovered distributions of crystal-field strength among $LiSc(WO_4)_2{:}Cr^{3+}$, defined as number of sites having specific energy of the $^4T_2$ state, is presented in Fig. 4.7 The analysis presented in paper [83]

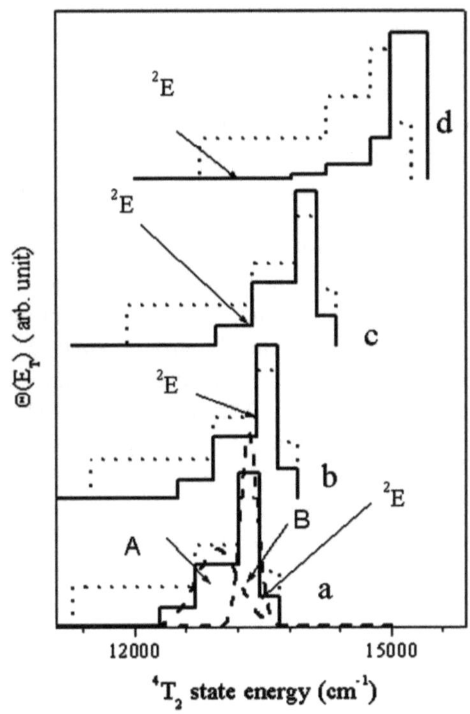

Fig. 4.7 Function describing the distribution energy of the $^4T_2$ state in the Cr^{3+} sites in $LiSc(WO_4)_2{:}Cr^{3+}$ at 146 (**a**), 157 (**b**), 181 (**c**), and 227 kbar (**d**). *Dashed histograms* correspond to the data obtained using a simplified model, and solid histograms correspond to the data obtained using a reliable model (for details see Ref. [83]). Energies of the ^2E state for different pressures are indicated by *arrows*. *Long dashed curves* in (**a**) correspond to the decomposition of recovered distribution into the two functions corresponding to the Cr^{3+}(**a**) site indicated as *A* and to the Cr^{3+}(**b**) site indicated as *B*. Reprinted from Ref. [83], copyright 2006, with permission from Elsevier

allowed distinguishing two distinct sites, which were tentatively attributed to Cr^{3+} replacing the Li^+ and Sc^{3+} in the lattice and labeled as Cr^{3+}(a) and Cr^{3+}(b), respectively. High pressure–tuning spectroscopy studies on Cr^{3+}-activated materials have clearly shown that pressure is an irreplaceable and powerful tool for quantifying the extent of electronic mixing between the ^2E_g and $^4T_{2g}$ states of the Cr^{3+} ion [88–90], for identifying the irresolvable multisite behavior of broadly emitting Cr^{3+} systems [91, 92], and for characterizing the inhomogeneous distribution of Cr^{3+} ions in disordered crystals [83, 86, 93, 99] or glass hosts [94].

The successful application of Cr^{3+}-doped materials stimulated interest in the research of other ions with $3d^3$ electronic configuration such as Mn^{4+}. High-pressure luminescence of Mn^{4+} was investigated in Al_2O_3 [95], gadolinium gallium garnet [96, 97], and $YAlO_3$ [98].

Pressure shifts of the energy of the $^4T_{2g}$ states in Cr^{3+} and Mn^{4+} ions in solid matrices are presented in Table 4.2. Equation (4.3.7) allows a qualitative and quantitative test of the dependence of the crystal-field strength, which is given by parameter Dq on the central ion–ligand distance R. Specifically, Eq. (4.3.7) allows calculation of the value of exponent n when the value of Dq, its dependence on pressure and bulk modulus of the material, are known. The data in Table 4.2 show that in all cases the value of n is much smaller than 5. This systematic deviation from the crystal-field model has been considered by means of increasing local compressibility caused by the impurity. A phenomenological model in which the reduction of local compressibility (increasing local stiffness) was related to the existence of occupied localized states, which is related to the impurity, is presented in the paper [100]. It is discussed therein that in the presence of the impurity the energy necessary for compression of the crystal volume is greater than in the case of the absence of the impurity because compression causes the increase in absolute energy of localized impurity states in the band gap.

The octahedrally coordinated V^{3+}, which contains two electrons occupying the 3d orbital, qualitatively has an energetic structure similar to that of Cr^{3+}. The ground state of V^{3+} is $^3T_{1g}$ (t_{2g}^2-electronic configuration). Like in Cr^{3+}, the first excited state can be either the $^1T_{2g}$ state, also belonging to the ground t_{2g}^2-electronic configuration, or the $^3T_{2g}$ state belonging to the exited electronic configuration $t_{2g}e_g$, separated from the ground state by energy equal to 10 Dq. As the result, in low-field materials the pressure-induced $^3T_{2g}$–$^3T_{1g}$ level crossing should be observed. Actually, Wegner and Güdel [111] observed a blue shift of the broadband luminescence related to the transition from the mixed state $^1T_{2g}$, $^3T_{2g}$ to the ground $^3T_{1g}$ state under pressure ≤ 26 kbar. The blue shift was accompanied by increasing the luminescence lifetime. Both of these effects were explained by different sensitivity of the $^1T_{2g}$ and $^3T_{2g}$ excited states to pressure.

4.4 Rare-Earth Ions

The spectral properties of RE ions, the lanthanides (Ln) and actinides (Ac), are determined by the localized states of the $4f^n$ and $4f^{n-1}5d$ and $5f^n$ and $5f^{n-1}6d$ electronic configurations, respectively. The Ln^{3+} and Ln^{2+} are frequently investigated because of their application as dopants in luminescence materials, and only some research has been devoted to Ac ions in solids.

When Ln appear in the form of atoms, their core electronic configuration is [Xe] and valence electrons are $4f^n\ 5d^m6s^2$ where n changes between 1 and 14, and m is 0 or 1. The ground electronic configuration of Ln^{3+} is always $\{[Xe]4f^n\}$, whereas for Ln^{2+} the ground state belong to either to the $\{[Xe]4f^n\}$ or to $\{[Xe]4f^n5d\}$ electronic configuration. Basic information on the energetic structure of Ln ions is summarized in [112].

4.4.1 Pressure Dependence of $4f^n$–$4f^n$ Transitions

The underlying phenomena associated with the energetic structure of the $4f^n$ electronic configuration and energies of the $4f^n \rightarrow 4f^n$ transitions are similar to those describing the $3d^n$ electronic configuration in TM ions. The principal difference between 4f and 3d orbitals is related to the special extension. In contras, to the 3d orbitals, the 4f orbitals in Ln ions are strongly localized in the Ln ion core and are additionally shielded by $5s^2d^6$ and $4d^{10}$ orbitals. Therefore, the 4f electrons interact only weakly with surrounding ligands.

To analyze the influence of high hydrostatic pressure on the emission of lanthanide ions, the semiempirical crystal-field approach is used where the total Hamiltonian of the $4f^n$ system is given by the sum of a free ion and the crystal-field terms as follows: [113].

$$H_F = \sum_i e_i E^{(i)} + H_{s-o} + \alpha L(L+1) + \beta G(G_2) + \gamma G(R_7) + \sum_{k,q} B_q^k(R) C_q^k(\theta, \phi) \tag{4.4.1}$$

Here $E^{(i)}$ are the Racah parameters describing the electron–electron interactions; and $H_{s\text{-}o}$ is a spin-orbit interaction Hamiltonian, which can be described by single constant ξ. Equivalently, instead of the Racah parameters, e_i and $E^{(i)}$, the f_k constants and the Slater integrals F_k can be used, e.g., $\sum_i e_i E^{(i)} = \sum_k f_k F_k$. The three constants α, β, and γ describe the interconfigurational interactions; and $G(G_2)$ and $G(R_7)$ are the eigenvalues of Casimirs' operator for the groups G_2 and R_7, respectively, used to classify the states of the $4f^n$ electronic configuration.

The last sum in Eq. (4.4.1) describes the interaction of electrons with the crystal field. The permitted values of subscripts k and q here are determined by the local point symmetry of Ln ions and vary from 0 to $6.C_q^k(\theta, \phi)$ are angular factors that can be evaluated exactly from the positions of ligands, and $B_q^k(R)$ are the radial crystal-field parameters, which depend on the 4f radial-wave functions. The values of parameters $B_q^k(R)$ are usually obtained by a fitting procedure. A useful single parameter that can be calculated and describes the interaction between $4f^n$ electrons and lattice is the crystal-field strength, Ξ [114]:

$$\Xi = \left[\frac{1}{3}\sum_k \frac{1}{2k+1}\left(B_{k0}^2 + 2\sum_{q>0} B_{kq}^2\right)\right]^{1/2} \quad (4.4.2)$$

Pressure affects the energetic structure of the Ln ion and the energies of the $4f^n$-$4f^n$ transitions in two ways: (1) the nephelauxetic [115] reduction of the Slater integrals; and (2) the spin-orbit coupling constant. This effect was broadly discussed by Shen and Holzapfel [116] in the framework of the covalence model. In this framework, the radial-wave functions of the Ln ion, when it is embedded in the lattice, expand compared with those of a free ion due to the penetration of the ligand electrons into the Ln ion space. Penetration increases when pressure decreases the Ln ion–ligand distances. The second effect is related to the pressure-induced increase of the crystal-field strength.

Usually, it is assumed that pressure effects under the hydrostatic limit are isotropic, which causes a proportional decrease of all distances in the lattice without a change in the local symmetry of the Ln ion. As the result, a standard assumption is that angular factors $C_q^k(\theta, \phi)$ in the crystal-field Hamiltonian are pressure independent and that the only pressure-sensitive parameters are the Slater integrals, spin-orbit coupling, and radial crystal-field parameters $B_q^k(R)$. Then, in most cases, pressure causes small linear shifts of the sharp-lines luminescence related to f-f transitions in the Ln^{3+} and Ln^{2+} ions.

One can find an extensive review on the research on spectroscopy and luminescence related to f-f transition under high hydrostatic pressure performed before 2001 in [2]. In this paper, a review on the research performed in the 21st century is provided. The most intensively investigated materials were those doped with Pr^{3+} ($4f^2$), Nd^{3+}($4f^3$), Eu^{3+} ($4f^6$), Eu^{2+} ($4f^7$), Tb^{3+} ($4f^8$), Ce^{3+} ($4f^1$), and Yb^{3+}($4f^{13}$).

4.4.1.1 Ce^{3+} and Yb^{3+}

The simplest structure of the ground electronic configuration characterize the Ce^{3+} and Yb^{3+} ions where the $4f^1$ and $4f^{13}$ electronic manifolds are split into the $^2F_{7/2}$ and $^2F_{5/2}$ states. In the case of Ce^{3+}, splitting usually does not exceed 3000 cm^{-1}, whereas splitting in the case of Yb^{3+} thief on the order of 10000 cm^{-1}. This is the reason why no reports on the pressure dependence of f-f luminescence in Ce^{3+} ions can be found. In the last decade, high-pressure luminescence of Yb^{3+} ions had been reported for InP [117, 118], GaN:Yb^{3+} [119], $LiNbO_3$ [120], and $Gd_3Ga_5O_{12}$ (GGG) [7]. In In P:Yb^{3+}, an increase of energy of the luminescence lines with pressure was observed [117, 118] and was related in Ref. [117] to the increase of the spin-orbit coupling constant, which was not been predicted by the model of Shen and Holzapfel [116]. The later ab initio calculations have allows relating this effect with the approach of the top of the valence band of the InP crystal host to the energies of the f-f Yb transitions [118]. In GaN:Yb^{3+} [119], a weak dependence of the luminescence lines on pressure was observed. The rates have been mostly negative and have varied from 0.06 to –0.75 cm^{-1}/kbar. All of these changes have been considered in the framework of the crystal-field approach, where pressure influences the crystal-field parameters, and under assumption that pressure does not influence the spin-orbit coupling. Similarly, very small pressure-induced shifts of the emission lines related to $^2F_{5/2} \rightarrow {}^2F_{7/2}$ luminescence in $LiNbO_3$:Yb^{3+} [120] and GGG:Yb^{3+} [7] have been shown. In $LiNbO_3$:Yb^{3+} [120], a strong decrease of the luminescence lifetime with an increase of pressure was observed. This result was explained by coupling of the ground electronic configuration $4f^{13}$ with the excited electronic configuration $4f^{12}5d^1$. The increase of the coupling coefficient with pressure is due to diminishment of the energetic distance between the emitting $^2F_{5/2}$ state and the lowest state of the $4f^{12}5d^1$ electronic configuration. Pressure dependence of the Yb^{3+} emission and luminescence decay has been analyzed in YPO_4:Yb^{3+} and $GdPO_4$:Yb^{3+} [121]. In both materials, small shifts of the emission lines with pressure have been observed. In $GdPO_4$:Yb^{3+}, pressure induced a small increase of energy of the emission line with the highest energy and negative shifts of the other lines, in which the values of the rates increase when the line energy decreases. This effect means that pressure caused an increase of the crystal-field splitting. A similar tendency was noticed for YPO_4:Yb^{3+}. In both materials, YPO_4:Yb^{3+} and $GdPO_4$:Yb^{3+}, pressure causes a decrease of the luminescence lifetimes, but this much smaller than in the case of $LiNbO_3$:Yb^{3+}.

4.4.1.2 Pr^{3+} Ions

The ground state of Pr^{3+} ($4d^2$) is the $^3H_4$ state. The host luminescence of Pr^{3+} consists of sharp lines related to the radiative transitions from the excited states $^1D_2$ and $^3P_0$ or only one of them. Pressure dependence of the energies of luminescence lines related to the radiative transition from the emitting $^1D_2$ state in Pr^{3+} in $LiNbO_3$:Pr^{3+} [122] and YVO_4:Pr^{3+} [123] was observed. Luminescence related to transitions from the $^1D_2$ and $^3P_0$ states in $Ca(NbO_3)_2$:Pr^{3+} [124], $LiTaO_3$:Pr^{3+} [125], $LiYF_4$:Pr^{3+} [126], $KLa(MoO_4)_2$:Pr^{3+} [127], and $CaWO_4$:Pr^{3+} [128], as well as the $^3P_0$ luminescence in $Gd_2(WO_4)_3$:Pr^{3+} [129], have been observed under pressure. The negative pressure shifts of all luminescence lines confirm the prediction of pressure-induced reductions of Slater integrals and spin-orbit coupling. Energies of the luminescence peaks and their pressure shifts are presented in Table 4.3. Quantitative calculations that obey the fitting of the pressure dependence of the Slater parameters and spin-orbit coupling was performed for $LiYF_4$:Pr^{3+} [126], $LiTaO_3$:Pr^{3+} [125], and $KLa(MoO_4)_2$:Pr^{3+} [127]. In the papers [125, 126], it was assumed that pressure does not changes the values α, β, and γ; therefore, the fitting

Table 4.3 Energies and pressure shifts of the Pr^{3+} luminescence lines in several materials

Material energy/shift [cm^{-1}/ $cm^{-1}kbar^{-1}$]	$YVO_4:Pr^{3+}$ [123]	$LiNbO_3:Pr^{3+}$ [122]	$CaWO_4:Pr^{3+}$ [128]	$Gd_2(WO_4)_3:Pr^{3+}$ [129]	$LiTaO_3:Pr^{3+}$ [125]	$KLa(MoO_4)Pr^{3+}$ [127]
$^1D_2 \rightarrow {}^3H_4$	16,180/−2.3	15,641/−2.34	16,120/−1.70		16,190/−2.13	16,750/−2.82
	16,455/−1.7	15,696/−2.35	16,190/−1.70		16,100/−2.31	
	16,500/−1.4	15,812/−2.35	16,515/−1.70		15,733/−2.15	
	16,800/−1.0	16,062/−2.35	16,590/−1.70			
		16,101/−2.35	16,790/−1.28			
		16,172/−2.35	16,830/−1.28			
$^1D_2 \rightarrow {}^3H_5$					14,016/−2.45	
					13,728/−2.61	
$^1D_2 \rightarrow {}^3H_6$					12,110/2.41	
					11,979/−2.63	
$^3P_0 \rightarrow {}^3H_4$					19,859/−2.45	21,200/−3.57
					19,664/−3.50	19,650/−3.74
					19,552/−2.40	
$^3P_0 \rightarrow {}^3H_4$						18,880/−2.44
						18,280/−2.88
						18,000/−2.87
$^3P_0 \rightarrow {}^3H_6$			15,920/−2.13		15,650/−2.41	16,210/−2.25
						16,200/−2.81
$^3P_0 \rightarrow {}^3F_2$			15,410/−2.65	15,300/−2.28	15,322/−2.68	15,450/−2.77
$^3P_0 \rightarrow {}^3F_4, {}^3F_3$					13,090/−2.71	14,600/−2.39
						13,700/−2.11

parameters were the Slater integrals, spin-orbit coupling, and crystal-field parameters. In addition, in paper [125] it was assumed that pressure does not change the ratios F_4:F_2 and F_6:F_2, so only the one Slater parameter F_2 was fitted. In both papers [126] and [125], it was shown that pressure diminishes the values of Slater integrals and spin-orbit coupling. Calculations performed in paper [126] were more extensive. The Slater parameters are considered to be independent of each other; additionally, the interaction between the $4f^2$ and 4f5d electronic configurations was explicitly taken into account. As far as spin-orbit coupling is concerned, both papers provide almost the same result: The relative change of spin-orbit coupling $\Delta\xi/\xi = -0.03$ %/kbar. Slater integrals have been found to diminish with increasing pressure, however, here the differences are larger: $\Delta F_2/F_2 = \Delta F_4 F_4 = \Delta F_6/F_6$ -0.02 %/kbar for $LiTaO_3$:Pr^{3+} [125], and $\Delta F_2/F_2 = -0.05$ %/kbar, $\Delta F_4 F_4 = -0.01$ %/kbar, and $\Delta F_6/F_6 - 0.05$ %/kbar for $LiYF_4$:Pr^{3+} [126]. In both materials, the values of B_q^k parameters have also been fitted.

The anisotropic effect of high hydrostatic pressure was observed for Pr^{3+}-doped $Y_3Al_5O_{12}$ (YAG) crystal by Turos-Matysiak et al. [12]. Ambient pressure emission of YAG:Pr^{3+} is dominated by two groups of lines related to ${}^1D_2 \rightarrow {}^3H_4$ and ${}^3P_0 \rightarrow {}^3H_4$ transitions with peaks at 16200–16300 and 20500–20600 cm^{-1}, respectively. Room temperature luminescence related to the ${}^1D_2 \rightarrow {}^3H_4$ and ${}^3P_0 \rightarrow {}^3H_4$ transitions obtained at different pressure are presented in Fig. 4.8a, b, respectively. In both cases,

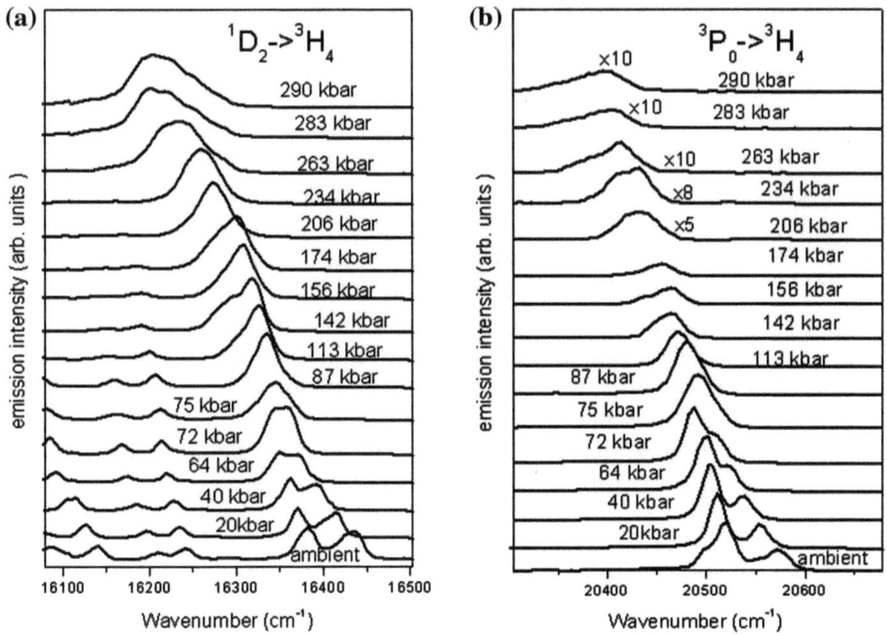

Fig. 4.8 Room-temperature luminescence spectra of YAG:Pr^{3+} at different pressures. **a** For the energy region corresponding to the ${}^1D_2 \rightarrow {}^3H_4$ transitions. **b** For the energy region corresponding to the ${}^3P_0 \rightarrow {}^3H_4$ transitions. Adapted from Ref. [12], copyright 2006, with permission from Elsevier

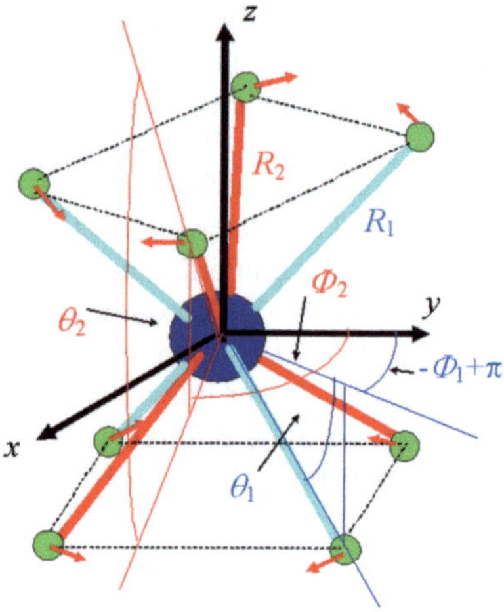

Fig. 4.9 The complex $(PrO_8)^{13-}$ in YAG. The directions of pressure-induced oxygen motion are indicated by *arrows*

energies of the spectral lines diminish with pressure. The number of observed lines representing the $^1D_2 \rightarrow {}^3H_4$ transition depends on pressure. Because the pressure shifts of lines with higher energy are smaller than those with lower energy, level crossings are seen at 87 and 206 kbar. Above 230 kbar, the lines become broader, and additional peaks appear at higher energies. The $^3P_0 \rightarrow {}^3H_4$ emission consists of two peaks, the energies of which also diminish with pressure with different rates. A level crossing occurs in the pressure range of 75–87 kbar.

Pr^{3+} replaces Y^{3+} in YAG and is surrounded by eight O^{2-} ligands (see Fig. 4.9). The energetic structure of the Pr^{3+} ions in YAG was calculated in crystal-field model using Hamiltonian (4.4.1) considering that the point symmetry of the $Pr^{3+}O_8$ cluster is D_2 [12]. The ground state $^3H_4$ splits into the $^3T_2$, $^3A_1$, $^3T_1$, and ^3E states due to the cubic field and additionally into nondegenerated states labeled by the irreducible representations of the D_2 point group. The lowest cubic field state, $^3T_2(^3H_4)$, splits into the Γ_3, Γ_1, and Γ_4 components, whereas the excited $^1E(^1D_2)$ state splits into the Γ_1 and Γ_2 components; the $^3P_0$ state is a nondegenerated (Γ_1) state. Ambient pressure–calculated energies are presented in Table 4.4.

To reproduce the specific dependence of the emission spectra on pressure, it was assumed that the Racah parameters, spin-orbit coupling, and crystal-field parameters depend on pressure, but the α, β, and γ parameters do not. To fit the experimental pressure shift of the $^1D_2 \rightarrow {}^3H_4$ and $^3P_0 \rightarrow {}^3H_4$ emission lines following the relations $\frac{dE^{(i)}}{dp} = -0.4 \cdot 10^{-4} E^{(i)}\ \mathrm{cm}^{-1}/\mathrm{kbar}$, for all (i) and $\frac{d\zeta}{dp} = -0.6 \cdot 10^{-4} \zeta\ \mathrm{cm}^{-1}/\mathrm{kbar}$ have been used [12]. The crystal-field effect was calculated using the superposition model [131] where parameters B_q^k are given by a superposition of k-rank contributions from the charged ligands. Considering that the Pr^{3+} ion occupies the origin of the coordinate system, the position of the L-th ligand is defined by distance R_L and angles: θ_L and ϕ_L.

Table 4.4 Energies of the $^3H_4$ and $^1D_2$ states in the crystal field of D_2 symmetry at ambient pressure. The data were obtained from diagonalization of the total Hamiltonian, which includes the cubic contribution to the crystal field described by $-B_q^k(cub)$ the and the D_2 symmetry crystal field described by B_q^k. Data are taken from [12]. Experimental data were taken from Ref. [130]

Free-ion	Cubic (O_h)		D_2		Experiment
	State	Energy (cm^{-1})	State	Energy (cm^{-1})	Energy (cm^{-1})
$^3H_4$	T_2	0	Γ_3	0	0
			Γ_1	20	9
			Γ_4	40	54
	A_1	565	Γ_2	494	
	T_1	595	Γ_1	506	
			Γ_3	506	
			Γ_4	532	
	E	670	Γ_1	755	
			Γ_2	759	
$^1D_2$	E	16,443	Γ_1	16,435	16,436
			Γ_2	16,443	
	T_2	17,073	Γ_1	16,906	
			Γ_3	17,117	
			Γ_4	17,150	
$^3P_0$	A_1	20,632	Γ_1	20,580	20,571

$$B_q^k = \sum_L \frac{A_k(R_L)}{\lambda_{k,q}} G_{kq}(\theta_L, \phi_L) \tag{4.4.3}$$

where coefficients $A_k(R_L)$ can be presented as the following integrals:

$$A_k(R_L) = \int \frac{\rho(R_L) r^k}{R_L^{k+1}} \mathrm{d}\tau \tag{4.4.4}$$

$G_{kq}(\theta_L, \phi_L)$ are functions depending on angles θ_L and ϕ_L; and $\lambda_{k,q}$ are constants. The Pr–ligand system is presented in Fig. 4.9. The system is described by two different distances R_1 (blue), R_2 (red), and two sets of angles θ_1 and ϕ_1 and θ_2 and ϕ_2, respectively. The ambient pressure values of these parameters are listed in Table 4.5. It was found in paper [12] that the assumption of isotropic pressure, which diminishes proportionally all distances in R_L, may not yield the level crossing seen in Fig. 4.8a, b.

To reproduce the level crossing, it was assumed that pressure continuously changes the symmetry of the Pr–ligand system by changing the angles θ_L and ϕ_L. Therefore, the following relations were used to describe the anisotropic effects of pressure:

Table 4.5 B_q^k and A_k coefficients for the PrO_8 system in YAG:Pr^{3+}. In the first column $B_q^k(cub)$, the cubic coefficients are presented. Data were taken from paper [12]

kq	$B_q^k(cub)$	B_q^k	dB_q^k/dp	A_k	λ_{kq}^{-1} dA_k/dp, $K = 0.58$
20	–	−391.1	−0.12	4471.1	1.39
22	–	−233.1	−0.072	−932.2	−0.29
40	−2961.5	−3023.1	−1.56	982.1	0.51
42	–	−428.3	−0.22	−3867.4	−2.00
44	1770.0	1268.2	0.66	921.6	0.48
60	1060.5	1024.5	0.74	531.3	0.38
62	–	334.6	0.24	−650.2	−0.47
64	1989.9	1584.6	1.15	661.6	0.48
66	–	7.2	0.0052	−96.01	−0.070

$$G_{kq}(\theta_L, \phi_L, p) = G_{kq}(\theta_L + \frac{d\theta_L}{dp}p, \phi_L + \frac{d\phi_L}{dp}p) \quad (4.4.5)$$

The isotropic reduction of volume was calculated by considering changes in parameters A_k

$$A_k(p) = A_k(p = 0) + \frac{dA_k}{dp} \cdot p \quad (4.4.6)$$

where

$$\frac{dA_k}{dp} = \frac{dA_k}{dR_L} \cdot \frac{dR_L}{dp} \approx \frac{A_k(k+1)}{3B_0}K \quad (4.4.7)$$

where coefficient K is smaller than unity and describes the increasing of local stiffness [12, 100]. The calculated positions of the maximums of the peaks versus pressure are presented by solid curves in Fig. 4.10a, b. Dashed-dotted curves represent calculated energies obtained under assumption of the isotropic effect of pressure. The solid curves represent the energies obtained when the anisotropic effect was considered using Eq. (4.3.5). The best fitting was obtained for values of the parameters listed in Tables 4.5 and 4.6. The calculated energies of states involved in the luminescence process are presented by solid curves in the insets in Fig. 4.10a, b.

In the last years, high-pressure spectroscopy has been used for the characterization of novel optical materials doped with Pr^{3+}. Mahlik et al. [133] investigated the luminescence and luminescence kinetics of β-SiAlON:Pr^{3+}under high hydrostatic pressures ≤ 260 kbar. It has been postulated [132] that Pr^{3+} in β-SiAlON can occupy not only the main crystallographic site (Wyckoff notation 2*b*), which

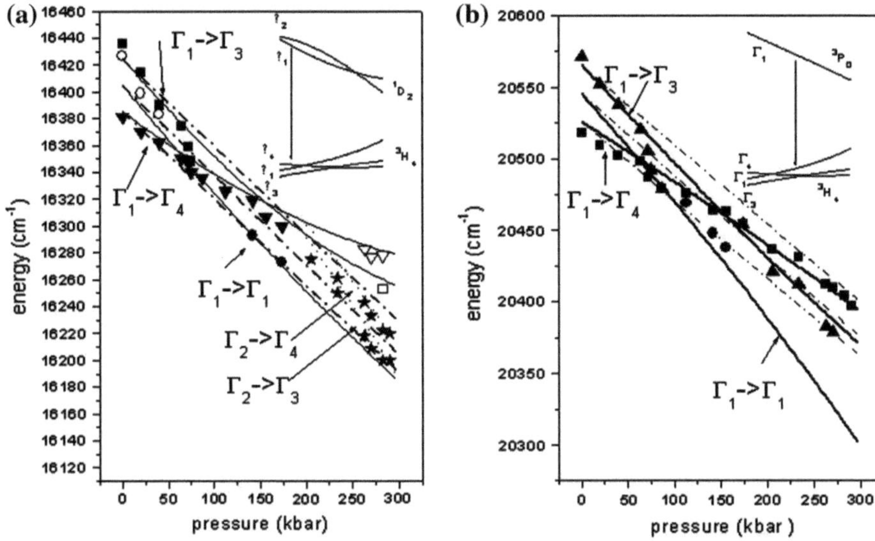

Fig. 4.10 Energies of the individual spectral lines versus pressure **a** corresponding to the $^1D_2$ (Γ_1, Γ_2) $\rightarrow$ $^3H_4$ (Γ_4, Γ_1, Γ_3) transitions and **b** corresponding to the $^3P_0$ (Γ_1) $\rightarrow$ $^3H_4$ (Γ_4, Γ_1, Γ_3) transition. Experimental positions are represented by experimental points (*circles*, *squares*, *asterisks*, and *triangles*). *Dashed-dotted curves* represent calculated energies obtained under assumption of the isotropic effect of pressure. *Solid curves* represent the calculated energies corresponding to respective transitions obtained under the assumption of anisotropic effect. In the insets of the figures are the calculated energies of $^1D_2$ (Γ_1, Γ_2), $^3H_4$ (Γ_4, Γ_1, Γ_3), $^3P_0$ (Γ_1), and $^3H_4$ (Γ_4, Γ_1, Γ_3)

Table 4.6 Parameters describing the PrO_8 system. Quantities in the first row represent the cubic symmetry, and quantities in the second row represent the D_2 symmetry. The third row contains the pressure coefficients of the respective angles. Data were taken from paper [12]

R_1 [Å]	R_2 [Å]	θ_1 (°)	θ_2 (°)	ϕ_1 (°)	$d\phi_1/dp$ (°/kbar)	ϕ_2 (°)	$d\phi_2/dp$ (°/kbar)
2.23115	2.23115	125.26	125.26	−180	−0.022	90	0.033
2.3030	2.4323	123.86	125.94	−192.52	−0.022	81.24	0.033

corresponds to the Pr^{3+} located at (0, 0, 0), but also the second site where Pr^{3+} is located at (0, 0, 1/4) position (Wyckoff notation 2*a*) and is coordinated by nine nitrogen/oxygen ions. High-pressure time-resolved spectroscopy has shown that energies of luminescence peaks related to the transitions $^1D_2 \rightarrow {}^3H_4$ and $^3P_0 \rightarrow {}^3H_4$ in Pr^{3+} ions in the 2*b* and 2*a* sites change monotonically and continuously with increasing pressure (see Fig. 4.11a). In addition, pressure dependence of the luminescence lifetime of the $^3P_0 \rightarrow {}^3H_4$ transition in Pr^{3+} ions in the 2*b* and 2*a* sites confirmed the existence of Pr^{3+} in the (0, 0, 1/4) lattice position.

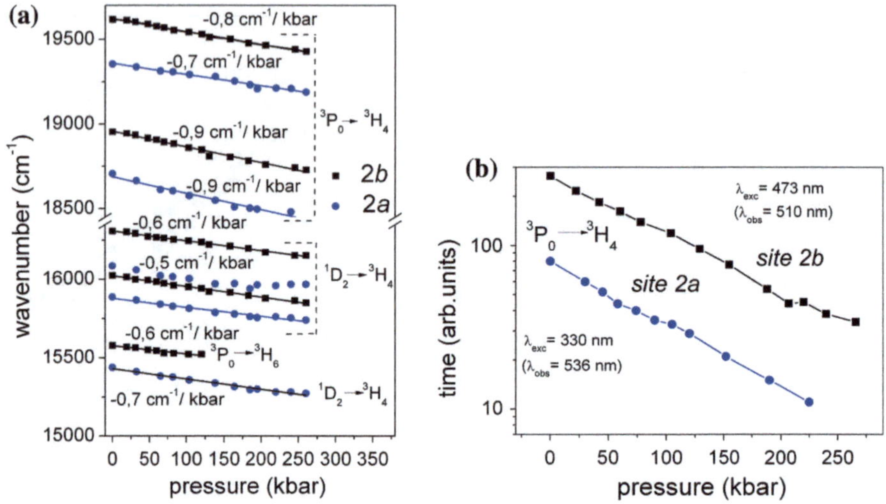

Fig. 4.11 **a** Positions of selected emission peaks related to luminescence of β-SiAlON:Pr^{3+} versus pressure. **b** Decay times of luminescence related to the $^3P_0 \rightarrow {}^3H_4$ transition in Pr^{3+} in $2a$ and $2b$ sites in β-SiAlON:Pr^{3+} Reprinted with the permission from Ref. [133]. Copyright 2013 American Chemical Society

4.4.1.3 Nd^{3+} Ion

Luminescence of Nd^{3+} ($4f^3$) ions under high hydrostatic pressure has been investigated in YAG [56], YVO_4 [134], $YLiF_4$ [136] and GGG [135]. The luminescence spectrum of Nd^{3+} is determined by the specific energetic structure of the ion. The metastable emitting state is $^4F_{3/2}$, which is split in the crystal field into two Kramer's doublets—one with lower energy (labeled R_1) and one with higher energy (labeled R_2)—separated by energy Δ. The final states are $^4I_{9/2}$, which is the ground state, and $^4I_{11/2}$, which is the first excited state. Both the $^4I_{9/2}$ and $^4I_{11/2}$ states are split by the crystal field, and their components are labeled X_i (i = 1,5) and Y_i(i = 1,6), respectively. At low temperature, only luminescence lines related to transitions from the R_1 state are observed. At higher temperatures, when the R_2 state is occupied, the emission from the R_2 state contributes to the luminescence by the spectral structure, which is repetition of the low-temperature spectrum shifted to higher energies by the value of the R_1–R_2 splitting Δ. In all materials mentioned above, pressure diminishes the energies of the $R_i \rightarrow X_i$ and $R_i \rightarrow Y_i$ transitions. Different pressure shifts of the lines related to transitions from the same initial state to different final components of the split $^4I_{9/2}$ and $^4I_{11/2}$ states, and between different initial states R_1 and R_2 and the same final state, indicate that pressure changes the crystal-field strength [134].

A spectacular effect was observed by Kaminska et al. [135] for luminescence lines related to the transitions from the R_1 and R_2 states to the highest component of the $^4I_{9/2}$ multiplet (X_5) in GGG:Nd^{3+}. The respective luminescence lines obtained

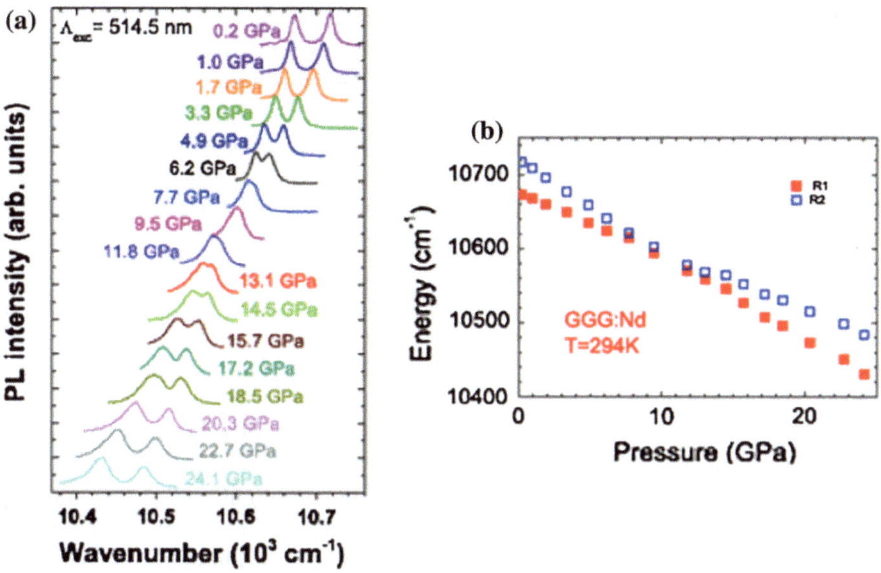

Fig. 4.12 **a** Pressure dependence of the luminescence spectra corresponding to the transitions from the R_1 and R_2 states to the highest component of the $^4I_{9/2}$ multiplet, **b** spectral positions of the luminescence lines as a function of hydrostatic pressure. Reprinted from Ref. [135] by permission of American Physical Society

for different pressures are presented in Fig. 4.12a. It is seen that the energy of the R_1–R_2 splitting Δ decreases with pressure and reaches almost zero at pressure of approximately 95 kbar (9.5 GPa) and then again increases when the pressure increases. The suggested interpretation is that the R_1–R_2 splitting energy Δ reaches the minimum at 95 kbar (9.5 GPa), which was confirmed by crystal-field calculations. However, the theoretically obtained minimal splitting is approximately three times greater than that provided by optical data [135]. It should be mentioned that diminishing of the splitting between the R_1 and R_2 states was also observed in YAG:Nd^{3+} [56] and YVO_4:Nd^{3+} [134] where the luminescence line related to the transition from the R_2 state to the X_5 state has a larger negative pressure shift than the line related to the transition from the R_1 state to the X_5 state.

4.4.1.4 Eu^{3+}, Tb^{3+} and Eu^{2+} Ions

Because the Eu^{3+} and Eu^{2+} ions are promising activators for red-, green-, and blue-emitting phosphors for light-emitting diodes (LED) and white light–emitting diodes (WLED), a large increase of interest in high-pressure spectroscopy of materials doped with Eu^{2+} and Eu^{3+} was observed. In this subsection, only research on the pressure dependence of f-f luminescence in Eu^{2+} and Eu^{3+} is reported. Dependence of the luminescence related to $^5D_0 \rightarrow {}^7F_J$ transitions, J = 0, 1, 2, 3, 4 in Eu^{3+}, on high hydrostatic pressure has been reported for $Ca_3Y_2(SiO_4)_3{:}Eu^{3+}$ [137] and $Gd_2(WO_3)_3{:}Eu^{3+}$ [129]. In both materials, most of the sharp emission lines shift to the lower energy with the absolute rates much <1 cm^{-1}/kbar. Pressure shifts of the emission lines related to $^5D_0 \rightarrow {}^7F_J$, $J = 0,1,2,3,4$ transitions in $Gd_2(WO_4)_3{:}Eu^{3+}$ [129] are much smaller than shifts of the emission related to the $^3P_0 \rightarrow {}^3F_2$ transition in Pr^{3+} in $Gd_2(WO_4)_3{:}Pr^{3+}$ [129], which is equal to –2.32 cm^{-1}/kbar, and the $^5D_4 \rightarrow {}^7F_5$ transition in Tb^{3+} in $Gd_2(WO_4)_3{:}Tb^{3+}$ [129], which is equal to –1.08 cm^{-1}/kbar. Pressure dependence of Tb^{3+} luminescence has been reported in mixed garnet $Y_{1.725}Tb_{0.575}Ce_{0.05}Gd_{0.65}Al_5O_{12}$ [138]. The pressure shifts are equal to 0.1, –1.0, –1.8, and –1.1 cm^{-1}/kbar for the $^5D_4 \rightarrow {}^7F_6$, $^5D_4 \rightarrow {}^7F_5$, $^5D_4 \rightarrow {}^7F_4$, and $^5D_4 \rightarrow {}^7F_3$, transitions, respectively [138]. This sequence of pressure rates appears due to the fact that pressure causes an increase of splitting between the ^7F_J states.

Eu^{2+} is widely known as an activator that provides broadband luminescence related to the $4f^65d \rightarrow 4f^7$ transitions [139]; however in some materials where the energy of the lowest state of the excited electronic configuration $4f^65d$ is higher than the energy of the $^6P_{7/2}$ state, the sharp line luminescence related to the $^6P_{7/2} \rightarrow {}^8S_{7/2}$ transition inside the $4f^7$ electronic configuration is observed. Dependence on pressure of the sharp line emission related to the $^6P_{7/2} \rightarrow {}^8S_{7/2}$ transition has been reported for $LiBaF_3{:}Eu^{2+}$ [140]. The luminescence consists of a single zero-phonon line and five phonon repetitions. In $LiBaF_3{:}Eu^{2+}$, the repetition lines are much more intensive that in the zero-phonon one. Pressure shift of the zero-phonon line was estimated to be equal to –0.95 cm^{-1}/kbar. The shifts of phonon lines were larger as a result of the increased photon energies with increasing pressure. Detailed analysis allows for calculation of the Grüneisen parameters for all phonon modes, which varied from 0.89 to 3.20 [140]. A different situation was found in the case of $KMgO_3{:}Eu^{3+}$ [141]. Although here the emission spectrum also consists of a single zero-phonon line and five phonon repetitions, the intensity of the zero-phonon line is much higher. The pressure shift of the zero-phonon line is equal to –0.61 cm^{-1}/kbar. Larger shifts of phonon-repetition lines are also caused by the increase of the energies of phonon modes with pressure. It is interesting that at temperature <100 K, under excitation with 340 nm, the emission is different and consists of several that which have much shorter lifetime and lower energy than the zero-phonon line [142]. These lines shift to the red region with approximately the same rate as $^6P_{7/2} \rightarrow {}^8S_{7/2}$ luminescence. Detailed analysis of the spectroscopic data allowed for attribution of this emission to the different Eu^{2+} site that occupies the K^+ ion position and is accompanied by the F^- vacancy [142].

4.4.1.5 Spectroscopic Evidence of Pressure-Induced Phase Transitions

Abrupt spectral changes of the f-f luminescence and discontinuity in the pressure shifts of the f-f luminescence lines of the Ln^{3+} ions are usually related to structural and phase transitions in the crystal host. Although the analysis of changes in the spectral structure of Ln^{3+} luminescence does not provide such definite information on the crystal structure as X-ray data, it provides qualitative (sometimes also quantitative [143]) information on phase transitions. Pressure dependence of the Nd^{3+} luminescence in $YLiF_4:Nd^{3+}$ [136] showed a subtle structural change of the scheelite phase at 55 kbar and a first-order structural phase transition at 100 kbar. Lazarowska et al. [143] observed a pressure dependence of the Pr^{3+} luminescence in $LiLuF_4$. The discontinuity in pressure behavior of the luminescence lines observed at 100 kbar has been attributed to the phase transition from a tetragonal to fergusonite structure. Crystal-field calculations have shown that this phase transition decreases the point-group symmetry of the Pr^{3+} site from the S_4 to C_2 point-group symmetry. The dependence of Pr^{3+} luminescence on pressure has shown the existence of phase transition in $CaMoO_4:Pr^{3+}$ [144] at pressure between 80 and 100 knar. Phase transition in $CaMoO_4$ at a pressure of 80 kbar has been confirmed by high-pressure Raman spectroscopy [145]. Pressure-induced ferroelectric phase transition has been reported in Yb^{3+}-doped congruent $Sr_{0.6}Ba_{0.4}Nb_2O_6$ [146]. It was shown that when the system undergoes a phase transition from a ferroelectric to a paraelectric phase, the relative contributions of the various Yb^{3+} centers to the emission spectrum is strongly modified and passes from four different emitting sites in the ferroelectric phase to only two emitting sites in the paraelectric phase.

4.5 Luminescence Related to the $4f^{n-1}5d \rightarrow 4f^n$ Transitions in Ln^{3+} and Ln^{2+} Ions

Luminescence related to parity allowed the radiative $4f^{n-1}5d \rightarrow 4f^n$ transitions in Ln^{3+} and Ln^{2+} ions, if they appear, to be bright and efficient. Compared with electrons belonging to the $4f^n$ electronic configuration, the 5d electron wave functions spread at the ligand space, the energy of the lowest state of the $4f^{n-1}5d$ electronic configuration, and the wavelength of the $4f^{n-1}5d \rightarrow 4f^n$ luminescence for a given lanthanide depend on the host.

Extended research on the energetic structure of the Ln^{3+} and Ln^{2+} ions in various lattices allow us to understand the performance of the system quite well. Energies of the lowest state of the $4f^{n-1}5d$ electronic configurations of Ln^{3+} and Ln^{2+} ions in different matrices were analyzed by Dorenbos [147–149]. The broad review of the archive literature allowed development the empirical model that relates the energy of the $4f^{n-1}5d$ state of the ion in the lattice, ΔE, to the respective free ion energy, ΔE_0, by the following relation:

$$\Delta E = \Delta E_0(n) - E_{\mathrm{depr}}(n) \tag{4.5.1}$$

where the depression energy E_{depr} is a sum of the energy of the centroid shift; and E_{centr} is related to the change in ligand polarization, accompanying the transition of an electron from f to d shell, and the energy of the crystal-field splitting E_{cr}. In the ligand-field polarization model [150] E_{centr} is proportional to R^{-6}, whereas in the framework of point-charge crystal-field approach [151] in cubic symmetry, E_{cr} is proportional to R^{-5}, where R is the central ion–ligand distance. However, it is useful to consider [100] that E_{depr} is proportional to R^{-n}, whereas n is an effective exponent usually <5 [67].

In contrast, Dorenbos [152–156] showed that the energies of Ln^{2+} and Ln^{3+} ions in the lattice can be schematically presented in the systematic alteration with respect to the vacuum level [154] and the energies of the host bands, which yields the following relation between the energies of the ground states of a given Ln ion with n electrons at the 4f shell existing as Ln^{3+} and with $n + 1$ electrons at 4f shell when it exists as Ln^{2+}:

$$E(\mathrm{Ln}^{2+}, n+1) = E(\mathrm{Ln}^{3+}, n) + \Delta E(n+1, n) \tag{4.5.2}$$

where $\Delta E(n+1, n)$ is the difference between the energies of the ground states of Ln^{2+} and Ln^{3+} in given lattices. $\Delta E(n+1, n)$ is equal to 6–7.5 eV for $1 \leq n \leq 6$ and $8 \leq n \leq 13$ and is approximately two times larger for $n = 7$ [154]. The alteration presented by Eq. (4.5.2) follows the dependence of the depression energy on n. As a result, the location of the $4f^{n-1}5d$ electronic manifold in a given lattice weakly depends on the electron number "n." For a given lattice, information on the energetic structure of lanthanides is summarized in the form of an energetic diagram presented in Fig. 4.13.

In Fig. 4.13, the situation for CaF_2 is presented; however, the presented scheme is qualitatively very similar in other crystals. Especially, it should be emphasized that the lowest state of the $4f^{n-1}5d$ electronic configuration for all lanthanides is located near the conduction band, and its energy with respect to the conduction band is approximately the same for all ions. Thus, different energies of the $4f^{n-1}5d \rightarrow 4f^n$ transitions actually result from different locations of the ground states of Ln^{3+} and Ln^{2+} ions in the band gap. From the scheme presented in

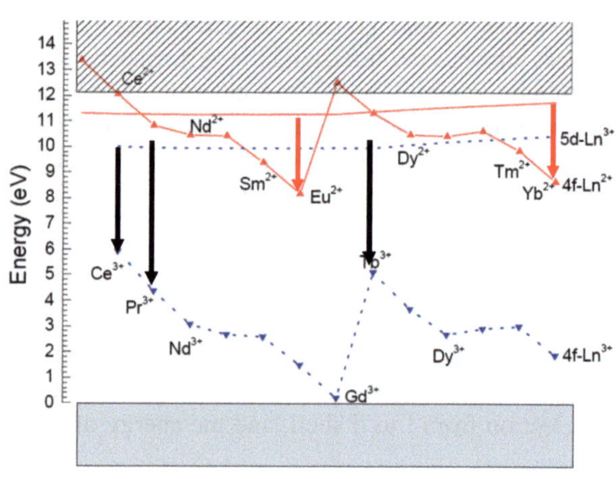

Fig. 4.13 Energy-level scheme of divalent and trivalent lanthanides in CaF_2. Energies corresponding to Ln^{3+} and Ln^{2+} are indicated by *blue dashed lines* and *red solid lines*, respectively. The lowest states of $4f^n$ electronic configurations are represented by *triangles*. Possible $4f^{n-1}5d \rightarrow 4f^n$ transitions are indicated by *vertical arrows*. Reprinted from Ref. [153], copyright 2004, with permission from Elsevier

Fig. 4.13, it is seen that only for a few ions the efficient luminescence, related to the $4f^{n-1}5d \rightarrow 4f^n$ transition, can be observed. The most promising are the Ce^{3+}, Eu^{2+}, and Yb^{2+} ions' where the luminescence varied from red to UV depending on the host. In the case of Pr^{3+}, UV luminescence is observed. No reports on $4f^75d \rightarrow 4f^8$ luminescence in Tb^{3+} are known, probably because the excited $^5D_4$ state from the $4f^8$ electronic configuration is usually located below the lowest state of the $4f^75d$ electronic configuration. It should also be emphasized that in some materials doped with Ce^{3+}, Eu^{2+}, and Yb^{2+}, $4f^{n-1}5d \rightarrow 4f^n$–related luminescence may not appear because the lowest state of $4f^{n-1}5d$ electronic configuration is degenerated with the conduction band [67].

Because pressure causes an increase of crystal-field strength and the centroid shift, pressure-induced red shifts of the luminescence related to the $4f^{n-1}5d^1 \rightarrow 4f^n$ transition have been observed in many lanthanide ions in different hosts (see data in Table 4.7).

4.5.1 5d → 4f Luminescence in Ce^{3+}

Ce^{3+} has the simplest energetic structure, in which a single electron can occupy the ground $4f^1$ and the excited $5d^1$ electronic configuration. The spin-orbit interaction splits the $4f^1$ state into the ground $^2F_{5/2}$ and excited $^2F_{7/2}$ electronic multiplets. Pressure changes the energy of the lowest state of the excited electronic configuration $5d^1$ with respect to the ground state due to the increase of the depression energy. As the result, pressure should diminish the energy of the d-f luminescence, ΔE. There are two important effects: (1) the increase of depression energy; and (2) the change of electron–lattice coupling, $S\hbar\omega$. Pressure dependence of the depression energy is discussed in [100] where it is described by the following relation

$$\frac{\mathrm{d}\Delta E}{\mathrm{d}p} = -n\frac{E_{\mathrm{depr}}}{3B_0} \tag{4.5.3}$$

where n is an exponent describing the dependence of depression energy on the central ion–ligand distance; and B_0 is bulk modulus of the material. The second effect, pressure dependence of the electron–lattice coupling, $S\hbar\omega$, is described by Eq. (4.3.8).

The measured quantities are the emission and absorption (or luminescence excitation) spectra. It is considered that the lowest band in the absorption spectrum is related to the energy difference between the ground state $4f^n$ and the excited state $4f^{n-1}5d$, $\Delta E = E_{\mathrm{abs}}$. Considering further that for the luminescence related to $4f^{n-1}5d \rightarrow 4f^n$, $E_{em} = E_{\mathrm{abs}} - 2S\hbar\omega$, one obtains following relation for the pressure rate of the emission energy:

Table 4.7 Pressure energies and pressure shifts of luminescence related to $4f^{n-1}5d \rightarrow 4f^n$ transitions

Material	E_{depr}, [148] (cm^{-1})	E_{em} (cm^{-1})	dE_{em}/dp (cm^{-1}/kbar)	E_{abs} (cm^{-1})	dE_{abs}/dp (cm^{-1}/kbar)	B_0 (kbar)	n
YAG:Ce^{3+} [160]	27570	19100 $^2F_{7/2}$	−12.5	21750	−12.5	1870 [106]	2.54
		17560 $^2F_{5/2}$	−11.8				
GSAG:Ce^{3+} [76]	27000	17340 $^2F_{7/2}$	−8.3	22320	−15.2	1916 [182]	2.67
		15250 $^2F_{5/2}$	−7.5				
LSO:Ce^{3+} [158]	21390	25480 $^2F_{7/2}$	−11.3	27930	−9.4	1350 [183]	1.80
		23300 $^2F_{5/2}$	−7.7				
$Y_{2.3}Ce_{0.05}Gd_{0.65}Al_5O_{12}$		18926$^2F_{7/2}$	−23				
		17435$^2F_{5/2}$	−20				
$Y_{1.725}Tb_{0.575}Ce_{0.05}Gd_{0.65}Al_5O_{12}$, [138].		18287 $^2F_{7/2}$	−17.3				
		16803 $^2F_{5/2}$	−17.3				
Lu_2S_3:Ce^{3+} [159]		16660	−30				
GGG:Ce^{3+} [161]		18500$^2F_{7/2}$	−12	23500	−21	1710 [109]	
		16500$^2F_{5/2}$		28600	−20		
YAG:Pr^{3+} [165]	27125	31484$^3H_4$	−14.1	34720 [147]	−14.1	1870 [106]	2.92
		29408$^3H_5$					
		27125$^3H_6$					
$La_2Be_2O_5$:Pr^{3+} [166]		33960$^3H_4$	−9.5				
		28840$^3F_3$, $^3F_4$					
$BaSO_4$:Pr^{3+} [167]		40000	7				
Ba_2SiO_4:Eu^{2+} [174]		19970	−12.6	28170			
Ca_2SiO_4:Eu^{2+} [174]		20060	−5.6	23000			
		16300	−7.8	20700			

(continued)

Table 4.7 (continued)

Material	E_{depr}, [148] (cm^{-1})	E_{em} (cm^{-1})	dE_{em}/dp (cm^{-1}/kbar)	E_{abs} (cm^{-1})	dE_{abs}/dp (cm^{-1}/kbar)	B_0 (kbar)	n
$Ca_3Y_2(SiO_4)_3:Eu^{2+}$ [137]		20500	−5.2	28570			
$BAM:Eu^{2+}$ [175]		23370	−6.71				
S0		22080	−9.3				
S1		20750	−9.4				
S2							
$SrF_2:Eu^{2+}$ [176], [179]	8344 [149]	24100	−17.3			690 [184]	4.3
p < 60 kbar		23060	0				
p > 60 kbar		23479	−14.5				3.7
$SrF_2:Eu^{2+}$ p ≤ 70 kbar [177]							
$BaF_2:Eu^{2+}$ kbar	7971 [149]	22300	−16.3 [176]			570 [185]	3.5
			−13.9 [179]				
			−17.7 [177]				3.7
$CaF_2:Eu^{2+}$ [177]		23870	−13.7			500 [177]	3.7
$Sr_2SiO_4:Eu^{2+}$ [168]		20400	13.9				
SI site p < 50 kbar		21100	0				
SI site p > 50 kbar		17540	−4.8				
SII site							
$Sr_2Si_5N_8:Eu^{2+}$ [181]		16260	−3.9				
Sr1		13990	−2.9				
Sr2							

$$\frac{dE_{em}}{dp} = -n\frac{E_{\rm depr}}{3B_0} + \frac{[2(n+1) - 6\gamma]\,2S\hbar\omega}{3B_0} \tag{4.5.4}$$

Then, when $\frac{dS\hbar\omega}{dp} \approx 0$, the luminescence pressure shift can be calculated as

$$\frac{dE_{em}}{dp} \approx -n\frac{E_{\rm depr}}{3B_0} \tag{4.5.5}$$

Pressure dependence of the Ce^{3+} luminescence spectrum has been observed in $Gd_3Sc_2Al_3O_{12}$ (GSAG) [76], Lu_2SiO_5 (LSO) [157] [158], Lu_2S_3 [159], $Y_3Al_5O_{12}$ (YAG) [160], $Y_{2.3}Ce_{0.05}Gd_{0.65}Al_5O_{12}$, and $Y_{1.725}Tb_{0.575}Ce_{0.05}Gd_{0.65}Al_5O_{12}$ [138]. Spectroscopic data related to the $5d^1 \rightarrow {}^2F_{5/2}$, ${}^2F_{7/2}$ transitions in Ce^{3+} are summarized in Table 4.7 where the ambient pressure energies of the $4f^{n-1}5d \rightarrow 4f^n$ transition and pressure shifts for different lanthanides are presented. For the materials where the bulk modulus and centroid shifts are known, the effective exponent that describes the influence of the lattice surrounding on depression energy is calculated using the Eq. (4.5.3) or (4.5.5). One notices that, such as in the case of d-d transitions in TM ions, the effective exponent is usually much smaller than 5. The reason may be the same, i.e., the increasing crystal stiffness in the vicinity of the RE dopant. In the case of GSAG:Ce^{3+} [76] and GGG:Ce^{3+} [161], the pressure shifts of the emission and absorption (excitation) energies have been measured. The absorption shift appeared two times larger than the emission shift, which corresponds to the strong pressure-induced diminishing of the Stokes shift. According to Eq. (4.3.8) the value of $S\hbar\omega$ diminishes with pressure under condition that $6\gamma > 2(n+1)$.

Because the lowest state of the $4f^{n-1}5d^1$ electronic configuration of Ln^{3+} is located near the conduction band edge, in some materials the $4f^{n-1}5d^1$ state is degenerated with the conduction band. In Ce^{3+}-doped materials, for which the $5d^1$ state is degenerated with the conduction band, the Ce^{3+} luminescence is quenched due to photoionization mechanism [159, 162]. No Ce^{3+} emission was observed in La_2O_2S:Ce [159, 162], X_2O_3:Ce (X = La, Y, and Lu) [163], and GGG:Ce [161]. Because pressure diminishes the energy of the $5d^1$ state, a pressure-induced dark-to-bright transition can be observed. The $5d^1 \rightarrow 4f^1$ Ce^{3+} luminescence was observed to appear in Lu_2O_3:Ce [163] at pressure >71 kbar and in GGG:Ce at pressure >30 kbar [161]. The systematic study of pressure dependence of the GGG: Ce^{3+} [161] emission and absorption has shown strong shifts toward lower energy of the two; the lowest absorption bands for all pressure range from ambient to 100 kbar (see Fig. 4.14a).

The absorption spectra of GGG:Ce^{3+} obtained for different pressure measured at RT are presented in Fig. 4.14a. Two absorption bands related to the transitions from the ground state to two first excited states of 5d electronic configuration are seen. Both absorption bands show a noticeable continuous decrease of peak energies with

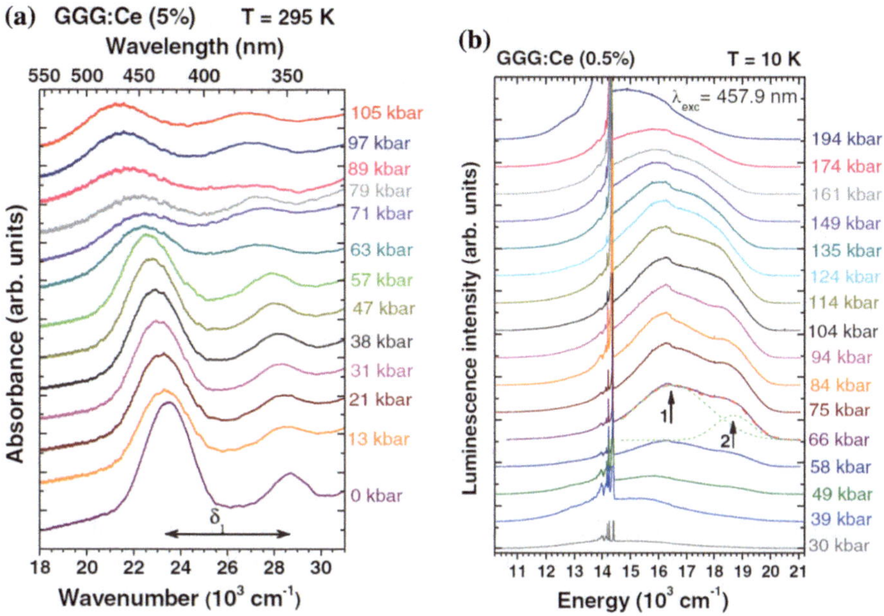

Fig. 4.14 **a** Pressure dependence of the absorption spectra of Ce^{3+} in GGG (Fig. 5 in Ref. [161]). **b** Pressure dependence of the luminescence spectra of Ce^{3+} in GGG. The *strong sharp lines* are related to ruby luminescence. Reprinted from Ref. [161] by permission of American Physical Society

increasing pressure. Despite the observed strong absorption of Ce^{3+} in GGG, no luminescence was observed under direct continuous-wave (cw) excitation into cerium absorption bands. After application of the pressure to this nonemitting system, at approximately 30 kbar, a weak broad luminescence appears in the spectral range between 10400 cm^{-1} (960 nm) and 20400 cm^{-1} (490 nm). The spectra consist of two overlapping broad bands corresponding to two transitions from the lowest sublevel of the excited 5d state to the $^2F_{7/2}$ (lower-energy band [peak labeled no. 1]) and $^2F_{5/2}$ (higher-energy band [peak labeled no. 2]) sublevels of the 4f ground state (see Fig. 4.14b). The pressure dependence of the luminescence is different under different pressures. At pressure <30 kbar, no luminescence was observed. At pressure between 30 and 60 kbar, broadband luminescence (much broader than the absorption band related to the $4f^1 \rightarrow 5d^1$ transition) that increases energy with pressure has been seen. Finally, for pressure >60 kbar, typical Ce^{3+} luminescence related to $5d^1 \rightarrow 4f^1$ ($^2F_{5/2}$, $^2F_{7/2}$) transitions has been observed. The emission at pressure >60 kbar is symmetrical to the respective absorption and presents shifts to the red spectral region with increasing pressure.

Apart from the already-discussed pressure-induced increase of depression energy, two additional effects that influence the phenomenon of pressure-induced crossover of the $5d^1$ and conduction band edge should be considered. The first effect is related to the fact that pressure can shift the energy of the ground state of

Ce^{3+} with respect to the conduction band (position of the $4f^1$ level in the band gap can depend on pressure). Actually, as will be discussed later in this review, this energy increases with increasing pressure. The second effect is the pressure-induced change of the value of the band gap. The compensation of these effects can cause the fact that although the dark-to-bright transition achieves the efficiency of luminescence related to the $5d^1 \rightarrow 4f^1$ transition is not very high. The nonradiative depopulation of the $5d^1$ state in Lu_2S_3:Ce^{3+} at high hydrostatic pressure is discussed in [159]. Although a large negative pressure shift of the energy of the $5d^1$ state was observed, the luminescence of the Ce^{3+} is thermally quenched, even at high pressure. This effect has been explained by the pressure-induced diminishing of the energy of the conduction band with respect to the ground state of Ce^{3+} [159].

4.5.2 *4f5d → 4f² Luminescence in Pr³⁺*

The energetic structure of Pr^{3+} (the $4f^2$ system) is presented in Fig. 4.15. The ground state is the $^4H_4$ state from the $4f^2$ electronic configuration. The highest state of the ground electronic configuration is $^1S_0$, the energy of which is close to that of the lowest state of the first excited electronic configuration 4f5d. The spectroscopic properties of Pr^{3+}, and their possible applications, are determined by the relative energy of the excited 4f5d and $^1S_0$ states [164]. When the $^1S_0$ is lower than 4f5d, after UV excitation to the 4f5d state the system relaxes nonradiatively to the $^1S_0$ state, and efficient luminescence related to transitions from $^1S_0$–$^1I_6$ and from $^3P_0$–^3H_J occurs. Such luminescence enables application of the material to the photon-cutting process where one UV photon is replaced by two red photons. In contrast, when the 4f5d state is lower than the $^1S_0$ state, the efficient UV or blue fast-decaying luminescence related to the parity-allowed 4f5d → $4f^2$ transition takes place. This luminescence enables application of the material as a scintillator. Because pressure diminishes the energy of the 4f5d state and does not influence the energy of the $^1S_0$ state, it can change the energy difference between the $^1S_0$ and 4f5d states. Especially when the $^1S_0$ state has lower energy at ambient pressure, the increase of pressure can qualitatively change the luminescence properties of the Pr^{3+} ion. Unfortunately, the effective excitation due to the $^3H_4$ → 4f5d transition requires the use of deep UV, which is not transmitted by diamonds. As a result, only a few papers are devoted to pressure dependence of the 4f5d → $4f^2$ luminescence in Pr^{3+}. Two-photon excitation is sometimes used to avoid absorption of the exciting light by diamond anvils. Using two-photon excitation, Meltzer et al. [165] investigated the high-pressure dependence of the luminescence related to the 4f5d → $^3H_4$, $^3H_5$, and $^3H_6$ transitions in YAG:Pr^{3+} and obtained negative pressure shifts equal to –14.1 cm^{-1}/kbar for all detected bands. More transparent in the UV region, a sapphire anvil cell (SAC) was used by Mahlik et al. [166] for the investigation of pressure dependence of the luminescence related to the 4f5d → $^3F_3$, $^3F_4$, and $^3H_4$ transitions in $La_2Be_2O_5$:Pr^{3+} under excitation of 250 nm in a pressure range ≤ 22 kbar. A negative pressure shift equal to –9.5 cm^{-1}/kbar

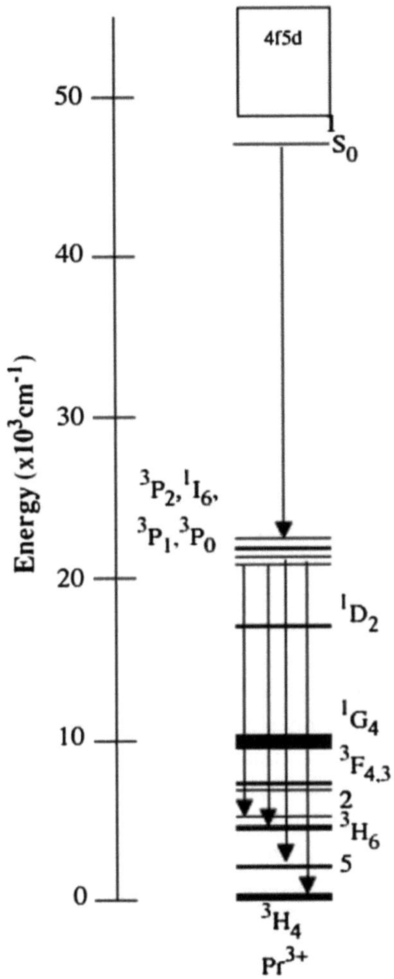

Fig. 4.15 Energetic structure and radiative transitions in Pr^{3+}

was detected. In the same paper [166], luminescence kinetics measured in DAC are presented. The luminescence lifetime of the 4f5d → $4f^2$ transition was found to be equal to 10.5 ns and stable against pressure ≤ 164 kbar [166].

Wang et al. [167] investigated the pressure dependence of the kinetics of luminescence related to transitions originating from the mixed 4f5d and $^1S_0$ states in $BaSO_4{:}Pr^{3+}$. They found that pressure causes an increase of the $^1S_0$ state lifetime from 82 ns at ambient pressure to 240 ns at 89 kbar. In addition, the broadband part of the emission spectrum peaking at 250 nm (40000 cm^{-1}) is quenched when the pressure is increased. To explain the increase of the luminescence lifetime, Wang et al. assumed that radiative transitions from the $^1S_0$ state are electric-dipole type and are allowed due to the interaction with the higher lying 4f5d state due to the spin-orbit interaction. Simple calculations have shown that the measured increase of lifetime with pressure takes place when the energetic distance between the 4f5d and $^1S_0$ states increases when pressure increases with a rate equal to 7 cm^{-1}/kbar. Although it may seem unexpected, the new results on high-pressure spectroscopy of Eu^{2+}-activated $Sr_2SiO_4:Eu^{3+}$ [168] have shown that the energy of $4f^65d \rightarrow 4f^7$ transitions can increase with pressure; therefore, the interpretation presented in paper [167] is reasonable.

4.5.3 *$4f^6$ 5d → $4f^7$ Luminescence in Eu^{2+} and $4f^{13}5d \rightarrow 4f^{14}$ Luminescence in Yb^{2+}*

Eu ions can exist in the divalent state when they are located in the cation sites in the lattices where they replace divalent ions such as Ba^{2+}, Sr^{2+}, and Ca^{2+} [112]. However, there are also reports on Eu^{2+} ions that replace univalent cations such as K^+ in $KMgF_3$ [169]. It has also been reported that Eu^{2+} ions might occupy empty lattices sites in β-SiAlON [170]. There are no reports on Eu^{2+} in trivalent cation sites. The ground electronic configuration of Eu^{2+} is $4f^7$ with the ground state $^8S_{7/2}$ and the first excited state $^6P_{7/2}$. Depending on the value of the depression energy, the energy of the lowest state of the $4f^65d$ electronic configuration can be either greater or smaller than the energy of the $^6P_{7/2}$ state. Thus, the luminescence of the Eu^{2+} can be either the sharp line(s) emission related to $^6P_{7/2} \rightarrow {}^8S_{7/2}$ transition, decaying with the lifetime of the order of milliseconds, or the broadband luminescence related to $4f^65d \rightarrow {}^8S_{7/2}(4f^7)$, decaying with the lifetime of the order of microseconds.

Because the increase of pressure causes an increase of the depression energy [see Eq. 5.3] and does not significantly change the energies of the states belonging to the $4f^7$ electronic configuration, it is expected that in the Eu^{2+}-doped materials, where the $^6P_{7/2} \rightarrow {}^8S_{7/2}$ transition is seen at ambient pressure, this luminescence can be replaced by the luminescence related to the $4f^65d \rightarrow {}^8S_{7/2}(4f^7)$ transition at high pressure. The pressure-induced level crossing for Eu^{2+} ion is presented in Fig. 4.16. Such a situation was shown for $LiBaF_3:Eu^{2+}$ by Mahlik et al. [140]. They measured the pressure dependence of the emission spectra and luminescence kinetics obtained at different temperatures for pressure ≤ 200 kbar. In Fig. 4.17a, b, luminescence spectra of $LiBaF_3:Eu^{2+}$ obtained at different pressures at ambient temperature of (a) the entire spectra and (b) only the region of the $^6P_{7/2} \rightarrow {}^8S_{7/2}$ emission are presented.

At ambient pressure, the luminescence spectrum of $LiBaF_3:Eu^{2+}$ consists of sharp lines at the spectral range 27,500–28,000 cm^{-1} related to the internal transitions $^6P_{7/2} \rightarrow {}^8S_{7/2}$ in the Eu^{2+}. Apart from these sharp lines, the broad emission band peaking at 24,500 cm^{-1} is seen. This broadband emission was earlier attributed to the emission from the excited electronic manifold of Eu^{2+}, $4f^65d^1$ [171–173]. It was shown latter by Mahlik et al. [140] that this band may not be related to the $4f^65d^1$ state and should be attributed to recombination of the Eu^{2+}-trapped exciton

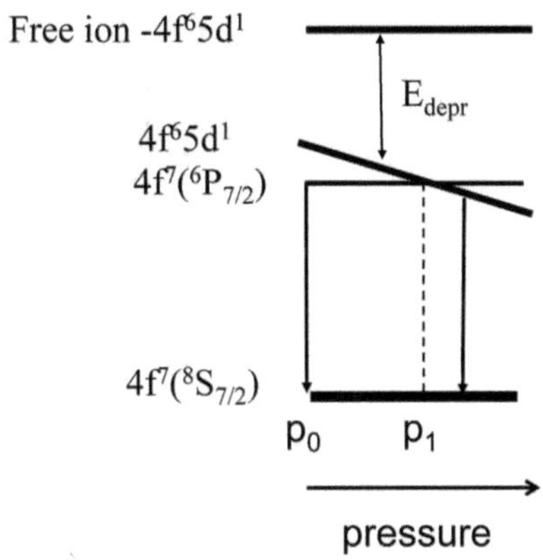

Fig. 4.16 Schematic diagram of the energies of the $4f^7$ and $4f^65d$ states of Eu^{2+} in dependence of pressure; p_0 and p_1 are ambient pressure and the pressure when the $4f^7$ ($^6P_{7/2}$) and $4f^65d$ crossover take place, respectively

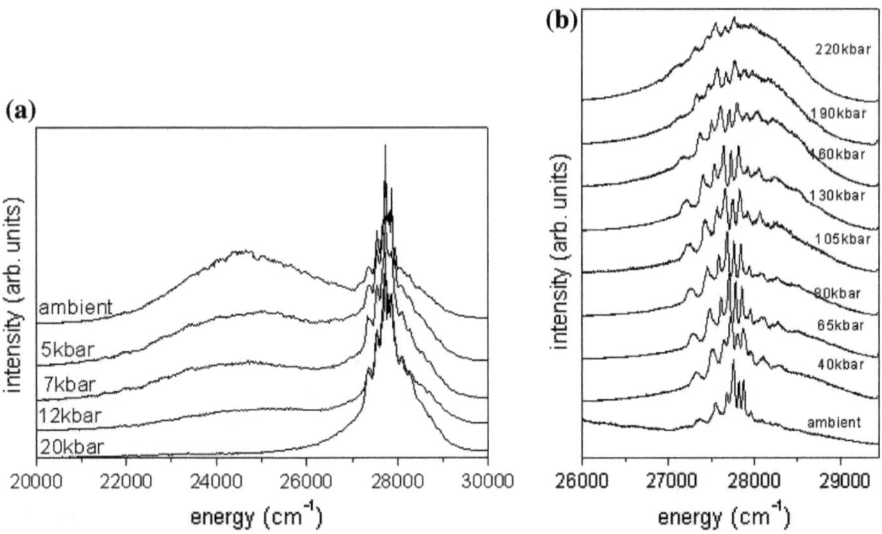

Fig. 4.17 Luminescence spectra of $LiBaF_3$:Eu^{2+} obtained at different pressures and at ambient temperature of (**a**) the entire spectra and (**b**) only the region of the $^6P_{7/2} \rightarrow {}^8S_{7/2}$ emission. The spectra are normalized with respect to the maximum intensity of the $^6P_{7/2} \rightarrow {}^8S_{7/2}$ emission. Reprinted from Ref. [140] by permission of IOP Publishing

(influence of pressure on impurity-trapped exciton states will be discuss later in this review). It is seen that the contribution from the impurity-trapped exciton (the broadband peaked at 24,500 cm^{-1}) diminishes with the increase of pressure and vanishes at 20 kbar (see Fig. 4.17a). This effect has been attributed to an increase of the energy of the Eu^{2+}-trapped exciton (ETE) state with respect to the $^6P_{7/2}$ state of

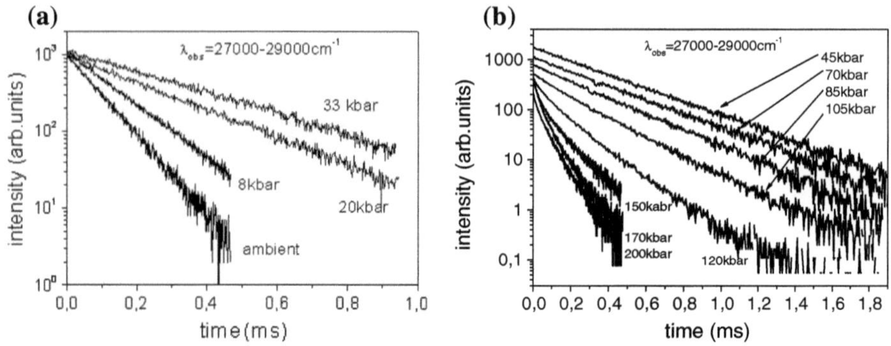

Fig. 4.18 Decays of the emission spectra of $LiBaF_3:Eu^{2+}$ monitored at the spectral range between 27,000 and 29,000 cm^{-1} obtained at different pressures at room temperature (**a**) ambient pressure (–33 kbar) and (**b**) 45 kbar (–200 kbar. Reprinted from Ref. [140] by permission of IOP Publishing

Eu^{2+} when the pressure increases. For pressure between 20 and 40 kbar, the $^6P_{7/2} \rightarrow {}^8S_{7/2}$ emission dominates the spectra at ambient temperature. At pressure >40 kbar, a new broadband luminescence peaking at approximately 28,000 cm^{-1} appears, shifts toward lower energy when pressure increases, and dominates the spectrum at 200 kbar (see Fig. 4.17b). Actually, this emission was attributed to the transition from the $4f^65d^1$ state to the ground state $^8S_{7/2}$ in Eu^{2+}. The luminescence decays of the emission measured in the spectral region of the $^6P_{7/2} \rightarrow {}^8S_{7/2}$ transition for different pressures are presented in Fig. 4.18a, b.

It is seen that when the pressure increases, the lifetime increases in the pressure range from ambient to 33 kbar, is stable at pressure between 33 and 45 kbar, and then, for pressure >45 kbar, begins to decrease. This effect was interpreted in [140] by the mixing of the $^6P_{7/2}$ state with europium-trapped exciton state and with $4f^65d^1$ state, which allowed the parity-forbidden $^6P_{7/2} \rightarrow {}^8S_{7/2}$ transition. At pressure range from ambient to 33 kbar, mixing with the trapped exciton state decreases with increasing pressure because of the increase of the energy of the exciton state; then at pressure >45 kbar, the mixing with the $4f^65d^1$ increases due to the decrease of the energy of the $4f^65d^1$ state.

In materials where the energy of the $4f^65d^1$ is lower than the energy of the $^6P_{7/2}$ state, pressure diminishes the energy of the emission but does not qualitatively change the spectrum. The ambient pressure energies of the $4f^65d \rightarrow {}^8S_{7/2}(4f^7)$ transition and respective pressure shifts of the Eu^{2+} emission in different materials are presented in Table 4.7.

Luminescence related to the $4f^65d \rightarrow {}^8S_{7/2}(4f^7)$ transition in Eu^{2+} was investigated in different pressures in $Ba_2SiO_4:Eu^{2+}$ and $Ca_2SiO_4:Eu^{2+}$ [174]. A single-broadband luminescence in $Ba_2SiO_4:Eu^{2+}$ and two-broadband luminescence in $Ca_2SiO_4:Eu^{2+}$ have shown red shifts with increasing pressure. Luminescence of $BaMgAl_{10}O_{17}$(BAM):Eu^{2+} [175] consists of three bands related to different Eu^{2+} sites in the BAM lattice. It was found that all bands shift to the red region when pressure increases [175]. A similar effect was observed in

$Ca_3Y_2(SiO_4)_3$ doped with Eu^{3+} and Eu^{2+} [137]. The Eu^{2+} provided broadband luminescence that showed a red pressure shift equal to –5.2 cm^{-1}/kbar. In SrF_2: Eu^{2+}, the influence of pressure on the single broad band, related to $4f^65d \rightarrow {}^8S_{7/2}(4f^7)$ transition peaking at 24,100 cm^{-1}, was measured [176, 177]. The linear red shift of this luminescence was obtained with increasing pressure ≤ 60 kbar; then at >60 kbar, the energy of the emission peak equal to 23,060 cm^{-1} became independent of pressure [176]. This specific behavior of the $4f^65d \rightarrow {}^8S_{7/2}(4f^7)$ luminescence has been attributed to the phase transition in SrF_2 from cubic to orthorhombic phase, which has been reported to occur at 50 kbar [178]. In contrast, luminescence related to the $4f^65d \rightarrow {}^8S_{7/2}(4f^7)$ transition in $BaF_2:Eu^{2+}$, peaking at 22,300 cm^{-1}, has appeared at pressure >30 kbar and thus presents a red shift equal to –16.3 cm^{-1}/kbar for all a pressure range from 30 to 90 kbar. Similar results were earlier obtained by Gath et al. [179] and by Su et al. [177], who measured pressure the dependence of the $4f^65d \rightarrow {}^8S_{7/2}(4f^7)$ emission in $MF_2:Eu^{3+}$ (M = Ca, Sr, Ba) powders (see data in Table 4.7). A very small pressure shift of the $4f^65d \rightarrow {}^8S_{7/2}(4f^7)$ luminescence was observed by Lazarowska et al. [181] in $Sr_2Si_5N_8:Eu^{2+}$. Here, the luminescence was a superposition of two bands related to the $4f^65d^1 \rightarrow 4f^7$ transitions in Eu^{2+} occupying a tenfold (Sr1) and an eightfold coordinated site (Sr2). It was found that pressure diminished the energies of the emission bands with rates equal to –3.9 and –2.9 cm^{-1}/kbar for Eu (Sr1) and Eu(Sr2) site, respectively.

An untypical positive pressure shift of Eu^{2+} luminescence was reported by Barzowska et al. [168] who investigated the influence of high hydrostatic pressure on luminescence related to the $4f^65d \rightarrow {}^8S_{7/2}(4f^7)$ transition in $Sr_2SiO_4:Eu^{2+}$ powder, a mixture of á′ and â phase, ≤ 280 kbar. In both phases, the Eu^{2+} ions substitute the Sr^{2+} ions in the Sr_2SiO_4 lattice in two inequivalent sites: tenfold coordinated SI and ninefold coordinated SII.

At ambient pressure, the luminescence of $Sr_2SiO_4:Eu^{2+}$ consists of two broad bands with maxima at 490 nm (20,400 cm^{-1}) and 570 nm (17,540 cm^{-1}). This feature of $Sr_2SiO_4:Eu^{2+}$ luminescence has been attributed to the interconfigurational parity allowed $4f^65d^1 \rightarrow 4f^7$ transitions in Eu^{2+} ions that occupy two different Sr^{2+} sites: (SI) and (SII). However, there was no agreement concerning the attribution of these two bands to the particular sites (see Discussion and the respective references in [168]). Pressure dependence of the luminescence of $Sr_2SiO_4:Eu^{2+}$ is presented in Fig. 4.19a through c.

Luminescence spectra of $Sr_2SiO_4:Eu^{2+}$ excited at 325 and 473 nm obtained at room temperature at pressure range ≤ 285 kbar are presented in Fig. 4.19a, b. Bands positions as a function of pressure are shown in Fig. 4.19c. It has been shown that pressure causes a red shift of the 17540 cm^{-1} emission equal to (–4.79 ± 0.24) cm^{-1}/kbar. A very similar pressure dependence of the orange band has been also observed for $Ca_2SiO_4:Eu^{2+}$ [174]. The influence of pressure on the bands with energy of 20,400 cm^{-1} is different. Surprisingly, when the pressure increases from ambient to 50 kbar, the emission band distinctly shifts to the higher energy with the rate of (13.9 ± 1.4) cm^{-1}/kbar; or pressure >50 kbar, the luminescence energy does not depend of pressure. This effect is especially curious

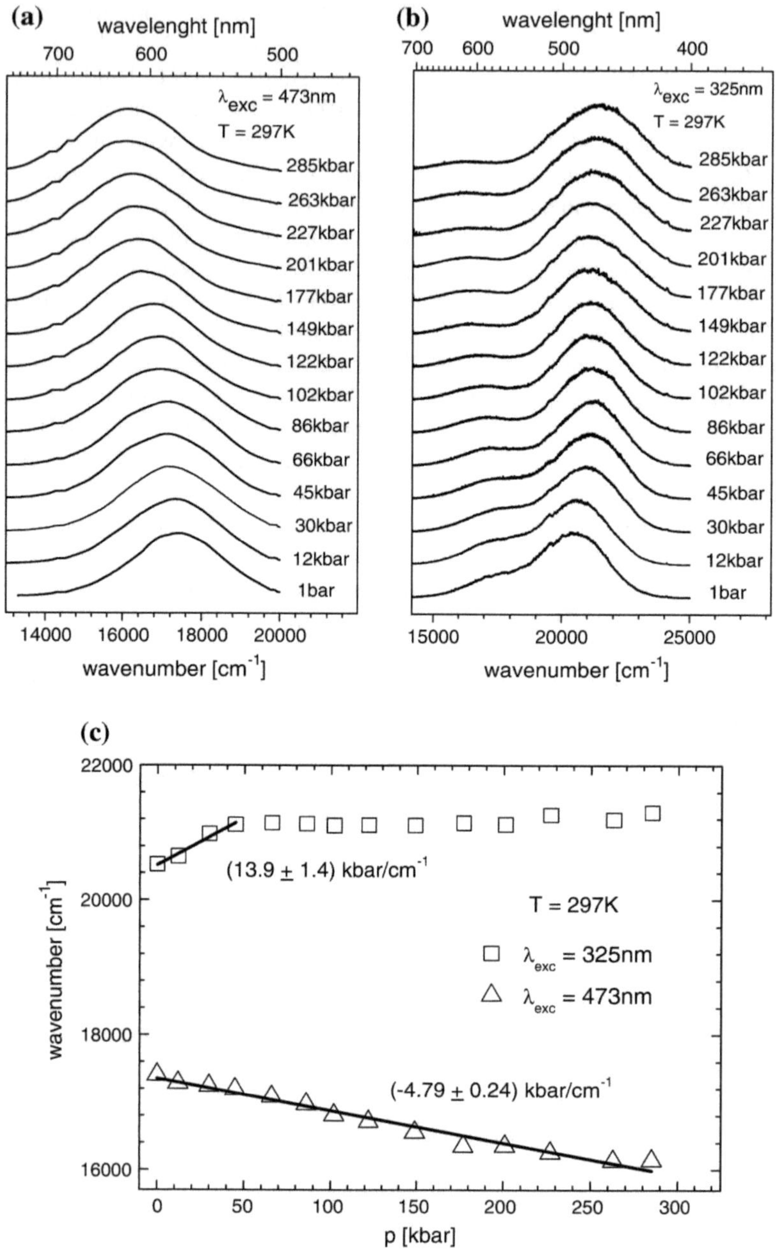

Fig. 4.19 Emission spectra of Sr_2SiO_4:Eu^{2+} obtained at different pressures: **a** excited with 473 nm, and **b** excited with 325 nm, **c** Influence of pressure on Sr_2SiO_4:Eu^{2+} emission band positions; the band corresponding to Eu^{2+} in SI site is represented by *squares*, and the band corresponding to SII site is represented by *triangles*. Reprinted from Ref. [168], copyright 2012, with permission from Elsevier

because in the case of very similar materials, such as $Ca_2SiO_4:Eu^{2+}$ and $Ba_2SiO_4:Eu^{2+}$, where the $4f^65d^1 \rightarrow 4f^4$ transition occurs, red shifts of the emissions have been observed [174].

Before we discuss the above-mentioned results it should be emphasized that Eqs. (4.5.3) and (4.5.5) describing the pressure shifts take into account changes of two contributions: the centroid shift and crystal-field splitting. The centroid shift should always increase with increasing pressure. Dependence of the crystal-field splitting on pressure can be different. It increases when the central ion is in a cubic field, and pressure does not change the symmetry of the center. For such a case, the emission and absorption bands related to $4f^65d \leftrightarrow {}^8S_{7/2}(4f^7)$ transitions in Eu^{2+} should shift to red when the pressure increases. The crystal-field splitting is equal zero in icosahedral coordination, and pressure does not change them. In specific cases, pressure can diminish crystal-field splitting. This can happen when pressure changes the symmetry of the crystal field from tetrahedral to octahedral or from octahedral to tetrahedral because such changes invert the sequence of split 5d states (see Fig. 4.3).

To explain the blue shift of Eu^{2+} luminescence with increasing pressure, it was considered that this emission is related to the $4f^65d^1 \rightarrow 4f^7$ transition in a tenfold coordinated Eu^{2+} (SI) site. The Eu^{2+} ion and its ligands in SI site are presented in Fig. 4.20. Among 10 oxygen ions, one can distinguish the 6 that form the octahedron (green balls) and the 4 that form the tetrahedron (red balls). This choice looks arbitrary, but in general one can distinguish two contributions to the crystal field—the tetragonal and the octagonal—independently of which ions are attributed to the specific component.

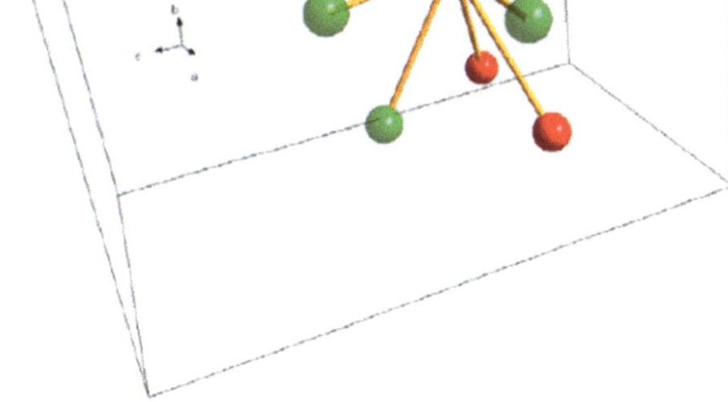

Fig. 4.20 Ligands position in tenfold coordinated Eu^{2+} sites: *green balls* indicate ligands assigned to octahedral symmetry, and *red balls* indicate ligands assigned to tetrahedral symmetry. Adapted from Ref. [168], copyright 2012, with permission from Elsevier

According to the crystal-field theory, in octahedral (O_h-point group) and tetrahedral (T_d-point group), the crystal field splits the excited electronic configuration of Eu^{2+}, $4f^65d^1$ into $4f^65d^1(e)$ and $4f^65d^1(t_2)$ states where the lower state is $4f^65d^1(e)$ and $4f^65d^1(t_2)$ in the octahedral and tetrahedral fields, respectively (see Fig. 4.3). Accordingly, a potential of the electrostatic field in a tenfold coordination site is a superposition of potentials of four tetrahedral ligands and six octahedral ligands

$$V_c(\boldsymbol{r}; R_4, R_6) = V_{\text{Tetr}}(\boldsymbol{r}, R_4) + V_{\text{Octa}}(\boldsymbol{r}, R_6) \tag{4.5.6}$$

The distance from the central ion to the tetrahedral and octahedral ligand is described by the R_4 and R_6 parameters, respectively (see Fig. 4.20). One notices that effective splitting depends on the contributions from the tetrahedral and octahedral fields, which can depend on pressure in a different way. In the paper [168], the energies of the $4f^65d^1(e)$ and $4f^65d^1(t_2)$ states represented by E and T levels were expressed as a function of R_4, R_6 as follows:

$$E = \left(\frac{4}{R_6^5} - \frac{16}{9R_4^5}\right) < r^4 >_{3d} T = \frac{8}{27}\left(-\frac{9}{R_6^5} + \frac{4}{9R_4^5}\right) < r^4 >_{3d} \tag{4.5.7}$$

In Fig. 4.21, one can see a dependence of energies of the E and T levels on distance (R_4, R_6). Point $(R_4 = 3.0\text{Å}, R_6 = 3.1\text{Å})$ can be approximately considered as the starting point in ambient pressure. The pressure should shrink the structure of the site. For example, shrinking the structure along parameter R_4 with holding $R_6 = 3.1\text{Å}$ constant causes an increase of the energy of the T level up to the crossing with the E level and then a flat plateau on the E level. Actually, the splitting of the $4f^65d^1$ electronic configuration of the Eu^{2+} is determined by the splitting of a single $5d^1$ orbital. Thus, such dependence of the energy of the E and T state splitting, as presented in Fig. 4.21, can explain the untypical pressure

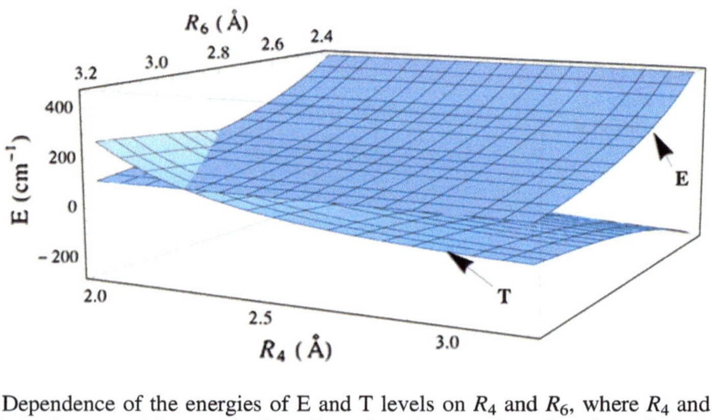

Fig. 4.21 Dependence of the energies of E and T levels on R_4 and R_6, where R_4 and R_6 are the distances from the central ion to the ligands assigned to tetrahedral and octahedral symmetry, respectively. Reprinted from Ref. [168], copyright 2012, with permission from Elsevier

dependence of the energy of the $4f^65d^1 \rightarrow 4f^7$ transition in the tenfold coordinated site. It should be mentioned that in ninefold coordinates, the decomposition of the total crystal field into almost equal contribution of tetrahedral and octahedral symmetry is probably impossible; therefore, the linear dependence on pressure by the energy of the $4f^65d^1 \rightarrow 4f^7$ transition is observed.

Materials doped with Yb^{2+} ($4f^{14}$) exhibit a strong absorption in the blue and UV region related to the spin and parity allowed $4f^{14}(^1S_0) \rightarrow 4f^{13}5d$ interconfigurational transitions. Depending on the host, the luminescence of the Yb^{2+} appears due to the $4f^{13}5d \rightarrow 4f^{14}$ transition, which is characterized by a medium Stokes shift; alternatively, the anomalous luminescence is observed (the influence of pressure on anomalous luminescence will be discussed latter in this review). When pressure is applied to the Yb^{2+} system, similar to the case of Eu^{2+}, red shifts of the absorption and emission spectra are expected, which can be described by Eqs. (4.5.3) and (4.5.4). The quantitative difference between the $4f^{13}5d \rightarrow 4f^{14}(^1S_0)$ transition in the Yb^{2+} and $4f^65d \rightarrow 4f^7(^8S)$ in Eu^{2+} results from the spin-selection rule.

In the case of Eu^{2+}, the ground state is the high-spin one, and the first excited state, according to the Hund rule, has the same multiplicity equal to 8. In the case of Yb^{2+}, the ground state is the spin singlet, whereas the first excited state, according the Hund rule, should be the spin triplet with multiplicity equal to 3. Therefore, compared with the $4f^65d \rightarrow 4f^7(^8S)$ transition in Eu^{2+}, the luminescence related to $4f^{13}5d \rightarrow 4f^{14}(^1S_0)$ transition is spin forbidden and has a long lifetime that reaches milliseconds; in addition, the luminescence kinetics strongly depends on temperature. Mahlik et al. [186] investigated the pressure and temperature dependence of the $4f^{13}5d \rightarrow 4f^{14}(^1S_0)$ luminescence in $MgF_2{:}Yb^{2+}$ in pressure ≤ 200 kbar and in the temperature range between 10 and 600 K. Positions of the emission bands versus pressure are presented in Fig. 4.22.

The results obtained confirmed the existence of the two phase transitions: one at 91 kbar from rutile to orthorhombic $CaCl_2$-type structure and at one approximately

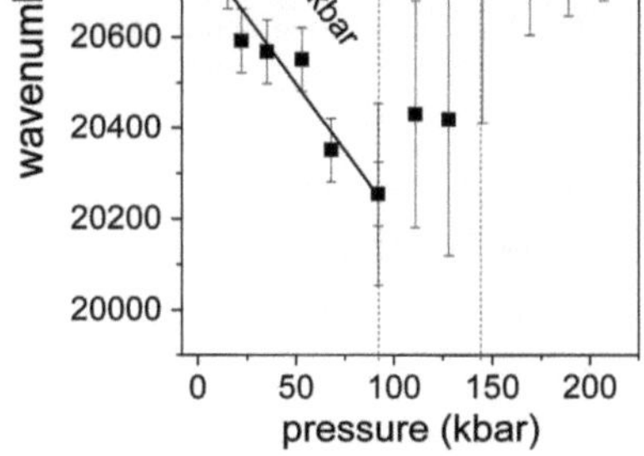

Fig. 4.22 Position of the $MgF_2{:}Yb^{2+}$ emission band versus pressure. Reprinted from Ref. [186], copyright 2015, with permission from Elsevier

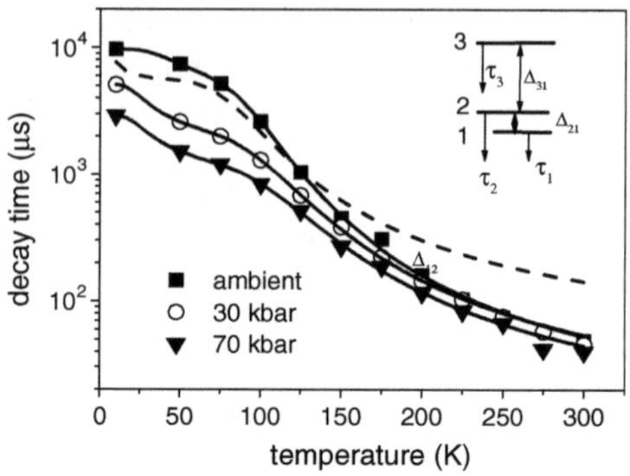

Fig. 4.23 Decay times of $MgF_2:Yb^{2+}$ luminescence at ambient pressure and 30 and 70 kbar as a function of temperature. Fitted dependences are presented by *solid curves*. The *dashed curve* represents the dependence obtained with data from the paper [180]. The energetic structure of the excited state is presented in the inset of the figure. Adapted from Ref. [186], copyright 2015, with permission from Elsevier

140 kbar where the MgF_2 orthorhombic phase is transformed into a modified fluorite-type structure. In these phases, the photoluminescence of $MgF_2:Yb^{2+}$ consisted of a broad band related to the $4f^{13}5d \rightarrow 4f^{14}$ transition in Yb^{2+}. Only in the low-pressure rutile phase was the red shift of $4f^{13}5d \rightarrow 4f^{14}$ luminescence equal to −5.9 cm^{-1}/kbar observed, and it was related to the increased energy of the splitting of the $4f^{13}5d$ electronic configuration. The luminescence decays versus temperature were measured for ambient pressure and 30 and 70 kbar. The decays were single exponential for all pressure and temperatures. The results are presented in Fig. 4.23.

The specific dependence of the lifetime on pressure was analyzed using a three-level scheme suggested by Lizzo et al. [180] where the lifetime was calculated using the following formula:

$$\frac{1}{\tau(T)} = \frac{\sum_{i=2}^{3} \frac{1}{\tau_i} \exp\left(\frac{-\Delta_{i1}}{kT}\right)}{\sum_{i=2}^{3} \exp\left(\frac{-\Delta_{i1}}{kT}\right)} \quad (4.5.8)$$

The fitted data are listed in Table 4.8. The high-pressure results showed that for the low-pressure rutile phase, pressure does not change the ordering of the first three excited states of Yb^{2+}, but it causes a decrease in the luminescence lifetimes of the

Table 4.8 Energetic structure and radiative lifetimes of the first excited states of $MgF_2:Yb^{2+}$

Pressure	Ambient	30 kbar	70 kbar	Data from the paper [180]
τ_1 (μs)	9670 ± 25	5115 ± 10	2915 ± 15	23400
Δ_{21} (cm^{-1})	97 ± 3	52 ± 2	53 ± 3	6.2
τ_2 (μs)	1500 ± 90	810 ± 20	480 ± 25	2900
Δ_{31} (cm^{-1})	520 ± 11	470 ± 10	473 ± 16	343 (494 exp.)
τ_3 (μs)	2.5 ± 0.4	3.0 ± 0.2	2.5 ± 0.4	13

two lowest excited states. This effect is accompanied by a significant decrease in total luminescence intensity and a small decrease of effective lifetime. This effect may be attributed to the nonradiative depopulation of these states. Because the lowest excited states are spin triplets, the probability of radiative transitions to the ground singlet state is small. As a result, the small but increasing probability of nonradiative depopulation of these sates can strongly quench the luminescence intensity.

4.5.4 *d-f Luminescence in Actinides*

The 5f6d → $5f^2$ luminescence of U^{4+} Cs_2NaYCl_6 under different pressures has been investigated by Behrendt et al. [187]. Three emission bands peaked at 27,200, 26,500, and 23,600 cm^{-1}, with respective pressure shifts of –22.6, –24.7, and –24.3 cm^{-1}/kbar, have been attributed to transitions from the $5f6d(t_{2g})$ to $^3H_4(^3T_{1g})$, $^3H_4(^3T_{2g})$, and $^3F_2$ + $^3H_5(^3T_{2g})$ states. The pressure shifts are much greater than those for lanthanides.

4.6 Influence of Pressure on Ionization and Charge-Transfer Transitions

Location of the localized states related to divalent and trivalent rare-earth ions relative to the valence and conduction bands of the host lattice is one of the most important factors that control the luminescence properties of rare-earth ions in solids. The location of the ground states of Ln^{2+} ions with respect to the valence band can be estimated from the energies of the charge-transfer transitions (CTT), which are responsible for the broad bands in the excitation spectra of Ln^{3+} ions. CTT is considered to be a transition of an electron from ligands to the Ln ion. In the energetic diagram, it corresponds to the transition from the top of the valence band to the Ln^{2+} level. The location of Ln^{3+} can be estimated if the energy of the ionization transition (IT) is known. The IT is the opposite process to the CTT and corresponds to the transition of an electron from the Ln ion to the conduction band.

The locations of Ln^{2+} and Ln^{3+} levels with respect to the bands are presented in Fig. 4.24a. It is visible that the energetic distance between the top of the valence band and the ground state of Ln^{2+} is relatively small for Sm^{2+}, Eu^{2+}, and Yb^{2+}. The respective CTTs in Sm^{3+}, Eu^{3+}, and Yb^{3+} are indicated by red arrows. As the result, the archival literature contains a large amount of information on CTT in Eu^{3+}, Sm^{3+}, and Yb^{3+}. In other ions, the CTT energies are rather large, usually larger that the band gap [188]. Srivastava and Dorenbos [188] analyzed the CTT and estimated absolute energy location of the ground states and $4f^{n-1}5d$ states in Ln^{2+} and Ln^{3+} ions in Ca-αSiAlON.

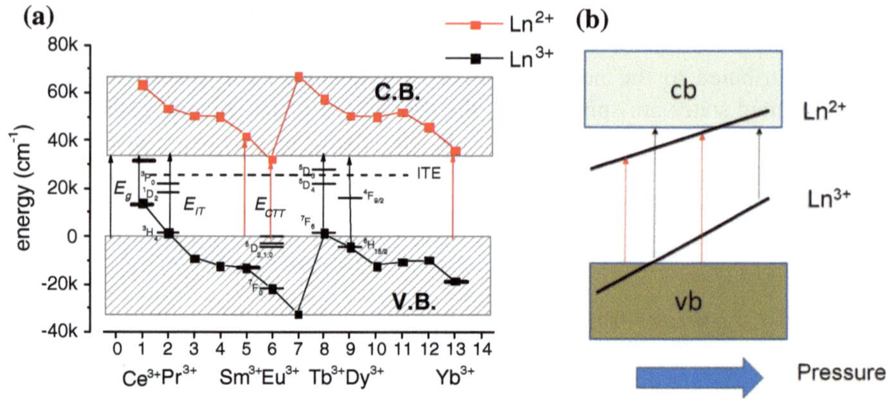

Fig. 4.24 **a** Diagram representing the energies of Ln^{3+} and Ln^{2+} with respect to the conduction and valence band edges. For Dy^{3+}, Eu^{3+}, Pr^{3+}, and Tb^{3+}, the energies of selected states of the $4f^n$ electronic manifold are also displayed; n is the number of electrons in the 4f shell of Ln^{3+}. The dashed line represents the energy of the ITE. **b** Expected dependence of energies of Ln^{2+} and Ln^{3+} ions on pressure. The CTT is represented by red arrows and the IT by *black arrows*

The energies of IT are the smallest for Ce^{3+}, Pr^{3+}, Tb^{3+}, and Dy^{3+} and are indicated by black arrows in Fig. 4.24a. A strong broadband related to the IT was observed by Boutinaud et al. [189] in several oxide lattices that contained metal cations M^{n+} with the empty d shell: Ti^{4+}, Zr^{4+}, V^{5+}, Nb^{5+}, and Ta^{4+} doped with Pr^{3+}. This band has been attributed to ionization of Pr^{3+} and creation of the excited state Pr^{4+} -$M^{(n-1)+}$ [190], which is called the "intervalence charge transfer state" (IVCTS) [189, 191]. It has bee shown that when the energy of IVCTS is low enough, it quenches the emission from the $^3P_0$ state of Pr^{3+} [189]. It should be emphasized here that after ionization of the Ln^{3+} ion that replaces the trivalent ion in the lattice, the Ln^{4+} and the electron create the bound-exciton states, independently if the lattice contains metal ions with empty d shell, which is called the "impurity-trapped exciton" (ITE). The ITE state (ITES) has bee observed in $LiNbO_3{:}Pr^{3+}$ by excited-state absorption (ESA) [193]. The luminescence of Pr^{3+} depends on the energy of the ITES. If this energy is greater than the energy of the $^3P_0$ state, the luminescence of Pr^{3+} contains lines related to the transitions from the $^3P_0$ and $^1D_2$ states; however, when the energy of the ITES is lower than the that of the $^3P_0$ state, the $^3P_0$ state–related emission is quenched [189, 191] and [193]. In contrast to the CTT, the IT often does not appear as a strong band in the Ln^{3+} luminescence excitation spectra [129]. In such a situation, estimating the location the Ln^{3+} ground state in the band gap is more difficult.

To predict the influence of pressure on the location of the levels of Ln^{2+} and Ln^{3+} with respect to the bands edges of the host, one should consider how the chemical environment of the lanthanide changes the $4f^n$ binding energy. It is known that the host lattice diminishes the binding energy of 4f electrons several times with respect to the free-ion energies (see Ref. [192]). Dorenbos [154] proposed a model in which binding energies of the free ions are diminished by an electrostatic Coulomb interaction of f electrons with the negative potential induced in the lattice by the lanthanide cation. In the framework of this approach, the binding energies of divalent at trivalent ions in the lattices, $E(\mathrm{Ln}^{2+}, n+1)$ and $E(\mathrm{Ln}^{3+}, n)$ (see Eq. (4.5.2)) are related to the respective binding energies of free ions in vacuum: $E_0(\mathrm{Ln}^{2+}, n+1)$ and $E_0(\mathrm{Ln}^{3+}, n)$ as follows [154]:

$$E(\mathrm{Ln}^{2+}, n+1) = E_0(\mathrm{Ln}^{2+}, n+1) + 1440\frac{2}{R_{2+}} \tag{4.6.1}$$

$$E(\mathrm{Ln}^{3+}, n) = E_0(\mathrm{Ln}^{3+}, n) + 1440\frac{3}{R_{3+}} \tag{4.6.2}$$

In formulas (4.6.1) and (4.6.2), the energies are expressed in eV and $R_{2+,}$ and R_{3+} are the ionic radii of Ln^{2+} and Ln^{3+} expressed in picamoles. The binding energies are calculated with respect to the vacuum level; therefore, the free-ion basic energies are $E_0(\mathrm{Ln}^{2+}, 7) = -24.92$ eV and $E_0(\mathrm{Ln}^{3+}, 6) = -42.97$ eV. Accordingly, when pressure compresses the system, the energies $E(\mathrm{Ln}^{2+}, n+1)$ and $E(\mathrm{Ln}^{3+}, n)$ increase with respect to the vacuum level. Another model has been proposed by Gryk et al. [125] and Grinberg [67] where the influence of a lattice on 4f electrons has been considered by the existence of a confined potential well crated by ligands. This is similar to the Dorenbos model, i.e., repulsing the influence of the confined potential well causes an increase of binding energy by a value proportional to $\frac{C}{R^m}$. The exponent m depends on the type of interaction; $m = 1$ corresponds to the electrostatic Coulomb repulsion proposed by Dorenbos [154]; $m = 2$ corresponds to the potential of the quantum well; and $m = 12$ corresponds to the short-distance Lennard Jones potential. Independently of the model, energies of the Ln^{2+} and Ln^{3+} ions should increase with pressure with respect to the energy of the free ion and with respect to the ambient pressure-binding energy. Assuming that the energies of the conduction and valence bands are not changed by pressure, pressure causes the increase of CTT energies and the decrease of IT energies. The situation is presented in Fig. 4.24b.

4.6.1 *Model of Impurity-Trapped Exciton States*

To consider the influence of pressure on the CTT and IT and on the luminescence and luminescence kinetics of Ln^{3+} and Ln^{2+} ions, a more convenient description called the ITE [67] [68] can be used that allows precise specification of the IT and CCT states. In this description, a perturbation approach is applied where the potential created by the Ln ion is the perturbation of the ideal crystal potential. A similar approach was successfully used in the 1950s to describe the "shallow" donor and acceptor states in semiconductors [196]; it is called the "effective mass approximation" (EMA).

Although in the dielectrics the mobilities of the carriers are too small and their effective masses too large for the formation of "shallow" donor and acceptor states, the initial Hamiltonian describing the system of the host and $Ln^{\alpha+}$ is the same as that for donor and acceptor impurities in semiconductors. In there is a one-electron approximation after the IT, the system consists of an unperturbed crystal with periodic potential—ionized lanthanide ion $Ln^{(\alpha+1)+}$—and a free electron (in the conduction band), whereas CTT creates a lanthanide ion with an extra electron $Ln^{(\alpha-1)+}$ and a hole in the valence band. To describe both cases, the following Hamiltonian has been proposed [125, 194, 195]

$$H = H_0 + H' \tag{4.6.3}$$

where

$$H_0 = \frac{-\hbar^2\nabla^2}{2m} + V_{cr}(r)_{r>R} \tag{4.6.4}$$

is the unperturbed Hamiltonian describing the ideal crystal without an Ln ion, and the perturbation potential is given as

$$H' = V_{\mathrm{Ln}}(\mathrm{r})|_{\mathrm{r}<R} - \frac{e^2}{\varepsilon r} + V_{\mathrm{latt}}(\mathbf{\Delta}, \mathbf{r}) \tag{4.6.5}$$

In Eqs. (4.6.3) through (4.6.5), V_{cr} is the lattice periodic potential; and V_{Ln} is the local potential of the $Ln^{(\alpha \pm 1)+}$ ion. Subscripts $r > R$ and $r < R$ denote the potential outside and inside of the first coordination sphere, respectively; R is the average distance between the $Ln^{\alpha+}$ ion and the ligands; $\frac{-e^2}{\varepsilon r}$ is a long-range Coulomb potential of $Ln^{(\alpha \pm 1)+}$; e is the electron charge; and ε is the static dielectric constant of the material. The last term represents the potential related to lattice relaxation, and Δ is the shift of the ligands, which appear as the response of the lattice on localization or delocalization of the electron.

The Hamiltonian (4.6.3) generates three following types of states:

(i) The first state is the band state related to the lattice potential $\boldsymbol{V}_{cr}(\mathbf{r})|_{\mathbf{r}>R}$, as described by Bloch functions φ_{nk}(r), where n is the number of the band and k is the wave vector: These states are the solutions of the unperturbed Hamiltonian (4.6.4) and describes the system far from the Ln ion.

(ii) The second sate is the localized state related to the local potential of $Ln^{(\alpha\pm1)+}$, $\boldsymbol{V}_{\mathrm{Ln}}(\mathbf{r})|_{\mathbf{r}<R}$ represented by atomic functions φ_{fd}(r), which are in fact the states of $Ln^{\alpha+}$ ($4f^n$ states are almost independent of the host and $4f^{n-1}5d$ i.e., weakly dependent on the host)

(iii) The third state is related to the long-range Coulomb potential and lattice-relaxation potential $\frac{-e^2}{\varepsilon r} + \boldsymbol{V}_{\mathrm{latt}}(\mathbf{\Delta}, \mathbf{r})$, which is represented by functions F(r). Obviously, φ_{fd}(r), and F(r) should be orthogonal; therefore, the state described by F(r) is expressed as a superposition of the band states φ_{nk}(r) rather than the φ_{fd}(r) states. It should be emphasized that Hamiltonian (4.6.3) describes the localized state after ionization as well as after CTT. Thus, the states described by F(r) are the ITE states independently of whether they were created by the IT or the CTT.

In the framework of EMA [196], in semiconductors such as Si, Ge, or GaAs, the binding energies of "shallow" donors and acceptors related to the potential $-\frac{e^2}{\varepsilon r} + V''_{\rm latt}(\Delta, {\rm r})$ are equal to 100–200 cm^{-1}; these states are directly related to the conduction and valence bands. The "shallow" states are delocalized because the electron (hole) spreads from the impurity on a distance many times larger than the lattice constant. Because it does not penetrate the core area of the impurity, the binding energy is not dependent on the type of impurity. Obviously, in dielectrics the binding energy if the ITE may not be calculated in the EMA. However, like in semiconductors in the dielectrics after the IT and CTT, respectively, the electron and the hole are bound by the Coulomb potential at the ITE states that are the excited states of the $Ln^{\alpha+}$. Because the bounded electron or hole is spatially located outside of the first coordination sphere of the dopant, the binding energy of the ITE, defined as the difference between the energy of the free carrier and the carrier trapped at the ITE state, $\Delta E_{\rm ITE}$ would depend rather on the lattice type rather the dopant lanthanide element and thus should behave with pressure as the conduction or valence band and not as the ground state of lanthanide.

The $V_{\rm latt}(\Delta,\mathbf{r})$ describes the lattice reaction to the transition of $Ln^{\alpha+}$ into $Ln^{(\alpha \pm 1)+}$ and includes the lattice-relaxation energy $S\hbar\omega$ as well as the electrostatic potential created by shifts of the lattice ions $V''_{\rm latt}(\Delta, {\rm r})$ [194].

Lattice relaxation $S\hbar\omega = K\frac{\Delta^2}{2}$ (K is the elastic constant) accompanies electronic transition in the covalence and ionic crystals. The physical situation is presented by the configurational coordinate diagram in Fig. 4.25. The IT is indicated in Fig. 4.25 by a blue arrow. The IT creates the system in the $Ln^{(\alpha+1)+}$ + e state (solid blue parabola). After IT the system diminishes the energy by value $K\frac{\Delta^2}{2}$ and relaxes to the bound ITE states (blue-dashed parabola) and additionally loses the energy $\Delta E_{\rm ITE}$. After the CTT (red arrow), a lanthanide ion with an extra electron and a hole in the conduction band $Ln^{(\alpha-1)+}$ + h is created (solid red parabola) and relaxes to the ITE state (red-dashed parabola), which diminishes the energy by $K\frac{\Delta^2}{2}$ and $\Delta E_{\rm ITE}$. Because in both cases—the free carrier and the ITE one—the carrier is outside the impurity, the electron–lattice interaction energy $S\hbar\omega$ is approximately the same in both states.

When the $Ln^{\alpha+}$ ion is a dopant in the covalence lattice, $V''_{\rm latt}(\Delta, {\rm r}) = 0$ and the perturbed potential consists of local potential $\boldsymbol{V_{Ln}}$ and Coulomb potential parts. The local part is responsible for the creation of localized states of the $4f^n$ and $4f^{n-1}5d$ electronic configuration, and the smooth Coulomb potential is responsible for the existence of the ITE. The cross-section of the potential is presented in Fig. 4.26a. Here the changes in ligand positions do not influence the potential, which is the same independently whether the electron occupied the localized $Ln^{\alpha+}$ state or the ITE state. When the $Ln^{\alpha+}$ ion is a dopant in the ionic lattice, additional potential

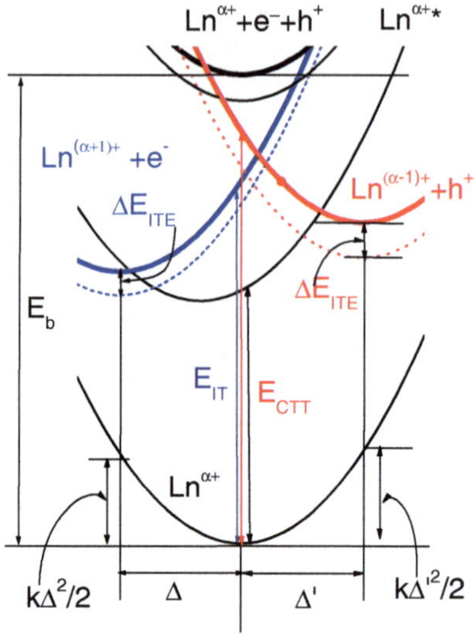

Fig. 4.25 Configuration-coordinate diagram representing the ITE states. The ITE states are represented by *dashed parabolas*

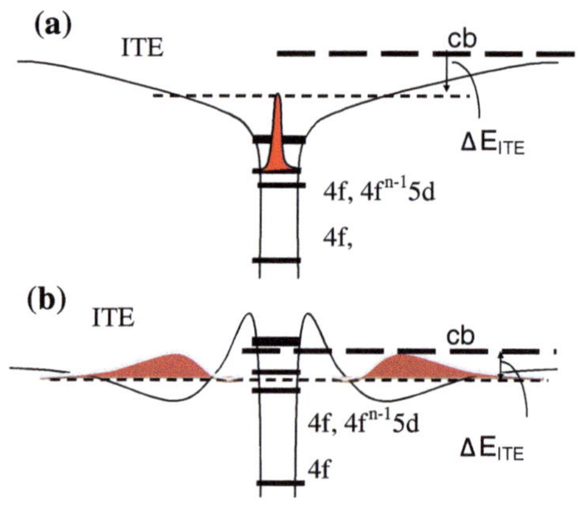

Fig. 4.26 The perturbation potential given by Eq. (4.6.5) for the ITE after the IT **a** in the absence of lattice distortion or in the case of a valence lattice **b** in the presence of lattice distortion in the ionic lattice

$V''_{\text{latt}}(\Delta, \text{r})$ appears after ionization as the result of the shift of negative ligands. This potential additionally diminishes the energy of the ITE state with respect to energy of the ground state of $Ln^{\alpha+}$ (or increases the energy of $Ln^{\alpha+}$ with respect to the ITE state) by a quantity approximately equal to $\Delta\frac{q}{R^2}$, for $r<R-\Delta$, $(R-\text{r})\frac{q}{R^2}$, for $R-\Delta<r<R$ and zero for $r>\text{R}$ [176] where q is the ligand charge. The situation is presented in Fig. 4.26b.

According to quantum mechanical rules, the binding energy of the ITE can be formally expressed as

$$\Delta E_{\mathrm{ITE}} = \int F(\mathbf{r})^{*}[T - \frac{e^2}{\varepsilon r} + V''_{\mathrm{latt}}(\Delta, \mathrm{r})]F(r)\,\mathrm{d}\mathbf{r} \tag{4.6.6}$$

where T represents the operator of kinetic energy of the electron in the conduction band or the hole in the valence band. Unfortunately, there is no unambiguous procedure for the quantitative calculation of ΔE_{ITE}.

The experimental data yield the binding energy ΔE_{ITE} quite large and equal to even a few thousand cm^{-1} [201]. In contrast, it can be small due to repulsive potential $V''_{\mathrm{latt}}(\Delta, \mathrm{r})$.

The diagram presented in Fig. 4.25 can be used to relate the energy of the IT and the $\mathrm{E_{IT}}$, the energy of the CTT (E_{CTT}), the energy of the ITE (E_{ITE}), and the lattice-relaxation energy $S\hbar\omega$. In both cases, one has

$$E_{\mathrm{ITE}} = E_{\mathrm{CTT/IT}} - S\hbar\omega - \Delta E_{\mathrm{ITE}} \tag{4.6.7}$$

4.6.2 High-Pressure Effect on an Anomalous Luminescence in Eu^{2+}- and Yb^{2+}-Doped Materials

Materials doped with Eu^{2+} and Yb^{2+} exhibit strong absorption in the blue and UV regions related to the spin and parity allowed $4f^{n} \rightarrow 4f^{n-1}5d$ interconfigurational transitions. The $4f^{n-1}5d \rightarrow 4f^{n}$ luminescence is observed when the lowest state of $4f^{n-1}5d$ electronic configuration is positioned below the conduction band. Otherwise, the $4f^{n-1}5d \rightarrow 4f^{n}$ emission is either (1) replaced by the red-shifted anomalous luminescence related to the recombination of the Ln^{2+}-trapped exciton (LnTE is the other name for the ITE) [67, 194, 197] or (2) no luminescence is observed. Luminescence of the Ln^{2+} (Ln = Yb, Eu) related to the $4f^{n-}5d \rightarrow 4f^{n}$ transition is characterized by a small or medium Stokes shift, whereas anomalous luminescence is characterized by the large Stokes shift [198, 199]. High-pressure behavior of the anomalous luminescence related to the recombination of the europium-trapped exciton (ETE) was observed in BaF_2:Eu^{2+} [176, 177, 179], mixed crystals $Ba_{0.3}Sr_{0.7}F_2$:Eu^{2+} [176, 179], $Ba_{0.5}Sr_{0.5}F_2$:Eu^{3+} [179], and $LiBaF_3$:Eu^{2+} [14]. The anomalous luminescence in BaF_2:Eu^{2+} and $Ba_{1-x}Sr_xF_2$:Eu^{2+} strongly depends on temperature. It is strong at helium temperature and is quickly quenched when temperature increases. In addition, BaF_2:Eu^{2+} luminescence is different in a single crystal [176, 179] versus in powder [177]. In the single crystal, the ambient pressure luminescence spectrum consists of a single band with a maximum at 16 700 cm^{-1} and is effectively quenched at 200 K [176, 179]. At temperature >200 K, an additional weak band appears in the luminescence spectrum at 24,000 cm^{-1}, which has been attributed to the $4f^{6}5d \rightarrow 4f^{7}$ transition from the thermally populated $4f^{6}5d$ in Eu^{2+} ions [179]. In powder, BaF_2:Eu^{2+} at a temperature range between 10 and 300 K, the luminescence spectrum is a superposition of two bands: the anomalous luminescence with maximum at 17,400 cm^{-1}, and the band with maximum at 22,600 cm^{-1}, which was attributed in [177] to self-trapped exciton (STE) emission but probably is related to the $4f^65d \rightarrow 4f^7$ transition in Eu^{2+}. The differences in the ambient pressure spectra result in different pressure dependence of the ETE emission in crystals and powders; however, the common feature is that at high pressure (>30 kbar for crystal and >47 kbar for powder), the ETE luminescence is replaced by luminescence related to the $4f^65d \rightarrow 4f^7$ transition in Eu^{2+}.

High-pressure luminescence spectra have been studied in detail for BaF_2:Eu^{2+} [177, 179, 176], mixed crystals $Ba_{0.3}Sr_{0.7}F_2$:Eu^{2+} [176, 179], and $Ba_{0.5}Sr_{0.5}F_2$:Eu^{3+} [179]. It was found that for all of these materials, at ambient pressure at low temperature, luminescence is dominated by the broadband emission related to the ETE, and with increasing temperature the relative contribution from the $4f^65d^1(e) \rightarrow 4f^7$ emission increases. In Fig. 4.27a and b, the luminescence spectra for $Ba_{0.5}Sr_{0.5}F_2$:Eu^{2+} obtained at different temperatures and pressures are presented. One concludes that at ambient pressure, the energy of the $4f^65d^1(e)$ state is greater than that of the ETE state. Pressure diminishes the energy of the $4f^65d^1(e)$ state, and at 30 kbar the $4f^65d^1(e)$ state is the first excited state in BaF_2:Eu^{2+}. This situation is similar for pressures <36 kbar in $Ba_{0.5}Sr_{0.5}F_2$:Eu^{2+} and <40 kbar in $Ba_{0.3}Sr_{07}F_2$:Eu^{2+}. For higher pressures, only the $4f^65d^1(e) \rightarrow 4f^7$ emission is observed in both materials.

The effect of pressure quenching of the ETE luminescence is also presented for $LiBaF_3$:Eu^{2+} [140] in Fig. 4.17a. It is seen that at ambient pressure the emission spectrum consists of a broad band with the maximum at 24500 cm^{-1}, which disappeared at pressure >20 kbar.

In Fig. 4.28, energies of the luminescence peaks (the energies of local maximums taken from the spectra) of $Ba_xSr_{1-x}F_2$:Eu^{2+} ($x = 1.0$, 0.5, 0.3, and 0) versus pressure are presented. In the case of BaF_2:Eu^{2+} and mixed $Ba_xSr_{1-x}F_2$:Eu^{2+} ($x = 0.5$ and $x = 0.3$) crystals, pressure causes a reduction in the energy of the anomalous luminescence as well as that of the $4f^65d^1(e) \rightarrow 4f^7$ transition in Eu^{2+}; at pressure lower <20 kbar, the energy difference between the $4f^65d^1(e)$ and ETE increases with pressure. At higher pressure, the energy of the anomalous

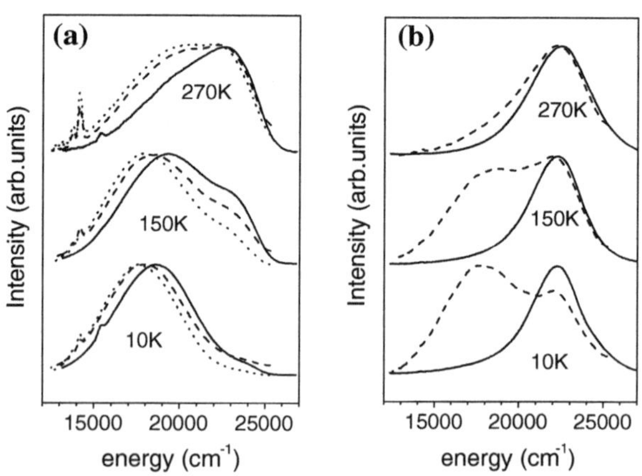

Fig. 4.27 Luminescence spectra for $Ba_{0.5}Sr_{0.5}F_2$:Eu^{2+} for different temperatures and pressures. All spectra are normalized to the maximum value. **a** Spectra for ambient pressure (*solid curves*), 11 kbar (*dashed curves*), and 21 kbar (*dotted curves*). **b** Spectra for 29 kbar (*dashed curves*) and 36 kbar (*solid curves*). Reprinted from Ref. [176] by permission of IOP Publishing

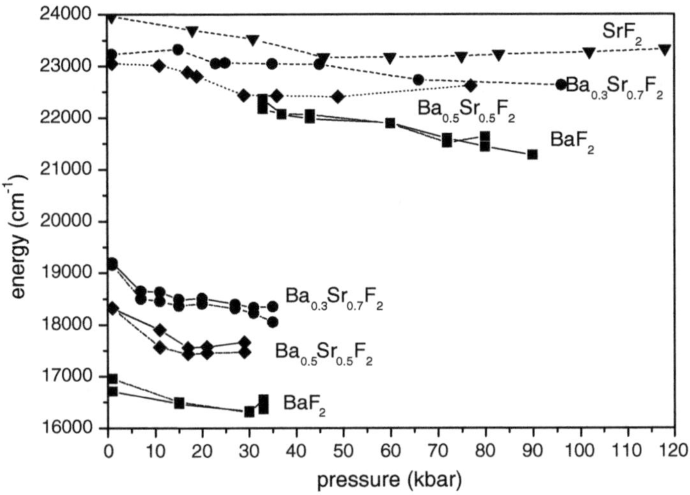

Fig. 4.28 Energies of the trapped exciton and $4f^65d^1(e) \rightarrow 4f^7$ luminescence for different pressure and temperatures. Symbols (*rectangle*), (*rhomb A*), (*circle*) and (*triangle B*) represent data obtained for $BaF_2:Eu^{2+}$ $Ba_{0.5}Sr_{0.5}F_2:Eu^{2+}$, $Ba_{0.3}Sr_{0.7}F_2:Eu^{2+}$, and $SrF_2:Eu^{2+}$, respectively. Solid, dashed-dotted, dotted, and dashed lines represent data obtained at 10, 50, and 250 K and room temperature, respectively. The data for $BaF_2:Eu^{2+}$ and $SrF_2:Eu^{2+}$ were taken from paper [179]. Reprinted from Ref. [176] by permission of IOP Publishing

luminescence decreases less rapidly and even starts to increase. In all materials at pressure >50 kbar, anomalous luminescence is replaced by normal luminescence. Although the $4f^65d^1(e) \rightarrow 4f^7$ transition energy decreases linearly with increasing pressure, anomalous luminescence energy changes are strongly nonlinear above a critical pressure. Such nonlinear behavior can be attributed to the phase transition observed in fluorides at pressures range between 30 and 50 kbar, but still one cannot unambiguously relate the type of luminescence to the specific phase. Considering $BaF_2:Eu^{2+}$ and mixed $Ba_xSr_{1-x}F_2:Eu^{2+}$ crystals, one notices that the simultaneous presence of anomalous and normal luminescence occurs in the cubic phase, and only normal luminescence occurs in the orthorhombic phase. In contrast, in $SrF_2:Eu^{2+}$, only normal luminescence is present regardless of phase, which is attributed to the fact that ETE state is in both phases above the $4f^65d^1(e)$. In the paper [176], it was shown that the nonlinear pressure shift of anomalous luminescence, especially the diminishing of the value of pressure rates of anomalous emission with increasing pressure, is the result of two effects: diminishing of the electron–lattice coupling energy $S\hbar\omega$ and quantum mechanical mixing of the $4f^65d^1(e)$ and ETE states. According to the proposed model [125, 194], the ETE system is formed by Eu^{3+} and an electron occupying the anti-bonding states (the conduction-band states). If the Eu^{2+} replaces divalent metal ions, the Eu^{3+} has additional Coulomb potential, which is in fact responsible for the formation of the ETE state. Because the electron is outside of the Eu ion (it does not matter where it actually is), the ligand anions respond to the ionization of Eu^{2+} to Eu^{3+} by a shift toward the Eu^{3+}. The lattice-relaxation energy is given by $S\hbar\omega = \frac{K}{2}\Delta^2$, which is much larger than that observed for the $4f^7$–$4f^65d^1$ transition. Actually, the large ion shift can generate electrostatic potential, which additionally diminishes the energy of the ETE with respect to the $4f^7$ and $4f^65d^1$ states [176] (see Fig. 4.26b). Pressure influences the system in three directions:

1. Due to compression, pressure causes an increase of $4f^7$ energy with respect to the band states, and the ETE that diminishes energy of the anomalous luminescence and stabilize ETE state.
2. The second effect is the diminishing of $4f^65d^1$ energy with respect to the ground state of Eu^{2+}.
3. The third effect is the influence of pressure on the local ions shifts Δ. If pressure diminishes the value of Δ by an increase of material stiffness, the energies of the $4f^7$ and $4f^65d^1$ states diminish with respect to the ETE state. The two latter effects are actually responsible for the replacement of ETE luminescence by normal luminescence related to the $4f^65d^1 \rightarrow 4f^7$ transition in $BaF_2:Eu^{2+}$ and mixed $Ba_xSr_{1-x}F_2:Eu^{2+}$ ($x = 0.5$ and $x = 0.3$).

Mahlik et al. [201] reported detailed spectroscopic investigation of CaF_2 doped with Yb^{2+} at high hydrostatic pressure. At ambient pressure and at temperatures <175 K, the luminescence consists of a single broad band with the maximum at 18500 cm^{-1}, which is attributed to the recombination of ytterbium-trapped excitons (YTE). This luminescence was found [201] to not depend of pressure but at 85 kbar it is blue shifted to 21630 cm^{-1}. This was attributed to the known phase transition of the CaF_2 crystal from the cubic to the orthorhombic phase [200]. As has been shown, the phase transition in $BaF_2:Eu^{2+}$ and mixed $Ba_xSr_{1-x}F_2:Eu^{2+}$ ($x = 0.5$ and x = 0.3) was accompanied by replacement of the anomalous luminescence by $4f^65d^1 \rightarrow 4f^7$; however, this is not the case with $CaF_2:Yb^{2+}$.

In the paper [201], the high-pressure luminescence properties of $CaF_2:Yb^{2+}$ have been discussed compared with the pressure dependencies observed for the luminescence of $BaF_2:Eu^{2+}$ and $CaF_2:Eu^{2+}$. The difference between the spectral properties of Eu^{2+} and Yb^{2+} CaF_2 lattice was attributed to the fact that the ground and $4f^65d$ state of Eu^{2+} are located deeper in the CaF_2 band gap than the ground and excited $4f^{13}5d$ state of Yb^{2+}, whereas the energies of the trapped exciton states for $Yb^{2+}$and Eu^{2+}, with respect to the conduction band, are approximately the same. Detailed analysis of the emission and excitation spectra has allowed us to conclude that the energy of the $4f^65d$ state in Eu^{2+} is lower than the energy of $4f^{13}5d$ state in Yb^{2+} by 970 cm^{-1}, whereas the $4f^65d$ state of Eu^{2+} is actually located deeper in the band gap than the $4f^{13}5d$ state of Yb^{2+} by 4970 cm^{-1}. The energy of YTE was found to be lower than the energy of the $4f^{13}5d$ state by 3800 cm^{-1}. If the location of the ITE state in the band gap is the same for Eu and Yb, in the $CaF_2:Eu^{2+}$ system the first excited state should be (and it is) the $4f^65d$ state. The scheme representing the energetic structure of $CaF_2:Eu^{2+}$ and $CaF_2:Yb^{2+}$ in real scale is presented in Fig. 4.29.

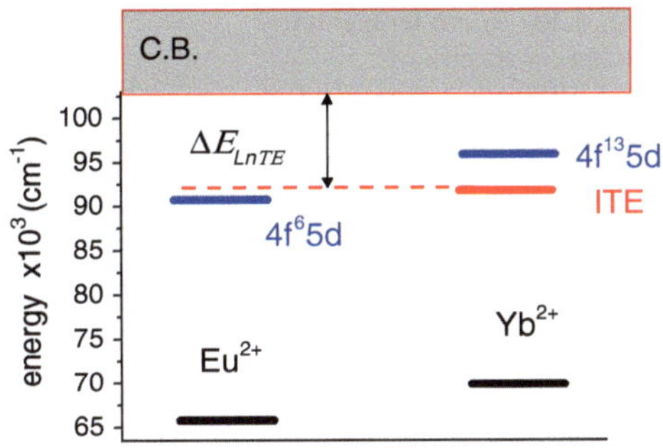

Fig. 4.29 The energetic structure of Eu^{2+} and Yb^{2+} in CaF_2 in cubic phase. Zero at the energetic scale corresponds to the energy of the top of valence band. Reprinted from Ref. [201] by permission of IOP Publishing

4.6.3 Pressure-Induced Luminescence Quenching in Pr^{3+}- and Tb^{3+}-Doped Materials

Considering the diagram in Fig. 4.25, one notices that the luminescence transition from the $Ln^{\alpha+*}$ to the ground state of $Ln^{\alpha+}$ can take place when the energy of the emitting state $Ln^{\alpha+*}$ is lower than energy of the ITE state. When the ITE has lower energy, the large lattice relaxation in the ITE state causes the luminescence quenching due to the nonradiative intersystem crossing the ITE and the ground state of $Ln^{\alpha+}$.

According to our earlier considerations, pressure can change the location of the localized states of the Ln^{3+} and Ln^{2+} ions with respect to the valence and conduction bands as well as the ITE states. Generally, it is expected that the energies of the $4f^n$ electronic configuration increase as presented in Fig. 4.24b. When we deal with two or more emitting excited states, increasing pressure can quench the emission in sequence from the highest to the lowest emitting state. Such pressure-tuning luminescence quenching effects allow obtaining the pressure shifts of the Ln^{3+} levels with respect to the ITE state and up to some extent with respect to the conduction band. The energetic structure of the Ln^{3+} ions presented in Fig. 4.24a shows that Pr^{3+} and Tb^{3+} ions are the best candidates to study pressure-induced luminescence quenching.

Mahlik et al. [128, 202] investigated the pressure dependence of Pr^{3+} luminescence in $CaWO_4:Pr^{3+}$ [128] and Tb^{3+} luminescence in $CaWO_4:Tb^{3+}$ [202]. The energetic structures of Pr^{3+} and Tb^{3+} are similar. The ground states, i.e., $^3H_4(4f^2)$ and $^7F_6(4f^8)$, are located approximately at the same energy in the band gap of the material (see Fig. 4.24a). Both ions have two emitting states. In $CaWO_4:Pr^{3+}$, the lowest emitting state is $^1D_2$ with energy approximately equal to 16,830 cm^{-1}, and the higher emitting state is $^3P_0$ with energy approximately equal to 20,480 cm^{-1} above the ground state. In $Tb^{3+,}$ the lowest emitting state is $^5D_4$ with energy equal to 20,510 cm^{-1}, and the higher emitting state is $^5D_3$ with energy approximately equal to 26,230 cm^{-1} above the ground state. The luminescence excitation spectra of $CaWO_4:Pr^{3+}$ and $CaWO_4:Tb^{3+}$ are presented in Figs. 4.30a, b, respectively. The excitation spectrum of $CaWO_4:Pr^{3+}$ monitored at 15,400 cm^{-1}, which corresponds to the $^3P_0 \rightarrow {}^3F_2$ transition in Pr^{3+}, consists of a broad band with maximum at

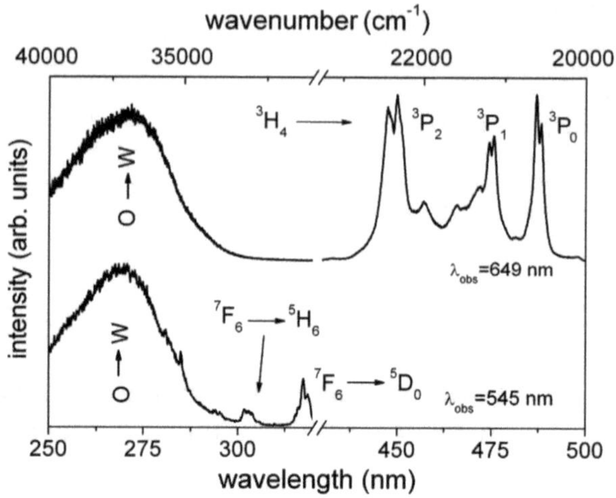

Fig. 4.30 Room-temperature excitation spectra of (**a**) $CaWO_4{:}Pr^{3+}$ monitored at 15 408 cm^{-1}, which is related to the $^3P_0 \rightarrow {}^3H_4$ Pr^{3+} emission, and (**b**) $CaWO_4$: Tb^{3+} monitored at 17391 cm^{-1}, which is related to the $^5D_4 \rightarrow {}^7F_5$ Tb^{3+} emission. Reprinted from Ref. [202], copyright 2013, with permission from Elsevier

approximately 37,000 cm^{-1}, related to a CT transition from O^{2-} to W^{6+} in the tungstate groups, and groups of sharp lines with peaks between 23,000 and 20,000 cm^{-1} corresponding to the $^3H_4 \rightarrow {}^3P_J$ transitions in Pr^{3+}. The luminescence-excitation spectrum of $CaWO_4{:}Tb^{3+}$ monitored at 17390 cm^{-1}, which corresponds to the $^5D_4 \rightarrow {}^7F_5$ transition in Tb^{3+} ion, consists of sharp lines in the spectral region from 30,000 cm^{-1} to 34,500 cm^{-1}, which are associated with the transitions from the ground state to the excited states of the $4f^8$ electronic configuration in Tb^{3+} ions, as well as a broad band with a maximum at approximately 37,000 cm^{-1}, which is assigned to a CT transition from O^{2-} to W^{6+}. In both spectra, the IT related to ionization of Pr^{3+} and Tb^{3+} is not observed.

The luminescence spectra of the $CaWO_4{:}Pr^{3+}$ excited at 488 nm, which corresponds to the $^3H_4 \rightarrow {}^3P_0$ transition in Pr^{3+}, and $CaWO_4{:}Tb^{3+}$ on excitation at 318 nm, which corresponds to the $^7F_6 \rightarrow {}^5D_0$ transition of Tb^{3+} obtained for different pressures, are shown in Figs. 4.31a, b, respectively.

The emission spectrum of $CaWO_4{:}Pr^{3+}$ at ambient pressure consists of a series of sharp lines related to the transitions from the $^3P_0$ and $^1D_2$ states. The most intense line, with a maximum at 649 nm, corresponds to the $^3P_0 \rightarrow {}^3F_2$ transition. The weaker lines, with peaks in the range 520 to 570 nm, are assigned to the $^3P_0 \rightarrow {}^3H_5$ transition, and the peaks in the 590–630-nm range are assigned to the $^3P_0 \rightarrow {}^3H_6$ and $^1D_2 \rightarrow {}^3H_4$ transitions. An evident decrease in the intensity of the emission from the $^3P_0$ state with increasing pressure and quenching of the $^3P_0$ emission at approximately 100 kbar was observed. Simultaneously, the intensity of emission from the $^1D_2$ state significantly increased. In the pressures range between 120 and 315 kbar, only the luminescence that originated from the $^1D_2$ state was observed. The intensity of this emission was almost constant up to 160 kbar, and then it decreased with further increasing pressure and was completely quenched at 315 kbar. Pressure-induced quenching of the emission from the $^3P_0$ and $^1D_2$ states was accompanied by significant shortening of the respective luminescence-decay times.

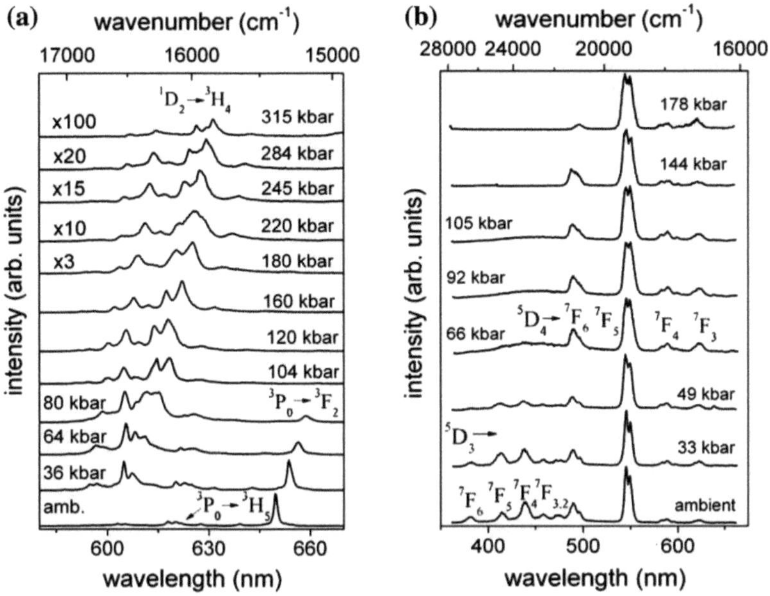

Fig. 4.31 The luminescence spectra of the **a** $CaWO_4:Pr^{3+}$ and **b** $CaWO_4:Tb^{3+}$ obtained at different pressures. The excitation wavelengths are 488 and 318 nm for (**a**) $CaWO_4:Pr^{3+}$ and (**b**) $CaWO_4:Tb^{3+}$, respectively. Reprinted from Ref. [202], copyright 2013, with permission from Elsevier

The luminescence spectra of $CaWO_4:Tb^{3+}$ obtained at different pressures are presented in Fig. 4.31b. Under ambient conditions, the emission consists of sharp lines related to transitions from the $^5D_3$ and $^5D_4$ excited states. The lines with peaks at 382, 414, 436, 458, and 470 nm correspond to the $^5D_3 \rightarrow {}^7F_J$ ($J = 2, 3, 4, 5$, and 6) transitions, respectively. The emission peaks at 488, 545, 587, and 622 nm correspond to the $^5D_4 \rightarrow {}^7F_6$, $^7F_5$, $^7F_4$, and $^7F_3$ transitions, respectively. Similar to the case of $CaWO_4:Pr^{3+}$, pressure-induced luminescence quenching of the $CaWO_4$: Tb^{3+} emission was observed. The intensity of luminescence from the $^5D_3$ state decreased with increasing pressure and was quenched at 70 kbar. This effect was accompanied by the shortening of the emission-decay time. Further increase in pressure caused a decrease in the intensity of $^5D_4$ luminescence, which was accompanied by a pronounced shortening of the luminescence-decay time. At pressures >190 kbar, the luminescence of Tb^{3+} was totally quenched.

The pressure effect on the energetic structure of $CaWO_4:Pr^{3+}$ and $CaWO_4:Tb^{3+}$ is presented in Fig. 4.32a, b, respectively. The energy of localized states of Pr^{3+} and Tb^{3+} increases with respect to the band states and the ITE state. In Fig. 4.32, the pressures P_1 and P_2 are the pressures at which the emission from the upper "*u*" and lower "*l*" emitting states is quenched. In the paper [202], values of P_1 and P_2 have been estimated for $CaWO_4:Pr^{3+}$ and $CaWO_4:Tb^{3+}$. The respective data are listed in Table 4.9. The values of P_1 and P_2 allow calculating of the pressure-induced shift

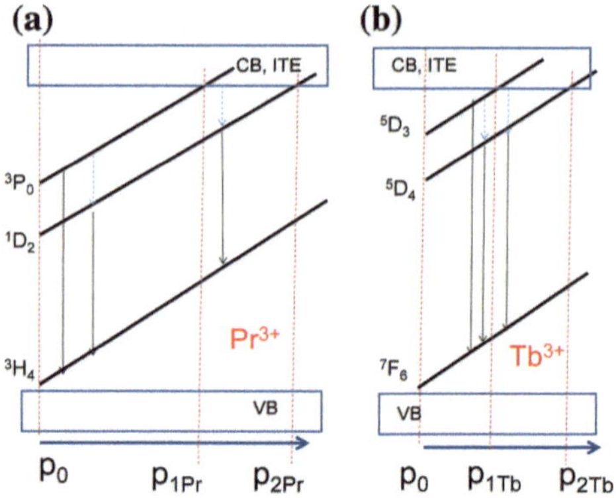

Fig. 4.32 The pressure dependence of the location of localized states of Pr^{3+} **a** and Tb^{3+} **b** with respect to the valence band and ITE states in $CaWO_4$

of ground states with respect to the conduction band (actually, with respect to the ITE state) as follows:

$$\frac{\mathrm{d}[E_{\mathrm{ITE}} - E(Ln^{3+})]}{\mathrm{d}p} = \frac{E_u - E_l}{P_1 - P_2} \tag{4.6.8}$$

The detailed mechanism of the nonradiative quenching of luminescence and the physical meaning of pressure P_1 and P_2 is presented in the configurational coordinate diagrams in Fig. 4.33a through c. In Fig. 4.33, the ambient-pressure situation is presented. After ionization, the system exists in the excited state, PTE, which relaxes nonradiatively to $^3P_0$ state and due to the large electron–phonon coupling to the $^1D_2$ state. As a result, the radiative transitions from $^3P_0$ and $^1D_2$ states are active. In Fig. 4.33b, the energy of the PTE state is diminished by pressure (equal to P_1),

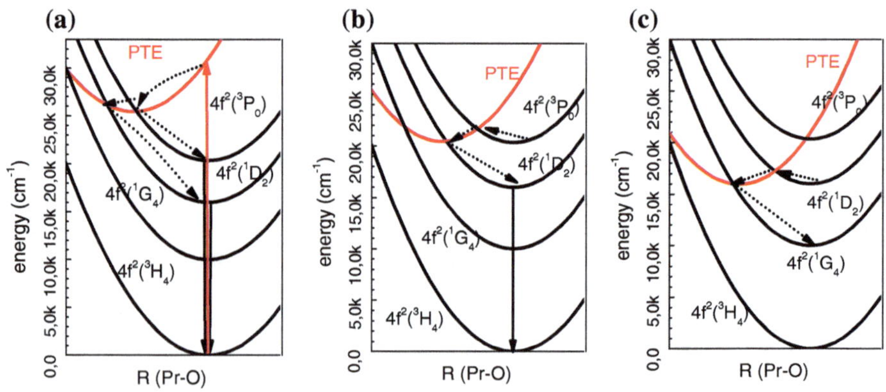

Fig. 4.33 Configurational coordinate diagram representing the structure of Pr^{3+} in the host, which is represented by the impurity (praseodymium)-trapped exciton PTE **a** represents the situation at ambient pressure, **b** at pressure P_1 (reprinted from Ref. [202], copyright 2013, with permission from Elsevier), and **c** at pressure P_2. The radiative and nonradiative transitions are indicated by *solid* and *dashed arrows*, respectively

Table 4.9 Critical pressures and relative energies of Ln^{3+} ions and their pressure shifts o with respect the ITE states for Pr^{3+} and Tb^{3+} in different lattices

Matrerial:dopand	P_1 (kbar)	P_2 (kbar)	$\frac{d[E_{ITE}-E(Ln^{3+})]}{dp}$ (cm^{-1}/kbar)	Estimated ambient pressure $E_{ITE}-E(Ln^{3+})$ (cm^{-1})	Pressure of phase transition (kbar)
$CaWO_4:Pr^{3+}$ [202]	105	315	−17.4	25360	100 [9]
$CaWO_4:Tb^{3+}$ [202]	50	190	−40	27100	100 [9]
$LiTaO_3:Tb^{3+}$ [125]	140	–			no
$LiNbO_3:Tb^{3+}$ [203]	–	135			no
$LiNbO_3:Pr^{3+}$ [144]	60	150	−40		80 [204]
β–SiAlON:Pr^{3+}, [133]	120–260	–			
$Ca(NbO_3)_2:Pr^{3+}$ [124]	58				
$Sr_xBa_{1-x}(NbO_2)_3:Pr^{3+}$ (x = 1/2) [205].	60	175	−22	21490	no
$Sr_xBa_{1-x}(NbO_2)_3:Pr^{3+}$ (x = 1/2) [205].	50	170	−23	21550	no
$Gd_2(WO_4)_3:Pr^{3+}$ [129]	90		−39	24000 (32000[a])	
$Gd_2(WO_4)_3:Tb^{3+}$ [129]		95	−37	24000 (32000[a])	

[a]In the paper [129] energy of the IT; $E_{IT}-E(Ln^{3+})$ have been calculated

and at this pressure it is equal to the energy of the $^3P_0$ state. The additional nonradiative pathway from the $^3P_0$ state to the $^1D_2$ state through the PTE is activated, which quenches the $^3P_0$ luminescence and increases the $^1D_2$ luminescence. In Fig. 4.33c, the energy of the PTE is equal to the energy of the $^1D_2$ state (pressure is equal to P_2). In this situation, the $^1D_2$ state is also nonradiatively depopulated through the PTE state to the $^1G_4$ state.

The calculated pressure shifts of the $E_{ITE}-E(Ln^{3+})$ energies for Tb^{3+} and Pr^{3+} are listed in Table 4.9.

It is seen that according to Eq. (4.6.8), the pressure shift of the ground state of Tb^{3+} is much larger than that of Pr^{3+}. This is, however, an artifact related to the phase transition in $CaWO_4$ lattice from tetragonal to monoclinic, which occurs at approximately 100 kbar [9]. Only the pressure shift of the ground state of Pr^{3+} is reliable because in this case both pressures P_1, which quenches the $^3P_0$ luminescence, and P_2, which quenches the $^1D_2$ luminescence, correspond to the high-pressure monoclinic phase of $CaWO_4$. In $CaWO_4:Tb^{3+}$ at pressure greater than P_1, which quenches the $^5D_3$ emission, and lower than P_2, which quenches the $^5D_4$ emission the phase transition takes place. During pressure-phase transition, the energetic distance between the ITE and the ground state of Tb^{3+}; $E_{ITE}-E(Tb^{3+})$, is additionally diminished. Under the assumption that the phase transition diminishes the energy of the ITE with respect to the energy of Tb^{3+} and Pr^{3+} by the same quantity (equal to 3200 cm^{-1}), the ambient pressure energies of the ground states $E_{ITE}-E(Tb^{3+})$ and $E_{ITE}-E(Pr^{3+})$ have been calculated in [202]. The respective data are presented in Table 4.9. Assuming that the binding energy of the ITE, ΔE_{ITE}, is the same for the PTE and the TTE, the ground state of Pr^{3+} is located in the band gap approximately 1740 cm^{-1} above the ground state of Tb^{3+}. Considering that the electron–lattice coupling of the system in the ITE state, $S\hbar\omega$, is approximately equal to 5000 cm^{-1}, one can expect the appearance of bands with maxima at 30,360 and 32,100 cm^{-1} related to ionization processes in the excitation spectra of $CaWO_4$:Pr^{3+} and $CaWO_4$:Tb^{3+}, respectively. No such bands can be seen in Fig. 4.30a, b.

In the paper [128], it was shown that the luminescence-decay time related to the transitions from the $^1D_2$ state in $CaWO_4$:Pr^{3+} is stable at pressure <150 kbar and then diminishes gradually with increasing pressure, whereas the luminescence-decay time related to $^3P_0$ diminishes rapidly when pressure increases above the ambient. At 100 kbar, where a phase transition takes place, no significant changes in the luminescence-decay times are observed. This effect supports the assumption that luminescence quenching is related to the pressure-induced increase in the energy of the ground states of Pr^{3+} and Tb^{3+} ions with respect to the conduction band and the ITE state and not to the phase transition.

Very reliable energies of the ground state of Pr^{3+} in the band gap have been obtained from the high-pressure luminescence spectroscopy of $Sr_xBa_{1-x}(NbO_2)_3$:Pr^{3+} (x = 1/2 and 1/3) [205]. Both materials at ambient pressure show luminescence that consists of lines related to transitions from the $^3P_0$ and $^1D_2$ states. With increasing pressure, $^3P_0$ and $^1D_2$ luminescence is gradually quenched. Pressures P_1 and P_2, which quench the $^3P_0$ and $^1D_2$ emission, are listed in Table 4.9. Because no phase transitions takes place in $Sr_xBa_{1-x}(NbO_2)_3$ in the pressure region between P_1 and P_2, one can estimate the relative energy of the Pr^{3+} ion, $E_{ITE} - E(Ln^{3+})$. It is equal to 21,550 cm^{-1} for $Sr_xBa_{1-x}(NbO_2)_3$:Pr^{3+} (x = 1/3) and 21,490 cm^{-1} for $Sr_xBa_{1-x}(NbO_2)_3$:Pr^{3+} (x = 1/2) [205]. Considering that the electron–lattice coupling $S\hbar\omega$ is approximately equal to 5000 cm^{-1}, one expects the appearance of a broad band in the excitation spectrum related to the IT at 26,500 cm^{-1}. As in the case of $CaWO_4$:Pr^{3+} and $CaWO_4$:Tb^{3+}, such a band is not seen in the excitation spectra of $Sr_xBa_{1-x}(NbO_2)_3$:Pr^{3+} [205].

The pressure dependence of the luminescence and the luminescence kinetics of Pr^{3+} was investigated in several other materials. Gryk et al. [125, 206], and Lazarowska et al. [203] investigated the spectral properties of $LiTaO_3$:Pr^{3+} and $LiNbO_3$:Pr^{3+} under high hydrostatic pressure. Ambient-pressure luminescence of $LiNbO_3$:Pr^{3+} consists of luminescence related to the transitions from the $^3P_0$ and $^1D_2$ states in Pr^{3+}, whereas the respective luminescence of $LiTaO_3$:Pr^{3+} consists of the emission related to transitions from the $^1D_2$ state only. In $LiNbO_3$:Pr^{3+} increasing pressure quenches emission from the higher excited state $^3P_0$, whereas luminescence related to the transition from $^1D_2$ is stable for pressure ≤ 200 kbar. In $LiTaO_3$:Pr^{3+}, the $^1D_2$-state luminescence intensity decreases gradually with pressure and is quenched at 135 kbar. The luminescence quenching is accompanied by diminishing of the luminescence lifetime [203]. Mahlik et al. [144] have investigated $CaMoO_4$:Pr^{3+}. Ambient-pressure luminescence spectra of $CaMoO_4$:Pr^{3+}

consist of lines related to transitions from the $^3P_0$ and $^1D_2$ states in Pr^{3+}. It was shown that, as in the case of $LiNbO_3:Pr^{3+}$, also in $CaMoO_4:Pr^{3+}$ the emission from $^3P_0$ is effectively quenched at 60 kbar, whereas the emission from $^1D_2$ is quenched at 150 kbar [144]. Similarly to the case of $CaWO_4:Tb^{3+}$, the pressure shift calculated using Eq. (4.6.8) is very large because at 80 kbar the phase transition of $CaMoO_4$ from the scheelite to fergusonite phase takes place [204]. As a result, the value of the pressure shift of Pr^{3+} energy includes the effect of phase transition. Quenching of $^3P_0$ luminescence under pressure was also observed in β–SiAlON: Pr^{3+} [133]. In this case, although the $^3P_0$ emission is strongly decreased at 120 kbar, it is still seen at the pressure 260 kbar. In the case of $Ca(NbO_3)_2:Pr^{3+}$, emission from the $^3P_0$ state is quenched at the pressure 58 kbar [124].

Pressure-induced quenching of luminescence related to the $^1D_2 \rightarrow {}^3H_4$ transition was observed in $YVO_4:Pr^{3+}$ [123]. Here, surprisingly, Pr^{3+} luminescence was replaced at 60 kbar by the intrinsic luminescence of the YVO_4 host. A similar effect was observed in Tb^{3+}-doped $CaMoO_4$ [145]. In contrast to $CaMoO_4:Pr^{3+}$, where intrinsic $CaMoO_4$ luminescence is not observed, the emission of $CaMoO_4:Tb^{3+}$ at ambient pressure is a superposition of Tb^{3+} luminescence related to the transitions from the lowest-emitting state $^5D_4$ and a broad band attributed to the host luminescence. The emission kinetics and intensity are independent of pressure ≤ 50 kbar, and then at higher pressure $^5D_4$ emission is quenched. Although the remaining host emission in $CaMoO_4:Tb^{3+}$ is observed for pressure >50 kbar, it is also quenched at 120 kbar. In contrast, in undoped $CaMoO_4$ host luminescence was observed at ≤ 250 kbar [145]. In the framework of the model where pressure causes an increase of energy of the Pr^{3+} and Tb^{3+} states, the effect of replacing the Pr^{3+} and Tb^{3+} luminescence by intrinsic luminescence of the host can be explained by pressure-induced degeneration of the emitting state of the Ln^{3+} with the conduction band. Specifically, when the ITE-binding energy ΔE_{ITE} is small or the system does not create a localized ITE state, after ionization the electron can be captured at self-trapped exciton states and recombine with the hole, thus providing the intrinsic luminescence [123].

In $Gd_2(WO_4)_3$ doped with Pr^{3+} [129] a strong emission from the $^3P_0$ state and a very weak emission from the $^1D_2$ state have been observed at ambient pressure. Increasing pressure quenches the $^3P_0$ luminescence, but it does not cause an increase of the $^1D_2$ emission, which remains very weak. Quenching of the $^3P_0$ luminescence is accompanied by shortening of its lifetime [129]. In $Gd_2(WO_4)_3$ doped with Tb^{3+} [129], only a strong emission from the $^5D_4$ state is observed. Cases of increasing pressure quenching this luminescence are accompanied by decreasing the luminescence lifetime [129]. The values of pressure that quench the $^3P_0$ luminescence in $Gd_2(WO_4)_3:Pr^{3+}$ and the $^5D_4$ luminescence in $Gd_2(WO_4)_3:Tb^{3+}$ are approximately the same (see Table 4.9). The fact that the energy of the $^5D_4$ state of Tb^{3+} is the same as the energy of $^3P_0$ state of Pr^{3+} and approximately equal to 20,500 cm^{-1} [129], very similar pressures quench this emission meaning that the ground states of the Pr^{3+} and Tb^{3+} ions are located approximately at the same energy in the $Gd_2(WO_4)_3$ band gap. In the paper [129], analysis of the excitation spectra allowed obtaining the energies of the ground states of Pr^{3+} and Tb^{3+} with respect to the conduction band (the ITE state) $E_{IT} - E(Ln^{3+})$. These energies are listed in parentheses in Table 4.9. The energies $E_{ITE} - E(Ln^{3+})$ were obtained from the spectroscopic data. There exist high-pressure data on a luminescence-quenching yield of $20500 < E_{ITE} - E(Ln^{3+}) < 26000\ cm^{-1}$. Assuming that $E_{ITE} - E(Ln^{3+}) = 24000\ cm^{-1}$, one obtains the energy $S\hbar\omega + \Delta E_{ITE} = 8000\ cm^{-1}$.

No evident quenching of luminescence related to the transitions from the $^3P_0$ and $^1D_2$ state of Pr^{3+} in $LiLuF_4$:Pr^{3+} in a pressure range ≤ 220 kbar [143] was observed. Only a weak dependence of luminescence kinetics on pressure was observed [143], specifically, the $^3P_0 \rightarrow {}^3H_4$ luminescence lifetime diminished from 35 μs at ambient pressure to 25 μs at 220 kbar. This effect can be easily related to the increase of the odd-parity lattice distortion without the necessity to engage the ITE state. Lack of pressure-induced luminescence quenching can be related to the fact that the localized states of Pr^{3+} ions are located in $LiLuF_4$ very deep in the band gap, far from the conduction band edge, or to the fact that the pressure shift of the localized states of Pr^{3+} with respect to the band states or ITE states is small.

Spectacular results have been obtained by Srivastava et al. [207] in cubic Y_2O_3:Pr^{3+}. Under ambient pressure, the luminescence spectrum of Y_2O_3:Pr^{3+} consists of sharp lines peaking at 600–700 nm related to the radiative transition $^1D_2 \rightarrow {}^3H_4$ in Pr^{3+} ion. No emission related to transition from the $^3P_0$ state was observed at ambient pressure. The excitation spectrum of this emission consists of sharp lines peaking in the range 450–500 nm related to $^3H_4 \rightarrow {}^3P_j$ transition and a broad band peaking at 300 nm attributed to the $^3H_4 \rightarrow$ 4f5d transition in Pr^{3+} ion. When pressure increases apart from the $^1D_2 \rightarrow {}^3H_4$ luminescence, additional sharp lines peaking at 500 to 520 nm appeared at 70 kbar (7GPa); these have been attributed to the $^3P_0 \rightarrow {}^3H_4$ transition in Pr^{3+} ion (see Fig. 4.34). This is quite the opposite situation to what was observed for other materials doped with Pr^{3+} where pressure successively quenches the $^3P_0$ and $^1D_2$ emission [124, 129, 133, 144, 202, 205] (see also the data in Table 4.9). It is important to note that C-Y_2O_3 retains its cubic structure for pressures smaller than approximately 110 kbar, and the first pressure-induced phase transition from the cubic to monoclinic form is completed at approximately 130 kbar at RT. [207].

For a better understanding of the luminescence pressure behavior in Pr^{3+}, Srivastava et al. [207] examined the experimental data that have been reported on the pressure dependence of the Ce^{3+} luminescence in C-Lu_2O_3 [163]. At ambient pressure, the Ce^{3+} ion is not luminescent at any temperature in this oxide, but it exhibits a red emission broad band peaking at 610 nm corresponding to the Ce^{3+} $5d^1 \rightarrow 4f^1$ transition at high pressures, even at room temperature. The explanation advanced for this result is given in Ref. [163] and is discussed in this review. At ambient pressure, the 5d state of Ce is degenerated with the conduction band; therefore, any excitation of the Ce^{3+} ion by way of the allowed $4f^1 \rightarrow 5d^1$ transition leads to formation of the cerium-trapped exciton (CTE) state. The subsequent nonradiative return of the CTE state to the ground state results in complete quenching of the Ce^{3+} luminescence in C-Lu_2O_3 and, more generally, in the C-Ln_2O_3 family of materials [208]. The observation of a Ce^{3+} $5d^1 \rightarrow 4f^1$ emission

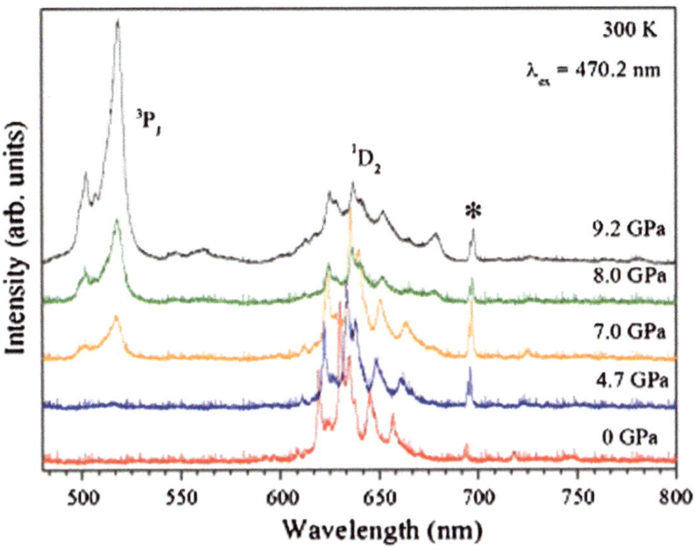

Fig. 4.34 Pressure dependence of the Pr^{3+} emission in C-$(Y_{0.995}Pr_{0.005})_2O_3$ at RT (λ_{ex} = 470.2 nm). Note the occurrence of the emission from the Pr^{3+} ^3P_J states near 500 nm (see text). The * indicates the ruby emission line. Reprinted from Ref. [207], copyright 2014, with permission from Elsevier

in C-Lu_2O_3:Ce^{3+} at high pressures is readily interpreted in terms of a pressure-induced crossover of the excited Ce^{3+} $5d^1$ state and the conduction band edge of the host lattice [159, 161–163]. The application of high pressure increases the band gap of Lu_2O_3 (it is a direct band-gap solid) and causes a red shift of the Ce^{3+} $5d^1$ state. Hence, when the pressure is high enough, there is a decrease in the overlap between the Ce^{3+} $5d^1$ and the host lattice–conduction band states. The resulting decreases in the photoionization probability results in the occurrence of Ce^{3+} luminescence. In C-Lu_2O_3:Ce^{3+}, the Ce^{3+} emission is generated at pressures of 70 kbar at RT [163]. Experimentally, the luminescence efficiency of C-Lu_2O_3:Ce^{3+} increases with increasing pressure and decreasing temperature. Sirvastava et al. [207] performed a comparative study between the pressure-dependent behavior of the luminescence of C-Lu_2O_3:Ce^{3+} and C-Y_2O_3:Pr^{3+} and concluded that the absence of Pr^{3+}, $^3P_0$ emission in C-Y_2O_3 can also be associated with the location of the lowest state of the electronic configuration $4f^15d^1$ of Pr^{3+} within the Y_2O_3 conduction band. Actually, at ambient pressure the PTE state in C-Y_2O_3:Pr^{3+} is located between the $^3P_0$ and $^1D_2$ states and therefore quenches the P_0 emission. Such a situation is presented in Fig. 4.33b and in Fig. 6 in Ref. [207]. Large lattice relaxation creates an additional potential that diminishes the energy of the PTE (the situation is presented in Fig. 4.26b). When pressure increases to 70 kbar, the energy of the 4f5d state diminishes, and the additional increase of the band gap removes its degeneration with the band states. Thus, at 70 kbar and after excitation to the 4f5d state, Pr^{3+} is not ionized, and the PTE cannot be formed as the stable state. The situation is presented in Fig. 4.26a and in Fig. 7 in Ref. [207].

It should be emphasized that in all other cases described in this subsection [124, 129, 133, 144, 202, 205], the 4f5d state of Pr^{3+} is also degenerated with the conduction band. In contrast to C-Y_2O_3:Pr^{3+}, pressure does not remove this degeneration; therefore, the dominating effect is the increase of the energy of the ground state of Pr^{3+} with respect to the relatively stable energy of the ITE and the conduction band.

4.6.4 Pressure Dependence of the Energy of CT Transitions

In the paper [129], the pressure dependence of the luminescence and luminescence kinetics of Tb^{3+}, Pr^{3+}, Eu^{3+}, and Dy^{3+} was measured. It was reported that compared with the luminescence related to Tb^{3+} and Pr^{3+}, which is quenched at pressures >95 and 90 kbar, the emission related to Eu^{3+} and Dy^{3+} is stable for all applied pressures ≤ 250 kbar. As discussed previously, Tb/Pr emission was quenched because the energy of the ITE $E_{\mathrm{ITE}} - E(\mathrm{Ln}^{3+})$ decreases with increasing pressure and at 100 kbar is lower than the energies of the emitting states. In the energy diagram presented in Fig. 4.24a, it can be seen that the ionization energies of Dy^{3+} and Eu^{3+} are much higher than those of Tb^{3+} and Pr^{3+}; therefore, the ITE is located high enough to not disturb the f-f emission in these ions. In contrast, the energy of CTT should increase with increasing pressure, therefore, they also do not influence the Eu^{3+} emission.

According to the diagram presented in Fig. 4.24, energy of CTT is the lowest for Eu^{3+}. Therefore, in the materials in which the energy of the Eu^{2+} is located deeply in the band gap, the ITE related to the CTT in Eu^{3+} can influence the internal radiative transitions in Eu^{3+}. Because pressure causes an increase of the energy of the Eu^{2+} state with respect to the valence, band pressure could recover the Eu^{3+} emission when it is quenched at ambient conditions by nonradiative processes through the ITE related to the CTT.

Only a few papers are available in the archival literature that discuss the pressure effect on energy of CTT in Eu^{3+}-doped materials. The most extensive research has been performed by Webster and Drickamer for La_2O_2S:Eu^{3+} and Y_2O_2S:Eu^{3+} [209]. Under ambient conditions, the luminescence spectrum of Eu^{3+} consists of lines related to the transitions from the three lowest-emitting states: $^5D_0$, $^5D_1$, and $^5D_2$. It was shown that applied pressure increases the energy of CT state with respect to the localized excited state ^5D_J of Eu^{3+}. As the result, the energy of the CT state becomes higher than the energy of the $^5D_3$ state, and emission from this state has appeared at 40 and 80 kbar for Y_2O_2S:Eu^{3+} and La_2O_2S:Eu^{3+}, respectively [209]. High-pressure spectroscopy of La_2O_2S:Eu^{3+} is also presented in Ref. [210]. In the paper [211], pressure dependence of excitation band edge was studied for La_2O_2S:Tb^{3+}, Gd_2O_2S:Tb^{3+}, and La_2O_2S:Eu^{3+}. It was found that pressure caused the quenching of Tb^{3+} luminescence and recovered emission from the $^5D_3$ state of Eu^{3+}.

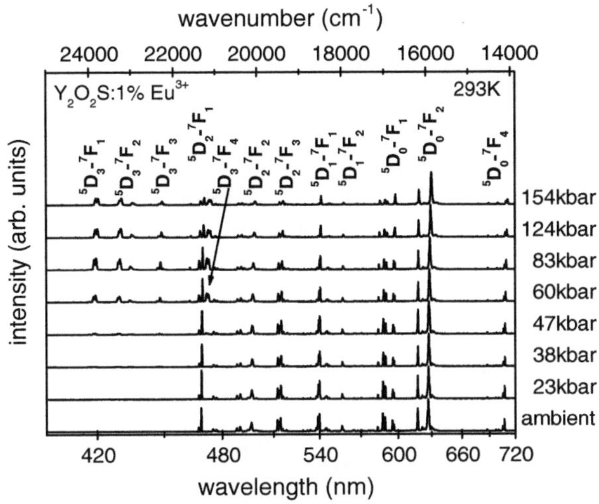

Fig. 4.35 Emission spectra of $Y_2O_2S:Eu^{3+}$ excited at 325 nm obtained on different pressures. Reprinted from Ref. [212], copyright 2014, with permission from Elsevier

Detailed study of the pressure dependence of $Y_2O_3S:Eu^{3+}$luminescence and luminescence kinetics was reported by Behrendt et al. [212]. The emission spectrum of Eu^{3+} in Y_2O_3S at ambient pressure consists of sharp lines related to radiative transitions from the three lowest emitting states: $^5D_0$, $^5D_1$, and $^5D_2$. At pressure >38 kbar, an additional emission from the highest-emitting state, $^5D_3$, appears. The luminescence spectra obtained at different pressure are presented in Fig. 4.35.

A similar effect was obtained by Webster and Drickamer [209] who measured the luminescence of $La_2O_2S:Eu^{3+}$ and $Y_2O_2S:Eu^{3+}$ at high hydrostatic pressure ranging from ambient to 120 kbar. Emission from $^5D_3$ appeared at approximately 20–40 kbar and 50–60 kbar for $Y_2O_2S:Eu^{3+}$and $La_2O_2S:Eu^{3+}$, respectively. The important contribution made in paper [212] results from the analysis of luminescence kinetics. It was shown that luminescence-kinetics profiles consist of two parts: the luminescence increase and the luminescence decay. Luminescence-kinetics profiles related to the $^5D_0 \rightarrow {}^7F_2$, $^5D_1 \rightarrow {}^7F_1$, and $^5D_2 \rightarrow {}^7F_3$ transitions are presented in Fig. 4.36a through c, respectively. The decay profiles of the $^5D_3 \rightarrow {}^7F_1$ emission are presented in Fig. 4.36d, e. It is seen that luminescence kinetics profiles are different for different transitions and depend on pressure. For all pressures, the kinetics of the $^5D_0 \rightarrow {}^7F_2$ emission can be divided into two processes. Just after excitation, the intensity of the emission increases and then decreases. In the case of the $^5D_1 \rightarrow {}^7F_1$ and $^5D_2 \rightarrow {}^7F_3$ transitions at ambient pressure, the luminescence appeared immediately after excitation (no luminescence increase is observed). The increased luminescence is manifested at pressure >40 kbar and 80 kbar for the $^5D_1 \rightarrow {}^7F_1$ and $^5D_2 \rightarrow {}^7F_3$ transitions, respectively. Emission related to the $^5D_3 \rightarrow {}^7F_1$ transition appears just after excitation without any delay for all considered pressures.

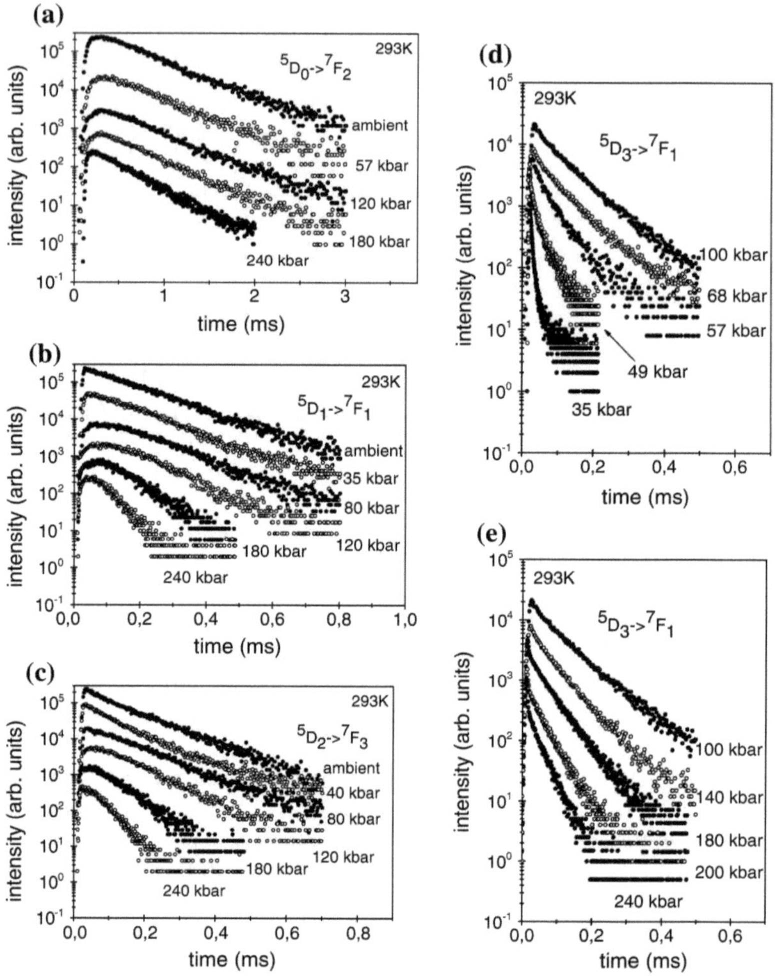

Fig. 4.36 Luminescence-decay profiles of $Y_2O_2S:Eu^{3+}$ monitored at a maximum of luminescence obtained under different pressures for the following transitions: $^5D_0 \rightarrow {}^7F_2$ (**a**), $^5D_1 \rightarrow {}^7F_1$ (**b**), $^5D_2 \rightarrow {}^7F_3$ (**c**), and $^5D_3 \rightarrow {}^7F_1$ (**d**, **e**). Reprinted from Ref. [212], copyright 2014, with permission from Elsevier

It should be emphasized that the increase in luminescence appears when the emitting state ^5D_J is populated through the muliphonon nonradiative process from the higher excited state $^5D_{J+1}$. If no increase in luminescence is observed, it means that the emitting state is populated just after the excitation. In the case of Y_2O_2S: Eu^{3+}, it corresponds to the situation when the excitation energy is transferred due to the intersystem crossing the ITE–^5D_J. It is seen that when the pressure increases, the intersystem crossing the ITE–^5D_J is replaced by nonradiative multiphonon transition $^5D_{J+1} \rightarrow {}^5D_J$. This is possible when the energy of the ITE state increases with

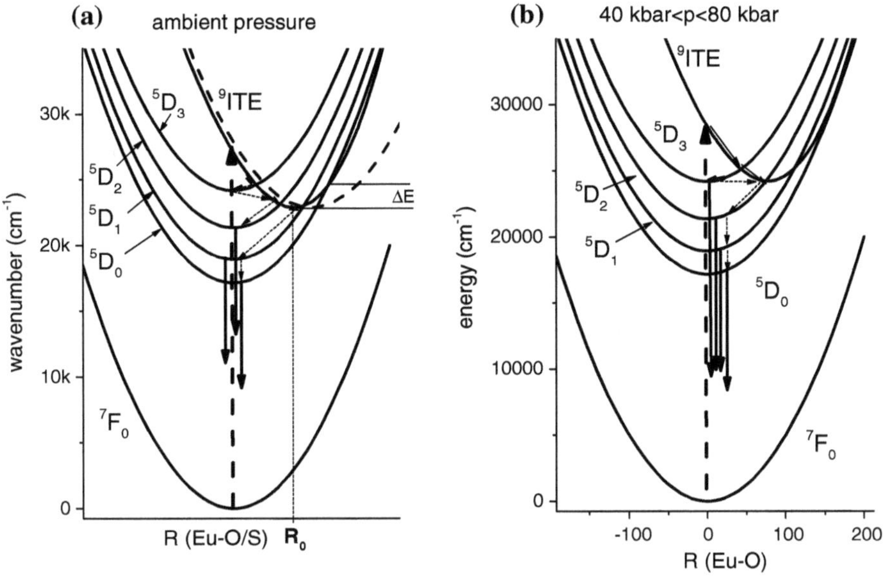

Fig. 4.37 **a** Configurational coordinate diagram of the $Y_2O_2S:Eu^{3+}$ system. The *dashed parabola* represents the ^{9}ITE state harmonic potential; and the *solid parabola* represents the electronic energy of the ^{9}ITE state, which includes an harmonicity. Possible transitions include excitation (*dashed arrow*), luminescence (*solid arrows*), and nonradiative intersystem crossing and nonradiative multiphonon relaxation (*dotted arrows*). **b** Configurational coordinate diagram of the $Y_2O_2S:Eu^{3+}$ system at pressure between 40 and 80 kbar.[9] The ITE parabola represents the electronic energy of the excitonic state, which includes an harmonicity. Reprinted from Ref. [212], copyright 2014, with permission from Elsevier

increasing pressure. Detailed analysis of pressure's influence on luminescence kinetics allowed development of a model of the ITE state that considers different spin states: the higher-energy ^{7}ITE state, which is seen in the excitation spectrum, and the lower-energy ^{9}ITE state, which is responsible for the de-excitation processes. In addition, to explain the pressure-induced appearance of the increased $^5D_1$ and $^5D_2$ luminescence, the anharmonic contribution to the potential electronic energy, representing the ITE state in the adiabatic approximation, has been considered. In the paper [212], the minimum energy of the ^{9}ITE electronic manifold has been located with respect to the ^5D_J states for ambient and high pressure. The configurational coordinate diagrams for ambient pressure and for pressure of 40–80 kbar are presented in Fig. 4.37a, b, respectively. In Fig. 4.37a, the energy barrier ΔE is the activation energy, which prevents the intersystem crossing the ITE- $^5D_0$ and causes the fact that the $^5D_0$ luminescence is always excited by multiphonon nonradiative-energy transfer from $^5D_1$ (see Fig. 4.36a) For the harmonic potential, this barrier does not exists, and no rise time would be observed for luminescence from the $^5D_0$ state.

4.7 Summary

In this chapter, we have summarized recent accomplishments in the area of high-pressure luminescence spectroscopy of phosphor materials. It was shown that the unique insight and understanding of phosphor performance is possible from high-hydrostatic study of luminescence centers created by TM and RE ions. The effect of pressure on luminescence related to f-f, d-d, and d-f ionization and charge-transfer transitions has been discussed. Several recent examples from the literature have been presented to illustrate the influence of pressure on luminescence energy, intensity, lineshape, and luminescence kinetics and efficiency.

The research presented showed the ability to use pressure to systematically and continuously study the static and dynamic luminescence properties of luminescence centers by the homogeneous decrease in bond length. In contrast, it was shown that hydrostatic pressure can be responsible for continuous changes of point symmetry of the dopant ion.

We also showed the way in which high hydrostatic pressure has been used for qualitative and quantitative verification of models and theories describing the energetic structure of and electronic transitions in solids such as the crystal-field model, adiabatic approximation, or quantum mechanical mixing. High pressure–tuning spectroscopy studies on Cr^{3+}-activated materials have shown that pressure is an irreplaceable and powerful tool for quantifying the extent of electronic mixing between the ^2E_g and $^4T_{2g}$ states of the Cr^{3+} ion and for characterizing the inhomogeneous distribution of Cr^{3+} ions in disordered crystals and glass hosts.

Especially, the unique ability of pressure to investigate the influence of impurity-trapped exciton states, which are created after ionization and charge transfer transitions, on the luminescence of TM and RE ions in solids was presented. The effect relies on pressure-induced luminescence quenching and recovery if the degeneration of the emitting level with the impurity-trapped exciton state is caused or removed, respectively.

In consequence, spectroscopy under high-hydrostatic pressure is one of the most efficient methods of phosphor characterization and provides much more information on the physical processes and performance of investigated materials than spectroscopy performed under ambient conditions. Simultaneously, this information is useful for the designation of novel phosphor materials.

Acknowledgments The author most gratefully acknowledges J. Barzowska, S. Mahlik, A. Lazarowska, and M. Behrendt for their contributions to the works presented in this review. A. Baran and A. Szymańska are acknowledged for their assistance in the preparation of the article. Special thanks go to A. Suchocki and A. Kamińska for enlightening discussion and to Ru-Shi Liu for encouragement. Some of the works presented in this review were supported by the POIG.01.01.02-02-006/09 project "New efficient phosphors for lighting and solar concentrators" cofunded by the European Regional Development Fund within the Innovative Economy Program. Priority I, Activity 1.1. Supporting Scientific Research to Build Knowledge-Based Economy, Sub-activity 1.1.2. Strategic Programmes for scientific research and development work, which is gratefully acknowledged. Author thanks Polish National Center for Research and Development for grants no PBS3/A5/48/2015 and PL-TWII/8/205.

References

1. Callen HB (1985) Thermodynamics and an introduction to thermostatistics. Wiley, New York
2. Bray KL (2001) High pressure probes of electronic structure and luminescence properties of transition metal and lanthanide systems. Top Curr Chem 213:1
3. Errandonea D, Pellicer-Porres J, Manjón FJ, Segura A, Ferrer-Roca Ch, Kumar RS, Tschauner O, Rodríguez-Hernández P, López-Solano J, Radescu S, Mujica A, Muñoz A, Aquilanti G (2005) High-pressure structural study of the scheelite tungstates CaWO4 and SrWO4. Phys Rev B 72: 174106
4. Birch F (1978) Finite strain isotherm and velocities for single-crystal and polycrystalline NaCl at high pressures and 300°K. J Geophys Res 83:1257
5. Errandonea D, Pellicer-Porres J, Manjón FJ, Segura A, Ferrer-Roca Ch, Kumar RS, Tschauner O, López-Solano J, Rodríguez-Hernández P, Radescu S, Mujica A, Muñoz A, Aquilanti G (2006) Determination of the high-pressure crystal structure of BaWO4 and PbWO4. Phys Rev B 73: 224103
6. Murnaghan FD (1944) The compressibility of media under extreme pressures. Proc Natl Acad Sci USA 30:244
7. Kamińska A, Biernacki S, Kobyyakov S, Suchocki A, Boulon G, Ramirez MO, Bausa L (2007) Probability of Yb^{3+} 4f-4f transitions in gadolinium gallium garnet crystals at high hydrostatic pressures. Phys Rev B 75:174111
8. Mujica A, Rubio A, Munoz A, Needs RJ (2003) High-pressure phases of group-IV, III–V, and II–VI compounds. Rev Mod Phys 75:863
9. Lacomba-Paralles R, Errandonea D, Segura A, Ruiz-Fuertes J, Rodríguez-Hernández P, Radescu S, López-Solano J, Mujica A, Muñoz A (2011) A combined high-pressure experimental and theoretical study of the electronic band-structure of scheelite-type AWO4 (A = Ca, Sr, Ba, Pb) compounds. J Appl Phys 110:043703
10. Henderson B, Imbush GF (1989) Optical spectroscopy of inorganic solids. Oxford Science Publications, Oxford
11. Kamińska A, Suchocki A, Arizmendi L, Calleyo D, Jaque F, Grinberg M (2000) Spectroscopy of near-stoichiometric LiNbO3:MgO, Cr crystals under high pressure. Phys Rev B 62:10802
12. Turos-Matysiak R, Zheng H, Wang JW, Yen WM, Meltzer RS, Łukasiewicz T, Świrkowicz M, Grinberg M (2007) Pressure dependence of the 3P0 → 3H4 and 1D2 → 3H4 emission in Pr3+:YAG. J Lumin 122–123:322
13. Pascual JL, Barandiáran Z, Seijo L (2007) Relation between high-pressure spectroscopy and $f^{n-1}d^1$ excited-state geometry: a comparison between theoretical and experimental results in SrF_2:Sm^{2+}. Phys Rev B 76:104109
14. Hozapfel WB, Isaacs NS (1997) High pressure techniques in chemistry and physics, a practical approaches. Oxford University Press, Oxford
15. Sherman WF, Stadtmuller AA (1987) Experimental techniques in high pressure research. John Wiley & Sons Ltd, London
16. Lawson AW, Tang TY (1950) A diamond bomb for obtaining powder pictures at high pressures. Rev Sci Instrum 21:815
17. Jamieson JC, Lawson AW, Nachtrieb ND (1959) New device for obtaining X-ray diffraction patterns from substances exposed to high pressure. Rev Sci Instrum 30:1016
18. Block S, Piermarini G (1976) The diamond cell stimulates high-pressure research. Phys Today 29(9):44
19. Jayaraman A (1983) Diamond anvil cell and high-pressure physical investigations. Rev Mod Phys 55:65
20. Jayaraman A (1986) Ultrahigh pressure. Rev Sci Instrum 57:1013
21. Dunstan DJ, Spain IL (1989) Technology of diamond anvil high-pressure cells: I. Principles, design and construction. J Phys E: Sci Instrum 22: 913

22. Spain IL, Dunstan DJ (1989) The technology of diamond anvil high-pressure cells: II. Operation and use. J Phys E: Sci Instrum 22: 923
23. Merrill L, Bassett WA (1974) Miniature diamond anvil pressure cell for single crystal x-ray diffraction studies. Rev Sci Instrum 45:290
24. Knoop F, Peters CG, Emerson WB (1939) A sensitive pyramidal-diamond tool for indentation measurements. J Res Natl Bur Stand 23(1):39
25. McSkimin HJ, Andreatch P (1972) Elastic moduli of diamond as a function of pressure and temperature. J Appl Phys 43(7):2944
26. Patterson DE, Margrave JL (1990) Use of gem-cut cubic zirconia in the diamond anvil cell. J Phys Chem 94:1094
27. Chervin JC, Syffose G, Besson JM (1994) Mechanical strength of sapphire windows under pressure. Rev Sci Instrum 65:2719
28. Takano KJ, Wakatsuki M (1991) An optical high pressure cell with spherical sapphire anvils. Rev Sci Instrum 62:1576
29. Piermarini GJ, Block S, Barnnett JS (1973) Hydrostatic limits in liquids and solids to 100 kbar. J Appl Phys 44:5377
30. Fujishiro I, Piermarini GJ, Block S, Munro RG (1982) Viscosities and glass transition pressure in the methanol–ethanol–water system. In: Backman CM, Johannisson T, Tagner L (eds) Proceedings 8 th AIRAPT conference Uppsala, vol II: 608
31. Ragan DD, Clark DR, Scheiferl D (1996) Silicone fluid as a high-pressure medium in diamond anvil cells. Rev Sci Instrum 67:494
32. Otto JW, Vassiliou JK, Frommeyer G (1998) Nonhydrostatic compression of elastically anisotropic polycrystals. I. Hydrostatic limits of 4:1 methanol-ethanol and paraffin oil. Phys Rev B 57:3253
33. Shen Y, Kumar RS, Pravica M, Nicol MF (2004) Characteristics of silicone fluid as a pressure transmitting medium in diamond anvil cells. Rev Sci Instr 75:4450
34. Tateiva N, Haga Y (2009) Evaluations of pressure-transmitting media for cryogenic experiments with diamond anvil cell. Rev Sci Instrum 80:123901
35. Jacobsen SD, Reichmann HJ, Kantor A, Spetzler H (2005) Advances in High-pressure technology for geophysical applications. In: Chen J, Wang Y, Duffy TS, Shen G, Dobrzhietskaya L (eds) (Elsevier, Amsterdam) 25
36. Varga T, Wilkinson AP, Angel RJ (2003) Fluorinert as a pressure—transmitting medium for high-pressure diffraction studies. Rev Sci Instrum 74:4564
37. Angel RS, Bujak M, Zhao J, Gatta GD, Jacobsen SD (2007) Effective hydrostatic limits of pressure media for high-pressure crystallographic studies. J Appl Cryst 40:26
38. Syassen K (2008) Ruby under pressure. High Pres Res 28:75 and references therein
39. Liebenberg DH (1979) A new hydrostatic medium for diamond anvil cells to 300 kbar pressure. Phys Lett A 73:74
40. LeSar R, Ekberg SA, Jones LH, Mills RL, Schwalbe LA, Schiferl D (1979) Raman spectroscopy of solid nitrogen up to 374 kbar. Solid State Commun 32(2):131
41. Nicol M, Hirsch KR, Holzapfel WB (1979) Oxygen phase equilibria near 298 K. Chem Phys Lett 68:49
42. Decker DL, Basset WA, Merrill L, Hall HT, Barnett JD (1972) High-pressure calibration: a critical review. J Phys Chem Ref Data 1:773
43. Ruoff AL, Lincoln RC, Chen YC (1973) A new method of absolute high pressure determination. J Phys D 6:1295
44. Li B, Kung J, Uchida T, Wang Y (2005) Pressure calibration to 20 GPa by use of simultaneous ultrasonic and X-ray techniques. J Appl Phys 98:013521
45. Decker DL (1966) Equation of state of sodium chloride. J Appl Phys 37:5012
46. Decker DL (1971) High-pressure equation of state for NaCl, KCl, and CsCl. J Appl Phys 42:3239
47. Brown JM (1999) The NaCl pressure standard. J Appl Phys 86:5801

48. Nellis WJ, Moriarty JA, Mitchell AC, Ross M, Dandrea RG, Ashcroft NW, Holmes NC, Gathers RG (1988) Metals physics at ultrahigh pressure: aluminum, copper, and lead as prototypes. Phys Rev Lett 60:1414
49. Schmidt SC, Schiferl D, Zinn AS, Ragan DD, Moore DS (1991) Calibration of the nitrogen vibron pressure scale for use at high temperatures and pressures. J Appl Phys 69:2793
50. Popov M (2004) Pressure measurements from Raman spectra of stressed diamond anvils. J Appl Phys 95:5509
51. Schiferl D, Nicol M, Zaug JM, Sharma SK, Cooney TF, Wang S-Y, Anthony TR, Fleischer JF (1997) The diamond $^{13}C/^{12}C$ isotope Raman pressure sensor system for high-temperature/pressure diamond-anvil cells with reactive samples. J Appl Phys 82:3256
52. Ono S, Mibe K, Ohishi Y (2014) Raman spectra of culet face of diamond anvils and application as optical pressure sensor to high temperatures. J Appl Phys 116:053517
53. Hess NJ, Schiferl D (1990) Pressure and temperature dependence of laser-induced fluorescence of Sm:YAG to 100 kbar and 700°C and an empirical model. J Appl Phys 68:1953
54. Wei Q, Dubrovinskakaia N, Dubrovinsky L (2011) Ruby and Sm:YAG fluorescence pressure gauges up to 120 GPa and 700 K. J Appl Phys 110:043513
55. Hua H, Vohra YK (1997) Pressure-induced blueshift of Nd^{3+} fluorescence emission in $YAlO_3$: Near infrared pressure sensor. Appl Phys Lett 71:2602
56. Kobyakov S, Kamińska A, Suchocki A, Galanciak D, Malinowski M (2006) Nd^{3+} doped Yttrium Aluminum Garnet crystal as a near-infrared pressure sensor for diamond anvil cells. Appl Phys Lett 88:234102
57. Mao H, Bell PM, Shaner JW, Steinberg DJ (1978) Specific volume measurements of Cu, Mo, Pd and Ag and calibration of the Ruby R1 fluorescence pressure gauge from 0.06 to 1 Mbar. J Appl Phys 49:3276
58. Mao HK, Xu J, Bell PM (1986) Calibration of the ruby pressure gauge to 800 kbar under quasi-hydrostatic conditions. J Geophys Res 91:4673
59. Forman RA, Piermarini GJ, Barnett JD, Block S (1972) Pressure measurement made by the utilization of ruby sharp-line luminescence. Science 176:284
60. Ragan DD, Gustavsen R, Schiferl D (1992) Calibration of the ruby R1 and R2 fluorescence shifts as a function of temperature from 0 to 600 K. J Appl Phys 72:5539
61. Kaplianskii AA, Przevluckij AK, Rozenbaum RB (1968) (in russian) Koncentrationnyj sdvig linni v opticheskicg spiektrach rubina. Fizika Tverdogo Tela (Physics of the Solid State) 10:2370
62. Galanciak D, Ph.D. Thesis (in polish), Wysokociśnieniowa spektroskopia chromu w materiałach laserowych Nichoals Copernicus University, Toruń, Poland (1996)
63. Yamaoka H, Zekko Y, Jarrige I, Lin J-F, Hiraoka N, Ishii H, Tsuei K-D, Mizuki J (2012) Ruby pressure scale in a low-temperature diamond anvil cell. J Appl Phys 112:124503
64. Powell RC, Dibartolo B (1972) Optical properties of heavily doped ruby. Phys Stat Sol (a) 10:315
65. Galanciak D, Legowski S, Meczynska H, Grinberg M (1994) High pressure spectroscopy of heavily doped ruby. J Phys IV France 04(C4):565
66. Trushkin S, Kamińska A, Trzeciakowski W, Hopkinson M, Suchocki A (2012) InAsP-based quantum wells as infrared pressure gauges for use in a diamond anvil cell. J Appl Phys 112:074504
67. Grinberg M (2011) Excited states dynamics under high pressure in lanthanide-doped solids. J Lumin 131:433
68. Grinberg M (2013) Impurity trapped excitons under high hydrostatic pressure. Opt Mater 35:2006
69. Grüneisen E (1912) Theorie des festen Zustandes einatomiger elemente. Ann Phys 12:257
70. Suski T (1998) High pressure in semiconductor physics I. In: Suski T, Paul W (eds) Academic Press, San Diego, 485

71. Panchal V, Errandonea D, Segura A, Rdríguez-Hernández P, Muñoz A, Lopez-Moreno S, Bettinelli M (2011) The electronic structure of zircon-type orthovanadates: effects of high-pressure and cation substitution. J Appl Phys 110:043723
72. Ibáñez J, Seguera A, García-Domene B, Oliva R, Manjón FJ, Yamaguchi T, Nanishi Y, Artús L (2012) High-pressure optical absorption in InN: Electron density dependence in the wurtzite phase and reevaluation of the indirect band gap of rocksalt InN. Phys Rev B 86:035210
73. Kaminska A, Franssen G, Suski T, Gorczyca I, Christiansen NE, Svane A, Suchocki A, Lu H, Schaff WJ, Dimakis E, Georgakilas A (2007) Role of conduction-band filling in the dependence of InN photoluminescence on hydrostatic pressure. Phys Rev 76:075203
74. Bersuker IB (1996) Electronic structure and properties of transition metal compounds, introduction to theory. Wiley, New York
75. Sugano S, Tanabe Y, Kamimura H (1970) Multiplets of transition metal ions in crystals. Academic Press, New York and London
76. Grinberg M, Barzowska J, Shen YR, Meltzer RS, Bray KL (2004) Pressure dependence of electron-phonon coupling in Ce^{3+}-doped $Gd_3Sc_2Al_3O_{12}$ garnet crystals. Phys Rev B 69:205101
77. García –Revilla S, Rodríguez F, Valiente R, Polnau M (2002) Optical spectroscopy of Al_2O_3: Ti^{3+} single crystal under hydrostatic pressure. The influence on the Jahn-Teller coupling. J Phys: Cond Matter 14: 447
78. Grinberg M, Mandelis A, Fieldsted K, Othonos A (1993) Spectroscopy and analysis of radiative and nonradiative processes in Ti^{3+}:Al_2O_3 crystals. Phys Rev B 48:5922
79. Grinberg M, Mandelis A, Phys Fieldsted K (1993) Theory of interconfigurational nonradiative transitions in transition-metal ions in solids and application to the Ti^{3+}:Al_2O_3 system. Rev B 48:5935
80. Gryk W, Grinberg M, Kukliński B, Shen YR, Bray KL (2003) High Pressure spectroscopy of Ti doped Al_2O_3 and $YAIO_3$ host crystals. Radiation Effect and Defects 158:141
81. Grinberg M, Suchocki A (2007) Pressure-induced changes in the energetic structure of the $3d^3$ ions in solid matrices. J Lumin 125:97
82. Wamsley PR, Bray KL (1994) The effect of pressure on the luminescence of Cr^{3+}: YAG. J Lumin 59:11
83. Shen YR, Grinberg M, Barzowska J, Bray KL, Hanuza J, Dereń PJ (2006) The effect of pressure on luminescence properties of Cr^{3+} ions in $LiSc(WO_4)_2$ crystals-Part I: Pressure dependent emission lineshape. J Lumin 116:1
84. Kamińska A, Kaczor P, Durygin A, Suchocki A, Grinberg M (2002) Low-temperature high-pressure spectroscopy of lanthanum lutetium gallium garnet crystals doped with Cr3 + and Nd3+. Phys Rev B 65:104106
85. Piermarini GJ, Block S, Barnett JD, Forman RA (1975) Calibration of the pressure dependence of the R_1 ruby fluorescence line to 195 kbar. J Appl Phys 46:2774
86. Grinberg M, Barzowska J, Shen YR, Bray KL, Hanuza J, Deareń PJ (2006) The effect of pressure on luminescence properties of Cr^{3+} ions in $LiSc(WO_4)_2$ crystals-Part II: Pressure-and temperature-dependent luminescence kinetics. J Lumin 116:15
87. Kamińska A (2006) Photoluminescence studies of TM and RE doped oxides using diamond anvil cell. Optica Applicata 36:163
88. Eggert JH, Goettel KA, Silvera IF (1989) Ruby at high pressure. II. Fluorescence lifetime of the R line to 130 GPa. Phys Rev B 40:5733
89. Galanciak D, Perlin P, Grinberg M, Suchocki A (1994) High pressure spectroscopy of LLGG doped with Cr^{3+}. J Lumin 60–61:223
90. Shen YR, Bray KL (1997) Effect of pressure and temperature on the lifetime of Cr^{3+} in yttrium aluminum garnet. Phys Rev B 56:10882
91. Basun SA, Kaplyanskii AA, Kutsenko AB, Dierolf V, Tröster T, Kapphan SE, Polgar K (2001) Optical characterization of Cr^{3+} centers in $LiNbO_3$. Appl Phys B 73:453
92. Shen Y, Bray KL, Grinberg M, Barzowska J, Sokólska I (2000) Identification of multisite behavior in a broadly emitting transition-metal system using pressure. Phys Rev B 61:14263

93. Kornyło A, Grinberg M, Kukliński B, Łukasiewicz T, Świrkowicz M (2004) Photoluminescence of LLGG:Cr^{3+} crystals under high pressure. Radiat Meas 38:579
94. Grinberg M, Barzowska J, Shen YR, Bray KL, Padlyak B, Buchynski PP (2002) High-pressure luminescence of Cr^{3+}-doped $CaO - Ga_2O_3 - GeO_2$ glasses. Phys Rev B 65:064203
95. Jovanić BR (1997) Shift under pressure of the luminescence transitions of corundum doped with Mn^{4+}. J Lumin 75:171
96. Galanciak D, Grinberg M, Gryk W, Kobyakov S, Suchocki A, Boulon G, Brenier A (2005) Influence of high pressure on the luminescence transitions of Mn^{4+}-doped gadolinium gallium garnet. J Phys: Cond Matter 17:7185
97. Brenier A, Suchocki A, Pedrini C, Boulon G, Madey C (1992) Spectroscopy of Mn^{4+}-doped Ca-substituted gadolinium gallium garnet. Phys Rev B 46:3219
98. Ya Zhydachevskii, Galanciak D, Kobyakov S, Berkowski M, Kamińska A, Suchocki A, Ya Zakharko, Durygin A (2006) Photoluminescence studies of Mn^{4+} ions in $YAlO_3$ crystals at ambient and high pressure. J Phys: Cond Matter 18:11385
99. Grinberg M, Barzowska J, Shen YR, Bray KL (2001) Inhomogeneous broadening of Cr^{3+} luminescence in doped $LiTaO_3$. Phys Rev B 63:214104
100. Grinberg M (2006) High pressure spectroscopy of rare earth ions doped crystals-new results. Opt Mater 28:26
101. Duclos SJ, Vohra YK, Ruoff AL (1990) Pressure dependence of the $^4T_2$ and $^4T_1$ absorption bands of ruby to 35 GPa. Phys Rev B 41:5372
102. Kim-Zajonz J, Werner S, Schulz H (1999) High pressure single crystal X-ray diffraction study on ruby up to 31 GPa. Zeitschrift fur Kristallographie 214:331
103. Smith RT, Welsh FS (1971) Temperature dependence of the elastic, piezoelectric, and dielectric constants of lithium tantalate and lithium niobate. J Appl Phys 42:2219
104. Liebermann RC, Jones LEA, Ringwood AE (1977) Elasticity of aluminate, titanate, stannate and germanate compounds with the perovskite structure. Phys Earth Planet Inter 14(2):165
105. Freire PTC, Pilla O, Lemos V (1994) Pressure-induced level crossing in $KZnF_3$:Cr^{3+}. Phys Rev B 49:9232
106. Kitaeva VF, Zharikov EV, Chistyi IL (1985) The properties of crystals with garnet structure. Phys Stat Sol (a) 92:475
107. Hommerich U, Bray KL (1995) High-pressure laser spectroscopy of Cr^{3+}:$Gd_3Sc_2Ga_3O_{12}$ and Cr^{3+}:$Gd_3Ga_5O_{12}$. Phys Rev B 51:12133
108. Sinkovits RS, Bartram RH (1991) Computer modeling of lattice dynamics in halide elpasolites. J Phys Chem Solids 52:1137
109. Graham LJ, Chang L (1970) Elastic Moduli of single-crystal Gadolinium Gallium Garnet. J Appl Phys 41:2247
110. Zneng W-C (1995) Determination of the local compressibilities for Cr^{3+} ions in some garnet crystals from high-pressure spectroscopy. J Phys: Cond Matter 7:8351
111. Wenger OS, Güdel HU (2002) Luminescence spectroscopy of V^{3+}-doped Cs_2NaYCl_6 under high pressure. Chem Phys Lett 354:75
112. Kano T (1999) Luminescence centers of rare-earth ions. In: Shionoya S, Yen WM (eds) Phosphor handbook, CRC Press Boca Raton, Boston, London, New York, Washington) 177
113. Goldschmidt ZB (1978) Gschneider KA, Eyring L (eds). Handbook on the physics and chemistry of rare earths. North-Holland, New York 1:1
114. Chang NC, Gruber JB, Leavitt RP, Morrison CA (1982) Optical spectra, energy levels, and crystal-field analysis of tripositive rare-earth ions in Y_2O_3. I. Kramers ions in C_2 sites. J Chem Phys 76:3877
115. Jorgensen CK (1962) The nephelauxetic series. Prog Inorg Chem 4:73
116. Shen YR, Holzapfel WB (1995) Nephelauxetic effects on Sm^{2+} and Eu^{3+} in ternary MYX compounds. Phys Rev B 52:12618
117. Kamińska A, Kozanecki A, Trushkin S, Suchocki A (2010) Spectroscopy of ytterbium-doped InP under high hydrostatic pressure. Phys Rev B 81:165209

118. Brik MG, Kaminska A, Suchocki A (2010) Ab initio calculations of structural, electronic, optical and elastic properties of pure and Yb-doped InP at varying pressure. J Appl Phys 108:103520
119. Kaminska A, Ma C-G, Brik MG, Kozanecki A, Boćkowski M, Alves E, Suchocki A (2012) Electronic structure of ytterbium-implanted GaN at ambient and high pressure: experimental and crystal field studies. J Phys: Cond Matter 24:095803
120. Ramirez M, Bausa L, Biernacki SW, Kaminska A, Suchocki A, Grinberg M (2005) Influence of hydrostatic pressure on radiative transition probability of the intrashell 4f transitions in Yb^{3+} ions in lithium niobate crystals. Phys Rev B 72:224104
121. Kaminska A, Duzynska A, Suchocki A, Bettinelli M (2010) Spectroscopy of f–f radiative transitions of Yb^{3+} ions in ytterbium doped orthophosphates at ambient and high hydrostatic pressures. J Phys: Cond Matter 22:225902
122. Gryk W, Kukliński B, Grinberg M, Malinowski M (2004) High pressure spectroscopy of Pr^{3+} in $LiNbO_3$. J Alloy Comp 380:230
123. Mahlik S, Grinberg M, Cavalli E, Bettinell M, Boutanaud P (2009) High pressure evolution of YVO_4:Pr^{3+} luminescence. J Phys: Cond Matter 21:105401
124. Mahlik S, Grinberg M, Kamiński AA, Bettinelli M, Boutanaud P (2009) Luminescence of Ca $(NbO_3)_2$:Pr^{3+} at ambient and high hydrostatic pressure. J Lumin 129:1219
125. Gryk W, Dyl D, Ryba-Romanowski W, Grinberg M (2005) Spectral properties of LiTaO3: Pr3 + under high hydrostatic pressure. J Phys: Cond Matter 17:5381
126. Tröster Th, Lavín V (2003) Crystal fields of Pr^{3+} in $LiYF_4$ under pressure. J Lumin 101:243
127. Galanciak D, Gryk W, Grinberg M, Macalik L, Mączka M (2006) High pressure spectroscopy of Pr^{3+} doped $KLa(MoO_4)_2$ crystal. Optica Applicata 36:443
128. Mahlik S, Cavalli E, Bettinelli M, Grinberg M (2013) Luminescence of $CaWO_4$:Pr^{3+} and $CaWO_4$:Tb^{3+} at ambient and high hydrostatic pressures. Radiat Meas 56:1
129. Mahlik S, Behrendt M, Grinberg M, Cavalli E, Bettinell M (2012) Pressure effects on the luminescence properties of CaWO4:Pr^{3+}. Optical Mater 34:2012
130. Carnall WT, Fields PR, Rajnak K (1968) Electronic energy levels in the trivalent lanthanide aquo ions. I. Pr^{3+}, Nd^{3+}, Pm^{3+}, Sm^{3+}, Dy^{3+}, Ho^{3+}, Er^{3+}, and Tm^{3+}. J Chem Phys 49:4424
131. Newman DJ, Ng BKC (2000) Crystal field handbook. Cambridge University Press, New York
132. Lazarowska A, Mahlik S, Grinberg M, Liu T-C, Liu R-S (2013) New Pr^{3+} sites in β-SiAlON red phosphor. Opt Mater 35:2001
133. Mahlik S, Lazarowska A, Grinberg M, Liu T-C, Liu R-S (2013) Luminescence spectra of β-SiAlON/Pr^{3+} under high hydrostatic pressure. J Phys Chem C 117:13181
134. Manjón FJ, Jandl S, Riou G, Ferrand B, Syassen K (2004) Effect of pressure on crystal-field transitions of Nd-doped YVO_4. Phys Rev B 69:165121
135. Kaminska A, Buczko R, Paszkowicz W, Przybylińska H, Werner-Malento E, Suchocki A, Brik M, Durygin A, Drozd V, Saksena S (2011) Merging of the $^4F_{3/2}$ level states of Nd^{3+} ions in the photoluminescence spectra of gadolinium-gallium garnets under high pressure. Phys Rev B 84:075483
136. Manjón FJ, Jandl S, Syassen K, Gesland YJ (2001) Effect of pressure on crystal-field transitions of Nd-doped $YLiF_4$. Phys Rev B 64:235108
137. Baran A, Mahlik S, Ginberg M, Zych E (2013) High pressure and time-resolved luminescence spectra of $Ca_3Y_2(SiO_4)_3$ doped with Eu^{2+} and Eu^{3+}. J Phys: Cond Matter 25:025603
138. Turos-Matysiak R, Gryk W, Grinberg M, Lin YS, Liu RS (2008) High pressure photoluminescence of Ce^{3+} doped $(Y_{1.725}\ Tb_{0.575}\ Ce_{0.05}Gd_{0.65})Al_5O_{12}$. Optical Materials 30: 722
139. Poort SH, Mayyerink A, Blasse G (1997) Lifetime measurements in Eu^{2+}-doped host lattices. J Phys Chem Solids 58:1451
140. Mahlik S, Grinberg M, Shi L, Seo HJ (2009) Pressure evolution of $LiBaF_3$:Eu^{2+} luminescence. J Phys: Cond Matter 21:235603

141. Wiśniewski K, Mahlik S, Grinberg M, Seo HJ (2011) Influence of high hydrostatic pressure on Eu^{2+}—luminescence in $KMgF_3$: Eu^{2+} crystal. J Lumin 131:306
142. Grinberg M, Mahlik S, Wiśniewski K, Seo HJ (2011) New Eu^{2+} sites in $KMgF_3$: Eu^{2+}. J Phys: Cond Matter 23:035404
143. Lazarowska A, Mahlik S, Krośnicki M, Grinberg M, Malinowski M (2012) Pressure-induced phase transition in $LiLuF_4$:Pr^{3+} investigated by an optical technique. J Phys: Cond Matter 24:115502
144. Mahlik S, Grinberg M, Cavalli E, Bettinelli M (2012) High pressure luminescence spectra of $CaMoO_4$:Pr^{3+}. J Phys: Cond Matter 24:215402
145. Mahlik S, Behrendt M, Grinberg M, Cavalli E, Bettinelli M (2013) High pressure luminescence spectra of $CaMoO_4$:Ln^{3+} (Ln = Pr, Tb). J Phys: Cond Matter 25:105502
146. Ramirez MO, Bausá LE, Garcia Solé J, Kaminska A, Kobyakov S, Suchocki A (2006) High-pressure-induced ferroelectric phase transition in the Yb^{3+}:$Sr_{0.6}Ba_{0.4}Nb_2O_6$ crystal at liquid helium temperature. Phys Rev B 74:174113
147. Dorenbos P (2000) The $4f^n \leftrightarrow 4f^{n-1}5d$ transitions of the trivalent lanthanides in halogenides and chalcogenides. J Lumin 91:91
148. Dorenbos P (2000) The 5d level positions of the trivalent lanthanides in inorganic compounds. J Lumin 91:155
149. Dorenbos P (2003) Energy of the first $4f^7 \rightarrow 4f^65d$ transition of Eu^{2+} in inorganic compounds. J Lumin 104:239
150. Morrison CA (1980) Host dependence of the rare-earth ion energy separation 4fn–4fn − 1 nl. J Chem Phys B 72:1001
151. Henderson B, Bartram RH (2000) Crystal—field engineering of solid state laser materials. Cambridge University Press, Cambridge
152. Dorenbos P (2003) Systematic behaviour in trivalent lanthanide charge transfer energies. J Phys: Cond Matter 15:8417
153. Dorenbos P (2004) Locating lanthanide impurity levels in the forbidden band of host crystals. J Lumin 108:301
154. Dorenbos P (2012) Modeling the chemical shift of lanthanide 4f electron binding energies. Phys Rev B 85:165107
155. Dorenbos P, Krumpel AH, van der Kolk E, Boutinaud P, Bettinelli M, Cavalli E (2010) Lanthanide level location in transition metal complex compounds. Opt Mater 32:1681
156. Dorenbos P (2005) The Eu^{3+} charge transfer energy and the relation with the band gap of compounds. J Lumin 111:89
157. Rodriguez-Mendoza UR, Cunningham GB, Shen YR, Bray KL (2001) High-pressure luminescence studies in Ce^{3+}:Lu_2SiO_5. Phys Rev 64:195112
158. Gryk W, Grinberg M, Joubert M-F, Dujardin C, Grosvalet L, Pedrini C (2006) High pressure photoluminescence study of cerium-doped Lu_2SiO_5. Optical Mater 28:115
159. Cunningham GB, Shen YR, Bray KL (2001) Effect of pressure on luminescence properties of Ce^{3+}:Lu_2S_3. Phys Rev 65:024112
160. Barzowska J, Grinberg M, Tsuboi T (2003) High pressure spectroscopy of Ce doped YAG crystal. Rad Effect Defects Solids 158:39
161. Kaminska A, Duzynska A, Berkowski M, Trushkin S, Suchocki A (2012) Pressure-induced luminescence of cerium-doped gadolinium gallium garnet crystal. Phys Rev B 85:155111
162. Yen WM (1999) Photoconductivity and delocalization in rare earth activated insulators. J Lumin 83–84:399
163. Shen Y-R, Gatch DB, Rodróguez Mendoza UR, Cunningham G, Meltzer RS, Yen WM, Bray KL (2002) Pressure-induced dark-to-bright transition in Lu_2O_3:Ce^{3+}. Phys Rev 65:212103
164. Srivastava AM (2009) Inter- and intraconfigurational optical transitions of the Pr^{3+} ion for application in lighting and scintillator technologies. J Lumin 129:1419
165. Meltzer RS, Zheng H, Wang JW, Yen WM, Grinberg M (2005) Pressure dependence of the $4f^15d^1 \rightarrow 4f^2$ emission of Pr^{3+}:YAG using excited state absorption. Phys Stat Sol (c) 2:284

166. Mahlik S, Malinowski M, Grinberg M (2011) High pressure luminescence and time resolved spectra of $La_2Be_2O_5:Pr^{3+}$. Opt Mater 34:164
167. Wang JW, Turos-Matysiak R, Grinberg M, Yen WM, Meltzer RS (2006) Mixing of the f2 (So) and 4f5d states of Pr^{3+} in $BaSO_4$ under high pressure. J Lumin 119–120:473
168. Barzowska J, Szczodrowski K, Krośnicki M, Kukliński B, Grinberg M (2012) Influence of high pressure on $Sr_2SiO_4:Eu^{2+}$ luminescence. Opt Mater 34:2095
169. Seo HJ, Moon BK, Tsuboi T (2000) Two-photon excitation spectroscopy of $4f^7 \rightarrow 4f^7$ transitions of Eu^{2+} ions doped in a $KMgF_3$ crystal. Phys Rev B 62:12688
170. Brgoch J, Gaultois MW, Balasubramanian M, Page K, Hong B-C, Seshadri R (2014) Local structure and structural rigidity of the green phosphor β-SiAlON:Eu^{2+}. Appl Phys Lett 105:181904
171. Altshuler NS, Korableva SL, Livanova LD, Stolov AL (1974) ESR and optical spectra of Eu^{2+} ion in $LiBaF_3$. Phys Solid State 15:2155
172. Sommerdijk JL, Versteegen JMPJ, Bril A (1975) On the luminescence of $LiBaF_3:Eu^{2+}$. J Lumin 10:411
173. Meijerink A (1993) Spectroscopy and vibronic transitions of divalent europium in $LiBaF^3$. J Lumin 55:125
174. Grinberg M, Barzowska J, Baran A, Kukliński B (2011) Characterization of various Eu^{2+} sites in $Ca_2SiO_4:Eu^{2+}$ and $Ba_2SiO_4:Eu^{2+}$ by high pressure spectroscopy. Mater Sci Pol 29:272
175. Turos-Matysiak R, Grinberg M, Wang JW, Yen WM, Meltzer RS (2007) Luminescence of BAM under high pressure: the Eu^{2+} Sites. J Lumin 122–123:107
176. Mahlik S, Wiśniewski K, Grinberg M, Meltzer RS (2009) Temperature and pressure dependence of the luminescence of Eu^{2+} doped fluoride crystals $Ba_xSr_{1-x}F_2$ (x = 0, 0.3, 0.5 and 1): experiment and model. J Phys: Cond Matter 21:245601
177. Su FH, Chen W, Ding K, Li H (2008) New Observations on the pressure dependence of luminescence from Eu^{2+}-Doped MF_2 (M = Ca, Sr, Ba) Fluorides. J Phys Chem A 112:4772
178. Francisco E, Blanco MA, Palacios P (2002) Atomistic simulation of the equation of state of SrF_2 using electron gas interionic potentials. Int J High Press Res 22:227
179. Gath DB, Boye DM, Shen YR, Grinberg M, Yen YM, Meltzer RS (2006) Pressure dependence of the impurity-trapped excition in BaF2: Eu and BaxSr1-xF2:Eu. Phys Rev B 74:195117
180. Lizzo S, Meijerink A, Driksen GJ, Blasse G (1995) Luminescence of divalent ytterbium in magnesium fluoride crystals. J Lumin 63:223
181. Lazarowska A, Mahlik S, Grinberg M, Yeh C-W, Liu R-S (2015) Pressure dependence of the $Sr_2Si_5N_8:Eu^{2+}$ luminescence. J Lumin 159:183
182. Xu Y-N, Ching WY, Brickeen BK (2000) Electronic structure and bonding in garnet crystals $Gd_3Sc_2Ga_3O_{12}$, $Gd_3Sc_2Al_3O_{12}$, and $Gd_3Ga_3O_{12}$ compared to $Y_3Al_3O_{12}$. Phys Rev B 61:1817
183. Ching WY, Ouyang L, Xu Y-N (2003) Electronic and optical properties of Y_2SiO_5 and $Y_2Si_2O_7$ with comparisons to α-SiO_2 and Y_2O_3. Phys Rev B 67:245108
184. Samara GA (1976) Temperature and pressure dependences of the dielectric properties of PbF_2 and the alkaline-earth fluorides. Phys Rev B 13:4529
185. Leger JM, Haines J, Atouf A, Schuete O, Hull S (1995) High-pressure x-ray- and neutron-diffraction studies of BaF_2: An example of a coordination number of 11 in AX_2 compounds. Phys Rev B 52:13247
186. Mahlik S, Lazarowska A, Grinberg M, Wells J-PR, Reid MF (2015) Luminescence properties of $MgF_2:Yb^{2+}$ at high hydrostatic pressure. J Lumin 169:788
187. Behrendt M, Kukliński B, Grinberg M, Karbowiak M (2009) $6d^15f^1 \leftrightarrow 5f^2$ transitions in U^{4+} in Cs_2NaYCl_6. Opt Mater 31:514
188. Srivastava AM, Dorenbos P (2009) Charge transfer transitions and location of the rare earth ion energy levels in Ca-α-SiAlON. J Lumin 129:634
189. Boutinaud P, Pinel E, Oubaha M, Mahiou R, Cavalli E, Bettinelli M (2006) Making red emitting phosphors with Pr^{3+}. Opt Mater 28:9

190. Reut EG, Ryskin AI (1973) Virtual recharge: Mechanism of radiationless transition in scheelite and fergusonite type crystals doped with rare-earth ions. Phys Stat Sol (a) 17:47
191. Boutinaud P, Sarakha L, Mahiou R, Dorenbos P, Inaguma Y (2010) Intervalence charge transfer in perovskite titanates $R_{1/2}Na_{1/2}TiO_3:Pr^{3+}$ (R = La, Gd, Y, Lu). J Lumin 130:1725
192. Dorenbos P, Rogers EG (2014) Vacuum referred binding energies of the lanthanides in transition metal oxide compounds. ECS J Sol State Sci Tech 3:R150
193. Cz Koepke, Wiśniewski K, Dyl D, Grinberg M, Malinowski M (2006) Evidence for existence of the trapped excition states in Pr^{3+} -doped $LiNbO_3$ crystal. Opt Mater 28:137
194. Grinberg M, Mahlik S (2008) Impurity trapped excitons: experimental evidences and theoretical concept. J Noncryst Solids 354:4163
195. Grinberg M (2010) High pressure study of localized states related to lanthanide ions in solids. J Electrochem Soc 157(4):G100
196. Kohn W (1957) Shallow impurity states in Si and Ge. Solid State Phys Adv Res Appl 5:257
197. Dorenbos P (2003) Anomalous luminescence of Eu^{2+} and Yb^{2+} in inorganic compounds. J Phys: Cond Matter 15:2645
198. Moine B, Courtois C, Pedrini C (1989) Luminescence and photoionization processes of Yb^{2+} in CaF_2, SrF_2 and BaF_2. J Phys France 50:2105
199. Jagannathan R, Kutty TRN (1997) Anomalous fluorescence features of Eu^{2+} in apatite-pyromorphite type matrices. J Lumin 71:115
200. Kanchana V, Vaitheeswaran G, Rajagopalan M (2003) Structural phase stability of CaF_2 and SrF_2 under pressure. J Phys B 328:283
201. Mahlik S, Lazarowska A, Grinberg M, Wells J-PR, Reid MF (2015) Pressure dependence of the emission in $CaF_2:Yb^{2+}$. J Phys: Cond Matter 27:305501
202. Mahlik S, Cavalli E, Bettinelli M, Grinberg M (2013) Luminescence of $CaWO_4:Pr^{3+}$ and $CaWO_4:Tb^{3+}$ at ambient and high hydrostatic pressures. Radiat Meas 56:1
203. Lazarowska A, Mahlik S, Grinberg M, Malinowski M (2011) High pressure luminescence and time resolved spectra of $LiNbO_3:Pr^{3+}$. Photon Lett Plans 3:67
204. Christofilos D, Kourouklis GA, Ves S (1995) A high pressure Raman study of calcium molybdate. J Phys Chem Solids 56:1125
205. Mahlik S, Lazarowska A, Speghini A, Bettinelli M, Grinberg M (2014) Pressure evolution of luminescence in $Sr_xBa_{1-x}(NbO_2)_3:Pr^{3+}$ (x = 1/2 and 1/3). J Lumin 152:62
206. Gryk W, Dujardin C, Joubert M-F, Ryba-Romanowski W, Malinowski M, Grinberg M (2006) Pressure effect on luminescence dynamics in Pr + 3 -doped LiNbO3 and LiTaO3 crystals. J Phys: Cond Matter 18: 117
207. Srivastava AM, Renero-Lacuna C, Santamaría-Pérez D, Rodríguez F, Valentine R (2014) Pressure-induced Pr^{3+} $^3P_0$ luminescence in cubic Y_2O_3. J Lumin 146:27
208. Yen WM, Raukas M, Basun SA, van Schaik W, Happek U (1996) Optical and photoconductive properties of cerium-doped crystalline solids. J Lumin 69:287
209. Webster G, Drickamer HG (1980) High pressure studies of luminescence efficiency and lifetime in La2O2S: Eu and Y2O2S:Eu. J Chem Phys 72:3740
210. Liu S, Chi Y, Gao C, Wang L, Zou G (1998) Influence of pressure on free-ion and crystal field parameters of Eu^{3+} in Y_2O_2S:Eu. Rev High Pressure Sci Technol 7:757
211. Gleason JK, Offen HW, Turley WD (1993) Pressure-induced luminescence quenching of terbium-doped oxide sulfides. Inorg Chem 32:639
212. Behrendt M, Szczodrowski K, Mahlik S, Grinberg M (2014) High pressure effect on charge transfer transition in $Y_2O_2S:Eu^{3+}$. Opt Mater 36:1616

Chapter 5
First-Principles Calculations of Structural, Elastic, Electronic, and Optical Properties of Pure and Tm^{2+}-Doped Alkali–Earth Chlorides MCl_2 (M = Ca, Sr, and Ba)

C.-G. Ma and M.G. Brik

Abstract The structural, elastic, and electronic properties of pure and Tm^{2+}-doped alkali–earth chlorides are studied in detail using the hybrid density-functional theory as implemented in the CRYSTAL09 code. The calculated local geometrical structures of the dopant site, Mulliken populations, electronic-band structures, and density of states for the pure and doped crystals are analyzed and compared to reveal the changes induced by the impurity ion. Additionally, the electronic and optical properties of the Tm^{2+}-ions doped into alkali–earth chlorides are modeled by employing the exchange-charge model of the semi-empirical crystal-field theory with the help of the optimized local coordination structures around impurities. The successful simulation of the 4f-5d transition spectrum of Tm^{2+} ions in $SrCl_2$ is demonstrated as an example to show the validity of the combined theoretical scheme of the first-principles and crystal-field models.

C.-G. Ma (✉) · M.G. Brik (✉)
College of Sciences, Chongqing University of Posts and Telecommunications, Chongqing 400065, People's Republic of China
e-mail: cgma.ustc@gmail.com

M.G. Brik
e-mail: mikhail.brik@ut.ee

M.G. Brik
Institute of Physics, University of Tartu, Ravila 14C, 50411 Tartu, Estonia

M.G. Brik
Institute of Physics, Jan Dlugosz University, Armii Krajowej 13/15, 42200 Czestochowa, Poland

R.S. Liu (ed.), *Phosphors, Up Conversion Nano Particles, Quantum Dots and Their Applications*, DOI 10.1007/978-3-662-52771-9_5

5.1 Introduction

Rare-earth ions (or lanthanides, chemical elements from La [No. 57] to Lu [No. 71]) have been actively used for more than several decades in numerous optical applications such as phosphor materials for displays and lighting, solid-state lasers operating in various spectral regions from infrared to ultraviolet, infrared to visible up-conversion materials, etc. [1 and references therein]. The physical reason behind such a wide range of applications is that all these elements have an unfilled 4f electron shell, which is shielded from the crystalline environment by the outer completely filled 5s and 5p electron shells. A large number of energy levels, which arises from various interelectron interactions within the 4f shell, can be additionally split by a crystal field. The electronic transitions between all those energy levels cover a wide energy range and can be employed for various applications.

In a vast majority of these applications, these ions are introduced into crystalline solids in their stable trivalent state. However, there are a few lanthanide ions whose divalent state is also stable, e.g., Eu^{2+}, Tm^{2+}. One of the main differences between the trivalent and divalent lanthanides (apart from the electrical charge) is that the charge-transfer transitions and the parity allowed 5d-4f transitions of the divalent ions are located at lower energies than those of the trivalent ions; because of this they can more easily studied experimentally. Regarding Tm^{2+} ions, it should be kept in mind that the $4f^{13}$ electron shell has only two energy levels: the ground state $^2F_{7/2}$ and the excited state $^2F_{5/2}$ located at approximately 9500 cm^{-1} [2]. As a result, the 4f-4f and 4f-5d transitions of the Tm^{2+} ions do not overlap and can be easily studied separately.

The MCl_2 (M = Ca, Sr, and Ba) crystals are ideal candidates for studies of the spectroscopic properties of Tm^{2+} ions. First, no charge-compensation defects are required after thulium ions are introduced into these hosts. Second, there is reliable experimental information about the absorption/emission spectra of these materials [2–8]. Moreover, the earth-alkaline halides are also important from an application point of view because the laser action has been achieved in CaF_2:Tm^{2+} [9]. The spectra of Tm^{2+} were modeled in our earlier publication [10].

Theoretical interpretation of peculiar features of impurity-ion spectra in crystals can be advanced by the combination of the semi-empirical crystal-field (CF) theory and the modern density-functional theory (DFT)–based computational methods. Such a combination highlights the strong features of both computational approaches and allows for creating a complementary picture of the electronic properties of doped optical materials. The short comparative list of the advantages and disadvantages of both methods is as follows: The CF theory can handle the multiplet structure of the impurity ion energy levels including low-symmetry splitting of the J manifolds and admixture of the electronic configurations of the opposite parity. These questions are beyond the current opportunities of currently used DFT-based computational packages because they are essentially based on the one-electron approximation. At the same time, the electronic-band structure of a solid simply cannot be obtained within the framework of a CF theory, whereas such calculations are already among the standard ones in the DFT-based methods. However, from calculations of the density of states of a doped material, it is possible to locate the ground state of an impurity ion in the band gap. Determination of the ground-state position in the host band gap is a crucial issue and has decisive importance for potential applications of a given crystal with intentionally introduced impurities. After that, the energy levels of an impurity ion, calculated with the help of the CF theory, can be superimposed onto the host's band structure starting from the ground state. Such an approach was used earlier for a number of crystals doped with transition-metal and rare-earth ions [11–13].

Recently we presented an example of such consistent joint application of these two different approaches to the description of the optical properties of Tm^{2+}ions in $SrCl_2$ [10], and the success of such an approach was demonstrated and confirmed by good agreement between the experimental and calculated optical properties of $SrCl_2$:Tm^{2+}. This system has not been thoroughly studied so far: To illustrate such a statement, we also mention that earlier the crystal-field parameters for the 4f electrons of the Tm^{2+} ions in similar compounds of CaF_2, SrF_2, and BaF_2 were analyzed in Refs. [14, 15], but they were not analyzed in $CaCl_2$, $SrCl_2$ and $BaCl_2$.

Based on the above-given consideration and examples, we decided to apply the DFT-based calculations to three alkali metal halide crystals, namely, $CaCl_2$, $SrCl_2$, and $BaCl_2$ doped with Tm^{2+} ions. The structural, electronic, optical, and elastic properties of these compounds were determined using first-principles calculation: The location of the 4f and 5d energy levels of the Tm^{2+} ions in the band gaps was identified. All obtained results are compared with the experimental data (if available), and agreement between the theory and the results of the experiment is discussed.

5.2 Crystal Structure

All three compounds studied here crystallize in different space groups. The corresponding structural data are summarized in Table 5.1, whereas Fig. 5.1 depicts unit cells for each of the studied crystals as well as the local coordination structures of alkaline-earth cations.

The Ca atoms in $CaCl_2$ are sixfold coordinated by the chlorine ions, and the Ba and Sr atoms in $BaCl_2$ and $SrCl_2$ are eightfold and ninefold coordinated by the chlorine atoms, respectively.

As a common feature of all these crystals, we note here that the metal–chlorine distance is rather large. In $CaCl_2$, there are two Ca–Cl distances of 2.704 Å and four distances of 2.765 Å. In $SrCl_2$, all eight Sr–Cl bonds are equal to 3.019 Å. Finally, Ba–Cl distances in $BaCl_2$ vary in a wide range from 2.864 to 3.579 Å.

Table 5.1 Calculated and experimental structural data for MCl_2 (M = Ca, Sr and Ba) crystals

	$CaCl_2$ (*Pnnm*, No. 58)		$SrCl_2$ ($Fm\bar{3}m$, No. 225)		$BaCl_2$ (*Pnma*, No. 62)	
	Calc.	*Exp.* [16]	*Calc.*	*Exp.* [17]	*Calc.*	*Exp.* [18]
a, Å	6.32847	6.24	6.97205	6.971	7.88907	7.865
b, Å	6.32626	6.43	6.97205	6.971	4.74148	4.731
c, Å	4.23826	4.2	6.97205	6.971	9.46948	9.421
$\alpha = \beta = \gamma$, °	90	90	90	90	90	90
V, Å^3	169.681	168.52	338.908	338.75	354.214	350.55
Z	2	2	4	4	4	4
M	0	0	0	0	0.251450	0.2514
	0	0	0	0	0.25	0.25
	0	0	0	0	0.117864	0.1209
Site	*2a*		*4a*		*4c*	
Cl1	0.303617	0.275	0.25	0.25	0.143961	0.1504
	0.305114	0.325	0.25	0.25	0.25	0.25
	0	0	0.25	0.25	0.429683	0.413
Site	*4g*		*8c*		*4c*	
Cl2					0.029152	0.029
					0.25	0.25
					0.829654	0.8392
Site					*4c*	

Coordinates of the crystallographic positions (in units of the lattice constants, (x, y, z) from *top* to *bottom*) of all ions are given. The symbol "Z" stands for the number of chemical formula units in one conventional unit cell

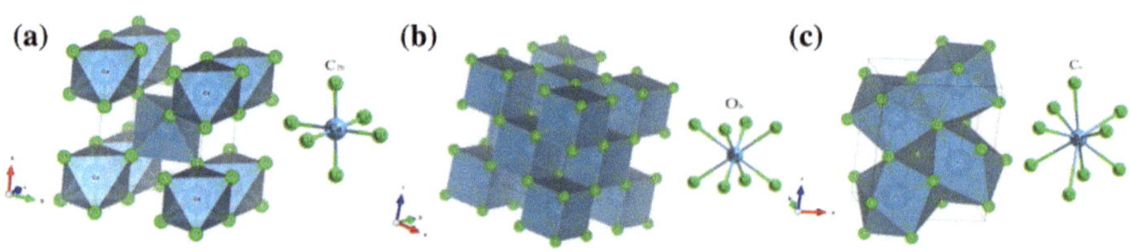

Fig. 5.1 Schematic representations for the unit cell of pure MCl_2 crystal and the local coordination structure around M cation (M = Ca [a], Sr [b], and Ba [c]). Drawn with VESTA [19]

5.3 Methods of Calculations

5.3.1 Ab Initio Calculations

The first-principles calculations of the structural, elastic, and electronic properties of pure and Tm^{2+}-doped hosts were performed using the periodic ab initio CRYSTAL09code [20] based on the linear combination of the atomic-orbitals method. Closed-shell and spin-polarized calculation forms were adopted for the pure and doped systems, respectively. To achieve better agreement between the calculated and experimental properties of the considered materials and improve the description of the strongly correlated lanthanide 4f orbitals [21], we employed the hybrid exchange–correlation (EXC) functional WC1PBE consisting of a PBE correlation part and a Wu-Cohen exchange part with a fractional mixing (16 %) of the nonlocal Hartree–Fock exchange [22]. The positive performance of such a hybrid EXC functional was confirmed by our recent works [23–25]. The local Gaussian-type basis sets (BS) were chosen as follows: the all-electron BSs in the form of 86-311G and 86-511d21G were used for the Cl [26] and Ca [27] atoms, respectively; the full-relativistic effective core pseudo-potentials (ECP), i.e., ECP28MDF and ECP46MDF—together with their corresponding valence BSs from the Stuttgart/Cologne group—were applied to Sr and Ba atoms [28], respectively; for Tm atom, the inner electrons $[Ar]3d^{10}$ were described by the scalar-relativistic ECP developed by Dolg et al. [29]; the other electrons $4s^2 4p^6 4d^{10} 5s^2 5p^6 4f^{13} 6s^2 5d^1$ were explicitly treated by the segmented-contraction type of the valence BS (14s13p10d8f)/[10s8p5d4f] [30] related to the ECP we used; and all the primitive functions with exponents less than 0.1 $bohr^{-2}$ were excluded because it is well-known that in the LCAO calculations for crystals, the BS of a free atom must be modified because the diffuse functions may cause numerical problems due to the large overlap with the core functions of the neighboring atoms in a densely packed crystal [31].

The Monkhorst–Pack schemes for the $8 \times 8 \times 8$ and $4 \times 4 \times 4$ k–point meshes in the Brillouin zone were applied to the pure and doped cases, respectively. The truncation criteria for bielectronic integrals (Coulomb and HF exchange series) were set to 8, 8, 8, 8, and 16, and a predefined "extra extra large" pruned DFT integration grid was adopted together with much higher DFT density and grid-weight tolerances (values 8 and 16). The tolerance of the energy convergence on the self-consistent field iterations was set to 10^{-8} Hartree. The Anderson scheme with 60 % mixing was used to achieve the converged solutions for all the calculation cases, and the total spin of the doped systems was locked to a value of 1/2 in the first 15 cycles due to the single positive hole occupation in the Tm^{2+}4f orbitals. In the geometry optimization with full structural relaxation, the convergence criteria of the root mean square of the gradient and the nuclear displacement were set to 0.00006 Hartree/bohr and 0.00012 bohr, respectively. For the elastic constant calculation, the default parameter values recommended in Ref. [32] were used.

5.3.2 *Crystal-Field Calculations and Exchange-Charge Model*

The parameterized effective Hamiltonians for the $4f^{13}$ and $4f^{12}5d$ configurations of Tm^{2+} ions can be respectively written as:

$$\boldsymbol{H}(4\mathrm{f}^{13}) = \boldsymbol{E}_{\mathrm{avg}} + \varsigma_{\mathrm{f}} A_{so}(\mathrm{f}) + \boldsymbol{H}_{\boldsymbol{CF}}(4\mathrm{f}) \tag{1}$$

$$\begin{aligned}\boldsymbol{H}(4\mathrm{f}^{12}5\mathrm{d}) = {} & \Delta_E(\mathrm{fd}) + \sum_{k=2,4} F^k(\mathrm{fd}) f_k + \sum_{j=1,3,5} G^j(\mathrm{fd}) g_j \\ & + \sum_{k=2,4,6} F^k(\mathrm{ff}) f_k + \varsigma_{\mathrm{f}}(\mathrm{ff}) A_{so}(\mathrm{ff}) + \alpha(\mathrm{ff}) L(L+1) + \beta(\mathrm{ff}) G(G_2) \\ & + \gamma(\mathrm{ff}) G(R_7) + \sum_{h=0,2,4} M^h(\mathrm{ff}) m_h + \sum_{k=2,4,6} P^k(\mathrm{ff}) p_k \\ & + \varsigma_{\mathrm{d}}(\mathrm{d}) A_{so}(\mathrm{d}) + \boldsymbol{H}_{\boldsymbol{CF}}(4\mathrm{f}) + \boldsymbol{H}_{\boldsymbol{CF}}(5\mathrm{d})\end{aligned} \tag{2}$$

where all notations and terms/operators/parameters are defined according to the standard practice [33]. The sum of all the interactions, except for the terms denoted as $\boldsymbol{H}_{\mathbf{CF}}(nl)(nl = 4\mathrm{f}$ or $5\mathrm{d})$, describes a quasi–free ion configuration Hamiltonian, the parameters of which can be taken as the free-ion ones with a reasonable reduction, caused by the nephelauxetic effect. Shortly, where E_{avg} represents the energy barycenter of the $4\mathrm{f}^{13}$ configuration. The term denoted as$\Delta_E(\mathrm{fd})$ is the (barycenter-to-barycenter) energy difference between both the $4\mathrm{f}^{13}$ and the $4\mathrm{f}^{12}5\mathrm{d}$ configurations; the ζ_{f}, $\zeta_{\mathrm{f}}(\mathrm{ff})$, and $\zeta_{\mathrm{d}}(\mathrm{d})$ entries are the spin-orbit coupling parameters for 4f electrons in the $4\mathrm{f}^{13}$ configuration, $4\mathrm{f}^{12}$ core and for the 5d single electron, correspondingly. The Slater radial integrals for the radial part of the electrostatic interaction among the 4f electrons or between the 4f and 5d electrons are denoted as $F^k(\mathrm{ff})(k = 2,4,6)$, $F^k(\mathrm{fd})(k = 2,4)$, and $G^j(\mathrm{fd})(j = 1,3,5)$. The two-body corrections terms describing the configuration interaction are denoted by $\alpha(\mathrm{ff})$, $\beta(\mathrm{ff})$, and $\gamma(\mathrm{ff})$. Finally, the $M^h(\mathrm{ff})(h = 0,2,4)$ and $P^k(\mathrm{ff})(k = 2,4,6)$ are related to the spin–spin and spin–other orbit relativistic interactions and the electrostatically correlated spin–orbit interactions between 4f elections, respectively (usually the ratios of M^0:M^2:M^4 and those of P^2:P^4:P^6 are kept fixed as described and explained in Ref. [34]). The last term, $\boldsymbol{H}_{\mathbf{CF}}(nl)$, is the CF interaction produced by the crystal lattice ions, which is felt by the 4f or 5d electrons of rare-earth ions and leads to the splitting of the J manifolds of an impurity ion. It is a standard practice to expand this term as follows:

$$\boldsymbol{H}_{CF}(nl) = \sum_{k,q=-k,\cdots k} B_q^k(nl) \boldsymbol{C}_{\boldsymbol{q}}^{(\boldsymbol{k})} \tag{3}$$

where $\boldsymbol{C}^k q$ are the spherical functions with rank k and order q defined by Wybourne [35]; and $B^k q(nl)$ are the crystal-field parameters (CFP). The former act on the angular parts of an impurity ion's wave functions, whereas the latter reflect the local symmetry of an impurity ion's site. The CFP values can be calculated using the crystal-structure data and the exchange-charge model (ECM) as shown below. More details related to this method of calculations can be found in our recent work [36].

The choice of the quasi–free -ion parameter values for the $4\mathrm{f}^{13}$ and $4\mathrm{f}^{12}5\mathrm{d}$ configurations of Tm^{2+} ions was performed according to data in the earlier literature. Thus, Pan et al. [37] reported the quasi–free ion parameter values for the $4\mathrm{f}^6 5\mathrm{d}$

configuration of the Eu^{2+} ions in $SrCl_2$. From their data, it is possible to obtain the reduction ratio between the free ion's values and those in the crystalline environment. The free ion–parameter values for the $4f^{N-1}5d$ configurations of $Eu^{2+}(N = 6)$ and $Tm^{2+}(N = 12)$ ions were calculated by the standard atomic-physics codes of Cowan [38, 39], and then the same reduction ratio as for the Eu^{2+} ions was directly applied to the Tm^{2+} ions. The reduced quasi–free ion parameter values for the $4f^{12}5d$ configuration of Tm^{2+} ions (using the standard notations in Ref. [33]) were obtained as follows: F^2(ff) = 84176, F^4(ff) = 59189, F^6(ff) = 45569, ζ_f(ff) = 2483, α(ff) = 17.26, β(ff) = –624.5, γ(ff) = 1820, M^0(ff) = 3.81, P^2(ff) = 695, ζ_d(d) = 1053, F^2(fd) = 11611, F^4(fd) = 5354, G^1(fd) = 5198, G^3(fd) = 4051, and G^5(fd) = 3056 (all in cm^{-1}). For the $4f^{13}$ configuration of the Tm^{2+} ions, the parameter ζ_f value can be approximately taken from that of ζ_f(ff). The parameters Δ_E(fd) and $\boldsymbol{E}_{avg}$ are respectively adjusted to 34756 and 3912 cm^{-1} to obtain the best agreement between the experimental and calculated energy levels.

The basic ideas of the exchange-charge model of the crystal field as outlined in Ref. [40] are as follows: The CFPs of any impurity ion, i.e., $B^kq(nl)$, are represented as a sum of two different contributions that arise from the point charges of all crystal lattice ions and introduced fictitious exchange charges (placed in the space between an impurity ion and ligands) due to the spatial overlap of the impurity-ion and ligand wave functions:

$$\begin{aligned} B_q^k(nl) = & -e^2\langle nl|r^k|nl\rangle \sum_i q_i\beta_k\cdot(-1)^k C_{-q}^k(\theta_i,\varphi_i)/R_i^{k+1} \\ & + e^2\frac{2(2k+1)}{2l+1}\sum_L S_k^{nl}(R_L)\cdot(-1)^k C_{-q}^k(\theta_L,\varphi_L)/R_L \end{aligned} \quad (4)$$

where the indexes i and L are, respectively, used to enumerate all the crystal lattice ions and the ligand ions in the nearest-neighbor coordination shell; (R_i, θ_i, φ_i) and (R_L, θ_L, φ_L) are the spherical coordinates of the ith crystal lattice ions and the Lth ligand ions in the reference system centered at the Tm^{2+} ions; and the ion charges q_i for Sr and Cl atoms can be taken as +2 and –1 as the first approximation. However, in our calculations we went further and took the values of these charges from the Mulliken population analysis as obtained by the performed ab initio calculations. The reduction factor β_k (k = 2, 4, …2 l) is defined as $1-\sigma_k$ (the so-called shielding constant [41]) for the 4f electrons, which accounts for the screening effects produced by the outer filled 5s and 5p electron shells. However, for the 5d electrons, it is set to 1 (no screening because this is an outer orbital). The $<nl|r^k|nl>$ entry is the radial integral of r^k (electron radial coordinate) between nl orbitals of the Tm^{2+} ions. The total overlap integral $S_k^{nl}(R_L)$ can be further expressed as:

$$S_k^{nl}(R_L) = G_s^{nl}S_s^{nl}(R_L)^2 + G_\sigma^{nl}S_\sigma^{nl}(R_L)^2 + \gamma_k^{nl}G_\pi^{nl}S_\pi^{nl}(R_L)^2 \quad (5)$$

where $\gamma_2^{4f} = 3/2, \gamma_4^{4f} = 1/3, \gamma_6^{4f} = -3/2, \gamma_2^{5d} = 1,$ and $\gamma_4^{5d} = -4/3$ [40]; $S_s^{nl}(R_L) = \langle nl0|3s0\rangle$, $S_\sigma^{nl}(R_L) = \langle nl0|3p0\rangle$, and $S_\pi^{nl}(R_L) = \langle nl\pm1|3p\pm1\rangle$; and G_s^{nl}, G_σ^{nl} and G_π^{nl} entries are the dimensionless adjustable parameters. They are determined by matching the calculated energy levels of an impurity ion to the experimental positions of its three lowest in energy absorption bands. For the sake of simplicity (and this turns out to be a very good approximation indeed in most cases), these three parameters can be approximated to a single value *G*(*nl*). In such a case, the value of this only parameter of the model can be easily found by matching the calculated position of the first excited state to the lowest absorption band in the experimental absorption/excitation spectrum. The moment of the *nl*-electron's radial coordinate $r \langle nl|r^k|nl\rangle$, as well as the overlap integrals between the *nl* orbitals of the Tm^{2+} ions and the outer 3s and 3p orbitals of Cl^- ions for various interionic distances, can be calculated using the corresponding numerical radial-wave functions provided by the Cowan code [38].

After all CFPs are determined, the "f-shell" program code written by M.F. Reid [33] can be employed to calculate the $4f^{13}$ and $4f^{12}5d$ CF energy levels and simulate the 4f-5d transition spectrum of Tm^{2+} ions in all considered cases. The results of these calculations are described below after the description of properties of pure $CaCl_2$, $SrCl_2$, and $BaCl_2$.

5.4 Ab Initio Calculations for Pure $CaCl_2$, $SrCl_2$, and $BaCl_2$ Crystals

The calculated structural parameters of pure MCl_2 (M = Ca, Sr, and Ba) crystals are listed in Table 5.1. As seen from that table, agreement between the experimental and calculated structural characteristics is very good, especially for $SrCl_2$ and $BaCl_2$, including not only the crystal lattice constants but also the fractional coordinates of all ions in a crystal lattice. The difference between the calculated and experimental structural parameters is somewhat greater in the case of $CaCl_2$, but this can be due to the fact that the corresponding experimental data are rather old [16], and the accuracy of those measurements might not be too high (as is evidenced by the number of digits after the decimal point in the corresponding structural data).

Good agreement between the optimized and experimental structural parameters allows calculations of the electronic, optical, and elastic properties of these compounds.

Figure 5.2 shows the calculated band structures of all three studied chlorides. The common feature is that for all compounds the valence band is rather narrow (3–4 eV), whereas the conduction band is rather wide. The upper states in the valence bands are very flat, which suggests a weak dispersion of holes in the reciprocal space. The lowest states of the conduction bands are also rather flat (low mobility of electrons), except for the vicinity of the Brillouin zone center. The band

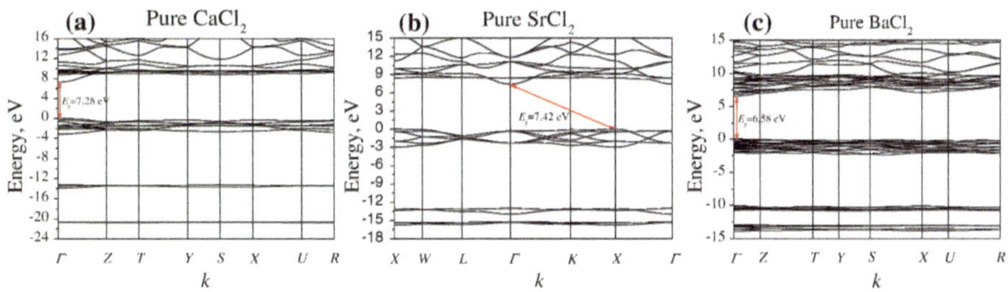

Fig. 5.2 Calculated band-structure diagrams for pure MCl_2 (M = Ca [**a**], Sr [**b**], and Ba [**c**]) crystals. In the cases of (**a**) and (**c**), the letters Γ, Z, T, Y, S, X, U, and R stand for the high-symmetry k points (0,0,0), (0,0,1/2), (0,1/2,1/2), (0,1/2,0), (1/2,1/2,0), (1/2,0,0), (1/2,0,1/2) and (1/2,1/2,1/2), respectively, whereas the high-symmetry k points (1/2,0,1/2), (1/2,1/4,3/4), (1/2,1/2,1/2), (0,0,0) and (3/4,3/8,3/8) corresponding to the letters X, W, L, Γ and K, respectively, are chosen in the case of (**b**). **b** Reproduced from Ref. [10] by permission of John Wiley & Sons Ltd

gaps are direct for $CaCl_2$ and $BaCl_2$ and is indirect for $SrCl_2$. The calculated band gaps are 7.28 eV ($CaCl_2$), 7.42 eV ($SrCl_2$) and 6.58 eV ($BaCl_2$), which is in good agreement with the experimental data of 6.9 eV ($CaCl_2$), 7.5 eV ($SrCl_2$), and 7 eV ($BaCl_2$), with all experimental data taken from Ref. [42].

Composition of the calculated electronic bands can be analyzed with the help of the density-of-states (DOS) and partial-density-of-states (PDOS) diagrams shown in Fig. 5.3. The valence band in all cases is formed by the 3p states of Cl ions, whereas the conduction band is basically composed of the d states of metals (3d states of Ca, 4d states of Sr, and 5d states of Ba). These are the most important states, which determine the optical properties of these crystals. The remaining DOS peaks are located deeply in energy; they come from the completely filled 3s states of Cl and p states of metal ions and do not manifest themselves in the optical spectra.

Table 5.2 lists the calculated elastic constants for all materials. All non-zero values of the elastic constants C_{ij} (the number of these non-zero components is

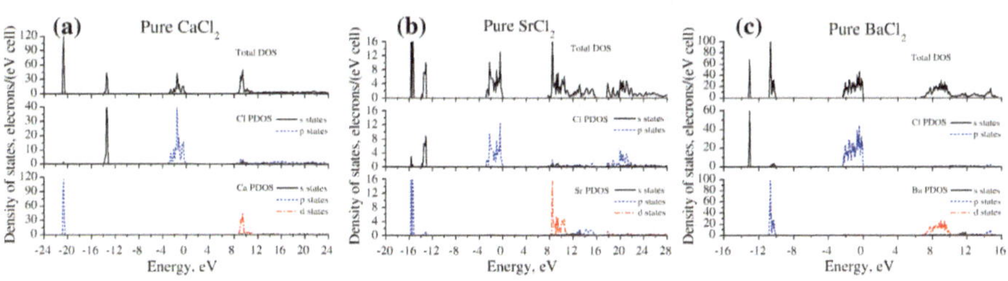

Fig. 5.3 Calculated PDOS/DOS diagrams for pure MCl_2 [M = Ca (**a**), Sr (**b**) and Ba (**c**)] crystals. **b** Reproduced from Ref. [10] by permission of John Wiley & Sons Ltd

Table 5.2 Calculated elastic constants C_{ij}(GPa), bulk moduli B (GPa), Poisson's ratios v, and longitudinal and transverse sound velocities $v_{l,t}$ (km/s) for MCl_2 (M = Ca, Sr and Ba) crystals

		$CaCl_2$ (*Pnnm*, No. 58)	$SrCl_2$ ($Fm\bar{3}m$, No. 225)			$BaCl_2$ (*Pnma*, No. 62)	
		This work	This work	*Exp.* [43]	*Calc.* [44]	This work	*Calc.* [45]
C_{11}		45.065	75.584	70.2	66	49.693	56.75
C_{12}		35.579	19.937	16.4	16	21.647	24.75
C_{13}		25.234	19.937	16.4	16	22.499	25.66
C_{22}		45.068	75.584	70.2	66	50.447	56.66
C_{23}		25.232	19.937	16.4	16	25.688	28.11
C_{33}		64.662	75.584	70.2	66	64.335	63.05
C_{44}		14.496	13.795	9.72	12	8.042	24.42
C_{55}		14.490	13.795	9.72	12	16.254	20.46
C_{66}		37.260	13.795	9.72	12	18.783	23.17
B		36.23	38.49	34.3	33	33.49	36.97
v		0.32	0.29	0.292	0.276	0.31	0.272
v_l	[001]	2.595	2.112			2.042	
	[010]	4.161	2.112			2.195	
	[100]	4.577	4.943			3.571	
	[111]	1.927	4.288			3.688	
v_{t1}	[001]	2.596	2.112			1.436	
	[010]	4.577	4.943			3.598	
	[100]	4.161	2.112			2.195	
	[111]	5.559	2.735			1.857	
v_{t2}	[001]	5.482	4.943			4.063	
	[010]	2.596	2.112			1.436	
	[100]	2.595	2.112			2.042	
	[111]	3.144	2.735			2.094	

The notation"[…]" represents the direction of sound wave propagation

determined by the crystal-lattice symmetry), bulk moduli, and Poisson's ratio—along with the sound velocities corresponding to different directions in the crystal lattices are listed in the table. Low values of the bulk moduli suggest that all three crystals are very soft substances.

Comparison of our calculated results with those available in the literature [43–45] yields good agreement between corresponding sets of data. We also went further with an analysis of the elastic anisotropy of these compounds, which can be visualized by plotting three-dimensional surfaces of the Young moduli along various directions in the crystal lattice. In the case of an an orthorhombic crystal (such as $CaCl_2$ and $BaCl_2$), such a three-dimensional surface is described by the following expression [46]:

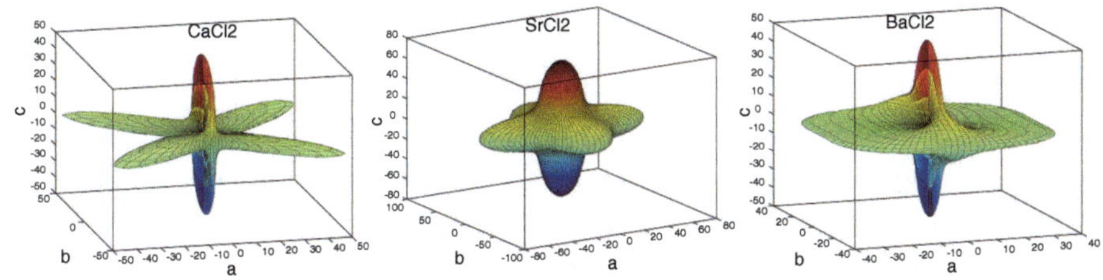

Fig. 5.4 Directional dependences of the calculated Young moduli for the MCl_2 (M = Ca, Sr, and Ba, from *left* to *right*) crystals (the axes units are GPa)

$$\frac{1}{E} = l_1^4 S_{11} + 2l_1^2 l_2^2 S_{12} + 2l_1^2 l_3^2 S_{13} + l_2^4 S_{22} + 2l_2^2 l_3^2 S_{23} + l_3^4 S_{33} + l_2^2 l_3^2 S_{44} + l_1^2 l_3^2 S_{55} + l_1^2 l_2^2 S_{66}, \quad (6)$$

where E is the value of the Young's modulus in the direction determined by the direction cosines l_1, l_2, l_3 ; and S_{ij} are the elastic compliance constants, which form the matrix inverse to the matrix of the elastic constants C_{ij}. The same surface in the case of a cubic crystal ($SrCl_2$) is given as

$$\frac{1}{E} = S_{11} - 2\left(S_{11} - S_{12} - \frac{1}{2}S_{44}\right)\left(l_1^2 l_2^2 + l_2^2 l_3^2 + l_1^2 l_3^2\right). \quad (7)$$

The distance from the center of such a surface to a point at its surface is equal to the Young modulus in a given direction determined by the direction cosines l_1, l_2, l_3. Figure 5.4 shows these surfaces plotted for the studied systems, and as seen from the surfaces, the anisotropy of the elastic properties is well-pronounced, especially for $CaCl_2$ and $BaCl_2$.

5.5 Ab Initio Calculations for Tm^{2+}-doped $CaCl_2$, $SrCl_2$, and $BaCl_2$ Crystals

Introduction of an impurity ion into a crystal can considerably change its optical properties. Even the color of a crystal can be changed because the purity-ion's energy levels, which appear in the band gap, can cause absorption in the visible range.

To model the geometrical structures of the Tm^{2+}-doped MCl_2 (M = Ca, Sr, and Ba) crystals, we constructed three super-cells from the optimized structural data of the pure host's primitive cells by respectively using the symmetrical super-cell transformation matrices (1 − 1 0; 1 1 0; 0 0 2), (2 0 0; 0 2 0; 0 0 2) and (1 − 1 0; 1 1 0; 0 0 1) [31]. Each studied super-cell contains eight MCl_2 chemical formula units, and one M^{2+}cation was replaced by the Tm^{2+} ion. In other words, the geometry optimizations were performed on the form of M_7TmCl_{16} for all cases. At first, due

Table 5.3 Calculated lattice parameters and volumes of the pure and Tm^{2+}-doped MCl_2 (M = Ca, Sr and Ba) supercells

	$CaCl_2$		$SrCl_2$		$BaCl_2$	
	Pure	Tm_{Ca}	Pure	Tm_{Sr}	Pure	Tm_{Ba}
a', Å	8.94825	8.96384	9.85997	9.82001	9.20429	9.13257
b', Å	8.94825	8.96384	9.85997	9.82001	9.20429	9.13257
c', Å	8.47652	8.50451	9.85997	9.82001	9.46948	9.35929
α', °	90	90	60	60	90	90
β', °	90	90	60	60	90	90
γ', °	89.9800	89.695	60	60	62.0134	62.3335
V', Å^3	678.724	683.332	677.816	669.608	708.428	691.351
		(+0.68 %)		(−1.21 %)		(−2.41 %)

The Tm^{2+}-induced volume changes with respect to the pure host cases are given in the parentheses

to the differences in the ionic radii of the Tm^{2+} and substituted metal ions (Ca^{2+}, Sr^{2+}, and Ba^{2+}), the lattice constants and interionic distances in the doped crystal will be somewhat different from those ones for the pure host. Table 5.3 illustrates this structural difference between the pure and thulium-bearing crystals.

As seen from Table 5.3, the volume of a unit cell of $CaCl_2$ doped with Tm^{2+} is slightly increased (by approximately 0.68 %), whereas the unit cells of $SrCl_2$ and $BaCl_2$ undergo a compression by 1.21 and 2.41 %, respectively. Such behavior can be explained by a simple comparison of the ionic radii of all ions involved. However, inspection of the database of the ionic radii given by Shannon [47] shows that the ionic radii of Tm^{2+} ions with eightfold and ninefold coordination numbers (CN) are absent. Due to the consideration of the chemical similarity across the whole lanthanide series, we can propose a linear ionic-radius dependence of divalent lanthanide ions on their atomic number Z and coordination number N to bypass the problem we encountered above as follows:

$$R(Ln^{2+}) = 0.0521N - 0.0176Z + 1.9382 \tag{8}$$

where the values of the two slopes and the intercept were obtained by fitting the experimental data set collected in Table 5.4 [47]. The calculated ionic radius of Tm^{2+}using Eq. (8) is slightly greater than that of Ca^{2+}, but slightly smaller than those of Sr^{2+} and Ba^{2+}, as shown in Table 5.4. This straightforward equation reveals the change reason for the super-cell volume with respect to the pure-host case after Tm^{2+}-doping.

Another geometrical manifestation of doping effects is the change of characteristic interionic distance. Comparison of the M-Cl (M = Ca, Sr, and Ba) and Tm–Cl distances in the pure and doped crystals is given by Table 5.5. Again, in accordance with the general trends in the change of the unit cell volumes from Table 5.3, the Tm–Cl distances are slightly increased, if the Tm^{2+} ions enter $CaCl_2$, and decreased when the Tm^{2+} ions occupy the Sr and Ba sites in $SrCl_2$ and $BaCl_2$, respectively.

Table 5.4 Effective ionic radii (in Å) of divalent lanthanides and M^{2+}(M = Ca, Sr and Ba) cations with different CNs

CN	Nd^{2+}	Sm^{2+}	Eu^{2+}	Dy^{2+}	Tm^{2+}	Yb^{2+}	Ca^{2+}	Sr^{2+}	Ba^{2+}
6	(1.197)	(1.162)	1.17	1.07	1.03	1.02	1.00	1.18	1.35
7	(1.249)	1.22	1.20	1.13	1.09	1.08	1.06	1.21	1.38
8	1.29	1.27	1.25	1.19	(1.143)	1.14	1.12	1.26	1.42
9	1.35	1.32	1.30	(1.248)	(1.195)	(1.177)	1.18	1.31	1.47
10	(1.405)	(1.370)	1.35	(1.300)	(1.247)	(1.229)	1.23	1.36	1.52

The values obtained from the linear fitting calculation are given in the parentheses

Table 5.5 Calculated distances (in Å) from the dopant site to the chlorine ligands in the first coordination shell before and after Tm^{2+}-doping in $CaCl_2$, $SrCl_2$ and $BaCl_2$ crystals

	$CaCl_2$			$SrCl_2$			$BaCl_2$		
M	Ca		Tm_{Ca}	Sr		Tm_{Sr}	Ba		Tm_{Ba}
Site	C_{2h}		C_{2h}	O_h		O_h	C_s		C_1
R	*Exp.*	*Calc.*	*Calc.*	*Exp.*	*Calc.*	*Calc.*	*Exp.*	*Calc.*	*Calc.*
Cl1	2.704	2.724	2.756	3.019	3.019	2.963	2.864	3.069	2.920
Cl2	2.704	2.724	2.756	3.019	3.019	2.963	3.167	3.081	2.829
Cl3	2.765	2.749	2.788	3.019	3.019	2.963	3.167	3.081	2.829
Cl4	2.765	2.749	2.788	3.019	3.019	2.963	3.154	3.139	2.924
Cl5	2.765	2.749	2.788	3.019	3.019	2.963	3.179	3.239	3.003
Cl6	2.765	2.749	2.788	3.019	3.019	2.963	3.256	3.275	3.026
Cl7				3.019	3.019	2.963	3.256	3.275	3.036
Cl8				3.019	3.019	2.9634	3.579	3.560	3.909
Cl9							3.579	3.560	3.914

The symbol "Cl*i*" stands for the *i*thchlorine ligands in the first coordination (see Fig. 5.1 for the definition of atomic labels). Those columns entitled as "*Exp.*" contain the experimental data of the cation-chlorine bond lengths in the pure hosts

Introduction of an impurity ion also produces local changes of the electron-density distribution, which can be evaluated by comparison of the calculated effective Mulliken charges of the crystal-lattice ions before and after doping (Table 5.6).

The Mulliken charge calculations indicate an increasing degree of covalency for MCl_2 crystals when M changes along the Ca–Sr–Ba series due to the decreasing effective charge of M cation. The Tm^{2+} effective charge is in all three hosts slightly greater than the charge of the substituted ion, which causes then a slight increase of the charge of the surrounding chlorine ions to maintain the local charge balance. It can be also noticed that the effective charges of all ions are different from their formal charges, which indicates formation of covalent bonds between the neighboring ions. Such a covalent chemical bonding behavior has been confirmed by the analysis of the bond populations between the dopant site and its nearest chlorine ligands before and after Tm^{2+}-doping. In addition, the electron-density distributions

Table 5.6 Effective Mulliken charge Q(e) of individual ion and "metal-chlorine" bond population P (milli e) in the local cluster composed of the dopant site and its nearest chlorine ligands before and after Tm^{2+}-doping in $CaCl_2$, $SrCl_2$ and $BaCl_2$ crystals

	$CaCl_2$		$SrCl_2$		$BaCl_2$	
M	Ca	Tm_{Ca}	Sr	Tm_{Sr}	Ba	Tm_{Ba}
$Q(M^{2+})$	1.630	1.658	1.498	1.738	1.403	1.739
$Q(Cl1^-)$	−0.815	−0.819	−0.749	−0.782	−0.692	−0.762
$Q(Cl2^-)$	−0.815	−0.819	−0.749	−0.782	−0.692	−0.751
$Q(Cl3^-)$	−0.815	−0.817	−0.749	−0.782	−0.692	−0.751
$Q(Cl4^-)$	−0.815	−0.817	−0.749	−0.782	−0.692	−0.783
$Q(Cl5^-)$	−0.815	−0.817	−0.749	−0.782	−0.711	−0.747
$Q(Cl6^-)$	−0.815	−0.817	−0.749	−0.782	−0.711	−0.773
$Q(Cl7^-)$			−0.749	−0.782	−0.711	−0.773
$Q(Cl8^-)$			−0.749	−0.782	−0.711	−0.699
$Q(Cl9^-)$					−0.711	−0.699
$P(M^{2+}–Cl1^-)$	23	13	25	12	−3	14
$P(M^{2+}–Cl2^-)$	23	13	25	12	−2	16
$P(M^{2+}–Cl3^-)$	23	14	25	12	−2	16
$P(M^{2+}–Cl4^-)$	23	14	25	12	5	15
$P(M^{2+}–Cl5^-)$	23	14	25	12	20	13
$P(M^{2+}–Cl6^-)$	23	14	25	12	21	13
$P(M^{2+}–Cl7^-)$			25	12	21	12
$P(M^{2+}–Cl8^-)$			25	12	15	2
$P(M^{2+}–Cl9^-)$					15	2

around the Ba^{2+} and Tm^{2+} ions occupying the same site in $BaCl_2$ are not similar but rather very different: The electrons lost by Ba atom trend to go to those chlorine ligands far away from the cation site, whereas most of the transferred electrons from the Tm atom stop in the coordination environment composed of the first seven chlorine ligands as indicated by the opposite dependences of the "metal–chlorine" bond populations on the chemical bond length for the pure and doped cases.

Finally, the changes of the electronic band structure that take place on doping are shown in Fig. 5.5.

The 4f states of the Tm^{2+} ions are clearly seen in the band gap. Because Tm^{2+} ions have 13 4f electrons, spin-polarized calculations were performed to distinguish between the spin-up and spin-down states. The band-structure diagrams from Fig. 5.5 allow for the following estimation of the position of the Tm^{2+} 4f ground state in the band gap above the top of the host's valence band: 1.05 and 2.91 eV (spin-up and spin-down states in $CaCl_2$, respectively), 1.37 and 3.20 eV (spin-up and spin-down states in $SrCl_2$, respectively), and 1.39 and 3.12 eV (spin-up and spin-down states in $BaCl_2$, respectively).

Certain asymmetry of the spin-up and spin-down state distributions is also clearly seen in Fig. 5.6, which presents the DOS/PDOS diagrams for the Tm^{2+}-doped crystals. In the case of $SrCl_2$ (cubic crystal and O_h local symmetry around the Tm^{2+}

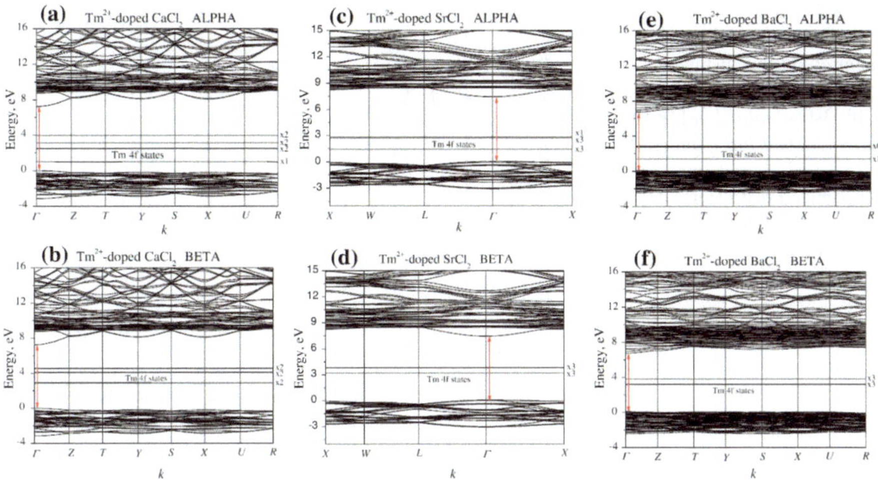

Fig. 5.5 Calculated band-structure diagrams for Tm^{2+}-doped MCl_2 [M = Ca (**a** and **b**), Sr (**c** and **d**), and Ba (**e** and **f**)] crystals. The *red two-way arrows* indicate the energy separation between the host valence and conduction bands. ALPHA and BETA represent the electronic pictures with the up and down spins, respectively. **c**, **d** Adapted from Ref. [10] by permission of John Wiley & Sons Ltd

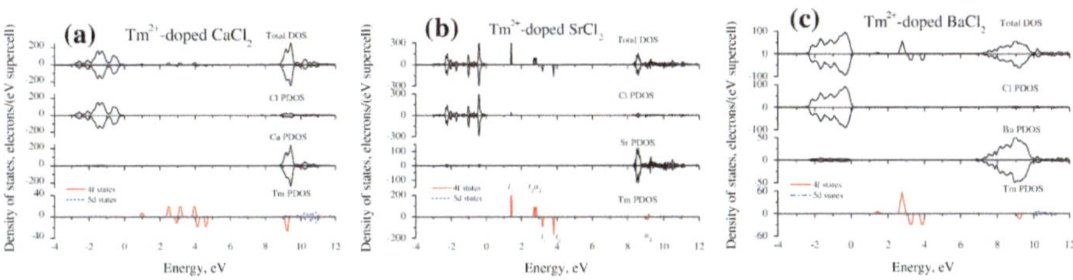

Fig. 5.6 Calculated PDOS/DOS diagrams for Tm^{2+}-doped MCl_2 [M = Ca (**a**), Sr (**b**), and Ba (**c**)] crystals. **b** Adapted from Ref. [10] by permission of John Wiley & Sons Ltd

site), it was also possible to determine the symmetry properties of the split electron states and assign the O_h group irreducible representations as shown in Fig. 5.6.

5.6 Crystal-Field Modeling of the Tm^{2+} Spectra in $SrCl_2$ Crystal

As described in Sect. 5.3.2, the overlap integrals between an impurity ion and the ligands are needed to apply the ECM. These integrals, in the case of the Tm^{2+} and Cl^- ions, were calculated numerically for various Tm^{2+}–Cl^- distances. They vary smoothly with the interionic separation as shown in Figs. 5.7 and 5.8.

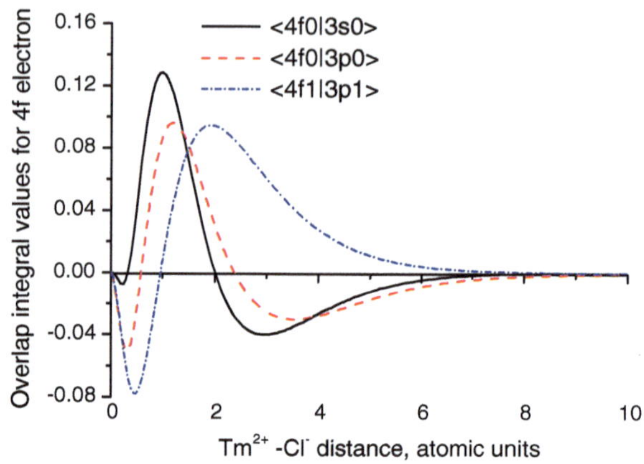

Fig. 5.7 Calculated dependence of overlap integrals on the distance between Tm^{2+} and Cl^- ions (<4f0|3s0> [*solid line*], <4f0|3p0> [*dashed line*], and <4f1|3p1> [*dash-dotted line*]). Reproduced from Ref. [10] by permission of John Wiley & Sons Ltd

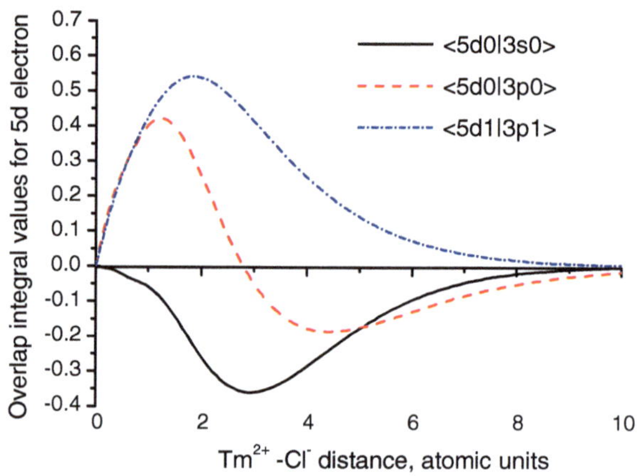

Fig. 5.8 Calculated dependence of overlap integrals on the distance between Tm^{2+} and Cl^- ions (<5d0|3s0> [*solid line*], <5d0|3p0> [*dashed line*], and <5d1|3p1> [*dash-dotted line*]). Reproduced from Ref. [10] by permission of John Wiley & Sons Ltd

Based on the structure of the Tm^{2+}-doped $SrCl_2$ optimized by the first-principles calculation, the non-zero 4f and 5d CFPs of Tm^{2+} ions in $SrCl_2$ can be estimated as follows—$B^4_0(5d) = -18944$, $B^4_4(5d) = B^4_{-4}(5d) = -11321$, $B^4_0(4f) = -567$, $B^4_4(4f)$ = (4f) = −339, $B^6_0(4f)$ 340, and $B^6_4(4f) = B^6_{-4}(4f) = -636$ (all in cm^{-1})—where the G(4f) and G(5d) parameter values can be fitted from the experimental 4f–4f and 4f–5d spectra [2] and are equal to 7.3 and 1.1, respectively.

With all of these parameters, it was possible to calculate the energy levels of the $4f^{13}$and $4f^{12}5d$ electron configurations of the Tm^{2+} ions in $SrCl_2$. In addition, probabilities of the electric dipole transitions between all possible states were also calculated. Finally, the simulated and measured 4f–5d spectra of $SrCl_2:Tm^{2+}$ are shown together shown in Fig. 5.9 where the pure 4f–5d transition lines are represented by the vertical bars, and the broad bands are reproduced by using the Gaussian-shaped curves with a full width at half maximum $E_{width} = 400$ cm^{-1},

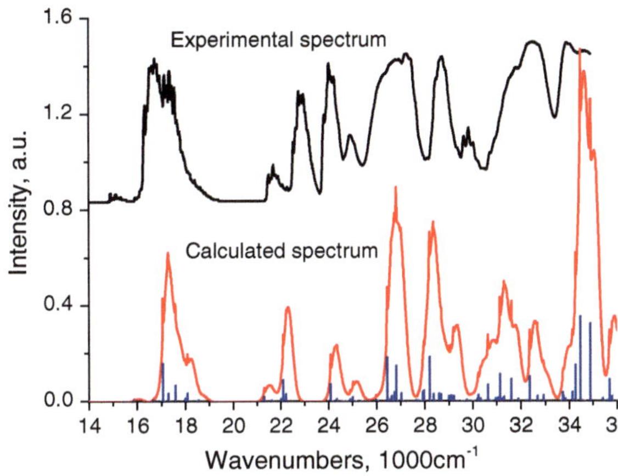

Fig. 5.9 Experimental absorption spectrum (*top*) and simulated $4f^{13}$–$4f^{12}5d$ transition spectrum (*below*) for $SrCl_2$:Tm^{2+}. The *vertical bars* show the calculated locations and relative intensities of pure electronic transitions. Reproduced from Ref. [10] by permission of John Wiley & Sons Ltd

which are displaced from the zero-phonon lines by E_{shift} = 200 cm^{-1}. It can be seen from Fig. 5.9 that the simulated spectrum is in good agreement with the experimental one including the positions of the absorption maxima and relative intensity of the absorption bands.

5.7 Summary

Detailed analysis of the structural and electronic properties of pure and Tm^{2+}-doped $CaCl_2$, $SrCl_2$, and $BaCl_2$ is presented in this chapter. First-principles calculations of the structural, electronic, and elastic properties were performed. Special emphasis was placed on variation of the structural and electronic properties on doping with Tm^{2+} ions. In particular, the position of the ground 4f level of thulium ions in the band gap of each considered material was determined. The calculated local structure of the impurity Tm^{2+} ions were applied to the understanding of the 4f–5d absorption spectrum of Tm^{2+} ions in $SrCl_2$ by employing semi-empirical CF theory. The successful theoretical simulation reveals that such a hybrid method is a good tool for describing the structural, electronic, and related optical properties of lanthanide defects in solids.

Acknowledgments C.-G. Ma acknowledges financial support from the National Natural Science Foundation of China (Grant No. 11204393), Scientific Research Foundation for the Selected Overseas Chinese Scholars, Ministry of Human Resources and Social Security of China (Grant No. 2014-167), National Training Program of Innovation and Entrepreneurship for Undergraduates (Grant No. 201410617001), Natural Science Foundation Project of Chongqing (Grant No. CSTC2014JCYJA50034), and Research Training Program for Undergraduates of Chongqing University of Posts and Telecommunications (Grant No. A2013-70). M. G. Brik acknowledges support from the European Regional Development Fund (Center of Excellence 'Mesosystems: Theory and Applications', TK114), Marie Curie Initial Training

Network LUMINET (Grant Agreement No. 316906), Chinese Recruitment Program of High-end Foreign Experts (Grant No. GDW20145200225), and the Programme for Foreign Experts offered by CQUPT and the Ministry of Education and Research of Estonia, Project PUT430.

References

1. Liu GK (2005) In: Liu GK, Jacquier B(eds) Spectroscopic properties of rare earths in optical materials. Tsinghua University Press & Springer (Berlin), Heidelberg, 2005, p 1
2. Grimm J, Wenger OS, Krämer KW, Güdek HU (2007) 4f-4f and 4f-5d excited states and luminescence properties of Tm^{2+}-doped CaF_2, $CaCl_2$, $SrCl_2$ and $BaCl_2$. J Lumin 126:590
3. Kiss ZJ (1962) Energy levels of divalent thulium in CaF_2. Phys Rev 127:718
4. McClure DS, Kiss ZJ (1963) Survey of the spectra of the divalent rare-earth ions in cubic crystals. J Chem Phys 39:3251
5. Loh E (1968) $4f^n \rightarrow 4f^{n-1}5d$ spectra of rare-earth ions in crystals. Phys Rev 175:533
6. Sabisky ES, Anderson CH (1967) Optically induced Tm^{2+} ions in trigonal sites in BaF_2. Phys Rev 159:234
7. Feofilov PP, Kaplyansky AA (1962) Spectra of divalent rare earth ions in the crystals of alkali-earth fluorides. 1. Samarium. Opt Spectr 12:493
8. Kaplyansky AA, Feofilov PP (1962) Spectra of divalent rare earth ions in the crystals of alkali-earth fluorides. 2. Europium and ytterbium. Opt Spectr 13: 235
9. Duncan RC, Kiss ZJ (1963) Continuously operating CaF_2:Tm^{2+} optical maser. Appl Phys Lett 3:23
10. Ma CG, Brik MG (2013) First-principles calculations of structural and electronic properties of pure and Tm^{2+}-doped $SrCl_2$. Phys Stat Solidi B 250:858
11. Brik MG, Sildos I, Kiisk V (2011) Calculations of physical properties of pure and doped crystals: Ab initio and semi-empirical methods in application to $YAlO_3$:Ce^{3+} and TiO_2. J Lumin 131:396
12. Brik MG, Kityk IV, Ozga K, Slezak A (2011) Structural, electronic and optical properties of pure and Ni^{2+}-doped CdI_2 layered crystals as explored by ab initio and crystal field calculations. Phys B 406:192
13. Nazarov M, Brik MG, Spassky D, Tsukerblat B, Nor Nazida A, Ahmad-Fauzi MN (2013) Structural and electronic properties of $SrAl_2O_4$:Eu2 + from density functional theory calculations. J Alloys Compds 573:6
14. Anderson CH, Call P, Stott J, Hayes W (1975) Correlation between superhyperfine constants and ef crystal field parameters of Tm^{2+} in CaF_2, SrF_2, and BaF_2. Phys RevB 11:3305
15. Voronko YK, Larionov AL, Malkin BZ (1976) Vibrational structure of optical spectra of CaF_2:Tm^{2+}, CaF_2:Yb^{3+} crystals. Optika i Spectroskopiya 40:86
16. Van Bever AK, Nieuwenkamp W (1935) Die Kristallstruktur von Calciumchlorid, $CaCl_2$. ZKristallographie 90:374
17. Brauer G, Müller O (1958) ZurKristallchemie des Strontiumchlorids. Z AnorgAllgChem 295:218
18. Brackett EB, Brackett TE, Sass RL (1963) The crystal structures of barium chloride, barium bromide, and barium iodide. J PhysChem 67:2132
19. Momma K, Izumi F (2011) VESTA 3 for three-dimensional visualization of crystal, volumetric and morphology data. J Appl Crystallogr 44:1272
20. Dovesi R et al (2009) CRYSTAL09 user's manual. University of Torino, Torino, Italy, p 76
21. Blanca-Romero A, Berrondo M, Rivas-Silva JF (2011) Kondo-like resonance in ZnO:Eu. IntJ Quant Chem 111:3831
22. Demichelis R, Civalleri B, Ferrabone M (2010) Dovesi R (2010) On the performance of eleven DFT functionals in the description of the vibrational properties of aluminosilicates. Int J Quant Chem 110:406

23. Li G, Lin CC, Chen WT, Molokeev MS, Atuchin VV, Chiang CY, Zhou W, Wang CW, Li WH, Sheu HS, Chan TS, Ma CG, Liu RS (2014) Photoluminescence tuning via cation substitution in oxonitridosilicate phosphors: DFT calculations, different site occupations, and luminescence mechanisms. Chem Mater 26:2991
24. Liu CM, Qi ZM, Ma CG, Dorenbos P, Hou DJ, Zhang S, Kuang XJ, Zhang JH, Liang HB (2014) High light yield of $Sr_8(Si_4O_{12})Cl_8:Eu^{2+}$ under X-ray excitation and its temperature-dependent luminescent characteristics. Chem Mater 26:3709
25. Basu M, Zhang ZW, Chen CJ, Chen PT, Yang KC, Ma CG, Lin CC, Hu SF, Liu RS (2015) Heterostructure of Si and $CoSe_2$: a promising photocathode based on a non-noble metal catalyst for photoelectrochemical hydrogen evolution. Angew Chem Int Ed 54:6211
26. Aprà E, Causà M, Prencipe M, Dovesi R, Saunders VR (1993) On the structural properties of NaCl: an ab initio study of the B1–B2 phase transition. J Phys Condens Matter 5:2969
27. Valenzano L, Torres FJ, Klaus D, Pascale F, Zicovich-Wilson CM, Dovesi R (2006) Ab initio study of the vibrational spectrum and related properties of crystalline compounds; the case of $CaCO_3$ calcite. Z Phys Chem 220:893
28. Lim IS, Stoll H, Schwerdtfeger P (2006) Relativistic small-core energy-consistent pseudo potentials for the alkaline-earth elements from Ca to Ra. J Chem Phys 124:034107
29. Dolg M, Stoll H, Preuss H (1989) Energy-adjusted ab initio pseudo potentials for the rare earth elements. J Chem Phys 90:1730
30. Cao XY, Dolg M (2002) Segmented contraction scheme for small-core lanthanide pseudo potential basis sets. J Mol Struct Theo Chem 581:139
31. Evarestov RA (2007) Quantum chemistry of solids. Springer-Verlag, Heidelberg
32. Perger WF, Criswell J, Civalleri B, Dovesi R (2009) Ab-initio calculation of elastic constants of crystalline systems with the CRYSTAL code. Comp Phys Comm 180:1753
33. Reid MF, van Pieterson L, Wegh RT, Meijerink A (2000) Spectroscopy and calculations for $4f^N \rightarrow 4f^{N-1}5d$ transitions of lanthanide ions in $LiYF_4$. Phys Rev B 62:14744
34. Duan CK, Tanner PA (2010) What use are crystal field parameters? A chemist's viewpoint. J Phys Chem A 114:6055
35. Wybourne BG (1965) Spectroscopic properties of rare earths. Interscience Publishers (Wiley, Inc., New York), p 164
36. Ma CG, Brik MG, Ryba-Romanowski W, Swart HC, Gusowski MA (2012) Spectroscopy and calculations for $4f^N \rightarrow 4f^{N-1}5d$ transitions of lanthanide ions in K_3YF_6. J Phys Chem A 116:9158
37. Pan ZF, Ning LX, Cheng BM, Tanner PA (2006) Absorption, excitation and emission spectra of $SrCl_2:Eu^{2+}$. Chem Phys Lett 428:78
38. Cowan RD (1981) The theory of atomic structure and spectra. University of California Press, Berkeley, p 456
39. Ma CG, Brik MG, Tian Y, Li QX (2014) Systematic analysis of spectroscopic characteristics of the lanthanide and actinide ions with the $4f^{N-1}5d$ and $5f^{N-1}6d$ electronic configurations in a free state. J Alloys Compds 603:255
40. Malkin BZ, Solovyev OV, Malishev AY, Saikin SK (2007) Theoretical studies of electron-vibrational $4f^N$-$4f^{N-1}5d$ spectra in $LiYF_4:RE^{3+}$ crystals. J Lumin 125:175
41. Edvardsson S, Klintenberg M (1999) The use of CI calculated polarizabilities to study rare-earth CFP dependencies in the laser host RE:YLF. Mater Sci Forum 315:407
42. Sugiura C (1974) Chlorine K-x-ray spectra and electronic band structure of $MgCl_2$, $CaCl_2$, $SrCl_2$ and $BaCl_2$. Phys Rev B 9:2679
43. Lauer HV Jr, Solberg KA, Kühner DH, Bron WE (1971) Elastic constants of EuF_2 and $SrCl_2$. Phys Lett A 35:219
44. Kanchana V, Vaitheeswaran G, Souvatzis P, Eriksson O, Lebegue S (2010) Density functional study of the electronic structure and lattice dynamics of $SrCl_2$. J Phys Condens Matter 22:445402

45. Kumar P, Vedeshwar AG (2015) DFT calculations of structural, electronic, optical and elastic properties of scintillator materials BaCl2 and BaBr 2. J Phys D Appl Phys 48:105301
46. Nye JF (1985) Physical properties of crystals. Clarendon Press, Oxford, p 143
47. Shannon RD (1976) Revised effective ionic radii and systematic studies of interatomic distances in halides and chalcogenides. Acta Crystallogr Sect A: Found Crystallogr 32:751

Chapter 6
First-Principles Calculation of Luminescent Materials

Bangfu Ding and Junying Zhang

Abstract Since the establishment of the well-known Kohn–Sham equation on the basis of the density functional theory (DFT), the DFT has been the most powerful tool of electronic-structure calculation in the field of condensed-matter physics. In luminous research regions, the DFT calculation finds wide application in revealing the luminescence mechanism and designing luminous materials. In this chapter, we first introduce the first-principles calculation foundation theory, related software, and the luminescence physics processes. Then various materials properties obtained using DFT are reviewed. Application of the first-principles calculation in exploring the intrinsic, native defect and dopant, as well as the doping-induced defect of photoluminescence origin, is illustrated by providing several examples. In addition, the excited-state calculation and band-gap correction approaches are presented. The calculation method and computation procedures must be improved and optimized in the future to make first-principles calculation play a more important role in luminescence-materials investigation.

6.1 The First-Principles Basic Theory, Related Software, and Luminescence Foundation

The solution of the Schrodinger equation only needs nuclear charge numbers and simulation environment parameters not involving any empirical parameters within the framework of the first-principles theory. The obtained results are eigenvalues and eigenfunctions of the systems studied, which can derive all properties of the system from theory. Moreover, the simulated results agree better with the experiment due to the gradual improvement of the models and the calculation procedure as well as the promotion of computation precision. Therefore, the reliability of the

B. Ding · J. Zhang (✉)
Key Laboratory of Micro-Nano Measurement, Manipulation and Physics (Ministry of Education), Department of Physics, Beihang University, Beijing 100191, China
e-mail: zjy@buaa.edu.cn

R.S. Liu (ed.), *Phosphors, Up Conversion Nano Particles, Quantum Dots and Their Applications*, DOI 10.1007/978-3-662-52771-9_6

first-principles calculation has been enhanced gradually. In this section, we first introduce the fundamental theory of the first-principles calculation and the related software, and then we present the basic physics process of luminescence.

6.1.1 Born–Oppenheimer Approximation

Generally, luminescence materials, such as crystals and nano-scale bodies, are many-electron systems. To acquire a system's electron-energy levels, the starting point is solving the Schrodinger equation of the multi-atom system, $H\psi(\vec{r}, \vec{R}) = E\psi(\vec{r}, \vec{R})$. Here $\vec{r}$ and $\vec{R}$ are sets of all electron coordinates $\{\vec{r}\}$ and atom coordinates $\{\vec{R}\}$. Without the external fields, the total Hamiltonians include the nuclear kinetic energy $T_{\rm N}(\vec{R})$, the electronic kinetic energy $T_{\rm e}(\vec{r})$, the internucleus interaction $V_{\rm N}(\vec{R})$, the interelectron interaction $V_{\rm e}(\vec{r})$, and the interaction $V_{\rm e-N}(\vec{r},\vec{R})$ between the electrons and the nuclei. Therefore, the total Hamiltonian can be written as $H = T_{\rm e}(\vec{r}) + V_{\rm e}(\vec{r}) + T_{\rm N}(\vec{R}) + V_{\rm N}(\vec{R}) + V_{\rm e-N}(\vec{r},\vec{R})$. Because the electron masses are far less than those of the nuclei, the electron thermal motion speeds are much faster than those of the nuclei, and they move at higher velocity. In contrast, the nuclei vibrate just near their equilibrium positions. That is to say, the electrons move adiabatically around the nuclei, whereas the nuclei can only move slowly to keep up with the electrons distribution change. According to this assumption, Born and Oppenheimer propose that electrons and nuclei motions are treated separately, thus forming the so called Born–Oppenheimer (adiabatic) approximation [1].

Under adiabatic approximation, the Schrodinger equation solution of multi-particle system can be written as the product of a nuclear wave function $\phi(\vec{R})$ and an electronic wave function $\psi(\vec{r},\vec{R})$. The electron wave function $\psi(\vec{r},\vec{R})$ is determined by many electrons Hamiltonian H_0 under the condition of instantaneous atom coordinates $\vec{R}$. Using $V_{\rm ext}(\vec{r},\vec{R})$ to express the external effect of the system containing the electron–nucleus interaction and other external potential field, we obtain

$$\begin{cases} H_0\psi_n(\vec{r},\vec{R}) = E_n\psi_n(\vec{r},\vec{R}) \\ H_0 = T_{\rm e}(\vec{r}) + V_{\rm e}(\vec{r}) + V_{\rm ext}(\vec{r},\vec{R}) \end{cases}. \tag{6.1}$$

The E_n and $\psi_n(\vec{r}, \vec{R})$ indicate the n-th eigenvalue and eigenfunction, respectively. We apply the $V(\vec{r}, \vec{R})$ representing the potential energy term in the Schrodinger equation, i.e., $V(\vec{r},\vec{R}) = \sum_p \sum_{q(q\neq p)} \frac{Z_pZ_q}{\vec{R}_{pq}} + \sum_i \sum_{k(k\neq i)} \frac{1}{\vec{r}_{ik}} - \sum_p \sum_j \frac{Z_p}{\vec{r}_{pj}}$. The $i(j)$ and $p(q)$ represent the $i(j)$-th electron and the $p(q)$-th nucleus, respectively. $Z_p(Z_q)$ is the nuclear charge of the $p(q)$-th nucleus. The symbol $\sum_{\dots}\cdots$ is the nuclei and electrons summation notation. Applying the separation-variation approach, the motion equations of the electrons and nuclei are derived as:

$$-\frac{1}{2}\sum_{i}\nabla_i^2\psi(\vec{r},\vec{R})+V(\vec{r},\vec{R})\psi(\vec{r},\vec{R})=E_i(\vec{R})\psi(\vec{r},\vec{R}), \tag{6.2}$$

$$-\frac{1}{2}\sum_{p}\nabla_p^2\phi(\vec{R})+E_i(\vec{R})\phi(\vec{R})=\varepsilon_i(\vec{R})\phi(\vec{R}). \tag{6.3}$$

The symbol $\nabla_i^2\left(\nabla_p^2\right)$ is expressed as $\frac{\partial^2}{\partial x_i^2}+\frac{\partial^2}{\partial y_i^2}+\frac{\partial^2}{\partial z_i^2}\left(\frac{\partial^2}{\partial x_p^2}+\frac{\partial^2}{\partial y_p^2}+\frac{\partial^2}{\partial z_p^2}\right)$. Equation (6.2) denotes electrons motion in case of fixing the nuclear positions. Equation (6.3) represents the nuclei motion. The $E_i(\vec{R})$ describes the electrons energy in (6.2) and the nuclei potential energy in (6.3). The $\varepsilon_i(\vec{R})$ is the sum of electronic total energy and nuclear repulsion energy, i.e., the system total energy. Because a solid system involves many interactions (such as electron–electron and nucleus–electron interaction) and huge numbers of electrons due to a very large volume, further approximations are needed to solve such complexity as seen in Eqs. (6.2) and (6.3).

6.1.2 Hartree–Fock Approximation

Through adiabatic approximation, motions of the electrons and the nuclei are handled separately. The crucial problem is that the many-electron wave functions are difficult to express and calculate. When a solid system's electron numbers exceed 1000 (the solid with volume 1 cm^3 contains 10^{29} electrons), the multi-electron wave function will be an unreasonable scientific concept [2, 3]. For many-electron system, Eq. (6.2) still cannot be solved strictly due to the electron–electron repulsion operator $\frac{1}{\overrightarrow{r_{ik}}}$ in terms of potential energy. To solve the variable-separation problem, Hartree and Fock present a further approximation method before the DFT formation.

From the Hartree point, every electron state can be described by a single electron wave function, $\psi_i(\overrightarrow{x_i})$, if the electron–electron interactions are negligible. The wave function of the whole system containing N electrons is written as $\psi(\overrightarrow{x_1},\overrightarrow{x_2},\ldots\overrightarrow{x_i},\ldots\overrightarrow{x_j},\ldots\overrightarrow{x_N})=\psi_1(\overrightarrow{x_1})\psi_2(\overrightarrow{x_2})\ldots\psi_i(\overrightarrow{x_i})\ldots\psi_j(\overrightarrow{x_j})\ldots\psi_N(\overrightarrow{x_N})$. Substituting $\psi(\overrightarrow{x_1},\overrightarrow{x_2},\ldots\overrightarrow{x_i},\ldots\overrightarrow{x_j},\ldots\overrightarrow{x_N})$ into (6.2) and applying the variation method to obtain the energy minimal value, the Hartree equation is derived as follows:

$$\left[-\nabla^2 + V(\vec{r}) + \sum_{i' \neq i} \int \frac{\left|\psi_{i'}\left(\vec{r'}\right)\right|^2}{\left|\vec{r} - \vec{r'}\right|} d\vec{r'}\right] \psi_i(\vec{r}) = E_i \psi_i(\vec{r}). \tag{6.4}$$

In fact, the Hartree approximation is the independent-particle model. However, every quantum state in the Hartree wave function satisfies the Pauli exclusion principle such as exchanging the i-th and j-th electrons

$$\begin{cases} \psi\left(\vec{x_1}, \vec{x_2}, \ldots \vec{x_i}, \ldots \vec{x_j}, \ldots \overrightarrow{x_N}\right) \overset{i \leftrightarrow j}{\Rightarrow} \\ \psi\left(\vec{x_1}, \vec{x_2}, \ldots \vec{x_i}, \ldots \vec{x_j}, \ldots \overrightarrow{x_N}\right) = \psi\left(\vec{x_1}, \vec{x_2}, \ldots \vec{x_j}, \ldots \vec{x_i}, \ldots \overrightarrow{x_N}\right). \end{cases} \tag{6.5}$$

The Hartree wave function lacks the anti-symmetry of electron-wave function exchange. Thereafter Fock uses the Slater determinant wave function like

$$\psi\left(\vec{x_1}, \vec{x_2}, \ldots \vec{x_i}, \ldots \vec{x_j}, \ldots \overrightarrow{x_N}\right) = \begin{vmatrix} \psi_1(\vec{x_1}) & \psi_1(\vec{x_2}) & \cdots & \psi_1(\vec{x_i}) & \cdots & \psi_1(\vec{x_j}) & \cdots & \psi_1(\overrightarrow{x_N}) \\ \psi_2(\vec{x_1}) & \psi_2(\vec{x_1}) & \cdots & \psi_2(\vec{x_i}) & \cdots & \psi_2(\vec{x_j}) & \cdots & \psi_2(\overrightarrow{x_N}) \\ \vdots & \vdots & \vdots & \vdots & \vdots & \vdots & \vdots & \vdots \\ \psi_i(\vec{x_1}) & \psi_i(\vec{x_2}) & \cdots & \psi_i(\vec{x_i}) & \cdots & \psi_i(\vec{x_j}) & \cdots & \psi_i(\overrightarrow{x_N}) \\ \vdots & \vdots & \vdots & \vdots & \vdots & \vdots & \vdots & \vdots \\ \psi_j(\vec{x_1}) & \psi_j(\vec{x_2}) & \cdots & \psi_j(\vec{x_i}) & \cdots & \psi_j(\vec{x_j}) & \cdots & \psi_j(\overrightarrow{x_N}) \\ \vdots & \vdots & \vdots & \vdots & \vdots & \vdots & \vdots & \vdots \\ \psi_N(\vec{x_1}) & \psi_N(\vec{x_2}) & \cdots & \psi_N(\vec{x_i}) & \cdots & \psi_N(\vec{x_j}) & \cdots & \psi_N(\overrightarrow{x_N}) \end{vmatrix}$$

to replace the Hartree multi-electron wave function. The Slater wave function satisfies the exchange anti-symmetry, i.e., the i-th and j-th electrons exchange

$$\begin{cases} \psi\left(\vec{x_1}, \vec{x_2}, \ldots \vec{x_i}, \ldots \vec{x_j}, \ldots \overrightarrow{x_N}\right) \overset{i \leftrightarrow j}{\Rightarrow} \\ \psi\left(\vec{x_1}, \vec{x_2}, \ldots \vec{x_i}, \ldots \vec{x_j}, \ldots \overrightarrow{x_N}\right) = -\psi\left(\vec{x_1}, \vec{x_2}, \ldots \vec{x_j}, \ldots \vec{x_i}, \ldots \overrightarrow{x_N}\right) \end{cases} \tag{6.6}$$

Applying the variation method, the Hartree–Fock (HF) equation is obtained as below:

$$\begin{aligned} &\left[-\nabla^2 + V(\vec{r})\right]\psi_i(\vec{r}) + \sum_{i' \neq i} \int \frac{\left|\psi_{i'}\left(\vec{r'}\right)\right|^2}{\left|\vec{r} - \vec{r'}\right|} \mathrm{d}\vec{r'}\, \psi_i(\vec{r}) \\ &- \sum_{i' \neq i, \parallel} \int \frac{\psi_{i'}^*\left(\vec{r'}\right)\psi_i\left(\vec{r'}\right)}{\left|\vec{r} - \vec{r'}\right|} \mathrm{d}\vec{r'}\, \psi_{i'}(\vec{r}) \\ &= E_i \psi_i(\vec{r}) \end{aligned} \tag{6.7}$$

with $\parallel$ denoting the spin parallel. The HF Eq. (6.7) adds a term to the Hartree equation, and this term is called an "exchange-interaction term". The advantage of this method is that its results are precise and the HF equation is an ab initio approach in strict meaning. This method also contains many drawbacks, e.g., the calculation quantity increases exponentially with the increase of electron numbers, thus requiring larger storage and more central processing units (CPU). Therefore, these disadvantages promote the DFT generation.

6.1.3 Density Functional Theory

In DFT, the multi-electron wave function and Schrodinger equation solution are replaced by the electronic density $\rho(\vec{r})$ and the corresponding computation procedure. Hence, the many-electron problem is simplified to a single–electron problem. This theory becomes a powerful tool for the computation of electronic structures as well as total energy. Therefore, the DFT is an important method to study the ground state of a multi-electron system.

Combining molecular dynamics method, the DFT obviously promotes progress in materials design, reaction-process investigation, structure simulation, and properties estimation. The DFT quickly becomes an important basis and core technology in computational materials science. Recently, the DFT work has increased dramatically and has greatly exceeded the HF method in research. The DFT method fits various types of applications because the relationship between the electronic ground-state energy and nuclear positions can be used to determine the molecular and crystal structures. When the atoms are not in their equilibrium positions, the DFT can also ascertain the stress of the nuclei. Therefore, the DFT can solve many problems in atomic and molecular physics such as the ionization-potentials calculation, vibration-spectrum study, chemical-reaction process investigation, biological molecule–structure analysis, and catalytic-activity characterization. In condensed-matter physics, the DFT can also gives rational conclusion on many properties such as material electronic and geometrical structures as well as metal-phase transition in solid or liquid. Another advantage of the DFT is that it provides the first-principles or ab initio computational framework and the development foundation of various energy band–calculation methods.

The electron-density functional theory proposed by Thomas and Fermi [4, 5] is applied to represent energy and thus can reduce the computation variables greatly. Due to nonprecise processing of kinetic energy, the early theory displays a little rough and is mainly applied in chemistry. Sometimes the Thomas–Fermi theory calculation result presents that a molecular does not bond. This conclusion hinders the further development of the Thomas–Fermi theory. In 1964, Hobenberg and Kohn proposed the well-known Hobenberg–Kohn theorem [6]. Applying this theorem and a series of mathematic derivation, the Kohn–Sham equation is obtained:

$$\left[-\nabla^2 + V_{\text{ext}}(\vec{r}) + \int d\vec{r}\frac{\rho\left(\vec{r}'\right)}{\left|\vec{r}-\vec{r}'\right|} + \frac{\delta E_{xc}[\rho]}{\delta\rho(\vec{r})}\right]\psi_i(\vec{r}) = E_i\psi_i(\vec{r})$$
$$\text{or } \left[-\frac{1}{2}\nabla^2 + V(\vec{r}) + \mu_{xc}(\vec{r})\right]\psi_i(\vec{r}) = E_i\psi_i(\vec{r}) \tag{6.8}$$

where $\rho(\vec{r}) = \sum_{i=1}^{N}|\psi_i(\vec{r})|^2$. It is the core to use a noninteracting model instead of corresponding-particles interaction in the Hamiltonian. In addition, all complexity containing any other particle interaction is included in the exchange correlation functional $E_{xc}[\rho]$. Unfortunately, all of the complex-interactions functional $E_{xc}[\rho]$ remains unknown. Therefore, various approximations—such as the local-density approximation (LDA) [5] and the generalized-gradient approximation (GGA)—were taken for the functional $E_{xc}[\rho]$ in later applications [7, 8]. The potential energy term treatment contains an all-electron full potential, all-electron muffin-tin, pseudo-potential, and jellium model. Similarly, the wave-function treatment includes plane waves, augmented plane waves, linearized augmented plane waves, scattering function, linear combination of atomic orbitals, and full numerical calculation. The processing method and the approximation of every term in Eq. (6.8) are shown in Fig. 6.1. The more detailed DFT introduction can be referred to in some monographs [9].

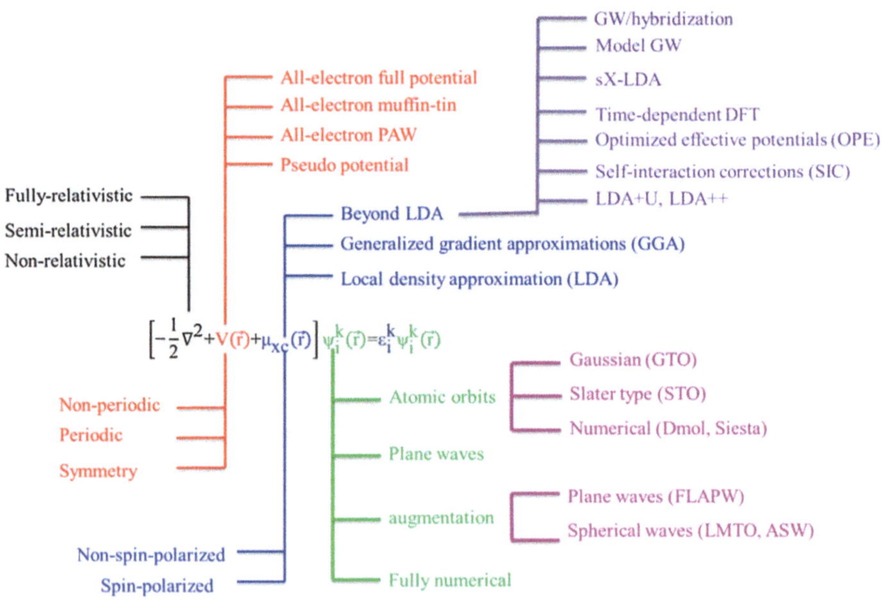

Fig. 6.1 Different approximation and processing methods for each term of the Kohn–Sham equation

6.1.4 Related Calculation Software

Based on the DFT, researchers have developed many open sources and commercial calculation software as shown in Table 6.1. In Table 6.1, we list software names, basis sets, potential-energy treatments, and Web sites. Moreover, there are many other similar first-principles calculation software that are not listed in Table 6.1.

In the following text, we introduce the Vienna Ab initio Simulation Package (VASP) utilization to consider self-activated luminescence $ZnMoO_4$ as an example. VASP calculations mainly have three procedures including structure relaxation, self-consistency, and calculation of properties. At each step of the calculation, the four files (INCAR, KPOINTS, POTCAR, and POSCAR) can be supplied. For the other files, one can refer to the Web page introduction. The INCAR file is a VASP-input document, which determines what to do and how to do it. The POSCAR file describes the system structure–containing unit or the super cell–basis vectors and atom positions. The KPOINTS file includes the k-point coordinates and weights as well as the mesh size for creating the k-point grid. The POTCAR file denotes the pseudo-potential for each atom species in the calculation.

Moreover, P4VASP software is mainly employed to deal with the VASP-calculation results to obtain the visualization data. Visualization for

Table 6.1 Some related software based on the DFT

Type	Code name	Basis set	Potential treatment	Web sites
Open source	ABINIT	Plane wave	Pseudo-potential, projection augmented wave	www.abinit.org
	Dacapo	Plane wave	Pseudo-potential	wiki.fysik.dtu.dk/dacapo
	FHImd	Plane wave	Pseudo-potential	th.fhi-berlin.mpg.de/th/fhi98md/
	PWscf	Plane wave	Pseudo-potential	www.pwscf.org/
	Quickstep	Gaussian and plane wave	Pseudo-potential	www.cp2k.org/about
	SIESTA	Local/numerical	Pseudo-potential	departments.icmab.es/leem/siesta/
	CPMD	Plane wave	Pseudo-potential	www.cpmd.org/
Commercial	Castep	Plane wave	Pseudo-potential	accelrys.com/products/materials-studio/
	VASP	Plane wave	Pseudo-potential	www.vasp.at/
	CRYSTAL	Local	All-electron	www.crystal.unito.it/index.php
	WIEN2 K	LAPW	All-electron	www.wien2k.at
	Gaussian	Local	All-electron	www.gaussian.com

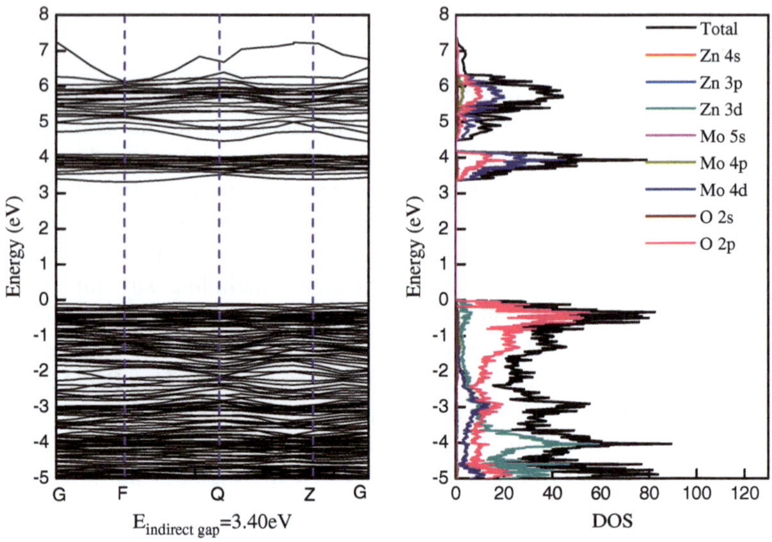

Fig. 6.2 Band structure and total and partial densities of states for $ZnMoO_4$ crystal with the Perdew–Burke–Ernzerhof (PBE) functional calculation

Electronic and Structural Analysis (VESTA) software can visualize the VASP POSCAR, CHARGE, and other files. Through the VASP calculation, the energy band and density of states (DOS) of $ZnMoO_4$ crystal are obtained in Fig. 6.2. The pure $ZnMoO_4$ crystal is an indirect band-gap semiconductor with a band-gap value of 3.40 eV, which is lower than the experimental band-gap value 4.50 eV [10]. The band-gap underestimation is a drawback of DFT formalism. The CB and VB are composed of Mo 4*d* state and O 2*p* state, respectively. The Zn 3*d* and O 2*p* states contribute to the VB and the CB with small quantity. Therefore, the luminous origin of $ZnMoO_4$ is ascribed to the MoO_4^{2-} anion group. Moreover, the electronic structure of molybdates can be tuned by the doping method, especially applied in luminescence materials [11]. The use of other software is referred to on the Web site mentioned in Table 6.1.

6.1.5 Luminescence Foundation

After absorbing external energy, such as light radiation, some materials can emit light. Why do they emit light? To explore the nature of luminescence, classical electrodynamical and quantum mechanical theories are applied. According to classical electrodynamics, the electromagnetic radiation vector A [12] for light absorption and emission is denoted as:

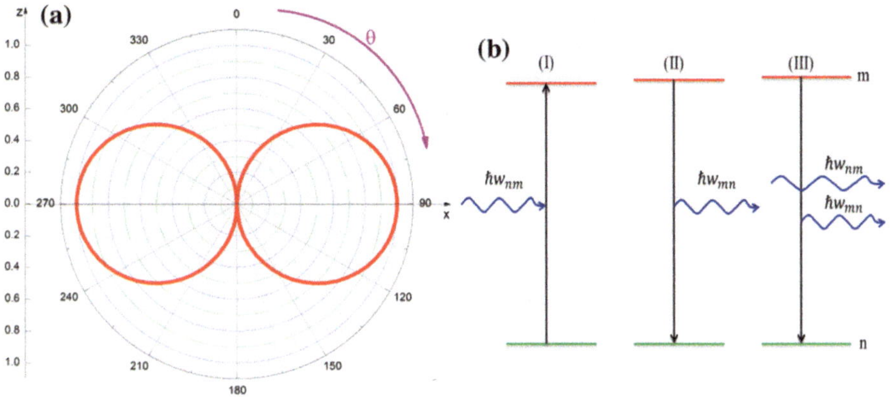

Fig. 6.3 **a** Angle-dependent linear dipole oscillation and **b** absorption I, self-emission II, and stimulated radiation III

$$A(\vec{r}) = \frac{\mu_0 e^{ikR}}{4\pi R} \int J\left(\vec{r'}\right)\left[1 - ik\vec{n}\cdot\vec{r'} + \frac{1}{2!}\left(ik\vec{n}\cdot\vec{r'}\right)^2 \ldots\right] d^3r' \tag{6.9}$$

with permeability of vacuum μ_0. The k and $\vec{n}$ are the wave vector and its unit vector, respectively. The $J\left(\vec{r'}\right)$ is current density, and R is the distance from the origin point to the field point $\vec{r}$, namely, $R = |\vec{r}|$. The first term on the right indicates the dipole transition, and the second one represents the magnetic and electronic quadrupole transitions. Generally the dipole transition intensity is stronger than that of the other two transitions, and thus the dipole transition is considered to be primary. If the dipole transition becomes forbidden, the other two transitions are also considered. The dipole emission can be described by linear dipole oscillation as displayed in Fig. 6.3a, and its intensity obviously depends on the oscillation direction denoted as θ. The oscillator intensity reaches the maximum and minimum values for $\theta = 0$ and $\theta = \frac{\pi}{2}$. The electric dipole momentum is expressed as $M = ez = M_0 e^{-iw_0 t}$ with the characteristic frequency w_0 and the radiation electric dipole moment amplitude M_0. The total energy containing the kinetic and potential energy is $\frac{m_e w_0^2 M_0^2}{2e^2}$ with electronic mass m_e. Therefore, the radiation electromagnetic energy per second is $\frac{w_0^4 M_0^2}{12\pi\varepsilon_0 c^3}$.

From quantum mechanical theory, the electron-transition description is shown in Fig. 6.3b containing the absorption, self-emission, and simulated emission processes [13]. The electron energy takes discrete values in a single atom as indicated by m and n. The light absorption and emission in these two quantum states can be described by statistical methods, and the energy relation between the m-th and n-th quantum states in the absorption and emission processes is expressed as $\hbar\omega_{mn} = E_m - E_n$. Under the light irradiation of energy-density $\rho(w_{mn})$, the emission probability is the sum of the self-emission and simulated emission probabilities. Their equilibrium relation is

$$\mathrm{N}_n B_{n\to m}\rho(w_{mn}) = \mathrm{N}_m[A_{m\to n} + B_{m\to n}(w_{mn})\rho(w_{mn})] \quad (6.10)$$

with the electron numbers N_m and N_n in the m-th and n-th states in atomic emission and absorption processes, respectively. The $\rho(w_{mn})$, $B_{n\to m}$, $A_{m\to n}$ and $B_{m\to n}$ are the excitation energy density, absorption, self-emission, and simulated emission coefficients, respectively. As we know, the absorption coefficient does not depend on excitation energy density and only relates to excitation frequency. The simulated-emission coefficient relies on excitation energy density. Assuming equality of the absorption and simulated emission constants, we can obtain $A_{m\to n} = \frac{\hbar w_{mn}^3}{\pi^2 c^3} B_{m\to n}$. Thus, the absorption probability can be calculated as $W_{mn} = B_{n\to m}\rho(w_{mn})$. Through quantum-mechanism derivation, the electron–dipole transition optic absorption is $W_{mn} = \frac{\pi}{3\varepsilon_0 c\hbar^3} I(w_{mn})|M_{mn}|^2$ with the excited state ψ_m^* and ground state ψ_n wave functions. $|M_{mn}|^2 = \left| \int \psi_m^* \left(\sum_i \mathrm{e}\vec{r_i} \right) \psi_n d\tau \right|^2$ is the electric dipole moment. Finally, we acquire the absorption, self-emission, and simulated emission constants $B_{n\to m} = B_{m\to n} = \frac{\pi}{3\varepsilon_0 \hbar^3} |M_{mn}|^2$ and $A_{m\to n} = \frac{w_{\mathrm{e.g.}}^3}{3\pi\varepsilon_0 \hbar c^3} |M_{mn}|^2$. The self-emission intensity $I(w_{mn})$ is proportional to $\hbar w_{mn} A_{m\to n} = \frac{w_{mn}^4}{3\pi\varepsilon_0 c^3} |M_{mn}|^2$. We can see that the self-emission intensity decays exponentially, and its life is inversely proportional to the absorption constant. The absorption cross-section, σ is defined as the probability of one atom absorbing one photon in unit section $\sigma_{mn} = \frac{\pi w_{mn}}{3\varepsilon_0 c\hbar} |M_{mn}|^2$. The oscillator intensity, f_{mn}, is a physical quantity to characterize absorption and emission intensity, namely $f_{mn} = \frac{2m_e w_{mn}}{3\hbar \mathrm{e}^2} |M_{mn}|^2$. Therefore, the emission life and absorption cross-section are expressed as a function of the oscillator strength $\tau_{mn}^{-1} = A_{m\to n} = \frac{\mathrm{e}^2 w_{mn}^2}{2\pi\varepsilon_0 mc^3} f_{mn}$ and $\sigma_{mn} = \frac{\pi \mathrm{e}^2}{2\varepsilon_0 mc} f_{mn}$.

In addition, the high-order perturbation is ignored in the electromagnetic-radiation vector. Considering these perturbations, the dipole moment is written as $|M_{mn}|^2 = |(\mathrm{e}\vec{r})_{mn}|^2 + |(\frac{\mathrm{e}}{2\mathrm{mc}} \vec{r} \times \vec{p})_{mn}|^2 + \frac{3\pi w_{mn}^2}{40c^2} |(\mathrm{e}\vec{r} \cdot \vec{r})_{mn}|^2$. If the initial and final wave functions have the same parity, $|M_{mn}|^2$ becomes zero for forbidden transition. The other two transitions become the main transition modes. For example, the *f*–*f* transition in the 4*f* shell isolated rare-earth ions, and the *d*–*d* transition in transition-metal ions belong to the parity-forbidden transition. From the quantum and classical description, the luminescence processes mainly contain absorption and emission. For the simplest isolated system, their intensity and probability can be calculated quantitatively. The luminescence materials are a many-body–interaction system, and thus their luminescence descriptions are more complicated. In the following section, we will introduce energy-band theory.

The luminous materials contain a large number of molecules, atoms, or ions forming a regular lattice structure. Thus, the electrons are located in a periodic

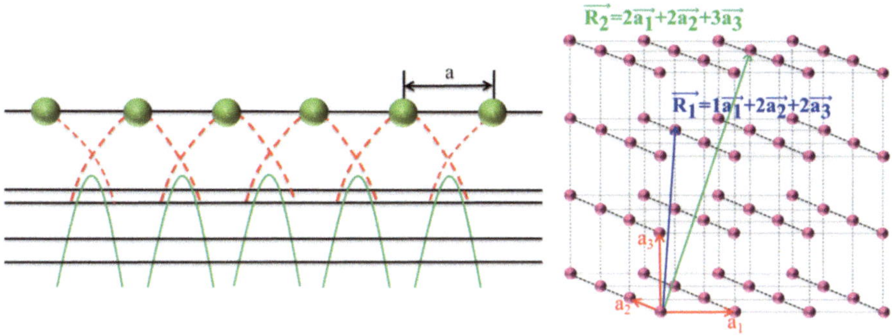

Fig. 6.4 Schematic one-dimensional periodic lattice structure potential and three-dimensional structure as well as two sets of lattice vectors $\overrightarrow{R_1} = 1\overrightarrow{a_1} + 2\overrightarrow{a_2} + 2\overrightarrow{a_3}$ and $\overrightarrow{R_2} = 2\overrightarrow{a_1} + 2\overrightarrow{a_2} + 3\overrightarrow{a_3}$

potential field as shown in Fig. 6.4. For a one-dimensional case, the electronic energy and wave function satisfy the Schrodinger equation $\left[-\frac{\hbar^2}{2M}\frac{d^2}{dx^2} + V(x)\right]\psi_k(x) = E(k)\psi_k(x)$. The solution is a Bloch function, namely, $\psi_k(x) = e^{ikx}u_k(x)$ with periodic function $u_k(x) = u_k(x + na)$. When the crystal has N atoms, every separate atom energy level become the close N level due to atomic interaction. Therefore, the energy band is formed as shown in Fig. 6.5. From Fig. 6.5, the energy band formed by the outer electrons is wider and the corresponding energy difference ΔE is larger. When the lattice spacing (a) is smaller, the band width and energy level difference ΔE are greater. In addition, the two energy bands may overlap. Generally speaking, the semiconductor and the insulator can be the luminator, whereas the metal materials cannot emit. Therefore, the VB and CB determine the luminous properties to a large extent. Similarly, for the three-dimensional case, we can also solve the Schrodinger equation to obtain the energy and satisfied Bloch theorem wave function.

The luminescence mainly originates from the inter-band transitions, which are divided into direct and indirect transitions according to the transition modes. If the electrons jump at the same point between the VBM (valence band maximum) and the CBM (conduction band minimum), this transition is direct. In contrast, there is indirect transition. The semiconductors silicon (Si) and gallium arsenide (GaAs) are typical examples as shown in Fig. 6.6. They have an indirect and direct band gap with the values 1.95 and 0.17 eV, respectively. When the crystal size becomes smaller, e.g., forming quantum dots, the Si becomes a better self-activated luminescence material.

Based on the electromagnetic quantum theory, the light absorption coefficient $\alpha(\hbar\nu)$ is equal to $A\sum p_{if}n_i n_f$ with A, p_{if}, n_i, and n_f being the coefficient, transition probability, initial, and final states density, respectively. The light absorption satisfies the energy and momentum conservation, i.e.,

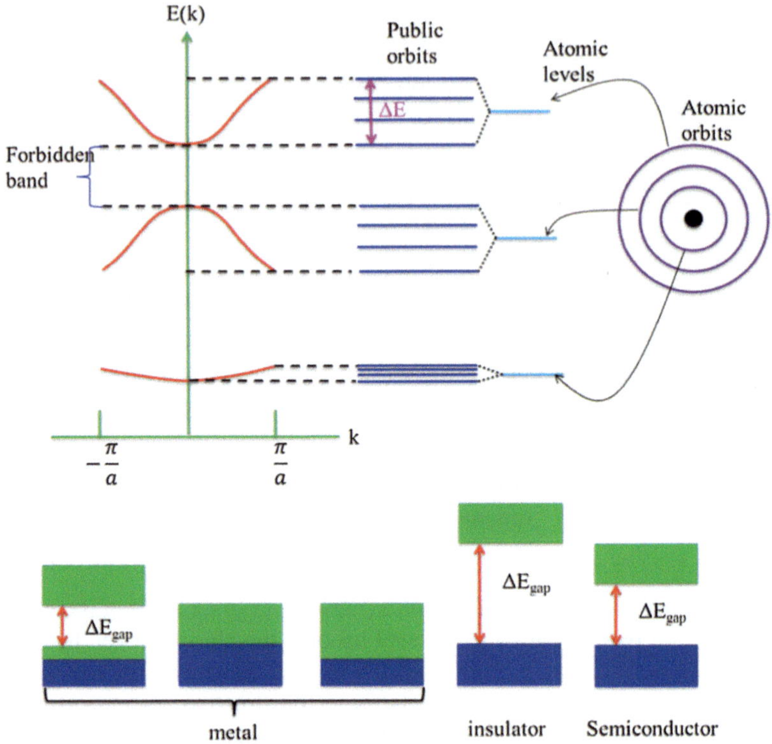

Fig. 6.5 The schematic energy-band formation and three types of energy-band structure

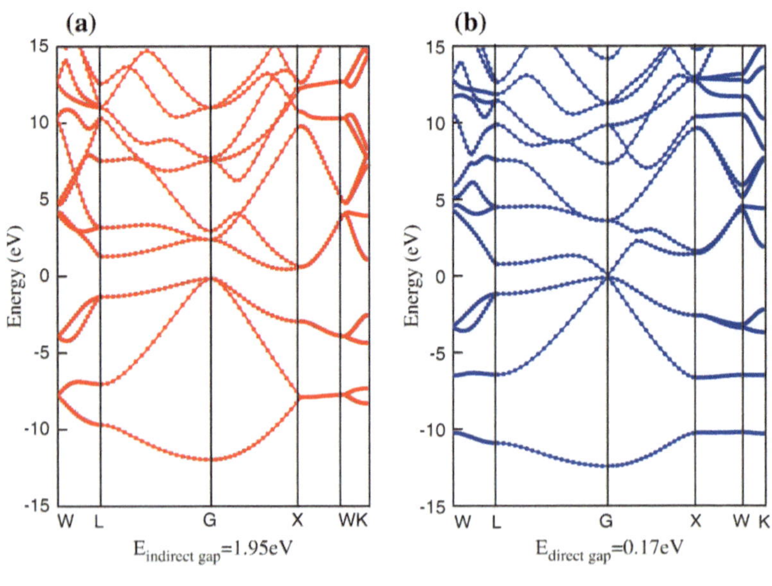

Fig. 6.6 The energy band structure of **a** Si and **b** GaAs crystals

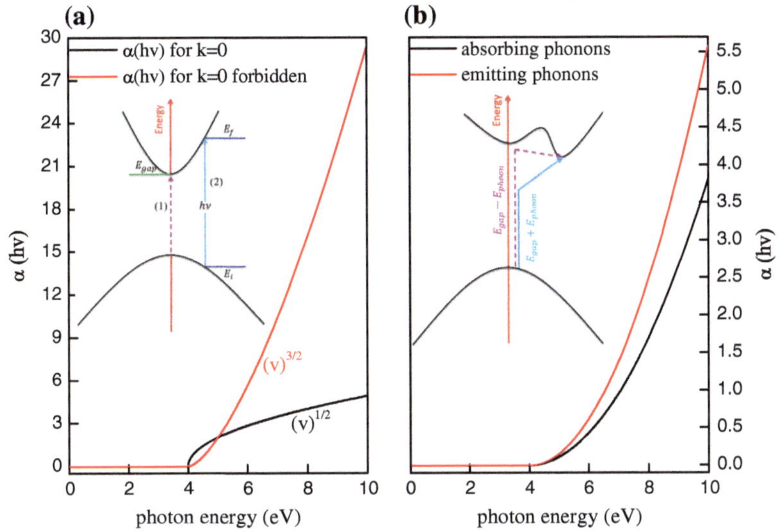

Fig. 6.7 The absorption constants change along with the photon energy in **a** direct and **b** indirect band-gap materials at room temperature. The *inset* pictures display the schematic electrons transitions

$$\begin{cases} \frac{\hbar^2}{2m^*} k_f^2 = \frac{\hbar^2}{2m^*} k_i^2 + h\nu \\ \hbar k_f = \hbar(k_i + q) \end{cases} \quad (6.11)$$

with $\hbar q$ being the photon momentum, and m^* being the effective mass of electrons. Because the photon momentum is smaller than the electrons-and-lattice momentum, an optics transition can only happen in the vertical direction of the energy-band dispersive curve. Therefore, the electron transfers directly from the VB to CB in direct gap materials, whereas this transition needs phonon association to meet the momentum conservation in indirect gap materials as shown in Fig. 6.7.

For direct band-gap materials, the absorption coefficients are derived as $A^*\left(h\nu - E_{\text{gap}}\right)^{1/2}$ for $k = 0$. When the transition in $k = 0$ becomes a forbidden transition, the absorption coefficient is $A'\left(h\nu - E_{\text{gap}}\right)^{3/2}$ for $k \neq 0$. For $k = 0$ and $k \neq 0$, the absorption constants are approximately proportional to $\sqrt{\nu}$ and $\sqrt{\nu^3}$, which is shown in Fig. 6.7a. In Fig. 6.7a, we assume that the host band-gap value is 4 eV. Thus, the absorption coefficient α is zero until the photon energy is larger than 4 eV. For the indirect-gap material, the absorption coefficient is $A\left(h\nu - E_{\text{gap}} + E_{\text{phonon}}\right)^2 \frac{1}{e^{\frac{E_{\text{phonon}}}{k_B T}} - 1}$ through absorbing phonons and $A\left(h\nu - E_{\text{gap}} - E_{\text{phonon}}\right)^2 \frac{1}{1 - e^{-\frac{E_{\text{phonon}}}{k_B T}}}$ via emitting phonons, respectively. The indirect transition probability is lower than that of the direct transition. The absorption coefficients of indirect gap materials are displayed in Fig. 6.7b. In Fig. 6.7b, the

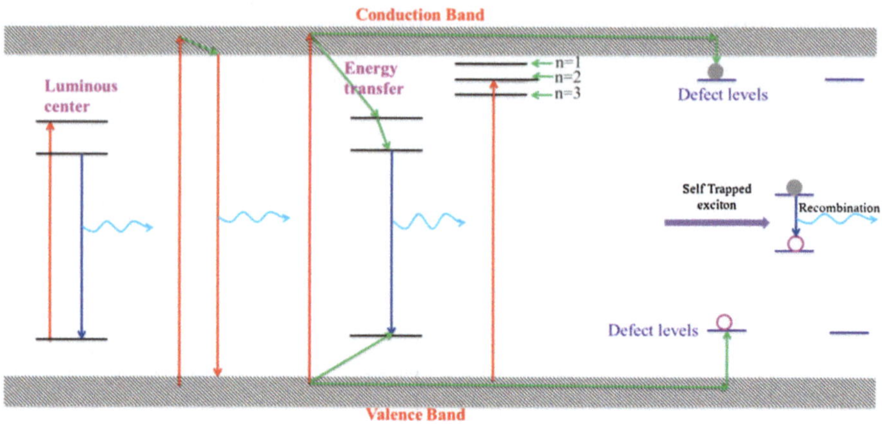

Fig. 6.8 The schematic band structure, defect, and exciton energy level as well as the luminous processes

band-gap and phonon energy values are 4 and 0.01 eV, respectively. For the indirect band-gap materials, the absorption coefficient is not zero under the photo energy slightly smaller than 4 eV. When the photon energy exceeds 4 eV, the absorption constants increase gradually.

Electrons in excited states are not stable; thus, they will return to the ground state releasing energy. The emission probability is $R = B\sum p_{ul}n_u n_l$ with the upper and lower energy state electron density n_u and n_l, respectively. In addition, the emission can also be completed by the phonon absorption and emission association $L = B'\left(h\nu - E_{\text{gap}} \pm E_{\text{phonon}}\right)^{1/2} e^{-\frac{h\nu - E_{\text{gap}}}{k_B T}}$, as shown in Fig. 6.8. From Fig. 6.8, the luminescent process can be differentiated into three types. Excitation energy directly excites the luminous centers or generates the exciton labeled n = 1, 2, 3 in the case of $E_{ex} < E_g$ (E_{ex} = excitation energy, E_g = host band gap). The exciton state is actually a neutral and nonconducting excited state that is divided into free and bound excitons. If the exciton is bound to defect states, it is called a "bound exciton." The self-trapped exciton emission is understood as explained later in the text. Some electrons are promoted to the CB by absorbing the photons to form small polarons. These polarons can interact with holes of the crystal defect or an impurity in the crystal to form self-trapped excitons (STE). The STE emission contributes to the PL. The STE can also form in a nondefective crystal through spontaneous symmetry breaking, for example, when an excited electron and hole lower their energy by creating a localized defect in the lattice. The polaron is defined as the electron and polarization field around the electron forming an interacting integral. If the effect range of the polarization field is larger than the lattice constants, it is called "large polarization". In contrast, it is "small polarization" when the effect range is smaller than the lattice parameters. When $E_{ex} > E_g$, the host first absorbs the energy and then transfers it to the luminous centers or emits the energy itself. White light is obtained by combining the luminous ions and host emission, which is commonly applied at present to realize the white light–emission diodes. When $E_{ex} \gg E_g$, the emission is a multi-stage and complex processes. In the next section, some concrete examples are cited to illustrate the luminous mechanism on the basis of the first-principles calculation.

6.2 Photoluminescence Mechanism Based on the First-Principles Calculation

Luminescence is commonly seen in nature and real life. For instance, all sorts of lamps, cathode ray tubes, and light-emitting diodes (LEDs) are applied to lighting, entertainment, and information displays [14]. The luminescence processes include excitation, emission, heat production, and energy transfer [15]. Luminescence can last for a period of time; thus it is different from thermal radiation, reflection, scattering, and Cherenkov radiation. Based on luminescent processes, luminous materials are divided into three categories, namely, self-activated matrixes, host + activator, and host + sensitizer + activator complex materials. Self-activated materials are a large family in which emission comes from the matrix or native defects. When exotic impurity is introduced in a matrix, it may also act as an activator or sensitizer to tune luminescence properties. Furthermore, more efficient use of irradiation can be realized by combing an activator and a sensitizer [16]. The self-activated host can absorb excitation energy and then emit radiation, e.g., tungstates, molybdates, vanadates, zirconates, titanates, niobates, tantalates, chromates, manganeses, metal oxides, and nano-size materials [17–19]. The other two types of luminescence matters include more examples such as rare-earth ion-doped oxides (Y_2O_3:Eu^{3+}) and co-doped silicates (Y_2SiO_3:Ce^{3+}, Tb^{3+}). To improve the luminescence properties, various preparation approaches are applied such as solid-phase reaction technique, polymeric precursor methods, and microwave-assisted hydrothermal/solvothermal route. Different morphologies are obtained including bulk powder, single crystal, and nano-scale particles with various size and shape [20]. Moreover, nonactivator or activator ion mono-doping and co-doping are also employed to tune luminescence. Therefore, various native or nonintrinsic defects grow in the matrix, which will substantially influence the luminescent properties. Although experimental characterization and the corresponding analysis can qualitatively or quantitatively identify different defects in the samples, microcosmic changes in the lattice structure induced by these defects and their formations mechanism are not clearly known.

To solve these problems, the first-principles calculation is employed to determine where the defects locate in the lattice sites and derive the defect-related electronic structure and optical properties. These properties typically include equilibrium crystal structure, total energy, charge density, energy-band structure, density of state (DOS), permittivity, reflectivity, and absorption spectra. From crystal structure, we can obtain the equilibrium lattice parameters, bond length, bond angle, and atom coordination or their variation caused by defects or impurities in the crystal. The probability of defect and dopant formation can be estimated through comparing formation energy based on the obtained total energy. For formation-energy calculation, we must consider different atmospheres, raw materials, stoichiometric ratio, etc. For example, the oxygen vacancy in ZnO can appear under oxygen-rich, oxygen-poor, and Zn-rich conditions. Its formation energy is denoted as

$$E_{\text{formation}}(\mathrm{V_O}, q) = E_T(\mathrm{ZnO{:}V_O}, q) - E_T(\mathrm{ZnO}) - E_T(\mathrm{O}) + qE_{Fermi} \tag{6.12}$$

with q being the charge of the oxygen vacancy such as −2, −1 and 0. E_{Fermi} is the Fermi energy level of the $\mathrm{V_O}$-containing system. The $E_T(\mathrm{ZnO{:}V_O}, q)$ and $E_T(\mathrm{ZnO})$ are the total energies of $\mathrm{V_O}$-containing and the pristine crystals, respectively. The oxygen atom single-point energy can be expressed as $E_T(\mathrm{O}) = \frac{E_T(\mathrm{O_2})}{2}$ under oxygen-rich and $E_T(\mathrm{O}) = E_T(\mathrm{ZnO}) - E_T(\mathrm{Zn})$ in oxygen-poor or Zn-rich conditions. $E_T(\mathrm{Zn})$ is the Zn atom single-point energy. $\mathrm{qE_{Fermi}}$ is related to the chemical potential of electrons. We assume that electron gains and losses are performed by exchanging them between the VB top and the Fermi sea. If the energy reference zero point is chosen as the VB top energy $E_{\text{VBM}}(\mathrm{ZnO{:}V_O}, q)$, qE_{Fermi} can be written as $q[E_{\text{Fermi}} + E_{\text{VBM}}(\mathrm{ZnO{:}V_O}, q)]$. Moreover, we define potential correction as $E_{\text{VBM}}(\mathrm{ZnO{:}V_O}, q) = E_{\text{VBM}}(\mathrm{ZnO}) + \Delta\mathrm{V}$ with $\Delta\mathrm{V} = \mathrm{V}_{\text{average}}(\mathrm{ZnO{:}V_O}, q) - \mathrm{V}_{\text{average}}(\mathrm{ZnO})$. $E_{\text{VBM}}(\mathrm{ZnO})$ is the VB top energy of pure ZnO, and ΔV is the average electrostatic potential difference between the $\mathrm{V_O}$-containing system and the pure supercells. The electron-exchange energy between the VBM and the Fermi sea is $E_{\text{VBM}}(\mathrm{ZnO}) = E_T(\mathrm{ZnO},0) - E_T(\mathrm{ZnO},q)$ with $E_T(\mathrm{ZnO},q)$ being the total energy of a perfect crystal with the charge q. Finally, the oxygen vacancy formation (6.12) becomes

$$\begin{aligned} E_{\text{formation}}(\mathrm{V_O}, q) = {} & E_T(\mathrm{ZnO{:}V_O}, q) - E_T(\mathrm{ZnO}) \\ & - E_T(\mathrm{O}) + q[E_{\text{Fermi}} + E_{\text{VBM}}(\mathrm{ZnO}) + \Delta V] \end{aligned} \tag{6.13}$$

The charge density can give the bonding and antibonding states of adjacent atoms. Moreover, the charge density has various derived forms such as deformation, difference, and spin-polarized charge density. The deformation-charge density indicates the charge redistribution after the atoms form a compound. The difference charge density represents charge redistribution after chemical composition or a geometrical configuration change in the systems. Through the deformation charge density, we can clearly see the bonding state of every atom. Moreover, we can see the strength of the bond polarity by way of the specific space distribution, e.g., charge accumulation or depletion. The charge distribution shape around certain lattice points can depict the bonding orbit (e.g., by analyzing d orbit).

The energy-band structure and DOS offer information on band gap, atomic energy level and constitution, defect or dopant energy-level position, as well as composition. Thus, we can judge whether this system is a metal, a semiconductor, or an insulator. The judgment standard is to look at the intersection between the Fermi energy level and CB. If they cross, the system is metal; otherwise it is a semiconductor or an insulator. For an intrinsic semiconductor, we can distinguish them as being a direct or indirect band-gap material.

In the calculation of luminescence material, the unit cell or supercell configuration, which contains dozens of atoms and hundreds of electrons is mostly used. Therefore, the band curves appear very flat and dense far below the VB. In principle, we are mainly concerned about the band structure around the Fermi level, namely, the VB and the CB. Band widths occupy an important role in band analysis. If the band is broadly fluctuated, the charge carrier effective mass is small, and the delocalization degree is large. In addition, the photoconductivity speed or charge mobility μ can be judged by the effective charge mass according to the simple formula $\mu = \frac{q}{m^*}\tau$. Here τ is the scattering time; q is the charge; and m* is the effective mass. Changes in effective mass can influence the charge mobility and further the conductivity. If τ does not change in the scattering of quasi-particles, μ is inversely proportional to the m*. When the semiconductor system is doped with different impurities, some new and narrower bands may occur in the band gap, namely, doping states. These doping states are called “donor states” or “acceptor states” depending on whether they give or gain electrons from the matrix, respectively. From the DOS graph, the composition of the VB, CB, and local state —as well as the hybridized peaks, which directly display the interaction of neighboring atoms—can be displayed.

The DFT calculation is employed widely to explore the defect-formation mechanism and the luminescence origin. There are a large number of reports studying the luminescence origin based on the DFT calculation. In what follows, we first introduce the self-activated photoluminescence (PL) mechanism investigation by way of the first-principles calculation and then demonstrate how the first-principles calculation has been used to treat the luminescent process related to intrinsic defects and dopants.

6.2.1 Intrinsic Luminescence

Initially, the PL mechanism is mainly studied by the molecular orbit theory, and this theory only treats some high-symmetry crystal. For intrinsic PL materials, first-principles calculations are used extensively to discuss the PL origin. From the calculation result, the fundamental crystal information and electronic properties can be obtained. The electronic-transition modes and their allowed or forbidden transition nature can be revealed. Thus, the theoretical results can predict the excitation and emission band positions approximately, which helps to perform the band assignment in the experimental spectra. After knowing the luminescent mechanism, we can modify the luminescence intensity and shift peak position as well as broaden the emission ranges by utilizing various experimental strategies.

Generally, self-activated luminescent tungstates and molybdates are divided into three classes: normal MeW(Mo)O_4, rare-earth Me_2W(Mo)O_6, and poly-tungstates/molybdates. Their luminescence origin is mainly charge-transfer transition between the O 2*p* orbits and W 5*d* or Mo 4*d* orbits. Moreover, the defect emission, such as vacancy or interstitial defects emission, and exciton recombination are also luminescence origin. In the following sections, several examples are provided to elaborate intrinsic luminous mechanism based on DFT.

According to the size of Me^{2+} radius, MeW(Mo)O_4 is divided into two types: wolframite monoclinic type with small ions—such as Mg, Fe, Mn, Zn, Cd, Cu—and scheelite tetragonal type with large ions such as Ca, Sr, Ba, and Pb [21]. $PbWO_4$ photoluminescence properties have been studied intensively for several decades because of plans to use it as a scintillation detector at the Large Hadron Collider in the European Center for Nuclear Research [22]. $PbWO_4$ luminescence contains one blue band and two green bands. Researchers believe that the blue band is ascribed to the charge-transfer transition of tungstate groups connected to Pb^{2+}. Zhang et al. calculated the electronic structures of $PbWO_4$, $CaWO_4$, $CaMoO_4$, and $PbMoO_4$ through the linearized augmented plane wave method [23]. In Fig. 6.9, we apply projection augment wave pseudo-potential method to recalculate their electronic properties. The results calculated using the two methods are consistent with each other. For the four materials, the VB and CB are mainly composed of O 2*p* and W 5*d* (Mo 4*d*), whereas the band gaps and band-to-band transition types are different. For instance, the Pb 6*s* states form the narrow band 1 eV below the bottom of the VB and thus hybridize with O 2*p* states. Thus, Pb^{2+} can also join in

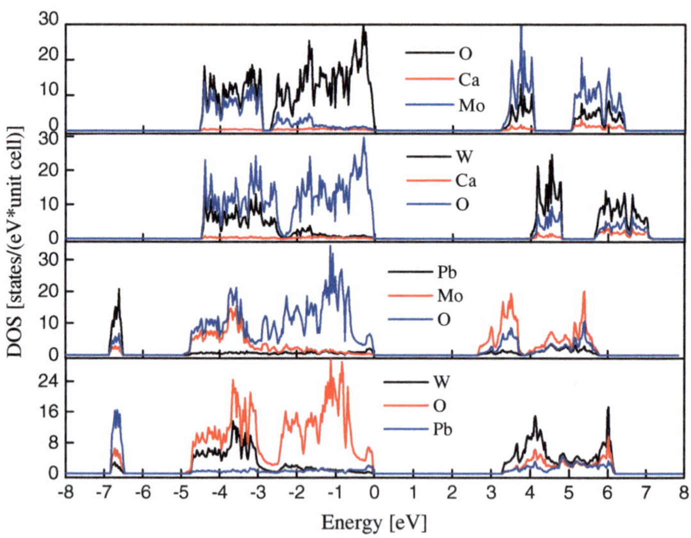

Fig. 6.9 The DOS of constituted elements in tungstates and molybdates without the spin-orbit coupling interaction

the luminous process. Therefore, $PbWO_4$ photoluminescence can be tuned by doping impurity ions in Pb sites [24].

For $MeWO_4$—like $CuWO_4$, $FeWO_4$, $CoWO_4$, $CdWO_4$, $ZnWO_4$, $BaWO_4$, $CaWO_4$, $CaMoO_4$, and $PbMoO_4$ [23, 25–28]—their electronic structures are also studied by the first-principles calculation, and the intrinsic luminescence is charge transfer transition between O and W or Mo. From these calculations, Me ions in most $MeW(Mo)O_4$ are found to be unable to participate in the luminescence and only serve as host element. For the other two types of tungstates/molybdates, the intrinsic luminescence is similar to the normal $MeW(Mo)O_4$ [29, 30]. For example, the monoclinic yttrium tungstate luminescence is ascribed to WO_6^{6-} based on the Perdew–Burke–Ernzerhof (PBE) functional calculation [31].

Scheelite-type $BaWO_4$ is a prospective material for application in Raman converters, lasers, and amplifiers. The noncrystalline $BaWO_4$ emission intensity is stronger than those of crystalline. Quantum mechanical theory calculations demonstrated that the amorphous $BaWO_4$ containing short-range disorder presented large trapped electronic holes, thus indicating that this structure is more beneficial to the photoluminescence. Large trapped holes are more favorable for attracting excited electrons to radiative recombination when the lattice is submitted to photoexcitation [32]. Other compounds—such as $BaTiO_3$ [33], $SrWO_4$ [34], α-Ag_2WO_4 [35], β-$ZnMoO_4$ [36], $CaWO_4$ [37], $CaMoO_4$ [38], and $SrZrO_3$ [39]—also present similar luminescence phenomenon due to local short-range disorder. The Ba (Sr, Ag, Zn, Ca) or Ba/Ti (Sr/W, Ag/W, Zn/Mo, Sr/Zr) local mobile defects along different directions have been applied to simulate experimental results through the first-principles calculation.

In addition, the vanadate, zirconate, titanate, niobate, tantalate, chromate, manganese, and gallate compounds are also self-activated materials, and their intrinsic luminescence is also the charge-transfer transition from V(Ti, Zr, Nb, Ta, Cr, Mn and Ga) *d* orbit to O 2*p* orbit. For example, undoped $YTaO_4$ is well known as a self-activated phosphor. Its band gap is measured to be 5.1 eV from absorption and luminescence spectra. This is close to the 5.14 eV calculated by means of a generalized gradient approximation (GGA) + U approach. The DOS shows that the VB of the tantalate system is mainly composed of O 2*p* states, and the lower CB is mainly composed of Ta 5*d* states. The PL excitation peak at 6 eV under ultraviolet (UV) and vacuum UV radiations can be associated with the absorption of TaO_4^{3-} group via O–Ta host charge-transfer transition. Similarly, the excitation bands peaking at 7 and 10 eV are also associated with the absorption of the host lattice because of the O–Ta and O–Y transitions [40]. In addition, the pure $ZnGa_2O_4$ nano-spheres excitation spectrum shows a broad band from 225 to 275 nm with a maximum at 245 nm by monitoring the 365 nm emission [41]. On 245-nm excitation, the emission spectrum consists of a broad band from 300 to 500 nm with the peak position at 365 nm. The DFT calculation shows that the VB and CB are mainly composed of O 2*p* and Ga 3*d* states. The calculated band gap is 3.15 eV, smaller than the experimental results due to the insufficient description of exchange correlation in DFT calculation. The intrinsic luminescence is ascribed to the self-activated center of the octahedral Ga–O group in the spinel lattice.

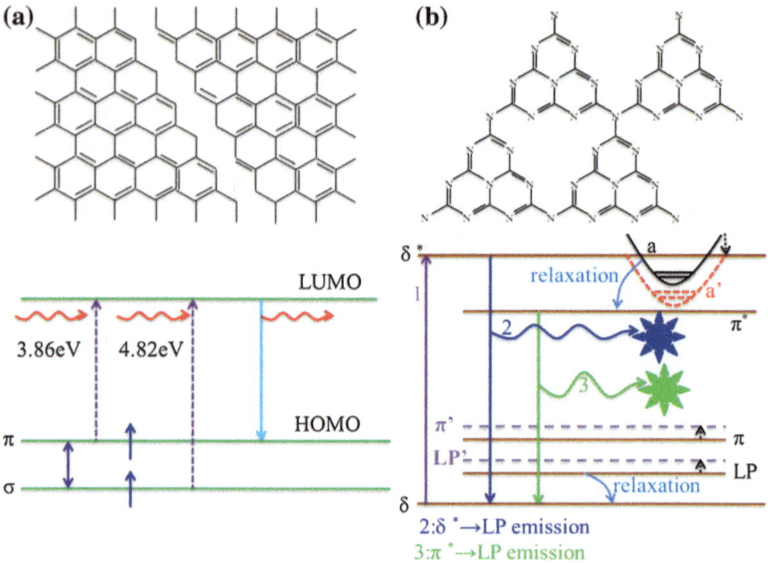

Fig. 6.10 Graphene quantum dots (**a**) and graphitic-phase C_3N_4 nano-structure (**b**) with their photoluminescence mechanism. The LUMO as π^* or δ^* and HOMO like π, σ, δ, as well as LP, are low and high occupied molecular orbits. **a** Adapted from Ref. [19] by permission of John Wiley & Sons Ltd. **b** Adapted from Ref. [44] by permission of Macmillan Publishers Ltd.

Furthermore, some oxides and nano-scale materials, such as ZnO and quantum dot, are also good self-activated luminescent materials [42, 43]. At room temperature, ZnO photoluminescence spectra typically consist of a UV emission and a series of visible bands. The UV emission originates from a Zn–O charge transfer emission. The lower-dimensional materials, such as graphene, graphitic-phase C_3N_4 quantum dots, and few-layer black phosphorus, can emit in visible and near infrared regions under UV irradiation. The ground-state configuration of graphene has two types, i.e., triplet state $\sigma^1\pi^1$ and singlet state σ^2. Two electronic transitions of 320 nm and 257 nm in graphene dots, respectively, can be regarded as transitions between the highest occupied π and σ orbits and the lowest unoccupied orbit based on the first-principles calculation. The g-C_3N_4 dots show two photoluminescence centers located at 405 and 480 nm under the 280-nm excitation. The photoluminescence peaks are assigned to the change of the optical band gap, which is related to the size of the sp^2 clusters [44, 45]. Their structures and PL mechanisms are shown in Fig. 6.10.

For bulk phosphorus with 10-nm thickness, no photoluminescence signal is observed within the detection spectrum range because its band gap is as low as 0.3 eV, thus falling in the infrared wave region. In contrast, an obvious photoluminescence signal centered at 1.45 eV is obtained on a single-layer phosphorene crystal. This PL peak is likely of exctionic nature and the peak position is slightly above single-shot GW (Green and screened Coulomb interaction)–BSE (Bethe–

Table 6.2 The band gap values of phosphorene by different calculation method in a DFT theory framework

Phosphorene	Exp[a]	PBE[b]	BSE-$G_0W_0^2$	BSE-$G_1W_1^2$	$G_0W_0^2$	$G_1W_1^2$
E_{gap} (eV)	1.45	0.82	1.20	1.40	2.00	2.45

[a]from Ref. [46]
[b]from Ref. [47]

Salpeter equation) calculation result. Self-consistently updating the Green's function spectrum and dielectric function using the G_1W_1 methodology yields an exciton energy of 1.4 eV as shown in Table 6.2. However, self-consistent G_1W_1 may not always give a better band gap. Therefore, additional experimental results must be assessed before further conclusions can be made because of the necessity of self-consistence [46, 47].

The DFT calculation can give the matrixes band gap as well as VB and CB compositions. From the band gap, we can decide the luminous centers in materials of self-activated luminescence. Comparing the theoretical and experimental band gap, we can estimate the accuracy of the calculation methods and improve the theoretical band gap by employing advanced methods. In Sect. 6.3, we will introduce this advanced DFT theory and the use on investigating luminescent materials. Moreover, we can also demonstrate the composition of the luminous center through the VB and CB DOS. Thus, based on the DFT calculation, we can modify the luminescence properties by changing the band gap and VB and CB edge compositions.

6.2.2 *Native Defect Luminescence*

The lattice defect is defined as irregularities or imperfections of the atom arrangement in the ideal crystal structure. It can be divided into point, line, and interface defects according to its geometrical morphology. For luminescence materials, most of the responsible defects are zero-dimensional point defects and two-dimensional interface defects. The native point defect results from crystal-structure imperfection including mainly vacancy, interstitial atoms, and dislocation defects. For instance, in silica-based luminescent materials, the main defects contain carbon-substitutional defects, oxygen vacancies, radical carbonyl defects, intrinsic diamagnetic centers (O–Si–O), and so on [48]. In phosphate luminescent materials, possible luminescence centers are carbon impurity, oxygen-related defects, and carbon dioxide radical defects. In oxide luminescence materials, the possible defects include surface defects, oxygen vacancies, interstitial carbon defects, interstitial oxygen ion, ionized oxygen vacancy, and Zn vacancy. Native defects are inevitable in the preparation of materials, which often creates luminescence. To understand the PL origin, in addition to experimental characterization, the first-principles calculation is often employed as a useful approach to revealing the nature of the defects. On the other hand, we can also predict whether or not the material can be luminescent and how the defect will influence the luminescence only through the first-principles calculation. By DFT calculation, the defect formation probability and the defect states positions can be obtained.

Various synthesis approaches lead to different luminescent phenomena in samples, and thus various models are established to explain the origins of luminosity. Zhang et al. prepared single-crystalline wolframite-type monoclinic structure $CoWO_4$ nano-wires with a width of 20 nm and a length of 200 nm by a solvothermal method [49]. In addition to the strong blue-green intrinsic emission at 10–250 K, a much stronger and broader near-infrared emission was found ranging from 700–1000 nm at approximately 300 K under the excitation wavelength of 325 nm. Through the first-principles calculation, the 700–1000 nm near infrared emission was mainly ascribed to the discrete Co^{3+} ions near the Co cationic vacancy. The carriers captured by the $(CoO_6)^{-10}$ and $(WO_6)^{-6}$ groups have a high probability to realize the local *d–d* transition to Co^{3+}. According to the above-mentioned model and using the VASP software, we recalculated the reference results as shown in Fig. 6.11. Similarly, a Cd vacancy in $CdWO_4$ would trap two holes to maintain local electrical neutrality [50]. The electronic-structure calculation reveals that the existent form of the hole in $CdWO_4{:}V_{Cd}$ should be oxygen

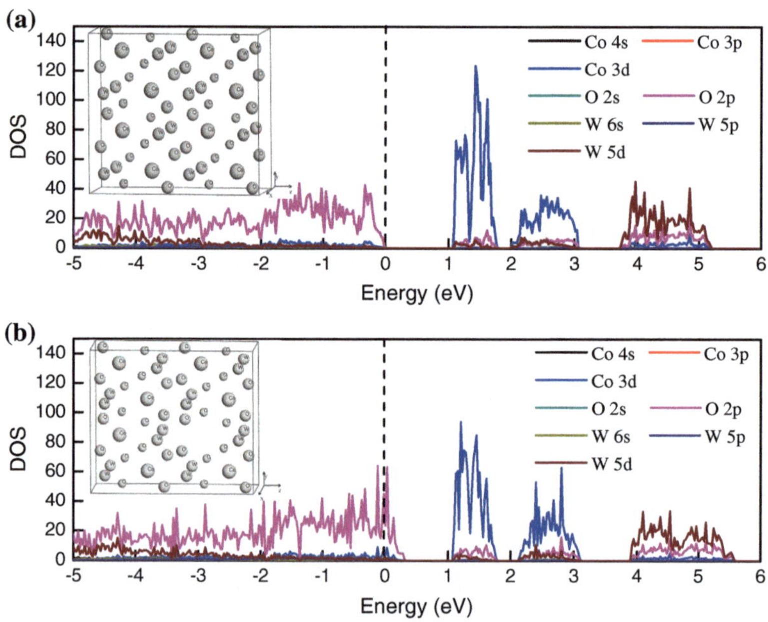

Fig. 6.11 The DOS of $CoWO_4$ (**a**) in the *upper* and $CoWO_4{:}V_{Co}$ (**b**) in the *lower*. The *dashed lines* represent the Fermi levels. The *inset* picture shows the model

molecular ions (O_2^{3-}). Therefore, an associated color center O_2^{3}–V_{Cd}–O_2^{3-}, named the "V_F center", would form. This V_F center causes an absorption band peaking at 650 nm with a shoulder at approximately 400 nm, which is consistent with the experimental optical absorptions of blue-grey-colored $CdWO_4$.

For rare-earth tungstate (Me_2WO_6) luminescence, there are also a large number of materials such as Y_2WO_6, La_2WO_6, and Lu_2WO_6 [51–54]. Due to the Me ions' radius equaling other rare ions' radius, rare-earth tungstates are often applied as hosts to hold luminescent rare-earth ions. In addition, there are some reports about exploration of matrix luminescence. Monoclinic Y_2WO_6 has a good performance as a negative thermal expansion and luminescent material [54]. Calcining at different atmospheres, Y_2WO_6 shows different excitation phenomena. According to Raman and synchrony-radiation analysis, the oxygen vacancy concentration in Y_2WO_6 increases when the calcination atmosphere changes from air to argon. Thus, single and twin oxygen vacancies are applied to simulate the air and Ar-calcined samples. First-principles calculation demonstrates that local states caused by the oxygen vacancy change their position if the oxygen vacancy concentration varies as shown in Fig. 6.12. Therefore, increasing the oxygen concentration induces the gradual diminishing of a long-wave excitation band and new visible excitation bands appear.

In addition, $SrTiO_3$ is a key material for oxide-based electronic device application [55]. The Ar^+-irradiated $SrTiO_3$ shows blue emission at room temperature. For irradiated $SrTiO_3$, the 430-nm blue-light intensity increases gradually from temperature 300–160 K. On further decreasing the temperature, this band

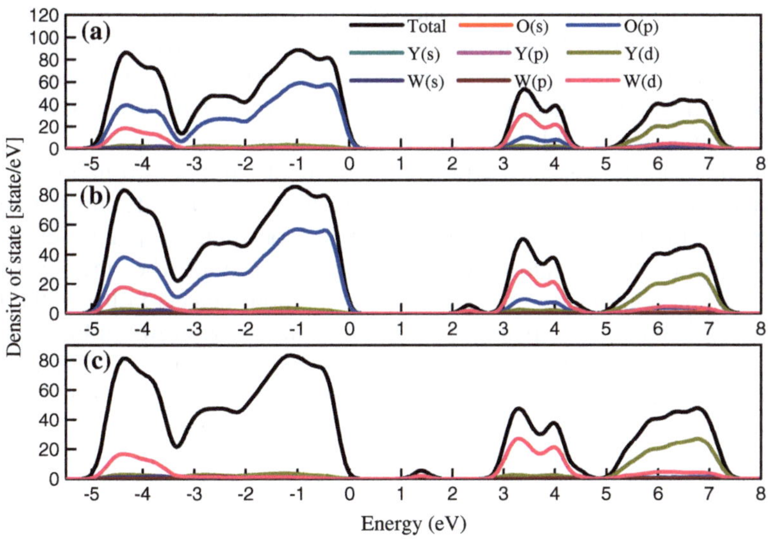

Fig. 6.12 The total DOS and partial DOS of constituted atoms for **a** super-cell Y_2WO_6, **b** $Y_2WO_{5.98}$, and **c** $Y_2WO_{5.95}$ from –5.5 to 8 eV. Reprinted from Ref. [54] by permission of American Chemical Society

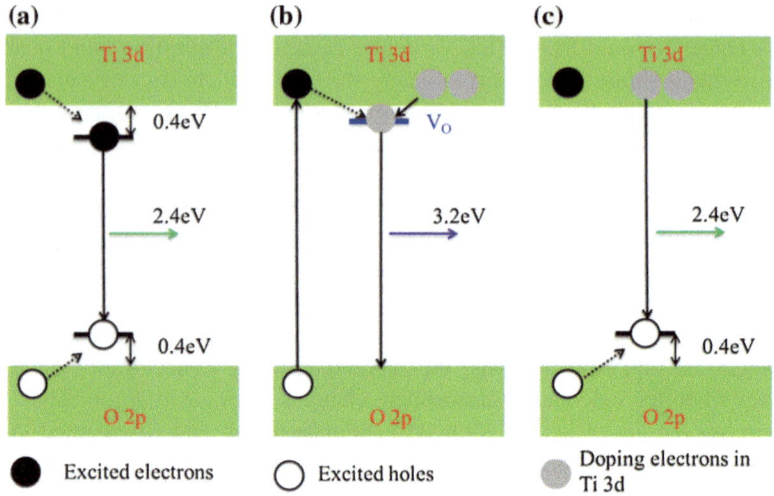

Fig. 6.13 **a** Self-trapped exciton emission, **b** oxygen-vacancy defect luminescence, and **c** recombination process between the conduction electron and the excited holes. Adapted from Ref. [55] by permission of Macmillan Publishers Ltd.

disappears, and two new peaks appear. The 550-nm green luminescence becomes prominent below 100-K, which originates from the recombination of excited electrons and holes that coupled in a self-trapped exciton state. Another 380-nm peak is originated from oxygen vacancy. The schematic luminescence processes of $SrTiO_3$ after Ar^+ irradiation are displayed in Fig. 6.13, which are further verified later by first-principles calculation [56]. For other compounds, the photoluminescence mechanism is studied by similar methods, and the luminescence contains an intrinsic luminescence and defect emission.

In addition to the vacancy defect, the interstitial atom is also important defect [57]. When an oxygen atom enters into the crystal to form an interstitial atom, such as sites A through F in Fig. 6.14, it will constitute a point defect. This defect-formation energy is lower than that of other positions when the oxygen atom exists near the tungsten atom in $PbWO_4$. In addition, defects in the A through D positions maintain the C_{3v} point symmetry invariability as shown in Fig. 6.14 [57]. The E and F sites are the possible positions of interstitial oxygen. The interstitial oxygen atom combines with one or two formal lattice oxygen ions forming oxygen molecular ions O_2^{2-} or O_3^{4-}, which is the 350-nm absorption source. Thus, the $PbWO_4$ green luminescence probably arises from $WO_4 + O_i$ centers [58]. Except for interstitial oxygen, C_i defect is important in enhancing phosphates luminescence. Zhang et al. [59] prepared BPO_4, BPO_4–$xSiO_2$, and BPO_4–xAl_2O_3 powders by the Pechini-type sol-gel process using glycerol and polyethylene glycol as additives. The pure BPO_4 shows a weak 413-nm blue emission, whereas the BPO_4–$xSiO_2$ and BPO_4–xAl_2O_3 samples exhibit bright bluish-white emission peaking at

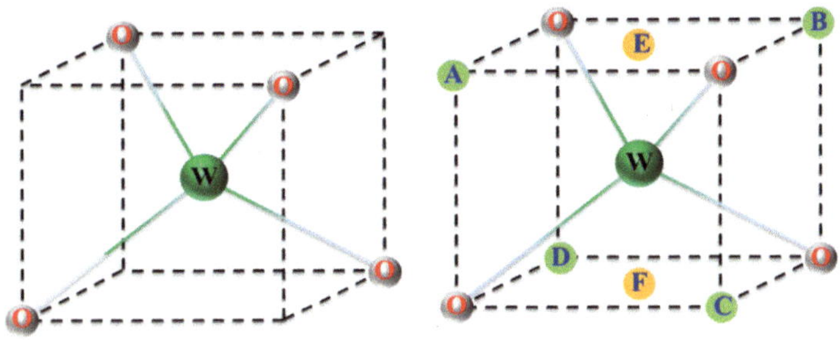

Fig. 6.14 The *left panel* shows the tungstate group, and the *right panel* presents the $WO_4 + O_i$ in $PbWO_4$. *A* through *F* sites are the most probable positions to form interstitial oxygen. Adapted from Ref. [57] by permission of John Wiley & Sons Ltd.

428 and 413 nm, respectively. Based on the first-principles calculation, the blue or bluish-white emission may be ascribed to the interstitial carbon impurities and the oxygen vacancies in the host lattice, respectively. Mixing SiO_2 and Al_2O_3 into BPO_4 results in the increase of an oxygen vacancy that enhances the luminous intensity.

Fabbri prepared ZnO nano-rods, nano-structure thin films, and nano-needles. Except for the emission near the band edge, ZnO nanostructures have weak or strong green luminescence. Based on the DFT calculation, Zn vacancy, O vacancy, H interstitial, and a combination of these in bulk and (100) surface are shown to explain how the presence of point-defect states affect the electronic properties of the ZnO [60]. The neutral oxygen vacancies in bulk generate a donor-like occupied state 1 eV below the CB bottom. This defect cannot be responsible for the observed green luminescence because the green luminescence corresponds to energy transition of 2.3 eV. The un-pair oxygen electrons due to the formation of Zn vacancies in bulk give rise to empty state at the Fermi level, close to the VB top. This state will act as an electrons acceptor in the emission process. As a consequence, the Zn vacancies are the most probable source for the green luminescence, which rules out the probability of the oxygen vacancy causing green emission. Similar consideration are applied to the electronic properties of (100) surface. The O vacancies at the non-polar (100) surface do not lead to significant change of the electronic properties, particularly close to the VB edges. When a complex Zn + O vacancy forms in this surface, a less-dispersion filled defect band appears on the valence band top, thus reducing the band gap. Therefore, green luminescence is ascribed to the transition from the defect state to CB. Moreover, the H_{int} in bulk and (100) surface containing Zn or O vacancies can increase the stability of a point defect. In short, the green emission is related to the surface state generated by Zn vacancies preferentially localizing at non-polar (100) surfaces. The assessment is supported by different cathode-excited green luminescence for nano-rods and nano-thin films or nano-needles as well as the increased stability of defects at surface. Indeed, the interface reconstruction at the grain boundaries can promote atomic rearrangements responsible for saturation of dangling bond states associated with V_{Zn} and finally

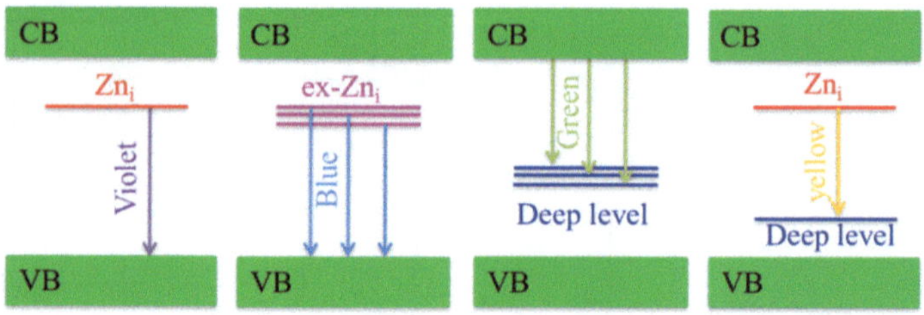

Fig. 6.15 The schematic photoluminescence origin for nano-scale ZnO. Adapted from Ref. [61] by permission of John Wiley & Sons Ltd

for the quenching of green luminescence. Zeng et al. synthesized nano-scale ZnO through non-equilibrium processes, analyzed resultant blue emission comprehensively, and focused on defect origins based on the previous theoretical calculation. The ZnO nanostructure shows blue and green emissions under excited energy larger than the band-gap value and has several fixed emission wavelengths at 415, 440, 455, and 488 nm. Combined with reported defect energy levels and formation thermodynamics, the luminescence origins are proposed based on the Zn_i defect and derivative extended states. In detail, the violet or blue emission peaking at 415 or 445 nm is ascribed to the transition between the Zn_i or ex-Zn_i state and VB. The 555-nm green luminescence originates from electron transition from CB to deep defect states. Finally, the 600-nm yellow emission is attributed to transition from the Zn_i to deep defect states as shown in Fig. 6.15 [61].

6.2.3 *Dopant or Doping-Induced Defect Luminescence*

In addition to the native defects, luminescence from dopants and doping-induced defects has also been studied in luminescence field. Dopants contain activator and nonactivator ions. The former are generally rare-earth ions with rich $4f^9$ luminescence configuration or sensitizer ions such as Bi^{3+}, Pb^{2+}, Sb^{3+} and Mn^{2+}. The latter include numerous metal or nonmetal ions. The doping can be divided into isovalent and nonisovalent doping according to charge-matching rules. DFT calculations verify that dopants can influence the local crystal environment and nonisovalent doping can produce a local state in the band gap because of local charge imbalance. These local states may be trapping or recombination centers. For these two types doping, DFT calculation gives the local coordination environment variation such as bond length or electronic distribution and local state positions of the impurity ions. Comparing the formation energy, we can tell what kind of sites the dopant can enter most easily. From the energy bands and DOS, the band gap, the energy level position, and the constitution of local states can be exhibited. Analysis of VB, CB edge, and local state charge density can determine whether or not the impurity forms a bond with the host lattice. Therefore, we can understand the luminescence origination from theoretical results, and the following examples are given to illustrate the application of first-principles calculation in dopant-containing luminescent materials.

Doping nonactivator or activator is an effective way to tune he photoluminescence properties. The main effect of doping is to tune the local environment of the luminescence group, broaden the luminous ranges, and enhance the output intensity or energy-transfer efficiency. Introducing La^{3+} in $PbWO_4$ could dramatically change the luminescence intensity. Light La-doping $PbWO_4$ shows lower luminescence intensity compared with the pure one. Heavy La-doping might cause a new effective recombination and thus severely degrade the luminescence [62]. Different charge compensation may result in the above-mentioned phenomena when La substitutes for Pb and W, respectively. To verify the experimental assumption, the first-principles calculation is carried out. The main compensating mechanism is $2La_{Pb}^{+} + V_{Pb}^{2-}$ for a light La-doped system, whereas doping brings interstitial oxygen ions to compensate for the positive electricity caused by La_{Pb}^{+}, forming defect cluster $2La_{Pb}^{+} + O_i^{2-}$ for heavy La doping [63]. Thus, the calculation result could reasonably explain the experimental results. Except for the nonactivator La ion, others, such as Gd, Y, Sb and Lu, are also employed to dope tungstates [64].

When the materials become solid solutions in the process of doping, some new luminescence phenomena can occur. In a solid solution, the dopants enter the crystal lattice without bringing variation to the whole crystal structure and symmetry, whereas the lattice size and composition change. The photoluminescence from each component can present in the solid solution. For instance, the photoluminescence band at 2.5 eV for nano-sized $ZnWO_4$ originates from radiative-electron transitions within the $WO_6{}^{6-}$ anions, and this PL band becomes modulated by the optical adsorption spectra of Ni ions in the $Zn_xNi_{1-x}WO_4$ solid solution [65].

By doping La in Y sites, oxygen-vacancy photoluminescence in the monoclinic Y_2WO_6 can be tuned [66]. Introducing the nonactivated La into Y_2WO_6 brings new excitation bands from violet to visible regions and strong near-infrared emission. Meanwhile, the band position and intensity depend on the doping concentration. By experimental analysis, La doping is considered as promoting the formation of oxygen vacancy, which is confirmed by the GGA + U calculation. As shown in Fig. 6.16, there are three type-Y sites. When La^{3+} occupies different Y sites, localized energy states, caused by the oxygen-vacancy pair, change their position in the forbidden band, thus inducing variation of the excitation and emission bands.

In addition to doping the metal ions, non-metal doping is also often used to tune the luminescence. Pristine and O-doped $SrSi_6N_8$ phosphors were successfully synthesized by gas-pressure sintering [67]. Both nitride-silicates $SrSi_6N_8$ and $SrSi_6N_{7.95}O_{0.05}$ exhibit a 452-nm blue emission under 370-nm UV excitation. Furthermore, $SrSi_6N_{7.95}O_{0.05}$ emits a 652-nm red emission under 460-nm blue excitation. From the calculated DOS plot, the $SrSi_6N_8$ unit cell shows a band gap of the matrix in the range 0–3.5 eV, and there is another band gap in the CB from 4 to 4.5 eV as displayed in Fig. 6.17a. The fundamental absorption edge of $SrSi_6N_8$ is estimated to be from 275 to 310 nm according to the energy difference between the

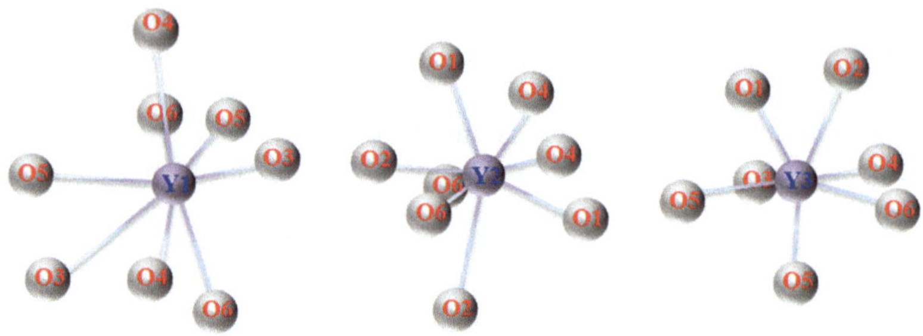

Fig. 6.16 The local ligand environment of three Y sites in monoclinic Y_2WO_6

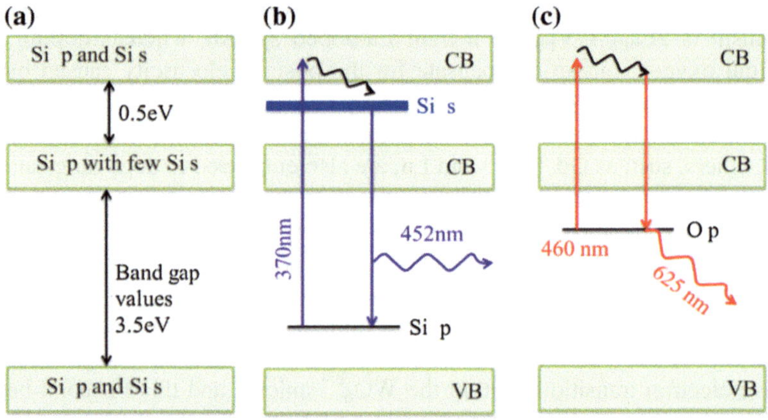

Fig. 6.17 The PL origins of $SrSi_6N_8$ phosphors based on different models according to calculation results in Ref. [54]: **a** $SrSi_6N_8$, **b** $SrSi_6N_8$ with nitrogen vacancy, and **c** $SrSi_6N_8$ with oxygen substituting nitrogen

VBM and CBM. The band gap of the ideal $SrSi_6N_8$ unit cell could not explain the luminescence under UV excitation, thus suggesting that it is not a self-activated luminescence host. From the experimental characterization, $SrSi_6N_8$ and $SrSi_6N_{7.95}O_{0.05}$ contain the N vacancy. According to first-principles calculation, an N vacancy in $SrSi_6N_8$ induces a local Si 3*p* state in the band gap, which is approximately 1.7 eV lower than the CB edge of the host. Therefore, a separation of 2.75 eV exists between the local Si 3*p* orbit and the Si 4*s* main peak in the CB, corresponding to a 452-nm emission. A new local O 2*p* state presents in the gap 1.4 eV below the CB when a nitrogen site is occupied by an oxygen atom. The separation value of >1.98 eV between the peaks of the O 3*s* orbit and the O 2*p* orbit is consistent with the experimentally observed emission wavelength of 625 nm [54]. In addition to nonactivated-ion doping, activated ion–doped silicon nitride compounds have also been investigated widely recently using theoretical calculation and experiment approaches [68–70].

Similarly, *N*-doped SnO_2 nano-crystals annealed in an O_2 atmosphere at different temperatures show two bands at 400 and 500–650 nm under 320-nm excitation [71]. The former consists of three sharp peaks, which are assigned to the band–acceptor and donor–acceptor pair transition as well as the phonon replica. For the latter, the intensity decreases, and the peak position shifts to low energy with the temperature increase. In addition, after annealing at 700 °C, the crystals also display excitation wavelength–dependent PL spectra. Based on the DFT calculation, three models are built (a) one shallow (relative to (110) surface) *N*-dopant (N_O) and one shallow oxygen vacancy (V_O); (b) one deep N_O and one shallow V_O; and (c) one deep N_O and one deep V_O. Different defect depths successfully explain the excitation wavelength–dependent PL spectra.

Compared with the doping of nonactivated ions, the doping of activated ions is a more common approach to tuning the luminescence. The dopant emission can broaden the luminescence range and increase the luminous intensity and energy-transition efficiency. The final aim is to obtain white-light emission. Kang et al. report a red-emitting $ScVO_4:Bi^{3+}$ phosphor that does not show excitation at energies <2.88 eV [72]. The incorporation of Bi^{3+} shifts the blue photoemission of the blank $ScVO_4$ matrix to red. The energy transfer from VO_4^{3-} groups to Bi^{3+} and the population redistribution of $^3P_1$ and $^3P_0$ of Bi^{3+} strictly depend on the testing temperature through luminescence analysis between 10 and 300 K. To explore the origin of red emission from Bi^{3+} in the $ScVO_4$ matrix, approximate calculation are performed to explore the influence of the replacement of Sc with Bi and the corresponding oxygen vacancy produced by this replacement using the DFT method. In bulk $ScVO_4$, as shown in Fig. 6.18a, every Sc atom has eight neighbors of O atoms with two kinds of Sc–O distances: 2.38 and 2.14 Å. When replacing Sc with large Bi, the distances of Bi–O bond are extended to 2.43 and 2.31 Å, which are slightly longer than those of Sc–O as seen in Fig. 6.18b. When an oxygen vacancy occurs around a Bi atom, the Bi–O bond lengths range from 2.34 to 2.95 Å as shown in Fig. 6.18c. Bi substitution induces distortion of the local structure and an important contribution at the VB maximum. Moreover, the formation energy of an

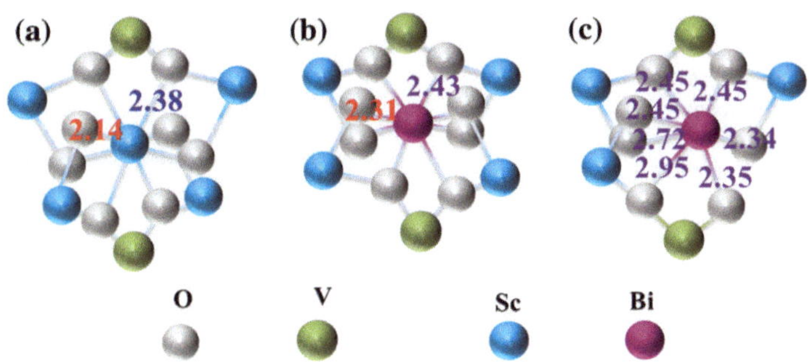

Fig. 6.18 The optimized structure of bulk $ScVO_4$ (**a**), $ScVO_4:Bi^{3+}$ (**b**), and $ScVO_4:Bi^{3+}$ with an oxygen vacancy around the Bi atom (**c**). Adapted from Ref. [72] by permission of American Chemical Society

oxygen vacancy defect is 0.8 eV higher in the pristine $ScVO_4$ than in the Bi-doped $ScVO_4$. The incorporation of Bi atom into $ScVO_4$ leads to oxygen loss more easily and then produces Bi_{Sc} + V_O complex defects. Accompanying an oxygen loss near the bismuth atom, a local state appears, and it lies 1.33 eV below the CB minimum. Therefore, the red emission of Bi^{3+} correlates with the formation of an oxygen vacancy.

When self-activated luminescence ions are doped into the host, the DFT calculation can also give the energy level of dopants to illustrate the luminous mechanism. $CaAl_2O_4$:Eu is a persistent luminescence material in the blue-light region, and the decay time can be lengthened with the doping of Nd [73]. These good properties closely relate with the electronic structure of dopants, defects, and the host material. According to the first-principles calculation, the Eu 4*f* level is located in the band gap below the CB minimum, and the Eu 5*d* level is located above the CB minimum, respectively. In addition, the Ca vacancy and Eu_{Ca} + V_{Ca} can also produce the local states in band gap above the VB maximum as shown in Fig. 6.19 [73]. The 4*f* electrons of Eu can move to the empty 5*d* impurity levels, and the electrons in the 5*d* levels can transfer back to the 4f levels, accompanying light emission. Some electrons also jump to the CB minimum, are trapped by the V_O states, and finally move back to the Eu 5*d* states. As a result, they may jump back to 4*f* levels, thus leading to persistent luminescence. Similarly, Nd doping can increase the persistent luminescence intensity and lengthen the luminous decay time because of electron trap-release between the CB and Nd 4*f* states.

In addition, co-doping the activator and the nonactivator is also an effective way to tune the luminescence intensity and coverage range. A new yellow-emitting γ-Ca_2SiO_4:Ce^{3+}, Li^+ phosphor was prepared by way of simple solid-phase reaction [74]. Its yellow luminescence is attributed to the direct absorption and emission of Ce ions. DFT calculation reveals that Ce ions at the Ca(1) sites in the matrix lattice are the primary contributors to the luminescence of the phosphors. Thus, by combing γ-Ca_2SiO_4:Ce^{3+}, Li^+ phosphors and blue LEDs, a white light was realized. Moreover, similar materials, i.e., Sr_2SiO_4:Eu^{2+} and *N*-modified Sr_2SiO_4:Eu^{2+} phosphors, were synthesized by a conventional solid-state reaction method [75]. The phosphors show a single and intense broadband emission peaking at 625 nm

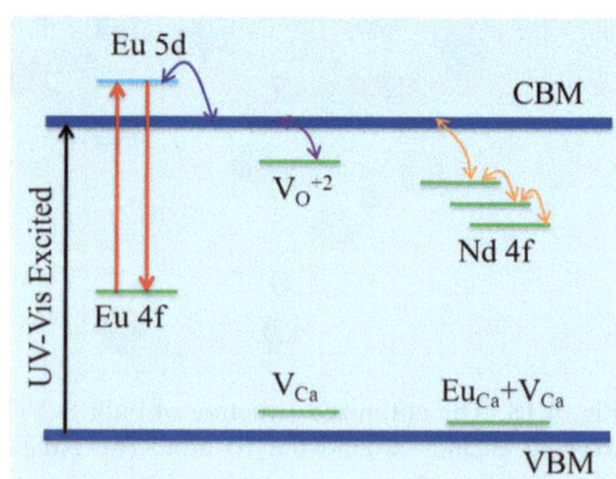

Fig. 6.19 The schematic persistent luminescence of $CaAlO_3$:Eu^{2+}, Nd^{3+} phosphors based on the DFT calculation. Adapted from Ref. [73] by permission of American Chemical Society

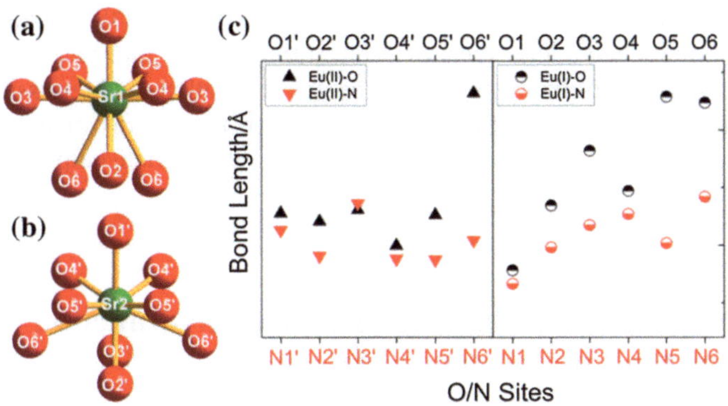

Fig. 6.20 The coordination environment of Sr(1) (**a**) and Sr(2) (**b**) as well as the bond length of Eu-O(O′) and Eu-N(N′) (**c**). Adapted from Ref. [75] by permission of Royal Society of Chemistry

under 453-nm blue-light excitation and 480-, 555-, and 625-nm emission bands under ultraviolet irradiation. Through the experimental characterization, 480- and 555-nm emissions were regarded as originating from Eu^{2+} ions occupying the Sr(1) and Sr(2) sites in the host, whereas 625-nm emission originates from the nitrogen-coordinated Eu^{2+} ions. Based on the DFT calculation, the nitrogen preferentially substitutes for the O5′ sites around Eu^{2+} in Sr(2) sites. In *N*-coordinated and O-coordinated Eu, the main difference is the bond-length change as shown in Fig. 6.20. Almost all the bond lengths become shortened after the oxygen is replaced by nitrogen. Nitrogen prefers to substitute the O5′ site around Eu in Sr(2) sites. The more covalent and shorter Eu–N bonds enable the broadening of Eu 4*f* energy states and the lower energy shifting of the Eu 5*d* energy band leading to the extra red emission.

In addition, novel high efficient red phosphors are obtained by doping Mo and Eu in the La_3BWO_9 [76]. When Mo ions are introduced in the host lattice, the Eu^{3+} characteristic excitation peak intensity are enhanced greatly. The DFT calculation demonstrates that the $La_3BW_{1-x}Mo_xO_9$ host gap becomes narrower than that of the pure La_3BWO_9, and the electron-transition or -transport efficiency from host to Eu^{3+} may become higher.

6.2.4 Conclusions

In summary, the origins of photoluminescence mainly stem from the matrix-, native-, and doping-induced defects. Based on the experimental characterization, the first-principles calculation gives an acceptable model to explain the luminescence mechanism. For intrinsic emission, different DFT methods are applied to modify the band gap to meet the experimental value. The intrinsic luminescence mechanism can be understood from the band structure and DOS constitution. For the defect luminescence, the vacancy, interstitial atoms, interface defects, and dopants are thought to be responsible for the excitation and emission bands. The DFT has become a powerful tool to investigate the origins of luminescence, and the advanced DFT theory is gradually developed, which helps advance the study of luminescence.

6.3 Calculation using the Advanced Density-Function Theory for Luminescence Materials

The DFT serves as the theoretical framework of ab initio or the first-principles calculation, and it gives a successful description for the many-electron system ground state. In solid or crystal form, the energy band structure, bulk modulus, lattice symmetry, lattice parameters, deformation potential, and lattice vibration–dispersion relation can be well described. For the multi-particle system, the band gaps of the semiconductor and the insulator are almost always underestimated, whereas the band gap for metals is often overestimated. The main reason is that the DFT is a ground-state theory, and the eigenvalues in LDA have only a Lagrange parameter without physical eigenvalue meaning. Thus, the DFT may give erroneous results. For example, many transition-metal oxides and high-temperature superconductors present metallic properties in DFT framework. To solve these problems, the solution method is to establish an excited-state (quasi-particles) motion equation such as a quasi-particle GW approximation. For example, the GaAs band gap value is shown in Table 6.3 through different theory approaches [77]. From Table 6.3, the calculated values are far smaller than the experimental values. When the lattice parameters have 2 % error, the band gap variation is approximately 0.5 eV. The LDA systematically underestimates the band gap of semiconductors and insulators. In the following section, we first roughly introduce the treatment approaches of the excited state and then present some advanced methods to improve the band gap of luminescent materials.

Molecules, atoms, clusters and solid excitation processes relate to the system perturbations induced by an external force such as interaction between photons or electrons and a solid [78]. The light-excitation process has been an important topic in the field of condensed matter physics and has not been solved fully. The simplest method is Hartree Fock (HF) approximation in linear response theory. Because of

Table 6.3 GaAs band gaps calculated by different approaches in DFT framework [77]

GaAs	Nonrelativistic	Scalar relativistic	Full relativistic	Core relaxation	Experimental
$E_{direct\ gap}\{[\Gamma]\}$ (eV)	1.1	0.6	0.5	0.2	1.5

Koopman's theorem, HF single-particle energy can represent the energy of the system by adding or subtracting a particle. At the same time, double-particle Hamiltonian eigenvalues can be regarded as the transition energy of quantum states. However, the electronic correlation effect is ignored in HF approximation, and thus the many-body theory method is necessary. The DFT-LDA theory proposed by Hohenberg–Kohn–Sham presents effective potential calculation of many bodies, and this theory is the ground-state approach. Thus, DFT-LDA cannot describe the excited state because it contains a static exchange–correlation effect. To represent a dynamical effect, we must adopt the many-body perturbation theory on the basis of the second quantization, the Green function, and the quasi-particle. When we employ a self-energy operator to replace the static exchange–correlation potential, the corresponding theories include the GW quasi-particle, time-dependent DFT, and sX-LDA approaches. These methods are the main tools to correct the band-gap value. To describe the exciton effect in a semiconductor, we must treat the electron–hole interaction and the Bethe–Salpeter equation suit for processing the exciton effect. For the specific implementation of the many-body theory method, the starting point can be an LDA solution of the Kohn–Sham equation, constructing various exchange-correlation potentials including the perturbation Green function, the screened Coulomb interaction, isolated quasi-particles polarizability, and different exchange-correlation kernel. In summary, the correlation schemes beyond LDA exchange-correlation potential have quasi-particle GW approximation and sX-LDA for the semiconductor excited state and the self-interaction correction, LDA + U and LDA++, for stronger correlation systems as shown in Fig. 6.1. For band-gap correction, currently most software can obtain good results comparable with the experimental results. However, software for the calculation of the excited state are few, and most can only tentatively calculate the electronic transition state.

For luminescence materials, calculation of the excited state is very important because emission originates from light irradiation. As we know, the electrons' motion-orbit radius becomes larger when they are excited from the ground to the excited states. In addition, the chemical bond in the excited state is weaker than that in the ground state. Therefore, the equilibrium position r_0 and elastic force constant k' are different from those of the ground states R_0 and k as shown in Fig. 6.21. The parabola of the excited state becomes flat with the decrease of k', and thus the gradient of potential energy U(e) is smaller than that of U(g). According to the Frank–Condon principle, the equilibrium position of U(e) and U(g) is the same, i.e., $r_0 = R_0$. Therefore, as described in Sect. 6.2, the first-principles ground state calculation can basically be consistent with the experimental results. However, in the excited-state condition, r_0 is different from R_0, and the electrons' population distribution and band gap can also change. In this section, we briefly cite some examples to illustrate calculation of the excited state in luminescent materials based on the first-principles theory. Then we introduce methods to treat band-gap underestimation of the semiconductor.

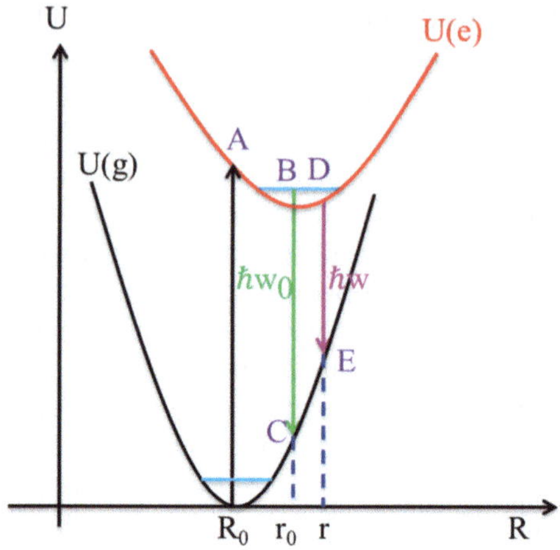

Fig. 6.21 The configuration coordinates the model and the schematic absorption-and-emission process

6.3.1 Excited-State Calculation

First, we start from the excited state studies of tungstates. $CaWO_4$ is a self-activated phosphor emitting blue or green light under ultraviolet or X-ray excitation and is a good laser host material. Gracia et al. [79] synthesized $CaWO_4$ crystals by means of microwave-assisted hydrothermal method, which allows a fast reaction rate and short reaction time, thereby leading to an overall reduction in energy consumption. $CaWO_4$ presents a broadband emission spectrum in the range from 350 to 800 nm. $CaWO_4$ emission is a typical multi-phonon and multi-level process, i.e., relaxation occurs by several paths involving the participation of numerous states within the band gap of the material. Based on calculations of the ground singlet (s) and excited singlet (s^*), the $CaWO_4$ band gap is 5.71 and 5.21 eV, respectively. By applying periodic models, a combination of different electronic states is to determine their electronic structure and specific atomic states that make up their corresponding energies. The calculation result shows that the excited state with triplet multiplicity is unsuccessful. The VBM of the ground singlet (s) and excited singlet (s^*) is similar in the DOS profiles, which are mainly composed by $2p_y$ and $2p_z$ orbits. The first CB in the ground state is mainly composed of $5d_{z^2}$ with smaller contribution from $5d_{x^2-y^2}$, and the second CB is composed of $5d_{\mathrm{xy}}$. In the singlet excited state, there is a first CB with dominance of W $5d_{x^2-y^2}$ over $5d_{z^2}$ contribution. The second CB is governed by W $5d_{yz}$ orbitals. Moreover, some new energy levels created by W $5d_{z^2}$ orbits appear lower than the first CB. These new energy levels can be understood as intermediate levels in the singlet excited DOS. During the excitation process, some electrons are promoted more feasibly from the oxygen $2p$ states to these intermediate W $5d_{z^2}$ states. The emission process occurs when an electron localized in a W $5d$ state decays into an empty O $2p$ state. The PL emission and these new energy levels can be attributed to the distorted $(WO_4)_d$ and $(CaO_8)_d$ pre-existing clusters [79].

In addition to self-activated hosts, rare-earth ion–doped materials are also studied by the excited-state calculation. Canning et al. [80] preformed excited-state calculation by manually setting the occupation of all the Ce 4*f* states to zero and filling the next highest state. For $Lu_2Si_2O_7$:Ce, the VB consists of O 2*p* states hybridized with Lu 4*f* states, and the CB consists of Lu 5*d* states. There are Ce 5*d* states below the Lu 5*d* states from the ground state DOS plot, and these Ce 5*d* states move to lower energy relative to the CB minimum in the excited-state plot. The excited-state Fermi level lies above the lowest Ce 5*d* level showing the filling of the lowest Ce 5*d* level. Thus, the lowest occupied excited Ce 5*d* levels can emit electrons to the Ce 4*f* states, which makes the Ce 5*d*–4*f* transition occur. In addition to rare-earth Ce luminescence, other luminescence ions—such as Sm and Eu—are also studied by way of excited-state calculation [81].

6.3.2 Band-Gap Correction for Luminescent Materials

In addition to excited-state calculation, the exact calculation of band gap is also important for luminescence-mechanism investigation. Currently, band-gap correction is performed by way of the DFT + U (U denotes the on-site *d*–*d* or *f*–*f* Coulomb interaction instead of the averaged Coulomb interaction), HSE (Heyd–Scuseria–Ernzerhof hybrid functional), and GW methods. DFT + U is generally applied in the systems containing *d* or *f* electrons due to the strong local interaction of these electrons. In the LDA + U method, the first step is to divide a system into two subsystems. The energy for a part of them is calculated by the normal DFT. For systems containing d or f electrons, the Hubbard model is used in calculating the energy band, whereas the other orbits can be treated according to the Kohn–Sham equation. The effective U value is applied to describe the orbit occupation as well as spin parameters. For U-value determination, the general principle refers to previous references [82, 83]. Another method is to test different U-values to observe the consistency of the experiment and calculation parameters such as band gap and local state position.

$YNbO_4$ has been known to be a self-activated phosphor with a band-gap value of 4.1 eV [84]. Under vacuum ultraviolet (UV) and UV excitation, the four excitation bands—labeled as A, B, C and D—are observed in the PL excitation spectrum. Based on the LDA + U approach, the calculated band gap is 4.28 eV, which agrees well with the experimental measurement value of 4.1 eV. Comparing the excitation spectra and DOS plots, bands B, C, and D are ascribed to the charge-transfer transition from O to Nb or Y. A and B originate from transition between the VB and defect state in the forbidden band.

Ji et al. prepared the pure $BaMgSiO_4:Eu^{2+}$ phosphor by solid-state reaction method under different atmospheres [85]. Under an N_2 atmosphere, the phosphor exhibits a strong green emission at 500 nm and a weak emission at 405 nm. After heat treatment under an NH_3 atmosphere, the 500-nm green emission intensity gradually decreases and completely disappears after heat treatment for 3 h, whereas

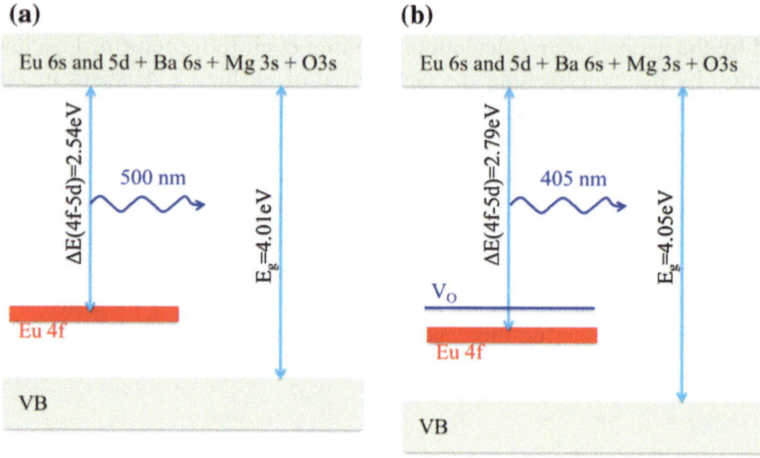

Fig. 6.22 Band-structure model of $BaMgSiO_4$:Eu with **a** $Eu_{Ba(3)}$ and **b** $Eu_{Ba(3)}$ + V_O to explain the luminous process of samples heat-treated in N_2 and NH_3, respectively, according to calculation results in Ref. [85]

a new blue emission peak at 445 nm appears and becomes very strong. According to the DFT + U calculation, Eu prefers to occupy Ba(3) sites coordinated with six oxygen atoms. The emission at 500 nm is attributed to the *4f–5d* transition of Eu in Ba(3) site as shown in Fig. 6.22a. Eu in the Ba(3) site has a *4f–5d* transition energy of 2.54 eV, which is ascribed to the emission peak at 500 nm. Eu in the Ba(1) and Ba(2) sites has a *4f–5d* transition energy of 2.90 and 2.80 eV, respectively, which are close to the 405-nm emission peak. The Eu doping in Ba sites explains the luminescence of the samples by N_2 heat treatment. For NH_3 heat-treated samples, the $Eu_{Ba(3)}$ + V_O model is applied to illustrate the emission process. The DFT + U calculation reveals that local electrons in the oxygen vacancy cause changes of the emission wavelength of $BaMgSiO_4$:Eu because of the down-shift of the Eu *4f* energy level as shown in Fig. 6.22b. Thus, the 500-nm emission weakens gradually, and new 445 nm luminous band occurs. In addition, the 405-nm luminescence gradually disappears due to the increasing overlap of emission and excitation bands over the increase of the heat-treatment time.

Moreover, on 396-nm near UV-light excitation, $Ba_3Eu(PO_4)_3$ and $Sr_3Eu(PO_4)_3$ exhibit an orange-red emission with two main peaks at 596 and 613 nm corresponding to the $^5D_0 \rightarrow {}^7F_1$ and $^5D_0 \rightarrow {}^7F_2$ transitions of Eu^{3+}. The 613-nm emission dominates the PL property because the Eu^{3+} ion occupies the 16c Wykoff position in the crystal, which has no inversion center. Electron-structure calculation verifies that the Eu *4f* orbital mainly occupies the VB top and the CB bottom. Electronic transition to the $^5D_0$ configuration dominates the PL properties [86]. From the above-mentioned three examples, we can see that the DFT + U corrects the band gap, and its calculation results explain the PL origins successfully. The DFT + U approach is widely applied in luminous region due to the smaller calculated quantity compared with other methods. For more details, please refer to some recent reports [87, 88].

The HSE function replaces the slowly decaying long-range part of the Fock exchange by the corresponding density functional counterpart [89, 90]. The resulting expression for the exchange-correlation energy is given by

$$E_{xc}^{\mathrm{HSE}} = \frac{1}{4}E_x^{\mathrm{HF.SR}}(\mu) + \frac{3}{4}E_x^{\mathrm{PBE.SR}}(\mu) + E_x^{\mathrm{PBE.LR}}(\mu) + E_c^{\mathrm{PBE}}(\mu). \tag{6.13}$$

The separation of the electron–electron interaction into a short- and long-rang part, labeled SR and LR, respectively, is realized only in the HF-exchange interaction. Electronic correlation is represented by the corresponding part of the PBE density function. μ is the parameter that defines the range of separation. Decomposition of the Couloumb kernel is obtained using the following construction $\frac{1}{r} = S_\mu(r) + L_\mu(r) = \frac{\mathrm{erfc}(\mu r)}{r} + \frac{erf(\mu r)}{r}$ with $r = \left|\vec{r} - \vec{r'}\right|$. If r is larger than characteristic distance $\frac{2}{\mu}$, the short-range interaction becomes negligible. From Eq. (6.13), we can see that the long-range term becomes zero for $\mu = 0$, and the short-range contribution then equals the full Coulomb operator. Consequently, the two limiting cases of the HSE03/HSE06 functional are a true PBE0 functional for $\mu = 0$ and a pure PBE calculation for $\mu = \infty$. To explore this topic clearly, we take some examples using HSE calculations [91–95].

The photoluminescence properties of $Ba_2MgSi_2O_7$:Eu^{2+} are investigated in the vacuum ultraviolet-to-visible–excitation energy range [91]. The band gap of the host is found to be approximately 7.44 eV. The excitation spectra exhibit two bands peaking at 314 and 348 nm and a shoulder at 263 nm by monitoring the Eu $5d \rightarrow 4f$ emission at 501 nm. Using the first-principles calculation, three configurations for Eu, i.e., Eu^{2+} ($4f^7$), Eu^{3+} ($4f^6$) and Eu^{2+} ($4f^65d^1$), are studied as shown in Fig. 6.23. The VB top is dominated by O 2*p*, and the CB bottom is formed by the

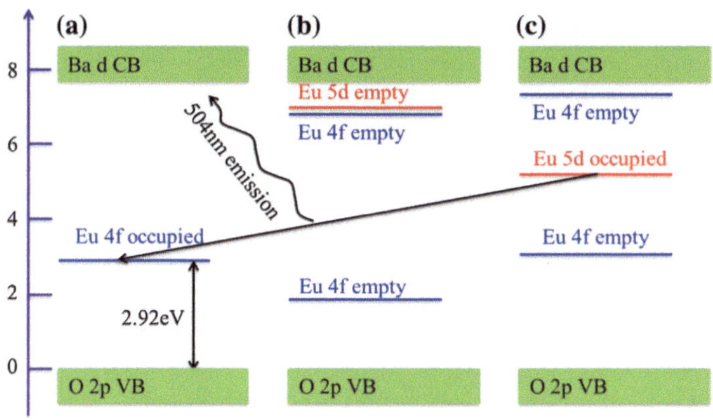

Fig. 6.23 The schematic band structures of $Ba_2MgSi_2O_7$ with **a** Eu^{2+} ($4f^7$), **b** Eu^{3+} ($4f^6$), and **c** Eu^{2+} ($4f^65d^1$) based on HSE and constraint excited-state calculation [91]

Ba 5*d* state with a small contribution from O 2*p* states through Ba–O bonding (Fig. 6.23a). Eu 4*f* and 5*d* states form local states in the band gap. Under high-energy excitation, the ground state Eu^{2+} ($4f^7$) may capture a hole from the VB to become Eu^{3+} ($4f^6$) as shown in Fig. 6.23b. In Eu^{3+} ($4f^6$), the six occupied 4*f* states are located deep inside the VB (not shown in Fig. 6.23) with a remaining empty 4*f* state in the gap. The three empty 5*d* states are positioned below the CB edge. In the lowest excited state of $Eu^{2+}(4f^6 5d^1)$, the $5d^1$ state can trap an electron from the CB to become the excited state $Eu^{2+}(4f^6 5d^1)$ from which the $5d \rightarrow 4f$ emission occurs [91]. The excitation electrons in occupied $Eu^{2+}(4f^6 5d^1)$ jumping to the empty ground state Eu^{2+} ($4f^7$) generate the strong 504-nm luminescence as shown in Fig. 6.23. Therefore, the position of the Eu^{2+}/Eu^{3+} 4*f* and 5*d* states relative to the host VB or CB depends on the electron occupation, which in turn determines the luminescence properties.

ZnO powders are heat treated at 700 °C in four atmospheres (air atmosphere, pure nitrogen gas flow, mixed gas flow of 5 % H_2, and 95 % N_2 and CO/CO_2 atmosphere) generated using activated carbon [92]. The ZnO UV luminescence intensity enhances in an H_2/N_2 reduction atmosphere and changes a little in air, pure nitrogen, and CO/CO_2 heat treatment. In contrast, the green emission intensity is greatly increased in H_2/N_2- and CO/CO_2-reduction atmospheres, and it strengthens slightly with heat treatment in pure N_2 flow. Therefore, the origins of the UV and green emission intensity boost are different. Through the DV-Xα method based on the linear combination of atomic orbits (LCAO) approximation, two cluster models of pure $(Zn_{29}O_{56})^{56-}$ and V_O-containing $(Zn_{29}O_{56})^{56-}$ are calculated. In a perfect crystal, the CB bottom mainly consists of the Zn 4*s* atomic orbits with some participation of O 2*p*. In the oxygen vacancy–containing ZnO cluster model, Zn 4*p* has considerable involvement in the CB bottom. The oxygen vacancy produces a deep donor level consisting of a mixture of Zn 4*s*, O 2*p*, and Zn 4*p* orbits. Figure 6.24 schematically shows the photoluminescence mechanism. In this diagram, the energy level of the VBM for both clusters is set at zero, and the band gap of the perfect cluster is set to the measured band gap of ZnO 3.4 eV. An oxygen vacancy makes a shallow localized well because hybridization of distort atom orbits produces empty levels whose energies are slightly below the CB minimum of the perfect crystal. The localization of electrons in this potential well seems to be the origin of thermal quenching for a green emission. The calculated potential well depth is 0.24 eV, which agrees well with the experimental fitted value 0.25 eV. The energy difference between the lowest unoccupied molecular orbit (LUMO) and the deep donor level for the V_O-containing cluster is calculated as 2.77 eV. The zero phonon line of the green emission appears at 438.5 nm corresponding to 2.83 eV. Broadband green luminescence can be simulated by setting the transition between the LUMO and the deep donor level to be the zero phonon emission and by constructing a phonon replica series with an interval of 72 meV. Thus, the DV-Xα method supports the conclusion that an oxygen vacancy acts as the luminescent center for green emissions [92]. In previous references, HSE methods also gave the same conclusion [93].

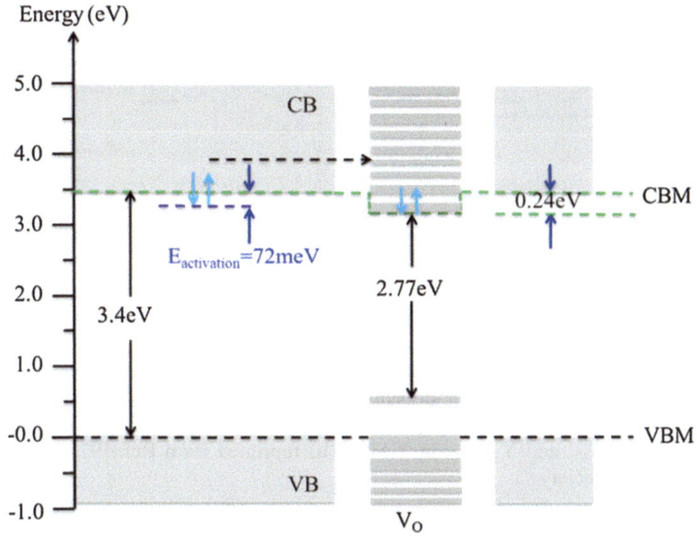

Fig. 6.24 Schematic energy band of a ZnO crystal with oxygen vacancies. The energy level was calculated by the DV-Xa molecular orbit method. The band gap is corrected on the basis of experimental measurement results. Reprinted from Ref. [92] by permission of Elsevier

Ning et al. calculated the electronic properties and $4f \rightarrow 5d$ transitions of Ce^{3+} substituted at the La^{3+} and Ca^{2+} sites of $La_2CaB_{10}O_{19}$ crystal using the hybrid function theory and wave function–based embedded cluster calculation [94]. The PBE, PBE0, and HSE06 functional calculation band gaps are 4.1, 8.1 and 7.3 eV, respectively. Thus, PBE0 result agrees well with the experimental value of 8.3 eV. The energy gap between the occupied Ce^{3+} $4f$ states and the VB maximum of the host is predicted to be 1.93 eV with a slight dependence on the local environment. Based on the PBE0-optimized supercell geometries, CASSCF/CASPT2 calculation is performed to derive the $4f^1$ and $5d^1$ energy levels of Ce^{3+} at the La^{3+} and Ca^{2+} sites. The experimental excitation bands are identified in association with the site occupations. The calculated energies and relative oscillator strengths of the $4f_i \rightarrow 5d_i$ $(i = 1\text{–}5)$ transitions in Ce_{La}- and Ce_{Ca}-doped crystal are shown in Fig. 6.25a [94]. Applying a similar method, the Ce^{3+} luminescence in the Sr_3AlO_4F [95] and Lu_2SiO_5 [96] hosts can be understood well.

Based on the hybrid DFT study, the formation energies of cation anti-site defects (Y_{Al} and Al_Y), oxygen vacancies (V_O), and nearest-neighbor complexes defect (Y_{Al}–Al_Y and Y_{Al}–V_O) in various charge states in $YAlO_3$ crystal are calculated [97]. The Y_{Al} formation is more energetically favorable than Al_Y under oxygen-poor conditions, which is consistent with the fact that the latter is not observed in experiments. With the optimized geometry, the band gap is predicted to be 8.13 eV with the standard PBE0, which contains an 0.25 HF exchange interaction, larger than the values of 5.58 and 7.37 eV obtained with PBE and HSE06 but smaller than the experimental value of 8.8 eV. By increasing the HF-exchange

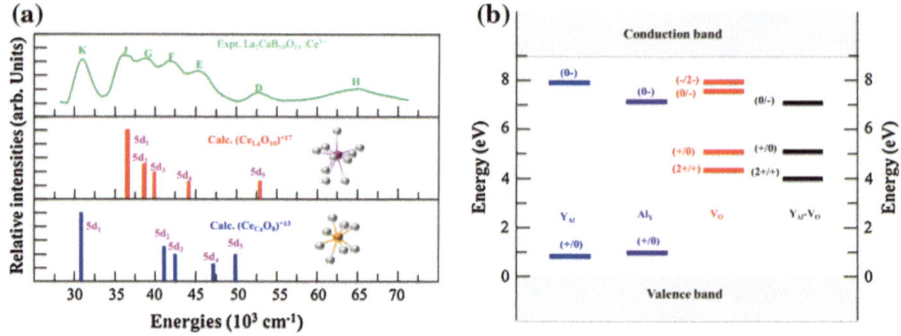

Fig. 6.25 The experimental excitation spectra, the cluster model calculation energy, and the relative oscillator strengths of Ce^{3+}-doped $La_2CaB_{10}O_{19}$ (**a**) reprinted from Ref. [94] by permission of American Chemical Society. Summary of derived thermodynamic transition-energy levels for Y_{Al}, Al_Y, V_O, and Y_{Al}-V_O in $YAlO_3$ (**b**) reprinted from Ref. [97] by permission of American Chemical Society

ratio to 32 %, the final band gap is 8.79 eV. On the basis of energetic calculations of the defects in different charge states, the optical transition energies of excitons trapped at Y_{Al}, V_O, Al_Y, and Y_{Al},-V_O are derived and compared with the experimental data observed under excitonic excitation at low temperature. The final local state positions for Y_{Al}, Al_Y, V_O, and Y_{Al}–V_O in $YAlO_3$ are displayed in Fig. 6.25b [97].

The Green function can describe the electron-add and -subtraction process; thus, the GW calculation can give the electron excitation spectra. In addition, a self-energy correction GW approximation includes primary scattering process. The mathematic description of quasi-particles is based on single-particle Green function, and it can fit strictly into the frame of quasi-particle self-energy. The self-energy operator is non-Hermitian, energy dependent, and a nonlocal operator that describes the exchange-and-correlation except for the Hartree approximation. The self-energy can be determined approximately and is the product of single-particle Green function and screened-interaction W.

Pure Ce and Ca + Ce-, Ce + Ba-, and Ce + B-doped $Gd_3Ga_3Al_2O_{12}$ crystals are grown by way of the Czochralski method [98]. Under identical excitation conditions, Ce^{3+} $4f$–$5d_{1/2}$ excitation peaks at 340 nm (3.65 eV) and 438 nm (2.83 eV) and the $5d$–$4f$ emission at 550 nm (2.25 eV) are very weak in 0.4 at.% Ca-doped $Gd_3Ga_3Al_2O_{12}$:Ce single crystal compared with $Gd_3Ga_3Al_2O_{12}$:Ce. Based on the DFT + G_0W_0 calculation, the positions of the lowest $4f$ and $5d$ levels of Ce relative to VB maximum and CB minimum for Ce^{3+} and Ce^{4+} are shown in Fig. 6.26 [98]. The calculated band gap using G_0W_0 is 6.63 eV, which is even closer to the experimental value compared with PBE0 hybrid function results (6.40 eV). As can be seen from Fig. 6.26, for the two samples containing Ce^{3+} and Ce^{4+}, the PL excitation peak at 3.65 eV originates from VB to $4f$ of Ce^{4+}. Similarly, 2.83 eV is ascribed to $4f \rightarrow 5d_1$. The theory calculation agrees well with the experimental data.

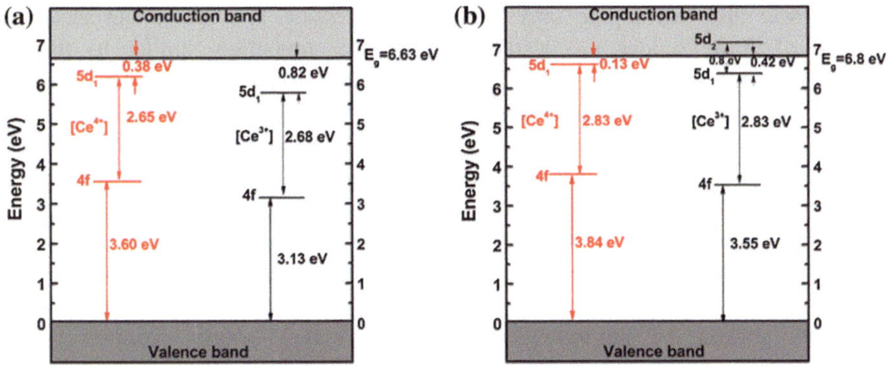

Fig. 6.26 Schematic energy-level diagram of $Gd_3Ga_3Al_2O_{12}$:Ce host obtained using DFT + G_0W_0 method (**a**) and the experiment (**b**). Reprinted from Ref. [98] by permission of American Physical Society

6.3.3 Conclusions

The band-gap values from the first-principles calculation agree well with the optical-measurement values by some advanced approaches such as LDA + U, HSE, and GW. They also have some disadvantages such as U-value uncertainty, the lengthy time needed for HSE and GW calculations, and consumption of storage. Therefore, new approaches and good computer technology must be developed in the future to precisely and rapidly calculate the electronic structure and understand the origin of luminescence.

6.4 Summary and Prospect

In this chapter, we introduce the DFT foundation theory, the first-principles calculation, related software, and the foundation of luminescence. For the isolated system, such as rare-earth ions or the simplest two-level atoms, we can calculate the energy level quantitatively. For complex systems, we can obtain electronic energy level through solving the Schrodinger equation based on the first-principles theory and various energy-band calculation theories. For example, transition-metal salt compounds—such as tungstates, molybdates, vanadates, zirconate, titanate, niobate, tantalate, chromate, manganese, gallate, etc.—are good self-activated luminescence materials. The first-principles calculation shows that the CB and VB are mainly composed of W(Mo, V, Zr, Ti, Nd, Ta, Cr, M, Ga)*d* and O 2*p* electrons. Thus, their intrinsic luminescence originates from the transition between the O 2*p* states and the *d* states. For quantum-dots luminescence, the cluster-model calculation shows that the native luminescence is ascribed to the transition between the HUMO and LUMO. In addition, the lone-pair electrons and occupied states below the HUMO also participate into luminous processes. The native luminescence origins of some oxides are also charge-transfer transition similar to that of transition-metal salt compounds. Except for the intrinsic luminescence, native defects can serve as the luminous centers. Based on DFT simulation, oxygen vacancy or interstitial oxygen forms local energy levels in a forbidden band. These defects levels are the luminescence center or quenching centers for PL. In the framework of DFT, the band-gap states induced by the vacancy of other elements or interstitial defects, such as C_i and Zn_i, can be calculated, thus revealing their contribution to the luminescence. For the nonactivator-doped materials, the first-principles calculation can reveal modification of the local coordination environments around the luminous centers such as bond-length variation or the appearance of new intrinsic defects. These changes lead to excitation and emission peak shift or intensity change as well as the appearance of new PL bands. Doping of rare-earth luminescence ions in matrixes can broaden the emission ranges and increase the utilization efficiency of radiation. The first-principles calculation shows that the rare ions 4*f* states locate in the VB or the band gap, and 5*d* states locate in the band gap or the CB. Therefore, the energy absorbed by the matrix can transfer to the rare earth ions. Combining the emissions from the host and the luminescent ions can generate white luminescence. Various methods beyond traditional DFT, such as LDA + U, HSE, and GW, are employed to improve the band gap. These methods have their respective advantages and shortcomings. Calculation of the excited state has not found wide application in luminescent materials, and only a few strategies have been proposed to perform this simple calculation.

In summary, the first-principles calculation has been playing significant roles in luminescent-mechanism investigation. Gradually, it demonstrates great power for designing luminescent materials. In the future, excited-state calculations become more important, and effective approaches to calculation should be developed.

References

1. Born M, Oppenheimer JR (1927) Zur Quanten Theorie der Molekeln. Ann Phys 389:457
2. Kohn W (1998) Nobel lecture: electronic structure of matter-wave functions and density functionals. Rev Mod Phys 71:1253
3. Jiang JF (2001) Wave function and density functional theory. Physics 23:549
4. Thomas LH (1927) The calculation of atomic fields. Proc Cambridge Philos Soc 23:542
5. Kohn W, Sham J (1965) Self-consistent equations including exchange and correlation effects. Phys Rev A 140:1133
6. Hohenberg P, Kohn W (1964) Inhomogeneous electron gas. Phys Rev B 136:864
7. Ceperley DM, Zunger A (1980) Ground state of the electron gas by a Stochastic method. Phys Rev Lett 45:566
8. Perdew TP, Zunger A (1981) Self-interaction correction to density-functional approximations for many-electron systems. Phys Rev B 23:5048
9. Engel E, Dreizler RM (2011) Density functional theory: an advanced course. Springer

10. Spassky DA, Vasilev AN, Kamenskikh IA, Mikhailin VV, Savon AE, Hizhnyi TA, Nedilko SG, Lykov PA (2011) Electronic structure and luminescence mechanism in $ZnMoO_4$ crystals. J Phys Condense Mat 23:365501
11. Blasses G (1988) Luminescence of inorganic solids: from isolated centres to concentrated systems. Prog Solid State Chem 18:79
12. Jackson JD (1975) Classical electrodynamics. Wiley, New York
13. Liu B (2010) Luminescence materials. Tongji University
14. Blasses G, Grabmaier BC (1994) Luminescence materials. Springer, Berlin
15. Blasses G (1984) The physics and chemistry of luminescence materials. Prog Phys 4:436
16. Shang MM, Li CX, Lin J (2014) How to produce white light in a single-phase host? Chem Soc Rev 43:1372
17. Blasses G (1980) The luminescence of close-shell transition-metal complexes: new development. Struct Bond 42:1
18. Ozgur U, Alivov YI, Liu C, Teke A, Reshchikov MA, Doan S, Avrutin V, Cho SJ, Morkoc H (2005) A comprehensive review of ZnO materials and devices. J Appl Phys 98:041301
19. Pan DY, Zhang JC, Li Zhen WuMH (2010) Hydrothermal route for cutting graphene sheets into blue-luminescent graphene quantum dots. Adv Mater 22:734
20. Arin P, Dumrongrojitanath P, Yayapao O, Phurangrat A, Thongtem S, Thongtem T (2014) Synthesis, characterization and optical activity of La-doped $ZnWO_4$ nanorods by hydrothermal method. Superlattices Microst 67:197
21. Zhang MR, Li PJ, Ying ZW (1998) Current status of R&D on $PbWO_4$ crystal-A new type of heavy scintillator. Chinese J Mater Res 12:113
22. Santos MA, Orhan E, Maurera MAMA, Simoes LGP, Souza AG, Pizani PS, Leite ER, Varela JA, Andres J, Beltran A, Longo E (2007) Contribution of structural order-disorder to the green photoluminescence of $PbWO_4$. Phys Rev B 75:165105
23. Zhang Y, Holzwarth NAW, Williams RT (1998) Electronic band structures of the scheelite materials $CaMoO_4$, $CaWO_4$, $PbMoO_4$, and $PbWO_4$. Phys Rev B 57:12738
24. Chen YH, Shi CH, Hu GQ (2000) Influence of Sb doping on the luminescent properties of $PbWO_4$ single crystals. J Appl Phys 87:1503
25. Kuyzhun OY, Bekenev VL, Solonin YM (2009) First-principles calculations and X-ray spectroscopy studies of the electronic structure of $CuWO_4$. J Alloy and Compd 480:184
26. Rajagopal S, Bekenev VL, Nataraj D, Mangalaraj D, Khyzhun OY (2010) Electronic structure of $FeWO_4$ and $CoWO_4$ tungstates: First-principles FP-LAPW calculations and X-ray spectroscopy studies. J Alloy Compd 496:61
27. Brik MG, Nagirnyi V, Kirm M (2012) Ab-initio studies of the electronic and optical properties of $ZnWO_4$ and $CdWO_4$ single crystals. Mater Chem Phys 134:1113
28. Tyagi M, Singh SG, Chauhan AK, Gadkari SC (2010) First principles calculation of optical properties of $BaWO_4$: A study by full potential method. Physica 405:4530
29. Zheng YH, You HP, Liu K, Song YH, Jia G, Huang YJ, Yang M, Zhang LH, Ning G (2011) Facile selective synthesis and luminescence behavior of hierarchical $NaY(WO_4)_2{:}Eu^{3+}$ and $Y_6WO_{12}{:}Eu^{3+}$. Cryst Eng Comm 13:3001
30. Huang MN, Ma YY, Huang XY, Ye S, Zhang QY (2013) The luminescence properties of Bi^{3+} Sensitized $Gd_2MoO_6{:}RE^{3+}$(RE = Eu or Sm) phosphors for solar spectral conversion. Spectrochim Acta A 115:767
31. Wang Q, Ci ZP, Zhu G, Xin SY, Zeng W, Que MD, Wang YH (2014) Multicolor bright Ln^{3+} (Ln = Eu, Dy, Sm) activated tungstate phosphor for multifunctional applications. Opt Mater Express 4:142
32. Marecos AS, Picon FC, Alves CN, Pizani PS, Varela JA, Longo E (2011) The role of short-range disorder in $BaWO_4$ crystals in the intensegreen photoluminescence. J Phys Chem C 115:12180
33. OrhanE Zenatti A (2005) Room-temperature photoluminescence of $BaTiO_3$: joint experimental and theoretical study. Phys Rev B 71:085112

34. OrhanE Santos M A, Maurera MAMA, Pontes FM, Santos COP, Souza AG, Varela JA, Pizani PS, Longo E (2005) Conditions giving rise to intense visible room temperature photoluminescence in $SrWO_4$Thin films: the role of disorder. Chem Phys 312:1
35. Cavalcante LS, Map Almeida, Jr WA, Tranquilin RL, Longo E, Batista NC, Mastelaro VR, Li MS (2012) Cluster coordination and photoluminescence properties ofα-Ag_2WO_4 microcrystals. Inorg Chem 51:10675
36. Cavalcante LS, Moraes E, Almeida MAP, Dalmaschio CJ, Bastista NC, Varela JA, Longo E, Li MS, Andres J, Beltran A (2013) A combined theoretical and experimental study of electronic structure and optical properties of β-$ZnMoO_4$ microcrystals. Polyhedron 54:13
37. CavalcanteLS Longo VM, Sczancoski JC, Almeida MAP, Batista AA, Varela JA, Orlandi MO, Longo E, Li MS (2012) Electronic structure, growth mechanism and photoluminescence of $CaWO_4$ Crystals. Cryst Eng Comm 14:853
38. Longo VM, Cavalcante LS, Paris EC, Sczancoski JC, Pizani PS, Li MS, Abdres J, Longo E, Varela JA (2011) Hierarchical assembly of CaMoO4 nano-octahedronsand their photoluminescence properties. J Phys Chem C 115:5207
39. Gupta SK, Ghosh PS, Pathak N, Ayra A, Natarajan V (2014) Understanding the local environment of Sm^{3+} in doped $SrZrO_3$ and energy transfer mechanism using time-resolved luminescence: a combined theoretical and experimental approach. RSC Adv 4:29202
40. Lim TL, Nazarov M, Yoon TL, Low LC, Fauzi MNA (2013) X-ray diffraction experiments, luminescence measurements and first-principles GGA+U calculations on $YTaO_4$. Comput Mater Sci 77:13
41. Zhang Y, Wu ZJ, Geng DL, Kang XJ, Shang MM, Li XJ, Lian HZ, Chen ZY, Lin J (2014) Full color emission in $ZnGa_2O_4$: simultaneous controlof the spherical morphology, luminescent, and electric properties via hydrothermal approach. Adv Funct Mater 10:1002
42. Aleksandra B, Yu HL (2006) Optical properties of ZnO nanostructures. Small 2:944
43. Baker SN, Baker GA (2010) Luminescent carbon nanodots: emergent nanolights. Angew Chem Int Ed 49:6726
44. Zhang YH, Pan WQ, Chai GQ, Liang MR, Dong GP, Zhang QY, Qiu JR (2013) Synthesis and luminescence mechanism of multicolor-emitting g-C_3N_4 nanopowders by low temperature thermal condensation of melamine. Sci Rep 3:1943
45. Zhang XD, Xie X, Wang H, Zhang JJ, Pan BC, Xie Y (2013) Enhanced photoresponsive ultrathin graphitic-phase C_3N_4 nanosheets for bioimaging. J Am Chem Soc 135:18
46. Liu H, Neal AT, Zhu Z, Luo Z, Xu XF, Tomanek D, Ye PD (2014) Phosphorene: an unexplored 2D semiconductor with a high hole mobility. ACS Nano 8:4033
47. Tran V, Soklaski R, Liang YF, Yang L (2014) Layer-controlled band gap and anisotropic excitons in few-layer black phosphorus. Phys Rev B 89:235319
48. Zhang CM, Lin J (2012) Defect-related luminescent materials: synthesis, emission properties and applications. Chem Soc Rev 41:7938
49. Zhang CL, Guo DL, Hu CG, Chen YX, Liu H, Zhang HL, Wang X (2013) Large-scale synthesis and photoluminescence of cobalt tungstate nanowires. Phys Rev B 87:035416
50. Zhou XW, Liu TY, Zhang QR, Cheng F, Qiao HL (2009) First-principles study of cadmium vacancy in $CdWO_4$ crystal. Solid State Sci 11:2071
51. Qian HJ, Zhang JY, Yin LQ (2013) Crystal structure and optical properties of white light Emitting Y_2WO_6:Sm^{3+} phosphor with excellent color rendering. RSC Adv 3:9029
52. Chen FM, Liu XH (2013) Structure and photoluminescence properties of $La_2Mo(W)O_6$:Eu^{3+} as red phosphors for white LED applications. Opt Mater 35:2716
53. Zhang ZJ, Zhang H, Duan CJ, Yuan JL, Wang XJ, Xiong DB, Chen HH, Zhao JT (2008) Structure refinement of Lu_2WO_6 and luminescent properties of Eu^{3+}, Pr^{3+} doped Lu_2WO_6. J Alloy Compd 466:258
54. Ding BF, Qian HJ, Han C, Zhang JY, Lindquist SE, Wei B, Tang ZL (2014) Oxygen vacancy effect on photoluminescence properties of self-activated yttrium tungstate. J Phys Chem C 118:25633

55. Kan D, Terashima T, Kanda R, Masuno A, Tanaka K, Chu S, Kan H, Ishizumi A, Kanemithsu Y, Shimakawa Y, Takano M (2005) Blue-light emission at room temperature from Ar^{+}-irradiated $SrTiO_3$. Nat Mater 4:816
56. Cuong DD, Lee B, Choi KM, Ahn HS, Han SW, Lee JC (2007) Oxygen vacancy clustering and electron localization in oxygen-deficient $SrTiO_3$: LDA+U study. Phys Rev Lett 98:115503
57. Chen T, Liu TY, Zhang QR, Li FF, Yi ZJ, Tian DS, Zhang XY (2007) First-principles study of $PbWO_4$ crystal with interstitial oxygen atoms. Phys Status Solidi A 204:776
58. Qi ZM, Shi CS, Zhou DF, Tang HG, Liu T, Hu TD (2001) The green emission and local structure of the scintillator $PbWO_4$. Phys B 307:45
59. Zhang CM, Liu CK, Li CX, Quan ZW, Liu XM, Lin J (2008) Enhanced luminescence of BPO_4 by mixing with SiO_2 and Al_2O_3. J Phys Chem C 112:2183
60. Fabbri F, Villani M, Catellani A, Calzolari A, Cicero G, Calestani D, Calestani G, Zappettini A, Dierre B, Sekiguchi T, Salviati G (2014) Zn vacancy induced green luminescenceon non-polar surfaces in ZnO nanostructures. Sci Rep 4:5158
61. Zeng HB, Duan GT, Li Y, Yang Sk XuXX, Cai WP (2010) Blue luminescence of ZnO nanoparticles based on non-equilibrium processes: defect origins and emission controls. Adv Mater 20:561
62. Huang YL, Zhu WL, Feng XQ, Man ZY (2003) The Effects of La^{3+}Doping on Luminescence Properties of $PbWO_4$ Single Crystal. J Solid State Chem 172:188
63. Chen T, Liu TY, Zhang QR, Li FF, Tian DS, Zhang XY (2007) First-principles study on the La^{3+} doping $PbWO_4$ crystal for different doping concentrations. Phys Lett A 363:477
64. Huang YL, Seo HJ, Feng Q, Yua SH (2005) Effects of trivalent rare-earth ions on spectral properties of $PbWO_4$ crystals. Mat Sci Eng B: Solid 121:103
65. Kalinko A, Kotlow A, Kuzmin A, Pankratov V, Popov AI, Shirmane L (2011) Electronic excitations in $ZnWO_4$ and $Zn_xNi_{1-x}WO_4$($x = 0.1$-0.9) using VUV synchrotron radiation. Cent Eur J Phys 9:432
66. Ding BF, Han C, Zheng LR, Zhang JY, Rm Wang, Tang ZL (2015) Tuning oxygen vacancy photoluminescence in monoclinic Y_2WO_6 by selectively occupying yttrium sites using lanthanum. Sci Rep 5:9443
67. Yeh CW, Liu YP, Xiao ZR, Wang YK, Hu SF, Liu RS (2012) Luminescence and density functional theory (DFT) calculation of undoped nitridosilicate phosphors for light-emitting diodes. J Mater Chem 22:5828
68. Liu TC, Cheng BM, Hu SF, Liu RS (2011) Highly stable red oxynitride β-SiAlON:Pr^{3+} phosphor for light-emitting diodes. Chem Mater 23:3698
69. Liu CC, Liu RS (2011) Advances in phosphors for light-emitting diodes. J Phys Chem Lett 2:1268
70. Chen WT, Shen HS, Liu RS, Attfield JP (2012) Cation-size-mismatch tuning of photoluminescence in oxynitride phosphors. J Am Chem Soc 134:8022
71. Zhou GX, Xiong SJ, Xl Wu, Liu LZ, Li TH, Chu PK (2013) N-doped SnO_2 nanocrystals with green emission dependentupon mutual effects of nitrogen dopant and oxygen vacancy. Acta Mater 61:7342
72. Kang FW, Yang XB, Peng MY, Wongraczek Ma ZJ, Zhang QY, Qiu JR (2014) Red photoluminescence from Bi^{3+} and the influence of the oxygen-vacancy perturbation in $ScVO_4$: a combined experimental and theoretical study. J Phys Chem C 118:7515
73. Qu BY, Zhang B, Wang L, Zhou RL, Zeng XZ (2015) Mechanistic study of the persistent luminescence of $CaAl_2O_4$:Eu,Nd. Chem Mater 27:2195
74. Jang HS, Kim HY, Kim YS, Lee HM, Jeon DY (2012) Yellow-emitting γ-Ca_2SiO_4:Ce^{3+}, Li^{+} Phosphor for solid-state lighting: luminescent properties, electronic structure, and white light-emitting diode application. Opt Express 20:2761
75. Ju LC, Xu X, Hao LY, Lin Y, Lee MH (2015) Modification of the coordination environment of Eu^{2+} in Sr_2SiO_4:Eu^{2+} phosphors to achieve full color emission. J Mater Chem C 3:1567
76. Huang JP, HouBh Ling HY, Liu J, Yu XB (2014) Crystal structure, electronic structure, and photoluminescence properties of $La_3BW_{1-x}Mo_xO_9$:Eu^{3+} red phosphor. Inorg Chem 53:8541
77. Huang MC (2003) The Density Functional Theory. Xiameng University

78. Huang MC (2005) Many body theory method of the excited state process. Chinese J Lumin 26:273
79. GraciaL Longo VM, Cavalcante LS, Beltran A, Avansi W, Li MS, Mastelaro VR, Varela JS, Longo E, Andres J (2011) Presence of excited electronic state in $CaWO_4$ crystals provoked by a tetrahedral distortion: an experimental and theoretical investigation. J Appl Phys 110:043501
80. Canning A, Chaudhry A, Boutchko R, Jensen NG (2011) First-principles Study of Luminescence in Ce-doped Inorganic Scintillators. Phys Rev B 83:125115
81. Chaudhry A, Boutchko R, Chourou S, Zhang G, Jensen NG, Canning A (2014) First-principles study of luminescence in Eu^{2+}-doped inorganic scintillators. Phys Rev B 89:155105
82. Solovyev IV, Dederichs PH (1994) Corrected atomic limit in the local-density approximation and the electronic structure of d impurities in Rb. Phys Rev B 50:16861
83. Anisimov VI, Zaanen J, Andersen OK (1991) Band theory and Mott insulators: Hubbard U instead of Stoner I. Phys Rev B 44:943
84. Lim TL, Nazarov M, Yoon TL, Low LC, Fauzi MNA (2014) First-principles LDA+U calculations and luminescence study of $YNbO_4$. Phys Scripta 89:095102
85. Ji WW, Lee MH, Hao LY, Xu X, Agathopoulos S, Zheng DW, Fang CH (2015) Role of oxygen vacancy on the photoluminescence of BaMgSiO4: Eu phosphors: experimental and theoretical analysis. Inorg Chem 54:1556
86. JiHP Huang Zh, Xia ZG, Molokeev MS, Jiang XX, Lin ZS, Atuchin VV (2015) Comparative investigations of the crystal structure and photoluminescence property of Eulytite-type $Ba_3Eu(PO_4)_3$ and $Sr_3Eu(PO_4)_3$. Dalton T 10:1039
87. Choi H, Cho SH, Khan S, Lee KR, Kim S (2014) Roles of an oxygen Frenkel pair in the photoluminescence of Bi^{3+}-doped Y_2O_3: computational predictions and experimental verifications. J Mater Chem C 2:6017
88. Spadavecchia F, Cappelletti G, Ardizzone S, Cetto M, Axxola MS, Presti LL, Cerrato G, Falciola L (2012) Role of Pr on the semiconductor properties of nanotitania: an experimental and first-principles investigation. J Phys Chem C 116:23083
89. Hummer K, Harl J, Kresse G (2009) Heyd-Scuseria-Ernzerhof hybrid functional for calculating the lattice dynamics of semiconductors. Pbys Rev B 80:115205
90. http://cms.mpi.univie.ac.at/vasp/vasp/vasp.html
91. Yan J, Ning LX, Huang YC, Liu CM, Hou DJ, Zhang BB, Huang Y, Tao Y, Liang HB (2014) Luminescence and electronic properties of$Ba_2MgSi_2O_7:Eu^{2+}$: a combined experimental and hybrid density functional theory study. J Mater Chem C 2:8328
92. Kim Y, Kang S (2011) Investigation of photoluminescence mechanisms of ZnO through experimental and first-principles calculation methods. Acta Mater 59:126
93. Hu J, Pan BC (2008) Electronic structures of defects in ZnO: hybrid density functional studies. J Chem Phys 128:154706
94. Ning LX, Wang ZC, Wang YF, Liu JX, Huang SZ, Duna CK, Zhang YF, Liang HB (2013) First-Principles study on electronic properties and optical spectra of Ce-Doped $La_2CaB_{10}O_{19}$ crystal. J Phys Chem C 117:15241
95. Ning LX, Wang YF, Wang ZC, Jin W, Huang SZ, Duan CK, Zhang YF, Chen WP, Liang HB (2014) First-principles study on site preference and 4f → 5d transitions of Ce^{3+} in Sr_3AlO_4F. J Phys Chem A 118:986
96. Ning LX, Lin LH, Li LL, Wu CB, Duan CK, Zhang YF, Seijo L (2012) Electronic properties and 4f → 5d transitions in Ce-doped Lu_2SiO_5: a theoretical investigation. J Mater Chem 22:13723
97. Ning LX, Cheng WP, Zhou CC, Duan CK, Zhang YF (2014) Energetic, optical, and electronic properties of intrinsic electron-trapping defects in $YAlO_3$: a hybrid DFT study. J Phys Chem C 118:19940
98. Wu YT, Meng F, Li Q, Koschan M, Melcher CL (2014) Role of Ce^{4+} in the scintillation mechanism of codoped $Gd_3Ga_3A_{12}O_{12}$:Ce. Phys Rev Appl 2:044009

Chapter 7
Color Tuning of Oxide Phosphors

Sun Woog Kim, Kenji Toda, Takuya Hasegawa, Kazuyoshi Uematsu and Mineo Sato

Abstract In white-LED application, the development of novel phosphors with high luminescence efficiency and high-emission color purity is extremely important because the performance of white LEDs is directly affected by the luminescence properties of the phosphors. In this chapter, therefore, we review some Eu^{2+}- and Ce^{3+}-activated phosphor materials and discuss the luminescence properties of these phosphors from the viewpoint of the relationship between the emission-band position and the crystallographic environment including the local site symmetry of the dopant site in the host lattice. We also propose a material-design concept to develop long wavelength-emission phosphors.

7.1 Introduction

The luminescence properties of phosphor materials are significantly important to achieve high-efficiency white-LED lamps and high-resolution flat panel displays. In particular, the color-rendering index (CRI) of white LEDs is considerably affected by the emission color of the phosphors. To improve the CRI of white LEDs, therefore, it is necessary to develop novel phosphors of high-emission intensities in the long-wavelength side and to exactly understand the terms of the luminescence mechanism of these phosphors. Among several phosphors, Eu^{2+}- and Ce^{3+}-activated phosphors are usually used in white LEDs because these phosphors show broad excitation and emission spectra due to the energy transitions between the 4*f* ground state and the 5*d* excited state of these ions [1]. It is well known that the position of the excitation and emission bands of Eu^{2+}- and Ce^{3+}-activated

S.W. Kim · K. Toda (✉) · T. Hasegawa
Graduate School of Science and Technology, Niigata University,
8050 Ikarashi 2-nocho, Niigata 950-2181, Japan
e-mail: ktoda@eng.niigata-u.ac.jp

K. Uematsu · M. Sato
Department of Chemistry and Chemical Engineering, Niigata University,
8050 Ikarashi 2-nocho, Niigata 950-2181, Japan

R.S. Liu (ed.), *Phosphors, Up Conversion Nano Particles, Quantum Dots and Their Applications*, DOI 10.1007/978-3-662-52771-9_7

phosphors is strongly dependent on the crystal-field strength around the activator. A strong crystal-field strength contributes to the larger energy splitting of the *5d* excited level [2], which result in the emission band of Eu^{2+}- and Ce^{3+}-activated phosphors shifting to the longer-wavelength side. The crystal-field strength is generally increased as the bond distance of activator—ligand and coordination number decrease [1, 3]. Furthermore, Van Uitert proposed that the emission-peak position (E) for Eu^{2+}- and Ce^{3+}-activated phosphors can be estimated approximately using the following equation [4]:

$$E = Q\left[1 - (V/4)^{1/V} \times 10^{-(n \times r \times Ea)/8}\right]$$

where Q is the position in energy for the lower *5d*-band edge data for the free Eu^{2+} ion (Q = 34,000 cm^{-1}) or Ce^{3+} ion (Q = 50,000 cm^{-1}); V is the valence of the activator; n and r are the coordination number and the ionic radius of cation, which is the dopant site of the activator, respectively; and Ea is the electron affinity of the atoms that form anions. This equation indicates that the emission-peak position of the Eu^{2+}- and Ce^{3+}-activated phosphors depends on the coordination number of the dopant site.

In other words, it is significantly important to investigate precisely the terms of the crystallographic environment of the dopant site to develop novel phosphors for white LEDs with high performance and high CRI. In addition, we previously proposed that for the dopant site to be symmetrical, the novel phosphors must emit long wavelengths. In this chapter, we will mainly discuss the effect of the crystallographic environment, including the local site symmetry, of the dopant site on the emission-band position of Eu^{2+}- or Ce^{3+}-activated phosphors. We also propose a novel design concept for further study on the development of novel longer-wavelength emission phosphors.

7.2 Eu^{2+}-Activated Phosphors

7.2.1 $Ba_9Sc_2Si_6O_{24}$:Eu^{2+}

Silicate materials have been widely investigated as host materials of phosphors for white LEDs owing to their excellent chemical and thermal stability, high luminescence efficiency, and low cost. Among them, Eu^{2+}-activated $Ba_9RE_2Si_6O_{24}$ (RE = rare earth) phosphors have been recently attracted considerable attention as novel phosphors for white LEDs because of their stable crystal structure [5–13]. Figure 7.1 shows the crystal structure of $Ba_9Sc_2Si_6O_{24}$, which was first determined using single-crystal X-ray diffraction by Wang et al. in 1994 [14]. The $Ba_9Sc_2Si_6O_{24}$ has a rhombohedral structure with the space group of *R*-3 (no. 148),

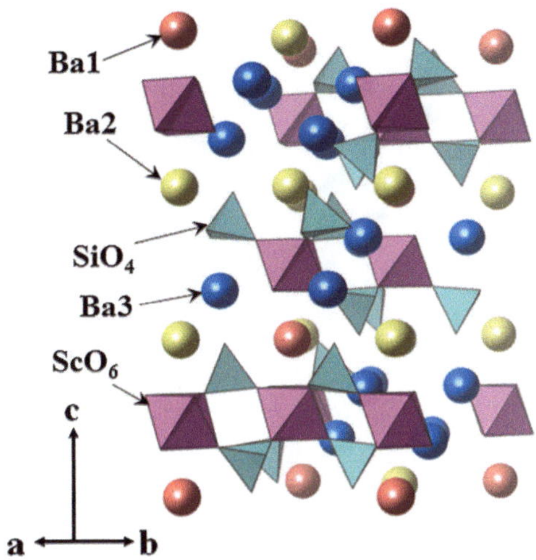

Fig. 7.1 Crystal structure of $Ba_9Sc_2Si_6O_{24}$

and it is built out of corner-connected ScO_6 octahedra and SiO_4 tetrahedra, which is similar to that of NASICON [15]. The bi-layers are comprised of SiO_4-ScO_6-SiO_4, in which the SiO_4 tetrahedra do not share oxide anions and stack along the c-axis direction, and the bi-layers are separated by Ba^{2+} ions. In addition, Ba^{2+} occupies three different sites having different crystallographic environments with BaO_9, BaO_{10}, and BaO_{12} polyhedra (Fig. 7.2). The BaO_9 and BaO_{12} polyhedra sites locate between the bi-layers, and the BaO_{10} polyhedra is coordinated by the ScO_6 octahedra and SiO_4 tetrahedra. The average bond distances of Ba–O in each Ba site are 0.2899, 0.2903, and 0.3047 nm for BaO_9, BaO_{10}, and BaO_{12} polyhedra, respectively.

The luminescence properties of the green-emitting phosphor based on $Ba_9Sc_2Si_6O_{24}$ were first reported in 2006 by Toda et al. [5] and was reported in detail in 2009 by Nakano et al. [6]. The excitation and emission spectra of the $Ba_9Sc_2Si_6O_{24}$:Eu^{2+} phosphor are shown in Fig. 7.3. The excitation spectrum of the $Ba_9Sc_2Si_6O_{24}$:Eu^{2+} phosphor consists of three broad bands from 280 to 480 nm with peaks at approximately 340, 380, and 440 nm. A weak absorption band of approximately 340 nm in the excitation spectrum represents the host absorption. Two strong broad bands centered at approximately 380 and 440 nm, which closely matches the emission wavelength of near-UV and blue LEDs, which correspond to the allowed transition from $4f^7$ to $4f^65d$ of Eu^{2+}. This result indicates the existence of two kinds of excitation centers, which is in accordance with the two different type of crystallographic environments around Eu^{2+} in the host $Ba_9Sc_2Si_6O_{24}$ lattice. As mentioned previously, the $Ba_9Sc_2Si_6O_{24}$ has three different Ba^{2+} sites with different crystallographic environments, and these sites are possible to substitute with Ce^{3+} ion. Because the average bond distance of Ba–O in each BaO_n ($n = 9$, 10, and 12) polyhedra are 0.2899, 0.2903, 0.3047 nm, respectively, and the crystal-field strength of O^{2-} around Eu^{2+} in each of the sites takes the order $BaO_9 > BaO_{10} \gg BaO_{12}$. The position of the excitation band of Eu^{2+}-activated

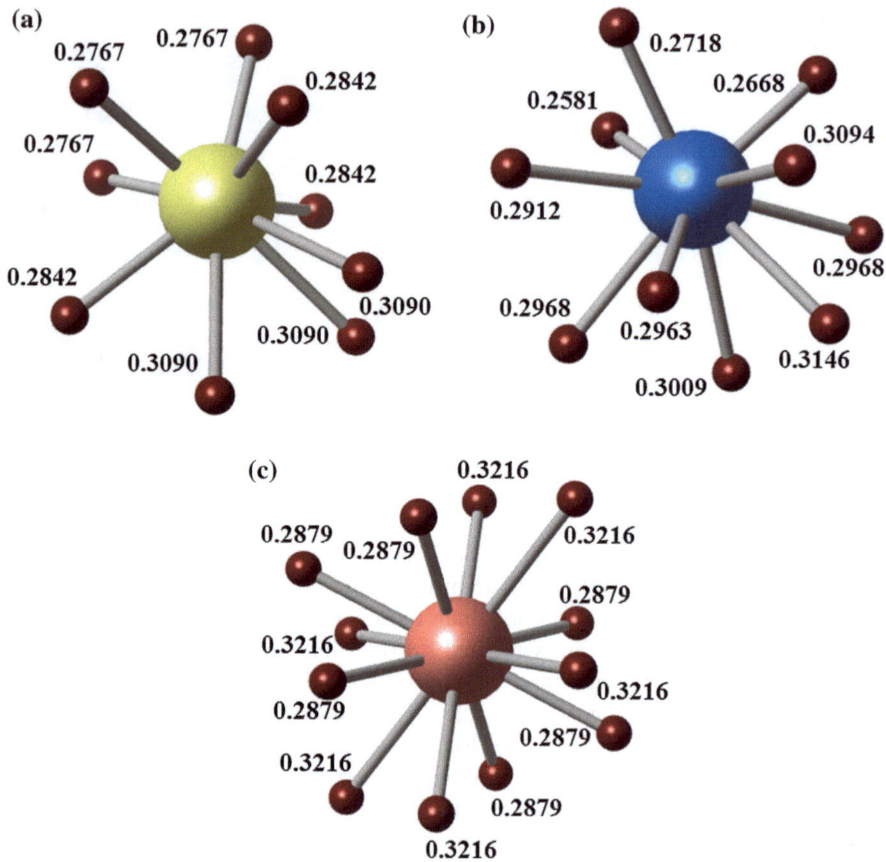

Fig. 7.2 Coordination environment of BaO_n (n = 9 (**a**), 10 (**b**), and 12 (**c**)) polyhedra

Fig. 7.3 Excitation and emission spectra of $Ba_9Sc_2Si_6O_{24}:Eu^{2+}$ phosphor

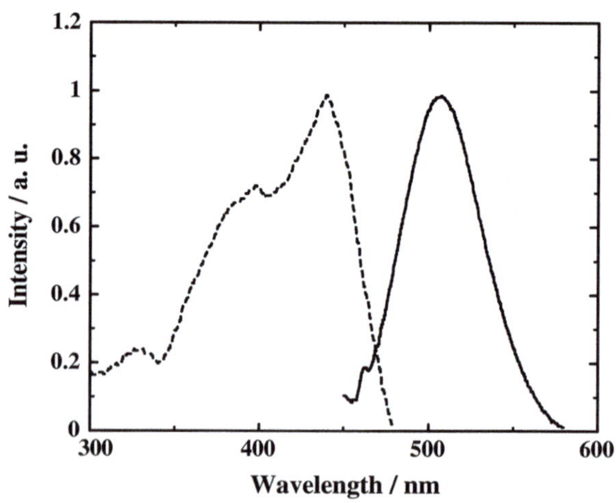

phosphors strongly depends on the crystal-field strength around Eu^{2+}; therefore, the excitation peaks at approximately 380 and 440 nm can be ascribed to BaO_{12} and BaO_{10} or BaO_9, respectively.

Furthermore, the emission spectrum exhibits a broad green emission band with a peak at 505 nm, and the emission band can be separated into two well-centered bands at approximately 480 and 540 nm, respectively, by Gaussian components. This result also supports that the $Ba_9Sc_2Si_6O_{24}:Eu^{2+}$ phosphor has two kinds of emission centers, which is in accordance with the two different types of crystallographic environments. The average bond distance of Ba–O in BaO_9 is similar to that of BaO_{10}, but it much smaller than that of the BaO_{12}. Therefore, crystal-field splitting of the 5*d* excited level of Eu^{2+} occupied at the BaO_9 and BaO_{10} polyhedra sites is nearly the same, and their emission bands could be distinguished clearly. This indicates that the two broad emission bands centered at approximately 480 and 540 nm are due to Eu^{2+} being occupied at the BaO_{12} and BaO_9 sites.

7.2.2 $AE_2SiO_4:Eu^{2+}$ (AE = Alkali Earth)

Eu^{2+}-activated phosphors based on alkaline earth orthosilicate, $AE_2SiO_4:Eu^{2+}$ (AE = alkaline earth), is one of the promising phosphors for use in blue LED–based white LEDs. Among them, Sr-rich $(Ba,Sr)_2SiO_4:Eu^{2+}$ (BOS or BOSE) has been usually used in white LEDs as a commercial yellow-emission phosphor. The crystal structure of alkaline earth orthosilicate, AE_2SiO_4 (AE = alkaline earth), was reported >30 years ago by Fields et al. [16], Catti et al. [17], and Barbier et al. [18]. AE_2SiO_4 (AE = Ba and Sr) has two crystallographic phases; monoclinic (β-type), which exists at low temperature <85 °C, and orthorhombic (α′-type), which exists at high temperature >85 °C. Both the β-type and α′-type structures consist of two cation sites with different crystallographic environments; the AE(I) and AE(II) sites are coordinated by 10 and 9 oxide anions, respectively. The $AE_2SiO_4:Eu^{2+}$ (AE = Ba and Sr) phosphors have orthorhombic α′-type structure. Figure 7.4 shows the crystal structure of AE_2SiO_4 and the crystallographic environment of AE in the structure. In the crystal structure, the AE(I) and AE(II) sites form a zig-zag chain structure along the b-axis and layering along the c-axis, i.e., AE(I) site layers are separated by AE(II) site layers as the layered structure. The average bond distances of AE–O in each AE(I) and AE(II) site are 0.2983 and 0.2824 nm, respectively, in the case of Ba_2SiO_4 and 0.2855 and 0.2701 nm, respectively, in the case of Sr_2SiO_4 [17].

The photoluminescence properties of Eu^{2+}-activated $(Ba,Sr)_2SiO_4$ phosphors were reported by Barry in 1968 [19] and recently have been begun to receive attention again as conversion phosphors in blue LED–based white LEDs. According to Barry's report [19], the excitation and emission bands of $(Ba,Sr)_2SiO_4:Eu^{2+}$ phosphors can be tuned by controlling the Ba/Sr ratios. Here the relationship between the crystallographic environment of AE sites and the

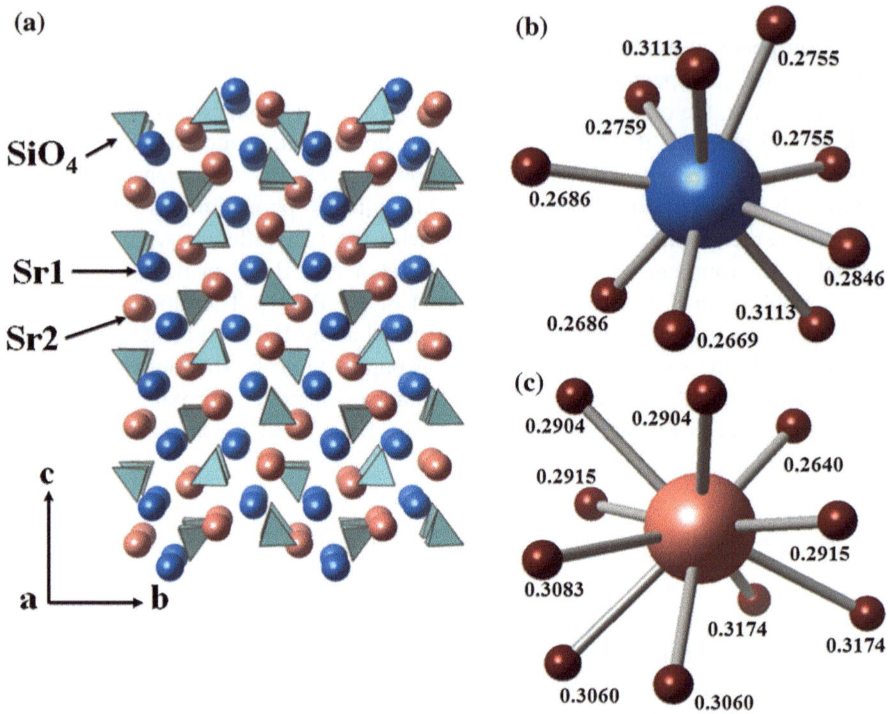

Fig. 7.4 Crystal structure of Sr_2SiO_4

luminescence properties of $Ba_2SiO_4{:}Eu^{2+}$ and $Sr_2SiO_4{:}Eu^{2+}$ is discussed in detail. Figure 7.5 shows the excitation and emission spectra of the $AE_2SiO_4{:}Eu^{2+}$ (AE = Ba and Sr) phosphors.

Both $AE_2SiO_4{:}Eu^{2+}$ (AE = Ba and Sr) phosphors show a broad excitation band covering the region from the UV to visible light part, and the photoluminescent

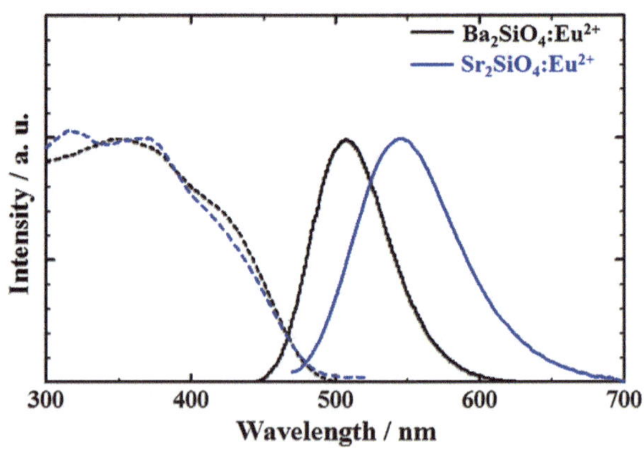

Fig. 7.5 Excitation and emission spectra of M_2SiO_4: Eu^{2+} (M = Ba and Sr)

emission spectra with excitation at 450 nm showed a strong broad emission band due to the $4f^65d^1 \rightarrow 4f^7$ transition of Eu^{2+}. The ionic radius of Sr^{2+} (0.131 and 0.136 nm for 9 and 10 coordination, respectively) [20] is smaller than that of Ba^{2+} (0.147 and 0.152 nm for 9 and 10 coordination, respectively) [20]. Because the average distance between AE^{2+} and O^{2-} of each site in Sr_2SiO_4 is shorter than that of Ba_2SiO_4, the crystal-field strength in alkali–earth orthosilicate, AE_2SiO_4, takes the order $Ba_2SiO_4 < Sr_2SiO_4$. The strong crystal-field strength leads to a shift of the emission band of Eu^{2+}-activated phosphors to the longer-wavelength side. These $Ba_2SiO_4:Eu^{2+}$ and $Sr_2SiO_4:Eu^{2+}$ phosphors exhibit broad green and yellow emissions with peaks at approximately 500 and 550 nm, respectively.

Because the $AE_2SiO_4:Eu^{2+}$ phosphors have two dopant sites with different crystallographic environments with AEO_{10} and AEO_9, the emission spectra of these phosphors can be well-separated into two spectra at shorter and longer wavelengths by Gaussian components [21, 22]. Although the difference between the size between the AE(I) and AE(II) sites in both Ba_2SiO_4 and Sr_2SiO_4 is nearly the same, the width of the emission band of the $Ba_2SiO_4:Eu^{2+}$ phosphor is narrower than that of the $Sr_2SiO_4:Eu^{2+}$ phosphor. Regarding this result, Hasegawa et al. showed that Eu^{2+} ions located only at the smaller site (CN = 9) in the $BaSrSiO_4:Eu^{2+}$ lattice and Sr^{2+} also located at the smaller site (CN = 9) based on result obtained by X-ray single-crystal diffraction analysis [23].

Compared with the $Ba_9Sc_2Si_6O_{24}:Eu^{2+}$ phosphor regarding the emission-band position of $Ba_2SiO_4:Eu^{2+}$ phosphor based on the result reported by Hasegawa et al., the emission band of $Ba_2SiO_4:Eu^{2+}$ is located at the shorter-wavelength side than in the $Ba_9Sc_2Si_6O_{24}:Eu^{2+}$ phosphor. The emission band corresponding to the BaO_9 polyhedra site of $Ba_9Sc_2Si_6O_{24}:Eu^{2+}$ phosphor shows the highest wavelength at 540 nm in the emission spectrum (Sect. 7.2.1). The emission-band position of the Eu^{2+}-activated phosphor strongly depends on the bond distance between Eu^{2+} and ligand, and the emission band generally shifts to the longer-wavelength side with decreasing bond distance of Eu–O in the oxide phosphors. In the case of Ba_2SiO_4: Eu^{2+} phosphor, the average bond distance of Ba–O in BaO_9 is 0.2824 nm, which is shorter than that of the $Ba_9Sc_2Si_6O_{24}:Eu^{2+}$ phosphor (0.2899 nm). These results indicate that it is necessary to consider other factors for determining the emission-band position in addition to the bond distance of the activator–ligand in the dopant site.

7.2.3 *$Li_2SrSiO_4:Eu^{2+}$*

In addition to the alkaline–earth orthosilicate phosphor, AE_2SiO_4 (AE = alkaline earth), orthosilicate materials have been widely investigated for white LEDs phosphors owing to their excellent chemical and thermal stability. The Li_2SrSiO_4: Eu^{2+} phosphor is also one of the novel orthosilicate phosphors for white LEDs [5, 24–29].

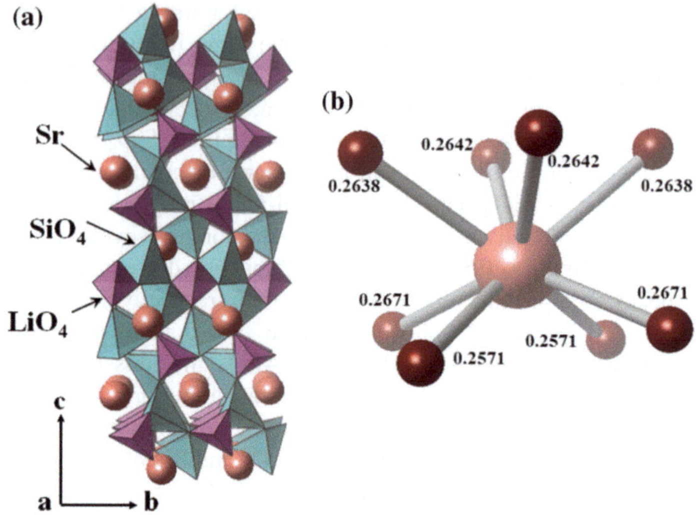

Fig. 7.6 Crystal structure of Li_2SrSiO_4

The crystal structure of Li_2SrSiO_4 is shown in Fig. 7.6. Li_2SrSiO_4 has a hexagonal crystal structure with the space group of $P3_121$, which is similar to that of Li_2EuSiO_4 [30]. In the crystal structure, SiO_4 and LiO_4 tetrahedral are connected by common corners and form a three-dimensional network, and Sr^{2+} ion locates at the outside space of the network. In addition, Sr^{2+} ions are surrounded by eight oxide ions, and the average bond distance of Sr-O in the SrO_8 dodecahedral site is 0.2633 nm. According to crystal-field theory [1, 2, 31], crystal-field splitting of the 5*d* excited level of Eu^{2+} became large when Eu^{2+} was doped into the small coordination site with short bonding distance. Therefore, the Li_2SrSiO_4:Eu^{2+} phosphor is expected to exhibit an emission band at longer wavelengths than the Sr_2SiO_4:Eu^{2+} phosphor.

Figure 7.7 shows the excitation and emission spectra of the Li_2SrSiO_4:Eu^{2+} phosphor. The phosphor has a strong optical absorption band in the near-UV and blue-light region, which is a good match the emission wavelength of near-UV and blue LED chips. The emission spectrum of the Li_2SrSiO_4:Eu^{2+} phosphor under excitation at 450 nm shows a strong broad reddish-yellow emission with a peak at approximately 580 nm, which is located at the longer-wavelength side than that of the Sr_2SiO_4:Eu^{2+} phosphor (550 nm). As shown in Fig. 7.6, the SrO_8 dodecahedron can be represented as a distorted cube in between a regular cube and an antiprism, likely YAG:Ce^{3+} phosphors. Furthermore, the shape of the SrO_8 dodecahedron is a little closer to the regular antiprism than YO_8 of YAG:Ce^{3+}. A lower symmetry into the dopant site usually results in the enhancement of the crystal-field strength around Eu^{2+}, and a strong crystal-field strength contributes to shift the emission band to the longer-wavelength side.

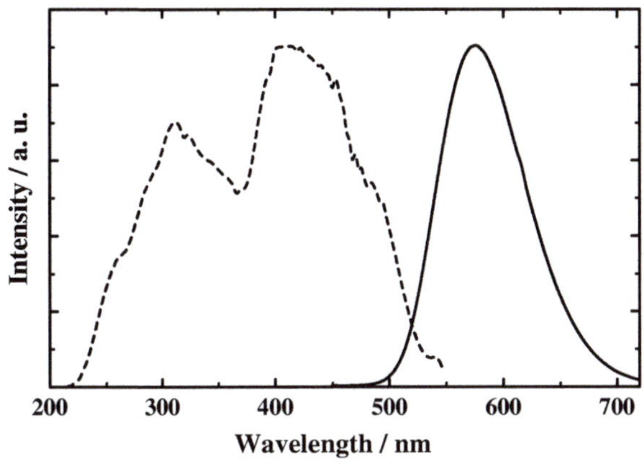

Fig. 7.7 Excitation and emission spectra of $Li_2SrSiO_4:Eu^{2+}$ phosphor

7.2.4 *$Sr_3B_2O_6:Eu^{2+}$*

Eu^{2+}-activated phosphors based on borate materials have been recently investigated and considered as a promising phosphor candidate for white LEDs owing to their broad excitation and emission spectra [32–36]. $Sr_3B_2O_6:Eu^{2+}$ is one of the borate phosphors for use in white LEDs [37–42]. The crystal structure of $Sr_3B_2O_6$ was first determined and reported by Richter and Muller in 1980 [43, 44], and $Sr_3B_2O_6$ has a rhombohedral structure with a space group of *R*3c. In addition, $Sr_3B_2O_6$ is also includes the SrO_8 dodecahedron in the crystal lattice, and the SrO_8 is close to the regular antiprism as shown in Fig. 7.8. The average bond distance of Sr–O in the SrO_8 dodecahedral site is 0.2635 nm, which is similar to that of $Li_2SrSiO_4:Eu^{2+}$ (0.2633 nm), thus indicating that Eu^{2+}-activated phosphor is also expected to exhibit emission bands at longer wavelengths.

Schipper et al. first reported in 1993 the luminescence properties of Eu^{2+}-activated $Sr_3B_2O_6$, and it presents reddish-yellow emission at 4.2 K [45], and it was shown by Song et al. that the $Sr_3B_2O_6:Eu^{2+}$ phosphor is a promising candidate for

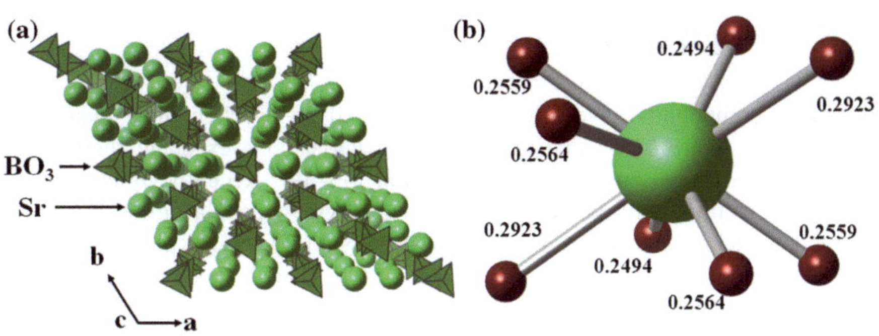

Fig. 7.8 Crystal structure of $Sr_3B_2O_6$

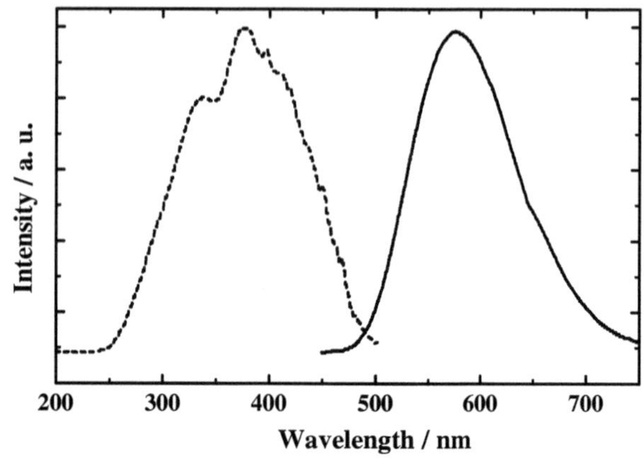

Fig. 7.9 Excitation and emission spectra of $Sr_3B_2O_6$: Eu^{2+} phosphor

white LEDs to yield warm white light [38]. The excitation and emission spectra of the $Sr_3B_2O_6:Eu^{2+}$ phosphor are given in Fig. 7.9. The excitation spectrum is consisted of a broad band covering the region from the part of UV to visible light indicating that the phosphor is suitable for a color converter using any excitation wavelength as the primary light source. The emission spectrum exhibits a broad reddish-yellow emission band centering at 574 nm attributed to the typical $4f^65d^1 \rightarrow 4f^7$ transition of Eu^{2+} under excitation at 450 nm. This is one of the important results to understand the purpose of this chapter. As pointed out in the introduction, the excitation and emission bands position of Eu^{2+}-activated phosphors strongly depends on the crystallographic environment of the cation site, which is substituted by the Eu^{2+} ions. Although the $Sr_3B_2O_6:Eu^{2+}$ phosphor is composed of different material compound than the $Li_2SrSiO_4:Eu^{2+}$ phosphor, both phosphors exhibit similar emission properties due to their similar crystallographic environment of the Sr^{2+} site. These results indicate that the excitation and emission bands are expected to exist at a similar wavelength when Eu^{2+} is doped into the cation site having similar crystallographic environment without reference to the material compound.

7.2.5 $NaMgPO_4:Eu^{2+}$

Phosphate compounds have also been considered an excellent host material of phosphor for white LEDs due to their low material cost, excellent thermal and hydrolytic stabilities, and charge stabilization [46–49]. Recently, blue light–excitable red-emitting phosphate phosphor, the olivine-type $NaMgPO_4:Eu^{2+}$, was reported by Kim et al. [50]. The olivine-type $NaMgPO_4:Eu^{2+}$ phosphor has an orthorhombic structure and a space group of *Pnma* with the lattice parameters a = 1.03701(8) nm, b = 0.61729(5) nm, and c = 0.49347(4) nm and the lattice volume V = 0.31284(4) nm^3. The structure of the olivine-type $NaMgPO_4$ is built out of PO_4 tetrahedra and MgO_6 octahedra connected to form a three-dimensional

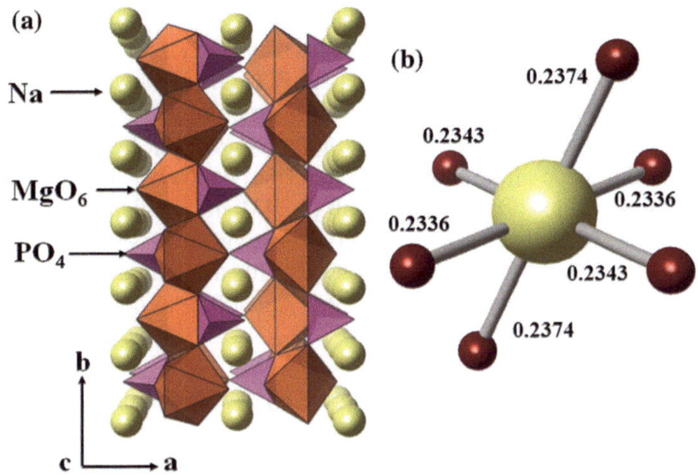

Fig. 7.10 Crystal structure of olivine-type $NaMgPO_4$

framework enclosing tunnels along the b-axis and Na^+ ions located in the tunnels as shown in Fig. 7.10. In the $NaMgPO_4:Eu^{2+}$ phosphor, from the viewpoint of the ionic radius, the Eu^{2+} ion is substituted into the Na^+ site in the $NaMgPO_4$ lattice. In the crystal structure, Na^+ ions are coordinated by six-oxide ions, and the average bond distance of Na–O is 0.2351 nm, which is much shorter than that of Li_2SrSiO_4 (0.2633 nm) and $Sr_3B_2O_6$ (0.2635 nm).

Figure 7.11 shows the photoluminescence excitation and emission spectra of the olivine-type $NaMgPO_4:Eu^{2+}$ phosphor. The phosphor exhibits a strong broad red-emission band centering at 628 nm, which is due to the $5d \rightarrow 4f$ transition of Eu^{2+} under excitation at 450 nm. In addition, the excitation spectrum of $NaMgPO_4$: Eu^{2+} phosphor shows a broad band covering the region from the part of UV to

Fig. 7.11 Excitation and emission spectra of olivine-type $NaMgPO_4:Eu^{2+}$ phosphor

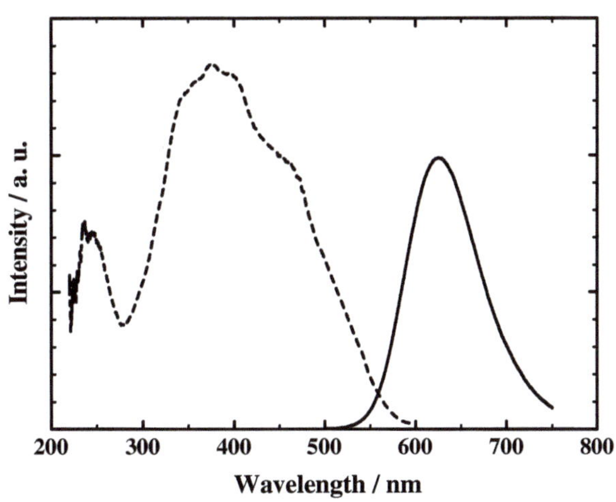

visible light indicating that the phosphor is suitable for a color converter using any excitation wavelength as the primary light source. It is well known that the orthophosphates with the general formula $ABPO_4$ (A: monovalent cations and B: divalent cations) have two different crystal structures from the glaserite- and olivine-type structure [51–54]. Glaserite-type $NaMgPO_4$ has an orthorhombic structure with the space group of $Pna2_1$ (No.33), in which Na^+ ions are coordinated by six-oxide ions with short bonding length (0.2431 nm), which is similar to that of the olivine-type $NaMgPO_4$. However, Eu^{2+}-activated glaserite-type $NaMgPO_4$ phosphor shows a broad blue-emission band with a maximum at approximately 437 nm under excitation at 332 nm [55, 56]. According to the following equation proposed by Van Uitert [4, 7]:

$$E = Q[1 - (V/4)^{1/V} \times 10^{-(n \times r \times Ea)/8}]$$

when Eu^{2+} ion is doped into crystallographically similar sites in the host lattice, the emission band of the phosphor is obtained at nearly the same wavelength.

In both the glaserite- and olivine-type $NaMgPO_4$ crystal structure, Na^+ ions, which are substituted by the Eu^{2+} ions, are coordinated by six-oxide ions even though the average bond distance of Na–O is different. In addition, the ionic radius (r) and the electron affinity (Ea) are also same in both crystal structures. In other words, the emission-band position of both the glaserite- and the olivine-type $NaMgPO_4:Eu^{2+}$ phosphors calculated by the above equation is nearly the same. In both the glaserite- and olivine-type crystal structure, furthermore, the average distance between Na^+ and O^{2-} is 0.2440 and 0.2360 nm, respectively, and the difference of the average bond distance is not greater. However, these phosphors exhibit blue emission peaking at 437 nm and red emission peaking at 625 nm, respectively. Therefore, we focused on the local site symmetry of the dopant site to determine the reason for the difference of emission-band position in the glaserite- and olivine-type $NaMgPO_4:Eu^{2+}$ phosphors. Figure 7.12 shows the Na^+

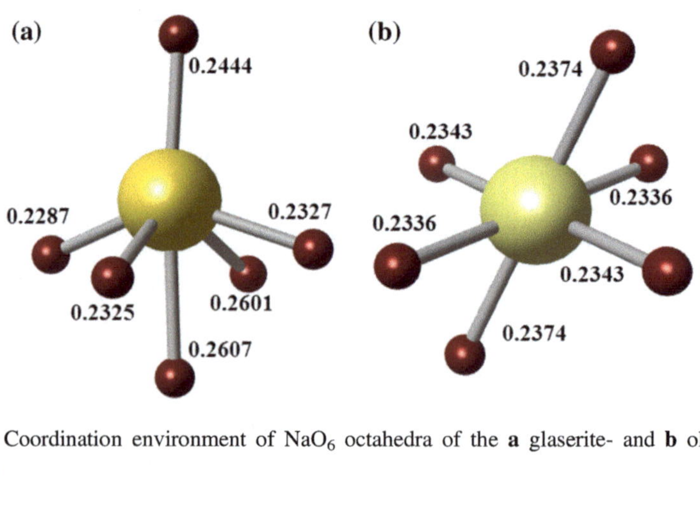

Fig. 7.12 Coordination environment of NaO_6 octahedra of the **a** glaserite- and **b** olivine-type $NaMgPO_4$

coordination environment of both the glaserite- and olivine-type $NaMgPO_4:Eu^{2+}$ phosphors. Both structures have an NaO_4 unit formed by the nearest four O^{2-} and Na^+ ions, and the average bond distances of Na–O in NaO_4 is 0.2385 nm for glaserite-type and 0.2340 nm for olivine-type phosphors, respectively. Furthermore, the shape of the NaO_4 unit for olivine-type has formed nearly closer to the plane-square, whereas the NaO_4 unit for glaserite-type is a pyramid shape. These results indicate that the interaction between the Eu^{2+} 5*d* orbital and the O^{2-} 2p orbital in olivine-type phosphors is stronger than that in glaserite-type phosphors, and strong interaction usually results in the enhancement of the crystal-field strength around Eu^{2+} in the crystal lattice.

Therefore, the olivine-type $NaMgPO_4:Eu^{2+}$ phosphors show the excitation and emission band at the longer-wavelength side compared with the glaserite-type $NaMgPO_4:Eu^{2+}$ phosphors.

7.2.6 $Ca_3Si_2O_7:Eu^{2+}$

$Ca_3Si_2O_7:Eu^{2+}$ phosphor is one of the famous materials that showed a broad red emission under blue-light excitation at 450 nm [5, 57–60]. The crystal structure of $Ca_3Si_2O_7$ and the crystallographic environment of the Ca^{2+} ion sites, which is substituted by Eu^{2+} ions, are shown in Fig. 7.13. The $Ca_3Si_2O_7$ crystallizer is well known to form a rankinite structure in the monoclinic crystal system with space group *P*21/*a* (no. 14) [57, 58]. The Ca^{2+} ions are coordinated by seven-oxide ions forming three different crystallographic coordination polyhedra. These CaO_7 polyhedra form a two-dimensional sheet along the a-axis and are separated by

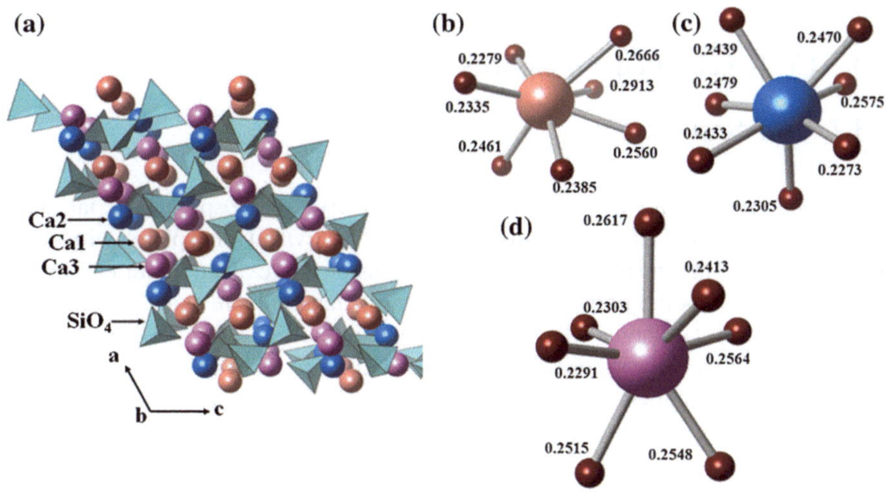

Fig. 7.13 Crystal structure of $Ca_3Si_2O_7$

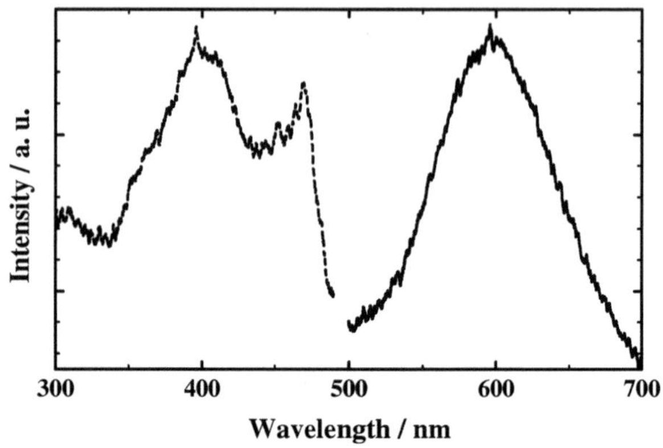

Fig. 7.14 Excitation and emission spectra of $Ca_3Si_2O_7$: Eu^{2+} phosphor

$[Si_2O_7]$ double-tetrahedra groups in the direction of the c-axis. The average bond distances of Ca–O in each CaO_7 polyhedra (Ca1, Ca2, and Ca3) are 0.2425, 0.2464, and 0.2511 nm, respectively.

The excitation and emission spectra of the $Ca_3Si_2O_7$:Eu^{2+} phosphor are shown in Fig. 7.14. A broad excitation band extending to 470 nm can be observed in the excitation spectrum of the $Ca_3Si_2O_7$:Eu^{2+} phosphor. When Eu^{2+} doped into host materials having different crystallographic coordination sites, the emission due to the Eu^{2+} generally shows the asymmetry broad band. Although $Ca_3Si_2O_7$ has three crystallographically nonequivalent Ca^{2+} sites, the emission spectrum of $Ca_3Si_2O_7$: Eu^{2+} phosphor shows a symmetric broad red-emission band with a peak at 600 nm due to the $4f^65d^1$–$4f^7$ transitions of Eu^{2+} [5, 57, 60]. As mentioned for the crystal structure of $Ca_3Si_2O_7$, three different Ca^{2+} sites have nearly same the neighboring environment, which usually results in a single symmetric emission band [58].

7.3 Ce^{3+}-Activated Phosphors

7.3.1 $Ba_3Sc_4O_9$:Ce^{3+}

Trivalence cerium ion Ce^{3+} has been also widely used in phosphor materials as an activator for white LEDs. Recently, many Ce^{3+}-activated phosphors with novel longer-wavelength emission over yellow emission have been reported. In this chapter, we discuss the phosphors including an octahedral site as a dopant site for Ce^{3+} in the crystal lattice. The luminescence properties of the $Ba_3Sc_4O_9$:Ce^{3+} phosphor were recently reported by Hasegawa et al. in 2014 [61].

The crystal structure of $Ba_3Sc_4O_9$ is given in Fig. 7.15. $Ba_3Sc_4O_9$ has a rhombohedral structure with the space group of *R*3 (no. 146), and it is composed of a BaO_n (n = 9 and 12) polyhedra and an ScO_6 octahedron [62]. Activator Ce^{3+} ions preferably substitute for the Sc^{3+} sites. The average bond distance of Sc-O in ScO_6 octahedra is 0.2143 nm, which is shorter than that of Na-O in the $NaMgPO_4$:Eu^{2+}

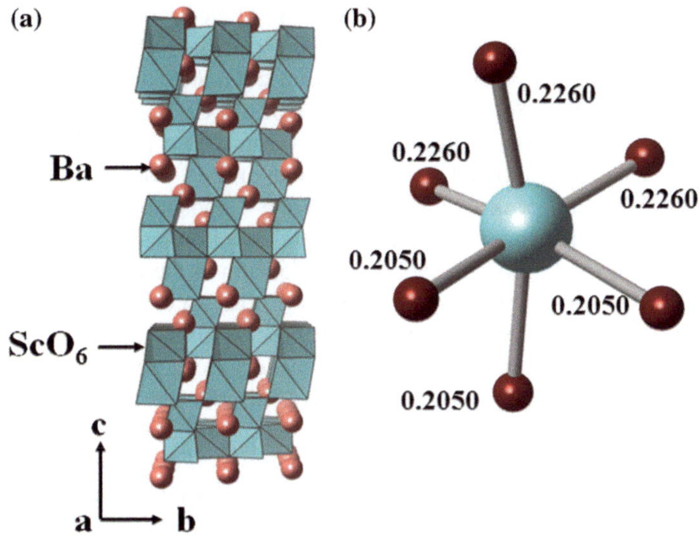

Fig. 7.15 Crystal structure of $Ba_3Sc_4O_9$

phosphor (0.2351 nm). In the ScO_6 octahedron, the Sc^{3+} ion and the nearest four O^{2-} ions form a distorted ScO_4 plane quadrilateral. In addition, the bond distance of Sc–O (0.2143 nm) is also shorter than that of Y–O (0.2435 nm) in the $Y_3Al_5O_{12}$:Ce^{3+} (YAG:Ce^{3+}) phosphor, which indicates that the $Ba_3Sc_4O_9$:Ce^{3+} phosphor is expected to exhibit an emission band at longer wavelength than the YAG:Ce^{3+} phosphor.

The excitation and emission spectra of the $Ba_3Sc_4O_9$:Ce^{3+} phosphor are shown in Fig. 7.16. The excitation spectrum consisted of a broad band covering the region from near-UV to visible light, which is a good match with the emission band of

Fig. 7.16 Excitation and emission spectra of $Ba_3Sc_4O_9$:Ce^{3+} phosphor

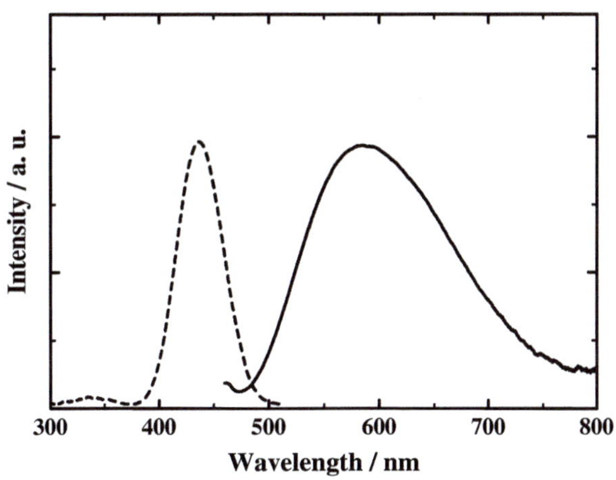

near-UV LED and blue LED chips. Because the 4*f* electron configuration of Ce^{3+} has two ground states ($^2F_{5/2}$ and $^2F_{7/2}$), the Ce^{3+}-activated phosphors are generally well known to exhibit a broader emission band compared with the Eu^{2+}-activated phosphors. The emission spectrum of $Ba_3Sc_4O_9$:Ce^{3+} also shows a broad reddish-yellow emission band from 480 to 750 nm with a peak at 583 nm, which can be ascribed to the allowed 5*d*-4*f* transition of Ce^{3+} ions under excitation at 437 nm. The emission peak of the $Ba_3Sc_4O_9$:Ce^{3+} phosphor was observed at the longer-wavelength side than that of a commercial yellow-emitting YAG:Ce phosphor, which shows the highest wavelength at 555 nm in the emission spectrum. As mentioned for the Eu^{2+}-activated phosphors, the position of the emission band due to the 5*d*-4*f* transition strongly depends on the crystallographic environment and site symmetry of the dopant site for the activator. Because the luminescence of Ce^{3+}-activated phosphors is also directly involved with the 5*d* orbit, the theory is considered to be suitable for explaining the luminescence behavior of the Ce^{3+}-activated phosphors. Although Y^{3+} ions are coordinated by eight-oxide ions in the $Y_3Al_5O_{12}$ structure, the YO_8 dodecahedra are composed of the intersected two different YO_4-plane quadrilaterals as presented in Fig. 7.17. In addition, $Ba_3Sc_4O_9$ also contains the ScO_4- plane quadrilateral in the ScO_6 octahedra and the bond distance of Sc–O in the ScO_4-plane quadrilateral, and ScO_6 octahedra are shorter than those of $Y_3Al_5O_{12}$. Consequently, the emission band of the $Ba_3Sc_4O_9$:Ce^{3+} phosphor was observed at a longer wavelength compared with the YAG:Ce^{3+} phosphor. These results support that the emission-band position of Ce^{3+}-activated phosphor can also be controlled by modifying the crystallographic environment and site symmetry of the dopant site. In addition, the Ce^{3+}-activated oxide phosphors can also serve as a red-emission phosphor for use in white LEDs.

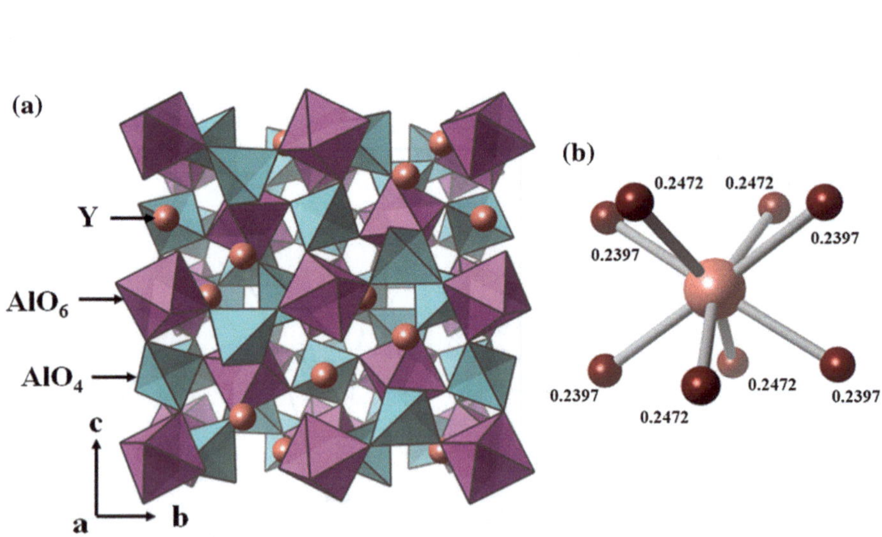

Fig. 7.17 Crystal structure of $Y_3Al_5O_{12}$

7.3.2 $Sr_6Y_2Al_4O_{15}$:Ce^{3+}

$Sr_6Y_2Al_4O_{15}$:Ce^{3+} phosphor is one of the recently reported novel Ce^{3+}-activated longer wavelength–emission phosphors [63]. $Sr_6Y_2Al_4O_{15}$ has a monoclinic structure with a space group of *C*2, which is first reported by Wang et al. [64]. The crystal structure of $Sr_6Y_2Al_4O_{15}$ is built out of corner-connected YO_6 octahedra and AlO_4 tetrahedra as well as Sr^{2+} ions located at two different crystallographic environment sites—SrO_9 and SrO_{10} polyhedra—as shown in Fig. 7.18. In addition, the ionic radii of Sr^{2+} (CN = 8: 0.1260 nm and CN = 9: 0.1310 nm) and Y^{3+} (CN = 6: 0.0900 nm) are similar to that of Ce^{3+} (CN = 6: 0.1010 nm, CN = 8: 0.1143 nm, and CN = 9: 0.1196 nm) [20]. This indicates that Ce^{3+} can possibly be doped into the three different sites in the $Sr_6Y_2Al_4O_{15}$ lattice having different crystallographic environments with SrO_n (n = 8 and 9) polyhedra and an YO_6 octahedron. As a result, the $Sr_6Y_2Al_4O_{15}$:Ce^{3+} phosphor shows the broad blue-emission band having two centers at approximately 396 and 450 nm, respectively, and an orange-red emission band with a peak at 600 nm. Because the average bond distances of M–O (M = Sr and Y) in SrO_n (n = 8 and 9) polyhedron and YO_6 octahedron are 0.2613, 0.2655, and 0.2338 nm, respectively, the crystal-filed strength takes the order $SrO_9 < SrO_8 < YO_6$. The increase of the crystal-field strength of O^{2-} around Ce^{3+} usually results in an emission-band shift to longer wavelength side. This suggests that the blue and red emissions of the

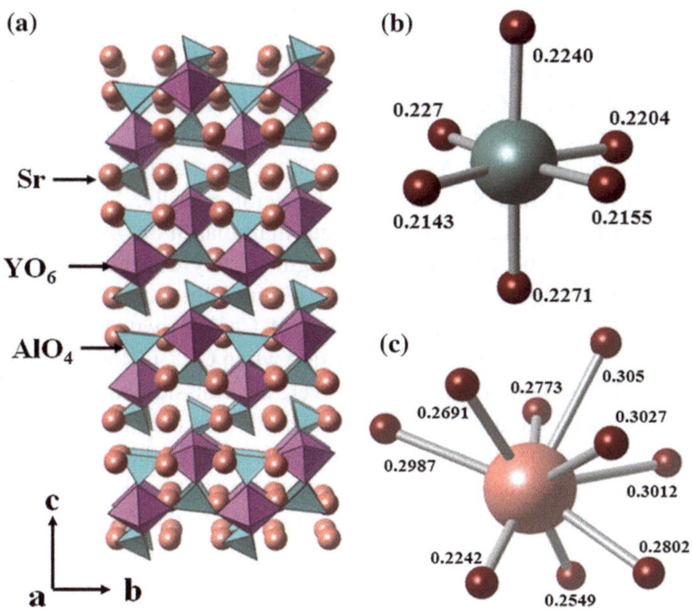

Fig. 7.18 Crystal structure of $Sr_6Y_2Al_4O_{15}$

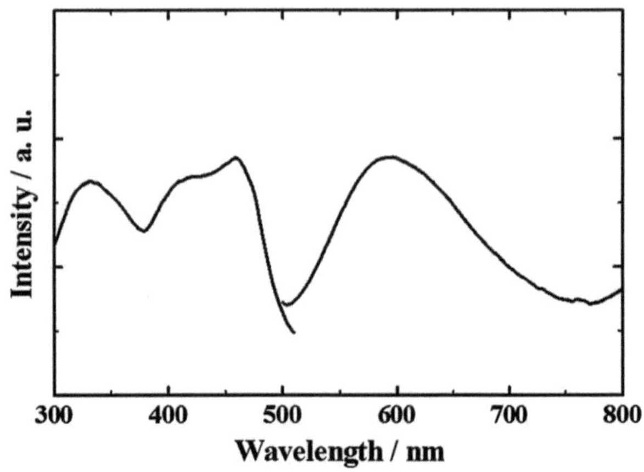

Fig. 7.19 Excitation and emission spectra of $Sr_6Y_2Al_4O_{15}:Ce^{3+}$ phosphor

$Sr_6Y_2Al_4O_{15}:Ce^{3+}$ phosphor correspond to the $5d \rightarrow 4f$ transition of Ce^{3+} doped into the SrO_n polyhedron and YO_6 octahedron sites, respectively.

Here we have mainly discussed the red emission of the $Sr_6Y_2Al_4O_{15}:Ce^{3+}$ phosphor from the viewpoint of the crystallographic environments of the YO_6 octahedra site. Figure 7.19 shows the excitation and emission spectra, which were recorded for emission at 600 nm and for excitation at 460 nm, of the $Sr_6Y_2Al_4O_{15}:Ce^{3+}$ phosphor. The phosphor shows two broad optical absorption bands from 280 to 500 nm with peaks at approximately 330 and 460 nm, respectively, and presents a broad orange-red emission band centered at approximately 600 nm due to the $5d \rightarrow 4f$ transition of Ce^{3+} under excitation at 460 nm. The emission band of this phosphor occupied the longer-wavelength side compared with $Ba_3Sc_4O_9:Ce^{3+}$ phosphor, which has maximum peak intensity at 580 nm. In both $Sr_6Y_2Al_4O_{15}:Ce^{3+}$ and $Ba_3Sc_4O_9:Ce^{3+}$, Ce^{3+} ions substituted into the YO_6 and ScO_6 octahedral sites, and the average bond distances of YO_6 and ScO_6 are 0.2287 and 0.2143 nm, respectively. In addition, both YO_6 and ScO_6 octahedra in each crystal structure have an MO_4 (M = Y and Sc) plane quadrilateral, and the Y–O distance in the YO_4-plane quadrilateral (0.2324 nm) is larger than that in ScO_4-plane quadrilateral (0.2155 nm). Because the crystal-field strength greatly depends on the bond distance between the activator and the ligands, it generally increases with decreasing bond distance. Therefore, the emission band of Ce^{3+}-activated phosphors is well known to appear on the longer-wavelength side when Ce^{3+} is doped into the smaller site having the shorter bond distance. However, the emission band of $Sr_6Y_2Al_4O_{15}:Ce^{3+}$ phosphors appeared at the longer-wavelength side than that of the $Ba_3Sc_4O_9:Ce^{3+}$ phosphors. It can be considered that the difference of the site symmetry of the YO_6 and ScO_6 octahedra. As pointed out in the olivine-type $NaMgPO_4:Eu^{2+}$ phosphor, the position of the emission band due to the $5d \rightarrow 4f$ transition is strongly affected by the local site symmetry of the dopant site in addition to the bond distance between the activator and the ligands. Although the Y–O distance in YO_4 is larger than that in ScO_4, the shape of the YO_4-plane quadrilateral is closer to a square compared with the ScO_4-plane quadrilateral, which indicates that the interaction between the Ce^{3+} 5*d* orbital and the O^{2-} 2p orbital in YO_6 octahedra is stronger than that in the ScO_6 octahedra. Strong interaction between the Ce^{3+} 5*d* orbital and the O^{2-} 2p orbital will contribute to enhance the crystal-field strength around the Ce^{3+} in the crystal lattice. As a result, the emission band of the $Sr_6Y_2Al_4O_{15}:Ce^{3+}$ phosphor appears on the longer-wavelength side than that of the $Ba_3Sc_4O_9:Ce^{3+}$ phosphors.

7.3.3 $LiSr_2YO_4:Ce^{3+}$

The $Sr_6Y_2Al_4O_{15}:Ce^{3+}$ and $LiSr_2YO_4:Ce^{3+}$ phosphors also show unreasonable luminescence behavior. Figure 7.20 shows the crystal structure of $LiSr_2YO_4$, which has a orthorhombic structure with a space group of *Pnma* (no. 62) [65].

The crystal structure of $LiSr_2YO_4$ is composed of an LiO_4 tetrahedron, SrO_n (n = 6 and 7) polyhedra, and distorted YO_6 octahedron. In the crystal lattice, the SrO_n (n = 6 and 7) polyhedral and distorted YO_6 octahedral sites are possible to

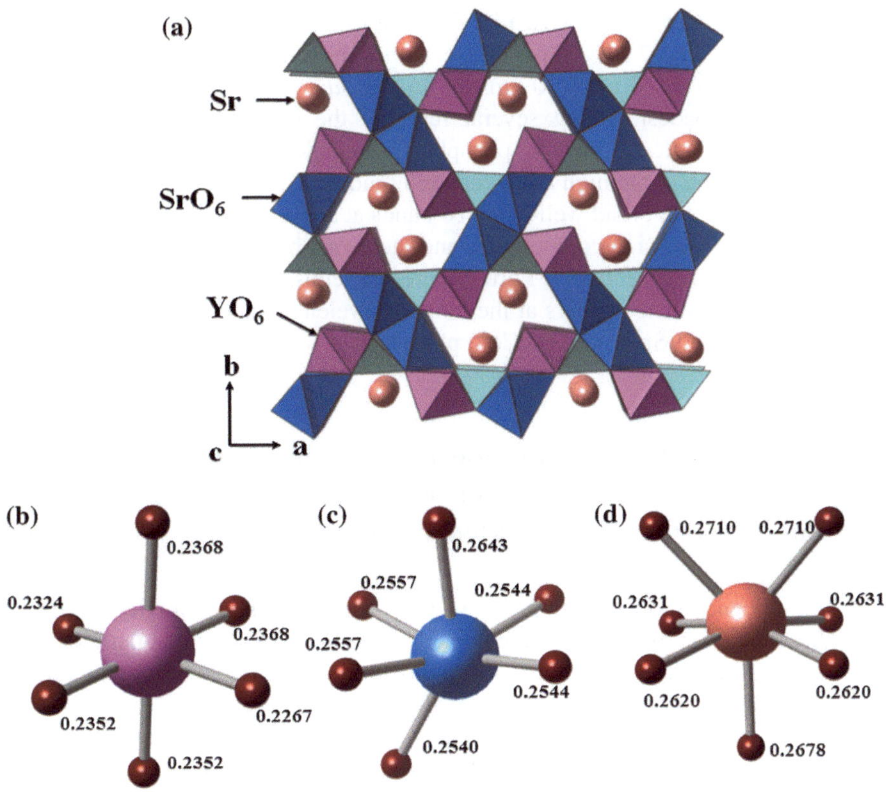

Fig. 7.20 Crystal structure of **a** $LiSr_2YO_4$, **b** YO_6, **c** SrO_6, **d** SrO_7

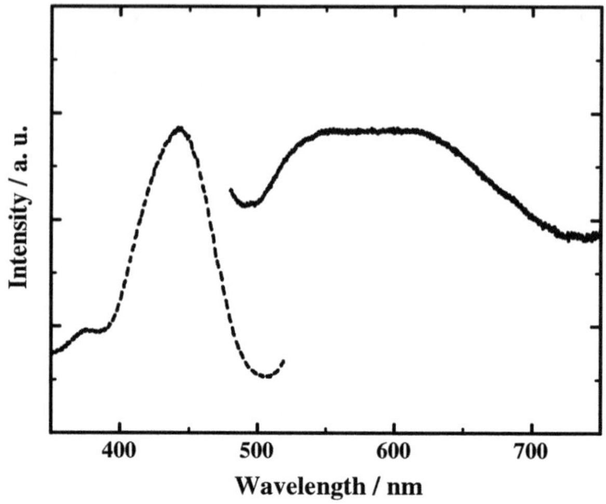

Fig. 7.21 Excitation and emission spectra of $LiSr_2YO_4$: Ce^{3+} phosphor

substitute with Ce^{3+} ion. This indicates the existence of three kinds of emission centers, which is in accordance with the three different types of crystallographic environments of Ce^{3+} in the host $LiSr_2YO_4$ lattice.

As shown in Fig. 7.21, $LiSr_2YO_4$:Ce^{3+} phosphor exhibits an extremely broad emission band from 500 to 720 nm due to the $5d \rightarrow 4f$ transition of Ce^{3+} [66], which is the broadest among several reported the conventional Ce^{3+}-activated phosphors. This indicates that the phosphor is suitable for use in blue LED-based white LEDs to obtain the high color-rendering index. The broad emission band can be separated into two broad well-centered bands at approximately 520 and 620 nm, respectively, by Gaussian components, and these two broad bands can be ascribed to SrO_6 and YO_6, respectively. Furthermore, the emission band corresponding to the YO_6 octahedral site appears at the longer-wavelength side compared with both $Ba_3Sc_4O_9$:Ce^{3+} and $Sr_6Y_2Al_4O_{15}$:Ce^{3+} phosphors.

7.4 Fundamental Data for the Relationship Between the Average Distance and the Emission Wavelength in Eu^{2+} or Ce^{3+} -Activated Phosphor

The luminescence properties of Eu^{2+}- or Ce^{3+}-activated phosphors depend on the crystallographic environments of the dopant site in the host lattice. In particular, the emission-band position due to $5d \rightarrow 4f$ transition Eu^{2+} or Ce^{3+} is strongly affected by the crystallographic environments of the dopant site. Table 7.1 lists the crystallographic data and the emission-band wavelength of the conventional Eu^{2+}-activated oxide phosphors. The emission-band wavelength can be calculated by following equation, which is proposed by Van Uitert [4]:

Table 7.1 Information of the emission band position and crystallographic environments of the conventional Eu^{2+}-activated phosphors

Host material	CN	Ionic radius (nm)	Average bond distance (nm)	Calculated λ_{em} (nm)	Measured λ_{em} (nm)	References
$Ba_9Sc_2Si_6O_{24}$	9	0.147	0.2899	424	505	[5, 6]
Ba_2SiO_4	9	0.147		424	500	[21]
Sr_2SiO_4	9	0.131	0.2824	442	550	[21]
$CaMgSiO_4$	8	0.118	0.2370	481	550	[67]
Li_2SrSiO_4	8	0.126	0.2633	469	580	[27]
$NaScSi_2O_6$	8	0.118		481	533	[68]
$Ba_2MgSi_2O_7$	8	0.142	0.2831	449	505	[5]
$NaAlSiO_4$	8	0.118	0.2563	481	556	[69]
$RbBaScSi_3O_9$	7	0.138	0.2837	477	500	[70]
$Ca_3Si_2O_7$	7	0.106	0.2467	527	600	[57]
Sr_3SiO_5	6	0.118	0.2601	536	570	[71]
$Na_2Mg_2Si_6O_{15}$	6	0.102	0.2636	566	430	
$SrCaP_2O_7$	9	0.131	0.2835	422	435	[72]
$KBaBP_2O_8$	8	0.142	0.2873	428	445	[73]
$Ca_9Gd(PO_4)_7$	8	0.112	0.2505	468	494	[74]
$Ba_7Zr(PO_4)_6$	7	0.138		454	585	[75]
SrP_2O_7	7	0.121	0.2637	478	420	[76]
$KBaPO_4$	6	0.138	0.2747	482	430	[77]
$NaMg_4(PO_4)_3$	6	0.102	0.2612	540	424	[78]
Glaserite-type $NaMgPO_4$	6	0.102	0.2413	544	433	[55]
Olivine-type $NaMgPO_4$	6	0.102	0.2351	544	625	[50]
$BaMgAl_{10}O_{17}$	8	0.142	0.2780	568	450	[79]
$SrMgAl_{10}O_{17}$	8	0.126	0.2634	518	469	[80]
$Sr_2Al_6O_{11}$	8	0.126	0.2661	529	460	[81]
$Sr_3B_2O_6$	8	0.126	0.2635	487	574	[38]
$LiSr_4(BO_3)_3$	6	0.118	0.2536	555	610	[82]
CaO	6	0.100	0.2405		738	[83]
SrO	6	0.118	0.2581		630	[84]

$$E = Q[1 - (V/4)^{1/V} \times 10^{-(n \times r \times Ea)/8}]$$

According to the equation, the coordination number (n) and ionic radius of the cation of dopant site are significantly important factors to determine the emission-band wavelength of the Eu^{2+}-activated phosphors. Figure 7.22 shows plots of the calculated and measured emission-band wavelengths of the

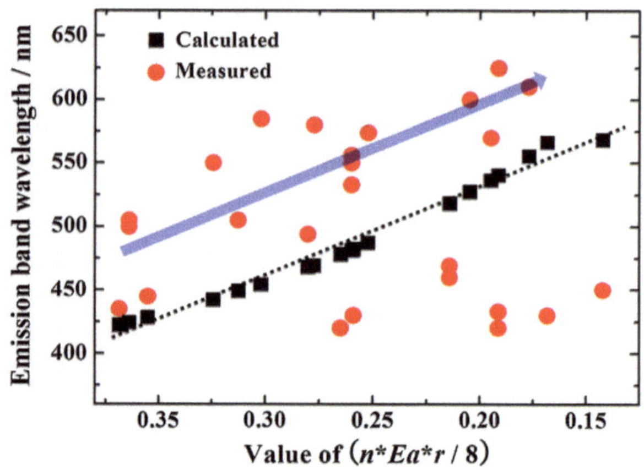

Fig. 7.22 Plots of calculated and measured emission-band positions of the Eu^{2+}-activated oxide phosphors as a function of ($n \times r \times Ea$)/8

Eu^{2+}-activated conventional oxide phosphors as a function of ($n \times r \times Ea$)/8. Although the calculated emission-band positions are underestimated compared with the measured one, both the calculated and measured emission-band wavelengths are shifted to the longer-wavelength side with increasing value of ($n \times r \times Ea$)/8. This result supports that the coordination number (n) and ionic radius (r) of the cation are key factors for estimating the emission-band wavelength of Eu^{2+}-activated phosphors. In addition, similar luminescence behavior is observed in the Eu^{2+}-activated (oxy)nitride phosphors [85].

In addition to the coordination number and ionic radius of the cation, it is well known that the emission-band wavelength due to the $5d \rightarrow 4f$ transition is also strongly affected by the bond distance between the cation and the anion in the dopant site. Figure 7.23 shows the dependence of the emission-band wavelength on the average bond distance of M–O in the Eu^{2+}-activated phosphors. The

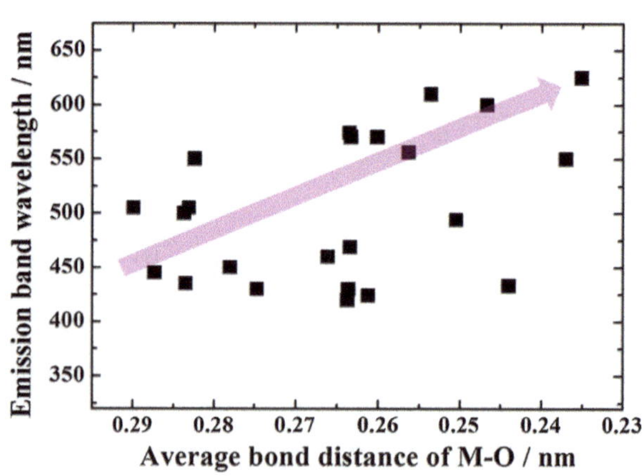

Fig. 7.23 dependence of the emission band wavelength on the average bond distance of M–O in the Eu^{2+}-activated oxide phosphors

emission-band wavelengths of the Eu^{2+}-activated phosphors shifted monotonically to the longer wavelength side with decreasing M–O bond distance in the dopant site. This result is reasonable because the crystal-field strength increases with decreasing bond distance in the dopant site.

As shown in Figs. 7.22 and 7.23; however, the crystallographic parameters of the dopant site—such as the coordination number, ionic radius, and bond distance—are not enough to explain the luminescence behavior of the some reported conventional Eu^{2+}- or Ce^{3+}-activated phosphors. As pointed out regarding the Sect. 7.2.5 ($NaMgPO_4:Eu^{2+}$) phosphor, although glaserite-type and olivine-type $NaMgPO_4$ have different space groups such as $Pna2_1$ and *Pnma*, both materials have the same orthorhombic structure. In both phosphors, an activator Eu^{2+} ion is doped into the Na^+ sites, which has same coordination number. As a result, the emission band of both phosphors, as calculated by the equation proposed by Van Uitert, are located at the same wavelength position. Furthermore, the difference of the Na–O bond distance of both glaserite-type (0.2440 nm) and olivine-type $NaMgPO_4$ (0.2360 nm) is not large. However, olivine-type $NaMgPO_4:Eu^{2+}$ shows a red emission with a peak at 625 nm, whereas the glaserite-type phosphor shows a blue emission with a peak at 433 nm (Table 7.1). In addition, Ce^{3+}-activated $Ba_3Sc_4O_9$ (Sect. 7.3.1), $Sr_6Y_2Al_4O_{15}$ (Sect. 7.3.2), and $LiSr_2YO_4$ (Sect. 7.3.3) phosphors also show unreasonable luminescence behavior compared with a conventional crystal-field theory. Although $Ba_3Sc_4O_9$, $Sr_6Y_2Al_4O_{15}$, and $LiSr_2YO_4$ all have different chemical composition, these materials have MO_6 octahedra as the dopant site for Ce^{3+} and the shorter M–O bond distance of M–O takes the order $LiSr_2YO_4 \rightarrow Sr_6Y_2Al_4O_{15} \rightarrow Ba_3Sc_4O_9$. According to conventional theory, the crystal-field strength around the Ce^{3+} ion in the host lattice takes the order $Ba_3Sc_4O_9:Ce^{3+} < Sr_6Y_2Al_4O_{15}:Ce^{3+} < LiSr_2YO_4:Ce^{3+}$. However, the emission bands of these phosphors are found at the longer-wavelength side on the order of $Ba_3Sc_4O_9:Ce^{3+} \rightarrow Sr_6Y_2Al_4O_{15}:Ce^{3+} \rightarrow LiSr_2YO_4:Ce^{3+}$. These results indicate that some other factors are required to determine the emission-band wavelength of the Eu^{2+}- or Ce^{3+}-activated phosphors. In this chapter, we focused on the local symmetry of the dopant site, as opposed to conventional crystal-field theory, to explain the unreasonable luminescence behavior.

As shown in Fig. 7.24, the MO_4 unit formed by the metal ion and the nearest four oxide ions becomes close to the plane square on the order of $LiSr_2YO_4 \rightarrow Ba_3Sc_4O_9 \rightarrow Sr_6Y_2Al_4O_{15}$. In the case of the octahedral site, high site symmetry of the MO_4 unit usually results in enhanced interaction between the Eu^{2+} or Ce^{3+} 5*d* orbital electron and the O^{2-} 2p orbital electron, thus leading to the strong crystal-field strength around the activator in the host lattice. In addition, the effect of local site symmetry on the luminescence behavior of phosphors can be described in some of the phosphors having an MO_8 dodecahedron as the dopant site [e.g., $Li_2SrSiO_4:Eu^{2+}$ (Sect. 7.2.3) and $Sr_3B_2O_6:Eu^{2+}$ (Sect. 7.2.4)], especially in the garnet-structure phosphors. These results support that local site symmetry of the dopant site is also significantly important to determine the emission-band wavelength of Eu^{2+}- or Ce^{3+}-activated phosphors. However, it is also difficult to conclude that the effect of local site symmetry is an appropriate approach to determine

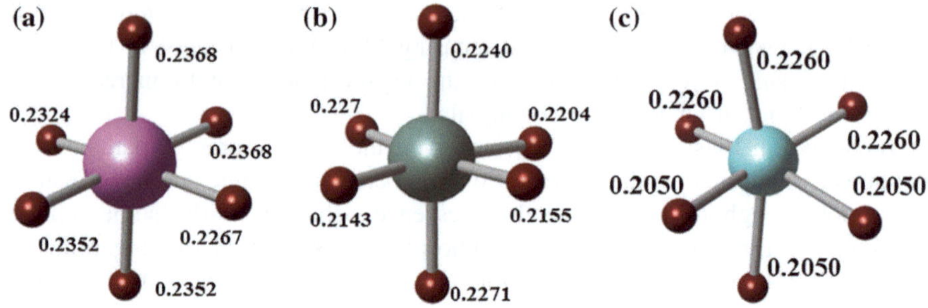

Fig. 7.24 Coordination environment of MO_6 octahedra of **a** $LiSr_2YO_4$, **b** $Sr_6Y_2Al_4O_{15}$, and **c** $Ba_3Sc_4O_9$

the luminescence behavior of all Eu^{2+}- or Ce^{3+}-activated phosphor materials. Therefore, further studies on the crystallographic environment of dopant sites are required to clearly determine luminescence behavior and to develop novel Eu^{2+}- or Ce^{3+}-activated phosphors for white LEDs with high performance.

References

1. Tamatani M (2006) Principal phosphor materials and their optical properties. In: Yen WM, Shionoya S, Yamamoto H (eds) Phosphor handbook, 2nd edn. CRC Press, Boca Raton, FL, pp 153–176
2. Dorenbos P (2006) Fundamentals of luminescence. In: Yen WM, Shionoya S, Yamamoto H (eds) Phosphor handbook, 2nd edn. CRC Press, Boca Raton, FL, pp 139–154
3. Henderson B, Imbush GG (1989) Optical spectroscopy of inorganic solids. Clarendon, Oxford, pp 67–79
4. Van Uitert L (1984) An empirical relation fitting the position in energy of the lower d-band edge for Eu^{2+} or Ce^{3+} in various compounds. J Lumin 29:1–9
5. Tada K, Kawakami Y, Kousaka S, Ito Y, Komeno A, Uematsu K, Sato M (2006) New silicate phosphors for a white LED. IEICE Trans Electron E89-C: 1406–1412
6. Nakano T, Kawakami Y, Uematsu K, Ishigaki T, Toda K, Sato M (2009) Novel Ba-Sc-Si-oxide and oxynitride phosphors for white LED. J Lumin 129:1654–1657
7. Bian L, Zhou TL, Yang JJ, Song Z, Liu QL (2012) Crystal structure and photoluminescence of $(Ba_{1-x-y}Sr_yEu_x)_9Sc_2Si_6O_{24}$. J Lumin 132:2541–2545
8. Bian L, Du F, Yang S, Ren Q, Liu QL (2013) Crystal structure and near-ultraviolet photoluminescence properties of $Ba_9Sc_2Si_6O_{24}:Ce^{3+}$, Na^+. J Lumin 137:168–172
9. Lee S, Park S (2013) Preparation and luminescent properties of Tb^{3+} and Tb^{3+}-Ce^{3+} doped $Ba_9Y_2Si_6O_{24}$ phosphors. J Lumin 143:215–218
10. Brgoch J, Borg CKH, Denault KA, DenBaars SP, Seshadri R (2013) Tuning luminescent properties through solid-solution in $(Ba_{1-x}Sr_x)_9Sc_2Si_6O_{24}:Ce^{3+}$, Li^+. Solid State Sci 18:149–154
11. Kim Y, Park S (2013) Preparation and luminescent properties of Eu-substituted barium-yttrium orthosilicate phosphors. Opt Mater 36:458–462
12. Kim Y, Park S (2014) Eu^{2+}, Mn^{2+} co-doped $Ba_9Y_2Si_6O_{24}$ phosphors based on near-UV-excitable LED lights. Mater Res Bull 49:469–474

13. Bian L, Wang T, Liu S, Yang SS, Liu QL (2015) The crystal structure and luminescence of phosphor $Ba_9Sc_2Si_6O_{24}:Eu^{2+}$, Mn^{2+} for white light emitting diode. Mater Res Bull 64:279–282
14. Wang LH, Schneemeyer LF, Cava RJ, Siegrist T (1994) A new barium scandium silicate: $Ba_9Sc_2(SiO_4)_6$. J Solid State Chem 113:211–214
15. Hasegawa Y, Tamura S, Imanaka N, Adachi G, Takano Y, Sekizawa K (2004) Trivalent praseodymium ion conducting solid electrolyte composite with NASICON type structure. J Alloys Compd 375:212–216
16. Fields JM, Dear PS, Brown JJ (1972) Phase equilibria in the system BaO-SrO-SiO_2. J Am Ceram Soc 55:585–588
17. Catti M, Gazzoni G, Ivaldi G (1983) Structures of twinned β-Sr_2SiO_4 and of α-$Sr_{1.9}Ba_{0.1}SiO_4$. Acta Cryst C 39:29–34
18. Barbier J, Hyde BG (1985) The structures of the polymorphs of dicalcium silicate, Ca_2SiO_4. Acta Cryst B 41:383–390
19. Barry TL (1968) Fluorescence of Eu^{2+}-activated phases in binary alkaline earth orthosilicate systems. J Electrochem Soc 115:1181–1184
20. Shannon RD (1976) Revised effective ionic radii and systematic studies of interatomic distances in halides and chalcogenides. Acta Cryst A 32:751–767
21. Kim JS, Jeon PE, Choi JC, Park HL (2005) Emission color variation of $M_2SiO_4:Eu^{2+}$ (M = Ba, Sr, Ca) phosphors for light-emitting diode. Solid State Commun 133:187–190
22. Zhang M, Wang J, Zhang Q, Ding W, Su Q (2007) Optical properties of $Ba_2SiO_4:Eu^{2+}$ phosphor for green light-emitting diode (LED). Mater Res Bill 42:33–39
23. Hasegawa S, Kim SW, Uematsu K, Ishigaki K, Toda K, Sato M, Masaki T, Yoon DH, Koide J, Toda M, Kudo Y (2014) Growth and structural analysis of silicate phosphor single crystal using gas phase method. In: Abstract of 226th meeting of The Electrochemical Society, Cancun, Mexico, 5–10 October 2014
24. Pardha Saradhi M, Varadaraju UV (2006) Photoluminescence studies on Eu^{2+}-activated Li_2SrSiO_4—a potential orange-yellow phosphor for solid-state lighting. Chem Mater 18:5267–5272
25. Zhang X, He H, Li Z, Yu T, Zou Z (2008) Photoluminescence studies on Eu^{2+} and Ce^{3+} -doped Li_2SrSiO_4. J Lumin 128:1876–1879
26. He H, Fu RL, Wang H, Song X, Pan Z, Zhao X, Zhang X, Cao Y (2008) $Li_2SrSiO_4:Eu^{2+}$ phosphor prepared by the pechini method and its application in white light emitting diode. J Mater Res 23:3288–3294
27. Kulshreshtha C, Sharma AK, Sohn KS (2009) Effect of local structure on the luminescence of $Li_2(Sr, Ca, Ba)SiO_4:Eu^{2+}$. J Electrochem Soc 156:J52–J56
28. He H, Fu R, Cao Y, Song X, Pan Z, Zhao X, Xiao Q, Li R (2010) $Ce^{3+} \rightarrow Eu^{2+}$ energy transfer mechanism in the $Li_2SrSiO_4:Eu^{2+}$, Ce^{3+} phosphor. Opt Mater 32:632–636
29. Levshov SM, Berezovskaya IV, Efryushina NP, Zadneprovskii BI, Dotsenko VP (2011) Synthesis and luminescence properties of Eu^{2+}-doped Li_2SrSiO_4. Inorg Mater 47:285–289
30. Haferkorn B, Meyer G (1998) Li_2EuSiO_4, an europium (II) litho-silicate:Eu[$(Li_2Si)O_4$]. Z Anorg Allg Chem 624:1079–1081
31. Dorenbos P (2000) The 5d level positions of the trivalent lanthanides in inorganic compounds. J Lumin 91:155–176
32. Yavetskiy RP, Dolzhenkova EF, Tolmachev AV, Parkhomenko SV, Baumer VN, Prosvirnin AL (2007) Radiation defects in $SrB_4O_7:Eu^{2+}$ crystals. J Alloys Compd 441:202–205
33. Yao SS, Xue LH, Yan YW (2011) Concentration quenching of Eu^{2+} in $Ba_2Mg(BO_3)_2:Eu^{2+}$ phosphor. Current Appl Phys 11:639–642
34. Wang R, Xu J, Chen C (2012) Luminescent characteristics of $Sr_2B_2O_5:Tb^{3+}$, Li+green phosphor. Mater Lett 68:307–309
35. Li X, Liu C, Guan L, Wei W, Yang Z, Guo Q, Fu G (2012) An ideal blue $Sr_3B_2O_6:Ce^{3+}$ phosphor prepared by sol-combustion method. Mater Lett 87:121–123

36. Zhang X, Chen Y, Zhou L, Gong M (2013) NUV chip based white LED using thermal stable Eu^{2+}-activated phosphate and borate phosphor. Mater Lett 93:390–392
37. Chang CK, Chen TM (2007) $Sr_3B_2O_6$:Ce^{3+}, Eu^{2+}: A potential single-phased white-emitting borate phosphor for ultraviolet light-emitting diodes. Appl Phys Lett 91:081902
38. Song WS, Kim YS, Yang H (2009) Yellow-emitting phosphor of $Sr_3B_2O_6$:Eu^{2+} for application to white light-emitting diodes. Mater Chem Phys 117:500–503
39. Xue Y, Xu X, Hu L, Fan Y, Li X, Li J, Mo Z, Tang C (2011) Synthesis and photoluminescence characteristics of $(Sr, Ca)_3B_2O_6$: Eu for application in white light-emitting diodes. J Lumin 131:2016–2020
40. Fan L, Zhao X, Zhang S, Ding Y, Li Z, Zou Z (2013) Enhanced luminescence intensity of $Sr_3B_2O_6$:Eu^{2+} phosphor prepared by sol-gel method. J Alloys Compd 579:432–437
41. Song Y, Liu Q, Zhang X, Fang X, Cui T (2013) The effect of Eu^{2+} doping concentration on luminescence properties of $Sr_3B_2O_6$:Eu^{2+} yellow phosphor. Mater Res Bull 48:3687–3690
42. Sekar S, Arunkumar P, Jeyakumar D, Lakshminarasimhan (2015) White Light emission in alkali metal ion co-doped single host lattice phosphor $Sr_3B_2O_6$:Ce^{3+}, Eu^{2+}, A^+[A = Li, Na, and K]. Ceram Int 41:3497–3501
43. Richter VL, Müller F (1980) On the structure of $Sr_3B_2O_6$. Z Anorg Allg Chem 467:123–125
44. Vegas A (1985) New description of the $Ca_3(BO_3)_2$ structure. Acta Cryst C 41:1689–1690
45. Schipper WJ, van der Voort D, van den Berg P, Vroon ZAEP, Blasse G (1993) The luminescence of europium in strontium borates. Mater Chem Phys 33:311–317
46. Huang CH, Chen TM, Liu WR, Chiu YC, Yeh YT, Jang SM (2010) A single-phased emission-tunable phosphor $Ca_9Y(PO_4)_7$:Eu^{2+}, Mn^{2+} with efficient energy transfer for white-light-emitting diodes. ACS Appl Mater Interf 2:259–264
47. Huang CH, Chen TM (2010) $Ca_9La(PO_4)_7$: Eu^{2+}, Mn^{2+}: an emission-tunable phosphor through efficient energy transfer for white light-emitting diodes. Opt Exp 18:5089–5099
48. Pang R, Li C, Shi L, Su Q (2009) A novel blue-emitting long-lasting proyphosphate phosphor $Sr_2P_2O_7$:Eu^{2+}, Y^{3+}. J Phys Chem Solids 70:303–306
49. Zhang M, Wang J, Ding W, Zhang Q, Su Q (2007) A novel white light-emitting diode (w-LED) fabricated with $Sr_6BP_5O_{20}$: Eu^{2+} phosphor. Appl Phys B 86:647–651
50. Kim SW, Hasegawa T, Ishigaki T, Uematsu K, Toda K, Sato M (2013) Efficient red emission of blue-light excitable new structure type $NaMgPO_4$:Eu^{2+} phosphor. ECS Solid State Lett 2: R49–R51
51. Yang Z, Yang G, Wang S, Tian J, Li X, Guo Q, Fu G (2008) A novel green-emitting phosphor $NaCaPO_4$:Eu^{2+} for white LEDs. Mater Lett 62:1884–1886
52. Zhang S, Huang Y, Seo HJ (2010) Luminescence properties and structure of Eu^{2+} doped $KMgPO_4$ phosphor. Opt Mater 32:1545–1548
53. Shinde KN, Dhoble SJ, Kumar A (2011) Photoluminescence studies of $NaCaPO_4$:RE (RE = Dy^{3+}, Mn^{2+} or Gd^{3+}). Phys B 406:94–99
54. Morimoto H, Awano H, Terashima J, Nakanishi S, Hirama Y, Ishikawa K, Tobishima S (2012) Charge-discharge properties of $LiCoO_2$ electrodes modified by olivine-type compounds of $LiMgPO_4$ for lithium secondary batteries. J Powder Sources 211:66–70
55. Tang W, Zheng Y (2009) Synthesis and luminescence properties of a novel blue emitting phosphor $NaMgPO_4$:Eu^{2+}. Luminescence 25:364–366
56. Tang W, Chen D, Yang H (2011) Luminescence characteristics of energy transfer between Ce^{3+} and Eu^{2+} in $NaMgPO_4$ phoshor. Appl Phys A 103:263–266
57. Park WJ, Jung MK, Kang SM, Masaki T, Yoon DH (2008) Synthesis and photoluminescence characterization of $Ca_3Si_2O_7$:Eu^{2+} as a potential green-emitting white LED phosphor. J Phys Chem Solids 69:1505–1508
58. Qian F, Fu R, Agathopoulos S, Gu X, Song X (2012) Synthesis and luminescence properties of a broad-band red phosphor $Ca_3Si_2O_7$:Eu^{2+} for warm white light-emitting diodes. J Lumin 132:71–75
59. Zhang X, Lu Z, Meng F, Lu F, Hu L, Xu X, Tang C (2012) A yellow-emitting $Ca_3Si_2O_7$:Eu^{2+} phosphor for white LEDs. Mater Lett 66:16–18

60. Ju H, Wang L, Wang B, Ma Y, Wang H, Chen S, Tao X (2013) Single-phased emission-tunable $Ca_3Si_2O_7$:Ce^{3+}, Eu^{2+} phosphors for white light-emitting diodes. Ceram Int 39:8001–8005
61. Hasegawa T, Kim SW, Ishigaki T, Uematsu K, Takaba H, Toda K, Sato M (2014) Novel reddish yellow-emitting Ce^{3+}-doped $Ba_3Sc_4O_9$ phosphors for blue-light-based white LEDs. Chem Lett 43:828–830
62. Spitsyn VI, Kovba LM, Peromova MV, Yudinskaya IV, Prokof'eva IG (1968) Dokl Akad Nauk SSSR 180:879
63. Kawano Y, Kim SW, Ishigaki T, Uematsu K, Toda K, Takaba H, Sato M (2014) Site engineering concenpt of Ce^{3+}-activated novel orange-red emission oxide phosphors. Opt Mater Express 4:1770–1774
64. Wang CH, Guo DF, Li ZF, Wang XM, Lin JH, Zeng ZZ, Jing XP (2012) Crystal structure of $Sr_6Y_2Al_4O_{15}$: XRD refinements and first-principle calculations. J Solid State Chem 192:195–200
65. Kubota S, Suzuyama Y, Yamane H, Shimada M (1998) Luminescence properties of $LiSr_2Y_{1-x}Ln_xO_4$ (Ln = Eu, Tb, Tm) ($0 \leq x \leq 1$). J Alloys Compd 268:66–71
66. Hasegawa T, Kawano Y, Kumagai S, Muto M, Kim SW, Uematsu K, Toda K, Sato M (2015) Design and synthesis of Ce^{3+}-activated longer wavelength side emission phosphor in oxide crystal. In: Abstract of the 32th annual meeting of the rare earth society of Japan, Kagoshima, Japan, 21–22 May 2015
67. Li YQ, Wang YH, Xu XH, Yu G, Zhang F (2010) Photoluminescence properties and valence stability of Eu in $CaMgSiO_4$. J Electrochem Soc 157:J39–J43
68. Xia Z, Zhang Y, Molokeev MS, Atuchin VV (2013) Structural and luminescence properties of yellow-emitting $NaScSi_2O_6$:Eu^{2+} phosphor: Eu^{2+} site preference analysis and generation of red emission by codoping Mn^{2+} for white-light-emitting diode applications. J Phys Chem C 117:20847–20854
69. Abe T, Kim SW, Ishigaki T, Uematsu K, Sato M, Masaki T, Yoon DH, Toda K (2014) Synthesis and the luminescent properties of $NaAlSiO_4$:Eu^{2+} phosphor using SiO powder as a silica source. J. Inf. Disp 15:31–35
70. Abe T, Kim SW, Ishigaki T, Uematsu K, Toda K, Sato M (2014) Synthesis of new green-emitting $KBa_{1-x}ScSi_3O_9$:Eu_x^{2+} phosphors for white LEDs. Opt Mater 38:57–60
71. Bhatkar VB, Omanwar SK, Moharil SV (2007) Combustion synthesis of Silicate phosphors. Opt Mater 29:1066–1070
72. Yu J, Zhang X, Hao Z, Luo Y, Wang X, Zhang J (2012) Blue emission of $Sr_{2-x}Ca_xP_2O_7$:Eu^{2+} for near UV excitation. J Alloy Compd 515:39–43
73. Sun J, Shen G, Wang X, Shen D (2013) $KBaBP_2O_8$:Eu^{2+}: A novel near-ultraviolet converting blue phosphor for white light-emitting diodes. Mater Lett 93:169–171
74. Huang CH, Liu WR, Chen TM (2010) Single-phased white-light phosphors $Ca_9Gd(PO_4)_7$: Eu^{2+}, Mn^{2+} under near-ultraviolet excitation. J Phys Chem C 114:18698–18701
75. Li C, Dai J, Deng D, Shen C, Xu S (2015) Synthesis, luminescent properties and white light emitting diode application of $Ba_7Zr(PO_4)_6$:Eu^{2+} yellow-emitting phosphor. Phys B 475:105–109
76. Ju G, Hu Y, Chen L, Wang X, Mu Z (2013) The influence of auxiliary codopants on persistent phosphor $Sr_2P_2O_7$:Eu^{2+}, R^{3+} (R = Y, La, Ce, Gd, Tb and Lu). Mater Res Bill 48:4743–4748
77. Poort SHM, Janssen W, Blasse G (1997) Optical properties of Eu^{2+}-activated orthosilicates and orthophosphate. J Alloys Compd 260:93–97
78. Lan L, Feng H, Tang Y, Tang W (2011) Novel blue-emitting phosphor $NaMg_4(PO_4)_3$:Eu^{2+}, Ce^{2+} with energy transfer. Opt Mater 34:175–178
79. Zhang Z, Feng J, Huang Z (2010) Synthesis and characterization of $BaMgAl_{10}O_{17}$:Eu^{2+} phosphor prepared by homogeneous precipitation. Particuology 8:473–476
80. Singh V, Gundu Rao TK, Zhu JJ (2008) A rapid combustion process for the preparation of $MgSrAl_{10}O_{17}$:Eu^{2+} phosphor and related luminescence and defect investigations. J Lumin 128:583–588

81. Zhong R, Zhang J, Zhang X, Lu S, Wang XJ (2006) Energy transfer and red phosphorescence in strontium aluminates co-coped with Ce^{3+}, Eu^{2+} and Dy^{3+}. J Lumin 119–120:327–331
82. Li G, Wang Y, Zeng W, Han S, Chen W, Li Y, Li H (2014) Enhanced long persistence of $LiSr_4(BO_3)_3$:Eu^{2+} orange phosphors by co-doping with Dy^{3+}. Opt Mater 36:1808–1813
83. Yamashita N (1994) Photoluminescence spectra of the Eu^{2+} center in SrO:Eu. J Lumin 62:187
84. Marí B, Singh KC, Moya M, Singh I, Om H, Chand S (2012) Characterization and photoluminescence properties of some CaO, SrO and $CaSrO_2$ phosphors co-doped with Eu^{3+} and alkali metal ions. Opt Mater 34:1267–1271
85. Xie RJ, Hirosaki N, Takeda T, Suehiro T (2013) On the performance enhancement of nitride phosphors as spectral conversion materials in solid state lighting. ECS J Solid State Sci Tech 2:R3031–R3040

Chapter 8
Oxide Phosphors

Weidong Zhuang, Yunsheng Hu, Ronghui Liu and Jiyou Zhong

Abstract In this chapter, the structures and luminescent properties of some widely used practical oxide phosphors for three-band fluorescent lamps, plasma displays, and white light-emitting diodes are briefly introduced. The problems of preparation, luminescent properties, and applications of these phosphors in industry are presented. Many efforts have been made to improve the performance of these phosphors mainly by composition adjustment, surface modification, and improvements in preparation methods. However, some of these problems in engineering still remain; therefore, developing new phosphors with superior stabilities and efficiency to replace traditional phosphors with poor performance is an urgent demand, and some newly developed promising phosphors are briefly introduced. Finally, future developments of oxide phosphors are briefly discussed.

8.1 Introduction

Various phosphors have been developed for different application systems, wherein the oxide phosphors are particularly noticed and mostly widely used because of their excellent properties such as high luminescence efficiency and thermal stability as well as the superiority of simple preparation methods and inexpensive raw materials. In this chapter, oxide phosphors for lighting and display will be introduced, and more attention will be focused on the performance improvement of practical oxide phosphors for three-band fluorescent lamps, plasma displays, and white light-emitting diodes. Due to the space limitation, the aim of this chapter is not to discuss all of the works involved in the progress of oxide phosphors but to introduce the efforts made by our research group in the General Research Institute for Nonferrous Metals, China.

W. Zhuang (✉) · Y. Hu · R. Liu · J. Zhong
General Research Institute for Nonferrous Metals, Beijing, China
e-mail: wdzhuang@126.com

R.S. Liu (ed.), *Phosphors, Up Conversion Nano Particles, Quantum Dots and Their Applications*, DOI 10.1007/978-3-662-52771-9_8

8.2 Phosphors for Three-Band Fluorescent Lamps

Nearly all the practical phosphors for three-band fluorescent lamps are either oxides or oxyacid salts (collectively called “oxides”), which can be efficiently excited by 254-nm ultraviolet light generated by low-pressure mercury vapor discharge. Those phosphors include simple oxide, aluminate, phosphate and borate, etc. Simple oxide combined with aluminate phosphors are the most widely phosphors used in three-band fluorescent lamps mainly due to their high emission efficiency as well as physical and chemical stability.

8.2.1 Simple Oxide Red-Emitting Phosphor

Here, “simple oxide phosphor” refers to Y_2O_3:Eu^{3+} red-emitting phosphor, which is the only available red-emitting oxide phosphor for three-band fluorescent lamps.

Y_2O_3:Eu^{3+}

Crystal structure: The crystal structure of Y_2O_3 belongs to a cubic system, and the Y^{3+} ions have two sites, which are sixfold coordinated sites with C_2 and S_6 symmetry.

Luminescent Properties: Y_2O_3:Eu^{3+} is an ionic crystal with strong ionic bonds, which could produce a strong crystal field. Perturbation of this crystal field weakens the intrinsic forbidden transitions of the Eu^{3+} ions, which is the main reason for the high luminescence efficiency of this phosphor. The excitation and emission spectra of Y_2O_3:Eu^{3+} are shown in Fig. 8.1. The excitation spectrum of Y_2O_3:Eu^{3+} is made up of a set of *f-f* transition lines and a strong charge transfer band covering 200 to approximately 300 nm, which fits the 254-nm ultraviolet light perfectly. The emission spectrum contains a very narrow and intense emission band centered at

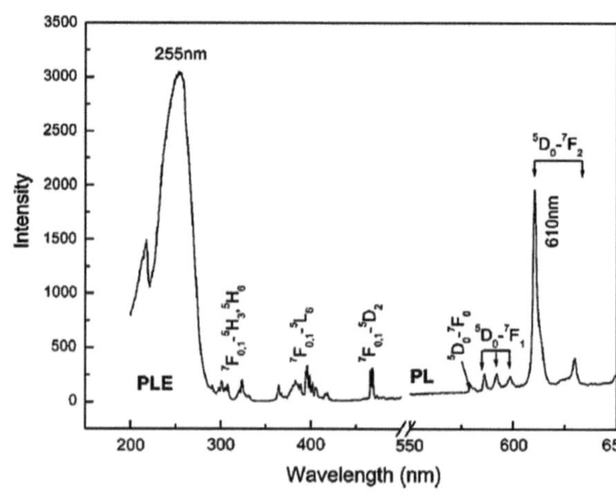

Fig. 8.1 Excitation and emission spectra of Y_2O_3:Eu^{3+}. Reprinted with the permission from Ref. [1]. Copyright 2008 American Chemical Society

610 nm (originating from the $^5D_0 \rightarrow {}^7F_2$ transition of Eu^{3+}) with a full width at half maximum (FWHM) of approximately 5 nm. This phosphor exhibits high color purity, high quantum efficiency (approaching 100 %), excellent thermal-quenching properties, and stable chemical properties.

Improvements: With the purpose of enhancing the emission efficiency, many chemical-composition adjustments were tried such as partially replacing *Y* by other rare-earth elements and adding small amount of Al_2O_3 or SiO_2, but the benefit was not so obvious, so much more attention has been focused on the improvements of crystalline structure, particle size, and morphology, which are important ways to enhance emission efficiency and helpful for the application of this phosphor.

Generally, $Y_2O_3{:}Eu^{3+}$ phosphor is prepared by high-temperature, solid-state reaction method in industry, but the products prepared by this method result in large particle size, broad particle distribution, and uncontrollable surface morphology. To improve this situation, Huang et al. [2–4] proposed a complex molten-salt system (Na_2CO_3 + S + NaCl) to prepare this phosphor based on the fact that the complex molten salt would provide a steady high-temperature liquid sintering environment for phase formation. The phosphor obtained in this way has higher emission intensity, finer and more uniform particles than the commercial phosphor, and a more spherical morphology. Ye et al. [5] successfully prepared the ultra-fine (nano- and sub-micron) $Y_2O_3{:}Eu^{3+}$ phosphor by a modified solution-combustion method. Ju et al. [6] made several physical treatments to improve the luminescent properties of commercial $Y_2O_3{:}Eu^{3+}$ phosphor.

Additionally, $Y_2O_3{:}Eu^{3+}$ is also an important phosphor used in the field emission display. However, $Y_2O_3{:}Eu^{3+}$ is an insulator and does not emit secondary electron when excited by low-energy electron beam. Thus, Hu et al. [7] improved the conductivity of $Y_2O_3{:}Eu^{3+}$ phosphor by coating with In_2O_3 on the surface, and made it more suitable for field emission display applications.

8.2.2 Aluminate Phosphors

The series of aluminate phosphors were first introduced for use in three-band fluorescent lamps in 1974, and even today they are still the most widely used due to their excellent stability and high luminous efficiency. Here, "aluminate phosphors" mainly refer to (Ce, Tb)$MgAl_{11}O_{19}$ (CTMA) green-emitting phosphor and $BaMgAl_{10}O_{17}{:}Eu^{2+}$ (BAM) blue-emitting phosphor. Generally, aluminate phosphors require high-temperature firing for synthesis; thus, many efforts have been made to reduce the synthesis temperature.

(Ce, Tb)$MgAl_{11}O_{19}$

Crystal structure: (Ce, Tb)$MgAl_{11}O_{19}$ belongs to a hexagonal system and has the same magnetoplumbite structure as the $PbFe_{12}O_{19}$ crystal, which has the symmetry of β-alumina compounds.

Luminescent Properties: The excitation and emission spectra of (Ce, Tb) $MgAl_{11}O_{19}$ are shown in Fig. 8.2. The emission spectrum shows the typical

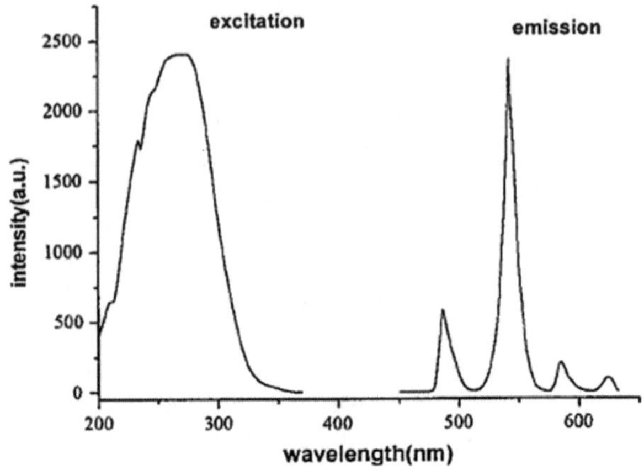

Fig. 8.2 Excitation and emission spectra of (Ce, Tb) $MgAl_{11}O_{19}$. Reproduced from Ref. [9] by permission of John Wiley & Sons Ltd

transitions of Tb^{3+} with a dominant emission peak at 545 nm. The excitation spectrum contains an intense and broad excitation band (with peak at approximately 285 nm) corresponding to Ce^{3+} absorption and several very weak absorption lines of Tb^{3+}; this implies that the energy absorbed by the Ce^{3+} ions can be efficiently transferred to Tb^{3+} ions. The energy-transfer mechanism were widely considered as multipolar interactions, mainly because several excitation lines (transitions of $^7F_6 \rightarrow {}^5G_2$, $^5D_1$, $^5H_1$) of Tb^{3+} locate near the emission peak of Ce^{3+} in this phosphor. Sun et al. [8] studied the luminescent properties of $CeMgAl_{11}O_{19}$, $TbMgAl_{11}O_{19}$, and (Ce, Tb)$MgAl_{11}O_{19}$ and proposed that the energy-transfer mechanism is dominated by resonance. The quantum efficiency of this phosphor has been improved as high as 97 %. The high durability and luminescent efficiency make it widely used as green-emitting phosphor for three-band fluorescent lamps.

Improvements: As mentioned previously, (Ce, Tb)$MgAl_{11}O_{19}$ phosphor has excellent luminescent properties and high durability; thus, much more attention has been focused on preparation technology, reduction of the reaction temperature, and improvements of particle morphology and size distribution etc. Kang et al. [10, 11] proposed that appropriate excessive Al_2O_3 could improve the emission intensity and that MgF_2 could take the place of conventional-flux H_3BO_3 because MgF_2 could stimulate crystal-lattice formation and accelerate grain growth. In addition, Feng et al. [12] studied the effects of adopting MgF_2, H_3BO_3, and MgF_2–H_3BO_3 as flux, respectively. This resulted phosphor, prepared with MgF_2–H_3BO_3 flux, displayed the highest emission intensity, and its brightness reached 167 % of the phosphor prepared without flux.

$BaMgAl_{10}O_{17}$:Eu^{2+}

Crystal structure: $BaMgAl_{10}O_{17}$:Eu^{2+} belongs to a hexagonal system and has a crystal structure similar to β-alumina.

Luminescent Properties: The excitation spectrum of $BaMgAl_{10}O_{17}$:Eu^{2+} has two broad bands in the ultraviolet region with peaks at approximately 270 and 300 nm, respectively. The emission spectrum mainly consists of a broad band with FWHM of approximately 55 nm and the emission peak approximately 450 nm, which corresponds to the $4f^65d \rightarrow 4f^7$ transition of the Eu^{2+} ion.

Improvements: $BaMgAl_{10}O_{17}:Eu^{2+}$ is not only a blue-emitting phosphor for three-band fluorescent lamps but also an important phosphor for plasma display. However, $BaMgAl_{10}O_{17}:Eu^{2+}$ blue-emitting phosphor has severe thermal quenching, and many efforts have been made to reduce its luminescence degradation and quenching.

Deng et al. [13] investigated the luminescent properties of $Ba_{x-0.05}MgAl_{10}O_{16+x}$: $Eu^{2+}_{0.05}$ ($0.88 \leq x \leq 1.10$) phosphors with different Ba^{2+} content, and the excitation and emission spectra are shown in Fig. 8.3. The optimized Ba^{2+} concentration is 0.94. In addition, the effects of a series of fluxes (BaF_2, MgF_2, AlF_3, $BaCl_2$, $MgCl_2$, $AlCl_3$, and H_3BO_3) on luminescent properties were studied [14], and the result showed that the fluoride fluxes have better effects than the chlorides and H_3BO_3. By replacing part of the Ba^{2+} with small amounts of Sr^{2+} or Ca^{2+}, Yu et al. [15] found that replacement of Ba^{2+} with 0.08 mol Sr^{2+} replacing has greater emission efficiency and better thermostability due to the reduction of defect concentration.

Liu et al. [16] investigated the thermal-degradation mechanism, and the results indicate that Eu^{2+} migration and aggregation should be considered for the thermal degradation of BAM. Teng et al. [17] proposed that coating the surface of phosphor with Y_2SiO_5 film could reduce the thermal degradation, but it would decrease the luminescent intensity. To improve the particle characteristics of this phosphor, Zhang et al. [18] developed the microemulsion method to synthesize the BAM phosphor and investigated the phase behavior in microemulsion system. With this method, the well-crystallized and pure BAM phosphor can be prepared at a relative lower temperature of approximately 1300 °C.

8.2.3 Phosphate Phosphors

Phosphate compounds are attractive to be used as hosts of phosphors because they are stable in composition, have inexpensive raw materials, and can be easily synthesized. Until today, various phosphate phosphors have been developed. In

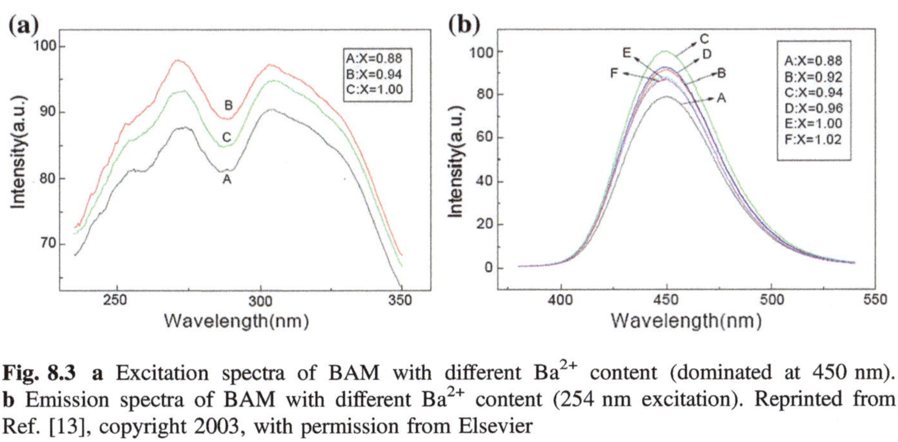

Fig. 8.3 **a** Excitation spectra of BAM with different Ba^{2+} content (dominated at 450 nm). **b** Emission spectra of BAM with different Ba^{2+} content (254 nm excitation). Reprinted from Ref. [13], copyright 2003, with permission from Elsevier

particular, (La, Ce, Tb)PO_4 green-emitting phosphate phosphor for three-band lamps was thought to be a competitive material to (Ce, Tb)$MgAl_{11}O_{19}$ aluminate phosphor because it can be combined with Y_2O_3:Eu^{3+} red-emitting phosphor and $Sr_5(PO_4)_3Cl$:Eu^{2+} blue-emitting phosphor to produce excellent white light for three-band lamps. The (La, Ce, Tb)PO_4 phosphate phosphor is introduced in the following text.

(La, Ce, Tb)PO_4

Crystal structure: This (La, Ce, Tb)PO_4 orthophosphate phosphor belongs to the monazite crystal group.

Luminescent Properties: The (La, Tb)PO_4 phosphor has very weak absorption lines at the ultraviolet region, but by adding Ce^{3+} this phosphor results in a remarkable increase in absorption and emission intensity because Ce^{3+} can act as an efficient sensitizer to Tb^{3+}. The absorption, excitation, and emission spectra are shown in Fig. 8.4a. As presented, the absorption and excitation spectrum shows a broad band in UV region with a dominant peak at approximately 280 nm. The emission spectrum remains the characteristic emission peak of Tb^{3+} with a dominant peak at 543 nm. Similar to (Ce, Tb)$MgAl_{11}O_{19}$, energy transfer from Ce^{3+} to Tb^{3+} can be explained in terms of a resonant-transfer process, and the Ce^{3+} concentration may greatly affect the thermal quenching properties, i.e., the quenching temperature of this phosphor decreases sharply with increasing Ce^{3+} concentration.

Improvements: As is known, the (La, Tb)PO_4 phosphor has many distinct advantages such as lower synthesis temperature and even higher luminescent properties compared with (Ce, Tb)$MgAl_{11}O_{19}$ green-emitting phosphor. However, one of the drawbacks of this phosphor is its relatively high sensitivity to ambient temperature, which may be the main limitation for its applications. To enhance the luminescence and thermal stability, He et al. [19] investigated the effect of alkali carbonate fluxes on the properties of (La, Ce, Tb)PO_4 phosphor. As shown in Fig. 8.4b, the luminance was remarkably enhanced by adding alkali carbonate fluxes. The thermal stability was also improved. This phosphor exhibited the best thermal stability with the addition of Li_2CO_3 as fluxes.

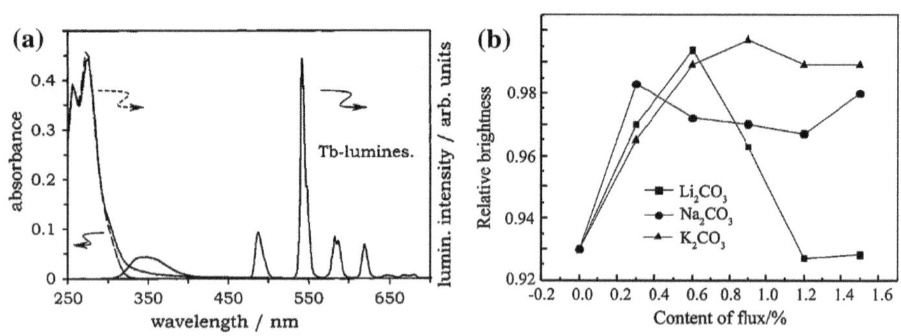

Fig. 8.4 **a** Absorption, excitation and emission spectra of (La, Ce, Tb)PO_4. Reprinted with permission from reference [20]. Copyright 2000 American Chemical Society. **b** Effects of fluxes on luminance of (La, Ce, Tb)PO_4 [19]

8.3 Oxide Phosphors for Plasma Display Panels

Generally, the term "plasma display panel" (PDP) refers to a kind of display device in which the pixels consist of small gas-discharge cells coated with phosphors. The phosphors can be excited by vacuum ultraviolet (VUV) light, which is produced by the gas discharge controlled individually in each cell. Thus, phosphors are very important to realize full-color PDPs. The phosphors for PDPs mainly include simple oxides, aluminates, silicates, and borates, etc. However, the practical phosphors usually require high luminescence efficiency and resistance to VUV radiation. The commonly used phosphors are Y_2O_3:Eu^{3+}, (Y, Gd)BO_3:Eu^{3+} red-emitting phosphors, $BaAl_{12}O_{19}$:Mn^{2+}, Zn_2SiO_4:Mn^{2+} green-emitting phosphors, and $BaMgAl_{10}O_{17}$:Eu^{2+} blue-emitting phosphors.

8.3.1 Aluminate Phosphors

Aluminate phosphors for PDPs include $BaAl_{12}O_{19}$:Mn^{2+} green-emitting phosphor and $BaMgAl_{10}O_{17}$:Eu^{2+} blue-emitting phosphor.

$BaAl_{12}O_{19}$:Mn^{2+}

Crystal structure: $BaAl_{12}O_{19}$:Mn^{2+} has a crystal structure similar to that of hexagonal β-alumina.

Luminescent Properties: $BaAl_{12}O_{19}$:Mn^{2+} has an intense and broad and excitation band in the region of 100 to approximately 200 nm. Under VUV excitation, this phosphor exhibits intense green-light emission with high luminescent color purity, and its emission spectrum shows a narrow band with peak wavelength at approximately 515 nm and FWHM approximately of 30 nm.

Improvements: To enhance the luminescence efficiency, Zhuang et al. [21] investigated the influence of doping with Mg^{2+}, Sr^{2+} and La^{3+} ions on the crystal structure and luminescence of $BaAl_{12}O_{19}$:Mn^{2+} phosphor. The results indicate that Mg^{2+} ions enter the sites of Al^{3+} ions, but Sr^{2+} and La^{3+} ions enter the sites of Ba^{2+} ions. The emission is enhanced with the optimum concentration range of 0.3 to approximately 0.9 mol Mg^{2+}, 0.24 to approximately 0.40 mol Sr^{2+}, and 0.05 to approximately 0.25 mol La^{3+}, respectively. Comparing the effects of doping other ions, doping with Mg^{2+} ion is much more effective. But some reports pointed out that doping with Mg^{2+} ion would result in the luminescence of $BaMgAl_{10}O_{17}$:Mn^{2+}. To enhance the luminescence of $BaMgAl_{10}O_{17}$:Mn^{2+}, Liao et al. [22] studied the effect of P^{5+} doping basing on the fact that the PO_4^{3-} may enhance excitation in the VUV region. The results indicated that the excitation (shown in Fig. 8.5a) is enhanced with doping with an appropriate P^{5+} concentration; the emission intensity (shown as Fig. 8.5b) is greatly improved and exceeds that of commercial Zn_2SiO_4:Mn^{2+} and $BaAl_{12}O_{19}$:Mn^{2+} phosphors. In addition, Xia et al. [23] developed a new green-emitting phosphor—$(Ba_{0.66}Sr_{0.33})$ $Mg_{0.8}Al_{11.47}O_{19}$:$Tm^{3+}_{0.01}$ (BSMA)—for PDPs. The excitation spectrum of BSMA shows broadband excitation with peaks at

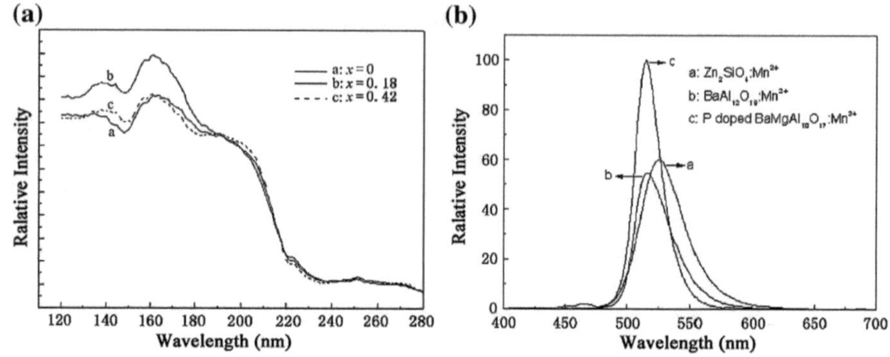

Fig. 8.5 **a** Excitation of $BaMgAl_{10}O_{17}:Mn^{2+}$ with different P^{5+} concentration (x) (λ_{em} = 515 nm); and **b** emission intensity of optimized P^{5+}-doped sample compared with commercial green-emitting phosphors (λ_{ex} = 147 nm) [22]

approximately 153 nm due to the host interband transition and peaks at approximately 186 and 193 nm corresponding to the absorption of Tm^{3+} ions. The emission spectrum shows a sharp line with peak at approximately 516 nm when excited by VUV light.

8.3.2 Silicate Phosphors

The silicate phosphors for PDPs mainly refer to $ZnSiO_4:Mn^{2+}$ green-emitting phosphor and $CaMgSi_2O_6:Eu^{2+}$ blue-emitting phosphor. The luminous efficacy of $CaMgSi_2O_6:Eu^{2+}$ is lower than that of $BaMgAl_{10}O_{17}:Eu^{2+}$ under 173-nm radiation; thus, $CaMgSi_2O_6:Eu^{2+}$ is not used widely. $ZnSiO_4:Mn^{2+}$ green-emitting phosphor is still used mainly because of its high luminous efficiency, inexpensive raw materials, and low synthesis temperature.

$Zn_2SiO_4:Mn^{2+}$

Crystal structure: $Zn_2SiO_4:Mn^{2+}$ is one member of the orthosilicate family belonging to a rhombohedral system.

Luminescent Properties: The excitation band of $Zn_2SiO_4:Mn^{2+}$ is very broad in the region of 100 to approximately 300 nm. Under VUV excitation, the emission spectrum shows an intense green emission with peak at approximately 525 nm, which corresponds to the $^4T_1 \rightarrow {}^6A_1$ transition of the Mn^{2+} ion.

Improvements: Although many series of green-emitting phosphors are available for PDPs, $Zn_2SiO_4:Mn^{2+}$ phosphor shows a more pure and high-efficiency green emission and lower cost among them. However, the long decay time of this phosphor is the biggest problem in application. To make the decay time shorter, higher Mn^{2+} concentrations are adopted at the expense of a decrease in luminous efficiency. In addition, the size and morphology of the particles are important for applications and usually can be improved by controlling the synthetic process. Xia et al. [24] synthesized this phosphor by precipitation-reaction method, and the products exhibit narrow size distribution with an average particle size of approximately 2 μm.

8.3.3 Borate Phosphors

Borates are excellent hosts for PDP phosphors due to their good transparency in the VUV region, and both Eu^{3+}- and Tb^{3+}-activated borate phosphors have high luminescence efficiency and excellent stabilities however; their fatal drawbacks are relative low color purity and nearly unadjustable color coordinates. Typically, the borate phosphors for PDPs mainly refer to the above-mentioned $YBO_3:Eu^{3+}$ red-emitting and $YBO_3:Tb^{3+}$ green-emitting phosphors. Generally, Gd^{3+} ions are used as sensitizer ions to enhance the absorption of light in the VUV region.

(Y, Gd)BO$_3$:Eu^{3+}

Crystal structure: The YBO_3 belongs to hexagonal system with space group $P6_3/m$. The Y^{3+} ions are eightfold coordinated, which are easy to replace by other rare-earth ions with no changes in the structure.

Luminescent Properties: The emission and excitation spectra of (Y, Gd)BO_3:Eu^{3+} are shown in Fig. 8.6a and b. The excitation spectrum in the VUV region mainly consists of two bands with peaks at approximately 166 and 220 nm, respectively, which is most probably due to the host sensitive band of BO_3^{3-} group. By 147-nm excitation, the emission spectrum shows intense orange-red emission with several sharp peaks. The 593-nm dominant emission peak corresponds to the magnetic dipole $^5D_0 \rightarrow {}^7F_1$ transition of Eu^{3+} ions. The other two emission peaks, 611 and 626 nm, are attributes to the electric dipole $^5D_0 \rightarrow {}^7F_2$ transition of Eu^{3+} ions.

Improvements: (Y, Gd)BO_3:Eu^{3+} is the most widely used red-emitting phosphor for PDPs mainly ascribing to its high efficiency, whereas its color purity is not

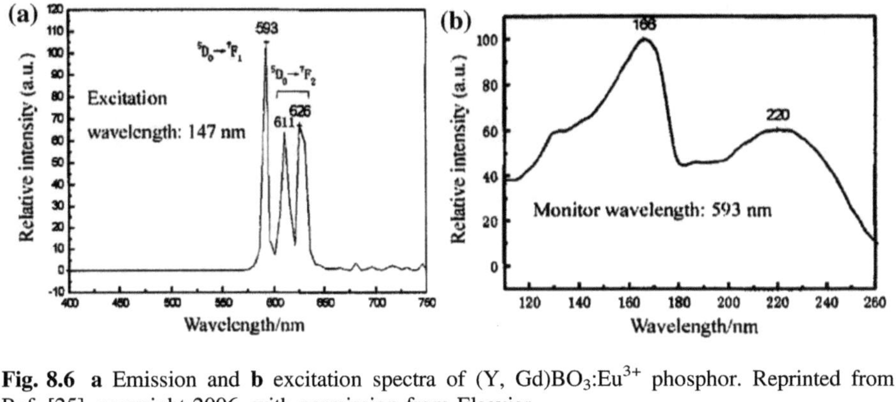

Fig. 8.6 a Emission and **b** excitation spectra of (Y, Gd)BO_3:Eu^{3+} phosphor. Reprinted from Ref. [25], copyright 2006, with permission from Elsevier

satisfactory. Many efforts have been devoted to improve the color purity; however, there is still no effective resolution. In addition, this phosphor is generally prepared by higher-temperature solid-state reaction method, and the resulting products are highly agglomerated and have a heterogeneous morphology and large particle sizes with wide range of distribution. Those disadvantages would lead to further difficulty in a screen-coating process and decrease the efficiency. To improve the particle characteristics of the phosphor, Cui et al. [25–27] and Yu et al. [28, 29] developed some soft chemistry synthesis methods, and high-quality phosphors have been successfully prepared.

(Y, Gd)BO_3:Tb^{3+}

Crystal structure: The crystal structure of (Y, Gd)BO_3:Tb^{3+} is the same as (Y, Gd)BO_3:Eu^{3+}, and it belongs to a hexagonal system.

Luminescent Properties: The excitation spectrum of (Y, Gd)BO_3:Tb^{3+} exhibits a broad band in the rage of 140 to approximately 200 nm, which corresponds to absorption of the BO_3^{3-} group. The emission spectrum shows typical transitions of Tb^{3+} ions with a dominant emission peak at 543 nm.

Improvements: Despite the deviation in chromaticity coordinates compared with those of NTSC green, (Y, Gd)BO_3:Tb^{3+} phosphor, with excellent stability and high luminescence efficiency is used to reduce the firing voltage of PDPs. Similar to (Y, Gd)BO_3:Eu^{3+}, the chromaticity coordinates of (Y, Gd)BO_3:Eu^{3+} are hard to adjust. Thus, much more attention has been focused on enhancing the emission efficiency. Xu et al. [30–32] studied the effect of doping with alkali metal ions on the properties of (Y, Gd)BO_3:Tb^{3+} phosphor. The results indicate that the addition of M_2CO_3 (M = Li, Na, K) can not only enhance the luminescence efficiency, this also promotes the quenching concentration of Tb^{3+}, which is clearly shown in Fig. 8.7a and b. Interestingly, structure analysis showed that the Li^+ ions prefer entering into interstitial sites rather than Y^{3+}-ion sites.

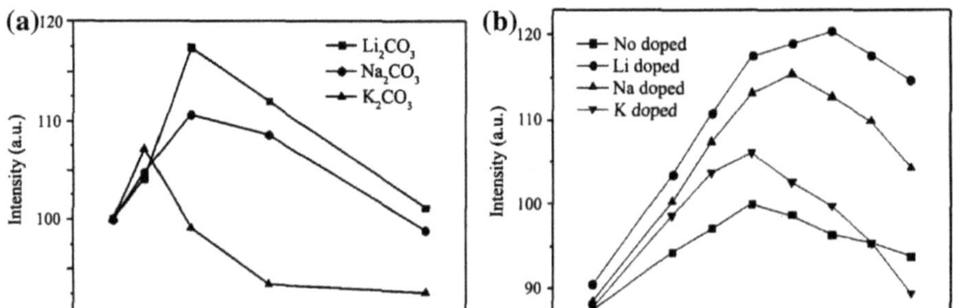

Fig. 8.7 **a** Effects of M_2CO_3 (M = Li, Na, K) on the brightness of (Y, Gd)BO_3:Tb^{3+}. **b** Effects of M_2CO_3 on the quenching concentration of (Y, Gd)BO_3:Tb^{3+} [31]

8.3.4 Other Phosphors

Many phosphors have been developed for PDPs, but those phosphors must be efficiently excited by VUV light, and the color purity, decay time, and stability are also very important parameters for the applications. In addition to the above-introduced phosphors, many other phosphors exist such as $Y(P, V)O_4:Eu^{3+}$ red-emitting phosphor and $Y_2SiO_5:Ce^{3+}$ blue-emitting phosphor, etc., which this chapter does not introduce but can be referred to in other literature.

8.4 Oxide Phosphors for White Light-Emitting Diodes

White light-emitting diodes (w-LEDs) are thought to have brought about a revolution in lighting due to their outstanding advantages such as high efficiency, energy savings, long lifetime, and environmental friendliness. Currently, w-LEDs are commonly generated by the combination of blue LED chips and yellow-emitting phosphors. However, the white light obtained by this approach would result in low color-rendering index and high correlated color temperature due to the red-emission deficiency in the visible spectrum. Generally, red-emitting phosphors are added to make up for red-light deficiency. In addition, w-LEDs fabricated using UV-LEDs, coupled with red-, green-, and blue- emitting phosphors, have the advantages of color stability and excellent color rendering and thus are a significant alternative for achieving high-quality white light. Based on different white light-generating systems, the phosphors developed for w-LEDs mainly include sulfide, oxide, and nitride systems. Aluminate and silicate phosphors are the most important phosphors in the oxide system and are introduced in the following text.

8.4.1 Aluminate Phosphors

Generally, aluminate phosphors with high luminescence efficiency and excellent thermal stabilities have attracted increasing attention. Typically, Ce^{3+}-doped yttrium aluminum garnet $Y_3Al_5O_{12}:Ce^{3+}$ ($YAG:Ce^{3+}$) yellow-emitting phosphor has excellent properties and is widely used to generate w-LEDs.

$Y_3Al_5O_{12}:Ce^{3+}$

Crystal structures: $Y_3Al_2Al_3O_{12}$ has the classic garnet structure represented by the general formula $A_3B_2X_3O_{12}$, where A, B, and X are eight, six, and four coordinated with the surrounding O, thus forming a dodecahedron, an octahedron, and a tetrahedron, respectively. The octahedron and tetrahedron do not share any edge among themselves, but they share edges with at least one dodecahedron.

Luminescent properties: The excitation and emission spectra of $YAG:Ce^{3+}$ are shown in Fig. 8.8. The excitation spectrum of $YAG:Ce^{3+}$ mainly consists of two

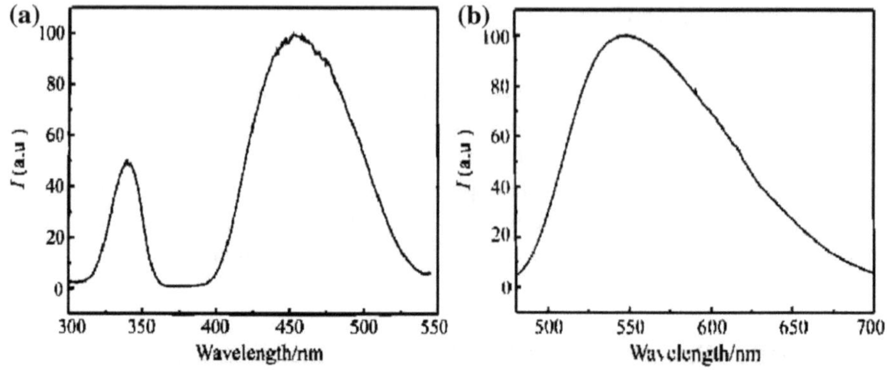

Fig. 8.8 **a** Excitation and **b** emission spectra of $(Y_{2.9}Ce_{0.1})Al_5O_{12}$ phosphor [33]

bands with excitation peaks located at approximately 340 and 460 nm, respectively. The emission spectrum shows a very broad band with emission peaks at approximately 535 nm and FWHM of approximately 100 nm.

Improvements: The external quantum efficiency of $YAG{:}Ce^{3+}$ is approximately 90 %, and the thermal-quenching properties are very good in the phosphors for w-LEDs. Especially, the strong absorption of blue light emitted by the blue LED chip made $YAG{:}Ce^{3+}$ one of the most important and widely used phosphors for white LEDs. However, $YAG{:}Ce^{3+}$ is actually a green-yellow-emitting phosphor, the emission spectrum in the red region is very weak, which leads to the production of white light with high color temperature and low color rendering. To improve this situation, La^{3+}, Gd^{3+} replacing part of Y^{3+}, increasing Ce^{3+} concentration, Si–N replacing part of Al–O, or even Mg–Si substituting Al–Al are adopted to turn the emission spectrum shift toward the longer-wavelength side, but then the quantum efficiency as well as the thermal-quenching properties decrease rapidly. In addition, for the purpose of obtaining green-emitting phosphors for w-LEDs, Lu^{3+} substituting Y^{3+} or Ga^{3+} replacing Al^{3+} are used to shift the emission spectrum toward the shorter-wavelength side.

Generally, $YAG{:}Ce^{3+}$ is synthesized by high-temperature solid-state reaction method in industry. The synthesis temperature is as high as 1600 °C. However, such a high temperature usually results in large particle size as well as nonhomogeneous and irregular morphology, which is harmful for luminescent properties and their applications. To improve this situation, Zhang et al. [34] investigated the influence of various flux on the properties of $YAG{:}Ce^{3+}$, and the results showed that the addition of BaF_2 and H_3BO_3 as flux is helpful to improve the luminescence efficiency and obtain a narrow particle-size distribution. In addition, Zhang et al. [35], Gao et al. [36], and Hou et al. [37] applied several soft chemical synthesis methods to lower the synthesis temperature and obtain high-quality $YAG{:}Ce^{3+}$ phosphors.

8.4.2 Silicate Phosphors

Here silicate phosphors for w-LEDs mainly refer to Eu^{2+}-activated metasilicate and orthosilicate and Ce^{3+}-activated garnet-type silicate.

Sr_3SiO_5:Eu^{2+}

Crystal structure: Sr_3SiO_5:Eu^{2+} has a space group of P4/ncc belonging to a tetragonal system. There are two kinds of sites for Sr^{2+} ions, and both of them are coordinated by six O^{2-} forming distorted octahedrals.

Luminescent properties: The excitation spectrum shows an intense and very broad band covering the spectral region of 250–550 nm. This phosphor can be efficiently excited by 440–480 nm of blue light, thus making it very suitable for white LEDs. The emission spectrum contains a broad emission band centered at approximately 580 nm with FWHM of approximately 90 nm.

Improvements: Sr_3SiO_5:Eu^{2+} is actually a yellow-orange-emitting phosphor. By combining this phosphor with blue-LED chips, amber lamps can be obtained, which emit very low color temperature and soft white light. The quantum efficiency of this phosphor can be improved up to 78 %; however, it then has poor thermal-quenching properties, which limit its applications. Generally, the thermal-quenching properties can be improved significantly by partial substitution of Ba^{2+} for Sr^{2+} or post-treatment using physical method. For this host matrix system, Ma et al. [38] introduced Al–F to partly replace Si–O, and the stability of the phosphor was improved to some extent and the emission spectrum shifted to shorter wavelength as well.

Sr_2SiO_4:Eu^{2+}

Crystal structure: Sr_2SiO_4 has two types of crystal structures, which are orthorhombic corresponding to α' phase and monoclinic corresponding to β phase. The α' and β phases of Sr_2SiO_4 are isostructural with Ba_2SiO_4 and β-Ca_2SiO_4, respectively. Generally, to stabilize the α'-Sr_2SiO_4 phase, Ba^{2+} ions are introduced to replace part of the Sr^{2+} ions. For α'-Sr_2SiO_4, the Sr^{2+} ions have two kinds of sites, Sr(1) and Sr(2), which coordinated by 10 and 9 oxygen atoms, respectively.

Luminescent properties: The excitation and emission spectra of Sr_2SiO_4:Eu^{2+} are shown in Fig. 8.9. The excitation spectrum exhibits a broad band covering the spectral region of 200–500 nm with an excitation peak at approximately 387 nm. Under blue or UV light excitation, the emission spectrum showed a broad band with an emission peak at approximately 560 nm.

Improvements: The luminescence efficiency of Sr_2SiO_4:Eu^{2+} is comparable to that of YAG:Ce^{3+} when combined with blue LED chips, although its thermal stability and structure stability pose limitations for its application. To improve the luminescent properties and thermal stabilities, Mg^{2+} and Ba^{2+} ions are generally used to replace part of the Sr^{2+} ions [39]. In addition, physical improvements are also important ways to improve the stability. To improve the particle performance, Hu et al. [40] successfully developed (Sr, Eu)CO_3@SiO_2 core shell-like precursor

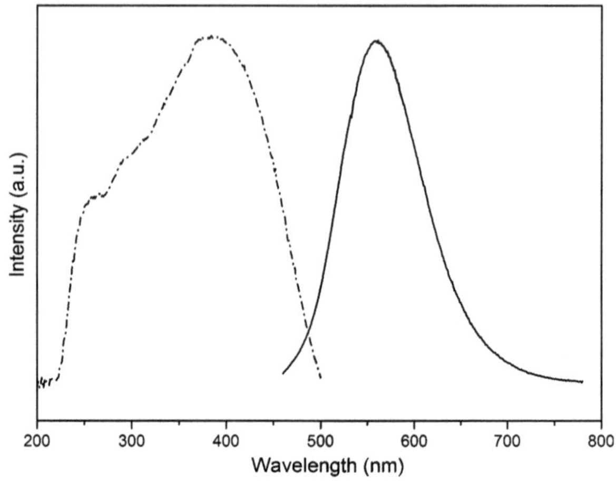

Fig. 8.9 Excitation and emission spectra of Sr_2SiO_4: Eu^{2+} phosphor (provided by Yunsheng Hu et al. with permission)

to prepare pure phosphor with small particle size, narrow distribution, and good morphology.

$Ca_3Sc_2Si_3O_{12}$:Ce^{3+}

Crystal structures: $Ca_3Sc_2Si_3O_{12}$ has the same crystal structure with $Y_3Al_5O_{12}$, corresponding to the general formula of $A_3B_2X_3O_{12}$. This means that Ca^{2+}, Sc^{3+}, and Si^{4+} ions occupy the dodecahedral, octahedral, and tetrahedral sites and are coordinated to eight, six, and four O^{2-}, respectively.

Luminescent properties: The excitation and emission spectra of $Ca_3Sc_2Si_3O_{12}$: Ce^{3+} are shown in Fig. 8.10. The excitation spectrum consists of several bands covering the range of 200 to approximately 500 nm with a dominant peak at approximately 455 nm, which is very suitable for blue LED chips. The emission spectrum presents one band with a dominant peak at approximately 505 nm.

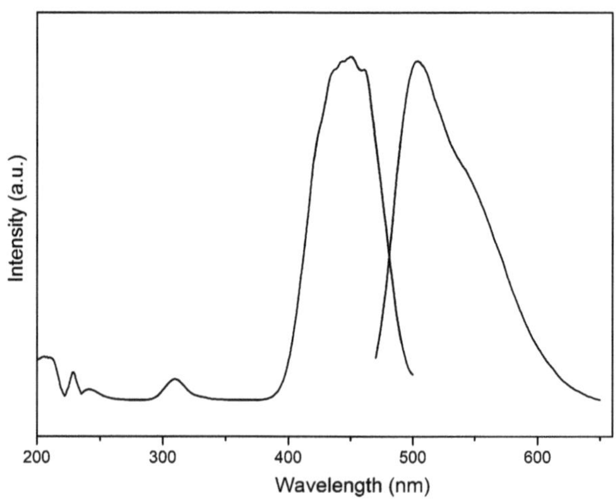

Fig. 8.10 Excitation and emission spectra of $Ca_3Sc_2Si_3O_{12}$:Ce^{3+} phosphor (provided by Yuanhong Liu, Wei Gao et al. with permission)

Improvements: The $Ca_3Sc_2Si_3O_{12}$:Ce^{3+} phosphor has high quantum efficiency and thermal stability. Especially, a broad and intense excitation band is located at the blue region, which is very suitable for the use as a green-emitting phosphor in white LEDs. However, it is difficult to avoid the Sc_2O_3 impurity phase when the phosphor is synthesized by high-temperature solid-state reaction method. As is known, Sc_2O_3 is an expensive raw material that is difficult to separate from the garnet phase. To solve this problem, Liu et al. [41–45] successfully applied gel-combustion method to obtain single-phase $Ca_3Sc_2Si_3O_{12}$:Ce^{3+} phosphor. In addition, to lower the consumption of Sc_2O_3 raw material, Mg^{2+} ion was introduced to replace part of Sc^{3+} ions, and the spectrum can be adjusted to shift toward the longer-wavelength side.

8.4.3 Other Phosphors

Many other phosphors have been developed for w-LEDs, but the phosphors with garnet, β-K_2SO_4 structures have received much more attention due to their similar structures with the efficient YAG:Ce^{3+} and orthosilicate phosphors. However, the luminescence efficiency of these phosphors reported by now is far less than those of YAG:Ce^{3+} and orthosilicate phosphors. Here, another promising system, molybdate phosphor, which was first developed by Hu et al. [46], is introduced in the following text.

$CaMoO_4$:Eu^{3+}

Crystal structures: $CaMoO_4$ has a tetragonal crystal structure with a space group of $I4_1/a$, Ca^{2+} ion with only one crystallography site, which can be occupied by Eu^{3+} ions.

Luminescent properties: The excitation and emission spectra of $CaMoO_4$:Eu^{3+} phosphor are shown in Fig. 8.11. The excitation spectrum consists of a broad band

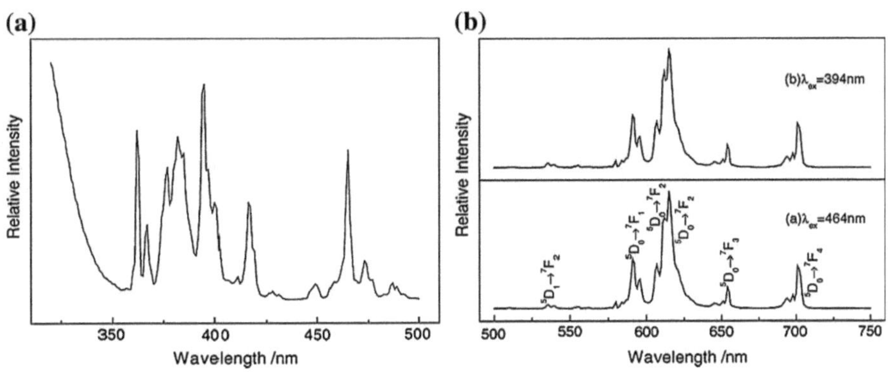

Fig. 8.11 **a** Excitation and **b** emission spectra of $CaMoO_4$:Eu^{3+} phosphor. Reprinted from Ref. [46], copyright 2005, with permission from Elsevier

and some sharp lines, corresponding to the charge-transfer band and transitions of Eu^{3+} ions, respectively. The phosphor can be well excited by ultraviolet (394 nm) and visible light (464 nm), which is nicely in agreement with the widely applied UV-LED or blue LED chips. The emission spectrum is composed of several sharp emission lines with a dominant peak at 612 nm. The emission intensity and stability of $CaMoO_4:Eu^{3+}$ phosphor is much better than those of commercial sulfide red-emitting phosphors.

Improvements: This phosphor shows intense luminescence when excited by 394- or 464-nm light, whereas the narrow excitation lines pose limitations for its application in w-LEDs. To broaden the excitation region, Zhao et al. [47] introduced Zn^{2+} and Sm^{3+} into this phosphor, and the excitation extended into the UV region. In addition, Hu et al. [48] investigated the luminescent properties of $CaMoO_4:Sm^{3+}$ and disclosed the energy transfer of $Sm^{3+} \rightarrow Eu^{3+}$.

8.5 Summary

This chapter focuses on the improvement of practical phosphors and the development of new phosphors in oxide systems. The problems of some practical phosphors are presented: Some of them can be solved by different means, whereas others still remain. To develop high-quality and inexpensive phosphors is an eternal pursuit for phosphor researchers, but the mechanism of investigation should receive more attention. Because there are almost no perfect phosphors, adjustments and improvements should be made according to the related mechanism. Finally, the relationship between structures and properties should be emphasized, which is helpful in understanding the luminescence mechanism and developing new materials.

Acknowledgments This work was financially supported by the Ministry of Science and Technology of China and the National Natural Science Foundation of China.

References

1. Li JG, Li X, Sun X, Ishigaki T (2008) Monodispersed colloidal spheres for uniform $Y_2O_3:Eu^{3+}$ red-phosphor particles and greatly enhanced luminescence by simultaneous Gd^{3+} doping. J Phys Chem C 112:11707
2. Huang Y, Hu YS, Teng XM, Long Z, Ye HQ, Zhuang WD (2007) Influence of molten salt on luminescent intensity and particle size of $Y_2O_3:Eu^{3+}$ phosphor. J Rare Earths 25:697
3. Huang Y, Ye HQ, Zhuang WD, Hu YS, Zhao CL, Li C, Guo SX (2007) Preparation of $Y_2O_3:Eu^{3+}$ phosphor by molten salt assisted method. Trans Nonferrous Met Soc China 17:644
4. Huang Y, Ye HQ, Zhuang WD, Hu YS (2007) Preparation of $Y_2O_3:Eu^{3+}$ red phosphor by firing assisted with (Na_2CO_3 + S+NaCl) compound molten salt. J Funct Mater 38:1056

5. Ye XY, Zhuang WD, Hu YS, He T, Huang XW, Liao CF, Zhong SW, Xu ZF, Nie HP, Deng GF (2009) Preparation, characterization, and optical properties of nano-and submicron-sized Y_2O_3:Eu^{3+} phosphors. J Appl Phys 105:064302
6. Ju RL, Hu XS, Lv ZG, Fang Y, Cui S, Yao P, Wang H, Zhuang WD (2007) Effect of physical treatment on luminescent properties of rare earth phosphor Y_2O_3:Eu^{3+}. J Chin Rare Earths Soc 25:412
7. Hu XS, Huang XW, Hu YS, Zhuang WD (2007) Research on Y_2O_3:Eu^{3+} phosphor coated with In_2O_3. J Rare Earths 25:11
8. Sun YM, Teng XM, Zhuang WD, Fang Y (2005) Luminescence and energy transfer in (Ce, Tb)$MgAl_{11}O_{19}$. J Anhui Normal Univ (Natural Science) 28:56
9. Zhang J, Zhang Z, Tang Z, Lin Y (2002) Synthesis and characterization of $(Ce_{0.67}Tb_{0.33})$ $Mn_xMg_{1-x}Al_{11}O_{19}$ phosphors derived by sol-gel processing. J Am Ceram Soc 85:998
10. Kang K, Zhuang WD, He DW, Deng CY, Huang XW (2004) Luminescent properties and preparation of $CeMgAl_{11}O_{19}$:Tb green phosphor. J Rare Earths Soc 22:206
11. Kang K, Zhuang WD, He DW, Deng CY, Huang XW (2004) Study on effect of flux on $CeMgAl_{11}O_{19}$:Tb^{3+} phosphor. J Rare Earths 22:114
12. Feng ZY, Zhuang WD, Huang XW, Wen XF, Hu YS (2010) Effect of MgF_2-H_3BO_3 flux on the properties of (Ce, Tb) $MgAl_{11}O_{19}$ phosphor. Rare Earths 28:351
13. Deng CY, Zhuang WD, He DW, Wang YS, Kang K, Huang XW (2004) Luminescence of $Ba_{x-0.05}MgAl_{10}O_{16+x}$:$Eu^{2+}_{0.05}$ with different Ba^{2+} content. Phys B 344:470
14. Deng CY, Zhuang WD, He DW, Wang YS, Kang K, Huang XW (2004) Preparation and luminescent properties of BAM blue phosphor for PDP and CCFL. J Rare Earths 22:108
15. Yu ZJ, Zhuang WD, Zhao CL, Chang YF, Huang XW (2001) Study on barium magnesium aluminate blue phosphors doped with Sr^{2+} and Ca^{2+}. J Rare Earths Soc 19:590
16. Liu RH, Zhuang WD, Huang XW, Hu YS, He HQ (2009) Thermal degradation properties of barium magnesium aluminate blue phosphor. J Chin Rare Earths Soc 27:178
17. Teng XM, Zhuang WD, Huang XW, Cui XZ, Zhang SS (2006) Luminescent properties of $BaMgAl_{10}O_{17}$:Eu^{2+} phosphors coated with Y_2SiO_5. J Rare Earths 24:143
18. Zhang XY, Zhuang WD, Cui XZ, He HQ, Huang XW (2006) Preparation of $BaMgAl_{10}O_{17}$:Eu^{2+} and its phase behavior in microemulsion system. J Rare Earths 24:736
19. He HQ, Zhao K, Hu XS, Zhang SS, Hu YS, Zhuang WD (2010) Effect of alkali carbonate fluxes on properties of (La, Ce, Tb)PO_4 green phosphor. Chin J Rare Met 34:717
20. Riwotzki K, Meyssamy H, Kornowski A, Haase M (2000) Liquid-phase synthesis of doped nanoparticles: colloids of luminescing $LaPO_4$:Eu and $CePO_4$:Tb particles with a narrow particle size distribution. J Phys Chem B 104:2824
21. Zhuang WD, Cui XZ, Yu ZJ, Huang XW, He HQ, Sheng ZK, Han JX (2001) Effect of doping on crystal structure and luminescence of $BaAl_{12}O_{19}$:Mn phosphor. J Chin Rare Earths Soc 19:586
22. Liao QR, Zhuang WD, Xia T, Liu RH, Teng XM, Liu YH (2009) Effect of phosphorus—doping on optical properties of $BaMgAl_{10}O_{17}$: Mn^{2+}. Acta Phys Sinica 58:2776
23. Xia T, Zhuang WD, Cui XZ, Zhao CL, Teng XM, Huang XW (2006) Low temperature luminescence properties of Tm^{3+} doped aluminate phosphor. J Rare Earths 24:141
24. Xia T, Zhuang WD, Cui XZ, Zhao CL, Huang XW (2007) Precipitation synthesis and low temperature luminescence prosperities of Zn_2SiO_4:Mn^{2+} green phosphor. Proc Asia Display 2:1723
25. Cui XZ, Zhuang WD, Zhang XY, Xia T, Long Z, Yu ZJ, Zhao CL, Huang XW (2006) Red emitting phosphor (Y, Gd)BO_3:Eu^{3+} for PDP prepared by complex method. J Rare Earths 24:149
26. Cui XZ, Zhuang WD, Yu ZJ, Xia T, Huang XW, Li HW (2008) Preparation of red phosphor (Y, Gd)BO_3:Eu by soft chemistry methods. J Alloys Compd 451:280
27. Cui XZ, Zhuang WD, Zhang XY, Xia T, Long Z, Yu ZJ, Zhao CL, Huang XW (2006) Synthesis of spherical (Y, Gd)BO_3:Eu^{3+} phosphor using W/O emulsion system. J Rare Earths 24:719

28. Yu ZJ, Huang XW, Zhuang WD, Cui XZ, Li HW (2005) Crystal structure transformation and luminescent behavior of the red phosphor for plasma display panels. J Alloys Compd 390:220
29. Yu ZJ, Huang XW, Zhuang WD, Cui XZ, He HQ, Li HW (2004) Co-precipitation preparation and luminescent behavior of (Y, Gd) BO_3:Eu phosphor. J Rare Earths 22:829
30. Xu HB, Zhuang WD, Wen XF, Liu RH, Hu YS, Xia T (2010) Effect of Li + Ions doping on structure and luminescence of (Y, Gd)BO3:Tb3+. J Rare Earths 28:701
31. Xu HB, Zhuang WD, Xia T, Liu RH, Tao Y, Huang Y (2010) Synthesis and properties of PDP green phosphor (Y, Gd)BO_3 Tb^{3+}. Chin J Rare Met 34:887
32. Xu HB, Zhuang WD, Hu YS, Xia T, Wen XF, Liu RH (2010) Influence of alkaline metal ions doping on properties of (Y, Gd)BO_3:Tb. J Chin Rare Earths Soc 28:268
33. Zhang SS, Zhuang WD, Zhao CL, Hu YS, He HQ, Huang XW (2004) Study on (Y, $Gd)_3$(Al, $Ga)_5O_{12}$:Ce^{3+} phosphor. J Rare Earths 22:118
34. Zhang SS, Zhuang WD, Zhao CL, He HQ, Huang XW (2002) Influence of flux on properties of $Y_3Al_5O_{12}$:Ce phosphor. J Chin Rare Earths Soc 20:605
35. Zhang SS, Zhuang WD, He T, Liu YH, Liu RH, Gao WG, Hu YS, Long Z (2010) Study on co-precipitation synthesized $Y_3Al_5O_{12}$:Ce yellow phosphor for white LED. J Rare Earths 28:713
36. Gao WG, Hu YS, Zhuang WD, Zhang SS, Liu YH, He HQ (2009) A Novel Method for the synthesis of YAG:Ce phosphor. J Rare Earths 27:886
37. Hou YK, Huang XW, Liu YH, Zhuang WD, Long Z, Hu YS (2008) Synthesis of rare earth oxide precursors and study on its effect on YAG:Ce phosphor. J Chin Rare Earths Soc 26:609
38. Ma XL, Zhuang WD, Guo HJ, Liu RH, Liu YH, He HQ, Peng P, Chen L (2013) Synthesis and luminescence properties of $Sr_{3-z}(Al_xSi_{1-x})O_{5-x}F_x$:z$Ce^{3+}$ phosphors. J Rare Earths 31:665
39. Fang Y, Xu HY, Ma HL, Zhuang WD (2009) Luminescent Properties of (Ba, $Sr)_2SiO_4$: Eu^{2+} Phosphor for LED. J Xinyang Normal Univ (Natural Science Edition) 22:256
40. Hu YS, Hao JH, Zhuang WD, Huang XW, He HQ (2011) Synthesis of (Sr,Eu)CO_3@SiO_2 core-shell-like precursor for alkali earth silicate phosphors. J Rare Earths 29:911
41. Liu YH, Hao JH, Zhuang WD, Hu YS (2009) Structural and luminescent properties of gel-combustion synthesized green-emitting $Ca_3Sc_2Si_3O_{12}$:Ce^{3+} phosphor for solid-state lighting. J Phys D Appl Phys 42:245102
42. Liu YH, Zhuang WD, Hu YS, Gao WG (2010) Improved photoluminescence of green-emitting phosphor $Ca_3Sc_2Si_3O_{12}$:Ce^{3+} for white light emitting diodes. J Rare Earths 28:181
43. Liu YH, Zhuang WD, Liu RH, Hu YS, He HQ, Zhang SS, Gao WG (2012) Spectral variations of $Ca_3Sc_2Si_3O_{12}$:Ce phosphors via substitution and energy transfer. J Rare Earths 30:339
44. Liu YH, Zhuang WD, Hu YS, Gao WG, Hao JH (2010) Synthesis and luminescence of sub-micron sized $Ca_3Sc_2Si_3O_{12}$:Ce green phosphors for white light-emitting diode and field-emission display applications. J Alloys Compd 504:488
45. Liu YH, Zhuang WD, Gao WG, Hu YS, He T, He HQ (2010) Effect of H_3BO_3 on preparation and luminescence properties of submicron green-emitting $Ca_3Sc_2Si_3O_{12}$:Ce phosphor. Acta Phys Sinica 59:8200
46. Hu YS, Zhuang WD, Ye HQ, Wang DH, Zhang SS, Huang XW (2005) A novel red phosphor for white light emitting diodes. J Alloys Compd 390:226
47. Zhao CL, Hu YS, Zhuang WD, Huang XW, He T (2009) Luminescence modification of Eu^{3+}-activated molybdate phosphor prepared via co-precipitation. J Rare Earths 27:758
48. Hu YS, Zhuang WD, Ye HQ (2004) Luminescent properties of samarium ion in calcium molybdate. J Rare Earths 22:821

Chapter 9
Categories of Oxide Phosphors

Dae-Ho Yoon and Young-Hyun Song

Abstract White light-emitting diodes (LEDs) are attractive as a solid-state lighting device. Especially, phosphor materials are fundamental component for white-light generation. For the generation of white LEDs, the InGaN blue LED chips and the yellow light from cerium-activated yttrium aluminum garnet phosphor ($Y_3Al_5O_{12}$: Ce^{3+}) have been used. These white LEDs usually have a highly correlated color temperature >5000 K and a low color-rendering index (CRI) value, which produces a cold white light. An alternative is to blend red, green, and blue (RGB) phosphors on a near-UV chips emitting chip. These oxide phosphors play a very important role. At a present time, remote phosphors are used as key materials on InGaN blue LED chips for solving the thermal problem. As the remote phosphors are applied to LEDs, the system efficiency could be increased ≤ 30 %. In this section, we will discuss the oxide phosphor including the new remote phosphor for the generation of white light-emitting diodes LEDs.

9.1 Introduction

Oxide phosphors are attractive as the key materials for the generation of white light with blue LEDs. Many phosphor compounds can be an adoptable host lattice with strong absorption and high quantum efficiency. According to their variable composition, oxide phosphors are divided into the followings categories:

- Garnet-type phosphor
- Silicate phosphor
- Remote phosphor with packaging

D.-H. Yoon (✉) · Y.-H. Song
School of Advanced Materials Science & Engineering,
Sungkyunkwan University, Suwon 440-746, Republic of Korea
e-mail: dhyoon@skku.edu

R.S. Liu (ed.), *Phosphors, Up Conversion Nano Particles, Quantum Dots and Their Applications*, DOI 10.1007/978-3-662-52771-9_9

9.1.1 *Garnet Phosphor*

9.1.1.1 YAG:Ce^{3+}

The composition of $3Y_2O_3$:$5Al_2O_3$ is an yttrium aluminum garnet (YAG) with a cubic garnet structure having a la3d (230) space group and a lattice parameter of 1.201 nm [1]. The garnet structure consist of Y–O dedecahedron, Al–O octahedron, and Al–O tetrahedron [2]. When doped with transition-metal and lanthanide components in the YAG host lattice, it is converted into luminescence materials for a variety of lamps. Generally, YAG ceramic is synthesized by a solid-state reaction method with Y_2O_3 and Al_2O_3 powder at >1600 °C [3]. Figure 9.1 shows the PL properties YAG:Ce^{3+} phosphor. The excitation spectrum mainly consists of two broad bands located at 340 and 460 nm, respectively [4]. The excitation spectrum is assigned to the electronic transition of the ground state of Ce^{3+} ($^2F_{5/2}$) to the crystal field-splitting bands of the excited 5*d* state of the Ce^{3+} ion [5]. It corresponds to the application of blue LEDs with YAG:Ce^{3+} phosphor for white-light generation because of the strong absorption in blue light. Generally, YAG:Ce^{3+} phosphor is a very important phosphor material for white LEDs. The emission spectrum of YAG: Ce^{3+} phosphor indicates the typically broad yellow band with full width at half maximum (FWHM) of 120 nm [6]. The emission light is attributed to the $5d \rightarrow 4f$ energy-level transition [7]. As a function of the host composition, the emission band can be shifted to the blue and red regions [8].

Generally, white LEDs with YAG:Ce^{3+} phosphor have presented a difficulty for the generation of warm white because of the high correlated color temperature (CCT) > 5000 K [9], which is derived from the lack of emission spectra in the red region. In case of the Gd^{3+} ion being substituted in the Y^{3+} site, the mission band is red-shifted [10]. Also, it is possible to tune the emission light to the red region with Pr^{3+} ions in YAG:Ce^{3+} phosphor [11]. It affects the high color-rendering index (CRI) and the low CCT value for the generation of warm white light.

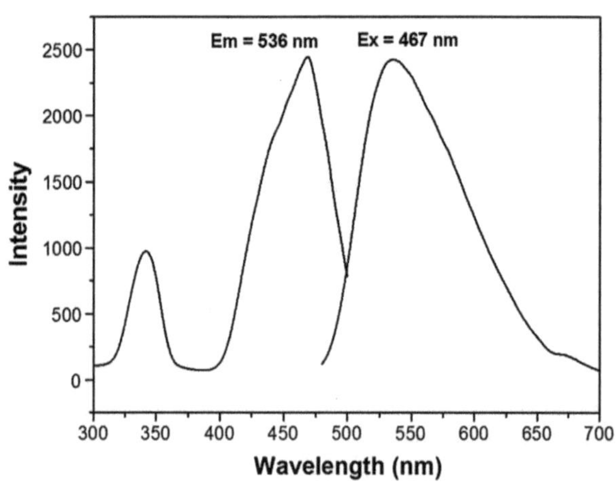

Fig. 9.1 Excitation and emission spectra of $Y_3AL_5O_{12}$:Ce^{3+}

9.1.1.2 $Ca_3Sc_2Si_3O_{12}$:Ce^{3+} Phosphor

Shimomura and Kijima reported a novel green phosphor that can be excited under excitation of the blue region [12]. It is synthesized by a solid-state reaction method at 1450 °C under a reducing atmosphere. The crystal structure of $Ca_3Sc_2Si_3O_{12}$:Ce^{3+} phosphor is a garnet-type structure with a lattice parameter of a = 12.25 Å; the Ca, Sc, and Si ions are occupied by the dodecahedral, octahedral, and tetrahedral sites and coordinated by eight, six, and four oxygen atoms, respectively [13]. Figure 9.2 shows the excitation and emission spectra [13]. The most excitation intensity is located at 400–500 nm, which is applicable to the intensity of white LEDs. It is attributed to the *f–d* electron transition of Ce^{3+} ions [14]. The emission spectra under 450 nm display the apparently broad emission band with a maxima peak at 505 nm, which corresponds to the $5d(^2D) \rightarrow 4f(^2F_{5/2}, {}^2F_{7/2})$ transitions of Ce^{3+} ions [15]. The decay curve of this phosphor is presented in Fig. 9.3 [14]. The

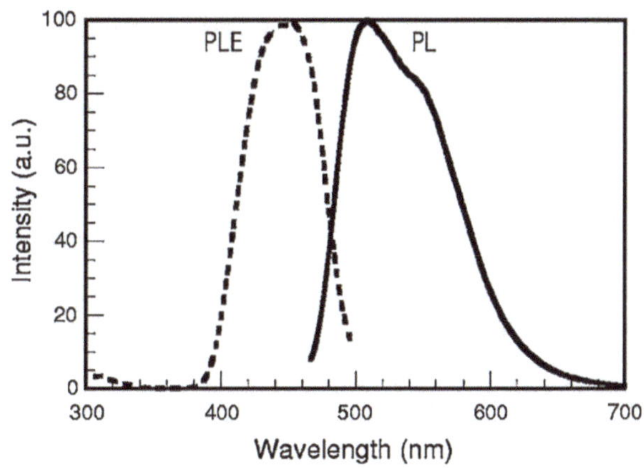

Fig. 9.2 The excitation and emission spectra of $Ca_3Sc_2Si_3O_{12}$:Ce^{3+} phosphor. Reprinted from Ref. [13] by permission of The Electrochemical Society

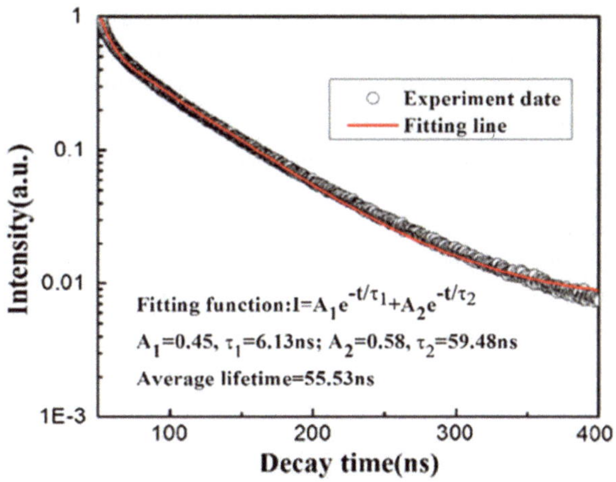

Fig. 9.3 PL decay curve of the transition of Ce^{3+} of annealed phosphors by solid-state reaction method. Reprinted from Ref. [14] by permission of IOP Publishing

plotting of the decay curve can be well-fitted. The average lifetime is 55.53 ns. These values are consistent with that of other Ce^{3+}-doped phosphors (approximately 60 ns).

9.1.1.3 $Ca_2GdZr_2(AlO_4)_3:Ce^{3+}$ Phosphor

The structure of CGZA crystal can be indexed to the cubic system. For the final acceptable refinement, which relates to a lower R factor, the Al^{3+} ions occupy the A sites, the larger Zr^{4+} ions occupy the D sites, and the largest Ca^{2+} and Gd^{3+} ions occupy the L sites with a ratio of 2:1 [16]. The crystallographic data and structure refinement for the CGZA crystal are listed in Table 9.1 [16]. The 24c, 16a, and 24d Wyckoff sites are fully occupied by Ca^{2+}/Gd^{3+}, Zr^{4+}, and Al^{3+}, respectively. Based on the effective ionic radii (r) of the cations, Ce^{3+} (r = 1.14 Å for CN = 8) is expected to preferably occupy the sites of Gd^{3+} [16].

Table 9.1 Crystallographic data and structure refinement for CGZA

Empirical formula	$Ca_2GdZr_2(AlO_4)_3$
Formula wt	692.79
Temp (K)	293 (2)
Wavelength (Å)	0.71073
Crystal system	Cubic
Space group	Ia $\bar{3}$ d
Unit-cell dimensions (Å)	a = 12.5057 (7)
Volume ($Å^3$)	1955.88 (19)
Z	8
Calcd density (g/cm^3)	4.706
Abs coeff. (mm^{-1})	10.184
F(000)	2552
Crystal size (mm)	0.2 × 0.1 × 0.05
è Range for data collection (deg)	3.99–27.5
Limiting indices	$-16 \leq h \leq 15, -16 \leq k \leq 15, -16 \leq l \leq 15$
Reflns collected/unique	6371–192 [R_{int} = 0.0431]
Completeness to = 27.47°	100 %
Maximum and minimum transmission	1.0000 and 0.5992
Abs correction	None
Refinement method	Full-matrix least-squares on F^2
Data/restraints/parameters	192/0/19
GOF on F^2	1.164
Final R indices [I > 2ó(I)]	R_1 = 0.0162, wR_2 = 0.0417
R indices (all data)	R_1 = 0.0162, wR_2 = 0.0417
Extinction coeff.	0.00087 (7)
Largest diff. peak and hole	0.309 and −0.400 e $Å^{-3}$

Figure 9.4 shows the PL and PLE spectra of the $Ca_2GdZr_2(AlO_4)_3:Ce^{3+}$ phosphor [16]. On excitation at 420 nm, the CGZA:Ce^{3+} phosphor exhibits an intense green luminescence, which is assigned to the electric dipole—allowed transition from the lowest level of the 5*d* excited state to the 4*f* ground state of the Ce^{3+} ions [17]. Figure 9.5 shows the normalized PL spectra of CGZA:$_xCe^{3+}$ with $x = 0.01$–0.12 under excitation at 415 nm [16]. The peak of PL band shifts to longer wavelength from 499 nm for $x = 0.01$–514 nm for $x = 0.12$. The emission intensity of samples obtained by integrating the relative emission bands for CGZA:$_xCe^{3+}$ ($x = 0.01$, 0.02, 0.04 ...0.12) measured under the same condition is shown in Fig. 9.5 [16]. It can be found that the maximum emission lies at $x = 0.02$.

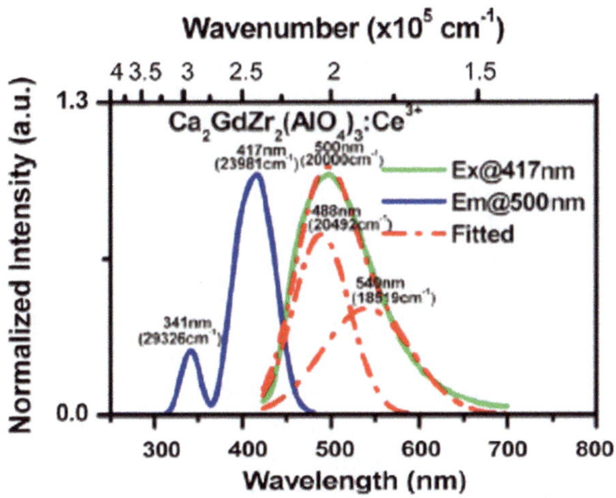

Fig. 9.4 Normalized PL and PLE spectra of $Ca_2GdZr_2(AlO_4)_3:Ce^{3+}$. Reprinted with the permission from Ref. [16]. Copyright 2014 American Chemical Society

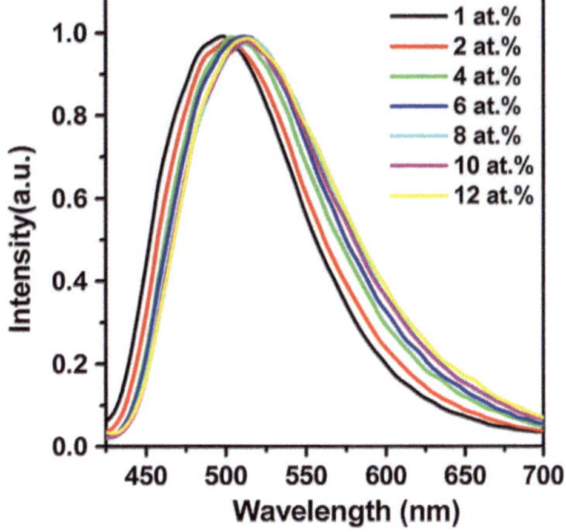

Fig. 9.5 Normalized PL spectra for $Ca_2Gd_{(1-x)}Zr_2(AlO_4)_3:xCe^{3+}$ (x = 0.01, 0.02, 0.04,...0.12). Reprinted with the permission from Ref. [16]. Copyright 2014 American Chemical Society

9.1.2 Silicate Phosphor

9.1.2.1 $NaAlSiO_4:Eu^{2+}$ Phosphor

$NaAlSiO_4$ host compounds with (Al, SiO_4) framework structure have different polymorphic forms including nepheline, low carnegieite, and high carnegieite [18]. The $NaAlSiO_4$ are tectosilicates that consist of the framework of an AlO_4 and SiO_4 tetrahedral with each oxygen atom surrounded by a tetrahedron, thus forming the 3D structure [19]. Nepheline is a stuffed derivative of tridymite, which has an Si:Al ratio of 1:1 and a decrease in symmetry from P63/mmc in tridymite to P63 in nepheline [20, 21]. Figure 9.6 indicates the excitation and emission spectra of $NaAlSiO_4:Eu^{2+}$ phosphor [22]. The typically broad emission band is attributed due to the $4f^65d^1 \rightarrow 4f^7$ transition of Eu^{2+} ions, which exhibits a greenish yellow emission centered at 551 nm [23]. Although the ionic radii (r) of Na^+ (r = 0.118 nm when CN 8, r = 0.124 nm when CN = 9) and Eu^{2+} (r = 0.125 nm when CN 8, r = 0.130 nm when CN = 9) cations differ, the substitution of Na^+ by Eu^{2+} ions will be readily carried out because enough interstices occur among the charge compensation of Al^{3+} and Si^{4+} ions in the aluminosilicate framework [24]. Toda et al. reported a $NaAlSiO_4:Eu^{2+}$ phosphor with an SiO powder as a silica source [24]. The XANES spectra of the $NaAlSiO_4:Eu^{2+}$ phosphor consisted of two strong absorption peaks at 6972 and 6980 eV, which correspond to the electron transitions of Eu^{2+} and Eu^{3+}, respectively [19, 25]. The intensity of the absorption peak (μt) at 6972 eV attributed to the Eu^{2+} ion increased with a small amount of SiO powder as a silica source, whereas that derived from the Eu^{3+} ion decreased [26]. These results indicate that the addition of SiO powder as a silica source in the preparation process significantly enhanced the Eu^{2+} contents of the $NaAlSiO_4:Eu^{2+}$ phosphors [27]. Compared with $YAG:Ce^{3+}$ phosphor, the $NaAlSiO_4:Eu^{2+}$ phosphors with SiO powder as a silica source show 68 % of the commercial $YAG:Ce^{3+}$ phosphor [22, 28]. In addition, this new phosphor is a promising candidate as a white-LED application.

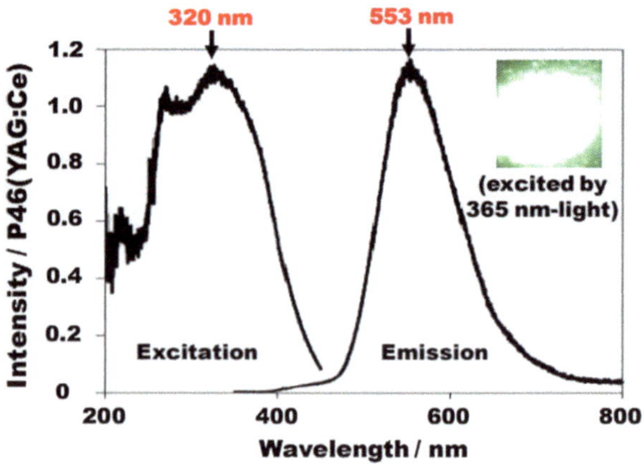

Fig. 9.6 Excitation and emission spectra of the $NaAlSiO_4:Eu^{2+}$. Reprinted from Ref. [22], Copyright © 2013 Masato Kakihana et al.

9.1.2.2 γ-Ca_2SiO_4:Ce^{3+} Phosphor

The Ce^{3+}-activated γ-Ca_2SiO_4 phosphors are an important luminescence material for white light-emitting diodes. Ce^{3+} activated host lattice is available to use the near-UV and visible regions because the Ce^{3+} in a crystal lattice lowers the energy of the 5d levels through to the centroid shift [29]. In yellow-to-green-emitting phosphors, where Ce^{3+} is excited by blue light, the dopant is commonly located on an eight-coordinate site [30]. The γ-phase is the stable, room temperature polymorph of the Ca_2SiO_4 host compounds [31]. However, the metastable β-phase can be stabilized at room temperature with various dopants [32]. Generally, it is in the Ce^{3+} ion–activated host that the β-phase is formed. For the stabilization of olivine-type γ-Ca_2SiO_4, a co-dopant is required, for example, Mg^{2+}, B^{3+}, and Al^{3+} [33]. In accordance with previous literature, γ-phase stabilization requires high synthesis temperatures and slow cooling rates. In the case of γ-Ca_2SiO_4:Ce,Al phosphor, two dopants are involved that appear to have opposing γ/β-stabilizing effects [34]. The PXRD patterns also demonstrate the purity of the γ-phase despite the Ce^{3+} and Al^{3+} doping [35]. Figure 9.7 indicates laboratory PXRD patterns for the yellow and blue Ca_2SiO_4:Ce,Al phosphors, showing the matching reference patterns [29]. The γ-Ca_2SiO_4:Ce,Al yellow phosphor is a promising competitor for YAG:Ce-based phosphors in SSL. The emission-band maximum of the silicate (575 nm) is red-shifted with respect to that of YAG:Ce (approximately 550 nm). In addition, the emission of the former tails into the red region. As a consequence, the emission band of γ-Ca_2SiO_4:Ce, Al can potentially be used to maximize the color rending of white LEDs or afford a warmer white light. The requirement of excess doping levels of Al^{3+} (or Ge^{4+}) undoubtedly causes a decrease in the quantum efficiencies [36]. This problem may be bypassed by modification of the synthetic procedures and exploration of other co-dopants. SSL applications aside, this phosphor, along with the other olivine-type oxide phosphors investigated, provides access to new information regarding Ce^{3+} emission in octahedral sites [37]. Such

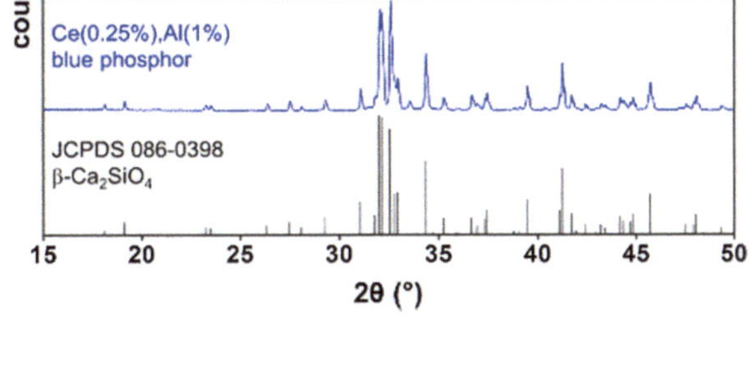

Fig. 9.7 Laboratory PXRD patterns for the yellow and blue Ca_2SiO_4:Ce, Al phosphors showing the matching reference patterns. Reprinted with the permission from Ref. [29]. Copyright 2014 American Chemical Society

information is invaluable for the design and optimization of novel phosphors for SSL. The results presented in this article have raised two important questions regarding the γ-Ca_2SiO_4:Ce,Al phosphor and other Ce^{3+}-doped olivine [29]. Why does the PL of γ-Ca_2SiO_4:Ce,Al differ so greatly from all the other olivine structures, and what doping site(s) does Ce^{3+} occupy in the olivine structure? There appears to be no doubt that these two questions are related. Of special interest is the broad yellow emission of γ-Ca_2SiO_4:Ce,Al that can potentially be used in SSL. This phosphor exhibits strong yellow PL under both blue and near-UV excitations. It is a promising candidate for suitable SSL application.

9.1.2.3 $NaBaScSi_2O_7:Eu^{2+}$ Phosphors

The new green-emitting Eu^{2+} phosphor was reported by Ray in 2012 [38]. In the case of $NaBaScSi_2O_7$ compounds, Na^{2+} and Ba^{2+} cations hold a position in void in an Na–Ba–Na–Ba sequence parallel to the c axis, and they consist of $[ScO_6]$ octahedral and $[SiO_2]$ tetrahedral ones [39]. The versatile structure consist of three types of sites: type 1 with trivalent Sc sites in ScO_6 polyhedral, type 2 with divalent Ba site in BaO_9 polyhedral, and type 3 with monovalent Na site consisting of eightfold coordination [40].

The ionic radii of Sc^{3+}, Ba^{2+}, and Na^+ are 0.74, 1.47, and 1.18 Å, and the ionic radii for the six-, eight-, and nine-coordinated Eu^{2+} are 1.17, 1.25 and 1.30 Å, respectively [41]. To the best our knowledge, regarding the effective ionic radii of cations with different coordination numbers, it is a possible for Eu^{2+}-doping ions to enter into the Sc^{3+}, Ba^{2+}, and Na^+ ion sites [42–44].

As seen in Fig. 9.8, ultraviolet-visible diffuse reflection (UV-Vis) spectra shows a strong absorption band from 275 to 500 nm, which is derived from the $4f \rightarrow 5d$ transition of Eu^{2+} ions [39]. In addition, Fig. 9.8 shows the photoluminescence excitation (PLE) and photoluminescence emission (PL) of $NaBaScSi_2O_7{:}0.1Eu^{2+}$ phosphor. The PLE spectrum indicates the typically broad

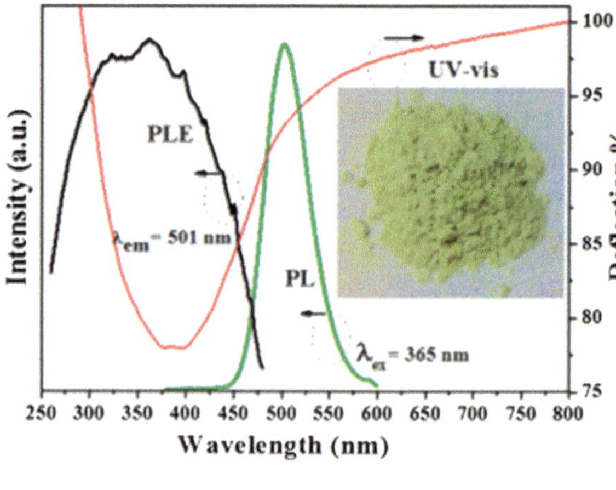

Fig. 9.8 UV-vis diffuse reflectance and the PLE (λ_{em} = 501 nm) and PL (λ_{ex} = 365 nm) spectra of the as-prepared $NaBaScSi_2O_{7:0.1}Eu^{2+}$ phosphor. The inset shows the image of the phosphor in the daylight. Reproduced from Ref. [39] by permission of The Royal Society of Chemistry

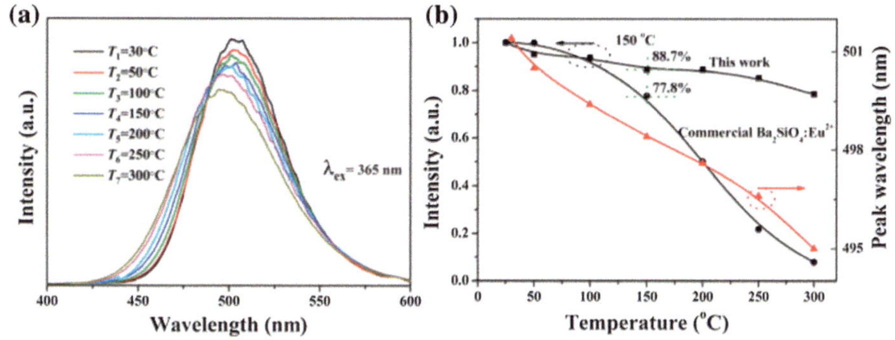

Fig. 9.9 **a** PL spectra (λ_{ex} = 365 nm) of the $NaBaScSi_2O_7{:}_{0.1}Eu^{2+}$ phosphor under different temperatures in the range of 30–300 °C. **b** Variations of the relative emission intensities and emission peaks as a function of temperature of the $NaBaScSi_2O_7{:}_{0.1}Eu^{2+}$ phosphor and data of the commercial phosphor are also given for comparison. Reproduced from Ref. [39] by permission of The Royal Society of Chemistry

band from 250 to 460 nm, which corresponds to the UV-Vis spectra. Moreover, the PL spectrum shows that $NaBaScSi_2O_7{:}0.1Eu^{2+}$ phosphors display the that broad-band emission at 501 nm and FWHM is 55 nm [45]. The inset of Fig. 9.8 shows a photograph of $NaBaScSi_2O_7{:}Eu^{2+}$ phosphors under daylight conditions, which is in agreement with the PL results. It is essential to make a thermally stable phosphor for application in white LEDs during the operation of the device [39]. Figure 9.9 shows the temperature-dependent emission spectra of $NaBaScSi_2O_7{:}Eu^{2+}$ phosphors from room temperature to 300 °C [39]. The PL intensity at 150 °C shows 88.7 % of the initial emission intensity at RT, but the commercial $BaSiO_4{:}Eu^{2+}$ phosphor maintained an emission intensity of 77.8 % at 150 °C [46]. The emission wavelength indicated the blue shift as a function of temperature, which can be explained by the active phonon-assisted excitation in the excited state of Eu^{2+} ions [47]. Figure 9.10 shows the electroluminescence (EL) spectra with the $NaBaScSi_2O_7{:}Eu^{2+}$, $BaMgAl_{10}O_{17}{:}Eu^{2+}$, and $CaAlSiN_3{:}Eu^{2+}$ phosphors as well as a near-UV LED under a driving current of 25 mA [39]. A color-rendering index (Ra) value of 86.5 at a correlated color temperature of 2528 K was found for the generation of warm white LEDs. This phosphor is a promising candidate for near UV-pumped white LEDs.

9.1.3 Remote Phosphor with Packaging

For white-light generation, the yellow-emitting phosphor ($YAG{:}Ce^{3+}$) was mixed with transparent encapsulated resin and then coated on the LED chip. However, the excitation wavelength exists as a function of forward-bias current during the operation of LEDs as well as affects the down-conversion process [48]. In case of

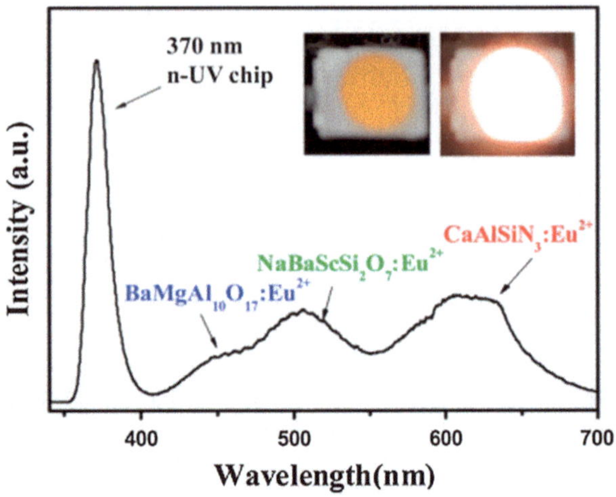

Fig. 9.10 EL spectrum of blending three blue-emitting $BaMgAl_{10}O_{17}:Eu^{2+}$, green-emitting $NaBaScSi_2O_7:Eu^{2+}$, and red-emitting $CaAlSiN_3:Eu^{2+}$ phosphors on a 370 nm-emitting n-UV chip. The *insets* show photos of the LED package. Reproduced from Ref. [39] by permission of The Royal Society of Chemistry

this type of white LEDs, the conversion efficiency suffers from energy loss as well as thermal properties, which are attributed to LED chips [49]. Therefore, the system efficiency of the device is decreased up to 60 % to approximately 70 %. The alternative is using a remote phosphor for solving the problems. Remote-phosphor LEDs, with a phosphor layer far from the chip, have been placed to improve the luminous efficiency of phosphor-converted LEDs [50]. As a results, color- and thermal-stable white LEDs can be achieved with a remote phosphor. In this section, on the basis of remote phosphors, these can be grouped into the following categories:

- Graphene-embedded remote phosphor
- Phosphor in glass
- Phosphor ceramic plate remote phosphor

9.1.3.1 Graphene-Embedded Remote Phosphor

Generally, white LEDs have drawbacks with respect of thermal stress due to the utilization of encapsulates including silicone [50], which brings about the unstable CCT and reduces the color quality. The alternative is a remote phosphor. However, remote phosphors with polymer-based materials also have several problems including thermal stress from heat occurrence during the operation of the device [51, 52]. By increasing the forward bias current, the performance of LEDs can increasingly degrade.

To overcome the thermal stress, instead of using a general remote phosphor, an rGO-embedded remote phosphor is a good candidate [53]. Graphene is a very attractive material because of its high transmittance, excellent thermal and electrical conductivity, and flexibility [54, 55]. In this section, the rGO-embedded remote

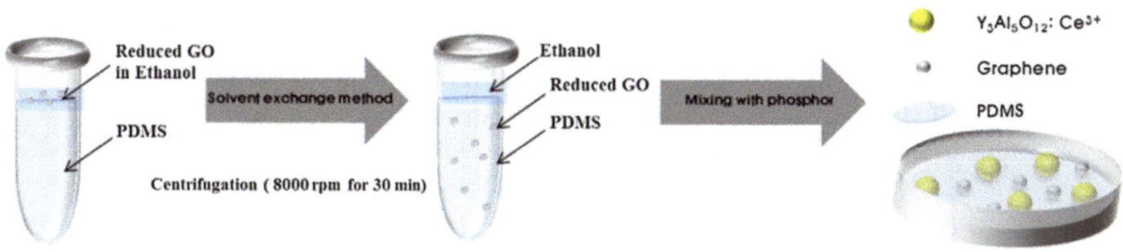

Fig. 9.11 Schematic diagram of rGO-embedded remote phosphor. Reproduced from Ref. [56] by permission of The Royal Society of Chemistry

phosphor is fabricated with polydimethylsiloxane (PDMS). The graphene oxide (GO) was synthesized by a modified Hummers' method, and then rGO was prepared with the addition of hydrazine to graphene oxide solution. This solution was stirred at 90 °C for 4 h. When it turned to black colour, the solution was filtered by vacuum filtration. The rGO-embedded remote phosphor was prepared with a solvent-exchange method as illustrated in Fig. 9.11 [56]. The electroluminescence spectrum was indicated in detail as shown in Fig. 9.12 [56]. The EL spectrum consisted of a 450-nm blue chip with a rGO-embedded remote phosphor. The colour-rendering index (CRI) value under a current forward bias of 20 mA was 64.38, and the QE was 32 %. For the confirmation of stable colour, the CCT was determined from the CIE colour coordinate as seen in Fig. 9.13 [56]. The calculated CCT value of the rGO-embedded remote phosphor was 5592 K, which agrees with daylight (5500–6000 K). In addition, by increasing the forward current bias from 20 to 50 mA, the colour point was shifted upward. It is assumed that the low variation of the colour point indicates a more stable remote phosphor than the conventional remote phosphor.

Fig. 9.12 The EL spectrum consists of a 450-nm blue chip with an rGO-embedded remote phosphor. Reproduced from Ref. [56] by permission of The Royal Society of Chemistry

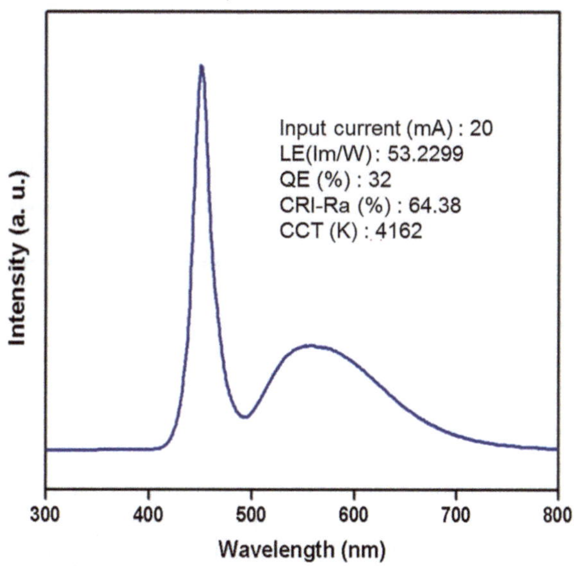

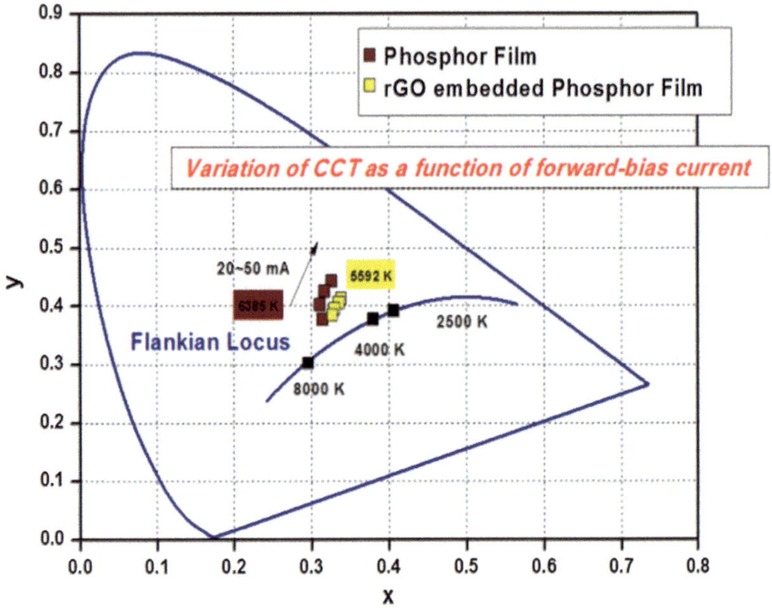

Fig. 9.13 CIE colour coordinate of an rGO-embedded remote phosphor compared with phosphor film. Reproduced from Ref. [56] by permission of The Royal Society of Chemistry

Fig. 9.14 Average temperature an rGO-embedded remote phosphor compared with phosphor film. Reproduced from Ref. [56] by permission of The Royal Society of Chemistry

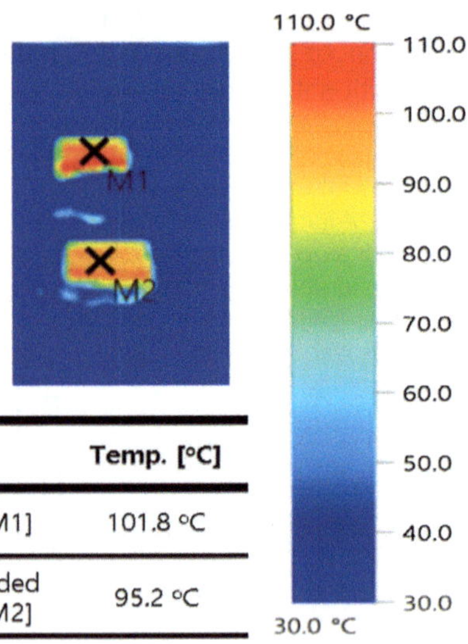

Sample	Temp. [°C]
Phosphor Film [M1]	101.8 °C
Graphene embedded Phosphor Film [M2]	95.2 °C

The temperature dependence of the rGO-embedded remote phosphor was compared with that of the conventional remote phosphor as seen in Fig. 9.14 [56]. The average temperature of the rGO-embedded remote phosphor was 83.2 °C. This indicates that the rGO-embedded remote phosphor is more stable than the conventional one. It is a promising candidate for high-power LEDs with respect to reliability and thermal stability.

9.1.3.2 Phosphor in Glass

Glass powders are used as a medium in which phosphors are spread so they maintain good optical properties of high transparency irrespective to an increase in temperature [57]. To fabricate phosphor-in-glass (PiG), glass powders (or frits), and phosphor powder mixtures are pelletized and sintered at the temperature <1000 °C [58]. The temperature is comparatively lower than that for phosphor ceramic plate (PCP) sintering [59]. Low-temperature process is beneficial to decrease the production cost. Moreover, phosphors can be dispersed in glass phase regardless of what kind of phosphor is used [60]. This enables the achievement of improved color rendering and correlated color temperature (CCT). Ji et al. reported an $Y_3Al_5O_{12}$:Ce^{3+} (YAG:Ce^{3+}) PiG [61]. Figure 9.15 shows the EL spectra of white LEDs of stacked YAG:Ce^{3+} PiG on blue InGaN/GaN blue LED chips of 450 nm [61]. The EL spectra of YAG:Ce^{3+} PiG exhibits a typical Ce^{3+}:5*d*–4*f* broad emission band, which is centered at 545 nm, under the excitation of 450 nm. The luminous efficacy was 78 lm/W, and the color rendering index (CRI) value under the input current of 350 mA was 61. CIE 1931 chromaticity coordinates in Fig. 9.16 show CCT changes as a function of temperature. Average CCT of YAG:Ce^{3+} PiG was approximately 5703 K and ranged from 5640 to 5800 K, whereas CCT changed from 5552 to 5778 K in powder [61]. CCT deviation of PiG has been proven to be smaller than that of powder, which indicates better thermal stability [62].

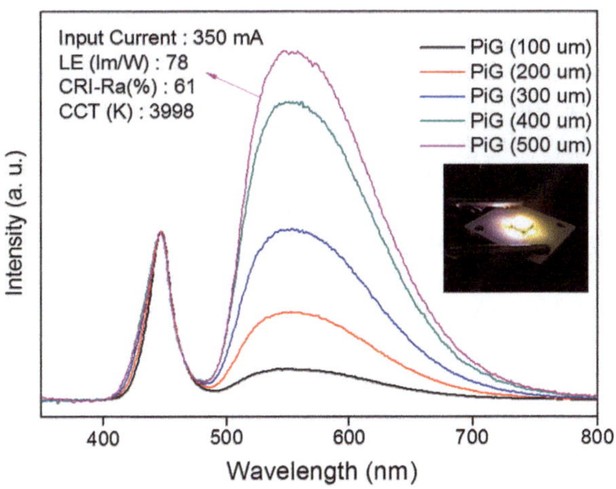

Fig. 9.15 Electroluminescence (EL) spectra of white LEDs stacked PiG using YAG:Ce^{3+} on a blue LED chip of 450 nm [61]

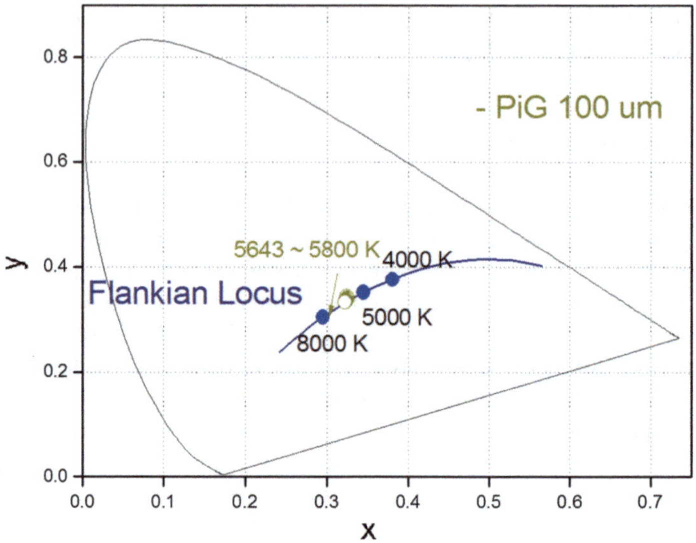

Fig. 9.16 Correlated color temperature (CCT) with color coordinates. Table 9.1 shows the range of CCT changes with temperature changes from 0 to 100 °C [61]

9.1.3.3 Plate Remote Phosphor

Polycrystalline YAG ceramics for application in high-power LEDs are attractive because of their excellent optical and thermal stability [63]. YAG was first doped with a neodymium single crystal was reported and was used as a laser material [64]. At present, polycrystalline YAG:Ce ceramics are placed far from the blue-emitting LEDs, and the optical properties can be tuned by the degree of porosity and grain structure. The advantages of polycrystalline YAG:Ce ceramic compared with silicone-based phosphor is the high thermal conductivity (>5 W/mK), which allows an excellent heat dissipation, thus affecting the optical performance [65]. Polycrystalline YAG:Ce ceramics are fabricated with using the nano-YAG:Ce phosphor in different from bulk YAG:Ce [66]. It is prepared at 1650 °C for 24 h under reducing atmosphere and then polished to the thickness of 100 μm. Figure 9.17 a shows the EL spectra of polycrystalline YAG:Ce ceramic on a blue LED chip (λ = 450 nm, P_{opt} = 1 W) [67]. The luminous efficacy was 74 lm/W under a current of 350 mA, and the CRI value was 73. The input current-luminous flux (I-L) was increased under a current from 10 to 350 mA. The luminous flux increased without any saturation effect because of the conversion of yellow light from blue light. Compared with phosphor-silicone, the I-L properties are increased due to their outstanding physical, chemical, and thermal stability. Temperature-luminous flux (T-L: 0 °C to approximately 100 °C@350 mA) was measured to the thermal stability as seen in Fig. 9.18 [67]. A decrease of T-L

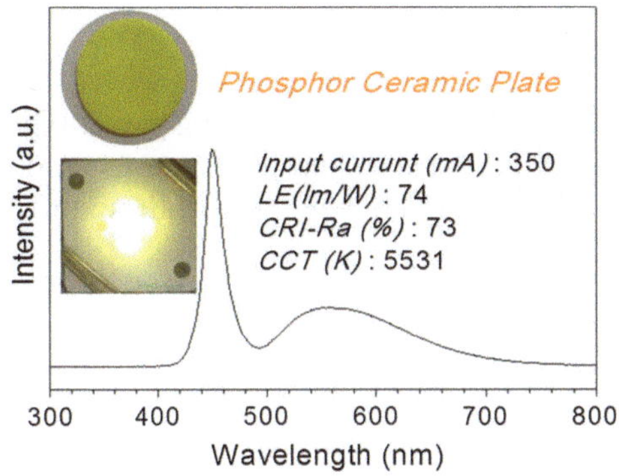

Fig. 9.17 EL spectra of polycrystalline YAG:Ce ceramic on a blue LED chip. Reproduced from Ref. [67] by permission of The Royal Society of Chemistry

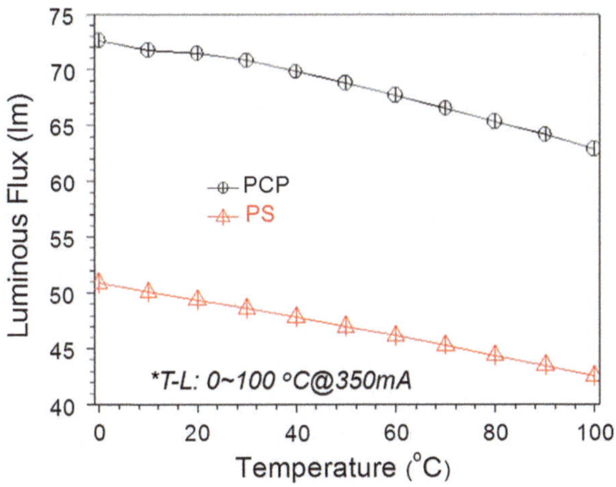

Fig. 9.18 Temperature-luminous flux (T-L = 0 to approximately 100 °C@350 mA) of PCP for thermal stability. Reproduced from Ref. [67] by permission of The Royal Society of Chemistry

property was observed as a function of temperature at <350 mA. The difference in luminous flux indicates that the change of polycrystalline YAG:Ce ceramic is lower than the phosphor–silicon value. Also, polycrystalline YAG:Ce ceramic displayed CCT variation from 5531 to 5463 K. The whole polycrystalline YAG:Ce ceramic demonstrated better colour stability and thermal stability than phosphor–silicon. From this result, polycrystalline YAG:Ce ceramic can be the best candidates for high-power LED applications.

9.2 Summary

This chapter focused on oxide phosphors including garnet-type, silicate, and remote phosphor with packaging [68]. These phosphors are focused on LED applications for the generation of white light. We hope that readers now understand these new types of phosphors.

References

1. Veith M, Mathur S, Kareiva A, Jilavi M, Zimmera M, Hucha V (1999) Low temperature synthesis of nanocrystalline $Y_3Al_5O_{12}$ (YAG) and Ce-doped $Y_3Al_5O_{12}$ via different sol-gel methods. J Mater Chem 9:3069
2. Saraswati V, Rao GVN, Ramarao GV (1987) Structural evolution in alumina gel. J Mater Sci 22:2529
3. Mukherjee S, Sudarsan V, Vatsa RK, Tyagi AK (2009) Luminescence studies on lanthanide ions (Eu^{3+}, Dy^{3+} and Tb^{3+}) doped YAG: Ce nano-phosphors. J Lumin 129:69
4. Mathur S, Shen H, Veith M, Rapalaviciute R, Agne T (2006) Structural and optical properties of highly Nd-doped yttrium aluminum garnet ceramics from alkoxide and glycolate precursors. J Am Ceram Soc 89(6):2027
5. Li Q, Gao L, Yan D (2000) The crystal structure and spectra of nano-scale YAG: Ce^{3+} materials. Chemistry and Physics 64:41
6. Zhang Y, LI L, Zhang X, Xi Q (2008) Temperature effects on photoluminescence of YAG: Ce^{3+} phosphor and performance in white light-emitting diodes. J Rare Earths 26:446
7. Jia D, Wang Y, Guo X, Li K, Zou YK, Jia W (2007) Synthesis and characterization of YAG: Ce^{3+} LED nanophosphors. J Electrochem Soc 154:J1
8. Tanner PA, Fu L, Ning L, Cheng BM, Brik MG (2007) Soft synthesis and vacuum ultraviolet spectra of YAG: Ce^{3+} nanocrystals: reassignment of Ce^{3+} energy levels. J Phys: Condens Matter 19:1
9. Xia Z, Zhuang J, Meijerink A, Jing X (2013) Host composition dependent tunable multicolor emission in the single-phase $Ba_2(Ln1-zTbz)(BO_3)^2Cl$: Eu phosphors. Dalton Trans 42:6327
10. Shen C, Chu J, Qian F, Zou X, Zhong C, Li K, Jin S (2012) High color rendering index white LED based on nanoYAG: Ce^{3+} phosphor hybrid with CdSe/CdS/ZnS core/ shell/shell quantum dots. J Mod Opt 59:1199
11. Gao H, Wang Y (2007) Preparation of (Gd, Y)AlO_3:Eu^{3+} by citric-gel method and their photoluminescence under VUV excitation. J Lumin 122:997
12. Hreniak D, Bettinelli M, Speghini A, Łukowiak A, Głuchowskil P, Wiglusz R (2009) The f–f Emission of Pr^{3+} Ion as an Optical Probe for the Structural Properties of YAG Nanoceramics. J Nanosci Nanotechnol 9:6315
13. Shimomura Y, Honma T, Shigeiwa M, Akai T, Okamoto K, Kijima N (2007) Photoluminescence and crystal structure of green-emitting $Ca_3Sc_2Si_3O_{12}$:Ce^{3+} phosphor for white light emitting diodes. J Electrochem Soc 154:J35
14. Liu Y, Hao J, Zhuang W, Hu Y (2009) Structural and luminescent properties of Gel-combustion synthesized green-emitting $Ca_3Sc_2Si_3O_{12}$:Ce^{3+} phosphor for solid-state lighting. J Phys D Appl Phys 42:1
15. Shimomura Y, Kurushima T, Shigeiwa M, Kijima N (2008) Redshift of green photoluminescence of $Ca_3Sc_2Si_3O_{12}$:Ce_{3+} phosphor by charge compensatory additives. J Electrochem Soc 155(2):J45

16. Gong X, Huang J, Chen Y, Lin Y, Luo Z, Huang Y (2014) Novel garnet-structure $Ca_2GdZr_2(AlO_4)_3:Ce^{3+}$ phosphor and its structural tuning of optical properties. Inorg Chem 53:6607
17. Lü W, Lv W, Zhao Q, Jiao M, Shao B, You H (2015) Generation of orange and green emissions in $Ca_2GdZr_2(AlO_4)_3:Ce^{3+}$, Mn^{2+}, Tb^{3+} garnets via energy transfer with Mn^{2+} and Tb^{3+} as acceptors. Journal of Materials Chemistry C 3(10):2334
18. Akaogi M, Tanaka A, Kobayashi M, Fukushima N, Suzuki T (2002) High-pressure transformations in $NaAlSiO_4$ and thermodynamic properties of jadeite, nepheline, and calcium ferrite-type phase. Phys Earth Planet Inter 130:49
19. Abe T, Toda K, Ishigaki T, Kim SW, Uematsu K, Sato M, Masaki T, Yoon DH (2013) Synthesis of yellow-emitting $NaAlSiO_4:Eu^{2+}$ phosphors using SiO powder as a silica source. Electrochem Soc 224:1
20. Jo DS, Kim BS, Masaki T, Yoon DH (2013) Luminescent properties of $(Na, M)AlSiO_4:Eu^{2+}$ phosphors (M = Li, K, Mg, Ca). J Ceram Process Res 14:s9
21. Hamilton DL (2015) To nephelines as crystallization temperature indicators. J Geol 69:321
22. Kakihana M, Kim JH, Komukai T, Kato H, Sato Y, Kobayashi M, Takatsuka Y (2013) Exploration of new phosphors using a mineral-inspired approach in combination with solution parallel synthesis. Opt Photonics J 3:5
23. Kim JH, Kato H, Kakihana M (2012) Control of $NaAlSiO_4:Eu^{2+}$ photoluminescence properties by charge-compensated aliovalent element substitutions. J Inform Display 13:97
24. Abea T, Kim SW, Ishigaki T, Uematsua K, Satoa M, Masaki T, Yoon DH, Toda K (2014) Synthesis and the luminescent properties of $NaAlSiO_4$: Eu^{2+} phosphor using SiO powder as a silica source. J Inform Display 15:31
25. Jo DS, Luo YY, Senthil K, Toda K, Kim BS, Masaki T, Yoon DH (2012) Synthesis and photoluminescence properties of new $NaAlSiO_4:Eu^{2+}$ phosphors for near-UV white LED applications. Opt Mater 34:696
26. Pang R, Zhao R, Jia Y, Li C, Su Q (2014) Luminescence properties of a new yellow long-lasting phosphorescence phosphor $NaAlSiO_4:Eu^{2+}$, Ho^{3+}. J Rare Earths 32:792
27. Bagga R, Achanta VG, Goel A, Ferreira JMF, Singh NP, Singh DP, Contini V, Falconieri M, Sharma G (2013) Luminescence study of mixed valence Eu-doped nanocrystalline glass–ceramics. Opt Mater 36:198
28. Guo Y, Yu X, Liu J, Yang X (2010) Photoluminescence of Eu^{2+}-activated $Na_{1-x}Al_{1-x}Si_{1+x}O_4$ upon UV excitation. J Rare Earths 28:34
29. Kalaji A, Mikami M, Cheetham AK (2014) Ce^{3+}-activated γ-Ca_2SiO_4 and other olivine-type $ABXO_4$ phosphors for solid-state lighting. Chem Mater 26:3966
30. Mao ZY, Zhu YC, Wang Y, Gan L (2014) Ca_2SiO_4: Ln (Ln = Ce^{3+}, Eu^{2+}, Sm^{3+}) tricolor emission phosphors and their application for near-UV white light-emitting diode. J Mater Sci 49:4439
31. Zadov AE, Gazeev VM, Pertsev NN, Gurbanov AG, Yamnova NA, Gobechiya ER, Chukanov NV (2008) Discovery and investigation of a natural analog of calcio-olivine (γ-Ca_2SiO_4). In Doklady Earth Sciences 423:1431
32. Ishida H, Mabuchi K, Sasaki K, Mitsuda T (1992) Low-temperature synthesis of β-Ca_2SiO_4 from hillebrandite. J Am Ceram Soc 75:2427
33. Wen J, Ning L, Duan CK, Zhan S, Huang Y, Zhang J, Yin M (2015) First-principles study on structural, electronic, and spectroscopic properties of γ-Ca_2SiO_4: Ce^{3+} phosphors. J Phys Chem A 119(29):8031
34. Jang HS, Kim HY, Kim YS, Lee HM, Jeon DY (2012) Yellow-emitting γ-Ca_2SiO_4: Ce^{3+}, Li^+ phosphor for solid-state lighting: luminescent properties, electronic structure, and white light-emitting diode application. Opt Express 20(3):2761
35. Gobechiya ER, Yamnova NA, Zadov AE, Gazeev VM (2008) Calcio-olivine γ-Ca_2SiO_4: I. Rietveld refinement of the crystal structure. Crystallogr Rep 53(3):404
36. Zadov AE, Gazeev VM, Pertsev NN, Gurbanov AG, Gobechiya ER, Yamnova NA, Chukanov NV (2009) Calcioolivine, γ-Ca_2SiO_4, an old and new mineral species. Geol Ore Deposits 51(8):741

37. Tanner PA, Mak CS, Edelstein NM, Murdoch KM, Liu G, Huang J, Barandiarán Z (2003) Absorption and emission spectra of Ce^{3+} in elpasolite lattices. J Am Chem Soc 125(43):13225
38. Ray S, Toda K, Kamei S, Ishigaki T, Uematsu K, Sato M, Kato H, Kobayashi M, Kakihana M (2012) The 12th international meeting on information display
39. Liu C, Xia Z, Lian Z, Zhou J, Yan Q (2013) Structure and luminescence properties of green-emitting $NaBaScSi_2O_7$: Eu^{2+} phosphors for near-UV-pumped light emitting diodes. J Mater Chem C 1(43):7139
40. Liu C, Xia Z, Lian Z, Zhou J, Yan Q (2014) Correction: Structure and luminescence properties of green-emitting $NaBaScSi_2O_7$: Eu^{2+} phosphors for near-UV-pumped light emitting diodes. J Mater Chem C 2(39):8433
41. Yang P, Rivers T (2000) Trace element partitioning between coexisting biotite and muscovite from metamorphic rocks, Western Labrador: Structural, compositional and thermal controls. Geochim Cosmochim Acta 64(8):1451
42. Zhu G, Shi Y, Mikami M, Shimomura Y, Wang Y (2014) Electronic structure and photo/cathodoluminescence properties investigation of green emission phosphor $NaBaScSi_2O_7$: Eu^{2+} with high thermal stability. Cryst Eng Comm 16(27):6089
43. Zhang X, Wang J, Huang L, Pan F, Chen Y, Lei B, Wu M (2015) Tunable luminescent properties and concentration-dependent site preferable distribution of Eu^{2+} ions in silicate glass for white LEDs applications. ACS Appl Mater Interfaces
44. Wierzbicka-Wieczorek M, Kolitsch U, Tillmanns E (2010) The crystal structures of three new complex silicates of scandium. Can Mineral 48(1):51
45. Li G, Wang Y, Zeng W, Chen W, Han S, Guo H, Wang X (2015) Luminescent properties of a new green afterglow phosphor $NaBaScSi_2O_7$: Eu^{2+}. Dalton Trans
46. Grinberg M, Barzowska J, Baran A, Kukliński B (2011) Characterization of various Eu^{2+} sites in Ca_2SiO_4: Eu^{2+} and Ba_2SiO_4: Eu^{2+} by high-pressure spectroscopy. Mater Sci Poland 29 (4):272
47. Kim JS, Park YH, Kim SM, Choi JC, Park HL (2005) Temperature-dependent emission spectra of M_2SiO_4: Eu^{2+}(M = Ca, Sr, Ba) phosphors for green and greenish white LEDs. Solid State Commun 133(7):445
48. Richards BS (2006) Luminescent layers for enhanced silicon solar cell performance: down-conversion. Sol Energ Mater Sol Cell 90:1189
49. Schlotter P, Schmidt R, Schneider J (1997) Luminescence conversion of blue light emitting diodes. Appl Phys A Mater Sci Process 64:417
50. Meneghini M, Trevisanello LR, Meneghesso G, Zanoni E (2008) A review on the reliability of GaN-based LEDs. IEEE Trans Device Mater Reliab 8(2):323
51. Kim JK, Luo H, Schubert EF, Cho J, Sone C, Park Y (2005) Strongly enhanced phosphor efficiency in GaInN white light-emitting diodes using remote phosphor configuration and diffuse reflector cup. Jpa J Appl Phys 44:L649
52. Liu Z, Liu S, Wang K, Luo X (2009) Status and prospects for phosphor-based white LED packaging. Front Optoelectron China 2:119
53. Kuo HC, Hung CW, Chen HC, Chen KJ, Wang CH, Sher CW, Cheng YJ (2011) Patterned structure of remote phosphor for phosphor-converted white LEDs. Opt Express 19(104):A930
54. Geim AK, Novoselov KS (2007) The rise of graphene. Nat Mater 6(3):183
55. Pei S, Zhao J, Du J, Ren W, Cheng HM (2010) Direct reduction of graphene oxide films into highly conductive and flexible graphene films by hydrohalic acids
56. Song YH, Han GS, Mang SR, Jung MK, Jung HS, Yoon DH (2015) Design of a thermally stable rGO-embedded remote phosphor for applications in white LEDs. J Mater Chem C 3:235
57. Saravanapavan P, Jones JR, Pryce RS, Hench LL (2003) Bioactivity of gel–glass powders in the $CaO\text{-}SiO_2$ system: a comparison with ternary ($CaO\text{-}P_2O_5\text{-}SiO_2$) and quaternary glasses ($SiO_2\text{-}CaO\text{-}P_2O_5\text{-}Na_2O$). J Biomed Mater Res, Part A 66(1):110
58. Lee YK, Lee JS, Heo J, Im WB, Chung WJ (2012) Phosphor in glasses with Pb-free silicate glass powders as robust color-converting materials for white LED applications. Opt Lett 37:3276

59. Fujita S, Sakamoto A, Tanabe S (2008) Luminescence characteristics of YAG glass–ceramic phosphor for white LED. Selected topics in quantum electronics. IEEE J Sel Top Quantum Electron 14(5):1387
60. Zhang R, Lin H, Yu Y, Chen D, Xu J, Wang Y (2014) A new-generation color converter for high-power white LED: transparent Ce^{3+}: YAG phosphor-in-glass. Laser Photonics Rev 8 (1):158
61. Ji EK, Song YL, Lee MJ, Song YH, Yoon DH (2015) Fabrication and analysis of luminous properties of ceramic phosphor plate for high-power LED. J Korean Cryst Growth Cryst Technol 25:35
62. Zhang R, Lin H, Yu Y, Chen D, Xu J, Wang Y (2014) A new-generation color converter for high-power white LED: transparent Ce^{3+}:YAG phosphor-in-glass. Laser Photon Rev 8:158
63. Ikesue A, Furusato I, Kamata K (1995) Fabrication of polycrystal line, transparent YAG ceramics by a solid-state reaction method. J Am Ceram Soc 78:225
64. Ikesue A (2002) Polycrystalline Nd:YAG ceramics lasers. Opt Mater 19:183
65. Yagi H, Yanagitani T, Numazawa K, Ueda K (2007) The physical properties of transparent $Y_3Al_5O_{12}$: elastic modulus at high temperature and thermal conductivity at low temperature. Ceram Int 33:711
66. Li X, Li Q, Wang J, Yang S, Liu H (2007) Synthesis of Nd^{3+} doped nano-crystalline yttrium garnet(YAG) powders leading to transparent ceramic. Opt Mater 29:528
67. Song YH, Han GS, Ji EK, Lee MJ, Song YL, Kong DS, Jung MK, Jeong BW, Jung HS, Yoon DH (2015) The novel design of a remote phosphor ceramic plate for white light generation in high power LEDs. J Mater Chem C 3:6148
68. Kim JS, Kwon OH, Jang JW, Lee SH, Han SJ, Lee JH, Cho YS (2015) Long-term stable, low-temperature remote silicate phosphor thick films printed on a glass substrate. ACS Comb Sci 17:234

Chapter 10
Crystal Structures and Luminescence Properties of Oxyhalide-Based Phosphors

Zhiguo Xia and Zihan Xu

Abstract This chapter focuses on the crystal structure and luminescence properties of oxyhalide-based phosphors for different optical applications, such as wavelength-conversion materials, in white light–emitting diodes (w-LEDs), long-lasting phosphorescence lighting signal indication, X-ray storage phosphor, etc. On the basis of the chemical compositions of the reported oxyhalide-based phosphors, we first summarize the binary oxyhalide phosphors, and then we focus on the ternary oxyhalide–based phosphors such as halo-silicate phosphors, halo-phosphate phosphors, halo-borate phosphors, halo-aluminate phosphors, halo-sulphate phosphors, and other systems. Finally we sum up the representative work on the oxyhalide-based phosphors in our group, which are concentrated on the structural design, new phase formation, and structure-property relationship of the oxyhalide-based phosphors. The proposed categories and the summary in this chapter will be important for the understanding of the oxyhalide-based phosphors and, give new insight into the development of novel oxyhalide-based phosphors; moreover, their potential optical applications can be expected.

10.1 Introduction

Halogen-bonding plays an important role in the construction of the new inorganic phosphor hosts and the modification of the crystal-field environment of the activators in phosphor materials [1, 2]. Moreover, such halogen-bonding will be related to the design of advanced functional materials. Therefore, the introduction of the halogen ions (such as F^-, Cl^-, Br^-, and I^-) into some conventional inorganic oxide compounds has great potential for the exploration of technologically important materials with special functional properties [3]. To explore suitable phosphors with excellent luminescence properties, it is significant to construct the rigid crystal

Z. Xia (✉) · Z. Xu
School of Materials Science and Engineering, University of Science and Technology Beijing, Beijing 100083, People's Republic of China
e-mail: xiazg@ustb.edu.cn

R.S. Liu (ed.), *Phosphors, Up Conversion Nano Particles, Quantum Dots and Their Applications*, DOI 10.1007/978-3-662-52771-9_10

structure of the host material because the luminescent properties of phosphors remarkably depend on interaction with the environment around the excited activators. Therefore, the introduction of the halogen ions offers additional opportunities for new phosphor materials compared with conventional oxide materials.

It is well-known that oxyhalide-based phosphors, such as halo-silicate, halo-phosphate, halo-borate, halo-aluminate, etc., have received remarkable attention in inorganic materials science because these compounds possess plenty of crystal-structure types as well as adjustable crystal-chemistry environment and exhibit particularly promising optical properties especially after the controlled rare-earth–doping experiments [4–7]. As we know, oxyhalide-based phosphors hosts belong to the active intermediate in between oxide systems (such as Y_2O_3, $Y_3Al_5O_{12}$) and halide systems including $NaYF_4$ and LaF_3. Hence, oxyhalide-based phosphors hosts have good stability, low phonon energy, and rich crystal-structure types, and plenty of new oxyhalide-based phosphors can be designed and excellent luminescence properties expected. Finally, the different optical applications—such as wavelength-conversion materials in white light–emitting diodes (*w*-LEDs), long-lasting phosphorescence lighting signal indication, X-ray storage phosphor, etc.—can be realized in such kinds of important and interesting oxyhalide-based phosphors systems [8–10].

In this chapter, we will first summarize the binary oxyhalide phosphors represented by Sr_4OCl_6:Eu^{2+} phosphor; then the main topic is focused on the ternary oxyhalide–based phosphors such as halo-silicate phosphors, halo-phosphate phosphors, halo-borate phosphors, halo-aluminate phosphors, and other oxyhalide-based phosphors systems. Finally, we sum up the representative work on crystal-structure and luminescence-property studies for the oxyhalide-based phosphors in our group.

10.2 Binary Oxyhalide Phosphors

10.2.1 Sr_4OCl_6:Eu^{2+} (SOC:Eu^{2+})

Im et al. first reported Eu^{2+}-doped Sr_4OCl_6 (SOC) phosphors for possible application in *w*-LEDs [11]. The unit cell representation of SOC and the existence of two crystallographic sites of strontium are depicted in Fig. 10.1a and b, respectively. The SOC structure shows two different strontium sites with high asymmetry with eight ($SrOCl_7$) and seven ($SrOCl_6$) coordinations. The SOC:Eu^{2+} phosphor exhibits broad excitation spectra ranging from 250 to 425 nm, and an intense broad blue emission band centered at 446 nm under λ_{ex} = 370 nm. The optimum concentration of Eu^{2+} in $Sr_{4-x}Eu_xOCl_6$ was found with x = 0.02 (0.5 mol%). The temperature-dependent PL studies show that the phosphor exhibits thermal quenching resistance, thus retaining the luminance of approximately 91 % at 200 °C.

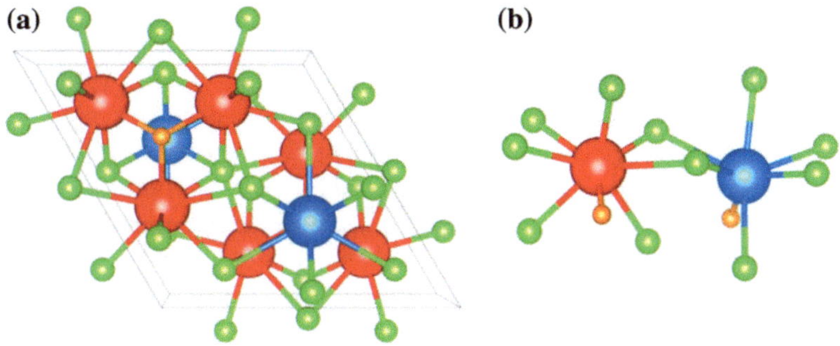

Fig. 10.1 **a** Unit-cell representation of the crystal structure of Sr_4OCl_6 with *red, blue, green* and *orange spheres* as Sr1, Sr2, Cl and O atoms, respectively. **b** Two different Strontium sites are depicted with eight- and seven-coordination with chlorine and oxygen atoms. Reprinted with permission from Ref. [11]. Copyright 2014 American Chemical Society

10.2.2 (La,Ca)OCl:Eu^{3+}

LnOX-type binary oxyhalide phosphors have drawn much attention in many different applications and different forms of crystalline [12–14]. As a typical example, red-emitting phosphors based on lanthanum–calcium oxychloride $(La1_{1-x}Ca_x)OCl$: $y\%Eu^{3+}$ ($0 \leq x \leq 0.17$, $1 \leq y \leq 5$) have been reported [13]. Oxychloride phosphors with stable tetragonal PbFCl-type structure were obtained in a single phase. The photoluminescence intensity was effectively enhanced by doping Ca^{2+} into the host LaOCl:Eu^{3+} lattice. By optimizing the composition, maximum emission intensity was obtained for $(La_{0.85}Ca_{0.15})OCl$:2 %Eu^{3+}. The relative emission intensity was 63 % compared with that of the commercial Y_2O_3:Eu^{3+} phosphor.

10.2.3 LaOBr:Er^{3+}

The LaOBr host has low phonon energy, which indicates that LaOBr:Er^{3+} may have high luminescent efficiency [14]. Under excitation into the $^4I_{11/2}$ level of Er^{3+} ions by 980-nm laser, the two- and three-photon upconversion luminescence of LaOBr:Er^{3+} was recorded. The most intense emissions come from the $^2H_{11/2}$, $^4S_{3/2} \rightarrow {}^4I_{15/2}$ transitions. Figure 10.2a shows the Stokes emission spectra of LaOBr:Er^{3+} samples under 261-nm excitation. The spectra exhibit six emission bands corresponding to the radiative transitions from some excited states of the Er^{3+} ions. According to the energy-level structure of Er^{3+} ions, the assignment can be easily made. The bands in the blue region of 400–412 and 463–482 nm are

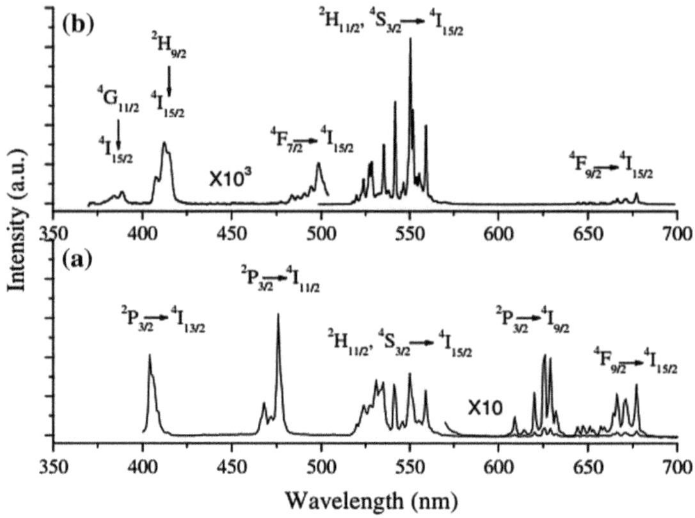

Fig. 10.2 Stokes (λ_{ex} = 261 nm) spectra of LaOBr:Er^{3+} (0.1 %) (**a**) and upconversion spectra of LaOBr:Er^{3+} (0.1 %) λ_{ex} = 980 nm (**b**). Reprinted from Ref. [14], Copyright 2007, with permission from Elsevier

assigned to the transitions $^2P_{3/2} \rightarrow {}^4I_{13/2}$ and $^2P_{3/2} \rightarrow {}^4I_{11/2}$ of Er^{3+} ions, respectively. The bands in the green region of 516–539 and 539–570 nm are associated with the transitions $^2H_{11/2} \rightarrow {}^4I_{15/2}$ and $^4S_{3/2} \rightarrow {}^4I_{15/2}$ of Er^{3+} ions, respectively. The transitions in the red region can be assigned as follows: $^2P_{3/2} \rightarrow {}^4I_{9/2}$ (604–638 nm) and $^4F_{9/2} \rightarrow {}^4I_{15/2}$ (642–683 nm) of Er^{3+} ions. Figure 10.3b gives the multi-photon upconversion emissions under 980-nm laser excitation. The bands in the region 516–539, 539–570, and 642–683 nm are assigned to the $^2H_{11/2} \rightarrow {}^4I_{15/2}$, $^4S_{3/2} \rightarrow {}^4I_{15/2}$, and $^4F_{9/2} \rightarrow {}^4I_{15/2}$ transitions of Er^{3+} ions, respectively, which is identical with the Stokes emission spectra.[3]

10.3 Halo-Silicate Phosphors

As discussed previously, ternary oxyhalide–based phosphors will be the main basis for the development of halo-containing phosphors, and the halo-silicate phosphors play an especially important role in such a field. There are many different chemical compositions, luminescence behaviors, and corresponding applications for halo-silicate phosphors. Here, we first give a summarization of the halo-silicate phosphors, including their host, activators, and corresponding excitation and emission peaks, as shown in Table 10.1. Then we will select some important halo-silicate phosphors for the detailed introduction.

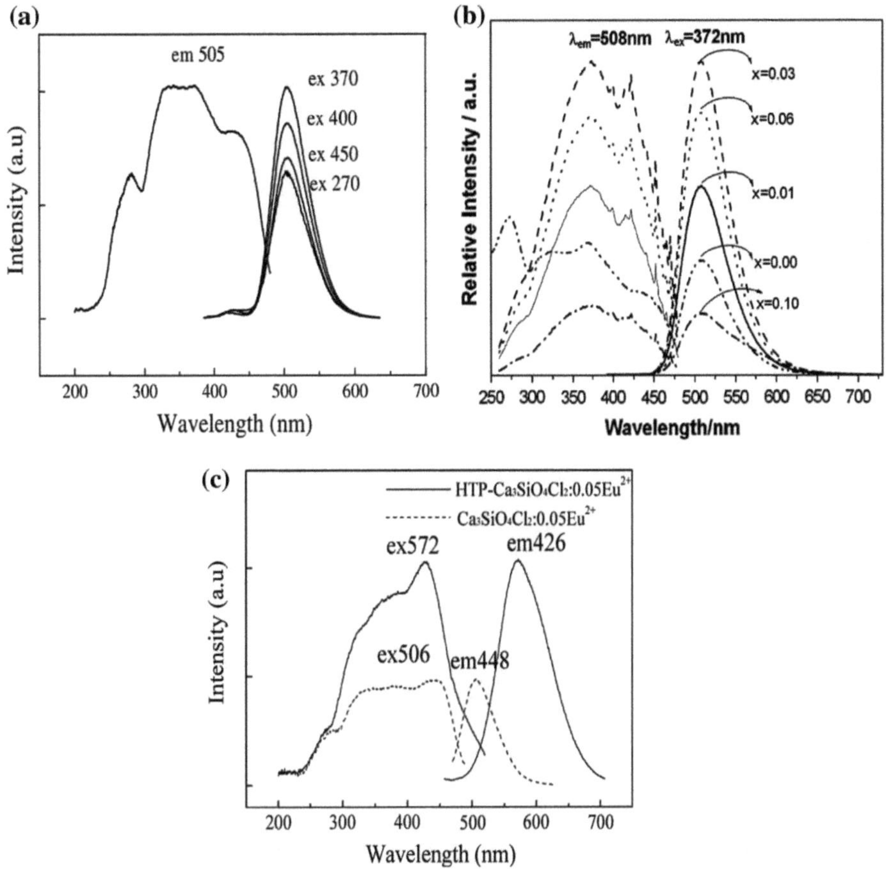

Fig. 10.3 PL and PLE spectra of LT phase $Ca_3SiO_4Cl_2$:Eu^{2+} (**a**) (Reproduced from Ref. [15] by permission of The Electrochemical Society) and $Ca_3SiO_4(Cl_{1}\_F_x)_2$:Eu^{2+} (x = 0.00, 0.01, 0.03, 0.06, 0.10) (**b**) (Reproduced from Ref. [17] by permission of The Electrochemical Society) phosphors. Comparison of PLE (*left*) and PL (*right*) spectra of LT phase and HT phase $Ca_3SiO_4Cl_2$:Eu^{2+}(**c**) (Reproduced from Ref. [15] by permission of The Electrochemical Society)

10.3.1 $Ca_3SiO_4Cl_2$

Among all of the reported halo-silicate phosphors, $Ca_3SiO_4Cl_2$:Eu^{2+} phosphors are a hot topic that has drawn much attention because of its adjustable crystal structure and interesting luminescence properties [15–19, 43]. $Ca_3SiO_4Cl_2$ was reported as early as the nineteenth century by Chaterlier, and investigations on $Ca_3SiO_4Cl_2$ phase transition were reported by Winkler in the 1990s [44–46]. Depending on the synthesis temperature, there are two phases—low temperature (LT)/high temperature (HT) phase—for $Ca_3SiO_4Cl_2$. LT-phase $Ca_3SiO_4Cl_2$ is composed of alternating layers of $CaCl_2$ and Ca_2SiO_4. The primary crystal structure of HT-phase

Table 10.1 Summarization of the host compositions, activators and the corresponding excitation and emission peaks of the selected halosilicate phosphors

No.	Phosphor		λ_{ex}(nm)	λ_{em}(nm)	References
	Host	Activator			
1	$Ca_3SiO_4Cl_2$ (Two phases)	Eu^{2+}	395 426	509 572	[15–19]
		Mn^{2+}	419	570	
2	$Ca_{10}(Si_2O_7)_3Cl_2$	Eu^2	395	508	[20]
3	$Ca_5(SiO_4)_2Cl_2$	Eu^{2+}	395	585	[21]
4	$Sr_4Si_3O_8Cl_4$	Eu^{2+}	370	484	[22]
3	$Sr_8(Si_4O_{12})Cl_8$	Eu^{2+}	320	495	[10]
4	$Ba_2SiO_3Cl_2$	Eu^{2+}–Mn^{2+}	365	425, 492 and 608	[23]
5	$Ba_5SiO_4Cl_6$	Eu^{2+}	405	440	[24]
16	$Ba_5SiO_4(F,Cl)_6$	Eu^{2+}	370	440	[25]
11	$Ba_5SiO_4Br_6$	Eu^{2+}	300	440	[26, 27]
6	$NaLa_4(SiO_4)_3F$	Eu^{3+}	254	610	[28]
7	$Ca_8Mg(SiO_4)_4Cl_2$	Eu^{2+}	381	507	[29–31]
		Dy^{3+}	350	492	[32]
8	$Ca_8Zn(SiO_4)_4Cl_2$	Eu^{2+}	450	505	[33]
9	$Ca_{12}Al_{14}O_{32}Cl_2$	Eu^{2+}	268, 324	442	[34]
10	$Ca_{12}Al_{10.6}Si_{3.4}O_{32}Cl_{5.4}$	Eu^{2+}	380	425, 500	[35]
13	$Y_3Si_2O_8Cl$	Ce^{3+}	278	358	[36]
		Tb^{3+}	234	542	
14	$La_3F_3(Si_3O_9)$	Ce^{3+}	280	309	[37]
18	Sr_2LiSiO_4F	Ce^{3+}–Li^+	340	445	[38]
		Eu^{2+}	350	510	[39–41]
12	$CaCl_2/SiO_2$	Eu^{2+}	338	427	[42]

$Ca_3SiO_4Cl_2$ is determined as monoclinic structure with a space group of *P21/c* [16]. As Liu reported in 2005, both of the $Ca_3SiO_4Cl_2$:Eu^{2+}phosphors with the LT/HT phases show intense broad absorption bands between 250 and 450 nm, which matches well with the near-ultraviolet (n-UV) (380–420 nm) emission band of InGaN-based chips. It also exhibits a green emission (510 nm) and a tunable yellow (551 nm) to orange emission (577 nm), which is ascribed to the allowed 5d → 4f transition of Eu^{2+} ion in the LT and HT phases, respectively [15, 16].

The PL and PLE spectra of LT-phase $Ca_3SiO_4Cl_2$:Eu^{2+} are shown in Fig. 10.3a. On 370-nm excitation, $Ca_3SiO_4Cl_2$:Eu^{2+} shows a very strong green emission band with a peak at 505 nm, a half width of 59 nm, and a weak blue band peaking at approximately 427 nm. With different excitation wavelengths of 270, 370, 400, and 450 nm, there are few changes in the green emission except for the emission intensity. Figure 10.3b shows the excitation and emission spectra of LT-phase $Ca_3SiO_4(Cl_{1-x}F_x)_2$:Eu^{2+} (y = 0.00, 0.01, 0.03, 0.06, and 0.10). There is no significant change in the excitation and emission spectra except for the luminescent intensity between F-doped samples and the un-doped sample. The green emission intensity of Eu^{2+} is considerably enhanced by the substitution of Cl by F ions. The emission intensity of Eu^{2+} in F-substituted samples is approximately 164 % ($x = 0.01$), 270 % ($x = 0.03$), 227 % ($x = 0.06$), and 67 % ($x = 0.10$) times stronger than that of the sample without F ions [19].

In comparison, Eu^{2+}-doped HT-phase $Ca_3SiO_4Cl_2$ synthesized at 1020 °C shows an orange emission. The excitation spectrum exhibits a broad band between 250 and 500 nm with three peaks of approximately 327, 371, and 426 nm, which are coupled well with the near-UV (n-UV) or the deep blue emission of GaN-based light-emitting diodes (LEDs). Its emission spectrum shows a single intense broad emission band centered at 572 nm. The measured chromaticity of yellowish-orange luminescence is $x = 0.52$, $y = 0.45$. Accordingly, Fig. 10.3c comparatively gives the PLE and PL spectra of the LT- and HT-phase $Ca_3SiO_4Cl_2:Eu^{2+}$. There are great differences between the excitation of LT-phase $Ca_3SiO_4Cl_2:Eu^{2+}$ and that of HT-phase $Ca_3SiO_4Cl_2:Eu^{2+}$. The excitation intensity in the violet-blue region (approximately 426 nm) of HT-phase $Ca_3SiO_4Cl_2:Eu^{2+}$ is obviously stronger than that at approximately 328 and 374 nm, which makes the excitation spectrum profile of HT-phase $Ca_3SiO_4Cl_2:Eu^{2+}$ have a greater gradient than that of LT-phase $Ca_3SiO_4Cl_2:Eu^{2+}$. As for the emission properties, the emission band o f$Ca_3SiO_4Cl_2:Eu^{2+}$ is centered at 506 nm with a FWHM of 61 nm, whereas that of HT-phase $Ca_3SiO_4Cl_2:Eu^{2+}$ is centered at 572 nm with a FWHM of 93 nm. As is well known, the luminescence of Eu^{2+}-activated phosphors usually results from the ground $4f^7(8S_{7/2})$ level to the $4f^65d$ excited configuration, which depends strongly on the nature of the host lattice. It has been suggested that the mean covalency of the dopant site in HT-phase $Ca_3SiO_4Cl_2$: Eu^{2+} would be higher than of LT-phase $Ca_3SiO_4Cl_2:Eu^{2+}$, and the center of gravity of the 5d states would be at the lower energy.

Green/yellow phosphors $Ca_3SiO_4Cl_2:Eu^{2+}$, Mn^{2+} were also reported by Wang et al. in 2008 [17]. These phosphors show intense broad absorption bands between 250 and 450 nm and exhibit two dominating bands situated at 512 and 570 nm, which are ascribed to the allowed 5d → 4f transition of the Eu^{2+} ion and the ${}^4T_{1g}({}^4G) \rightarrow {}^6A_{1g}({}^6S)$ transition of the Mn^{2+} ion, respectively. The lifetime of the Eu^{2+} ion decreases with increasing concentration of the Mn^{2+} ion, thus strongly supporting an efficient energy transfer from Eu^{2+} to Mn^{2+}. For example, $Ca_3SiO_4Cl_2$ possess the LT phase. Figure 10.4 shows the excitation and emission spectra of $Ca_3(SiO_4)Cl_2$:$0.09Eu^{2+}$, xMn^{2+}($x = 0, 0.06, 0.09, 0.12, 0.15, 0.18$). For $x = 0$, $Ca_3(SiO_4)Cl_2$:$0.09Eu^{2+}$ shows an intense green broadband emission at approximately 509 nm on 395-nm excitation. With the introduction and increasing content of Mn^{2+}, the sample shows an additional strong yellow broad-band emission at approximately 568 nm in addition to the green one. The new yellow-emission band is considered to be associated with the Mn^{2+} ion. Monitoring the yellow emission at 568 nm, the Mn^{2+} ion shows the same excitation features as the Eu^{2+} ion as shown in curves 2 and 3 of Fig. 10.4. With increasing Mn^{2+} content, the yellow-emission intensity enhances, whereas that of the green one decreases gradually as shown in Fig. 10.4. These results support the efficient energy transfer from Eu^{2+} to Mn^{2+} ions.

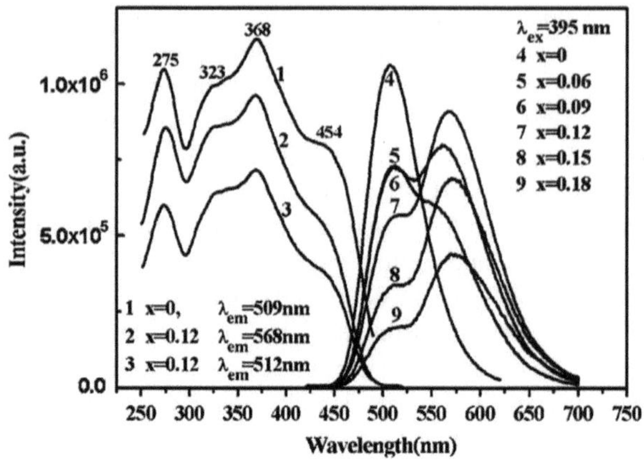

Fig. 10.4 Excitation and emission spectra of $Ca_3(SiO_4)Cl_2$:0.09Eu^{2+}, $x$$Mn^{2+}$ (x = 0, 0.06, 0.09, 0.12, 0.15, 0.18). Reproduced from Ref. [17] by permission of The Electrochemical Society

10.3.2 $Ca_{10}(Si_2O_7)_3Cl_2$

Eu^{2+}-activated $Ca_{10}(Si_2O_7)_3Cl_{12}$was reported as a kind of green-emitting phosphor [20]. It can be efficiently excited by the incident lights of 280–420 nm, perfectly matching with the emissions wavelength of near-UV LEDs. The PL and PLE spectra of $Ca_{9.7}Eu_{0.3}(Si_2O_7)_3Cl_2$ are shown in Fig. 10.5a. It is observed that the excitation spectrum consists of three broad absorption bands at approximately 276, 323, and

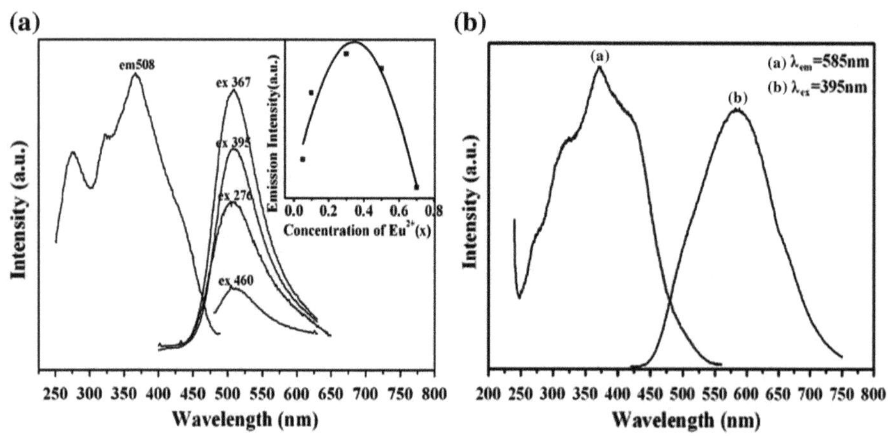

Fig. 10.5 **a** PL and PLE spectra of $Ca_{9.7}Eu_{0.3}(Si_2O_7)_3Cl_2$ (λ_{em} = 508 nm for excitation and λ_{ex} = 276, 367, 395, and 460 nm for emission). The *inset* represents the concentration dependence of the emission intensity of $Ca_{10-x}Eu_x(Si_2O_7)_3Cl_2$($\lambda_{ex}$ = 395 nm). Reprinted from Ref. [20], Copyright 2006, with permission from Elsevier. **b** Photoluminescence excitation spectra ((**a**) λ_{em} = 585 nm) and emission spectrum ((**b**) λ_{ex} = 395 nm) of $Ca_5(SiO_4)_2Cl_2$:Eu^{2+}. Reprinted from Ref. [21], Copyright 2006, with permission from Elsevier

367 nm with a shoulder at approximately 440 nm. As also given in Fig. 10.5a, the emission spectrum under 395-nm excitation consists of a single broad band with a maximum at approximately 508 nm, and the FWHM of the emission band is approximately 79 nm. With different excitation wavelengths of 276, 367, 395, and 460 nm, there are no significant changes in the emission spectra except for the emission intensity. Moreover, $Ca_{10}(Si_2O_7)_3Cl_2:Eu^{2+}$ has a higher quenching temperature >50 K. It can be expected that $Ca_{10}(Si_2O_7)_3Cl_2:Eu^{2+}$ is a promising green-phosphor candidate for creating white light in phosphor-converted white LEDs.

10.3.3 $Ca_5(SiO_4)_2Cl_2$

In fact, $Ca_5(SiO_4)_2Cl_2$ is not the real chemical formula, which belongs to the nominal chemical composition, and its crystal structure is not yet determined [21]. The as-prepared $Ca_5(SiO_4)_2Cl_2$:Eu^{2+} phosphor can be efficiently excited by incident light of 300–450 nm, which is well matched with the emission band of a 395-nm–emitting InGaN chip, and emits an intense orange light peaking at 585 nm. The excitation and emission spectra of $Ca_5(SiO_4)_2Cl_2$:Eu^{2+} phosphor are shown in Fig. 10.5b. The excitation spectrum exhibits a broad band between 250 and 500 nm with four peaks at approximately 275, 320, 372, and 426 nm. For the emission spectrum, an asymmetric broad band predominating at approximately 585 nm, with an FWHM of approximately 153 nm, is ascribed to the 5d → 4f allowed transition of Eu^{2+} ions. Furthermore, a shoulder at approximately 505 nm is also observed owing to the emission of Eu^{2+} in small amount of calcium orthsilicate impurity phase.

10.3.4 $Sr_4Si_3O_8Cl_4$

$Sr_4Si_3O_8Cl_4:Eu^{2+}$was the first reported strontium chlorosilicate phosphor that showed bright blue-green emission. This kind of phosphor was also reported by our group, which will be discussed in detail later. Excited by 370-nm near-ultraviolet light, the phosphors show a broadband bluish-green emission centering at 484 nm, which originates from the $4f^5d^1 \rightarrow 4f^7$ transition of Eu^{2+} ion. The excitation spectra of the phosphors are also a broad band extending from 250 to 400 nm. It was also reported that Mg^{2+}-codoping in $Sr_4Si_3O_8Cl_4:Eu^{2+}$ phosphor greatly enhanced the bluish–green emission of the phosphors [22].

10.3.5 $Sr_8(Si_4O_{12})Cl_8$

Liu et al. recently reported the $Sr_8(Si_4O_{12})Cl_8:Eu^{2+}$ phosphor with high light yield under X-ray excitation [10]. As shown in Fig. 10.6a, XRD plots on the Rietveld

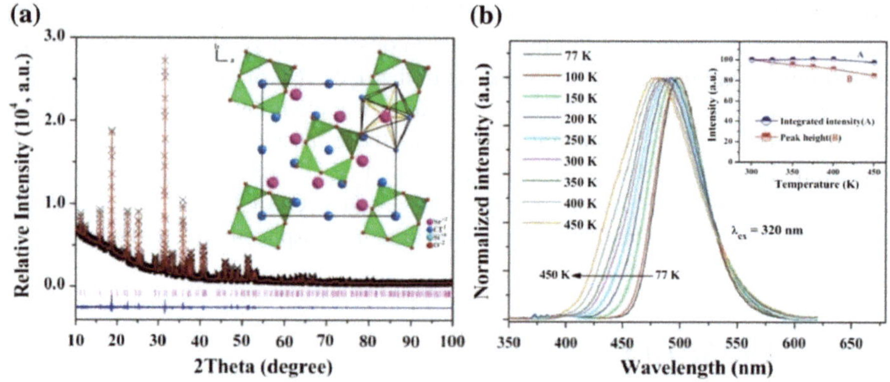

Fig. 10.6 **a** XRD plots on the Rietveld refinement of the as-prepared $Sr_{7.7}Eu_{0.3}(Si_4O_{12})Cl_8$, and the *inset* shows the projection of $Sr_8(Si_4O_{12})Cl_8$ along the c-axis. **b** Normalized emission spectra of $Sr_{7.7}Eu_{0.3}(Si_4O_{12})Cl_8$ different at temperatures from 77 to 450 K under excitation at 320 nm; the *inset* shows the temperature dependence of the integrated emission intensity (*curve A*) and the emission peak height (*curve B*). Reprinted with the permission from Ref. [10]. Copyright 2014 American Chemical Society

refinement of the as-prepared $Sr_{7.7}Eu_{0.3}(Si_4O_{12})Cl_8$ are given showing the pure phase. The inset in Fig. 10.6a demonstrates the crystal structure of $Sr_8(Si_4O_{12})Cl_8$. The structure of $Sr_8(Si_4O_{12})Cl_8$ consists of a three-dimensional framework made from $[SiO_4]$ tetrahedrons as well as irregular eightfold polyhedra of strontium, in which four $[SiO_4]$ tetrahedra are linked by corner-sharing to form a ring, and the rings are separated by an irregular $[SrO_4Cl_4]$ polyhedron. There is only one Sr site with *C*1 symmetry. The coordinating environments for all Sr^{2+} ions are the same. Therefore, when Eu^{2+}ions are doped into the $Sr_8(Si_4O_{12})Cl_8$ host lattice, substituting for Sr^{2+} ions, we expect only one broad peak in the emission spectra. Figure 10.6b gives the normalized emission spectra of $Sr_{7.7}Eu_{0.3}(Si_4O_{12})Cl_8$ at different temperatures from 77 to 450 K under excitation at 320 nm. The emission peak shifts to shorter wavelengths with increases in temperature, and the emission spectra at room temperature show a 58-nm FWHM broad band extending from 400 to 600 nm and peaking at approximately 487 nm. The temperature dependence of the integrated emission intensity and the emission peak height are also given in the inset of Fig. 10.6b.

10.3.6 $Ba_2SiO_3Cl_2$

$Ba_2SiO_3Cl_2:Eu^{2+}$, Mn^{2+} phosphor was synthesized by high-temperature solid-state reaction [23]. The emission band consists of three peaks located at 425, 492, and 608 nm, respectively. The emission peaks at 425 and 492 nm originate from the transition 5d → 4f of Eu^{2+} ions that occupy the two Ba^{2+} sites in the host of $Ba_2SiO_3Cl_2$, whereas the 608-nm emission is attributed to the Mn2 + emission;

there is efficient energy transfer from $Eu^{2+}$to Mn^{2+} ions. White light can be obtained by mixing the three emission colors of blue (425 nm), green (492 nm, and red-orange (608 nm) in such a single host. When the concentrations of the Eu^{2+} ions and Mn^{2+} ions were 0.02 and 0.004 mol, respectively, the sample presented intense white-light emission.

10.3.7 $Ba_5SiO_4(F,Cl)_6$

An intense blue emission of Eu^{2+}-doped $Ba_5SiO_4Cl_6$ under vacuum ultraviolet and near-ultraviolet excitation was observed [24]. The material exhibits an emission peak at 440 nm with relative intensity 50 % under 147-nm, 90 % under 254-nm, 120 % under 366-nm, and 220 % under 405-nm excitation compared with the commercial blue phosphor $BaMgAl_{10}O_{17}$:Eu^{2+}. It is a very promising candidate as a blue-emitting phosphor for potential applications in display and light-emitting diode devices. It was also found that the introduction of F into the Cl sites can modify the luminescence properties. Therefore, a series of halosilicate phosphors, $Ba_5SiO_4(F,Cl)_6$:Eu^{2+}, was synthesized. When excited by 370-nm light, $Ba_5SiO_4Cl_6$:Eu^{2+} exhibits a broad emission band peaking at 440 nm. The emission band seems to be asymmetric in the long-wavelength region, which indicates the existence of different luminescent centers. Partial substitution of Cl with F in the host lattice leads to a red shift in the emission band with centering wavelength from 440 to 503 nm suggesting that the second emission centers dominate [25]. This series of phosphors is considered to be a promising blue and green component used in the fabrication of near-UV–based white LEDs.

10.3.8 $Ba_5SiO_4Br_6$

It was reported that $Ba_5SiO_4Br_6$:Eu^{2+}could act as a kind of photostimulable X-ray storage phosphor [26], and the introduction of Nb^{5+} can also enhance such an effect [27]. The phosphor is also thermally stimulable. After X-ray irradiation, the EPR spectrum of Nb^{4+} is observed. UV irradiation at wavelengths between 250 and 340 nm has the same effect. It is concluded that an electron is trapped at the niobate group. The phosphor is optically stimulable between 300 and 650 nm. The corresponding absorption band is tentatively ascribed to the $NbO_4{}^{4-}$ group. In the region between 300 and 340 nm, a competition between photoionization and stimulation takes place resulting in characteristic thermoluminescence behavior.

10.3.9 $Ca_8Mg(SiO_4)_4Cl_2$

Calcium magnesium chlorosilicate ($Ca_8Mg(SiO_4)_4Cl_2$) (CMSC) belongs to another important group of chlorosilicate phosphors [29–32]. The phosphors show intense absorption in the range of 375–450 nm, which makes them potential candidates of green-emitting phosphor used for near-UV or blue-light–excited white LEDs, which can be prepared by sol-gel method and solid-state reaction. $Ca_8Mg(SiO_4)_4Cl_2:Eu^{2+}$ shows efficient green emission at 509 nm and is approximately 2.8 times higher than a commercial $(Y,Gd)_3(Al,Ga)_5O_{12}:Ce^{3+}$ in PL peak intensity under 450-nm light excitation. Moreover, the chlorosilicates show less thermal quenching than the reference YAG in terms of PL-integrated intensity. These results imply that some Eu^{2+}-doped alkaline–earth chlorosilicates are promising candidates as phosphors for near-UV and blue LEDs.

As shown in Fig. 10.7a, the PL intensity increases gradually with increasing of $CaCl_2$ amount and reaches the maximum at the excessive amount, which is approximately 100 %, when it is prepared by sol-gel method, whereas, the PL intensity decreases with further increase of the $CaCl_2$ amount. The change trend is also similar for the PLE intensity; moreover, the relative intensity of excitation bands in the blue region gradually increases with increasing $CaCl_2$ amount. Thus, the sample $Ca_8Mg(SiO_4)_4Cl_2:Eu^{2+}$ prepared with 100 % excess of $CaCl_2$ is a more suitable green-emission phosphor for the *n*-UV and blue LED chip.

Trivalent dysprosium ion (Dy^{3+}) can be also doped into the CMSC host. The as-prepared phosphor consists of two emission bands, blue and yellow, and the emission intensity of the former is stronger than that of the latter. Blue emission between 470 and 500 nm is assigned to the transition of Dy^{3+} ion from the $^4F_{9/2}$ excited state to the $^6H_{15/2}$ ground state, and it consists of two emission peaks at 480 and 492 nm. Yellow emission is observed between 560 and 600 nm corresponding to the $^4F_{9/2} \rightarrow {}^6H_{13/2}$ transition, which constitutes two emission peaks at 573 and

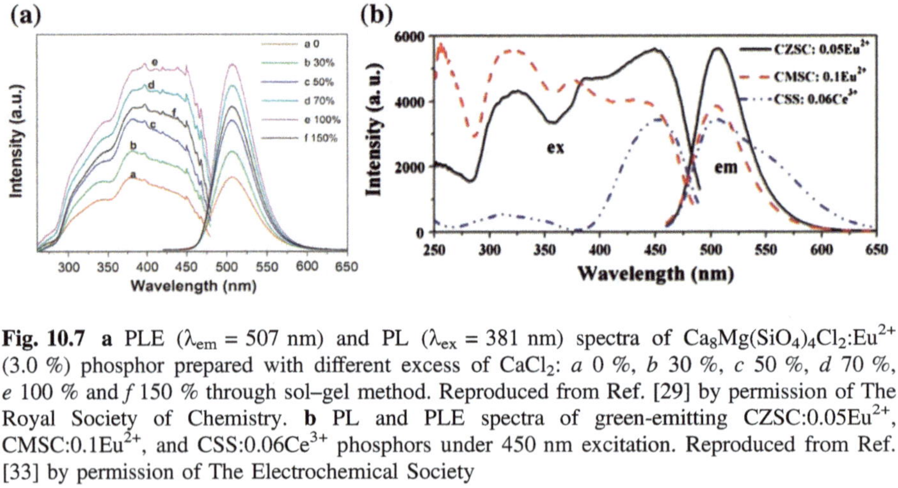

Fig. 10.7 **a** PLE (λ_{em} = 507 nm) and PL (λ_{ex} = 381 nm) spectra of $Ca_8Mg(SiO_4)_4Cl_2:Eu^{2+}$ (3.0 %) phosphor prepared with different excess of $CaCl_2$: *a* 0 %, *b* 30 %, *c* 50 %, *d* 70 %, *e* 100 % and *f* 150 % through sol–gel method. Reproduced from Ref. [29] by permission of The Royal Society of Chemistry. **b** PL and PLE spectra of green-emitting CZSC:0.05Eu^{2+}, CMSC:0.1Eu^{2+}, and CSS:0.06Ce^{3+} phosphors under 450 nm excitation. Reproduced from Ref. [33] by permission of The Electrochemical Society

587 nm. Dy^{3+} concentration not only affects the emission intensity, it also influences the color coordination and color temperature of the samples. The preferentially selected site occupied by Dy^{3+} is the six-coordinated Ca^{2+} site.

10.3.10 *$Ca_8Zn(SiO_4)_4Cl_2$*

Calcium zinc chlorosilicate ($Ca_8Zn(SiO_4)_4Cl_2$) (CZSC), similar to host CMSC, can be prepared by high temperature solid-state reaction with nominal composition of $(2 - x)CaO$-xZnO–SiO_2–$CaCl_2$ ($0 \leq x \leq 1.1$) [33]. The phase gradually converts from $Ca_3SiO_4Cl_2$:Eu^{2+}(CSC) to $Ca_8Zn(SiO_4)_4Cl_2$:Eu^{2+} followed by improved luminescent properties with increasing x. The two phosphors both emit with a maximum at 505 nm. Significantly, CZSC:Eu^{2+} shows a more intense excitation band in the blue region centered at 450 nm. On 450-nm excitation, the integrated-emission intensity of CZSC:Eu^{2+} is 1.3 times stronger that of CMSC: Eu^{2+} and nearly the same as the $Ca_3Sc_2Si_3O_{12}$:Ce^{3+} phosphors. PLE and PL spectra of CZSC:0.05Eu^{2+} are presented in Fig. 10.7b along with those of CMSC:0.1Eu^{2+} and CSS:0.06Ce^{3+} phosphors for comparison. The PL-integrated intensity of CZSC:0.05Eu^{2+} is 1.3 times stronger than that of CMSC:0.1Eu^{2+} and nearly the same as that of the CSS:0.06Ce^{3+} phosphor under the same excitation condition at 450 nm. Compared with the PLE spectrum of CMSC:0.1Eu^{2+}, the intensity of the PLE band in the range of 400–450 nm for the CZSC:0.05Eu^{2+} sample is much higher than CMSC:0.1Eu^{2+}due to a high 5d state density distributed nearby the lowest 5d state in CZSC:0.05Eu^{2+}, which might be related to a smaller cell volume in CZSC than CMSC because the effective ionic radius for Zn (coordination number = 4) is 0.60 nm smaller than 0.71 nm for Mg (coordination number = 4).

10.3.11 *$Ca_{12}Al_{14}O_{32}Cl_2$*

The crystal structure of $Ca_{12}Al_{14}O_{32}Cl_2$ is cubic (space group *I*43*d*, Z = 2) with lattice dimensions of a = 1.200950(5) nm and V = 1.73211(1) nm^3 [34]. In the structural model of $Ca_{12}Al_{14}O_{32}Cl_2$, there is one Ca site, two Al sites, two O sites, and one Cl site. This compound is isomorphous with $Ca_{12}Al_{10.6}Si_{3.4}O_{32}Cl_{5.4}$. The excitation spectrum of $Ca_{12}Al_{14}O_{32}Cl_2$:Eu^{2+} consisted of two wide bands, which were located at approximately 268 and 324 nm. The emission spectrum resulted in indigo light with a peak at approximately 442 nm when excited at 324 nm.

10.3.12 *$Ca_{12}Al_{10.6}Si_{3.4}O_{32}Cl_{5.4}$*

As mentioned previously, $Ca_{12}Al_{10.6}Si_{3.4}O_{32}Cl_{5.4}$ is isostructural with $Ca_{12}Al_{14}O_{32}Cl_2$. In such a structure, the Cl anions fill the large cages surrounded by eight Al (Si) tetrahedrons and two Ca ions in $Ca_{12}Al_{10.6}Si_{3.4}O_{32}Cl_{5.4}$. Cl anions are tightly bound in these cages, and do not randomly move because the radius of Cl anions ($r = 0.181$) is very close to that of the cages ($r = 0.171$). Eu^{2+}-activated $Ca_{12}Al_{10.6}Si_{3.4}O_{32}Cl_{5.4}$ can act as a novel bluish green phosphor for white light–emitting diodes. The phosphor shows broad and strong absorption in the region (320–450 nm). When excited at 380 nm, the phosphor shows two emission bands at approximately 425 and 500 nm. The main emission peak of Eu^{2+}-activated $Ca_{12}Al_{10.6}Si_{3.4}O_{32}Cl_{5.4}$ exhibits a red shift compared with that of Eu^{2+}-activated $Ca_{12}Al_{14}O_{33}$, which is due to the introduction of Si and Cl ions. The results show $Ca_{12}Al_{10.6}Si_{3.4}O_{32}Cl_{5.4}$ is a promising host candidate for the phosphors [35].

10.3.13 $Y_3Si_2O_8Cl$

The luminescence properties of Ce^{3+} and Tb^{3+} in $Y_3Si_2O_8Cl$ have been investigated [36]. The Ce^{3+} excitation bands in the region from 220 to 360 nm are attributed to the transitions from the 4f level to the crystal-field splitting levels of 5d. The Tb^{3+} excitation bands in the region from 220 to 280 nm are due to the 4f → 5d transitions. The spectral-energy distributions of Tb^{3+} emission strongly depend on the Tb^{3+} concentration. The $^5D_3$ emissions of Tb^{3+} decrease with increasing Tb^{3+} concentration, and this phenomenon is due to the cross-relaxation between Tb^{3+} ions. Efficient energy transfer from Ce^{3+} to Tb^{3+} in the $Y_3Si_2O_8Cl$ is observed.

10.3.14 $La_3F_3(Si_3O_9)$

The luminescence properties of Ce^{3+} in $La_3F_3(Si_3O_9)$ have been reported [37]. Excitation and emission bands corresponding to the $4f^1 \rightarrow 5d^1$ transitions of Ce^{3+} were identified. The center of gravity of the 5d states lies at remarkably high energy (43.2×10^{-3} cm^{-1}) for Ce^{3+} in a silicate compound, which is attributed to the combined oxygen/fluoride coordination of the Ce^{3+} ion. Emission from the lowest 4f5d level to the $^2F_{5/2}$ and $^2F_{7/2}$ levels was found at 32.4×10^{-3} and 30.4×10^{-3} cm^{-1}. These results are compared with the literature data on silicates and fluorides.

10.3.15 Sr_2LiSiO_4F

The compound Sr_2LiSiO_4F crystallizes with monoclinic structure with space group P21/m (no. 11). A view of the unit-cell diagram of Sr_2LiSiO_4F is given in Fig. 10.8a. The structure contains two types of 10-coordinate Sr atoms with different coordination environment, a 5-coordinate Li atom, and a 4-coordinate Si atom. The Li atom occupies a distorted square pyramid and the Si atom a slightly

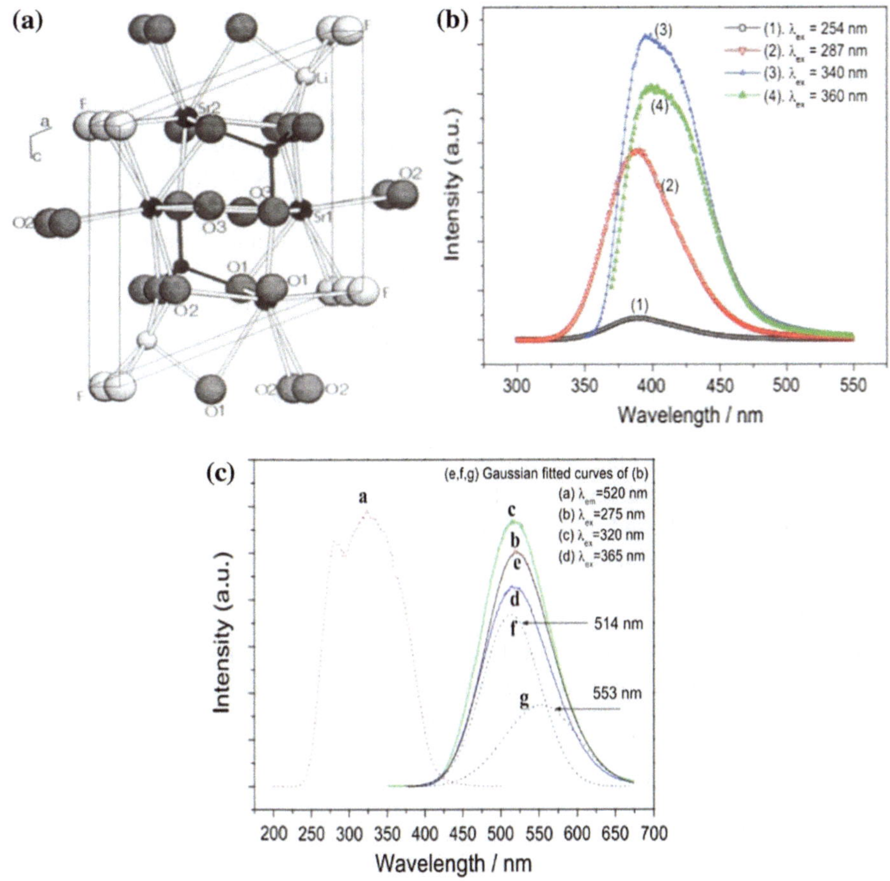

Fig. 10.8 **a** Unit cell diagram of Sr_2LiSiO_4F as viewed along the b axis. Reprinted with the permission from Ref. [39]. Copyright 1995 American Chemical Society. **b** PL emission spectra of $Sr_{1.98}LiSiO_4F$:$0.01Ce^{3+}$, $0.01Li^+$ excited with different UV light. Reproduced from Ref. [38] by permission of John Wiley & Sons Ltd. **c** PL emission and excitation spectra of $(Sr_{1.99}Eu_{0.01})$ $LiSiO_4F$ phosphor. Reprinted from Ref. [40], with kind permission from Springer Science +Business Media

distorted tetrahedron. As shown in Fig. 10.8a, the three-dimensional framework of this structure can be described as follows: The 10-coordinated Sr-containing polyhedra share faces and vertices with LiO_3F_2 square pyramids and SiO_4 distorted tetrahedra, respectively. In this structure, SiO4 groups are isolated from one another, whereas the LiO_3F_2 square pyramids are connected through F atoms. The two crystallographically distinct Sr sites can be described in detail as follows: A distorted bicapped square anti-prism [Sr(I)] is isformed by one O1, four O2, three O3, and two F atoms, whereas a distorted bicapped dodecahedron [Sr(II)] is formed by three O1, four O2, one O3, and two F atoms [38, 39].

$Sr_2LiSiO_4F{:}Ce^{3+}$, Li^+ phosphors have been reported. The PL emission spectra of $Sr_{1.98}LiSiO_4F{:}0.01Ce^{3+},0.01Li^+$ under 254-, 287-, 340-, and 360-nm excitation are shown in Fig. 10.8b. It can be seen that the peak of emission spectra shifts under different excitation light. The structure of Sr_2LiSiO_4Fcontains two crystallographically distinct 10-coordinate Sr atoms. Considering the radius of Sr^{2+}, Li^+, Si^{4+}, and Ce^{3+} ($Sr^{2+}r$ = 0.135 nm, CN = 10; $Ce^{3+}r$ = 0.125 nm, CN = 10; Li^+r = 0.07 nm, CN = 5; $Si^{4+}r$ = 0.026 nm CN = 4), we think it is reasonable that the dopant Ce^{3+} ions substitute on the Sr sites and that this emission shift may be attributed to the Ce^{3+} occupying different Sr sites in Sr_2LiSiO_4F [38].

Eu^{2+} luminescence in Sr_2LiSiO_4F is also studied and the emission spectrum shows an intense and broad green emission band [39]. As shown in Fig. 10.8c, the emission spectrum (curve b) can be well deconvoluted into two sub-bands with the maxima at 514 and 553 nm, respectively. The bands can be ascribed to the emission of Eu^{2+} ions occupied at different sites. The FWHM of the emission band is approximately 100 nm. The excitation curve covers a broad range from 250 to 400 nm, which is consistent with the diffuse reflection spectra of Eu^{2+}-doped Sr_2LiSiO_4F sample. Based on the PL emission and excitation spectra, Sr_2LiSiO_4F: Eu^{2+} phosphor is a candidate for the green-emitting component in UV chip–excited white LEDs.

It is also reported that he co-doping of Ce^{3+} can enhance the absorption of Eu^{2+} in the near-UV region. Ce^{3+}-co-doped Sr_2LiSiO_4F samples show broadband emission ranging from 360 to 620-nm in the blue to green–yellow region that results in white emission. This indicates partial energy transfer from Ce^{3+} to Eu^{2+}. Selected compositions of $Sr_2LiSiO_4F{:}Ce^{3+}$, Eu^{2+} phosphors can find potential application in solid-state lighting [41].

10.3.16 ($CaCl_2/SiO_2$)

($CaCl_2/SiO_2$) belongs to s kind special halo-silicate material, which is a heterogeneous luminescent material host [42]. XRD patterns of ($CaCl_2/SiO_2$) show only lines corresponding to the α-quartz of SiO_2. Whatever else is present must be either amorphous or in amounts too small to be detected by routine x-ray analysis. It was also found that the ratio of $CaCl_2/SiO_2$ will affect the luminescence behavior, and the optimum ratio has been determined by Hao et al. An intense violet–blue emitting ($CaCl_2/SiO_2$):Eu^{2+} phosphor with a composition of 25 % $CaCl_2$ and 75 % SiO_2 is prepared by a solid-state reaction. The phosphor emits at 427 nm with a narrow bandwidth of 21 nm. ($CaCl_2/SiO_2$):Eu^{2+} would be a promising new phosphor for converting near-ultraviolet radiation to violet–blue emission for a novel phototherapy illuminator using a near-ultraviolet (approximately 395 nm) light–emitting diode as the excitation source.

10.4 Halo-Phosphate Phosphors

Study on the halo-phosphate phosphors should originate from the apatite-type alkaline–earth halophosphates with the general formula of $M_5(PO_4)_3X$ (M = Ca, Sr, Ba; X = F, Cl, Br) [46, 47]. Such kinds of halo-phosphate as hosts of luminescence materials have been widely investigated because of the flexibility of the host lattice, good thermal stability, inexpensive raw materials, and simple synthesis conditions. Apatite-type phosphors also play an important role in solid-state lighting and the display industry, especially the fluorapatites, which have drawn much attention; they are good host lattices for luminescence and laser materials because fluorine atoms have the greatest electronegativity and exhibit the strongest electron-accepting ability. Thus, phosphors that contain fluorine atoms usually have good luminescence properties.

As for the apatite-type $M_5(PO_4)_3X$ structure, there are two types of cationic sites (M_1 and M_2), which are the sevenfold coordinated 6 h sites and ninefold coordinated 4f sites, respectively. As an example, Fig. 10.9a gives the unit-cell diagram of a complex apatite-type compound, $Ca_6Y_2Na_2(PO_4)_6F_2$. We can find that the Ca, Y,

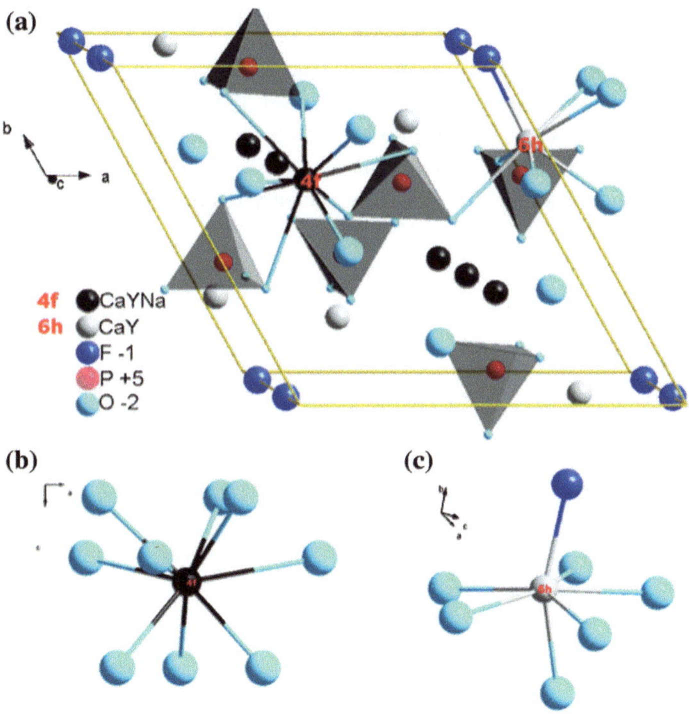

Fig. 10.9 **a** Unit cell diagram of the apatite $Ca_6Y_2Na_2(PO_4)_6F_2$ and the coordination atoms representation of the 4f site (CaYNa). **b** and the 6 h site (CaY) (**b**). Reproduced from Ref. [46] by permission of The Royal Society of Chemistry

Na occupy the M sites, and there are two different sites for the cations. It is known that all of the cations' sites at M positions are suitable and easily accommodate a great variety of foreign cations, for example, rare-earth ions, transition-metal ions, and alkaline–earth ions with different ionic radii, which make it possible to design a novel luminescence material. Except for the apatite-type halophosphates phosphors, many other halo-phosphate phosphors have been also explored in these years, and many different applications, such as w-LEDs, long-lasting phosphorescence lighting signal indication, biological fluorescence label, etc., have been studied. Here, we will select some important halo-phosphates phosphors for detailed introduction.

10.4.1 $Ca_5(PO_4)_3X$ (X = F, Cl)

$Ca_5(PO_4)_3X$ (X = F, Cl) belongs to a series of typical halophosphates phosphors, and several different activator ions can be doped in such a host showing efficient luminescence [48–52]. The luminescence properties of Ce^{3+} ions in $Ca_5(PO_4)_3X$ (X = F, Cl) have been studied. The excitation bands of $Ca_{5-2x}Ce_xNa_x(PO_4)_3Cl$ are shifted to lower energies compared with $Ca_{5-2x}Ce_xNa_x(PO_4)_3F$. Co-doping with Na^+ ions results in a remarkable increase of the luminescence intensity of Ce^{3+}-doped $Ca_5(PO_4)_3X$ apatites due to the charge compensation. Figure 10.10(i) shows the excitation and emission spectra of $Ca_{5-2x}Ce_xNa_x(PO_4)_3Cl$. The excitation spectrum consists of three peaks at 235, 273, and 327 nm. The large emission band is split into two bands at 365 and 386 nm. With increasing Ce^{3+} concentration, the emission intensity of the samples increases and reaches to a maximum at $x = 0.006$. Beyond this concentration, the emission intensity decreases [48].

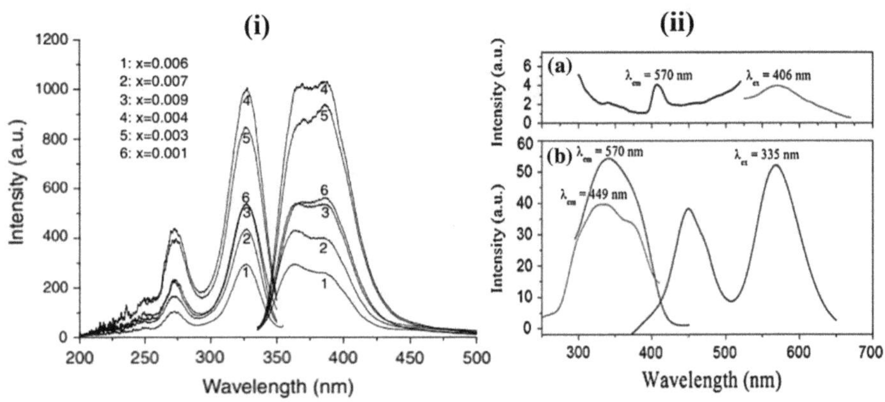

Fig. 10.10 **i** Excitation (*left*) and emission (*right*) spectra of $Ca_{5-2x}Ce_xNa_x(PO_4)_3Cl$. Reprinted from Ref. [48], Copyright 2004, with permission from Elsevier. **ii** PL and PLE spectra of phosphors CPF:0.08Mn^{2+} (**a**) and CPF:0.025Eu^{2+},0.08Mn^{2+} (**b**). Reproduced from Ref. [49] by permission of John Wiley & Sons Ltd

Eu^{2+}/Mn^{2+}-co-doped $Ca_5(PO_4)_3F$ phosphors have been reported. Their luminescence properties show that the developed $Ca_5(PO_4)_3F:Eu^{2+}/Mn^{2+}$ phosphors can efficiently convert UV light in a broad range from 250 to 420 nm into tunable white emission [49]. Figure 10.10(ii) shows clear evidence for energy transfer in the $Ca_5(PO_4)_3F:0.025Eu^{2+},0.08Mn^{2+}$ sample. The PLE spectrum of CPF:$0.025Eu^{2+},0.08Mn^{2+}$ monitored at 570 nm (Mn^{2+}emission) is similar to that of CPF:$0.025Eu^{2+}$ monitored at 449 nm. The presence of the broadband transition from the Eu^{2+} ions in the excitation spectrum, monitored at the $4f^65d^1 \rightarrow 4f^7$ transition of Mn^{2+}, strongly proves the occurrence of energy transfer from Eu^{2+} to Mn^{2+}. The excitation of Eu^{2+} ions into the PLE band yields emission from both Eu^{2+} and Mn^{2+} ions, which consists of a blue band attributed to the $4f^65d^1 \rightarrow 4f^7$ transition of the Eu^{2+} ions and a yellow band that corresponds to the $^4T_1(^4G) \rightarrow {}^6A_1(^6S)$ transition of the Mn^{2+} ions, thereby demonstrating the existence of efficient energy transfer from the Eu^{2+} to Mn^{2+} ions in the $Ca_5(PO_4)_3F$ lattice. The result verified that these phosphors could be used as blue and yellow double-color–emitting phosphors in n-UV white LEDs. The Eu^{2+}–Tb^{3+} couple can be also co-doped into the $Ca_5(PO_4)_3Cl$host, and the process of energy transfer from Eu^{2+} to Tb^{3+} is observed, the mechanism of which has been assigned to a balance of nonradiative resonant energy transfer caused by the exchange interaction and the phonon-assisted nonradiative process [50].

Dy^{3+} can be also introduced into such a host. Ultrafine $M_5(PO_4)_3F:Dy^{3+}$ (M = Ca, Ba) phosphors were prepared by way of combustion process. The PL excitation spectra show the excitation peaks observed at 250–400 nm due to the $f \rightarrow f$ transition of Dy^{3+} ion, which is useful for solid-state lighting purposes (i.e., mercury-free excitation). The PL emission of Dy^{3+} ion by 348-nm excitation gave an emission at 489 (blue), 582 (yellow), and 675 nm (red) [52, 53].

10.4.2 $Sr_5(PO_4)_3X$ (X = F, Cl, Br)

$Sr_5(PO_4)_3X$ (X = F, Cl, Br) acts as another important apatite-type halo-phosphate phosphor host. As shown in Fig. 10.11a, the photoluminescence emission spectra of $Sr_{10}(PO_4)_6X:Eu^{2+}$ (X = F_2, Cl_2, Br_2 and O) under 365-nm excitation are demonstrated. All of the samples can show violet-blue (when X = O) or blue emission peaking near 450 nm (X = F_2, Cl_2, Br_2). The results verified that Eu^{2+} can show efficient emission in such a system [54].

In addition to Eu^{2+}, Pb^{2+} can also show characteristic emission when Pb-doped $Sr_5(PO_4)_3Cl$ phosphors were prepared [55]. The red emission peaking at 725 nm was observed to be excited by the wavelength at 486 nm, which was ascribed to the Pb^{2+}-phosphate charge-transfer transition. The optimum concentration for Pb^{2+} luminescence was 1 %. At > 1 %, luminescence quenching occurred. Because the radius of Pb^{2+} ion (1.19 Å) is almost the same as that of Sr^{2+} ion (1.21 Å), Pb^{2+} ions can easily substitute for Sr^{2+} ions and act as a luminescence center. The luminescence of Pb^{2+} ion is quite diverse and depends strongly on the host lattice due to the

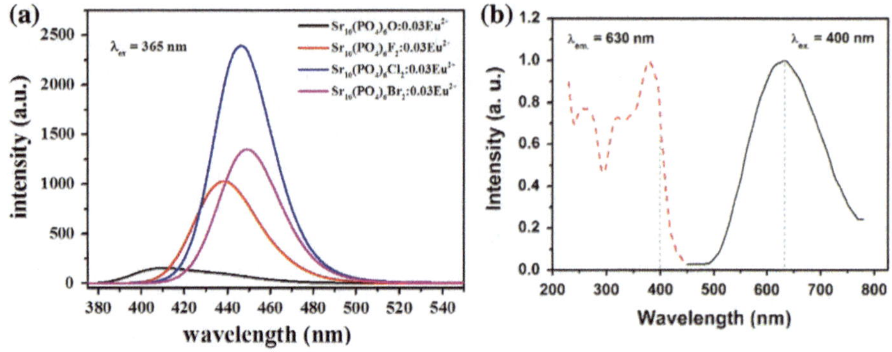

Fig. 10.11 **a** Photoluminescence-emission spectra of $Sr_{10}(PO_4)_6X$ ($X = F_2, Cl_2, Br_2$ and O) under 365-nm excitation. **b** PLE and PL spectra of $Ba_{4.95}Yb_{0.05}(PO_4)_3Cl$ phosphor (**b**). Reprinted from Ref. [58], Copyright 2009, with permission from Elsevier

well-known 6 $s^2 \rightarrow$ 6s6p transitions on the s^2 ion itself or due to a ligand-to-metal charge-transfer state.

Mn-doped $Sr_5(PO_4)_3Cl$ nanorods were also reported [56]. A red-emission band centered at 695 nm corresponding to the $^4T_1 \rightarrow {}^6A_1$ transition of Mn^{2+} ion was seen from the photoluminescence spectra. The optimum concentration for Mn^{2+} luminescence in the $Sr_5(PO_4)_3Cl$ host is approximately 1.5 mol%. Usually the two kinds of ligand environments around Mn^{2+} are proposed as a tetrahedral site and an octahedral site. The tetrahedrally coordinated Mn^{2+} gives a green emission by forming a weak crystal field, whereas the octahedrally (or higher) coordinated Mn^{2+} gives an orange to red emission by forming a strong crystal field. $Sr_5(PO_4)_3Cl$ belongs to the apatite structure, and two different crystallographic sites for the cations are available. Whichever sites the Mn^{2+} ions locate on, they all have a high coordination number (7 or 9) and a strong crystal field. Thus, Mn^{2+} in $Sr_5(PO_4)_3Cl$ host shows red emission.

10.4.3 $Ba_5(PO_4)_3X$ (X = F, Cl)

As discussed previously, Eu^{2+}-doped $M_5(PO_4)_3X$ (M = Ca, Sr, Ba; X = F, Cl, Br) phosphors generally show efficient blue emission, which can be used in tricolor fluorescent lamps and LEDs [53]. Except for this, it was interestingly found that $Ba_5(PO_4)_3Cl:Eu^{2+}$and $Ba_5(PO_4)_3Cl:Eu^{2+}$, Ce^{3+} can be used as long-persistent phosphors, but their afterglow intensity is very weak. The persistent luminescence of $Ba_5(PO_4)_3Cl:Eu^{2+}$, R^{3+} was enhanced with co-dopants Y^{3+}, La^{3+}, Ce^{3+}, Gd^{3+}, Tb^{3+}, and Lu^{3+}.

Barium haloapatite ($Ba_5(PO_4)_3X$, X = F, Cl) nanocrystals doped with Ce^{3+} ions were synthesized by a precipitation method. The emission intensity of Ce^{3+}-doped $Ba_5(PO_4)_3Cl$ was strongly dependent on its concentration. The optimum concentration for Ce^{3+} luminescence was 0.4 mol%. The sites of Ce^{3+} located in $Ba_5(PO_4)_3F$ and $Ba_5(PO_4)_3Cl$ are different, so their PL properties were also very different. It was found that the emission spectra excited by 270 nm exhibits the presence of three bands located at 345, 376, and 404 nm. It is well-known that the Ce^{3+} emission band consists of a doublet corresponding to the splitting between the $^2F_{5/2}$ ground state and $^2F_{7/2}$ state due to spin-orbit coupling. For Ce^{3+}-doped $Ba_5(PO_4)_3Cl$ nanocrystals, the two emission bands correspond to a splitting > or < 2000 cm^{-1} and must correspond to different Ce^{3+} sites [57].

Yb^{2+} ion-doped $Ba_5(PO_4)_3Cl$ phosphor was also reported. Four distinct absorption bands were observed in the UV-light region due to the electronic transitions of Yb^{2+} ion from the $^1S_0$ ground state to the $^2F_{5/2}(t_{2g})$, $^2F_{5/2}(e_{.g.})$, $^2F_{7/2}(t_{2g})$, and $^2F_{7/2}(e_{.g.})$ excited states [58]. Figure 10.11b gives the PLE and PL spectra of $Ba_{4.95}Yb_{0.05}(PO_4)_3Cl$ phosphor. The main emission wavelength of the phosphor was approximately 630 nm. The optimized Yb^{2+} ion concentration was 0.2 mol%, and concentration quenching was observed above this value due to the electric dipole–dipole interaction.

10.4.4 *$(Sr_3,Ca,Ba)(PO_4)_3Cl$*

$(Sr_3,Ca,Ba)(PO_4)_3Cl$ are isostructural to the apatite-type $Sr_5(PO_4)_3Cl$ compounds, which has cell parameters of a = 9.859 Å, c = 7.206 Å, V = 606.58 Å^3, and N = 2. Single-phased $(Sr_{3-x},Ca_{1-y-z},Ba)(PO_4)_3Cl{:}xEu^{2+}$, yTb^{3+}, zMn^{2+} phosphors were synthesized by high-temperature solid-state reaction, and their luminescent properties were investigated. Under UV excitation, white-light emission was obtained from the as-prepared triactivated phosphors by way of combining three emission bands centered at 450, 543, and 570 nm contributed by Eu^{2+}, Tb^{3+}, and Mn^{2+}, respectively. Moreover, both spectral overlapping and lifetime-decay analyses suggest that dual-energy transfers, that is, $Eu^{2+}\rightarrow Tb^{3+}$ and $Eu^{2+}\rightarrow Mn^{2+}$, play key roles in obtaining the white emission [59].

10.4.5 *$Ca_{2.6}Sr_{2.4}(PO_4)_3Cl$*

$Ca_{2.6}Sr_{2.4}(PO_4)_3Cl$ is also isostructural to the apatite-type $Ca_5(PO_4)_3Cl$ phosphor [60]. PL measurement showed that the $Ca_{2.6}Sr_{2.4}(PO_4)_3Cl{:}Eu^{3+}$phosphor exhibited bright orange emission. The broad excitation band from 370 to 500 nm is ascribed to the 4f → 4f transitions of Eu^{3+} in the host lattices and the strong excitation bands at 396 and 441 nm attributed to the $^7F_0 \rightarrow {}^5L_6$ and $^7F_0 \rightarrow {}^5D_3$ transitions of Eu^{3+}, respectively. On excitation with 396-nm UV irradiation, the phosphor has a sharp red emission bands at 593 nm. As we know, the emission spectra are described by the well-known $^5D_0 \rightarrow {}^7F_J$ (J = 1, 2, 3) emission lines of the Eu^{3+} ions with a strong emission for J = 1 at 593 nm, which allows the Eu^{3+} to occupy a center of symmetry in the host lattice.

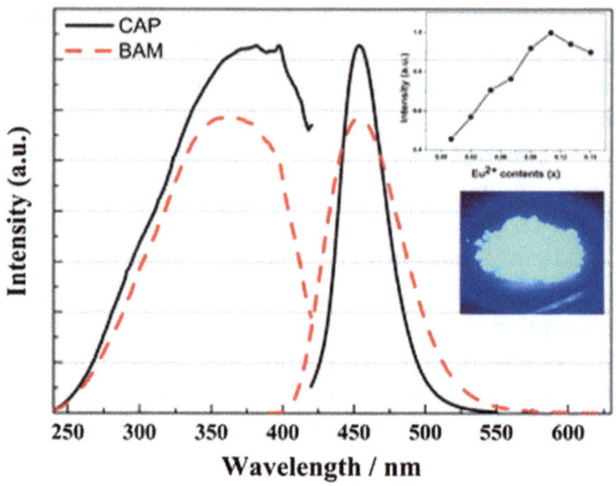

Fig. 10.12 PL and PLE spectra of Ca_2PO_4Cl:11 % Eu^{2+} compared with those of the commercial phosphor BAM. *Inset* PL intensity of $(Ca_{1-x}Eu_x)_2PO_4Cl$ as a function of Eu^{2+} content. Reproduced from Ref. [62] by permission of The Royal Society of Chemistry

10.4.6 Ca_2PO_4Cl

The crystal structure of Ca_2PO_4Cl does not belong to the apatite family. Greenblatt et al. first reported the X-ray single crystal structure of Ca_2PO_4Cl, which crystallizes in the orthorhombic system with a space group of *Pbcm* and with four formula units per unit cell [61]. The crystal structure consists of discrete and distorted $PO_4{}^{3-}$ tetrahedra, which are held together primarily by Ca^{2+} ions. Two different crystallographic sites are available for the divalent Ca^{2+} ions, one with site symmetry C_2 and the other with site symmetry C_S. In both sites, the Ca^{2+} is coordinated by six oxide ions and two chloride ions. A blue phosphor Ca_2PO_4Cl:Eu^{2+} was reported as a candidate for white LEDs by Chen et al. in 2010 [62]. Figure 10.12 shows a comparison between the PL/PLE spectra of Ca_2PO_4Cl:Eu^{2+} phosphors and those of the blue-emitting $BaMgAl_{10}O_{17}$:Eu^{2+} (BAM) commodity phosphor under the same measurement conditions. The samples presented a blue-emitting band peaking at 454 nm under optimal excitation at 370 nm, and the Stokes shift was estimated to be 5000 cm^{-1}. The luminescent intensity of Ca_2PO_4Cl:Eu^{2+} was found to be 128 % under excitation at 380 nm, 149 % under 400 nm, and 247 % under 420 nm, which is as high as that of BAM.

10.4.7 $Na_2M(PO_4)F$ (M = Sr, Ca)

$Na_2M(PO_4)F$ (M = Sr, Ca) belongs to another typical non–apatite family halo-phosphate phosphor host [63, 64]. The novel green-emitting phosphor Na_2CaPO_4F:Eu^{2+} was synthesized, which has broad excitation bands between 250 and 450 nm with the green emission centered at 506 nm. The optimum concentration of Eu^{2+} in Na_2CaPO_4F:Eu^{2+} is determined to be 0.02 mol. Moreover, Dy^{3+}

can be doped into the $Na_2Sr(PO_4)F$ host, whereupon the PL emission of the Dy^{3+} ion on the 348-nm excitation gives emission at 482 (blue), 575 (yellow), and 670 nm (red). These three different emission bands came from the one origin owing to their having the same excitation wavelength. The transitions involved in the blue, yellow, and red bands of Dy^{3+}ion are well known and have been identified as $^4F_{9/2} \rightarrow {}^6H_{15/2}$, $^6H_{13/2}$, $^6H_{11/2}$ transitions, respectively.

10.5 Halo-Borate Phosphors

Borate compounds have a versatile crystal structure. Boron-containing polyhedra can be formed in different forms, and the connections of the polyhedra will in turn lead to many possibilities to obtain new phases [79]. Furthermore, the introduction of halogen atoms will produce more chances to discover many new halo-borate compounds, of which many will act as the excellent phosphor hosts [80, 81]. Herein, we first summarizes the reported halo-borate phosphors including their host, activators, and the corresponding excitation and emission peaks as shown in Table 10.2. Then we will select some important halo-borate phosphors for detailed introduction.

Table 10.2 Summarization of the host compositions, activators and the corresponding excitation and emission peaks of the selected haloborate phosphors

No.	Phosphor		λ_{ex} (nm)	λ_{em} (nm)	References
	Host	Activator			
1	Ca_2BO_3Cl	Eu^{2+}	413	573	[65–67]
		Ce^{3+}	360	422	
		Ce^{3+}–Tb^{3+}–Yb^{3+}	369	420, 542, 978	[68]
2	$Sr_2B_5O_9Cl$	Ce^{3+}	308	320, 343	[69, 70]
	$Sr_2B_5O_9Br$		314	326, 349	
3	$Ca_2B_5O_9Cl$	Eu^{2+}	–	454.5	[71]
	$Ca_2B_5O_9Br$			455.2	
	$Sr_2B_5O_9Cl$			424.1	
	$Sr_2B_5O_9Br$			428.9	
	$Ba_2B_5O_9Cl$			416.9	
	$Ba_2B_5O_9Br$			425.4	
4	$Sr_2B_5O_9Br$	Sm^{2+}	405	420	[72, 73]
	$Sr_2B_5O_9Cl$			700, 685	
5	$Sr_2B_5O_9X$ (X = Cl, Br)	Yb^{2+}	340	420	[74]
6	$Sr_2B_5O_9X$(X = Cl, Br)	Pb^{2+}	250	278	[75]
7	$Ba_2Y(BO_3)_2Cl$	Ce^{3+}–Tb^{3+}–Eu^{3+}	360	592, 612, 624	[76]
		Eu^{2+}	365	526	[3]
		Ce^{3+}	355	417	[77]
8	$Sr_5(BO_3)_3Cl$	Eu^{2+}	380	615	[78]

10.5.1 Ca_2BO_3Cl

The crystal structure of the calcium borate chloride Ca_2BO_3Cl phosphor was first reported to crystallize in the monoclinic system with a space group of $P2_1/c$ by Majling et al. [82]. However, in 2007, the researchers reported that a bright yellow emission originated from Eu^{2+} in Ca_2BO_3Cl and that such a phosphor show great potential as an appropriate yellow-emitting phosphor for white LEDs [65]. Since then, studies on halo-borate phosphors have drawn much attention.

$Ca_2BO_3Cl:Eu^{2+}$ was first reported as an efficient yellow-emitting phosphor under near-ultraviolet and blue excitation. As shown in Fig. 10.13(i), the emission spectrum shows a single intense broad emission band centered at 573 nm, which corresponds to the allowed *f–d* transition of Eu^{2+}. The excitation spectrum is very broad extending from 350 to 500 nm, which is coupled well with the emission of UV LED (350–410 nm) and blue LED (450–470 nm) [65]. After that, $Ca_2BO_3Cl:Ce^{3+}$ was also reported, and Fig. 10.13(ii) shows the PLE and PL spectra monitored by different wavelengths [66, 67]. The dominant emission band from Ce (1) is located at approximately 435 nm (excited at 350 nm) with the weaker emission band from Ce (2) at 405 nm (Fig. 10.13(ii)a). Figure 10.13(ii)b shows the dominant emission band at 470 nm (excited at 375 nm) from Ce (2). As shown in Fig. 10.13 (ii), the intensity of the emission band excited at 375 nm is greater than the band excited at 350 nm, and the absorption at 375 nm matches well with the commercially available near-UV LED chip.

Because Eu^{2+} and Ce^{3+} can show corresponding yellow and blue emission in the Ca_2BO_3Cl host, as expected, the emission spectrum of Ce^{3+}/Eu^{2+}-co-doped Ca_2BO_3Cl shows not only a strong blue band of Ce^{3+} but also a strong yellow band of Eu^{2+}. Figure 10.13(iii)a shows the PL spectra for phosphors Ca_2BO_3Cl:6 % Ce^{3+}, nEu^{2+} with different Eu^{2+} concentrations, which are excited at 349 nm corresponding to the optimal excitation wavelength of Ce^{3+}. With increasing Eu^{2+} content (n), the PL intensity of Eu^{2+} is observed to increase, whereas that of Ce^{3+} is simultaneously found to decrease gradually, which confirms the existence of energy transfer between Ce^{3+} and Eu^{2+} in the host Ca_2BO_3Cl. Moreover, Fig. 10.13(iii)b illuminates the PLE and PL spectra for the typical sample of Ca_2BO_3Cl:6 %Ce^{3+}, 1 % Eu^{2+}. It is found that two broad emission bands centering at 421 and 575 nm in the PL spectrum are attributed to the emission of Ce^{3+} and Eu^{2+}, respectively. The inset herein shows a schematic of the energy-level system, which describes the energy transfer from Ce^{3+} to Eu^{2+} [67].

Except for the Eu^{2+} and Ce^{3+} luminescence in the Ca_2BO_3Cl host, near-infrared (NIR) quantum-cutting (QC) $Ca_2BO_3Cl:Ce^{3+}$, Tb^{3+}, Yb^{3+} phosphor is promising for the development of luminescent solar concentrators (LSC) with Si solar cells [68]. The NIR emissions of approximately 978, 1012, and 1067 nm due to the $^2F5/2 \rightarrow {}^2F_{7/2}$ transitions of Yb^{3+} ion can be observed in such a system. The NIR emission intensity of $Ca_2BO_3Cl:Ce^{3+}$, Tb^{3+}, Yb^{3+} on broadband excitation of Ce^{3+}

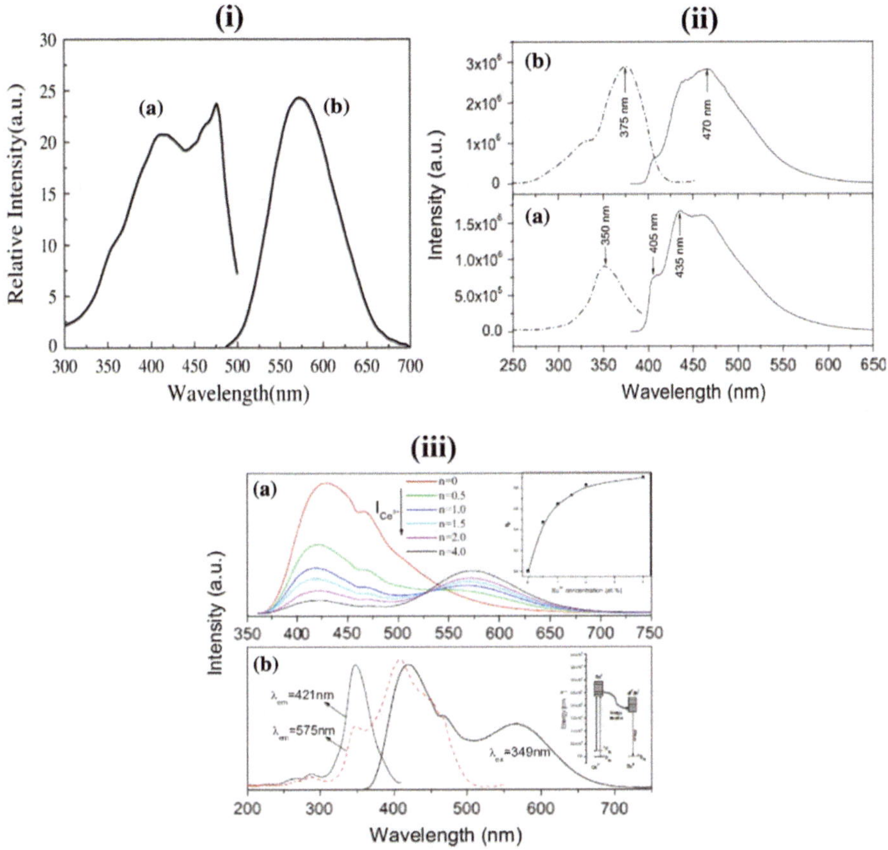

Fig. 10.13 **i** The excitation (*a*, $\lambda_{em} = 573$ nm) and emission (*b*, $\lambda_{ex} = 413$ nm) spectra of $Ca_2BO_3Cl:Eu^{2+}$. Reprinted from Ref. [65], Copyright 2007, with permission from Elsevier. **ii** Emission (*solid lines*) and excitation (*dashed lines*) of $Ca_2BO_3Cl:Ce^{3+}$. **a** Site 1, $\lambda_{ex} = 350$ nm for the emission spectrum and $\lambda_{em} = 405$ nm for the excitation spectrum. **b** Site 2, $\lambda_{ex} = 375$ nm for the emission spectrum and $\lambda_{em} = 470$ nm for the excitation spectrum. Reproduced from Ref. [66] by permission of The Electrochemical Society. **iii** (**a**) PL spectra for Ca_2BO_3Cl:6 % Ce^{3+}, nEu^{2+} phosphors excited at 349 nm. *Inset* Dependence of the energy transfer efficiency on Eu^{2+} concentration. (**b**) PL and PLE spectra for Ca_2BO_3Cl:6 %Ce^{3+},1 %Eu^{2+}. *Inset* Schematic energy level describing energy transfer in the $Ca_2BO_3Cl:Ce^{3+}$, Eu^{2+} phosphor. Reproduced from Ref. [67] by permission of The Electrochemical Society

ion at 288, 315, and 369 nm is approximately 9.1, 10, and 2.8 times as intense as that of $GdBO_3:Tb^{3+}$, Yb^{3+} with the highest NIR quantum efficiency of approximately 182 % on narrow excitation of Tb^{3+} ion at 488 nm. This demonstrates that Ce^{3+} ion can be an efficient sensitizer to harvest UV photons and greatly enhances the NIR emission of Yb^{3+} ion through efficient energy feeding by the allowed 4f–5d absorption of Ce^{3+} ion with high oscillator strength.

10.5.2 $Sr_5(BO_3)_3Cl$

In 1992, Alekel et al. reported the crystal structure of $Sr_5(BO_3)_3Cl$ for the first time and found that it had a structure of stoichiometrically mimicked apatite compounds [78]. In fact, $Sr_5(BO_3)_3Cl$ shows an orthorhombic structure (space group $C222_1$). The inset of Fig. 10.16 shows the crystal structure of $Sr_5(BO_3)_3Cl$, and there are four Sr-atom coordination types: $Sr(1)O_5Cl_2$, $Sr(2)O_7Cl_2$, $Sr(3)O_8$, and $Sr(4)O_8$. The latter two sites are crystallographically distinct SrO_8 square anti-prisms but have a close bond distance. The PL and PLE spectra of $Sr_{4.955}Eu_{0.045}(BO_3)_3Cl$ at room temperature are shown in Fig. 10.16. It is observed that four broad bands at approximately 270, 332, 380, and 432 nm can be distinguished in the excitation spectra (monitored at 615 nm). The low-energy bands (290–490 nm) are due to the 4f → 5d transitions of Eu^{2+}, and the high-energy band (250–280 nm) is assigned to the host absorption. It is expected that $Sr_{4.955}Eu_{0.045}(BO_3)_3Cl$ can match with UV (350–410 nm) and blue (450–470 nm) lights from GaN- and InGaN-based LEDs. The PL spectrum consists of a broad asymmetric emission band at approximately 615 nm due to the 5d → 4f transitions of Eu^{2+} ion in$Sr_5(BO_3)_3Cl$. This asymmetric band can be deconvoluted into two sub-bands at approximately 600 and 639 nm, respectively.

10.5.3 $M_2B_5O_9X$ (M = Ca, Sr, Ba; X = Cl, Br)

During the development of X-ray storage phosphor or the low γ-ray sensitive storage phosphor materials for use in position sensitive thermal neutron detection, the halo-borate phosphors represented by $Sr_2B_5O_9Cl:Eu^{2+}$ have attracted much attention [69–75]. Researchers developed a series of $M_2B_5O_9X$ (M = Ca, Sr, Ba; X = Cl, Br)–based phosphors. The Eu^{2+} emission has been extensively used in various applications such as fluorescent and EL lamps, color television, X-ray imaging using photo-stimulated luminescence, etc. The blue emission of Eu^{2+} can also be useful in obtaining thermosluminescence dosimetry (TLD) phosphors for ionising radiations. Therefore, $Sr_2B_5O_9Cl:Eu^{2+}$ was first reported, which is shown to be at least 70 % more sensitive than the conventional $CaSO_4$:Dy phosphor used in the TLD of ionising radiations. Hence, $Sr_2B_5O_9Cl$:Eu should be suitable as a phosphor in personal dosimetry of ionising radiations using thermoluminescence (TL) technique. Figure 10.14(i) shows the PL emission spectra of $Sr_2B_5O_9Cl$:Eu. For a Eu concentration of 0.1 mol%, a prominent emission at 368 nm is obtained along with a weak emission in the region of 410 ± 430 nm. As for $Sr_2B_5O_9Cl$:Eu with 2 mol% Eu, we observed both bands at 368 and 419 nm.

Except for $Sr_2B_5O_9Cl:Eu^{2+}$, researchers have conducted some detailed study on $M_2B_5O_9X$-based phosphors. In the crystal structure of $M_2B_5O_9X$ (M = Ca, Sr, Ba; X = Cl, Br), boron atoms coordinate with three or four oxygen atoms to form [BO_3] and [BO_4] such that these infrastructures comprise the (B_5O_9) network. In contrast,

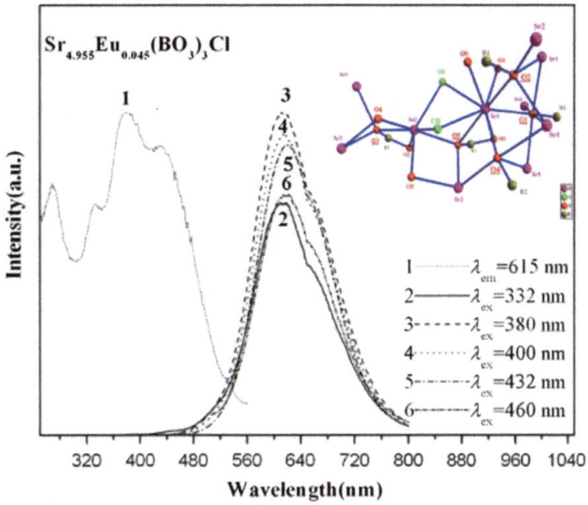

Fig. 10.14 PL (λ_{ex} = 332, 380, 400, 432, and 460 nm) and PLE (λ_{em} = 615 nm) spectra of $Sr_{4.955}Eu_{0.045}(BO_3)_3Cl$ phosphor at room temperature. The *inset* shows the schematic coordinate environment of O atoms in $Sr_5(BO_3)_3Cl$. Reproduced from Ref. [78] by permission of The Electrochemical Society

M atoms occupy two different lattices in $M_2B_5O_9X$. Every M atom has two X coordinating atoms and seven O coordinating atoms. Two different deformed bi-pyramidal polyhedra are formed, in which M is at the center; two X atoms are at the opposite vertex; and oxygen atoms are at the joint of the bi-pyramids. In the Eu-doped $M_2B_5O_9X$ compounds, Eu^{2+} ion substitutes for the M^{2+}ion, and therefore Eu^{2+} occupies two different lattices in $M_2B_5O_9X:Eu^{2+}$ phosphor. To give a representative description on the luminescence properties of $M_2B_5O_9X:Eu^{2+}$ phosphor, Fig. 10.14(ii) shows the typical PLE and PL spectra of Eu^{2+} in $M_{1.95}Eu_{0.05}B_4O_9X$: Eu^{2+} (M = Ca, Sr, Ba; X = Cl, Br). It was believed that all of the $M_{1.95}Eu_{0.05}B_4O_9X:Eu^{2+}$ phosphors have similar excitation and emission spectra; generally, blue emission can be observed from such a system. However, the specific wavelengths of emission peaks are different [80].

The luminescence properties of Ce^{3+}-activated $M_2B_5O_9X$ have been also studied. For example, the luminescence properties of Ce^{3+} ions in strontium haloborates were studied on excitation in the 3.5- to 22-eV region. In $Sr_{2(1-x)}Ce_{2x}B_5O_9R$ (R = Cl, Br; x $\leq$ 0.01) solid solutions, two principal Ce^{3+} centers are observed. The dominant center was found to be produced by the direct substitution of the dopant ion for Sr^{2+} without a local charge compensation. The other center is ascribed to an association of a Ce^{3+} ion and a cation vacancy. The charge-compensation mechanism on the cation sub-lattice includes the formation of one vacancy per two Ce^{3+} ions incorporated [81].

The storage phosphor properties of compounds $M_2B_5O_9X:Eu^{2+}$ (M = Ca, Sr, Ba; X = Cl, Br) have been studied. The best compounds, $Sr_2B_5O_9Br:Eu^{2+}$ and $Ca_2B_5O_9Br:Eu^{2+}$, show a photo-stimulated luminescence yield of, respectively, 21 and 14 photons per incident thermal neutron. They contain natural boron. With these yields, image plates with detective quantum efficiency values >40 % are envisaged when using enriched ^{10}B, but the stimulation energy at 633 nm is too high.

Radiation-induced defects in polycrystalline pure and Ce^{3+}-doped $Sr_2B_5O_9Br$ storage phosphors have been investigated using EPR and optical-absorption methods. The EPR of irradiated pure $Sr_2B_5O_9Br$ represents two overlapping spectra in the 335- to 340-mT range from paramagnetic electron- and hole-trapping centres, which are stable at room temperature. These centers are attributed to $F(Br^-)$ and O^-_{Br} centres, i.e., an electron trapped in a bromine vacancy and a hole trapped on an oxygen ion at a bromine site. The 1 s → 2p transitions of $F(Br^-)$ centres cause the optical absorption band at 560 nm. Another observed absorption band at 365 nm was attributed to the transitions of O^- centers. Thus, electron- and hole-trapping in pure $Sr_2B_5O_9Br$ occurs in the V_{Br} and O^{2-}_{Br} aggregates created during the synthesis. The treatment of reflection spectra of $Sr_2B_5O_9Br$ doped with Ce^{3+} is more complicated than in the case of pure material. During the measurements, a sample is excited with the whole spectrum of the xenon lamp (270–820 nm). Therefore, in reflection curves the bands due to Ce^{3+} absorption and Ce^{3+} luminescence at 330 and 350 nm are very intense. As can be seen in the inset of Fig. 10.14(iii), the reflectivity is >100 % at a particular wavelength, which is caused by Ce^{3+} luminescence, which in turn can be explained by the partial absorption of Ce^{3+} luminescence in the material after irradiation. Therefore, Ce^{3+} emission intensities in reflection curves measured before and after irradiation are different. Furthermore, a correct differential spectrum is also shown in Fig. 10.14(iii). The spectrum was fitted with two Gaussians with maxima at 380 and 580 nm.

10.5.4 $M_2B_5O_9X{:}Sm^{2+}$

The luminescence properties of Sm^{2+}-activated stoichiometric and non-stoichiometric $Ba_2B_5O_9Cl$ were investigated from 16 to 450 K. In stoichiometric $Ba_2B_5O_9Cl$, the emission spectra of Sm^{2+} are composed of $4f^6 \rightarrow 4f^6$ transitions over the 16- to 450-K temperature range studied. Luminescence was observed from Sm^{2+} in four inequivalent cationic sites at 16 K and two inequivalent sites at room temperature. In the non-stoichiometric compound, the emission is characterized by the $4f^6 \rightarrow 4f^6$ transitions at low temperature and $4f^55d^1 \rightarrow 4f^6$ transitions at high temperature. Sm^{2+} doped into the Ca^{2+} and Sr^{2+} analogues in the $M_2B_5O_9Cl$ series show the broadband $4f^55d^1 \rightarrow 4f^6$ luminescence in both the stoichiometric and non-stoichiometric compounds due to the increased ligand field caused by the smaller ionic radii of the metal ions [72].

The luminescence properties of Sm^{2+} in $Sr_2B_5O_9R$ (R = Cl, Br) have been studied and compared with those in SrB_4O_7. In the range from 80 to 300 K, the emission of Sm^{2+} in $Sr_2B_5O_9R$ is predominantly due to the $4f^55d \rightarrow 4f^6$ transition, whereas $SrB_4O_7{:}Sm^{2+}$ shows a very efficient emission of the $4f^6 \rightarrow 4f^6$ type. The excitation spectrum of $Sr_2B_5O_9Cl{:}Sm^{2+}$ consists of a series of intense broad bands. The spectrum of $Sr_2B_5O_9Cl{:}Sm^{2+}$ at 120 K differs from the room-temperature emission spectrum and consists of a structural band at approximately 700 nm and a relative intense line at 681.9 nm (14665 cm^{-1}). Assuming that this line is the zero-phonon line of the emission from $4f^55d$ configuration, one can conclude that the energy mismatch between the lowest $4f^55d$ crystal-field component and the $^5D_0$ ($4f^6$) level is at least 75 cm^{-1} [73].

10.5.5 $Sr_2B_5O_9X:Yb^{2+}$ and $Sr_2B_5O_9X:Pb^{2+}$

The luminescence properties of ytterbium ions in strontium haloborates were studied under optical, X-ray, and synchrotron excitation. The coexistence of Yb ions in two valence states (divalent and trivalent) was detected in the $Sr_2B_5O_9X$: Yb (X = Cl, Br) powder materials prepared in a slightly reducing (H_2/N_2) or an oxidizing atmosphere. Figure 10.15a shows the emission and UV excitation spectra of a $Sr_{2(1-x)}Yb_{2x}B_5O_9Cl$ ($x = 0.01$) sample in the as-prepared state. The emission spectrum consists of a single band with a maximum at approximately 420 nm. In the excitation spectrum of this emission, several bands at 267, 305, 338, and 374 nm are observed. These bands must be ascribed to transitions between the $4f^{14}$ and $4f^{13}5d$ configurations of the Yb^{2+} ion [74].

The luminescence properties of Pb^{2+} in $Sr_2B_5O_9X$ (X = Cl, Br) are also reported. The emission is situated at very short wavelength (approximately 280 nm). In these compounds, the position of the energy levels of the Pb^{2+} ion is determined by the oxygen ions of the borate groups surrounding the Pb^{2+} ion. The influence of the two surrounding halide ions appears in the Stokes shift, which is smaller for the Pb^{2+} luminescence in bromoborate as for the most diluted sample of $(Sr,Pb)_2B_5O_9Br$,viz. In $Sr_{1.98}BPb_{0.02}B_5O_9Br$, more than one excitation band can be observed at 4.2 K (Fig. 10.15b). There is a stronger band at approximately 254 nm and a weaker band with a maximum at 264 nm. The emission band has its maximum at 278 nm. A second emission band seems to be hidden on the longer-wavelength side of this emission band [75].

10.5.6 $Ba_2Ln(BO_3)_2Cl$ (Ln = Rare Earth)

Chloroborate phosphors $Ba_2Ln(BO_3)_2Cl:Eu^{2+}$ (Ln = rare-earth element) were first mentioned by Schipper in 1994 [83]. After that, our group published a pioneering study on the crystal-structure analysis, luminescence-properties study and related mechanism and theory of luminescence ions [3, 84]. The successful isomorphic substitution for Ln^{3+} sites in the $Ba_2Ln(BO_3)_2Cl$ host can be realized by different trivalent rare-earth ions, such as Y, Sm, Eu, Gd, Tb, Dy, Ho, Er, Tm, Yb, and Lu, which confirms that the compounds can only accommodate the smaller lanthanide ions [3]. $Ba_2Ln(BO_3)_2Cl$ phases are crystallized in a monoclinic unit cell with the space group $P2_1/m$. The structure unit of $Ba_2Ln(BO_3)_2Cl$ is constructed by isolated BO_3 triangles—(Ba-O, Cl) and (Ln-O) polyhedra—and there are two Ba sites and on Ln site, which appear as $Ba(1)O_5Cl_4$, $Ba(2)O_8Cl_2$, and LnO_7 polyhedra. The

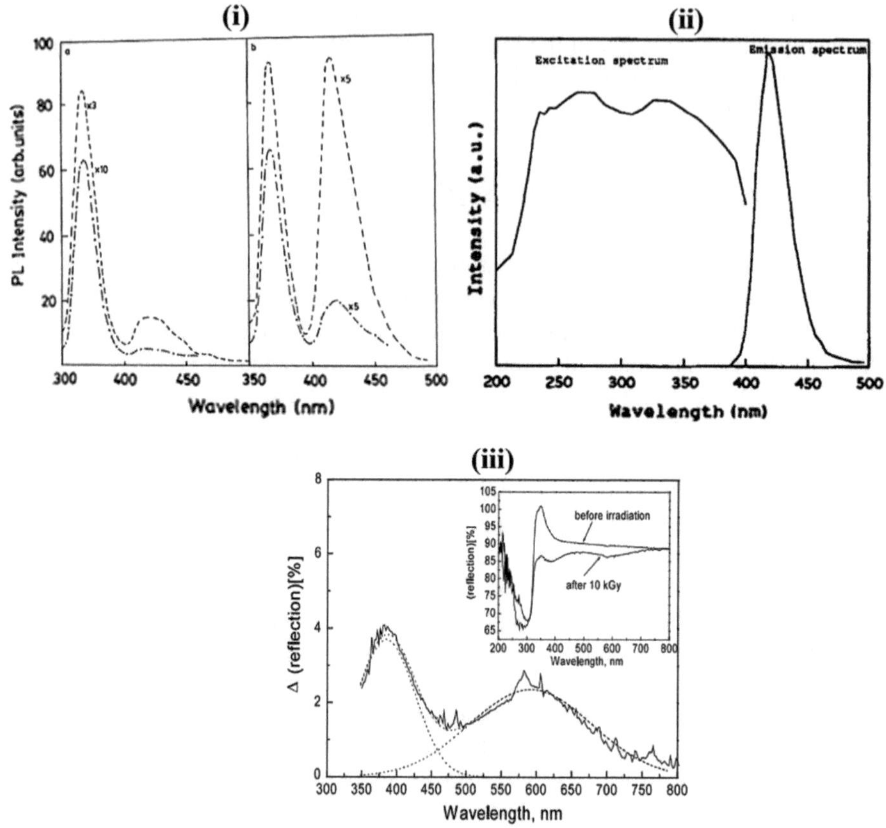

Fig. 10.15 **i** PL emission spectra of $Sr_2B_5O_9Cl$:Eu phosphor **a** 0.1 mol% Eu and **b** 2.0 mol% Eu (λ_{ex} = 335 nm). Reprinted from Ref. [80], Copyright 2000, with permission from Elsevier. **ii** Typical PLE and PL spectra of Eu^{2+} in $M_{1.95}Eu_{0.05}B_4O_9X:Eu^{2+}$ (M = Ca,Sr,Ba; X = Cl, Br). Reprinted from Ref. [79], Copyright 1995, with permission from Elsevier. **iii** Reflection spectrum of $Sr_2B_5O_9Br$:1 %Ce^{3+} recorded before irradiation minus the reflection spectrum recorded after irradiation with a ^{60}Co gamma source. The *dotted curves* represent a fit of the spectrum with two Gaussian peaks. *Inset* shows the reflection spectra recorded before and after irradiation (10 kGy with a ^{60}Co gamma source) are plotted. Reprinted from Ref. [81] by permission of IOP Publishing

detailed luminescence properties on such a host will be discussed later based on the research in our group.

In addition to our work, several other groups have also made some progress in this system. Guo's group studied the energy-transfer process of Ce^{3+}/Tb^{3+} in $Ba_2Ln(BO_3)_2Cl$, and they also studied the preferred site of Eu^{3+} in such a host [85, 86]. Recently, they also studied $Ce^{3+}/Tb^{3+}/Yb^{3+}$ tri-doped $Ba_2Y(BO_3)_2Cl$ phosphors with intense near-infrared (NIR) emission and broadband absorption in the near-ultraviolet region. Zhang et al. reported a series of $Ce^{3+}/Tb^{3+}/Eu^{3+}$ tri-doped $Ba_2Y(BO_3)_2Cl$ red-emitting phosphor. The $Ce^{3+} \rightarrow Tb^{3+} \rightarrow Eu^{3+}$ energy-transfer

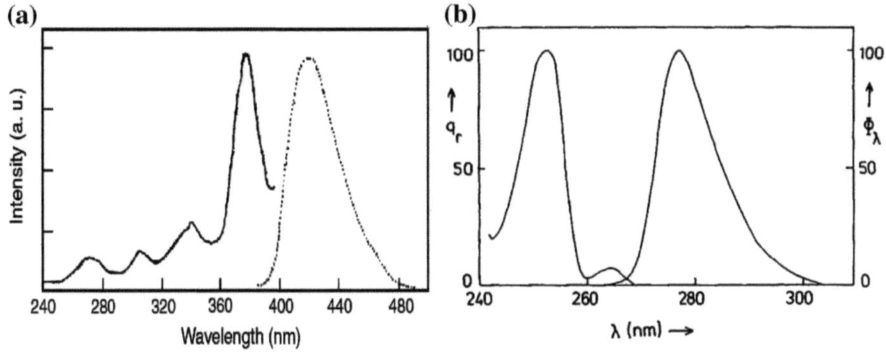

Fig. 10.16 **a** Emission and excitation spectra of $Sr_{2(1-x)}Yb_{2x}B_5O_9Cl$ ($x = 0.01$) at 300 K ($\lambda_{ex} = 340$ nm, $\lambda_{em} = 420$ nm). Reprinted from Ref. [74], Copyright 2002, with permission from Elsevier. **b** Excitation and emission spectra ($\lambda_{em} = 271$ nm, $\lambda_{ex} = 250$ nm) of the luminescence of $Sr_{1.98}BPb_{0.02}B_5O_9Br$ at 4.2 K. Reprinted from Ref. [75], Copyright 1988, with permission from Elsevier

scheme has been proposed to realize the sensitization of Eu^{3+} ion emission by Ce^{3+} ions. Energy transfers from Ce^{3+} to Tb^{3+}, and Tb^{3+} to Eu^{3+}, as well as the corresponding energy-transfer process, have been proposed. The combination of narrow-line red emission and near-UV broadband excitation makes $Ba_2Y(BO_3)_2Cl$:Ce^{3+}/Tb^{3+}/Eu^{3+} a novel and efficient red phosphor for NUV LED applications [87].

10.6 Halo-Aluminate Phosphors

There are two typical halo-aluminate phosphors: (1) $Sr_3Al_2O_5Cl_2$:Ce^{3+}/Eu^{2+} phosphors and their derivatives e.g., $Sr_3(Al,Ga)_2O_5Cl_2$:Ce^{3+}/Eu^{2+} and (2) Sr_3AlO_4F:Ce^{3+} and its isostructural solid-solution compounds [88–94]. Here we will give a detailed description of the two series of halo-aluminate phosphors.

10.6.1 $Sr_3Al_2O_5Cl_2$ and $Sr_3Ga_2O_5Cl_2$

$Sr_3Al_2O_5Cl_2$:Eu^{2+} should be the first reported halo-aluminate phosphor that can show orange-yellow emission on NUV excitation at 400 nm [88]. The compound $Sr_3Al_2O_5Cl_2$ crystallizes in an orthorhombic structure with the space group $P2_12_12_1$. Orange-yellow phosphors $Sr_{3-x}(Al_2O_5)Cl_2$:Eu_x in pure phase were synthesized and characterized that were as being packaged in a yellow-emitting LED with a NUV chip with correlated color temperature (CCT) = 2300 K and CRI = 74.

As for $Sr_3Al_2O_5Cl_2$:Eu^{2+}, one broadband emission centered at 609 nm is observed under the excitation of 350 nm, which is attributed to the typical $4f^65d^1 \rightarrow 4f^7(^8S_{7/2})$ transition of the Eu^{2+} ions. In the excitation spectra, a broadband consisting of unresolved bands from 200 to 450 nm is due to the *4f5d* multiplet excited states of the Eu^{2+} ions. As for $Sr_3Al_2O_5Cl_2$:Ce^{3+}, Li^+, it belongs to a novel blue-emitting phosphor. The excitation spectrum shows a broad band extending from 300 to 400 nm, and the emission spectrum shows abroad blue band peaking at 450 nm with a half width of approximately 100 nm. Moreover, a conspicuous spectral overlap between the emission band of the Ce^{3+} ions and the excitation band of the Eu^{2+} ions can be found. Therefore, it is expected that inefficient energy transfer can occur from the Ce^{3+} to the Eu^{2+}ions [89].

In addition to the $Sr_3Al_2O_5Cl_2$ host, investigators have also discovered the $Sr_3Ga_2O_5Cl_2$ matrix, which is not isotopic to $Sr_3Al_2O_5Cl_2$. The $Sr_3Ga_2O_5Cl_2$ phase has a monoclinic structure with the space group $P2_1$, and its lattice parameter is a = b = c = 0.9569 nm. The phase contains two eight-coordinated (Sr–O_5Cl_3), four nine-coordinated (Sr–O_5Cl_4), and two tetrahedral (Ga–O_4) sites. $Sr_3Ga_2O_5Cl_2$:Eu^{2+} can be used as a kind of blue-emitting phosphor for white LEDs [93]. Figure 10.17 gives the excitation and emission spectra of $Sr_3Ga_2O_5Cl_2$:Eu^{2+} phosphor. It shows an efficient broad absorption band from 275 to 410 nm, and an efficient blue emission peaking at 460 nm. It is believed that $Sr_3Ga_2O_5Cl_2$:Eu^{2+} phosphor is a promising blue-emitting component for UV chip–excited white light–emitting diodes.

Moreover, long-persistent luminescence (LPL) phosphors can be also found in such a $Sr_3Al_2O_5Cl_2$system [94–96]. A novel orange-yellow–emitting $Sr_3Al_2O_5Cl_2$:Eu^{2+}, Tm^{3+} LPL phosphor has been recently developed. The incorporation of Tm^{3+} into $Sr_3Al_2O_5Cl_2$:Eu^{2+} as an auxiliary activator dominates its long-persistent luminescence. The emissions in $Sr_3Al_2O_5Cl_2$:Eu^{2+}, Tm^{3+} for both fluorescence and LPL are due to the 5d → 4f transitions of Eu^{2+}. Figure 10.18(i) shows the LPL decay curve of $Sr_3Al_2O_5Cl_2$:Eu^{2+}, Tm^3, and the afterglow can persist for nearly 220 min at

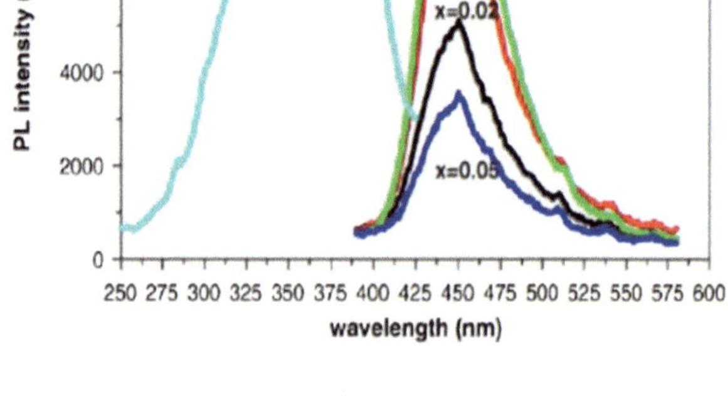

Fig. 10.17 Excitation and emission spectra of $Sr_3Ga_2O_5Cl_2$:Eu^{2+} phosphor. The *inset* shows the Eu^{2+} content–dependent emission intensities. Reprinted from Ref. [93], Copyright 2010, with permission from Elsevier

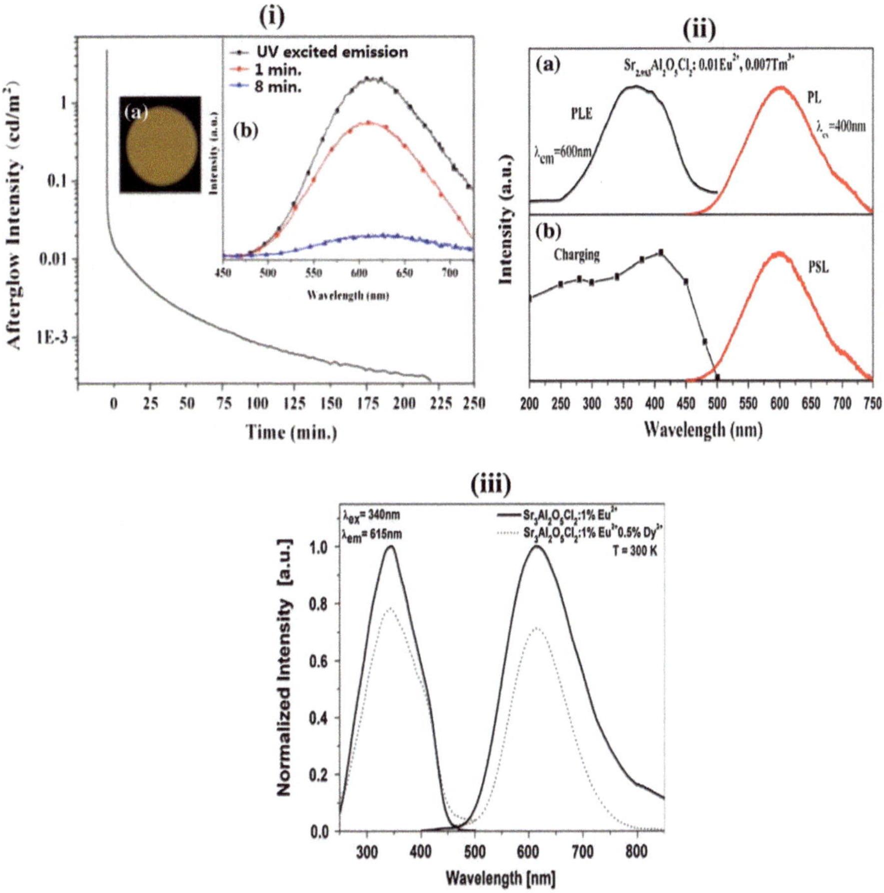

Fig. 10.18 **i** The LPL decay curve of $Sr_3Al_2O_5Cl_2$:Eu^{2+}, Tm^{3+}; *Inset A* Photos of typical sample in the *dark* after the removal of the 365 nm UV light. *Inset B* the LPL spectra measured at different times (1 and 8 min) after removal of the excitation source. Reprinted from Ref. [94] by permission of OSA Publishing. **ii** PL and PLE spectra of $Sr_3Al_2O_5Cl_2$:0.01Eu^{2+}, 0.007Tm^{3+} (**a**) and PSL emission spectrum and PSL response spectrum for charging wavelengths (**b**). Reprinted from Ref. [95], Copyright 2014, with permission from Elsevier. **iii** Excitation and emission spectra of $Sr_3Al_2O_5Cl_2$:1 %Eu^{2+} and $Sr_3Al_2O_5Cl_2$2:1 %Eu^{2+},0.5 %Dy^{3+}. Reprinted from Ref. [96], Copyright 2013, with permission from Elsevier

a recognizable intensity level (>0.32 mcd/m^2). As can be seen in Fig. 10.18(i), when exposed to a 365-nm UV lamp, both $Sr_{2.985}Al_2O_5Cl_2$:$Eu^{2+}_{0.015}$ and $Sr_{2.955}Al_2O_5Cl_2$:$Eu^{2+}_{0.015}$, $Tm^{3+}_{0.03}$ emit intense orange-yellow light; however, after the excitation source is switched off, only $Sr_{2.955}Al_2O_5Cl_2$:$Eu^{2+}_{0.015}$, $Tm^{3+}_{0.03}$ yields orange-yellow LPL [94]. Strong and long-lasting photo-stimulated luminescence (PSL) is also observed in $Sr_3Al_2O_5Cl_2$:Eu^{2+}, Tm^{3+} under 980-nm continuous stimulation after UV- and blue-light pre-irradiation. Compared with the Eu^{2+} singly doped one, the PSL intensity and energy storage capacity in the doubly doped material are enhanced by a factor of 400 and 91, respectively. Figure 10.18(ii)a shows the PLE and PL spectra of $Sr_3Al_2O_5Cl_2$:0.01Eu^{2+}, 0.007Tm^{3+} phosphor. The phosphor exhibits a typical orange-yellow emission band peaking at 600 nm, and the PLE spectrum covers a broad spectral region from 200 to 500 nm as preciously mentioned. Figure 10.18(ii) b presents the PSL emission spectrum and PSL charging spectrum, which gives the initial intensity of the PSL as a function of the excitation wavelengths over the 200- to 500-nm spectral range. The PSL spectrum is almost identical to the PL spectrum, indicating that the PL and PSL spectra have the same emission centers. However, there is a little difference between the PLE and PSL charging spectra. In the PLE spectrum, the photoluminescence cannot be effectively excited by UV irradiation between 200–300 nm. However, the PSL can be effectively achieved by a wide range of wavelengths (200–450 nm). The may perhaps be the reason that the energy of UV light in the 200–300-nm region is high enough to promote electrons in the conduction band and get trapped. Thus, a strong PSL phenomenon can be observed even after 5 min of irradiation by UV light ranging from 200–300 nm [95].

The luminescence and persistent luminescence properties of $Sr_3Al_2O_5Cl_2$:$Eu^{2+}$and $Sr_3Al_2O_5Cl_2$:Eu^{2+}, Dy^{3+} phosphors have been also comparatively investigated. The persistent luminescence of $Sr_3Al_2O_5Cl_2$:$Eu^{2+}$is rather weak and lasts only for only a few seconds, whereas the same material co-doped with Dy^{3+} shows a longer and stronger afterglow, which is similar to the effect of Tm^{3+} doping. The emission and excitation spectra of $Sr_3Al_2O_5Cl_2$:Eu^{2+} and $Sr_3Al_2O_5Cl_2$:Eu^{2+}, Dy^{3+} at 300 K show the same position of maxima (Fig. 10.18(iii)). The emission spectra of both phosphors consist of a broad band peaking at 615 nm, which is attributed to overlapping emission bands for Eu^{2+} ions in three different strontium sites. No additional peaks from Dy^{3+} were observed in the emission spectrum of the sample co-doped withDy^{3+}. The intensity of the emission band of $Sr_3Al_2O_5Cl_2$:1 %Eu^{2+}, Dy^{3+} is weaker compared with that of $Sr_3Al_2O_5Cl_2$:Eu^{2+}. No significant differences were noticed in the excitation spectra of $Sr_3Al_2O_5Cl_2$:Eu^{2+}, Dy^{3+} and $Sr_3Al_2O_5Cl_2$:Eu^{2+} except for different intensity. The excitation spectra of these two phosphors recorded for 615-nm emission consist of a broad band peaking at 340 nm [96].

10.6.2 Sr_3AlO_4F

Sr_3AlO_4F phase belongs to the A_3BX_5 family, which is isostructural to the Sr_3SiO_5 and $LaSr_2AlO_5$ compounds [97]. As for such a host, it contains the Sr_3AlO_4F and Sr_3GaO_4F phase, and these compounds comprise the Sr_2F sheets held together by isolated MO_4 tetrahedra and additional Sr^{2+} ions, which are also isostructural with Sr_2EuAlO_5 and Sr_2GdGaO_5. The luminescence properties of $Sr_{3-2x}Ce_xNa_xAlO_4F$ phosphors were the first investigated [98]. Ce^{3+} ions occupy two different Sr^{2+} sites in these phosphors, and the different emissions derived from the two luminescent centers are observed. It was also found that Ba substitution into the Sr sites can

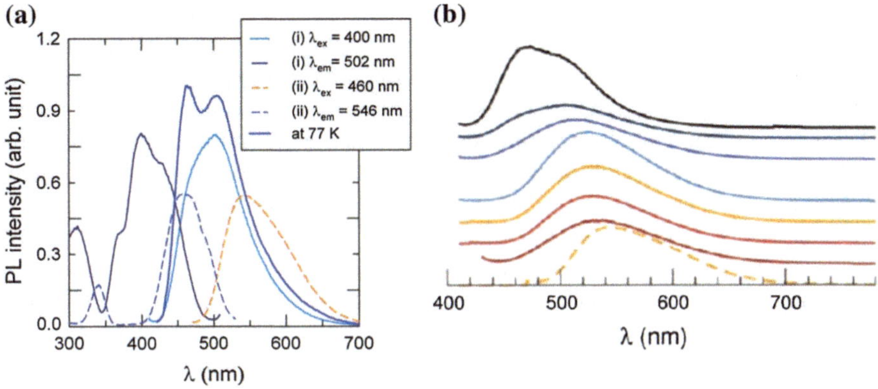

Fig. 10.19 **a** Excitation and emission spectra of **i** $(Sr_{1-x}Ba_x)_3AlO_4F{:}Ce^{3+}(x = 1.0)$ at room temperature compared with **ii** a commercial $YAG{:}Ce^{3+}$phosphor. The PL emission spectra of $SBAF{:}Ce^{3+}(x = 1.0)$ at 77 K are also displayed. Reprinted from Ref. [97], 2010 American Chemical Society. **b** Emission spectra of the solid-solution phosphors $Sr_{2.975}Ce_{0.025}Al_{1-x}Si_xO_{4+x}F_{1-x}$ with the various x values indicated. For comparison, the emission spectrum of $YAG{:}Ce^{3+}$phosphor are also shown as the *dotted line*. Reproduced from Ref. [99] by permission of John Wiley & Sons Ltd

enhance the emission intensities in such a host. For $x = 0.33$ and $y = 0.008$, $(Sr_{0.992-x-y}Ba_xCe_y)_3AlO_4F$ has a broad excitation band centered at approximately 400 nm and emission peak at 502 nm with QY = 95 % [97]. Figure 10.19a shows a comparison of the excitation and emission spectra of $(Sr_{1-x}Ba_x)_3AlO_4F{:}Ce^{3+}$ and $YAG{:}Ce^{3+}$phosphor, and the results shown here suggest a PL intensity (λ_{ex} = 400 nm) that is approximately 150 % that of commercial YAG: Ce^{3+}(λ_{ex} = 460 nm; P46-Mitsubishi Chem). Unfortunately, these powders have poor chemical stability and degrade on contact with moisture.

Solid solutions of Ce^{3+}-activated $(Sr_3AlO_4F)_{1-x}$–$(Sr_3SiO_5)_x$ have been prepared and the luminescence properties have been studied [99]. The emission spectra for $[Sr_{2.975}Ce_{0.025}Al_{1-x}Si_xO_{4+x}F_{1-x}]{:}Ce^{3+}$are shown in Fig. 10.19b. As x increases, the maximum emission peak wavelengths increase. The emission peak positions gradually move toward longer wavelengths in the range from 474 nm for $Sr_3AlO_4F{:}Ce^{3+}(x = 0)$ to 537 nm for $Sr_3SiO_5{:}Ce^{3+}(x = 1.0)$. This 63-nm shift allows for highly color-tunable phosphors, and the QY decreases and then increases with increasing x, which has a maximum of 70 % and a minimum of 54 %. It is believed that solid-solution phosphors have wider emission peaks, which provide enhanced color-rendering properties when incorporated into an LED chip.

In such a Sr_3AlO_4F host, many people have conducted a series of works to develop the phosphors for different applications [100]. It is found that some defect structures can be developed in the present host during exposure to reducing gases, and these defects in anion-deficient nonstoichiometric $(Sr_{3-x}A_x)_{1-\alpha-2\delta}MO_{4-\alpha}F_{1-\delta}$ (A = Ca, Ba) can be monitored by broad-band photoluminescence emissions centered near 500 nm when excited with 254-nm UV light. Samples of $Sr_{2.85}Eu_{0.1}Al_{1-z}In_zO4F$

(z = 0, 0.01, 0.1) were prepared, and the near-UV excited line and broadband photoluminescence can be observed [101]. $Sr_{3-2x}Eu_xNa_xAlO_4F$ phosphors were prepared, and their photoluminescence and cathodoluminescence properties investigated [102]. $Sr_{2.80}Eu_{0.10}Na_{0.10}AlO_4F$ shows a bright red luminescence with a dominant $^5D_0 \rightarrow {}^7F_2$ emission at 618 nm, thus showing that the phosphor may be a potential candidate for field emission display application. Under the excitation of UV light, $Sr_3AlO_4F{:}Tm^{3+}$ and $Sr_3AlO_4F{:}Tb^{3+}$ exhibit the characteristic emissions of $Tm^{3+}(^1D_2 \rightarrow {}^3F_4$, blue) and Tb^{3+} ($^5D_4 \rightarrow {}^7F_5$, green). In addition, the tunable luminescence in Bi^{3+}/Eu^{3+}-co-doped Sr_3AlO_4F have been also observed [103]. Except for this, novel $Ce^{3+}/Tb^{3+}/Yb^{3+}$ tri-doped Sr_3AlO_4F phosphors were synthesized, and the energy-transfer processes from $Ce^{3+} \rightarrow Tb^{3+} \rightarrow Yb^{3+}$ was observed; thus, near-infrared luminescence can be realized in such a system [104].

10.7 Other Oxyhalide-Based Phosphors

Except for the previously mentioned halo-silicate, halo-phosphate, halo-borate, and halo-aluminate phosphors, there are also other interesting halo-containing phosphors such as the following halo-sulphate [105–107] and halo-nitride [108, 109] compounds. Study of the crystal chemistry, synthesis technique, and luminescence properties will be very important.

10.7.1 *KMgSO₄Cl and KZnSO₄Cl*

UV-excited luminescence from Cu-doped halo-sulphate phosphors has been investigated, and they were synthesized by wet-chemical method [105]. The intense emission of the spectrum is assigned to the electronic transitions $3d^94\,s^1 \rightarrow 3d^{10}$ in Cu^+ ions. Figure 10.20(i) shows the PL emission spectra of $KMgSO_4Cl{:}Cu$; with increasing concentration of Cu^+ ions, the peak intensity increases, and maximum intensity is observed for 0.5 mol% of Cu^+ ion. The results suggest that Cu plays an important role in PL emission in this matrix. However, limited hosts with Cu^+ as luminescence center were studied due to difficulties in incorporating it in luminescence form.

Moreover, $KMgSO_4Cl{:}Ce^{3+}$, $KMgSO_4Cl{:}Ce^{3+}$, Dy^{3+}, and $KMgSO_4Cl{:}Ce^{3+}$, Mn^{2+}, were synthesized. Energy transfer from $Ce^{3+}/Dy^{3+}$and Ce^{3+}/Mn^{2+} results in an increase in PL peak intensity, thus suggesting that Ce^{3+} plays an important role in PL emission in the present matrix [107]. The PL emission spectra have two peaks at 482 and 571 nm and a single peak at 564 nm, which could be attributed to the $Ce^{3+}/Dy^{3+}$and Ce^{3+}/Mn^{2+} emissions, respectively. Figure 10.20(ii) gives the excitation and emission spectra of $KZnSO_4Cl{:}Ce_{10}$, $Dy_{0.1}$. A very strong PL emission of Dy^{3+} ions was observed in $KZnSO_4Cl{:}Ce$, Dy phosphor due to the presence of

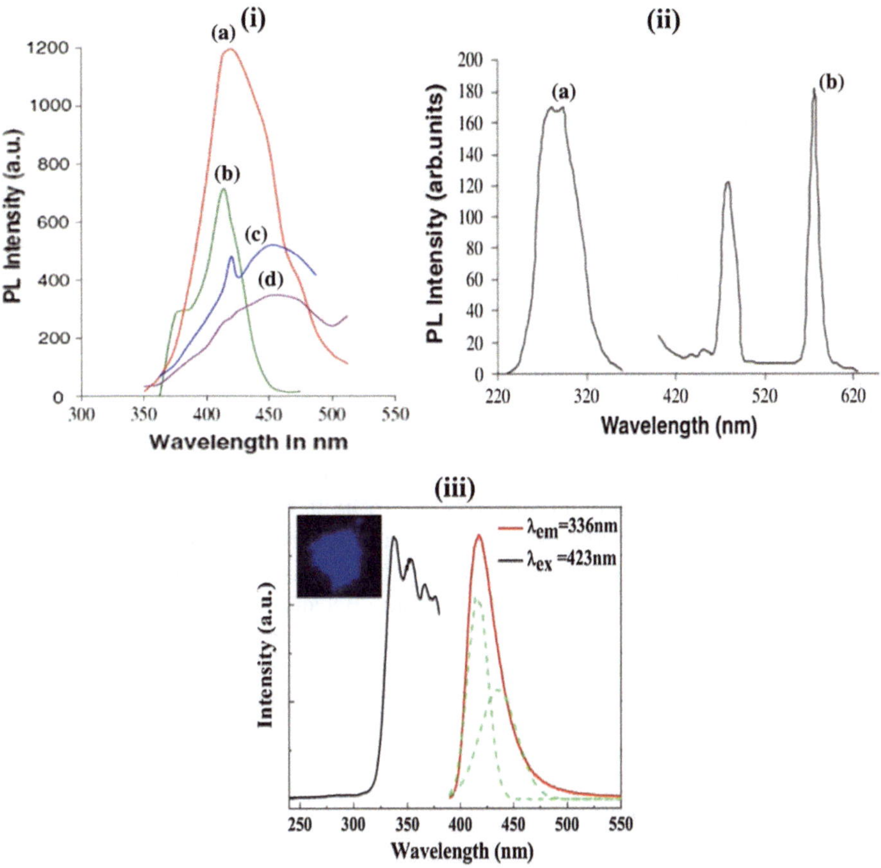

Fig. 10.20 **i** PL emission spectra of $KMgSO_4Cl$:Cu. **a** 0.5 mol%, **b** 0.3 mol%, **c** 0.2 mol%, and **d** 0.1 mol% (λ_{ex} = 257 nm). Reprinted from Ref. [105], with permission from Springer Science +Business Media. **ii** Excitation (**a**) and emission (**b**) spectra of $KZnSO_4Cl:Ce_{10},Dy_{0.1}$. Reprinted from Ref. [107], Copyright 2005, with permission from Elsevier. **iii** Excitation and emission spectra of Ca_2BN_2F:1 %Eu^{2+} phosphor. The *inset* shows the phosphor under 365-nm excitation in a UV box. Reprinted from Ref. [108] by permission of OSA Publishing

Ce^{3+} ion as a sensitizer. The PL emission spectra of $KZnSO_4Cl$: Ce(10 mol%), Dy (0.1 mol%) phosphors show theCe^{3+} emission at 327 and 340 nm due to 5d → 4ftransition of Ce^{3+} ion as Dy^{3+} emission at 475 and 575 nm due to the $^4F_{9/2} \rightarrow {}^6H_{15/2}$ and $^4F_{9/2} \rightarrow {}^6H_{13/2}$ transitions of Dy^{3+} ion, respectively (under excitation wavelength of 254 nm). In particular the PL characteristics of phosphors do not show individual Dy^{3+} or Mn^{2+} emission, whereas the presence of a co-activator ion of Ce^{3+} shows Dy^{3+} and Mn^{2+} emission in $KZnSO_4Cl$ halosulfate phosphor due to the energy transfer of $Ce^{3+}\rightarrow Dy^{3+}$ and $Ce^{3+}\rightarrow Mn^{2+}$ ions. The emission of $Dy^{3+}$and Mn^{2+} in $KZnSO_4Cl$ phosphor may be useful for TL dosimetry and the lamp industry, respectively.

10.7.2 Ca_2BN_2F

Nitride/oxynitride compounds—such as $M_2Si_5N_8$ (M = Ca, Sr, Ba), $MAlSiN_3$ (M = Ca, Sr), and $MSi_2O_2N_2$ (M = Ca, Sr, Ba)—have attracted much attentions because of their excellent luminescence properties as white-LED s phosphors. Thus, new nitride-based phosphors for normal pressure are worth developing. Ca_2BN_2Fcrystallizes in the orthorhombic system with space group *Penm*an with four formula units per unit cell [108]. The dimensions of the unit cell are a = 9.182 Å, b = 3.649 Å, and c = 9.966 Å. The crystal structure consists of tri-BN_2^{3-} anions and tetra-Ca1$(Ca2)_3$F units. There are two different crystallographic sites for the Ca^{2+} ions. The Cal site is coordinated by five nitrogen ions and one fluorine ion, and another Ca2 site is coordinated by three nitrogen and three fluorine ions. A blue-emitting $Ca_2BN_2F:Eu^{2+}$phosphor was synthesized. Figure 10.20(iii) shows the PL and PLE spectra of Ca_2BN_2F:1 %Eu^{2+}. The excitation spectrum consisted of four absorption bands peaking at 337, 352, 365, and 376 nm, which correspond to the transition from the ground state of Eu^{2+} to its field-splitting levels at the 5d1 state. The sample exhibited a blue-emitting band peaking at 423 nm under optimal excitation at 336 nm. A broad asymmetric band was observed in the emission spectrum for wavelength range 390–530 nm. Because Eu^{2+} ions occupied two different Ca^{2+} sites, the emission band deconvoluted into two individual emission bands centered at 415 and 435 nm. The inset demonstrates the phosphor excited at 365 nm in the UV box, which gives an intense blue color.

10.7.3 $Ba_3P_5N_{10}Br$

Recently, Schnick et al. reported the novel nitridophosphate$Ba_3P_5N_{10}Br$ with a zeolite-type structure [109]. Doped with Eu^{2+}, $Ba_3P_5N_{10}Br$ exhibits natural white-light luminescence on excitation by near-UV light, thus making this compound a promising candidate for use as a single white emitter in pcLEDs. The crystal structure of $Ba_3P_5N_{10}Br$ consists of a network of all-side vertex-sharing $[PN_4]$ tetrahedra leading to three-dimensional achter-ring channels, which contain alternating Ba and Br atoms. The crystal structure of $Ba_3P_5N_{10}Br$ contains five different crystallographic Ba positions. When Eu^{2+} was doped into the $Ba_3P_5N_{10}Br$ host, the excitation spectrum reveals a maximum at λ = 396 nm. Thus, this compound can be effectively excited by near-UV light. Excitation at λ = 400 nm results in an emission spectrum with two bands: one in the blue spectral region centered at λ = 472 nm and the other in the yellow spectral region centered at λ = 582 nm. The full–width-at–half maximum (FWHM) values are approximately 1502 cm^{-1} for the band at λ = 472 nm and 2374 cm^{-1} for the band at λ = 582 nm. The presence of two emission bands can be explained by the occupation of Eu^{2+} ions on different Ba lattice sites.

Based on the above study, Schnick et al. believe that the observed feature makes $Ba_3P_5N_{10}Br:Eu^{2+}$ a promising candidate for use as a single white emitter in pcLEDs. Moreover, the development of$Ba_3P_5N_{10}Br:Eu^{2+}$ demonstrated that nitridic zeolites are well suited as host lattices for high-performance luminescence materials. Consequently, the synthesis of isotypic compounds $M_3P_5N_{10}X:Eu^{2+}$ by the exchange of halide ion X or alkaline–earth ion M may lead to further interesting luminescent properties.

10.8 The Work on Oxyhalide-Based Phosphors in Our Group

As demonstrated in the previous section, we can find that the oxyhalide-based compounds demonstrated important chances to create new inorganic phosphors for different optical applications. It is believed that the more ionic nature and bonding state of the metal–halide bond was shown to decrease the synthesis and sintering temperatures of the resulting compounds Moreover, more flexible and alternative sites that can be occupied by the doped rare-earth ions can be formed, which will in turn induce multi-color PL emission originating from different crystal-field environments. Therefore, we have conducted a series of work on the development of the new oxyhalide-based phosphors in our group, and the studied phosphor system including the halo-silicate, halo-borate, halo-phosphate, halo-aluminate, and halo-vanadate, etc., phosphors. Here, we present some representative work to give an introduction.

10.8.1 Our Work on Halo-Silicate Phosphors

As we know, alkali–earth halide and silicate compound are both effective phosphor hosts. Therefore, as multiplex compounds, studies on halo-silicate materials, such as flurosilicate, chlorsilicate, and bromosilicate matrix, have attracted increasing attention. The first studied halo-silicate phosphors in our group is $Sr_4Si_3O_8Cl_4$: Eu^{2+}. We began this research in 2004, and some systematic work on the halo-silicate phosphors has been conducted from then on [110]. Figure 10.21a gives the PLE and PL spectra of $Sr_4Si_3O_8Cl_4:Eu^{2+}$ phosphor. The excitation spectra of $Sr_4Si_3O_8Cl_4:Eu^{2+}$ show the characteristic excitation spectra band of Eu^{2+} ion, which is broad in the range from 250 to 400 nm. Under 365-nm UV light, $Sr_4Si_3O_8Cl_4:Eu^{2+}$ shows broad emission, related to the 5d → 4f transition, and gives a strong blue-green luminescence with a peak wavelength at 496 nm and an FWHM of 70 nm. It is interestingly found that the substitution of Mg^{2+} or Ca^{2+} for Sr^{2+} in the $Sr_4Si_3O_8Cl_4$ host will modify the luminescence properties. By changing the dopant amount of Mg2 + ions, a series of powder materials represented by the

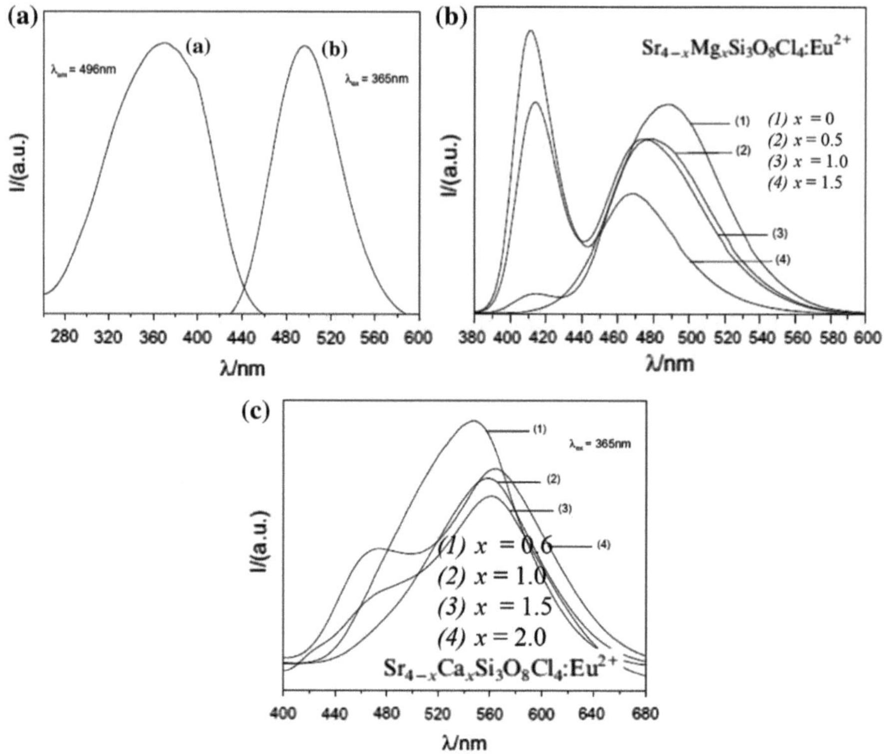

Fig. 10.21 **a** PL and PLE spectra of $Sr_4Si_3O_8Cl_4:Eu^{2+}$, **b** PL emission spectra of $Sr_{4-x}Mg_xSi_3O_8Cl_4:Eu^{2+}$ depending on Mg content. **c** PL emission spectra of $Sr_{4-x}Ca_xSi_3O_8Cl_4:Eu^{2+}$ depending on Ca content. Reprinted from Ref. [111] by permission of Elsevier

chemical formulas $Sr_4Si_3O_8Cl_4:Eu^{2+}$, $Sr_{3.5}Mg_{0.5}Si_3O_8Cl_4:Eu^{2+}$, $Sr_{3.0}MgSi_3O_8Cl_4:Eu^{2+}$, and $Sr_{2.5}Mg_{1.5}Si_3O_8Cl_4:Eu^{2+}$ were successfully synthesized, and their comparative emission spectra are given in Fig. 10.21b. It was found that a new blue-violet–emission center at 411 nm appears with the introduction of Mg^{2+}, and the peak intensities at 496-nm decrease, whereas those at 411 nm increase with increasing Mg^{2+} content. The observed emission peaks should be ascribed two different cation sites in the host lattice because of the dopant of Mg^{2+}. When the doping amounts of calcium (x) in $Sr_4Si_3O_8Cl_4$, ranged from 0 to 0.5, all of the samples showed blue-green emission similar to $Sr_4Si_3O_8Cl_4:Eu^{2+}$, and it is possible that Ca^{2+} ions are embedded in the host lattice of $Sr_4Si_3O_8Cl_4$ with little change. However, $Sr_{4-x}Ca_xSi_3O_8Cl_4:Eu^{2+}$ shows greenish-yellow luminescence when the range of x is between 0.5 and 2. Their emission spectra are shown in Fig. 10.21c. We can find two emission peaks in such a case: one near 470 nm and the other near 560 nm. With increasing Ca content, the emission peak disappear, and the intensities of the emission peak at a 560-nm increase. Thus, we can find an emission-color evolution from blue green to yellow [110, 111].

The second example on the development of the halo-silicate phosphor is $Ca_3SiO_4Br_2{:}Eu^{2+}$ [112–116]. On the basis of the formula of efficient green-emitting phosphor $Ca_3SiO_4Cl_2{:}Eu^{2+}$, we designed the chemical composition of the Eu^{2+}-doped $CaO–CaBr_2–SiO_2$ phosphor system as $Ca_3SiO_4Br_2{:}Eu^{2+}$. That is to say, $Ca_3SiO_4Br_2$ belongs to the nominal chemical composition when we proposed and designed such a host. To get the real crystal structure information of the nominal $Ca_3SiO_4Br_2$ phase, single crystals of new bromosilicate compound $Ca_3SiO_4Br_2$ were grown out of high-temperature $Ca_2SiO_4/CaBr_2$ melts by slow-cooling process. As shown in Fig. 10.22a, colorless, transparent, and flaky crystals of $Ca_3SiO_4Br_2$ were directly isolated from the final solid coagulation [114]. Under a 365-nm UV lamp, all of the as-prepared melts and the crystals in it show a bright blue emission (Fig. 10.22b). The structure of the new $Ca_3SiO_4Br_2$ compound was determined by single-crystal X-ray diffraction, and the depiction of the framework of the $Ca_3SiO_4Br_2$ compound emphasizing the layer structure is demonstrated in Fig. 10.22c. $Ca_3SiO_4Br_2$ phase crystallized in the triclinic system, space group *P–1* with unit cell parameters $a = 8.0051(18)$ Å, $b = 8.720(3)$ Å, $c = 11.749(3)$ Å, $\alpha = 69.07(0)^\circ$, $\beta = 89.98(0)^\circ$, $\gamma = 75.46(0)^\circ$, and $V = 737.88(196)$ Å^3, Z = 3. Structure analysis reveals that $Ca_3SiO_4Br_2$ is composed of alternating layers of $CaBr_2$ and Ca_2SiO_4, and all Si atoms are tetrahedrally coordinated to O atoms, thus forming two Si–O–Si bonds and two Si–O–Ca bonds. Of note, the Br atoms exist in the interlayer structure of two adjacent Ca_2SiO_4 layers, which are bonded with Ca atoms in the two layers. Figure 10.22d presents the excitation [photoluminescence excitation (PLE); $\lambda_{em} = 469$ nm] and emission (PL; $\lambda_{ex} = 365$ nm) spectra of Eu^{2+}-doped $Ca_3SiO_4Br_2$ phosphor powder. The PLE spectra have a broad band between 260 and 430 nm, and there are two obvious excitation peaks at approximately 291 and 358 nm. Interestingly, Eu^{2+}-doped $Ca_3SiO_4Br_2$ phosphor shows a strong blue emission band centered at 469 nm under 365-nm UV light, which demonstrates that $Ca_3SiO_4Br_2{:}Eu^{2+}$ is a promising blue phosphor for use in white light LEDs [115].

Interestingly, similar as the $Ca_3SiO_4Cl_2$ phase, there are also two phases for $Ca_3SiO_4Br_2$: the low-temperature (L) phase (a) and the high-temperature (H) phase. We also found different luminescence properties for the L and H phase [116]. Except for this, $Ca_3SiO_4Br_2{:}Eu^{2+}$phosphor not only emits greenish-yellow light, it also shows strong long-lasting phosphorescence after it is excited with 365-nm UV light. The phosphorescence lasts for >1 h in the light perception of the dark-adapted human eye (0.32 mcd/m^2) [112]. Figure 10.23 shows a comparison of the normalized emission spectra of L and H phase of $Ca_3SiO_4Br_2{:}Eu^{2+}$phosphor on 365-nm UV excitation, and the afterglow spectrum of H-phase is also given. It is obvious that the three emission spectra are all asymmetric in the spectra profiles. A solid arrow marked at 498, 474, and 540 nm in Fig. 10.23 indicates that the PL spectrum of L-phase exhibits a clear bluish-green color with peak wavelength at 498 nm; the PL spectrum of H-phase gives a clear blue color with a peak wavelength at 474 nm; and the LLP spectrum of H-phase shows a greenish-yellow color with the main peak wavelength at 540 nm. This also shown by their respective digital photographs for photoluminescence and phosphorescence for the corresponding phosphors.

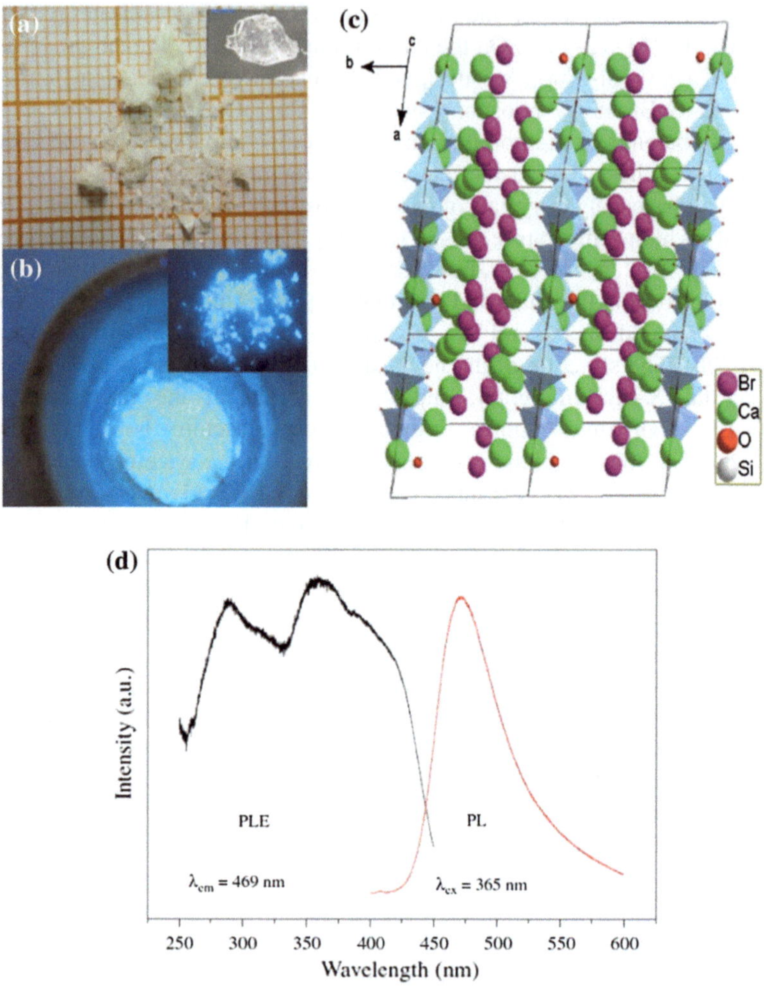

Fig. 10.22 As-grown Eu^{2+}-doped $Ca_3SiO_4Br_2$ crystals in the daylight (**a**) and under a 365 nm UV lamp showing *bright blue* emission (**b**). **c** Depiction of the framework of $Ca_3SiO_4Br_2$ compound emphasizing the layer structure. **d** PLE (*left*) and PL (*right*) spectra of Eu^{2+}-doped $Ca_3SiO_4Br_2$ crystals. **a**, **b**, **d** Reprinted from Ref. [114], Copyright 2010, with permission from Elsevier. **c** Reprinted from Ref. [115], Copyright 2013, with permission from Elsevier

The third example on the halo-silicate phosphor is $Ca_3GeO_4Cl_2:Eu^{2+}$ [117]. As we summarized previously, excellent photoluminescence properties can be realized in the calcium chlorosilicate $Ca_3SiO_4Cl_2$ phosphors, which show green or orange emission when the Eu^{2+} was doped. Because Si and Ge elements belong to the same group, Si/Ge substitution can form possible solid-solution phases. Therefore, we believe that $Ca_3GeO_4Cl_2$ may be also a kind of potential phosphor host for Eu^{2+} doping, and some controlled luminescence can be realized in such a phosphor system. However, $Ca_3GeO_4Cl_2$ crystallizes in the orthorhombic space group *Pnma*

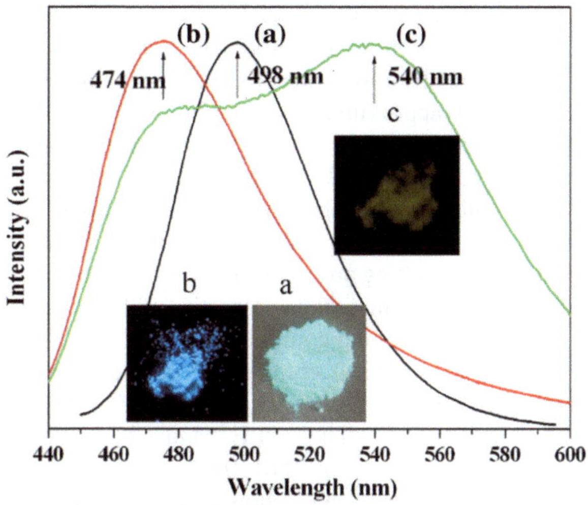

Fig. 10.23 Comparison of the normalized emission spectra of low-temperature (L) phase (**a**) and high-temperature (H) phase (**b**) on 365-nm UV excitation as well as the afterglow spectrum of H-phase (**c**) $Ca_3SiO_4Br_2$:Eu^{2+}phosphor. *Inset* Digital photographs of photoluminescence and phosphorescence for the corresponding phosphors. Reprinted from Ref. [116], copyright 2009, with permission from Elsevier

with Z = 4 (single-crystal XRD), which is different from the $Ca_3GeO_4Cl_2$ phase. As given in Fig. 10.24a, it represents the unit cell of $Ca_3GeO_4Cl_2$ phase with an orthorhombic structure, which consists of one distinct Ge site on the special position 4c, site symmetry m, three different O sites, one distinct Cl position and two different Ca sites, Ca1 and Ca2 (one on the general position 8d, site symmetry 1, and the other on the special position 4c). As given in Fig. 10.24b, two calcium atoms show the same sixfold coordination. Both Ca1 and Ca2 have four oxygen and two chlorine neighbors; however, the bond length and the coordination environment should be

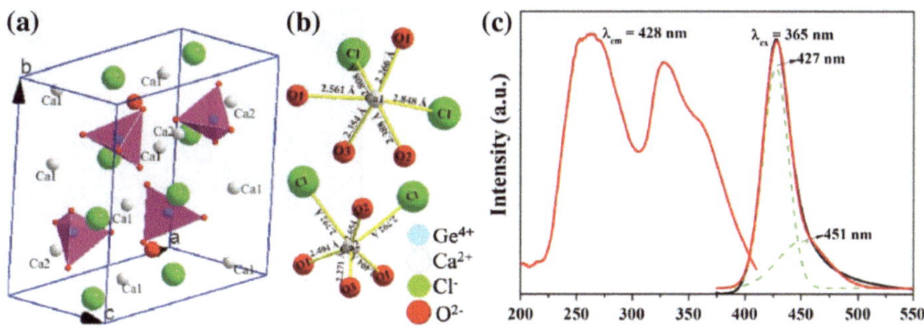

Fig. 10.24 **a** Crystal structure of $Ca_3GeO_4Cl_2$ compound. **b** Coordination spheres of the two different Ca^{2+} sites in $Ca_3GeO_4Cl_2$. **c** PLE and PL spectra of $Ca_3GeO_4Cl_2$:Eu^{2+} phosphors. Reproduced from Ref. [117] by permission of The Royal Society of Chemistry

different. Figure 10.24c gives the PLE and PL spectra of the $Ca_3GeO_4Cl_2:Eu^{2+}$ phosphors. The excitation spectrum monitored at 428 nm, ranging from 200 to 415 nm, exhibits one strong absorption peak located approximately 263 nm and two weak absorption peaks at approximately 327 and 359 nm, respectively. Under excitation at 365 nm, a blue emission with a typical asymmetrical band extending from 375 to 550 nm with a peak at 428 nm can be seen.

The last example of a halo-silicate phosphor is $Ca_{5.45}Li_{3.55}(SiO_4)_3O_{0.45}F_{1.55}:Ce^{3+}$ [4]. It is accepted that the low phonon energy in the fluoride anion causes efficient emission due to the minor quenching process from multi-phonon relaxation realized in this host, and thus interesting luminescence properties will be found. In 2010, Krüger first synthesized the new F-containing compound, $Ca_{5.42}Li_{3.55}(SiO_4)_3O_{0.45}F_{1.55}$ single crystal, and showed its chemical composition and detailed crystal structure. Therefore, we investigated blue-emitting $Ca_{5.45}Li_{3.55}(SiO_4)_3O_{0.45}F_{1.55}:Ce^{3+}$ phosphor for *w*-LED application. The PLE and PL spectra of $Ca_{5.45-x}Li_{3.55}(SiO_4)_3O_{0.45}F_{1.55}:xCe^{3+}$ ($x = 0.005, 0.01, 0.03, 0.05, 0.08, 0.1, 0.15$) phosphors are presented in Fig. 10.25(i). As shown in Fig. 10.25(i), $Ca_{5.42}Li_{3.55}(SiO_4)_3O_{0.45}F_{1.55}:0.03Ce^{3+}$ (monitored at 470 nm) has a broad PLE spectrum, and it covers the spectral region from 200 to 450 nm quite well. Under 365-nm UV-light irradiation, $Ca_{5.45-x}Li_{3.55}(SiO_4)_3O_{0.45}F_{1.55}:xCe^{3+}$ exhibits blue emissions peaking at 470 nm with typical unsymmetrical doublet bands that extend from 400 to 600 nm. As shown in the inset of Fig. 10.25(i), the emission band can be deconvoluted on an energy scale into two bands centered at 21413 cm^{-1} (467 nm) and 19762 cm^{-1} (506 nm). Ce^{3+} concentration x dependence of the emission intensities is also shown

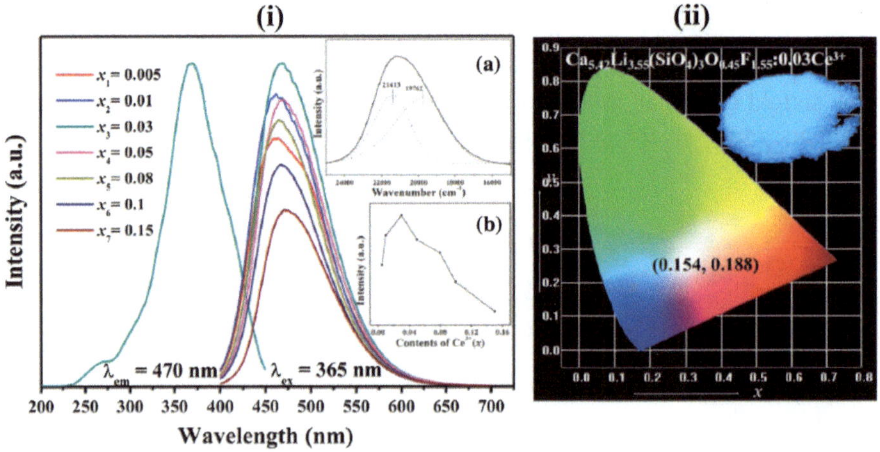

Fig. 10.25 **i** PLE and PL spectra of $Ca_{5.45-x}Li_{3.55}(SiO_4)_3O_{0.45}F_{1.55}:xCe^{3+}$ phosphors. *Inset* (**a**) shows the deconvolution of PL spectrum by Gaussian fitting, and *Inset* (**b**) shows the dependence of PL intensity on the Ce^{3+} doping concentration. **ii** Color coordinates of $Ca_{5.42}Li_{3.55}(SiO_4)_3O_{0.45}F_{1.55}:0.03Ce^{3+}$ in a CIE chromaticity diagram; the *inset* shows a digital photograph of the *blue*-emitting phosphor $Ca_{5.42}Li_{3.55}(SiO_4)_3O_{0.45}F_{1.55}:0.03Ce^{3+}$. Reproduced from Ref. [4] by permission of The Royal Society of Chemistry

in the inset of Fig. 10.25(i). We can see that the optimal doping concentration was observed to be approximately 3 mol%. The CIE chromaticity diagram for the $Ca_{5.42}Li_{3.55}(SiO_4)_3O_{0.45}F_{1.55}$:0.03$Ce^{3+}$ sample under 365-nm UV excitation is shown in Fig. 10.25(ii). The color coordinate is shown as (x = 0.154, y = 0.188), which means that this phosphor can be used as a blue-emitting phosphor for *w*-LED application. The inset of Fig. 10.25(ii) also shows a digital photograph of $Ca_{5.42}Li_{3.55}(SiO_4)_3O_{0.45}F_{1.55}$:0.03$Ce^{3+}$ phosphor, and its intense blue emission can be clearly observed.

10.8.2 Our Work on Halo-Borate Phosphors

In 1994, Schipper et al. reported the luminescence of Ce^{3+}, Eu^{2+}, Eu^{3+} and Gd^{3+} in the compound $Ba_2Ln(BO_3)_2Cl$ (*Ln* = Gd, Y, Lu) [27]. They found that Eu^{2+} ion in $Ba_2Y(BO_3)_2Cl:Eu^{2+}$ shows a strong luminescence at 4.2 K under 340-nm excitation, which consists of two emission bands. However, the shorter-wavelength emission band was quenched at room temperature. Therefore, we proposed and designed a series of $Ba_2Ln(BO_3)_2Cl$ compounds (Ln = rare-earth (RE) ions) and studied the structure and luminescence properties of Eu^{2+}-doped $Ba_2Ln(BO_3)_2Cl$ (Ln = Y, Gd and Lu) phosphors under *n*-UV light. Rietveld analysis reveals that the successful isomorphic substitution for Ln^{3+} sites in the $Ba_2Ln(BO_3)_2Cl$ host can be implemented only for the smaller lanthanide ions (such as Y, Sm, Eu, Gd, Tb, Dy, Ho, Er, Tm, Yb, and Lu), and they all crystallize in a monoclinic $P\ 2_1/m$ space group [3]. Figure 10.26 (left) gives the Rietveld analysis patterns for X-ray powder diffraction data of typical $Ba_2Y(BO_3)_2Cl$ sample. From this, we can find that the measured XRD pattern is identical with the calculated one. In the structure of Ba_2Ln $(BO_3)_2Cl$ (*Ln* = Y, Gd and Lu), the Ba atoms occupy two different sites–Ba(1) and Ba(2) $\leq$ and they have a complex coordination: Ba(1) is ninefold coordinated by 5O and 4C1, whereas Ba(2) is tenfold coordinated by 8O and 2C1 as shown in the inset of Fig. 10.26 (left). PLE and PL spectra and digital photographs under 365-nm UV light (b) of $Ba_2Ln(BO_3)_2Cl$:0.03Eu^{2+} (*Ln* = Y, Gd and Lu) phosphors are also shown in Fig. 10.26. These compounds display interesting and tunable luminescence properties after doping with Eu^{2+} $Ba_2Ln(BO_3)_2Cl:Eu^{2+}$ phosphors exhibit bluish-green/greenish-yellow light with peak wavelengths at 526, 548, and 511 nm under 365-nm UV light excitation for $Ba_2Y(BO_3)_2Cl:Eu^{2+}$, $Ba_2Gd(BO_3)_2Cl$: Eu^{2+}, and $Ba_2Lu(BO_3)_2Cl:Eu^{2+}$, respectively [3].

Except for the $Ba_2Ln(BO_3)_2Cl$:0.03Eu^{2+} (*Ln* = Y, Gd, and Lu) phosphors [118], we also found another very interesting phosphor system in the $Ba_2Ln(BO_3)_2Cl$ host, i.e., $a_2Tb(BO_3)_2Cl$:Eu phosphors [119, 120]. Novel red-emitting $Ba_2Tb(BO_3)_2Cl$: Eu phosphor possessed a broad excitation band in the near-ultraviolet region, and the coexistence of Eu^{2+}/Eu^{3+} and the red-emitting luminescence from Eu^{3+} with the aid of efficient energy transfer of $Eu^{2+}\rightarrow Eu^{3+}(Tb^{3+})$ and $Tb^{3+}\rightarrow Eu^{3+}$ have been investigated. As shown in Fig. 10.27, this phosphor gives PLE and PL spectra of $Ba_2Tb(BO_3)_2Cl$ (a) and $Ba_2Tb_{0.995}(BO_3)_2Cl$:0.005Eu (b) phosphor. Ba_2Tb

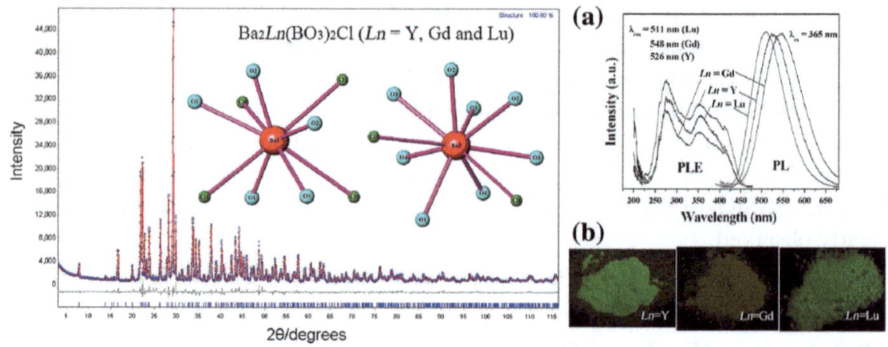

Fig. 10.26 (*Left*) Rietveld analysis patterns for X-ray powder diffraction data of a typical $Ba_2Y(BO_3)_2Cl$ sample. The coordination *spheres* of the two different Ba^{2+} positions in $Ba_2Ln(BO_3)_2Cl$ compound are given in the inset. PLE and PL spectra (**a**) and digital photographs under 365-nm UV light (**b**) of $Ba_2Ln(BO_3)_2Cl{:}0.03Eu^{2+}$ (Ln = Y, Gd, and Lu) phosphors. Reprinted with permission from Ref. [3]. Copyright 2011 American Chemical Society

$(BO_3)_2Cl$ emits green emission with the main peak at approximately 543 nm, which originates from the $^5D_4 \rightarrow {}^7F_5$ transition of Tb^{3+}. $Ba_2Tb(BO_3)_2Cl{:}Eu$ shows a bright red emission from Eu^{3+} with peaks at approximately 594, 612, and 624 nm under n-UV excitation (350–420 nm) [119]. The excitation spectrum of $Ba_2Tb(BO_3)_2Cl$ exhibits two obvious broad bands from 200 to 300 nm, with two peaks at 248 nm and 283 nm, and some weak transitions from 300 to 400 nm. In comparison, $Ba_2Tb(BO_3)_2Cl{:}Eu$ phosphor possesses the same excitation band in the short-wavelength region coming from the Tb-containing host, as shown in Fig. 10.27b. However, the intensity of the sharp lines corresponding to the transition of Tb^{3+} has been enhanced obviously, and the excitation band edge has been broadened to 450 nm showing a broad-band character. This new finding in $Ba_2Tb(BO_3)_2Cl{:}Eu$ phosphor on the broadband absorption in the *n*-UV region, as well as the realization of red-emitting luminescence ascribed to the energy transfer, will provide a new way to explore novel red-emitting phosphors for use in white LEDs.

Eu ions can be incorporated as Eu^{2+}on Ba^{2+} sites for the $Ba_2Ln(BO_3)_2Cl$ (Ln = Y, Gd and Lu) system and as Eu^{3+} on the Ln^{3+} sites in the $Ba_2Tb(BO_3)_2Cl$ compound. Considering the versatile nature of the $Ba_2Ln(BO_3)_2Cl$ compounds, we propose a new strategy based on controlling the valence state of Eu and the energy transfer to Eu^{3+} by varying the Tb content in the host composition. Accordingly, tunable multicolor emission changes from green to red. Even full-color-emission in $Ba_2Ln_{0.97-z}Tb_z(BO_3)_2Cl{:}0.03Eu$ (Ln = Y, Gd, and Lu, z = 0 to approximately 0.97) phosphors can be realized on *n*-UV excitation by using the allowed Eu^{2+}-absorption transitions to sensitize Tb^{3+} and Eu^{3+} emission by way of ET of $Eu^{2+} \rightarrow (Tb^{3+})_n \rightarrow Eu^{3+}$. Figure 10.27c gives the emission spectra for 377-nm excitation of $Ba_2(Lu_{0.97-z}Tb_z)(BO_3)_2Cl{:}0.03Eu$ phosphors with different Tb content. When the Tb content is increased in the $Ba_2(Lu,Tb)(BO_3)_2Cl{:}0.03Eu$ phosphor, the relative intensities of the green Tb^{3+} emission lines increases gradually, whereas the broadband emission

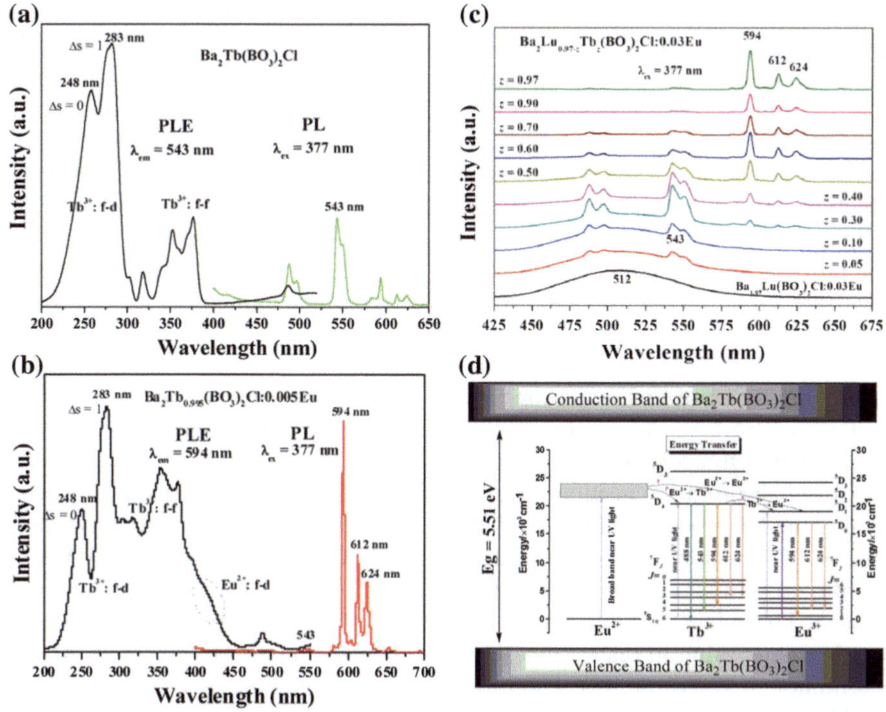

Fig. 10.27 PLE and PL spectra of $Ba_2Tb(BO_3)_2Cl$ (**a**) and $Ba_2Tb_{0.995}(BO_3)_2Cl{:}0.005Eu$ (**b**) phosphor. (**c**) Emission spectra for 377-nm excitation of $Ba_2(Lu_{0.97-z}Tb_z)(BO_3)_2Cl{:}0.03Eu$ phosphors with different Tb content (z values). **d** Schematic energy-level diagram showing the energy transfer among Eu^{2+}, Tb^{3+} and Eu^{3+} in $Ba_2(Lu_{0.97-z}Tb_z)(BO_3)_2Cl{:}0.03Eu$ phosphor (**a**, **b**, **d**). Reprinted with the permission from Ref. [119]. Copyright 2012 American Chemical Society. (**c**) Reproduced from Ref. [120] by permission of The Royal Society of Chemistry

decreases accordingly. Furthermore, when the Tb content reaches $z = 0.3$, emission peaks located at approximately 594, 612, and 624 nm (assigned to $^5D_0$–$^7F_1$ and $^7F_2$ transitions of Eu^{3+}) appear accompanied by a decrease of the green Tb^{3+} emission at 543 nm indicating the efficient ET between Tb^{3+} and Eu^{3+}. Therefore, Fig. 10.27d displays the proposed diagram of energy transfer between $Tb^{3+}{\rightarrow}Eu^{3+}$ and $Eu^{2+}{\rightarrow}Eu^{3+}(Tb^{3+})$ in the $Ba_2Tb(BO_3)_2Cl{:}Eu$ phosphors. It is believed that the excitation lines of $Ba_2Tb(BO_3)_2Cl{:}Eu$ phosphors in the n-UV region come from the absorption of combined effect of $Tb^{3+}/Eu^{3+}/Eu^{2+}$ such that a wide and strong absorption band can be observed in the PLE spectrum (Fig. 10.27b). The absorbed energy of Tb^{3+} could be transferred from the $^5D_4$ level of Tb^{3+} to the $^5D_0$ ($^5D_1$) level of Eu^{3+}; accordingly, the absorbed energy of Eu^{2+} can also be completely transferred from the $4f^65d^1$ band to the $^5D_1$ level of Eu^{3+} and the $^5D_4$ level of Tb^{3+}.

To further understand the present $Ba_2(Lu,Tb)(BO_3)_2Cl{:}0.03Eu$ phosphor, excitation spectra for emission at 512, 543, and 594 nm and emission spectrum under 377-nm excitation of the selected $Ba_2Lu_{0.57}Tb_{0.4}(BO_3)_2Cl{:}0.03Eu$ phosphor are shown in Fig. 10.28a. The PL spectrum shows band emission of Eu^{2+} and line emissions of Tb^{3+} and Eu^{3+}. The 512-nm band emission is due to the $4f^6 5d$-$4f^7$ ($^8S_{7/2}$) transition of Eu^{2+}, whereas the 488 and 543-nm peaked line emissions are due to the $^5D_4$-$^7F_{6,5}$ transitions of Tb^{3+} and the 594, 612 and 618-nm peaked line-emissions to the $^5D_0 \rightarrow {}^7F_{1,2}$ transitions of Eu^{3+}. The excitation spectra monitoring the Eu^{3+} emission at 594 and the Tb^{3+} emission at 543 nm give similar spectral profiles, which is a direct evidence of a $Tb^{3+} \rightarrow Eu^{3+}$ ET. More importantly, it is found that there are strong and broad absorption bands in the *n*-UV and violet regions (300–450 nm) for the excitation spectra monitoring both the Tb^{3+} and Eu^{3+} emissions, which are similar to the excitation bands observed while monitoring Eu^{2+} emission at 512 nm. The broad bands in the 300- to 450-nm range are assigned to the $4f^7 \rightarrow 4f^6 5d$ transitions on Eu^{2+} and the observation of these bands in the excitation spectra for Tb^{3+} and Eu^{3+} emission provide evidence for energy transfer from Eu^{2+} to Tb^{3+} in $Ba_2Lu_{0.57}Tb_{0.4}(BO_3)_2Cl$:0.03Eu phosphors. This series of excitation spectra demonstrates that $Tb^{3+} \rightarrow Eu^{3+}$ ET occurs and also shows that the allowed Eu^{2+} $4f$-$5d$ absorption transition sensitizes Tb^{3+} and Eu^{3+} line emissions, thus confirming the ET process of $Eu^{2+} \rightarrow (Tb^{3+})_n \rightarrow Eu^{3+}$. Therefore, the photographs in Fig. 10.28b illustrate the color tenability under excitation at 365 nm with a UV lamp for $Ba_{1.97}Ln(BO_3)_2Cl$:0.03Eu and $Ba_2(Ln_{0.97-z}Tb_z)(BO_3)_2Cl$:Eu phosphors with different Tb content, and the variation in color can be clearly observed by the naked eye. It is interesting to realize that the color-tunable emissions by way of the host substitution of Tb/*Ln* (*Ln* = Y, Gd, Lu) occurs for a fixed Eu-concentration of 3 mol %. The underlying mechanism originated from $Eu^{2+} \rightarrow (Tb^{3+})_n \rightarrow Eu^{3+}$ is also shown in Fig. 10.28b [120].

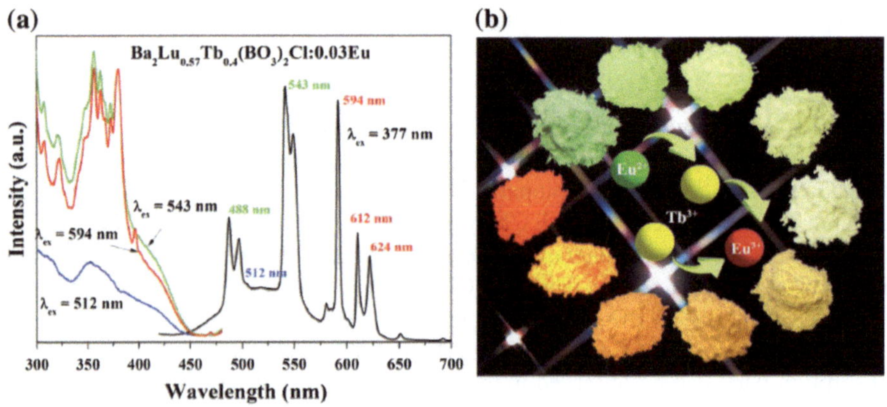

Fig. 10.28 **a** Excitation spectra for emission at 512, 543 and 594 nm and emission spectrum under 377-nm excitation of the selected $Ba_2Lu_{0.57}Tb_{0.4}(BO_3)_2Cl$:0.03Eu phosphor. **b** Photographs under 365-nm excitation of $Ba_{1.97}Ln(BO_3)_2Cl$:0.03Eu and $Ba_2(Ln_{0.97-z}Tb_z)(BO_3)_2Cl$:Eu phosphors with different Tb content: The underlying mechanism originated from $Eu^{2+} \rightarrow (Tb^{3+})_n \rightarrow Eu^{3+}$ energy transfer. Reproduced from Ref. [120] by permission of The Royal Society of Chemistry

10.8.3 Our Work on Halo-Aluminate Phosphors

This review has demonstrated that oxyhalides, in particular oxyfluorides, have recently become the subject of renewed focus of exploratory synthesis in the search for new compounds with intriguing luminescence properties since the introduction of rare-earth ions. One recent example of the interest in oxyfluoride-based phosphors is $Sr_3AlO_4F:Ce^{3+}$, which shows efficient blue-green Ce^{3+} emission under *n*-UV excitation, and some modified phosphors based on Sr_3AlO_4F have also been developed. The structural composition for the new Sr_3AlO_4F host comes from the substitution of F^- for O^{2-} and compensated by Al^{3+} for Si^{4+} substitution in the Sr_3SiO_5. Recent work reported by Im et al. also shows that a solid solution is possible between the nearly isostructural Sr_3AlO_4F and Sr_3SiO_5 compounds, and some excellent luminescence properties have also been reported. Accordingly, we noticed that $Ca_2Al_2SiO_7$ is a famous phosphor host with a tetragonal structure of melilite type. In this structure, if the chemical substitution of F^- for O^{2-}, compensated by Al^{3+} for Si^{4+}, is also effective, then a new compound with the chemical composition of $Ca_2Al_3O_6F$ can be obtained. On the basis of the search in the JCPDS database, the preliminary structural data of $Ca_2Al_3O_6F$ can be found, which is completely different from $Ca_2Al_2SiO_7$ phase. However, it clearly gives us an effective strategy when screening for new phosphor compositions from the well-known phosphor host. Herein, we propose that $Ca_2Al_3O_6F$ phase will be an excellent host as the phosphor materials [7, 121, 122].

To conduct such research, an understanding of the crystal structure of $Ca_2Al_3O_6F$ is very important. Structural analysis reveals that the $Ca_2Al_3O_6F$ crystal is trigonal with unit-cell parameters a = 17.305 Å, c = 6.9925 Å. The most probable space groups are $R3$, $R\overline{3}$, $R3\ m$, $R32$, and $R\overline{3}m$. The program TOPAS 4.2 was used to solve and refine the crystal structure. Figure 10.29a illustrates the Rietveld refinement analysis results for the X-ray diffraction pattern of $Ca_2Al_3O_6F$. The refinement of this crystal structure, with isotropic thermal parameters of all atoms, was stable and gave a low R_B factor and R_{wp} = 7.38 %. The crystal structure of

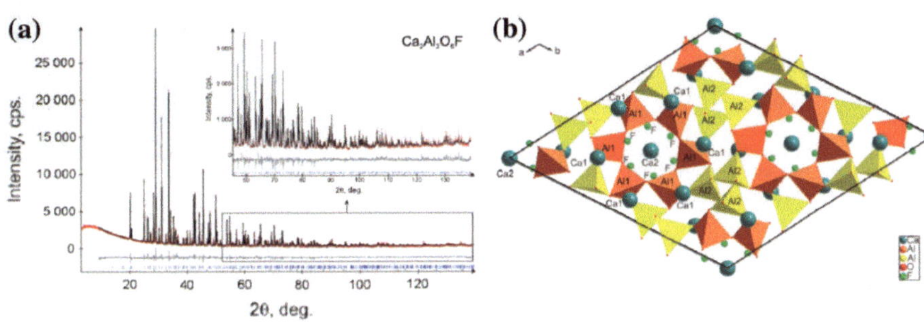

Fig. 10.29 **a** Rietveld refinement for the X-ray diffraction pattern of $Ca_2Al_3O_6F$. **b** Crystal structure of $Ca_2Al_3O_6F$, with the unit cell is outlined. Reproduced from Ref. [121] by permission of The Royal Society of Chemistry

$Ca_2Al_3O_6F$ is shown in Fig. 10.29b in the projection of the unit cell on the *ab* plane. The structure consists of an almost ideal AlO_4 tetrahedrons linked through corners, Ca^{2+} ions in voids, and F^- ions disordered over one site and six sites around the Ca1 and Ca2 ions, respectively. Only two tetrahedrons are linked in each node [121].

A novel green-emitting phosphor $Ca_2Al_3O_6F:Eu^{2+}$ was synthesized, and its luminescence properties were investigated for application in white light–emitting diodes (w-LEDs). This phosphor shows good absorption ranging from the ultraviolet to blue region [7]. Figure 10.30a gives the PLE (λ_{em} = 502 nm) and PL (λ_{ex} = 312, 350, and 400 nm) spectra of the selected $Ca_{1.95}Al_3O_6F:0.05Eu^{2+}$ phosphor. The PLE spectra have a broad band between 260 and 440 nm, and there are three excitation peaks at approximately 312, 350, and 400 nm. The emission spectra are broad with a maximum at approximately 502 nm (green emission) under different excitation wavelengths of 312, 350, and 400 nm showing the green emission, to which the emission band corresponds regarding the transition from the $4f^65d$ excited state to the $4f^7$ ground state of a Eu^{2+} ion. Figure 10.30b shows the emission spectra of $Ca_2Al_3O_6F:0.08Ce^{3+}$, yTb^{3+} phosphors. Except for $Ca_2Al_3O_6F:0.08Ce^{3+}$, the emission spectra (λ_{ex} = 350 nm) in Fig. 10.30a contain a blue emission band of Ce^{3+} ions and a series of strong emission lines at 488, 542, 584, and 619 nm due to the $^5D_4 \rightarrow {}^7F_J$ (J = 6, 5, 4, and 3) characteristic transitions of Tb^{3+} ions and a green emission line at 542 nm from the $^5D_4 \rightarrow {}^7F_5$ transitions dominates the entre spectrum. Accordingly, we can observe strong a green emission in this series of samples. The CIE chromaticity diagram of $Ca_2Al_3O_6F:0.08Ce^{3+}$, yTb^{3+} phosphors on 350-nm excitation is shown in Fig. 10.30c. The color tone can be tuned from blue (0.160, 0.047) to blue-greenish (0.219, 0.271) and from green (0.287, 0.455) to greenish yellow (0.287, 0.480), respectively. The inset in Fig. 10.30c also shows a series of

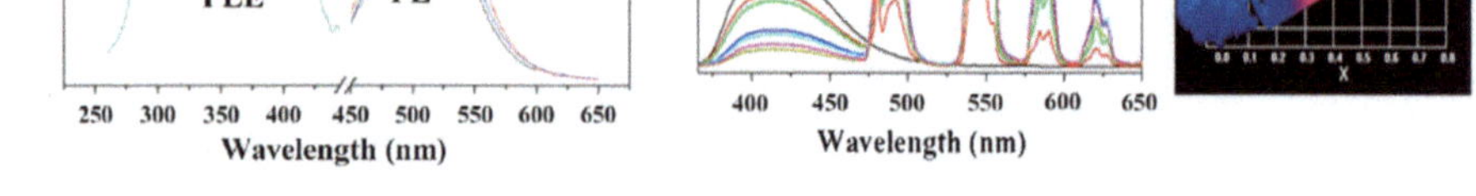

Fig. 10.30 **a** PLE (λ_{em} = 502 nm) and PL (λ_{ex} = 312, 350 and 400 nm) spectra of the selected $Ca_{1.95}Al_3O_6F:0.05Eu^{2+}$ phosphor. Reproduced from Ref. [7] by permission of The Royal Society of Chemistry. **b** PL spectra for $Ca_2Al_3O_6F:0.08Ce^{3+}$, yTb^{3+}phosphors on Tb^{3+} doping content (*y*). **c** Representation of the CIE chromaticity coordinates for $Ca_2Al_3O_6F:0.08Ce^{3+}$, yTb^{3+} phosphors on 350-nm excitation as well as digital photos of the selected samples under 365 nm UV lamp excitation (**b**, **c**) Reprinted with permission from Ref. [122]. Copyright 2012 American Chemical Society

digital photos of the selected $Ca_2Al_3O_6F:0.08Ce^{3+}$, yTb^{3+} phosphors under 365-nm UV-lamp excitation. It is found that $Ca_2Al_3O_6F:0.08Ce^{3+}$,0.05 Tb^{3+} shows intense green light, thus suggesting that it can be used as a potential green emitting candidate for *n*-UV LEDs [122].

10.9 Some Challenges and Future Work

Based on a review of recent papers on the main research topics and the important considerations in the design of new oxyhalide hosts phosphors, the following advantages or challenges can be expected in such a system:

1. First, the exploration of the oxyhalide phosphors expands the phase spaces to include oxyhalide hosts, thus offering additional opportunities for new phosphor materials. The crystal chemistry study verified that the lower electronegativity of Cl^- versus F^- leads to lower energy $4f^N \rightarrow 4f^{N-1}5d^1$ transitions through a larger $5d^1$centroid shift, which will help in discovering some new phosphor systems. The appearance of the new phase, such as$Ba_3P_5N_{10}Br$, will further enrich such a system.
2. Several oxyhalide phosphors— e.g., $Ca_3SiO_4Cl_2:Eu^{2+}$, $Ca_8Mg(SiO_4)_4Cl_2:Eu^{2+}$, $Ca_2BO_3Cl:Eu^{2+}$, $Sr_3Al_2O_5Cl_2:Eu^{2+}$, and $Sr_3AlO_4F:Ce^{3+}$—have been proposed for white LED applications.
3. Different applications can be realized based on the oxyhalide host phosphors, such as white LEDs, long-lasting phosphorescence, X-ray storage phosphors, γ-ray–sensitive storage phosphor, etc.
4. The synthesis temperature, by using the high temperature solid-state reaction, is relatively low compared with that of the normal silicate, borate, phosphate, etc.

However, some future work is still needed to further research on oxyhalide host phosphors as follows:

1. In general, nonsilicate oxychlorides have shown less potential due to their lower efficiencies at high temperature.
2. From a fundamental point of view, in hosts with multiple centers one can typically assign the various emission/excitation bands to specific sites from bond lengths and/or coordination arguments. However, the variety of structures and coordination environments in oxyhalide can make it difficult to draw further structure–property conclusions toward the emission, excitation, and thermal quenching of these materials.
3. Within the current set of oxyhalide phosphors, these materials generally have less stability toward moisture and could also require coatings to prevent hydrolysis reactions. Moreover, their thermal stability is also relatively poor, and it should be improved by some chemical composition and crystal engineering modification or the chemical coating on the surface.

References

1. Setlur AA, Lyons RJ, Murphy JE, Kumar NP, Kishore MS (2013) Blue Light-Emitting Diode Phosphors Based upon Oxide, Oxyhalide, and Halide Hosts. J Electrochem Soc 2: R3059
2. George NC, Denault KA, Seshadri R (2013) Phosphors for solid-state white lighting. Annu Rev Mater Res 43:481
3. Xia Z, Wang X, Wang Y, Liao L, Jing X (2011) Synthesis, structure, and thermally stable luminescence of Eu^{2+}-doped $Ba_2Ln(BO_3)_2Cl$ (Ln = Y, Gd and Lu) host compounds. Inorg Chem 50:10134
4. Zhou J, Xia Z, Yang M, Shen K (2012) High efficiency blue-emitting phosphor: Ce^{3+}-doped $Ca_{5.45}Li_{3.55}(SiO_4)_3O_{0.45}F_{1.55}$ for near UV-pumped light-emitting diodes. J Mater Chem 22:21935
5. Xia Z, Liao L, Zhang Z, Wang Y (2012) Combustion synthesis and luminescence properties of yellow-emitting phosphors $Ca_2BO_3Cl:Eu^{2+}$. Mater Res Bull 47:405
6. Mi R, Zhao C, Xia Z (2014) Synthesis, structure and tunable Luminescence properties of novel $Ba_3NaLa(PO_4)_3F:Eu^{2+}$, Mn^{2+} phosphors. J Am Cera Soc 97:1802
7. Xia Z, Liu R-S, Huang K-W, Drozd V (2012) $Ca_2Al_3O_6F:Eu^{2+}$: a green-emitting oxyfluoride phosphor for white light-emitting diodes. J Mater Chem 22:15183
8. Zhang X, Chen H, Ding W, Wu H, Kim J (2009) $Ca_2B_5O_9Cl:Eu^{2+}$, A Suitable Blue-Emitting Phosphor for n-UV Excited Solid-State Lighting. J Am Ceram Soc 92:429
9. Dutczak D, Ronda C, Meijerink A, Jüstel T (2013) Red luminescence and persistent luminescence of $Sr_3Al_2O_5Cl_2:Eu^{2+}$, Dy^{3+}. J Lumin 141:150
10. Liu C, Qi Z, Ma C-G, Dorenbos P, Hou D, Zhang S, Kuang X, Zhang J, Liang H (2014) High Light Yield of $Sr_8(Si_4O_{12})Cl_8:Eu^{2+}$ under X-ray Excitation and Its Temperature-Dependent Luminescence Characteristics. Chem Mater 26:3709
11. Gwak SJ, Arunkumar P, Im WB (2014) A New Blue-Emitting Oxohalide Phosphor Sr_4OCl_6: $Eu^{2+}$for Thermally Stable, Efficient White-Light-Emitting Devices under Near-UV. J Phys Chem C 118:2686
12. Xia Z, Li J, Luo Y, Liao L, Varela J (2012) Comparative Investigation of Green and Red Upconversion Luminescence in Er^{3+}Doped and Yb^{3+}/Er^{3+}Codoped LaOCl. J Am Ceram Soc 95:3229
13. Kim SW, Jyoko K, Masui T, Imanaka N (2010) A New Type of Red-emitting (La, Ca)OCl: Eu^{3+} Phosphors. Chem Lett 39:604
14. Guo H (2007) Two- and three-photon upconversion of $LaOBr:Er^{3+}$. Opt Mater 29:1840
15. Liu J, Lian H, Shi C, Sun J (2005) Eu^{2+}-Doped High-Temperature Phase $Ca_3SiO_4Cl_2$. J Electrochem Soc 152:G880
16. Liu J, Lian H, Sun J, Shi C (2005) Characterization and Properties of Green Emitting $Ca_3SiO_4Cl_2:Eu^{2+}$ Powder Phosphor for White Light-emitting Diodes. Chem Lett 34:1340
17. Ding W, Wang J, Liu Z, Zhang M, Su Q, Tang J (2008) An Intense Green/Yellow Dual-Chromatic Calcium Chlorosilicate Phosphor $Ca_3SiO_4Cl_2:Eu^{2+}–Mn^{2+}$ for Yellow and White LED. J Electrochem Soc 155:J122
18. Baginskiy I, Liu RS (2009) Significant Improved Luminescence Intensity of Eu^{2+}-Doped $Ca_3SiO_4Cl_2$ Green Phosphor for White LEDs Synthesized Through Two-Stage Method. J Electrochem Soc 156:G29
19. Wang J, Liu Z, Ding W, Su Q (2009) Luminescent Properties of $(Ca_{1-x}Sr_x)_3SiO_4(Cl_{1-y}F_y)_2$: Eu^{2+}Phosphors and Their Application for White LED. Int J Appl Ceram Tec 6:447
20. Ding W, Wang J, Zhang M, Zhang Q, Su Q (2007) Luminescence properties of new $Ca_{10}(Si_2O_7)_3Cl_2:Eu^{2+}$ phosphor. Chem Phys Lett 435:301
21. Ding W, Wang J, Zhang M, Zhang Q, Su Q (2006) A novel orange phosphor of Eu^{2+}-activated calcium chlorosilicate for white light-emitting diodes. J Solid State Chem 179:3582
22. Zhang X, Zhou F, Shi J, Gong M (2009) $Sr_{3.5}Mg_{0.5}Si_3O_8Cl_4:Eu^{2+}$ bluish–green-emitting phosphor for NUV-based LED. Mater Lett 63:852

23. Wang J, Shen C, Yang Y, Jin S, Lee C, Wang H (2007) Synthesis and luminous characteristics of $Ba_2SiO_3Cl_2$:Eu^{2+}, Mn^{2+} phosphor for white LED. Proc of SPIE 6828:682815
24. Zeng Q, Tanno H, Egoshi K, Tanamachi N, Zhang S (2006) $Ba_5SiO_4Cl_6$:Eu^{2+}: An intense blue emission phosphor under vacuum ultraviolet and near-ultraviolet excitation. Appl Phys Lett 88:051906
25. Zhang X, Wang X, Huang J, Shi J, Gong M (2009) Near UV-based LED fabricated with $Ba_5SiO_4(F, Cl)_6$:Eu^{2+} as blue- and green-emitting phosphor. Opt Mater 32:75
26. Meijerink A, Blasse G, Struye L (1989) A new photostimulable phosphor: Eu^{2+} -activated barium bromosilicate ($Ba_5SiO_4Br_6$). Mater Chem Phys 21:261
27. Schipper W, Blasse G, Leblans P (1994) Nb (V) as an Electron Trap in the Storage Phosphor Ba5SiO4Br 6: Eu (II), Nb (V). Chem Mater 6:1784
28. Feng G, Jiang W, Chen Y, Zeng R (2011) A novel red phosphor $NaLa_4(SiO_4)_3F$: Eu^{3+}. Mater Lett 65:110
29. Guo C, Li M, Xu Y, Li T, Ren Z, Bai J (2011) A potential green-emitting phosphor $Ca_8Mg(SiO_4)_4Cl_2$:Eu^{2+} for white light emitting diodes prepared by sol–gel method. Appl Surf Sci 257:8836
30. Koo HY, Hong SK, Han JM, Kang YC (2008) Eu-doped $Ca_8Mg(SiO_4)_4Cl_2$ phosphor particles prepared by spray pyrolysis from the colloidal spray solution containing ammonium chloride. J Alloys Compd 457:429
31. Koo HY, Han JM, Kang YC (2008) $Ca_{7.97-x}Mg(SiO_4)_4Cl_2$:$Eu_{0.03}$, D_x(D = Y, Gd, Mn) Phosphor Particles Prepared by Spray Pyrolysis. Jpn J Appl Phys 47:163
32. Fang Y, Zhuang W, Hu Y, Ye X, Huang X (2008) Luminescent properties of Dy^{3+} ion in $Ca_8Mg(SiO_4)_4Cl_2$. J Alloys Compd 455:420
33. Lü W, Hao Z, Zhang X, Liu Y, Luo Y, Liu X, Wang X, Zhang J (2011) Eu^{2+}-Activated $Ca_8Zn(SiO_4)_4Cl_2$: An Intense Green Emitting Phosphor for Blue Light Emitting Diodes. J Electrochem Soc 158:H124
34. Iwata T, Haniuda M, Fukuda K (2008) Crystal structure of $Ca_{12}Al_{14}O_{32}Cl_2$ and luminescence properties of $Ca_{12}Al_{14}O_{32}Cl_2$:Eu^{2+}. J Solid State Chem 181:51
35. Ju H, Su X, Xu S, Zhang Y, Deng D, Zhao S, Wang H, Wang B, Sun L (2009) Luminescence properties of Eu^{2+}-activated $Ca_{12}Al_{10.6}Si_{3.4}O_{32}Cl_{5.4}$: A promising phosphor for solid state lighting. Mater Lett 63:1275
36. You H, Wu X, Cui H, Hong G (2003) Luminescence and energy transfer of Ce^{3+} and Tb^{3+} in $Y_3Si_2O_8Cl$. J Lumin 104:223
37. Oskam K, Kaspers K, Meijerink A, Müller-Bunz H, Schleid T (2002) Luminescence of $La_3F_3[Si_3O_9]$:Ce^{3+}. J Lumin 99:101
38. Zhang X, Seo HJ (2010) Photoluminescence properties of blue-emitting Sr_2LiSiO_4F:Ce^{3+}, Li^{+}phosphor. physica status solidi (a) 207:428
39. Akella A, Keszler DA (1995) Sr_2LiSiO_4F: Synthesis, Structure, and Eu^{2+} Luminescence. Chem Mater 7:1299
40. Zhang X, Kim JS (2009) Intense green-emitting Sr_2LiSiO_4F:Eu^{2+} phosphor for n-UV white LEDs. Appl Phys A 97:549
41. Sivakumar V, Varadaraju UV (2009) Eu^{2+}, Ce^{3+} Luminescence and $Ce^{3+} \rightarrow Eu^{2+}$ Energy-Transfer Studies on Sr_2LiSiO_4F: A White Light-Emitting Phosphor. J Electrochem Soc 156:J179
42. Hao Z, Zhang J, Zhang X, Ren X, Luo Y, Lu S, Wang X (2008) Intense violet-blue emitting ($CaCl_2/SiO_2$): Eu^{2+} phosphor powders for applications in UV-LED based phototherapy illuminators. J Phys D Appl Phys 41:182001
43. Okamoto S, Yamamoto H (2009) Photoluminescent Properties of Eu^{2+}-Doped Alkaline-Earth Chlorosilicate Phosphors for LED-Based Solid-State Lighting. Electrochem Solid-State Lett 12:J112
44. Winkler A, Wieker W (1982) Zum Ablauf der Hydrothermal reaktion von CaO und Quarz in Suspension bei 190 °C. Z Anorg Allg Chem 22:52

45. Miskiewicz K, Pyzalske M (1984) Polymorphism of $Ca_3SiO_4Cl_2$-DTA studies. Cem Concr Res 18:819
46. Guo N, You H, Jia C, Oouyang R, Wu D (2014) A Eu^{2+} and Mn^{2+}-coactivated fluoro-apatite-structure $Ca_6Y_2Na_2(PO_4)_6F_2$ as a standard white-emitting phosphor via energy transfer. Dalton Trans 43:12373
47. Zhu H, Xia Z, Liu H, Mi R, Hui Z (2013) Luminescence properties and energy transfer of Bi^{3+}/Eu^{3+}-codoped $Ca_{10}(PO_4)_6F_2$ phosphors. Mater Res Bull 48:3513
48. Yang P, Yao G-Q, Lin J-H (2004) Photoluminescence of Ce^{3+} in haloapatites $Ca_5(PO_4)_3X$. Inorg Chem Commun 7:302
49. Wanjun T, Fen Z (2014) A Single-Phase Emission-Tunable $Ca_5(PO_4)_3F:Eu^{2+}$, Mn^{2+} Phosphor with Efficient Energy Transfer for White LEDs. Eur J Inorg Chem 2014:3387
50. Zhang Z, Wang J, Zhang M, Zhang Q, Su Q (2008) The energy transfer from Eu^{2+} to Tb^{3+} in calcium chlorapatite phosphor and its potential application in LEDs. Appl Phys B 91:529
51. Wang W-N, Iskandar F, Okuyama K, Shinomiya Y (2008) Rapid Synthesis of Non-Aggregated Fine Chloroapatite Blue Phosphor Powders with High Quantum Efficiency. Adv Mater 20:3422
52. Shinde K, Nagpure I, Dhoble S, Godbole S, Bhide M (2009) Dy^{3+} emission in $M_5(PO_4)_3F$ (M = Ca, Ba) phosphor. In. J. Phys. 83:503
53. Nagpure IM, Shinde KN, Dhoble SJ, Kumar A (2009) Photoluminescence characterization of Dy^{3+} and Eu^{2+} ion in $M_5(PO_4)_3F$ (M = Ba, Sr, Ca) phosphors. J Alloys Compd 481:632
54. Song Y, You H, Yang M, Zheng Y, Liu K, Jia G, Huang Y, Zhang L, Zhang H (2010) Facile synthesis and luminescence of $Sr_5(PO_4)_3Cl:Eu^{2+}$ nanorod bundles via a hydrothermal route. Inorg Chem 49:1674
55. Xiu Z, Liu S, Xu F, Ren M, Pan J, Cui X, Yu W, Yu J (2007) Synthesis and optical properties of Pb-doped $Sr_5(PO_4)_3Cl$ nanorods. J Alloys Compd 441:219
56. Liu S, Xiu Z, Xu X, Pan J, Cui X, Yu W, Yu J (2007) Synthesis and photoluminescence of $Sr_5(PO_4)_3Cl$: Mn nanorods. Mater Lett 61:1775
57. Xiu Z, Lü M, Liu S, Zhou G, Zhang H (2006) Synthesis and photoluminescence of Ce^{3+}-doped barium haloapatites nanocrystals. J Alloys Compd 416:236
58. Yoo HS, Vaidyanathan S, Kim SW, Jeon DY (2009) Synthesis and photoluminescence properties of Yb^{2+} doped $Ba_5(PO_4)_3Cl$ phosphor for white light-emitting diodes. Opt Mater 31:1555
59. Chen X, Dai P, Zhang X, Li C, Lu S, Wang X, Jia Y, Liu Y (2014) A highly efficient white light (Sr_3, Ca, Ba)$(PO_4)_3Cl:Eu^{2+}$, Tb^{3+}, Mn^{2+} phosphor via dual energy transfers for white light-emitting diodes. Inorg Chem 53:3441
60. Ren FQ, Chen DH (2009) Photoluminescence and structural characteristics of a new orange phosphor: $Ca_{2.6}Sr_{2.4}(PO_4)_3Cl:Eu^{3+}$. Appl Phys B 98:159
61. Greenblatt M, Banks E, Post B (1967) The crystal structures of the spodiosite analogs, Ca_2CrO_4Cl and Ca_2PO_4Cl. Acta Crysta 13:166
62. Chiu Y, Liu W, Chang C, Liao C, Yeh Y, Jang S, Chen T (2010) $Ca_2PO_4Cl:Eu^{2+}$: an intense near-ultraviolet converting blue phosphor for white light-emitting diodes. J Mater Chem 20:1755
63. Nagpure IM, Pawade VB, Dhoble SJ (2010) Combustion synthesis of $Na_2Sr(PO_4)F:Dy^{3+}$ white light emitting phosphor. Luminescence 25:9
64. Huang C-H, Chen Y-C, Kuo T-W, Chen T-M (2011) Novel green-emitting $Na_2CaPO_4F:Eu^{2+}$ phosphors for near-ultraviolet white light-emitting diodes. J Lumin 131:1346
65. Yang Z, Wang S, Yang G, Tian J, Li P, Li X (2007) Luminescent properties of Ca_2BO_3Cl: Eu^{2+} yellow-emitting phosphor for white light-emitting diodes. Mater Lett 61:5258
66. Guo CF, Luan L, Shi L, Seob HJ (2010) Photoluminescence Properties and Crystallographic Sites of Ce^{3+} in Ca_2BO_3Cl. Electrochem Solid-State Lett 13:J28
67. Guo C, Luan L, Shi FG, Ding X (2009) White-Emitting Phosphor $Ca_2BO_3Cl:Ce^{3+}$, Eu^{2+} for UV Light-Emitting Diodes. J Electrochem Soc 156:J125

68. Zhang Q, Wang J, Zhang G, Su Q (2009) UV photon harvesting and enhanced near-infrared emission in novel quantum cutting $Ca_2BO_3Cl:Ce^{3+}$, Tb^{3+}, Yb^{3+} phosphor. J Mater Chem 19:7088
69. Sidorenko A, Bos A, Dorenbos P, Rodnyi P, van Eijk C, Berezovskaya I, Dotsenko V (2003) Luminescence and thermoluminescence of $Sr_2B_5O_9X:Ce^{3+}$, A^+ (X = Cl, Br, A = Na^+, K^+) phosphors. J Phys: Condens Matter 15:3471
70. Dotsenko V, Berezovskaya I, Efryushina N, Voloshinovskii A, Dorenbos P, Van Eijk C (2001) Luminescence of Ce^{3+} ions in strontium haloborates. J Lumin 93:137
71. Knitel MJ, Hommels B, Dorenbos P, vanEijk CWE, Berezovskaya I, Dotsenko V (2000) The feasibility of boron containing phosphors in thermal neutron image plates, in particular the systems $M_2B_5O_9X:Eu^{2+}$ (M = Ca, Sr, Ba; X = Cl, Br) Part II: experimental results. Nuclear Instruments and Methods in Physics Research A 449:595
72. Zeng Q, Kilah N, Riley M, Riesen H (2003) Luminescence properties of Sm^{2+}-activated barium chloroborates. J Lumin 104:65
73. Dotsenko V, Radionov V, Voloshinovskii A (1998) Luminescence of Sm^{2+} in strontium haloborates. Mater Chem Phys 57:134
74. Dotsenko VP, Berezovskaya IV, Pyrogenko PV, Efryushina NP, Rodnyi PA, van Eijk CWE, Sidorenko AV (2002) Valence States and Luminescence Properties of Ytterbium Ions in Strontium Haloborates. J Solid State Chem 166:271
75. Meijerink A, Jetten H, Blasse G (1988) Luminescence and energy transfer in lead-activated strontium haloborate $(Sr, Pb)_2B_5O_9X$(X = Cl, Br). J Solid State Chem 76:115
76. Zhang X, Zhou L, Pang Q, Gong M, Setlur A (2014) Novel Broadband Excited and Linear Red-Emitting $Ba_2Y(BO_3)_2Cl:Ce^{3+}$, Tb^{3+}, Eu^{3+}Phosphor: Luminescence and Energy Transfer. J Am Ceram Soc 97:2124
77. Jing H, Guo C, Zhang G, Su X, Yang Z, Jeong J (2012) Photoluminescentproperties of Ce^{3+} in compounds $Ba_2Ln(BO_3)_2Cl$ (Ln = Gd and Y). J Mater Chem 22:7
78. Zhang Q, Wang J, Yu R, Zhang M, Su Q (2008) Luminescence Properties of Eu^{2+} Ions in $Sr_5(BO_3)_3Cl$. Electrochem Solid-State Lett 11:H335
79. Pu Z, Su M, Ruan S (1995) Studies on the luminescent properties of europium doped alkaline earth haloborates. Mater Res Bull 30:1227
80. Dhoble S, Moharil S (2000) Preparation and characterisation of Eu^{2+} activated $Sr_2B_5O_9Cl$ TLD phosphor. Nucl Instrum Methods Phys Res, Sect B 160:274
81. Sidorenko AV, Bos AJJ, Dorenbos P, Eijk CWEv, Rodnyi PA, Berezovskaya IV, Dotsenko VP, Guillot-Noel O O, Gourier D (2004) Radiation induced defects in $Sr_2B_5O_9Br:Ce^{3+}$ storage phosphor. J Phys: Condens Matter 16:4131
82. Majling J, Figusch V, Corba J, Hanic F (1974) Crystal data on calcium borate chloride, Ca_2BO_3Cl. 7: 402
83. SchipperWJ Blasse G (1994) Luminescence in the chloroborates $Ba_2Ln(BO_3)_2Cl$. J Alloys Compd 203:267
84. Xia Z, Zhuang J, Liao L, Liu H, Luo Y, Du P (2011) Synthesis and Luminescence Properties of $Ba_2Gd(BO_3)_2Cl:Eu^{2+}$ Phosphor. J Electrochem Soc 158:J359
85. Zhang N, Guo C, Jeong J (2013) Color tunable emission in Ce^{3+} and Tb^{3+} co-doped $Ba_2Ln(BO_3)_2Cl$ (Ln = Gd and Y) phosphors for white light-emitting diodes. Spectrochim acta A 116:556
86. Jing H, Guo C, Zhang N, Ren Z, Bai J (2013) Effects of Eu^{3+} Sites on Photoluminescence in $Ba_2Ln(BO_3)_2Cl$ (Ln = Gd, Y) Hosts. ECS J Solid State Sci Techno 2:R1
87. Zhao J, Guo C, Li T (2015) Enhanced near-infrared emission by co-dopingCe^{3+} in $Ba_2Y(BO_3)_2Cl:Tb^{3+}$, Yb^{3+} phosphor. RSC Adv 5:28299
88. Tang Y-S, Hu S-F, Ke W-C, Lin CC, Bagkar NC, Liu R-S (2008) Near-ultraviolet excitable orange-yellow $Sr_3(Al_2O_5)Cl_2:Eu^{2+}$ phosphor for potential application in light-emitting diodes. Appl Phys Lett 93:131114
89. Song Y, Jia G, Yang M, Huang Y, You H, Zhang H (2009) $Sr_3Al_2O_5Cl_2:Ce^{3+}$, Eu^{2+}: A potential tunable yellow-to-white-emitting phosphor for ultraviolet light emitting diodes. Appl Phys Lett 94:091902

90. Zhang X, Park B, Choi N, Kim J, Kim GC, Yoo JH (2009) A novel blue-emitting $Sr_3Al_2O_5Cl_2$:Ce^{3+}, Li^+ phosphor for near UV-excited white-light-emitting diodes. Mater Lett 63:700
91. Zhang X, Choi N, Park K, Kim J (2009) Orange emissive phosphor for warm-white light-emitting diodes. Solid State Commun 149:1017
92. Zhang X, Park B, Kim J, Lee J, Choi J (2010) Orange emission enhancement by energy transfer in $Sr_3Al_2O_5Cl_2$:Ce^{3+}, Eu^{2+} phosphor for solid-state lighting. J Lumin 130:117
93. Lee J, Choi J, Zhang X, Lee J, Park K, Kim J (2010) Novel blue-emitting $Sr_3Ga_2O_5Cl_2$:Eu^{2+} phosphor for UV-pumped white LEDs. Mater Lett 64:768
94. Li Y, Wang Y, Gong Y, Xu X, Zhou M (2010) Design, synthesis and characterization of an orange-yellow long persistent phosphor: $Sr_3Al_2O_5Cl_2$: Eu^{2+}, Tm^{3+}. Opt Express 18:24853
95. Dong X, Zhang J, Zhang X, Hao Z, Liu X (2014) Photostimulated luminescence of $Sr_3Al_2O_5Cl_2$:Eu^{2+}, Tm^{3+} with a large energy storage capacity. J Lumin 145:657
96. Dutczak D, Ronda C, Meijerink A, Jüstel T (2013) Red luminescence and persistent luminescence of $Sr_3Al_2O_5Cl_2$:Eu^{2+}, Dy^{3+}. J Lumin 141:150
97. Im WB, Brinkley S, Hu J, Mikhailovsky A, Denbaars SP, Seshadri R (2010) $Sr_{2.975-x}Ba_xCe_{0.025}AlO_4F$: a Highly Efficient Green-Emitting Oxyfluoride Phosphor for Solid State White Lighting. Chem Mater 22:2842
98. Chen W, Liang H, Ni H, He P, Su Q (2010) Chromaticity-Tunable Emission of Sr_3AlO_4F: Ce^{3+} Phosphor: Correlation with Matrix Structure and Application in LEDs. J Electrochem Soc 157:J159
99. Im WB, George N, Kurzman J, Brinkley S, Mikhailovsky A, Hu J, Chmelka B, Denbaars SP, Seshadri R (2011) Efficient and Color-Tunable Oxyfluoride Solid Solution Phosphors for Solid-State White Lighting. Adv Mater 23:2300
100. Im WB, Fellows NN, Denbaars SP, Seshadri R (2009) $La_{1-x-0.025}Ce_{0.025}Sr_{2+x}Al_{1-x}Si_xO_5$ solid solutions as tunable yellow phosphors for solid state white lighting. J Mater Chem 19:1325
101. ParkS Vogt T (2010) Near UV excited line and broad band photoluminescence of an anion-ordered oxyfluoride. J Am Chem Soc 132:4516
102. Chen W, Liang H, Xie M, Su Q (2010) Photoluminescence and Low Voltage Cathodoluminescence of Sr_3AlO_4F:Eu^{3+}. J Electrochem Soc 157:J21
103. Shang M, Li G, Kang X, Yang D, Geng D, Lin J (2011) Tunable Luminescence and Energy Transfer properties of Sr_3AlO_4F:RE^{3+} (RE = Tm/Tb, Eu, Ce) Phosphors. ACS Appl Mater Interf 3:2738
104. Sun J, Sun G, Sun Y, Han L (2014) Luminescence properties and energy transfer investigations of tri-doped Sr_3AlO_4F:Ce^{3+}, Tb^{3+}, Yb^{3+} phosphors. Opt Mater 36:1097
105. Dhoble SJ, Gedam SC, Nagpure IM, Godbole SV, Bhide MK, Moharil SV (2008) Luminescence of Cu^+ in halosulphate phosphor. Journal of Materials Science 43:3189
106. Gedam SC, Dhoble SJ, Moharil SV (2007) $Dy^{3+}$and Mn^{2+} emission in $KMgSO_4Cl$ phosphor. J Lumin 124:120
107. Gedam SC, Dhoble SJ, Moharil SV (2006) Synthesis and effect of Ce^{3+} co-doping on photoluminescence characteristics of $KZnSO_4Cl$: M (M = Dy^{3+} or Mn^{2+}) new phosphors. J Lumin 121:450
108. Chiu Y-C, Liu W-R, Huang C-H, Yeh Y-T, Jang S-M (2012) Luminescence properties, crystal structure and LED package of potential blue-emitting phosphors-Ca_2BN_2F:Eu^{2+}. Opt Express 20:27361
109. Marchuk A, Schnick W (2015) $Ba_3P_5N_{10}Br$:Eu^{2+}: A Natural-White-Light Single Emitter with a Zeolite Structure Type. Angew Chem 127:2413
110. Xia Z, Sun J, Du H (2004) Study on luminescence properties and crystal-lattice environmental of Eu^{2+} in $Sr_{4-x}Mg_xSi_3O_8Cl_4$:Eu^{2+} phosphor. J Rare earth 22:370
111. Xia Z, Sun J, Du H, Zhou W (2006) Luminescence of Eu^{2+} in alkali earth chlorosilicate phosphor and their color-tunable properties. Opt Mater 28:524
112. Xia Z, Li Q, Sun J (2006) Greenish-yellow Light-emitting, Long-lasting Phosphorescence in Eu^{2+} -doped $CaO–CaBr_2–SiO_2$ Phosphor System. Chem Lett 35:764

113. Xia Z, Liu J, Li Q, Sun J (2007) Comparative Investigation of Photoluminescence and Long-Lasting Phosphorescence Properties in the ($CaO–CaBr_2–SiO_2$):Eu^{2+} Phosphor System. Electrochem Solid State Lett 10:J4
114. Xia Z, Li Q, Li G, Xiong M, Liao L (2011) Crystal Growth of $Ca_3SiO_4Br_2$: New Photoluminescence Bromosilicate Host. J Crystal Growth 318:958
115. Xia Z, Liao L, Xiong M, Li G (2013) Synthesis, structure and Eu^{2+} -doped luminescence properties of bromosilicate compound $Ca_3SiO_4Br_2$. J Lumin 134:227
116. Xia Z, Li G, Chen D, Xiao H (2009) Synthesis and calcination temperature dependent photoluminescence properties of novel bromosilicate phosphors. Mater Lett 63:2600
117. Chen X, Xia Z, Liu Q (2014) Synthesis, structure and luminescence properties of new chloro-germanate phosphors $Ca_3GeO_4Cl_2$:Eu^{2+}. Dalton Trans 43:13370
118. Xia Z, Zhuang J, Liu H, Liao L (2012) Photoluminescence properties and energy transfer of $Ba_2Lu(BO_3)_2Cl$: Eu^{2+}/Eu^{3+}, Tb^{3+} phosphors. J Phys D Appl Phys 45:015302
119. Xia Z, Zhuang J, Liao L (2012) Novel red-emitting $Ba_2Tb(BO_3)_2Cl$: Eu phosphor with efficient energy transfer for potential application in white light-emitting diodes. Inorg Chem 51:7202
120. Xia Z, Zhuang J, Meijerink A, Jing X (2013) Host composition dependent tunable multicolor emission in the single-phase $Ba_2(Ln_{1-z}Tb_z)(BO_3)_2Cl$: Eu phosphors. Dalton Trans 42:6327
121. Xia Z, Molokeev MS, Oreshonkov AS, Atuchin VV, Liu RS, Dong C (2014) Crystal and local structure refinement in $Ca_2Al_3O_6F$ explored by X-ray diffraction and Raman spectroscopy. Phys Chem Chem Phys 16:5952
122. Xia Z, Liu R-S (2012) Tunable Blue-Green Color Emission and Energy Transfer of $Ca_2Al_3O_6F$:Ce^{3+}, Tb^{3+} Phosphors for Near-UV White LEDs. J Phys Chem C 116:15604

[illegible]

Chapter 11
Tuning Luminescence by Varying the O/N or Al/Si Ratio in Some Eu-Doped Nitride Phosphors

Quanlin Liu and Ting Wang

Abstract Silicon-(oxy)nitrides ((oxy)nitridosilicates) have rich structures based on $[SiN_4]$ or $[(Si, Al)(N, O)_4]$ tetrahedra, which can accommodate various cations and form complicated solid–solutions. These structures can adjust 5*d* energy levels of rear-earth ions and their luminescent features, which have been currently adopted to develop novel silicon-(oxy)nitride phosphors for white LEDs. In the structures of some silicon-nitrides, the N can be partially replaced by O with Al substitutions for Si. This substitution generally gives rise to a large change in the nephelauxetic and crystal-field splitting effects on Ce^{3+} and Eu^{2+} ions; thus, they can modify excitation and emission wavelengths for application in LEDs. In this chapter, we focus on the several typical nitride phosphors, i.e., $CaAlSiN_3$-, $M_2Si_5N_8$ (M = Ca, Sr, Ba)-, and sialon-based phosphors. We describe their structures in details and discuss the relationships of lattice parameters, compositions and O-preferential occupancy. We also show how to adjust their photoluminescence spectra by changing the rations of N/O or Si/Al. Finally, the relationship of 5d level energy of Ce^{3+}/Eu^{2+} with structure and composition in nitrides is presented and discussed.

11.1 Introduction

White light–emitting diodes (LEDs) have recently been receiving attention as light sources because of their energy efficiency, long lifetime, compactness, and environmental friendliness [1, 2]. The white light can be generated by using an LED chip–phosphor system, which is termed “phosphor-converted LED” (pc-LED), in which the phosphor strongly absorbs blue light (approximately 460 nm) from LED chips and efficiently re-emits in the red, green, or yellow parts of the visible spectrum. This revolution has led to the identification and development of numerous potential phosphor materials for further improvement of light quality and

Q. Liu (✉) · T. Wang
School of Materials Science & Engineering, University of Science and Technology Beijing, Beijing 100083, China
e-mail: qlliu@ustb.edu.cn

R.S. Liu (ed.), *Phosphors, Up Conversion Nano Particles, Quantum Dots and Their Applications*, DOI 10.1007/978-3-662-52771-9_11

extending LED applications. Typical commercially available white LED features a combination of blue InGaN chips with $Y_3Al_5O_{12}$:Ce^{3+} (YAG:Ce)-based phosphors [3, 4]. This type of white light has high correlated color temperature (CCT) and poor color-rendering index (CRI) because it lacks a red component, thus giving "cool"-white light, and cannot meet the requirements of some special fields such as medical and architectural applications. This prevents the use of these lamps in many general lighting applications, especially as replacements for lamps with lower CCTs, which make up the majority of the lighting market. The need to fabricate "warm"-white LEDs with CCT of 2700–3500 K and high CRI > 80 has driven LED phosphor development toward novel compositions and structure-types [5].

It is necessary to combine suitable phosphors with an LED chip to produce white light. The general requirements for LED phosphors are listed as follows [6]: (1) broad excitation spectrum matching well with the emission spectrum of blue LED chips; (2) suitable emission spectrum such as the emission position; (3) high quantum efficiency; (4) small thermal quenching; and (5) high chemical stability. Phosphors play key roles in determining the optical properties of white LED lamps such as luminous efficiency, chromaticity coordinate, correlation color temperature (CCT), and lifetime. Excitation and emission spectra are first and most important criterion, which allow one to judge the value of the phosphors at first.

Generally, the luminescent properties of RE ions usually involve two electric transitions: intra-$4f^n$ (*f-f*) transition and $4f^n$–$4f^{n-1}5d$ (*f-d*) transition (transition between a $4f^{n-1}5d^1$ excited state and a $4f^n$ ground state [such as from Eu^{2+} or Ce^{3+}]). The emission wavelength from the *f-f* transition of trivalent RE ions is relatively insensitive to the host and the temperature because the 4*f* shell is shielded from its surroundings by the filled 5*s* and 5*p* shells. The *f-d* transition is a parity-allowed electric dipole transition; therefore, broad emissions can be achieved. In addition, the wavelengths of absorption and emission can be tuned by the host lattice because of the strong interaction of the 5*d*-electron with the neighbouring anion ligands in the compounds. As a free ion, Ce^{3+} or Eu^{2+} has a large energy gap between the 4*f* ground state and the 5*d* excited state; for Ce^{3+} this gap is approximately 6.2 eV (50,000 cm^{-1}) [7]. This gap is decreased substantially by crystal-field splitting and a downward shift of the 5*d* centroid leading to a 4*f*-to-5*d* separation at approximately 4000–2000 cm^{-1}, which enables the absorption of blue light and the emission of yellow or red light. Because the 5*d* orbitals are strongly affected by the surrounding crystal field and by the polarizability of the host [8, 9], the excitation and emission energies can be tuned through the host crystal composition and structure. Therefore, Eu^{2+}/Ce^{3+}-doped phosphors have attracted increasing attention in recent years. Dorenbos [10, 11] carried out a systematic study of the energies of the lowest $4f^{n-1}5d^1$ states of Eu^{2+}/Ce^{3+} ions. The effect of the host crystal on the energy difference of *fd*-absorption is expressed by the energy red-shift (D) of 5*d* levels. This red shift is a combined effect of the centroid shift ε_c and the crystal-field splitting ε_{cfs}. The centroid shift ε_c is the downward shift of the average energy of the five 5*d* levels of Eu^{2+} relative to the energy of the first *fd* transition in the free Eu^{2+} ion. ε_c represents the nephelauxetic effect, which is often attributed to covalency between the lanthanide ion and the anion ligands of host lattice [12]. The crystal-field splitting ε_{cfs} is the difference in energy between the energy of the lowest and highest 5*d* levels.

The much lower electronegativity of nitrogen (χ(N) approximately 3.0 versus χ(O) approximately 3.4) could make it more possible to find new nitride phosphors that can effectively absorb blue-LED radiation. Although nitrides are less numerous than oxides due to the thermodynamic factors (e.g., larger triple-bond energy of 941 kJ/mol dinitrogen vs. 499 kJ/mol oxygen), they have a fairly large number of compounds, which leads to a rich nitride chemistry and structure. Currently one important focus has been on Eu^{2+}/Ce^{3+}-doped nitride phosphors because they have broad excitation and emission extending from UV to the visible spectral region, and thus they are suitable for use as downconversion phosphors in white LEDs [6].

The most diverse class of nitrides is nitridosilicates and their derived compounds, oxonitridosilicates, and oxonitridoaluminosilicates. These compounds are excellent host materials for phosphors due to their high strength and outstanding thermal and chemical stability, which had been previously studied as structural materials [13]. The structures of nitridosilicates typically consist of SiN_4 tetrahedra. The partial nitrogen/silicon of SiN_4 tetrahedra can be replaced by oxygen/aluminum to form $Si/Al[O/N]_4$ tetrahedra. Similar to oxosilicates, SiN_4 or $Si/Al[O/N]_4$ units are stacked together by sharing their corners to form a condensed framework, thus resulting in excellent thermal and chemical stability of nitridosilicates and oxonitridosilicates. Furthermore, SiN_4 tetrahedra can connect with each other by sharing edges to form more complex structures. The rare-earth or other metal ions can be accommodated in the voids of the network. These various structures can provide Ce^{3+}/Eu^{2+} with rich and different ligand environments as well as adjust the emission and absorption wavelengths.

Photoluminescence of rare earth–doped nitridosilicates was first reported in 1997 for $CaSiN_2:Eu^{2+}$ [14]. Later, van Krevel et al. [15] reported the long-wavelength Ce^{3+} emission in Y–Si–O–N materials. Uheda [16] and Höppe [17] investigated the photoluminescence of $LaSiN_5:Eu^{2+}$ and $Ba_2Si_5N_8:Eu^{2+}$, respectively. Since the interesting work by Krevel et al. [18] and Xie et al. [19], the rare earth–doped nitride phosphors have received increasing attention. In particular, α-SiAlON [20], β-SiAlON [21], $M_2Si_5N_8:Eu^{2+}$ (M = Ca, Sr, Ba) [22], and $Sr_xCa_{1-x}AlSiN_3:Eu^{2+}$ [23] phosphors have been proved to be good downconversion phosphors for white LED combined with a blue InGaN LED chip.

In the structures of nitridosilicates, N can be partially replaced by O, with Al substitutions for Si, because the mean bond lengths Si–N, Si–O, Al–N, and Al–O are similar, i.e., 0.174, 0.164, 0.187, and 0.175 nm, respectively. This substitution generally give rise to a large change in the nephelauxetic and crystal-field splitting effects on Ce^{3+} and Eu^{2+} ions and thus can modify the excitation and emission wavelengths for application in LEDs. In the next sections, we will show how to adjust photoluminescence spectra of some typical phosphors by changing the rations of N/O or Si/Al.

11.2 $CaAlSiN_3:Eu^{2+}$ and $(Sr, Ca)(Al, Si)_2(N, O)_3:Eu^{2+}$ Phosphors

The host compound, $CaAlSiN_3$, was reported to be formed in the $CaO–AlN–Si_3N_4$ phase diagram [24]. Ottinger [25] analyzed the crystal structure of $CaAlSiN_3$ using single-crystal diffraction data. Uheda et al. [23, 26] first reported the photoluminescence of Eu^{2+}-doped $CaAlSiN_3$, which has a broad excitation band from the UV region to 590 nm and gives a broadband emission with a maximum between 630 and 680 nm. This phosphor exhibits high quantum efficiency, small thermal quenching, and high chemical stability; therefore, it has attracted increasing attention as an important red phosphor for white LEDs.

11.2.1 $CaAlSiN_3$-Type Structure

The atomic coordinates and crystallographic data of $CaAlSiN_3$ were reported by Uheda et al. [26]; herein they are listed in Table 12.1. The $CaAlSiN_3$ crystallizes in the orthorhombic structure with the space group of Cmc21 (no. 36). There are four crystallographic sites: Ca atoms occupy the 4a site, Al/Si atoms take up the 8b site randomly, whereas N atoms locate at two crystallographic site, i.e., 8b and 4a site, denoted by N^I and N^{II}, respectively. The multiplicity of N^I and N^{II} are 8 and 4, respectively, so the amount of N^I is twice as that of N^{II}. Due to the difference of the site symmetry of N^I and N^{II}, the crystal-chemical formula of $CaAlSiN_3$ can be represented as $Ca(Al/Si)_2N^I_2N^{II}$ [27].

The structure of $CaAlSiN_3$ consists of highly dense and corner-sharing MN_4 ($M = Al^{3+}$, Si^{4+}) tetrahedral network. Six MN_4 tetrahedra form a six-membered ring by sharing corners, and the rings link together parallel to the (001) plane to form a sheet A. Another sheet, B, which is equivalent to sheet A rotated by 180°, is overlaid on sheet A to form a rigid three-dimensional framework. The Ca atoms are located in the large channels along the c-axis, which are coordinated by five nitrogen atoms. N^I atoms are coordinated with three Si/Al atoms, whereas N^{II} are only bonded to Si/Al atoms. The N^I and N^{II} are very different from each other from the view point of structural chemistry. The $CaAlSiN_3$ structure can be regarded as consisting of puckered hexagonal layers containing equal numbers of silicon and N^I atoms, and the layers are connected together along the *a*-axis by N^{II} atoms, which complete the tetrahedral around the Al/Si atoms. Half of the Al/Si atoms within a layer are, by way of N^{II} atoms, linked to the layer above and the other half to the layer below.

Many nitrides adopt the $CaAlSiN_3$-type structure, and the general formula can be expressed as AB_2X_2X', such as $LiSi_2N_3$ and $NaSi_2N_3$ (with A = Li or Na, B = Si and X = X′ = N), Si_2N_2O (A = vacancy, B = Si, X = N, X′ = O), $SrAlSiN_3$, and $MgAlSiN_3$ (A = Sr/Mg, B = Al/Si). A considerable variety of solid solutions can be formed between these compounds. Li, Na, Mg, Ca, and Sr ions can occupy A-sites; or a large amount of vacancies exist in A-sites. There are different atomic

Table 11.1 Crystallographic data and atomic coordinates of $CaAlSiN_3$

	$CaAlSiN_3^a$				$Sr_{0.99}Eu_{0.01}AlSiN_3^b$			
Site	Atom	*x*	*y*	*z*	Atom	*x*	*y*	*z*
4a	Ca	0.0	0.3146	0.4952	Sr/Eu	0.0	0.3153	0.5086
8b	Al/Si	0.1741	0.1570	0.0291	Al/Si	0.1777	0.1518	0.0256
8b	N^I	0.2126	0.1300	0.3776	N^I	0.2286	0.1177	0.3617
4a	N^{II}	0.0	0.2446	0.678	N^{II}	0.0	0.2081	0.015
Space group		$Cmc2_1$				$Cmc2_1$		
Lattice Parameters (Å)		a = 9.8020(4), b = 5.6506(2) c = 5.0633(2)				a = 9.843(3), b = 5.760(2) c = 5.177(2)		

[a]Data from Uheda et al. [26]. [b]Data from Watanabe et al. [28]

ratios of Al/Si in B-sites and various atomic ratios of N/O in X′-sites. For comparison, crystallographic data of $Sr_{0.99}Eu_{0.01}AlSiN_3$ are also listed in Table 11.1.

It is difficult to distinguish O/N or Al/Si sites based on X-ray diffraction data because their atomic scattering factors for X-ray are very similar, respectively. Generally it needs neutron-diffraction data to determine the O and N (or Al and Si) sites in the structures. Piao et al. [29] reported that $M–N^I$ is shorter than $M–N^{II}$, indicating that three-fold bridged N^I (X-site) contributed dominantly to the rigidness of the structure. Furthermore, in other compounds sharing the isotropic space groups, the N^{II} sites can be replaced by other anion or anion groups such as O^{2-}, NH^{2-}, and $-N = C{=}N^{-2-}$. Wang et al. [27] confirmed that O atoms preferentially occupy N^{II}-sites (X'-site) through the composition dependence of lattice parameters. This method is also valid for distinguishing O and N in $Sr_2(Si, Al)_5(N, O)_8{:}Eu^{2+}$ phosphors [30].

The lattice parameter ***a*** of Eu-doped $Ca(Al/Si)_2N_2(N_{1-x}O_x)$ shows a clear decrease with higher oxygen concentrations (Fig. 11.1) as expected given the

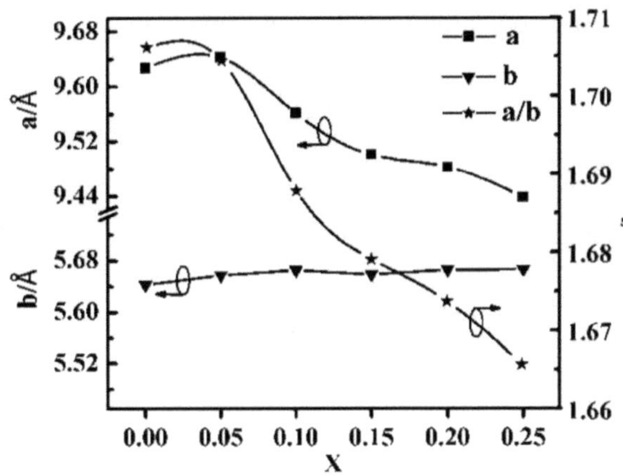

Fig. 11.1 Lattice parameters of $Ca_{0.875-0.5x}[Al_{0.75}Si_{1.25}]N_2(N_{1-x}O_x){:}0.02Eu$. Reprinted from Ref. [27], copyright 2013, with permission from Elsevier

smaller ionic radii of O^{2-} versus N^{3-}. However, the lattice parameter ***b*** does not follow a clear decrease with O^{2-} concentration as expected by Vegard's law. Thus, ***a*/*b*** ratio has a large change with x, resulting in separation of each pair of peaks [27]. Substitutions of O atoms for N atoms generally give rise to the lattice contraction because the mean bond lengths of Si–N, Si–O, Al–N, and Al–O are 0.174, 0.164, 0.187, and 0.175 nm, respectively. The observation that the crystal structure has a larger shrinkage along the [100] direction than the [010] direction with the oxygen concentration increasing indicates that oxygen atoms preferentially substitute for nitrogen atoms only at one of two crystallographic sites. In the crystal structure (Fig. 11.2), N^{I} atoms are linked with three Si/Al atoms, whereas N^{II} atoms are chain-linked with two Si/Al atoms. In fact, a large amount of structural data in oxygen-containing silicates or alumina silicates demonstrate that divalent O atoms can either occupy terminal sites of Si/Al–O tetrahedra or bridge two Si/Al–O tetrahedral. An O atom that takes up the site linking three Si/Al–O tetrahedra is not observed, which is consistent with oxosilicate chemistry and confirmed by the "Pauling's second rule" model for anion distributions [31]. This order was also confirmed in a $Ba_{1.95}Eu_{0.05}Si_4AlN_7O$ sample determined from the neutron diffraction by Chen et al. [32]. Therefore, only N^{II} atoms can be substituted by O atoms in $CaAlSiN_3$-based crystal structure, and the crystal-chemical formula can be represented as $Ca(Al/Si)_2N_2^{I}(N_{1-x}^{II}O_x)$. Along the [010] direction (the orange chain in Fig. 11.2b), the combination of $-N^{I}$–Si/Al–$N^{I}-$ without substitution of O contributed dominantly to the rigidness of the structure, which caused the ***b***-axis length to almost keep constant. However, along the [100] direction, the replacement of N^{II} by O forms the $-N^{I}$–Si/Al–N^{II}/O– chains leading to a rapid decrease of the lattice parameter of this direction, which agreed well with the results of XRD analysis (Fig. 11.1).

Li et al. [33] addressed that fact that Al/Si is found to be occupied at the 8b site with a ratio of about 1/2 by a solid-state reaction method in all cases, not 1/1 as previously proposed for both powder [23] and single-crystal [25] compound. The investigators proposed that an approximate composition can be formulated as

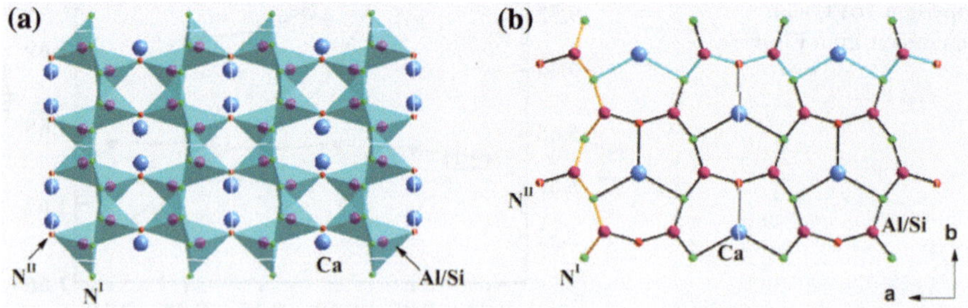

Fig. 11.2 Crystal structure of $Ca_{0.875-0.5x}[Al_{0.75}Si_{1.25}]N_2^{I}(N_{1-x}^{II}O_x)$:0.02Eu along the [001] direction. **a** Polyhedral view of the structure and **b** single layer of the structure. Reprinted from Ref. [27], copyright 2013, with permission from Elsevier

$Ca_{1-\delta/2}Al_{1-\delta}Si_{1+\delta}N_3$ or $CaAl_{1-4\delta/3}Si_{1+\delta}N_3$ ($\delta = 0.3$–0.4) instead of an ideal $CaAlSiN_3$. Li et al. [34] also reported that all Eu-doped $CaAlSiN_3$ samples from alloy at low temperatures showed a large deficiency in Ca and Al. From the viewpoint of structural chemistry, $Ca_{1-\delta/2}Al_{1-\delta}Si_{1+\delta}N_3$ (i.e., Ca vacancy) is better for structural stability than $CaAl_{1-4\delta/3}Si_{1+\delta}N_3$ (i.e., Al/Si vacancy) because the lattice can be regarded as (Si, Al)N_4 tetrahedra being stacked together by sharing their corners to form a condensed framework with Ca locating in the voids of the network. Vacancies can be easily generated in Ca sites due to the substitution of Si^{4+} for Al^{3+} or O^{2-} for N^{3-}. As an extreme example, the structure of Si_2N_2O is stable when a complete vacancy (no atom) exists in Ca sites. The concentration of cations in Ca sites can be changed according to electroneutrality requirement.

In $CaAlSiN_3$ structure, it is difficult to determine the ordered or disordered arrangement of Al^{3+} and Si^{4+} by X-ray diffraction method because of there being only a small difference in their atomic-scattering factors. Harris et al. [35] tried to measure the arrangement of Al^{3+} and Si^{4+} in $MgAlSiN_3$ by ^{15}N-NMR, but could not obtain a definitive answer as to whether both ions have an ordered or a disordered structure. Uheda et al. [36] reported that Al^{3+} and Si^{4+} occupied randomly at the same site with vacancies in the Ca site based on the analysis on data of the convergent-beam electron diffraction (CBED). Although Al^{3+} and Si^{4+} occupied randomly at the same site from diffraction data, Lee [37] and Jung [38] confirmed that there exist different local environments around the Eu^{2+} activators in the structure of $CaAlSiN_3$:Eu^{2+}, i.e., Si-rich and Al-rich environment, thus causing an inhomogeneous broadening and deconvoluted Gaussian peaks. Yang et al. [39] investigated the crystal structures and photoluminescence properties of Eu-doped $CaAlSiN_3$ prepared using an alloy–nitridation method in different Eu concentrations and in different Al/Si molar ratios. They gave a linear relationship between lattice volumes and Al/Si molar ratios indicating that $CaAlSiN_3$ structure allows Al/Si ratio having a large changes, i.e., not limited to 1:1.

11.2.2 Luminescence

The properties of a substance, to a large extent, rely on its composition and crystal structure. As mentioned previously, $CaAlSiN_3$-type structure can accommodate various ions and form complicated solid solutions. These changes in composition and local structures can adjust the 5d energy levels of Eu^{2+}/Ce^{3+} and their luminescent features.

$CaAlSiN_3$: Eu^{2+}, which has wide absorption bands from the UV to the visible range, is one of the most promising phosphors because of its chemical stability, low thermal quenching, and high quantum efficiency [23]. Despite the significant potential in the application, which requires red components, it is found that most of the emission peaks of $CaAlSiN_3$:Eu^{2+} are <640 nm and can hardly meet the required wavelengths (approximately 620 nm) of high-efficacy wLED due to effects of the relatively sensitive functioning of the human visual system. For lighting applications that require both high rendering and efficiency, it is desirable to shift the deep red emission of $CaAlSiN_3:Eu^{2+}$ to shorter wavelengths without loss of quantum output. The required emission wavelength is approximately 620 nm for high-efficacy wLED. A cationic substitution method has been used to make up this deficiency. Watanabe et al. [28, 40] reported that replacement of Ca^{2+} by Sr^{2+} can shift the emission maxima from 650 to 610 nm. They addressed the fact that the structures of $CaAlSiN_3$ and $SrAlSiN_3$ are closely related, and complete solid solutions are formed between the end members. Watanabe et al. [40] reported that Sr^{2+} with a larger ionic radius than that of Ca^{2+} was introduced to substitute for Ca^{2+}. Replacement of Ca^{2+} by larger Sr^{2+} expands the crystal lattice, which weakens the crystal-field strength around Eu^{2+} and results in the blue shift of emission without much losing quantum output. However, in this process, $Sr_{1-x}Ca_xAlSiN_3:Eu^{2+}$ phosphor was synthesized under an unapproachable pressure of 190 MPa.

Wang et al. [27] reported that emission peaks of $CaAlSiN_3:Eu^{2+}$ can be tuned from 650 to 610 nm by oxygen substitutions for nitrogen (Fig. 11.3). They confirmed that oxygen preferentially substitutes for nitrogen in 4a site, N^I-site. Incorporating more O to substitute for N atoms reduces the nephelauxetic effect around the Eu^{2+} and elevates the lowest 5*d* energy level of Eu^{2+} in the crystal because O^{2-} has higher electronegativity than N^{3-}, and Eu–O has lower covalency than Eu–N.

Due to oxygen substitution, two luminescent centers corresponding to two types of Eu^{2+}-coordination modes, i.e., $EuN^I_2N^{II}_3$ and $EuN^I_2N^{II}_2O$, are formed in the structure of $Ca(Al/Si)_2N_2(N_{1-x}O_x):Eu^{2+}$ (Fig. 11.4). To further clarify the

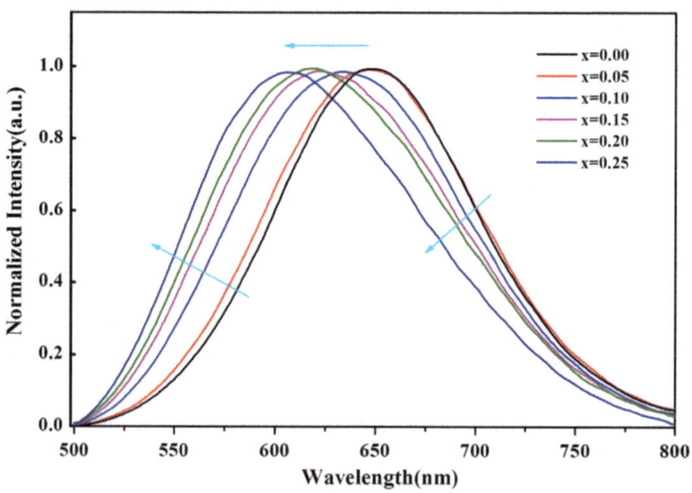

Fig. 11.3 Normalized emission spectra of $Ca_{0.875-0.5x}[Al_{0.75}Si_{1.25}]N_2(N_{1-x}O_x)$:0.02 Eu with various O content, *x*, excited at 450 nm. *x* indicates O concentration. The *arrows* illustrate the band shift and change in the spectra with increasing *x*. Adapted from Ref. [27], copyright 2013, with permission from Elsevier

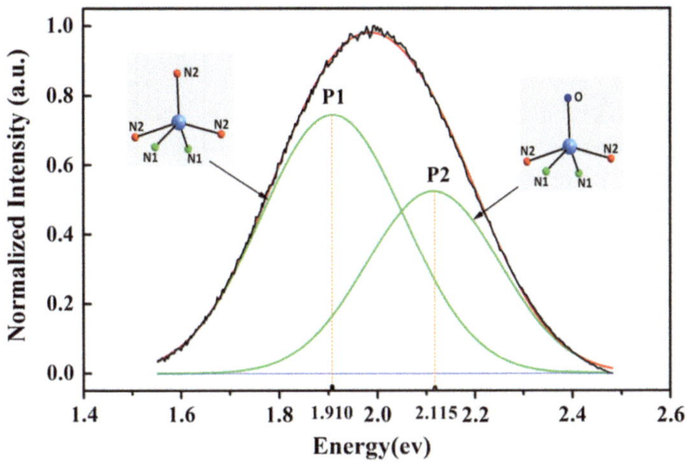

Fig. 11.4 Fitting curves by bi-Gaussian method of $Ca_{0.625}[Al_{0.75}Si_{1.25}]N_2(N_{0.85}O_{0.15})$:0.02Eu. The *inset* figures show $EuN^{I}2N^{II}3$ and $EuN^{I}2N^{II}2O$ molds corresponding to P1 (1.910 eV) and P2 (2.115 eV), respectively. Reprinted from Ref. [27], copyright 2013, with permission from Elsevier

luminescent feature of Eu^{2+} in the two coordinating environments, two fitting curves were estimated using bi-Gaussian functions to fit emission bands of samples with different O contents on an energy scale (Fig. 11.5). One type of luminescence center peaking at 1.910 eV (650 nm-P1) corresponds to $EuN_2^{I}N_3^{II}$ coordination mode, and another peaking at 2.115 eV (587 nm-P2) relates to $EuN_2^{I}N_2^{II}O$. With increasing O concentration P1 intensities decrease, whereas P2 intensities increase, thus indicating that more $EuN_2^{I}N_2^{II}O$ models form with more $EuN_2^{I}N_3^{II}$ models dismissing. A continuous blue shift of emission is due to the change in the numerical ratio of the two luminescent centers.

As temperature was increased from 15 to 300 K, the emission peak position of $Ca(Al/Si)_2N_2(N_{0.8}O_{0.2})$:0.02Eu shifted from 611 to 622 nm [27]. This behavior is against the law that increasing temperature generally leads to blue shift of emission band. This phenomenon can also be explained by the change of relative intensity of P1 and P2 with temperature. By fitting the curves of emission spectra using bi-Gaussian curves, the P1 intensities keep almost constant, whereas P2 intensities decrease and their full width half maximum (FWHM) widens with increasing temperature. This suggests that the effects of temperature on Eu^{2+} in the $EuN_2^{I}N_2^{II}O$ models are stronger than those in $EuN_2^{I}N_3^{II}$ models, which is in accordance with the fact that nitride phosphors pose a lower thermal quenching effect than most oxide phosphors [41].

Wang et al. [42] analyzed the time-resolved photoluminescence of Eu^{2+}-doped Ca $(Al/Si)_2N_2(N_{1-x}O_x)$ (CASNO:Eu^{2+}) and proved the presence of energy transfer among Eu^{2+} ions in two P1 ($EuN_2^{I}N_3^{II}$) and P2 ($EuN_2^{I}N_2^{II}O$) luminescent centers. As shown in Fig. 11.6, the relative P1/P2 ratio remains almost constant for CASNO:0.0005Eu^{2+},

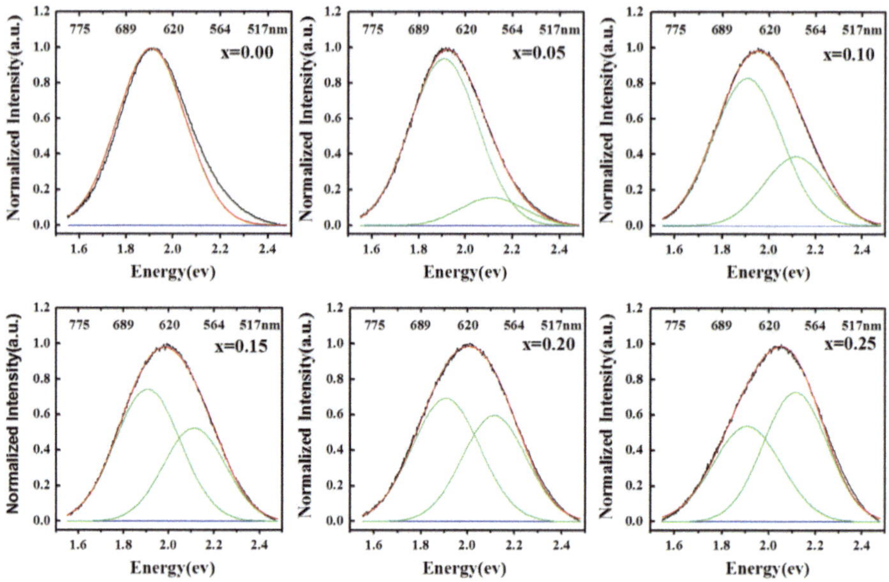

Fig. 11.5 Emission spectra ($\lambda_{ex} = 450$ nm) by bi-Gaussian fitting of $Ca(Al/Si)_2N_2(N_{1-x}O_x)$:0.02Eu. The figures indicate O concentration. The peak positions of P1 and P2 almost keep constant, i.e., 1.910 eV for P1 and 2.115 eV for P2. P1 intensities decrease, whereas P2 intensities increase with greater amount of oxygen. Reprinted from Ref. [27], copyright 2013, with permission from Elsevier

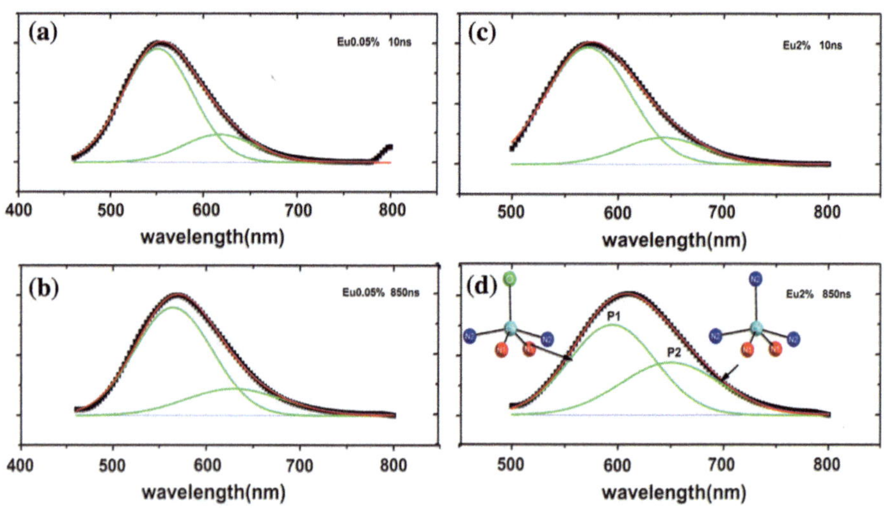

Fig. 11.6 Time-resolved emission spectra of CASNO $0.0005Eu^{2+}$ (**a** and **b**) and CASNO:$0.02Eu^{2+}$ (**c** and **d**). Fitted Gaussian emission bands are shown, and their corresponding environments around Eu^{2+} ions are designated in the *inset*. Reproduced with permission from Ref. [42], Copyright 2013, the electrochemical society

which suggests that P1 keeps almost the same time evolvement with P2 in CASNO:0.0005Eu^{2+}. Figures 11.6c, d show that the relative P1/P2 ratio decreases dramatically as the decay time increases in CASNO:0.02Eu2 + resulting from the relatively decreased P1 and increasing P2 intensity. These suggest that P1 has a faster time evolution than that of P2 as well as the energy transfer from Eu^{2+} ions at EuN_4O environments to those at pure N environments. According to the various time evolutions of P1 and P2 in two different CASNO hosts, the energy transfer hardly occurs in CASNO:0.0005Eu^{2+} phosphor, whereas it occurs obviously in CASNO:0.02Eu^{2+} because the higher Eu^{2+} concentration decreases the distance for energy transmission and allows more chances for the energy transfer. The energy transfer appears sequentially from the high-energy Eu^{2+} ions in the EuN_4O environments to the low-energy Eu^{2+} ions in the EuN_5 environments and leads to decreased P1 emission and simultaneous 9increase of P2 emission in CASNO:0.02Eu^{2+}. The active energy transfer eventually controls the spectral distribution. Therefore, the peak positions of emission spectra scarcely changes with the delay time for CASNO:0.0005Eu^{2+}, whereas the peak position shifts dramatically to the low-energy side for CASNO:0.02Eu^{2+}.

Jung et al. [38] investigated non-stoichiometric $CaAl_xSi_{(7-3x)/4}N_3{:}Eu^{2+}$ phosphors with $x = 0.7$ to approximately 1.3 to identify the origin of the deconvoluted Gaussian components that constitute the emission spectra of stoichiometric $CaAlSiN_3{:}Eu^{2+}$ phosphors. The Al/Si molar ratio around the Eu^{2+} activator caused the deconvoluted Gaussian peaks. The Eu^{2+} activator sites in Al-rich environments gave rise to the lower-energy emission peak, whereas those in Si-rich environments were related to the higher-energy emission peaks. Active energy transfer from the

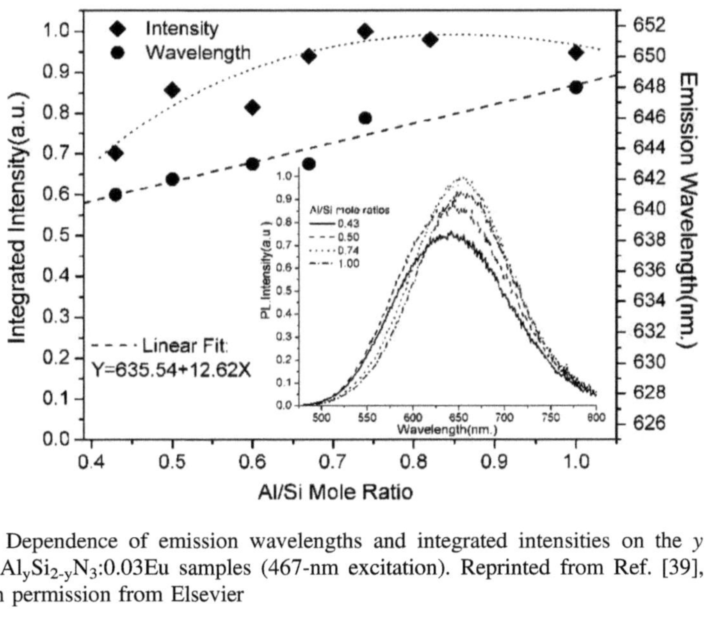

Fig. 11.7 Dependence of emission wavelengths and integrated intensities on the *y* values of $Ca_{0.47+0.5y}Al_ySi_{2-y}N_3$:0.03Eu samples (467-nm excitation). Reprinted from Ref. [39], copyright 2012, with permission from Elsevier

Eu^{2+} activator site in the Si-rich environment to that in the Al-rich environment was confirmed.

Yang et al. [39] investigated the dependence of emission wavelengths and integrated intensities on the y values of $Ca_{0.47+0.5y}Al_ySi_{2-y}N_3$:0.03Eu samples ($y$ = 0.43–1.0, 467-nm excitation) (Fig. 11.7). As mentioned previously, derivations of the charge from the change of Al/Si ratio are compensated by Ca vacancies. The emission wavelengths show a closely linear increase with the Al contents, which is in accordance with the trend of lattice volume. This phenomenon can be explained be the larger centroid shift of *5d* level of Eu^{2+} ions with increasing Al contents. Dorenbos [12] addressed the spectroscopic polarizablility that α_{sp} should be regarded as a phenomenological parameter representing the effects of ligand polarization, covalency, and possible charge-cloud expansion, which are all potential factors responsible for centroid shift. According to the conclusion drawn by Dorenbos [43], the α_{sp} value scales with the types of cations are approximately present in the compound as follows—$S^{6+} < C^{4+} < P^{5+} < B^{3+} < Si^{4+} < Al^{3+} < Mg^{2+}$ cation vacancy—where small cations are in the beginning of the series and large cations at the end of the series. The fact that Al^{3+} cations substitute the Si^{4+} increases the average α_{sp} value with increasing y value, so the centroid shift (ε_c) becomes larger, the energy level of *5d* configuration decreases, and the emission wavelength shifts to red.

11.3 $M_2Si_5N_8:Eu^{2+}$ and $M_2(Si, al)_5(N, O)_8:Eu^{2+}$ (M = Ca, Sr, Ba) Phosphors

The crystal structures of the host compounds, $M_2Si_5N_8$ (M = Ca, Sr, Ba), were investigated by Schlieper et al. [44, 45]. Höppe et al. [17] first reported the photoluminescence of $Ba_2Si_5N_8:Eu^{2+}$. Li et al. [22] investigated the photoluminescence properties of $M_2Si_5N_8:Eu^{2+}$ (M = Ca, Sr, Ba), which shows a long-wavelength broadband emission between 575 and 680 nm under the excitation band from the near UV to blue range depending on the type of the M ion and the Eu concentration. Therefore, red-emitting $M_2Si_5N_8:Eu^{2+}$ (M = Ca, Sr, Ba) phosphors are attracting extensive attention due to their potential application in warm-white LEDs.

11.3.1 Crystal Structure of $M_2Si_5N_8$ (M = Ca, Sr, Ba)

Schlieper et al. [44, 45] analyzed the crystal structure information of $M_2Si_5N_8$ (M = Ca, Sr, Ba), which is summarized in Table 11.2. $Ca_2Si_5N_8$ has a monoclinic structure with the space group of Cc, whereas $Sr_2Si_5N_8$ and $Ba_2Si_5N_8$ have an orthorhombic structure, which belongs to the space group $Pmn2_1$. In the crystal structure of the three compounds, $[SiN_4]$ tetrahedrons are combined to form a three-dimensional rigid network structure by sharing the corner. Ca^{2+} ions are

Table 11.2 Comparison of crystallographic data between $Ca_2Si_5N_8$ and $Sr_2Si_5N_8$

	$Ca_2Si_5N_8$[a]						**$Sr_2Si_5N_8$[b]**				
Site	Atom		*x*	*y*	*z*	**Site**	Atom		*x*	*y*	*z*
4a	Ca(1)		0	0.7637	0	**2a**	Sr(1)		0.0	0.3153	0.5086
4a	Ca(2)		0.6112	0.7457	0.2000	**2a**	Sr(2)		0.1777	0.1518	0.0256
4a	Si(1)		0.0581	0.8055	0.3539	**4b**	Si(1)		0.2518	0.6669	0.6836
4a	Si(2)		0.7557	0.2097	0.3182	**2a**	Si(2)		0	0.0549	0.6771
4a	Si(3)		0.7545	0.4966	0.0631	**2a**	Si(3)		0	0.4196	0.4619
4a	Si(4)		0.3627	0.2078	0.3681	**2a**	Si(4)		0	0.4014	0.9023
4a	Si(5)		0.8552	0.0027	0.1264						
4a	N^{II}	N(1)	0.9866	0.638	0.4289	**2a**	N^{II}	N(1)	0	0.191	0.520
4a		N(2)	0.1286	0.009	0.9959	**4b**		N(2)	0.2478	0.9122	0.6728
4a		N(5)	0.9798	0.999	0.2178	**2a**		N(5)	0	0.171	0.835
4a		N(7)	0.6309	0.157	0.2732						
4a	N^{I}	N(3)	0.7959	0.2424	0.1702	**4b**	N^{I}	N(3)	0.2488	0.4443	0.0105
4a		N(4)	0.8027	0.7484	0.1752	**2a**		N(4)	0	0.5871	0.7735
4a		N(6)	0.8335	0.0145	0.9349	**2a**		N(6)	0	0.4270	0.2722
4a		N(8)	0.7960	0.4826	0.4161						
Space group	Cc					Pmn2$_1$					
Lattice parameters (Å)	a = 14.352(3), b = 5.610(1) c = 9.689(2), β = 112.03					a = 5.710(2), b = 6.822(2) c = 9.341(2)					

[a]Data from Schlieper et al. [45]. [b]Data from Schlieper et al. [44]

contained in the structure along the channels of [010] direction, whereas Sr^{2+} and Ba^{2+} ions are located in cavities formed by Si_6N_6 rings parallel along [100] direction (Figs. 11.8 and 11.9).

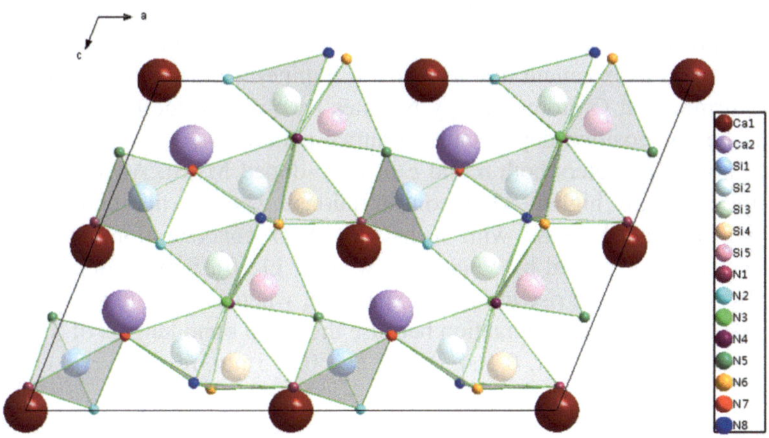

Fig. 11.8 Crystal structure of $Ca_2Si_5N_8$ along [010] direction

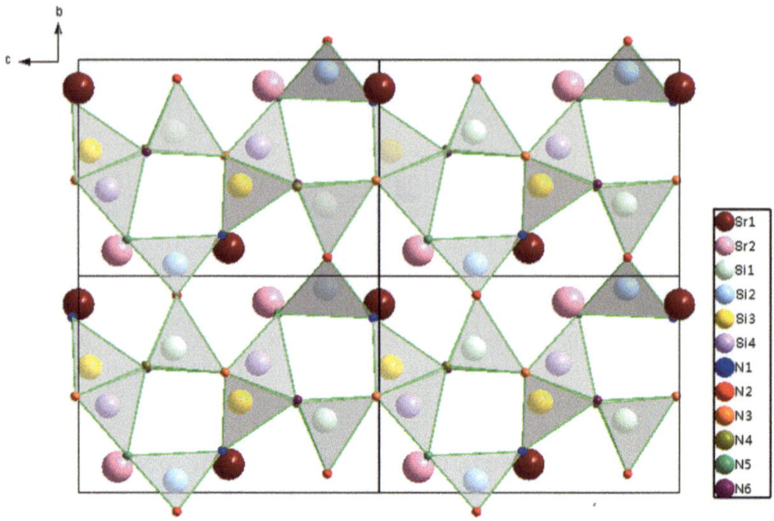

Fig. 11.9 Crystal structure of $Sr_2Si_5N_8$ along [100] direction

Interestingly, although the two compounds $Ca_2Si_5N_8$ and $Sr_2Si_5N_8$ each have a different crystal system (i.e., monoclinic versus orthorhombic), the crystal structure of $Ca_2Si_5N_8$ and $Sr_2Si_5N_8$ is very similar. The numerical relationship of the lattice parameters between $Ca_2Si_5N_8$ and $Sr_2Si_5N_8$ are found as $a \approx 2b'$, $b \approx a'$, $c \approx c'$ where m is the value of $Ca_2Si_5N_8$; and m' is the value of $Sr_2Si_5N_8$. Therefore, the unit cell volume of $Ca_2Si_5N_8$ is almost twice of that of $Sr_2Si_5N_8$. As listed in Table 11.2, the atom sites show a corresponding relationship of the crystal structure in $Ca_2Si_5N_8$ and $Sr_2Si_5N_8$. Similar with $CaAlSiN_3$ structure, there are two sorts of N atoms occupied at different crystallographic sites: Half of the N atoms are connected with three Si atoms, and the other half are linked with two Si atoms, which are herein represented as N^{I} and N^{II} (Table 11.2), respectively. To develop outstanding optical properties of $Sr_2Si_5N_8:Eu^{2+}$, O is considered to add into the crystal because it plays an excellent role in terms of luminescence in silicates. Wang et al. [42] confirmed that O atoms occupy N^{II}-sites (twofold bridge $[SiN_4]$ tetrahedron) through the composition dependence of lattice parameters.

The lattice parameters both ***a*** and ***c*** of Eu^{2+}-doped $M_2(Si, Al)_5(N, O)_8$ show a clear uptrend with higher Al/O concentrations, whereas the lattice parameter ***b*** follows a clear decrease with increasing Al/O concentration rather than an increase trend as expected by Vegard's law (Fig. 11.10). Substitutions of Al^{3+} for Si^{4+} ions generally give rise to lattice expansion, whereas substitutions of O atoms for N atoms cause lattice contraction because the mean bond lengths of Si–N, Si–O, Al–N, and Al–O are 0.174, 0.164, 0.187, and 0.175 nm, respectively. Obviously,

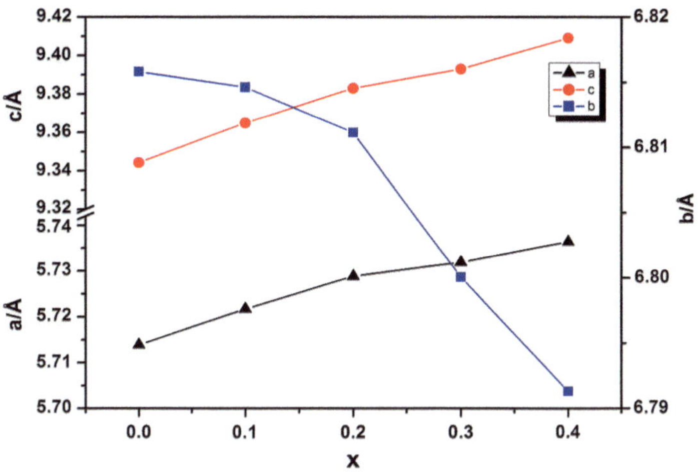

Fig. 11.10 Lattice parameters of $Sr_2Si_{5-x}Al_xN_{8-x}O_x{:}Eu^{2+}$. Reprinted from Ref. [30], copyright 2013, with permission from Elsevier

the extension of ***a*** and ***c*** is mainly attributed to the replacement of Si by Al atoms, and the shrinkage along ***b*** axis is predominantly ascribed to the substitutions of O atoms for N atoms. Accordingly, the large different transformations among three directions suggest that preferential occupation occurred in the evolution of the crystal structure.

Figure 11 shows the crystal structure of the crystal structure along the [001] (a) and [100] (b) directions, respectively. The atom labels were marked according to their own atomic sites; notably the N3, N4, and N6 atoms connecting three Si atoms belong to N^I, whereas N1, N2, and N5 atoms bonding two Si atoms are represented as N^{II}. For the same situation with N^I and N^{II} in $CaAlSiN_3$, the selective substitution of O for N contributed to the various changes of lattice parameters.

Along the [100] orientation (a), the skeleton structure is based on the condensed N^I (N3, N4, and N6) atoms, which are not replaced by O atoms and thus contribute dominantly to the rigidness of the structure. Therefore, the lattice parameter ***a*** shows a clear increase with higher x content due to the larger ionic radii of Al^{3+} versus Si^{4+}. Similarly, along the [001] direction (Fig. 11.11b), the $-N^I$–Si–N^I– chains protect themselves against the effects of replacement of N^{3-} by O^{2-} and the substitution of Al^{3+} for Si^{4+}, thus making the principle contribution to lattice expansion. However, along the [010] direction, the framework is connected by N^{II} atoms, i.e., N1, N2, and N5 atoms. The substitutions of O for N^{II} lead to a rapid decrease of the lattice parameter on this direction, which matches well with the results of XRD analysis as shown in Fig. 11.10.

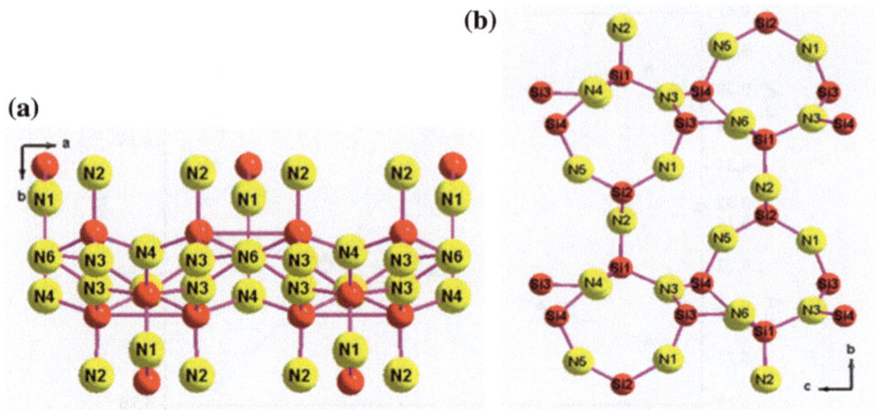

Fig. 11.11 Crystal structure of $Sr_2Si_{5-x}Al_xN_{8-x}O_x:Eu^{2+}$. **a** and **b** along the [001] and [100] direction, respectively, wherein Sr atoms were removed for clear expression of the skeleton pattern. Reprinted from Ref. [30], copyright 2013, with permission from Elsevier

11.3.2 Luminescence

$M_2Si_5N_8:Eu^{2+}$ phosphors were first investigated in the series of $Ba_2Si_5N_8:Eu^{2+}$ compounds by Höppe et al. [17]. They reported that the fluorescence emission of $Ba_2Si_5N_8:Eu^{2+}$ was observed at 600 nm due to two-photon excitation. The emission wavelength at approximately 590 nm could last for even 15 min in the dark, as seen with the naked eye, without an activating lamp, which is caused by the recombination effect of holes and traps consisting of nitrogen vacancies formed by reducing synthesis conditions. Li et al. [22] reported the influence of the type of the alkaline–earth ion and Eu^{2+} concentration on the luminescence properties of $M_2Si_5N_8:Eu^{2+}$ (M = Ca, Sr, Ba) phosphors. The maximum solubility of Eu^{2+} in the M = Ca series is approximately 7 mol%, whereas Eu^{2+} can be completely incorporated into the M = Sr and Ba lattice due to the isostructural of $M_2Si_5N_8$ compounds. The $Eu_2Si_5N_8$. $M_2Si_5N_8:Eu^{2+}$ phosphors show a typical broad emission band in the range of 570 to approximately 680 nm under a spectral excitation band from 370 to 460 nm depending on the type of M ion and the Eu concentration, which is found to be proportional to the changes of the unit cell volume for M = Sr and Ba. The red-shift behavior of the emission band in $Sr_2Si_5N_8:Eu^{2+}$ phosphors was discussed on the basis of the configuration coordination model by Piao et al. [46]. The luminescence of $(SrCa)_2Si_5N_8:Eu^{2+}$ solid solutions was investigated by Piao et al. [47] and Li et al. [48]. These phosphors showed intense orange-red emission bands under the broad excitation spectra of 380–470 nm corresponding to the output lights of near-UV or blue LEDs. Sohn et al. [49] examined the two-peak emission of $Sr_2Si_5N_8:Eu^{2+}$ phosphors by analyzing the time-resolved photoluminescence spectra and observed the energy transfer between Eu^{2+} activators at two different sites in the $Sr_2Si_5N_8:Eu^{2+}$ structure.

Chen et al. [32] reported a cation-size-mismatch mechanism to tune the luminescence property in $M_{1.95}Eu_{0.05}Si_{5-x}Al_xN_{8-x}O_x$ lattice. The Eu^{2+} dopants, which are larger than the host cations in the M = Ca series but smaller than the Ba^{2+} host cations, attract the increased oxide coordination in the Ca series and more nitride coordination in the Ba series. Size-mismatch between the host and the dopant cations tunes photoluminescence shifts systematically, which will lead to a blue shift when M is smaller than the Eu^{2+} ion in the M = Ca host but a red shift when M is larger in the M = Sr and Ba host.

Wang et al. [30] revealed that more Al–O pairs incorporated into lattice result in a small red shift of emission spectra in the $Sr_2(Si, Al)_5(N, O)_8$:Eu^{2+} phosphors (Fig. 11.12). According to Dorenbos' work, spectroscopic polarizability α_{sp} should be regard as a phenomenological parameter representing the effects of ligand polarization, covalence, and possible charge-cloud expansion, which are all potential factors responsible for centroid shift. By calculating the α_{sp} value of various cations present in some compounds, it was found that the α_{sp} value of Si is smaller than that of Al. The substitution of Al for Si increases the average α_{sp} value, thus resulting in the larger centroid shift (ε_c) and lower energy level of *5d* configuration with increasing Al content eventually leading to red shifts of the emission wavelength. On the other hand, incorporating more O to substitute for N atoms reduces the nephelauxetic effect around the Eu^{2+} and elevates the lowest *5d* energy-level of Eu^{2+} in the crystal because O^{2-} has higher electronegativity than N^{3-} and Eu–O has lower covalency than Eu–N. These ultimately result in a blue shift of emission band. The resultant red shifts can be attributed to the subtle competition

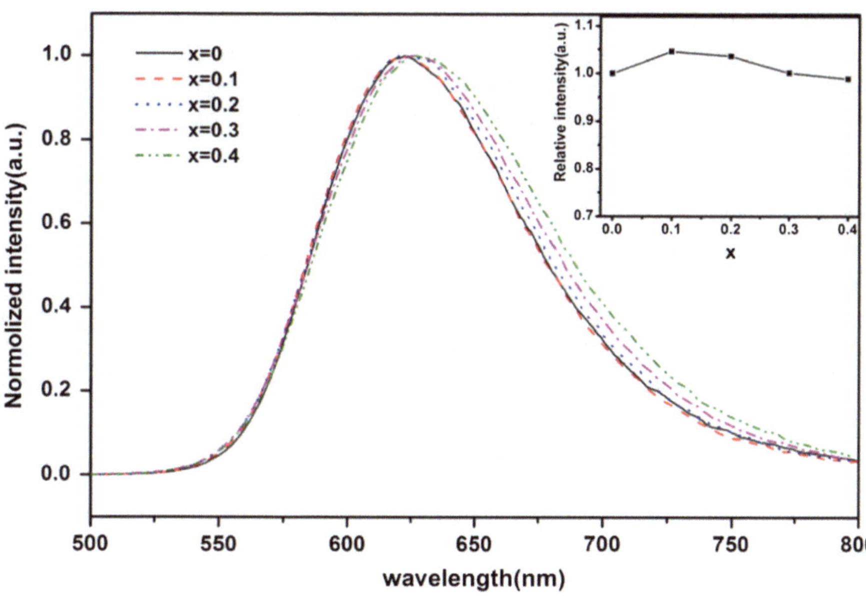

Fig. 11.12 Photoluminescence spectra of $Sr_2Si_{5-x}Al_xN_{8-x}O_x$:$Eu^{2+}$ with various *x* values excited at 400 nm. Reprinted from Ref. [30], copyright 2013, with permission from Elsevier

between red-shift effects from Al substitutions for Si and blue-shift effects from Al replacement by O. The resultant red shifts demonstrate that Eu^{2+} ions are more sensitive to the change of local structure caused by Al.

11.4 Sialon:Eu^{2+} Phosphors

Sialons constitute an important family of materials with unique structural features and various crystal chemistries. Because of their excellent mechanical properties and high temperature stability, sialon ceramics have been developed for structural engineering applications. Since the pioneering work of Krevel et al. [18] and Xie et al. [19], the rare earth–doped sialon phosphors have received increasing attention. In particular, Ca-α-sialon:Eu [50] and β-SiAlON [21] phosphors have been proved to be good downconversion phosphors for white LED combined with a blue InGaN LED chip.

11.4.1 Structures

Sialons are solid solutions of the (M–)Si–Al–O–N system, the structure of which is derived from Si_3N_4. Cao and Metselaar [51] overviewed the structures of Si_3N_4 and sialon. Si_3N_4 occurs in two different crystalline phases, denoted α and β, the crystallographic data of which are listed in Tables 11.3 and 11.4. The β-Si_3N_4 structure can be described as a stacking of Si–N layers in an ABAB… sequence, which gives long continuous channels parallel to the hexagonal c-direction as shown in Fig. 11.13. In the α-Si_3N_4 structure (Fig. 11.14), Si–N layers stack in

Table 11.3 Comparison of structural data between α-Si_3N_4 and Ca-α-sialon

		α-Si_3N_4[a]				**Ca-α-sialon**[b]		
Site	Atom	*x*	*y*	*z*	atom	*x*	*y*	*z*
2a	Null				Ca	1/3	2/3	0.228
6c	Si	0.5106	0.0815	0.1904	Si/Al	0.5090	0.0820	0.207
6c	Si	0.1658	0.2520	−0.0168	Si/Al	0.1681	0.2512	−0.003
2a	N	0.0	0.0	−0.029	N/O	0.0	0.0	0.0
2b	N	1/3	2/3	0.634	N/O	1/3	2/3	0.642
6c	N	0.3488	−0.0383	−0.022	N/O	0.3477	−0.0441	−0.012
6c	N	0.3159	0.3165	0.241	N/O	0.3164	0.3150	0.246
Space group and lattice parameter (Å)		P31c a = 7.650, c = 5.6275				P31c a = 7.8383, c = 5.7033		

[a]Data from Yang et al. [54]. [b]Data from Izumi et al. [55]

Table 11.4 Comparison of structural data between β-Si_3N_4 and β-sialon

		β-Si_3N_4[a]				β-sialon[b]		
Site	Atom	*x*	*y*	*z*	Atom	*x*	*y*	*z*
6 h	Si	0.1742	−0.2322	1/4	Si/Al	0.1686	−0.2355	1/4
2c	N	1/3	2/3	1/4	N/O	1/3	2/3	1/4
6 h	N	0.329	0.039	1/4	N/O	0.3320	0.0301	1/4
Space group and lattice parameter (Å)		$P6_3/m$ a = 7.6093, c = 2.9079				$P6_3/m$ a = 7.664, c = 2.963		

[a]Data from Yang et al. [54]. [b]Data from Gillott et al. [56]

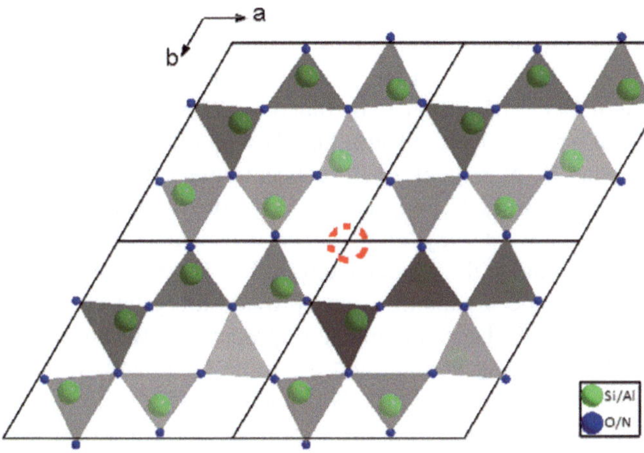

Fig. 11.13 Crystal structure of β-sialon. The red-dash ball represents positions of the rare-earth ions doped into the crystal

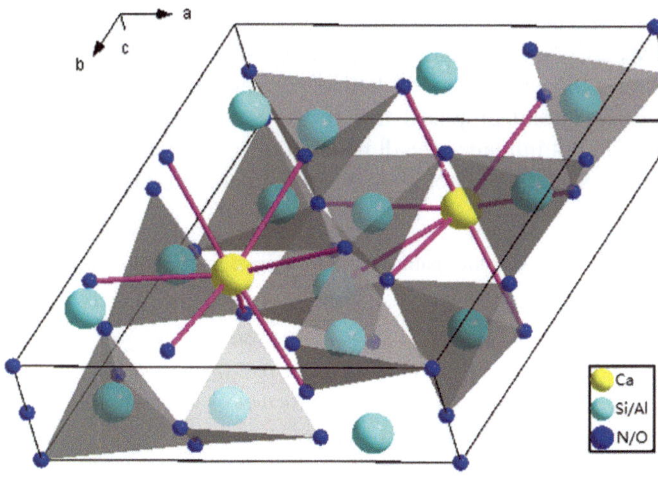

Fig. 11.14 Crystal structure of Ca-α-sialon

ABCD…sequence where the c-glide plane relates the layers CD with AB, thus forming two large closed interstitial sites per unit cell instead of the continuous channels [52]. The α-sialons are solid solutions of the M–Si–Al–O–N system, whose structure is derived from α-Si_3N_4 and can be represented as $M_xSi_{12\text{-}(m+n)}Al_{(m+n)}O_nN_{16-n}$ where m (Al–N) pairs and n (Al–O) pairs replace $m + n$ (Si–N) pairs in each unit cell; $x = m/v$; and v is the valence of the M ion. M is a modifying cation, typically of Li, Mg, Ca, Y or rare earth with Z > 60. In the α-sialon structure, there are two interstitial sites per unit cell that can accommodate M cations; thus, M content (x) is ≤ 2, and the values of m and x are taken to meet electroneutrality requirement, i.e. $x = m/v$. For comparison, crystallographic data of Ca-α-sialon are also listed in Table 11.3. In the structure of $CaAlSiN_3$ and $Ca_2Si_5N_8$, some of the N atoms connect three Si/Al atoms, and some others bond to two Si/Al atoms. But in the α- and β-sialon structures, all N atoms bond to three Si/Al atoms. Thus, the substitutions of O for N and Al for Si are random for different crystallographic sites, and the number of substitutions is limited.

The emission of Ca-α-sialon:Eu can be tuned by changing the composition, e.g., by changing the m and n values, substituting Ca with other metals such as Li, Mg and Y, and adjusting O, N content, or Eu concentration [20, 48]. The pure-nitride α-sialon:Eu (i.e., $Ca_{m/2}Si_{12\text{-}m}Al_mN_{16}$:$Eu^{2+}$, $n = 0$) may give excellent photoluminescence properties, e.g., it may give longer wavelength emission vs. oxygen contained α-sialon:Eu due to the nephelauxetic effect. Furthermore, the perfect rigid framework of pure-nitride α-sialon:Eu may provide a crystal chemistry basis for good concentration- and thermal-quenching properties. Because the scattering factors of Al and Si are very similar, and the same for O and N, it is difficult to determine the ratio of Al to Si, and O to N. Yang et al. [53] addressed that the fact that the composition dependence of lattice parameter is useful to determine ratio of Al to Si.

There exists a relationship between the lattice parameters and m/n values (i.e., Al/Si ratios and O contents) of the compositions $M_xSi_{12\text{-}(m+n)}Al_{m+n}O_nN_{16-n}$ (Table 11.5). Due to the differences in the bond lengths of Si–N (0.174 nm), Al–O (0.175 nm), and Al–N (0.187 nm), for a general composition of $M_xSi_{12\text{-}(m+n)}Al_{m+n}O_nN_{16-n}$, substitutions of m (Si–N) by m (Al–N) and n (Si–N) by n (Al–O) bring a linear change of lattice parameters with m and n. Jack [57] reported the Eqs. (1) and proposed that the size and solubility of the modifying cations, M, should not influence the dimension of α-sialon lattice, possibly because the interstitial holes are large enough to accommodate those cations [51]. Nordberg

Table 11.5 Relations between lattice parameters and compositions of $M_xSi_{12\text{-}(m+n)}Al_{(m+n)}O_nN_{16-n}$

No.	Equations	Reference
1	$\Delta a = 0.045m + 0.009n;\ \Delta c = 0.040m + 0.008n$	[57]
2	$a = 7.752 + 0.036m + 0.02n;\ c = 5.620 + 0.031m + 0.04n$	[58]
3	$a = 7.752 + 0.036m + 0.02n;\ c = 5.620 + 0.048m + 0.009n$	[59]
4	$a = 7.752 + 0.031m;\ c = 5.620 + 0.026m$ for $n = 0$	[53]

[58] gave the Eqs. (2) based on data of Y-α-sialon. Sun et al. [59] supplied the Eqs. (3), which nearly conform to Eqs. (1). Yang et al. [53] gave the Eq. (4) based on the data of $(Ca_{0.995}Eu_{0.005})_{m/2}Si_{12-m}Al_mN_{16}$ and summarized the equations and crystallographic data reported by different research groups and plotted them as shown in Fig. 11.15.

Yang et al. [53] confirmed that $(Ca_{0.995}Eu_{0.005})_{m/2}Si_{12-m}Al_mN_{16}$ samples have composition with very little oxygen content, i.e., the so-called Eu-doped pure-nitride α-sialons, according to relations in Fig. 11.15, e.g., the line of the lattice parameters (*a, c*) of the samples with *m* are nearly located under the line of $n = 0$.

β-sialon is a solid solution of the Si–Al–O–N system, the structure of which is derived from β-Si_3N_4 by the simultaneously equivalent substitution of Al–O for Si–N [62]. Its general formula can be expressed as $Si_{6-z}Al_zO_zN_{8-z}$ ($0 < z \leq 4.2$) [63]. Their crystallographic data are listed in Table 11.4 for convenient comparison. The β-Si_3N_4 structure can be described as a stacking of Si–N layers in an ABAB… sequence, which gives long continuous channels parallel to the hexagonal c-direction. Unlike α-sialon, metal ions cannot enter into the lattice of β-sialon because charge balance is not required. However, the investigators did observe the green emission of Eu^{2+} in β-sialon and reported that Eu^{2+} was homogenous in a whole particle [21, 64]. Li et al. [48] proposed that Eu^{2+} ions locate in the continuous channels along the *c*-axis, and gave the formula $Eu_xSi_{6-z}Al_{z-x}O_{z+x}N_{8-z-x}$ for Eu^{2+}-doped β-sialon, and that Eu^{2+} is directly connected by six adjacent nitrogen/oxygen atoms ranging from 2.4850 to 2.5089 Å. Kimoto et al. [65] directly observed a single Eu dopant atom in β-sialon using scanning transmission electron microscopy, thus revealing that a Eu dopant exists in a continuous atomic channel in the structure.

11.4.2 Luminescence

Krevel et al. [18] and Xie et al. [19] reported the luminescence properties of the Eu^{2+}-doped Ca-α-sialon, which have received increasing attention. The α-sialon:Eu^{2+} phosphors exhibit yellow emission (approximately 570–600 nm) and a strong absorption in the UV to blue spectral range (approximately 300–470 nm), and they have been proved to be a good downconversion phosphor for white LED combined with a blue InGaN LED chip [50]. Xie et al. [20] investigated the effects of activator concentration and overall composition of the host lattice on the luminescence of α-sialon:Eu^{2+} phosphors with the formula $(Ca_xEu_y)Si_{12-2x-3y}Al_{2x+3y}O_yN_{16-y}$. Luminescence concentration quenching occurs at a Eu^{2+} concentration of 0.075, and the strongest luminescence is achieved at $x = 1.4$ as the concentration of Eu^{2+} is fixed at $y = 0.075$. The absorption and emission bands shift to longer wavelengths with increasing Eu^{2+} concentration and Al content. Li et al. [66] reported the luminescence of α-sialon:Eu^{2+} phosphors with the target compositions of $Ca_{0.95\times m/2}Eu_{0.05\times m/2}Si_{12-(m+n)}Al_{m+n}O_nN_{16-n}$ ($m = 1.0$–4.5, $n = 0$–1.2), thus

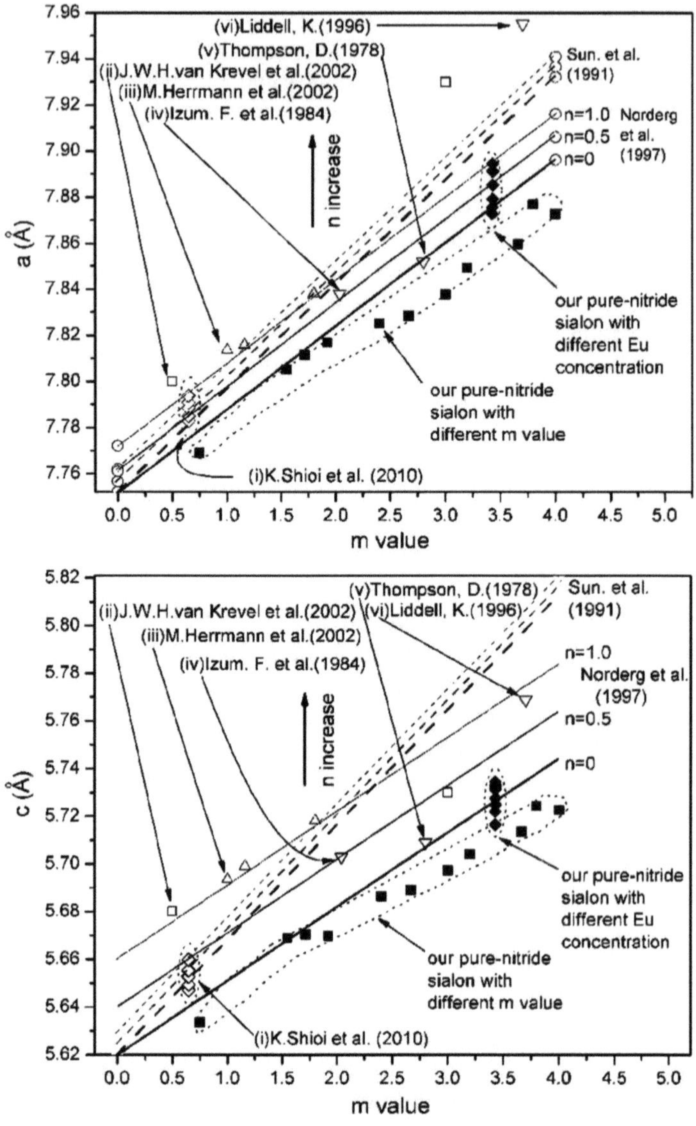

Fig. 11.15 Relationship of lattice parameters (*a, c*) with *m* values. The solid lines of $n = 0$, 0.5, and 1 are plotted according to the Eqs. (2) reported by Norderg et al. [58]; the dash lines of $n = 0$, 0.5, and 1, are plotted according to the Eqs. (3) reported by Sun et al. [59]. The different solid symbols are from our two series of specimens. The different hollow symbols are from the data in some literature, i.e., (i) Eu-α-sialon ($n = 0$ to approximately 0.325) from Shioi et al. [60]; (ii) Ca-α-sialon (*n* value unknown) from Krevel et al. [18]; (iii) Nd/Y-α-sialon ($n = 1.26$ to approximately 1.4) from Herrmann et al. [61]; (iv) $Ca_{0.68}Si_{9.96}Al_{2.04}O_{0.68}N_{15.32}$ from JCPDS No.: 42-0252 (1984); (v) $Ca_{0.8}Si_{9.2}Al_{2.8}O_{1.2}N_{14.8}$ from ref.: JCPDS No.: 33-0261 (1978); (vi) $Ca_{1.8}Si_{8.2}Al_{3.7}N_{16}$ from ref.: JCPDS No.: 48–1586 (1996). Reprinted from Ref. [53], copyright 2012, with permission from Elsevier

indicating that the strongest luminescence is achieved at approximately m = 3.5 and n = 0–0.2.

Yang et al. [53] investigated the luminescence properties of Eu^{2+}-doped pure-nitride α-sialons, nominal formula $(Ca_{0.995}Eu_{0.005})_{m/2}Si_{12-m}Al_mN_{16}$ with different m values, synthesized through solid-state reactions in a nitrogen atmosphere with CaAl, EuSi, AlN, and α-Si_3N_4 as starting materials. It was demonstrated that the phosphors with $2.4 \leq m \leq 4.0$ have the strongest emission (Fig. 11.16) and that the emissions shift to the red region with increasing m values as well as increasing Eu contents. They also reported the dependence of emission wavelengths and integrated intensities of $Ca_{1.71-x}Si_{8.57}Al_{3.43}N_{16}$:xEu samples on Eu concentrations indicating that the strongest luminescence is obtained at x = 0.5–1.0.

Hirosaki et al. [21] reported a β-sialon:Eu^{2+} green phosphor with the composition $Eu_{0.00296}Si_{0.41395}Al_{0.01334}O_{0.0044}N_{0.56528}$, which can be excited efficiently over a broad spectral range between 280 and 480 nm and has an emission peak at 535 nm with a full width at half maximum of 55 nm. This green phosphor has potential applications in phosphor-converted white LEDs. Xie et al. [64] conducted a comprehensive investigation of the β-sialon:Eu^{2+} phosphor, $Si_{6-z}Al_zO_zN_{8-z}$ with z = 0.1–2.0, and dopant Eu concentration varying in the range of 0.02–1.5 mol%. They demonstrated that concentration quenching occurs at 0.5 to 0.7 mol% for z = 0.1, and 0.5, and at 0.3 mol% for z = 1.0–2.0, thus indicating that the concentration quenching occurring at higher Eu^{2+} concentration in samples with low z values could be due to the high solubility of Eu^{2+} in β-sialon with smaller

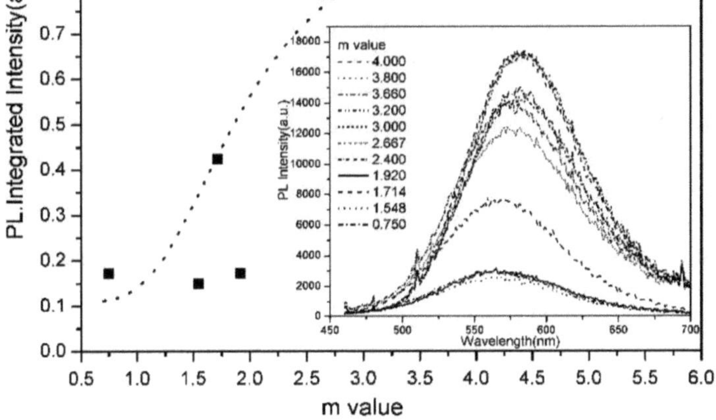

Fig. 11.16 Relationship of integrated emission intensities with m values of $(Ca_{0.995}Eu_{0.005})_{m/2}Si_{12-m}Al_mN_{16}$ samples. The *inset* shows the emission spectra by 450-nm excitation. Reprinted from Ref. [53], copyright 2012, with permission from Elsevier

z values. With increasing z value and Eu^{2+} concentration, the emission band shifts toward the long-wavelength side, i.e., red shift, especially for high z-value samples and in the region of low Eu^{2+} concentration. Zhu et al. [67] reported the z-value dependence of photoluminescence in Eu^{2+}-doped β-sialon ($Si_{6-z}Al_zO_zN_{8-z}$) phosphors with $1 \leq z \leq 4$ and demonstrated that they exhibit two broad emission bands with maxima at approximately 415 nm (violet) and 540 nm (green) under UV excitation. The violet emission becomes dominant with increasing z value, whereas the green emission eventually disappears with z value increasing to 4. The results suggest that there are probably two different kinds of N/O coordinations around Eu^{2+} ions in β-sialon: One is oxygen-rich, and the other is nitrogen-rich.

11.5 Relationship of *f-d* Transition with Structure and Composition in Nitrides

As described in the previous three sections, substitution of Si with Al in nitride phosphors generally leads to a red shift in emission spectra, whereas replacement of N by O leads to a blue shift. This can be understood by the fact that the f-d transition is related to the structure and composition of the host materials.

Dorenbos [12] addressed the fact that spectroscopic polarizablility α_{sp} should be regarded as a phenomenological parameter representing the effects of ligand polarization, covalency, and possible charge-cloud expansion, all of which are potential factors responsible for centroid shift of the 5*d* energy level. The α_{sp} value scales with the types of cations present in the compound is approximately as follows: $Si^{4+} < Al^{3+}$. The substitution of Al^{3+} for Si^{4+} elevates the average α_{sp} value, so the centroid shift (ε_c) becomes larger and the energy level of 5*d* declines, resulting in red-shifts of the absorption and emission bands [39]. Another possible reason for the emission red-shift can be attributed to an enhanced Stokes-shift resulting from the lattice expansion with Al-substituted Si [20].

The structure and composition of a compound play an important role in tuning its luminescence properties. Dorenbos related the energy of the 5*d* configuration of Ce^{3+} with anion polarizability and cation electronegativity and obtained the relationship in oxides and fluorides. Recently, the energy of 5*d* levels of Ce^{3+} in nitrides has become available due to the development of nitride phosphors. Wang et al. [68] collected information on 5*d* levels of Ce^{3+} in nitrides and obtained a linear relationship between the anion polarizability and the inverse square of the average electronegativity of the cations based on the Dorenbos model as shown in Fig. 11.17. Wang et al. [68] selected Ce^{3+}-doped β-sialon as an ideal compound to verify that the centroid shift due to the nephelauxetic effect can be predicted not only in nitrides but also in oxynitrides. For the 5*d* levels of Ce^{3+} doped β-sialon, one can obtain the centroid shift energy of 2.58 eV based on the excitation spectra. With the linear relationship mentioned previously, one obtains $\varepsilon_c = 2.50\text{eV}$, which is in excellent agreement with the observed value of 2.58 eV. This example

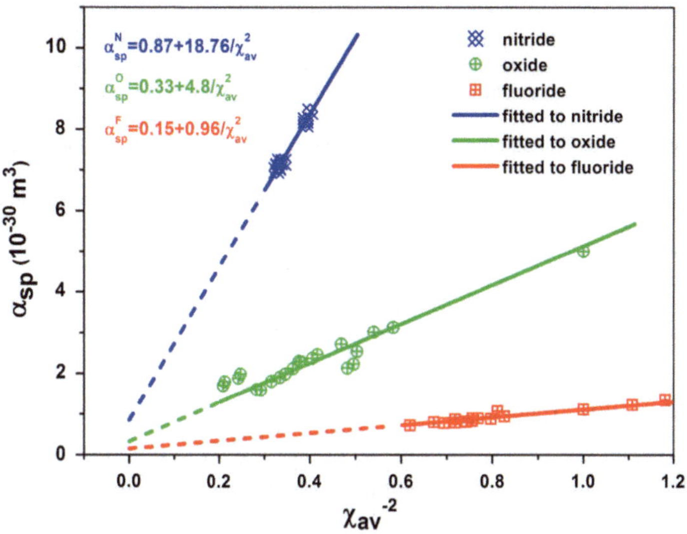

Fig. 11.17 Spectroscopic polarizability against the inverse square of the average cation electronegativity. The *solid lines* are least-squares linear fits through the data. The data of fluoride and oxide compounds are derived from the work reported by Dorenbos. The short *dotted lines* are extension of *solid lines* for a clear view. Reprinted from Ref. [68], copyright 2015, with permission from Elsevier

demonstrates that the linear relationship obtained in Fig. 11.17 is reliable and that this relationship is of value for the interpretation and prediction of spectroscopic properties involved in the *f-d* transition in nitrides and oxynitrides. Furthermore, one can predict the local structure features of Ce^{3+} based on the luminescence spectra according to the relationship in Fig. 11.17. As an example, it was testified that Ce^{3+} ions occupy the Sr^{2+} site rather than the Y^{3+} site in $SrYSi_4N_7$: Ce compound [68].

References

1. Malakoff D (2002) Materials science-lighting initiative flickers to life. Science 296:1782
2. Nakamura S, Fasol G (1997) The blue laser diode: GaN based light emitters and lasers. Springer, Berlin
3. Nakamura S (2009) Current status of GaN-based solid-state lighting. MRS Bull 34:101
4. Schubert EF, Kim JK (2005) Solid-state light sources getting smart. Science 308:1274
5. Setlur AA, Heward WJ, Hannah ME, Happek U (2008) Incorporation of Si^{4+}–N^{3-} into Ce^{3+}-doped garnets for warm white LED phosphors. Chem Mater 20:6277
6. Xie RJ, Li YQ, Hirosaki N, Yamamoto H (2011) Nitride phosphors and solid-state lighting. CRC Press
7. Lang RJ (1936) The spectrum of trebly ionized cerium. Can J Res 14:127

8. Dorenbos P (2002) Relating the energy of the [Xe]5d^1 configuration of Ce^{3+} in inorganic compounds with anion polarizability and cation electronegativity. Phys Rev B 65:235110
9. Dorenbos P, Andriessen J, van Eijk CWE (2003) $4f^{n-1}5d$ centroid shift in lanthanides and relation with anion polarizability, covalency, and cation electronegativity. J Solid State Chem 171:133
10. Dorenbos P (2000) The 5d level positions of the trivalent lanthanides in inorganic compounds. J Lumin 91:155
11. Dorenbos P (2003) Energy of the first $4f^7 \rightarrow 4f^6 5d$ transition of Eu^{2+} in inorganic compounds. J Lumin 104:239
12. Dorenbos P (2000) 5d-level energies of Ce^{3+} and the crystalline environment. I. Fluoride compounds. Phys Rev B 62:15640
13. Zeuner M, Pagano S, Schnick W (2011) Nitridosilicates and oxonitridosilicates: from ceramic materials to structural and functional diversity. Angew Chem Int Edit 50:7754
14. Lee SS, Lim S, Sun SS, Wager JF (1997) Proceedings of the society of photo-optical instrumentation engineers (SPIE). 3242:75
15. van Krevel JWH, Hintzen HT, Metselaar R, Meijerink A (1998) Long wavelength Ce3 + emission in Y–Si–O–N materials. J Alloy Compd 268:272
16. Uheda K, Takizawa H, Endo T, Yamane H, Shimada M, Wang CM, Mitomo M (2000) Synthesis and luminescent property of Eu^{3+}-doped $LaSi_3N_5$ phosphor. J Lumin 87–89:967
17. Höppe HA, Lutz H, Morys P, Schnick W, Seilmeier A (2000) Luminescence in Eu^{2+}-doped $Ba_2Si_5N_8$: fluorescence, thermoluminescence, and upconversion. J Phys Chem Solids 61:2001
18. van Krevel JWH, van Rutten JWT, Mandal H, Hintzen HT, Metselaar R (2002) Luminescence properties of terbium-, cerium-, or europium-doped α-sialon materials. J Solid State Chem 165:19
19. Xie R, Mitomo M, Uheda K, Xu F, Akimune Y (2002) Preparation and luminescence spectra of calcium- and rare-earth (R = Eu, Tb, and Pr)-codoped α-sialon ceramics. J Am Ceram Soc 85:1229
20. Xie R, Hirosaki N, Mitomo M, Yamamoto Y, Suehiro T, Sakuma K (2004) Optical properties of Eu^{2+} in α-SiAlON. J Phys Chem B 108:12027
21. Hirosaki N, Xie R, Kimoto K, Sekiguchi T, Yamamoto Y, Suehiro T, Mitomo M (2005) Characterization and properties of green-emitting β-SiAlON:Eu^{2+} powder phosphors for white light-emitting diodes. Appl Phys Lett 86:211905
22. Li YQ, van Steen JEJ, van Krevel JWH, Botty GB, Delsing ACA, DiSalvo FJ, de With G, Hintzen HT (2006) Luminescence properties of red-emitting $M_2Si_5N_8$:Eu^{2+} (M = Ca, Sr, Ba) LED conversion phosphors. J Alloy Compd 417:273
23. Uheda K, Hirosaki N, Yamamoto Y, Naito A, Nakajima T, Yamamoto H (2006) Luminescence properties of a red phosphor, $CaAlSiN_3$:Eu^{2+}, for white light-emitting diodes. Electrochem Solid St 9:H22
24. Huang Z, Sun W, Yan D (1985) Phase relations of the Si_3N_4-AlN-CaO system. J Mater Sci Lett 4:255
25. Ottinger F (2004) Synthese, struktur und analytische detailstudien neuer stickstoffhaltiger silicate und aluminosilicate. Ph.D. Thesis. Universität Karlsruhe
26. Uheda K, Hirosaki N, Yamamoto H (2006) Host lattice materials in the system Ca_3N_2–AlN–Si_3N_4 for white light emitting diode. Phys Status Solid(a) 203:2712
27. Wang T, Yang J, Mo Y, Bian L, Song Z, Liu QL (2013) Synthesis, structure and tunable red emissions of $Ca(Al/Si)_2N_2(N_{1-x}O_x)$:$Eu^{2+}$prepared by alloy-nitridation method. J Lumin 137:173
28. Watanabe H, Yamane H, Kijima N (2008) Crystal structure and luminescence of $Sr_{0.99}Eu_{0.01}AlSiN_3$. J Solid State Chem 181:1848
29. Piao X, Machida K, Horikawa T, Hanzawa H, Shimomura Y, Kijima N (2007) Preparation of CaAlSiN:Eu phosphors by the self-propagating high-temperature synthesis and their luminescent properties. Chem Mater 19:4592
30. Wang T, Zheng P, Liu X, Chen H, Bian L, Liu QL (2014) Effects of replacement of AlO^+ for SiN^+ on the structure and optical properties of $Sr_2Si_5N_8$:Eu^{2+} phosphors. J Lumin 147:173

31. Fuertes A (2006) Prediction of anion distributions using pauling's second rule. Inorg Chem 45:9640
32. Chen W, Sheu H, Liu R, Attfield JP (2012) Cation-size-mismatch tuning of photoluminescence in oxynitride phosphors. J Am Chem Soc 134:8022
33. Li YQ, Hirosaki N, Xie RJ, Takeda T, Mitomo M (2008) Yellow-orange-emitting $CaAlSiN_3$: Ce phosphor: structure, phostoluminescence, and application in white LEDs. Chem Mater 20:6704
34. Li J, Watanabe T, Sakamoto N, Wada H, Setoyama T, Yoshimura M (2008) Synthesis of a multinary nitride, Eu-doped CaAlSiN3, from alloy at low temperatures. Chem Mater 20:2095
35. Harris RK, Leach MJ, Thompson DP (1992) Nitrogen-15 and oxygen-17 NMR spectroscopy of silicates and nitrogen ceramics. Chem Mater 4:260
36. Uheda K, Yamamoto H, Yamane H, Inami W, Tsuda K, Yamamoto Y, Hirosaki N (2009) An analysis of crystal structure of Ca-deficient oxonitridoaluminosilicate, $Ca_{0.88}Al_{0.91}Si_{1.09}N_{2.85}O_{0.15}$. J Ceram Soc Jpn 117:94
37. Lee S, Sohn K (2010) Effect of inhomogeneous broadening on time-resolved photoluminescence in $CaAlSiN_3$:Eu^{2+}. Opt Lett 35:1004
38. Jung YW, Lee B, Singh SP, Sohn K (2010) Particle-swarm-optimization-assisted rate equation modeling of the two-peak emission behavior of non-stoichiometric $CaAl_xSi_{(7-3x)/4}N_3$:Eu^{2+} phosphors. Opt Express 18:17805
39. Yang J, Wang T, Chen D, Chen G, Liu Q (2012) An investigation of Eu^{2+}-doped $CaAlSiN_3$ fabricated by an alloy-nitridation method. Mater Sci Eng, B 177:1596
40. Watanabe H, Wada H, Seki K, Itou M, Kijima N (2008) Synthetic method and luminescence properties of $Sr_xCa_{1-x}AlSiN_3$:Eu^{2+} mixed nitride phosphors. J Electrochem Soc 155:F31
41. Piao X, Machida K, Horikawa T, Yun B (2010) Acetate reduction synthesis of $Sr_2Si_5N_8$:Eu^{2+} phosphor and its luminescence properties. J Lumin 130:8
42. Wang T, Zheng P, Liu X, Chen H, Yang S, Liu QL (2014) Decay behavior analysis of two-peak emission in $Ca(Al/Si)_2N_2(N_{1-x}O_x)$:$Eu^{2+}$ phosphors. J Electrochem Soc 161:H25
43. Dorenbos P (2001) 5d-level energies of Ce^{3+} and the crystalline environment. III. Oxides containing ionic complexes. Phys Rev B 64:125117
44. Schlieper T, Millus W, Schnick W (1995) Nitrido-silicate II, Hochtemperatur-Synthese und Kristallstruktur von $Sr_2Si_5N_8$ und $Ba_2Si_5N_8$. Zeitschrift für anorganische und allgemeine Chemie 621:1380
45. Schlieper T, Schnick W (1995) Nitridosilicate I, Hochtemperatur-Synthese und Kristallstruktur von $Ca_2Si_5N_8$. Zeitschrift für anorganische und allgemeine Chemie 621:1037
46. Piao X, Horikawa T, Hanzawa H, Machida K (2006) Characterization and luminescence properties of $Sr_2Si_5N_8$:Eu^{2+} phosphor for white light-emitting-diode illumination. Appl Phys Lett 88:161908
47. Piao X, Horikawa T, Hanzawa H, Machida K (2006) Preparation of $(Sr_{1-x}Ca_x)_2Si_5N_8$:Eu^{2+} solid solutions and their luminescence properties. J Electrochem Soc 153:H232
48. Li YQ, Hirosaki N, Xie RJ, Takeda T, Mitomo M (2008) Crystal and electronic structures, luminescence properties of Eu^{2+}-doped $Si_{6-z}Al_zO_zN_{8-z}$ and $M_ySi_{6-z}Al_{z-y}O_{z+y}N_{8-z-y}$ (M = 2Li, Mg, Ca, Sr, Ba). J Solid State Chem 181:3200
49. Sohn K, Lee S, Xie R, Hirosaki N (2009) Time-resolved photoluminescence analysis of two-peak emission behavior in $Sr_2Si_5N_8$:Eu^{2+}. Appl Phys Lett 95:121903
50. Xie R, Hirosaki N, Sakuma K, Yamamoto Y, Mitomo M (2004) Eu^{2+}-doped Ca-α-SiAlON: a yellow phosphor for white light-emitting diodes. Appl Phys Lett 84:5404
51. Cao GZ, Metselaar R (1991) α'-sialon ceramics: a review. Chem Mater 3:242
52. Hampshire S, Park HK, Thompson DP, Jack KH (1978) α′-sialon ceramics. Nature 274:880
53. Yang JJ, Song Z, Bian L, Liu QL (2012) An investigation of crystal chemistry and luminescence properties of Eu-doped pure-nitride α–sialon fabricated by the alloy-nitridation method. J Lumin 132:2390
54. Yang P, Fun H, Rahman IA, Saleh MI (1995) Two phase refinements of the structures of α-Si_3N_4 and β-Si_3N_4 made from rice husk by Rietveld analysis. Ceram Int 21:137

55. Izumi F, Mitomo M, Bando Y (1984) Rietveld refinements for calcium and yttrium containing alpha-sialons. J Mater Sci 19:3115
56. Gillott L, Cowlam N, Bacon GE (1981) A neutron diffraction investigation of some β′-sialons. J Mater Sci 16:2263
57. Jack KH (1983) Progress in nitrogen ceramics. E65: 45
58. Nordberg LO (1997) Doctorial Dissertation, Stockholm
59. Sun WY, Tien TY, Yen TS (1991) Solubility limits of α′-SiAlON solid solutions in the system Si, Al, Y/N, O. J Am Ceram Soc 74:2547
60. Shioi K, Hirosaki N, Xie R, Takeda T, Li YQ (2010) Synthesis, crystal structure and photoluminescence of Eu-α-SiAlON. J Alloy Compd 504:579
61. Herrmann M, Kurama S, Mandal H (2002) Investigation of the phase composition and stability of the α-SiAlONs by the Rietveld method. J Eur Ceram Soc 22:2997
62. Jack KH (1976) Review. Sialons and related nitrogen ceramics. J Mater Sci 11:1135
63. Ekstrom T, Nygren M (1992) SiAlON ceramics. J Am Ceram Soc 75:259
64. Xie RJ, Hirosaki N, Li HL, Li YQ, Mitomo M (2007) Synthesis and photoluminescence properties of β-sialon:Eu^{2+} ($Si_{6-z}Al_zO_zN_{8-z}$:Eu^{2+}). J Electrochem Soc 154:J314
65. Kimoto K, Xie RJ, Matsui Y, Ishizuka K, Hirosaki N (2009) Direct observation of single dopant atom in light-emitting phosphor of β-SiAlON:Eu^{2+}. Appl Phys Lett 94:41908
66. Li HL, Xie RJ, Hirosaki N, Suehiro T, Yajima Y (2008) Phase purity and luminescence properties of fine Ca-α-SiAlON: Eu phosphors synthesized by gas reduction nitridation method. J Electrochem Soc 155:J175
67. Zhu XW, Masubuchi Y, Motohashi T, Kikkawa S (2010) The z value dependence of photoluminescence in Eu^{2+}-doped β-SiAlON ($Si_{6-z}Al_zO_zN_{8-z}$) with $1 \leq z \leq 4$. J Alloy Compd 489:157
68. Wang T, Xia Z, Xiang Q, Qin S, Liu QL (2015) Relationship of 5d-level energies of Ce^{3+} with structure and composition of nitrides. J Lumin 166:106

Chapter 12
Characteristics and Properties of A(I,II)M(IV)F_6 Fluoride Phosphors

Hoang-Duy Nguyen and Ru-Shi Liu

Abstract Since the invention of high-brightness InGaN blue light-emitting diode (LEDs), most white LEDs have been fabricated using a combination of blue LED chips and yellow-emitting phosphors. Red-emitting materials improve the color reproducibility of white LEDs. Fluoride phosphors doped with Mn^{4+} have gained significant attention because of their excellent optical properties, thermal stability, and feasible preparation. This chapter presents recent studies on non–rare earth luminescent A(I,II)M(IV)F_6:Mn^{4+} materials for solid-state lighting. These studies have focused on synthesis approaches, morphologies, electronic structure, optical characterizations, waterproof properties, and device performance.

12.1 Introduction

A light-emitting diode (LED) is a semiconductor light source that provides many advantages compared with incandescent light sources including lower energy consumption, longer lifetime, and environment friendliness. Significant developments in increasing the luminescence efficiency and the band gap of LEDs were achieved in the 1990s [1–8]. The first high-brightness blue LED based on indium–gallium–nitride (InGaN) was manufactured by Isamu Akasaki and Hiroshi Amano Shuji Nakamura [1–3] who won the 2014 Nobel Prize in Physics for developing bright and

H.-D. Nguyen · R.S. Liu (✉)
Department of Chemistry, National Taiwan University, Taipei 106, Taiwan
e-mail: rsliu@ntu.edu.tw

H.-D. Nguyen
Institute of Applied Materials Science, Vietnam Academy of Science and Technology, Hochiminh City, Vietnam

R.S. Liu
Department of Mechanical Engineering and Graduate Institute of Manufacturing Technology, National Taipei University of Technology, Taipei 106, Taiwan

R.S. Liu (ed.), *Phosphors, Up Conversion Nano Particles, Quantum Dots and Their Applications*, DOI 10.1007/978-3-662-52771-9_12

energy-saving white-light sources. Since the production of high-brightness blue LEDs, suitable and intense light-emitting materials have been researched and developed to complement blue emission and yield an ideal white light. The combination of luminescent materials and blue LED chips are generally called "phosphor-converted white LEDs" (WLEDs). The performance of WLEDs, including their long-term reliability, strongly depends on the emitting materials and the package WLED method [9]. The blue-light extraction from LEDs should be enhanced, and the phosphors should absorb the blue light (440–470 nm) from LED emission to obtain maximum efficiency of the proper white light. $Y_3Al_5O_{12}:Ce^{3+}$ (YAG:Ce), $Tb_3Al_5O_{12}:Ce^{3+}$ (TAG:Ce), and silicate compounds doped with Eu^{2+}exhibiting broadband emission in the visible spectral region (500–700 nm), efficient absorption of blue light (420–480 nm), and favorable chemical and thermal stabilities have attracted interest as yellow–green phosphors with blue LEDs to make white light [10–13]. However, this approach limits the application range to cool white light with correlated color temperature (CTT approximately 4000–8000 K) and restricts the color-rendering index (CRI < 75) because of the lack of red emission in the luminescent spectra [14, 15] as presented in Fig. 12.1a. Recent advances in

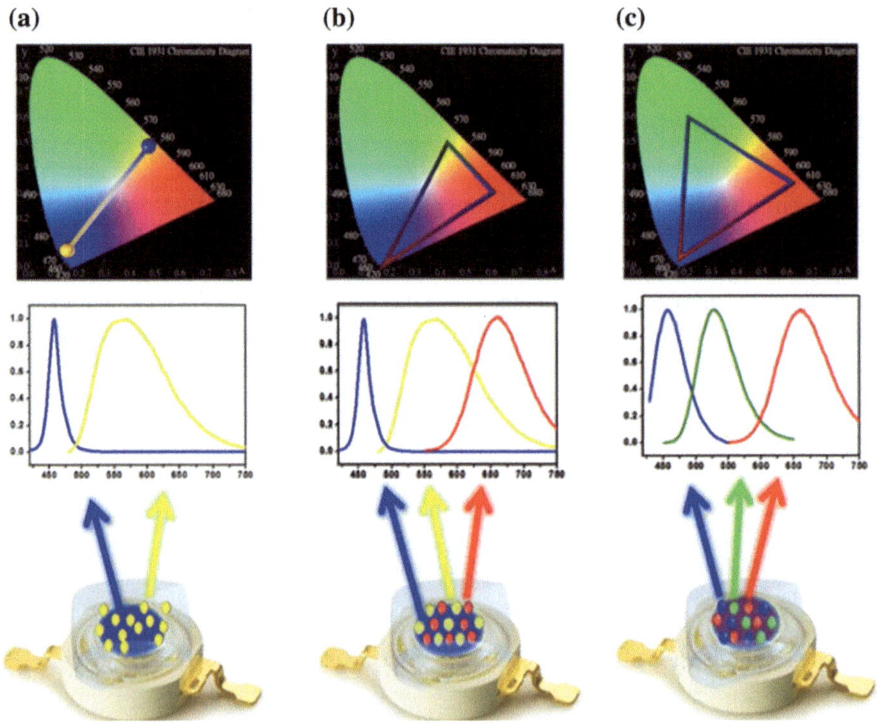

Fig. 12.1 Chromaticity coordinate and photoluminescence spectra of white LEDs including blue LED chip and **a** yellow phosphor, **b** yellow phosphor and red phosphor, and **c** green phosphor and red phosphor

red-emitting phosphors—such as rare earth-activated sulfide, molybdates/tungstate, and nitride red phosphors—have yielded warm WLEDs with a high CRI value of 80 and low CTT values of 2700–4000 K [15–17]. These improvements in color point and reproducibility properties (Fig. 12.1b, c) have rendered WLEDs superior to other traditional light sources such as incandescent and halogen bulbs. However, these red-emitting materials have inherent drawbacks. Given their attractive luminescence, sulfide phosphors strongly quench emission with temperature and are highly sensitive to hydrolysis because of the ionic nature of binary sulfides. Eu^{3+}-activated phosphors possess high CRI values but low absorption in blue or near-UV light, which can be attributed to parity-restricted *4f–4f* transitions. Nitride phosphors are commercially applied because of their sufficient chemical strength and efficient luminescence, but their broad emission bands significantly limit the maximum achievable luminous efficacies of high-quality warm WLEDs [17, 18]. Recently, narrow-band red-emitting fluoride phosphors doped with manganese (IV) (Mn^{4+}) have become extremely popular because of their favorable thermal stability and potential applications in solid-state light sources. In this chapter, we discuss the recent advancements in preparation routes, structure and optical characterizations, and waterproof properties of red fluoride phosphors A_2MF_6:Mn^{4+} (A = K, Na, Cs, or NH_4; M = Si, Ge, Zr, Sn, or Ti) and $BSiF_6$:Mn^{4+} (B = Ba or Zn).

12.2 Inorganic Luminescence Materials

Luminescence materials, also called "phosphors," are composed of an inorganic material (host lattice) and an activator (a small amount of impure atoms are distributed in the host crystal). Either the host lattice or the activator absorbs energy to excite the luminescence.

Self-absorbance and luminescence of the host lattice: Phosphors containing luminescence centers composed of complex ions have been well known since the beginning of this century. $CaWO_4$ has been recognized as a practical phosphor, and this compositere presents the Scheelite compound. The luminescence center is the WO_4^{2-} complex ion where W metal ions coordinate with four O^{2-} ions in a tetrahedral symmetry (T_d). Other analogous T_d complexes are MoO_4^{2-} and VO_4^{3-} [19]. In general, Sheelite phosphors assume the form A_pMO_4, where A is a mono-valence alkaline, di-valence alkaline earth, or tri-valence lanthanide metal ion; p is the number of ions; and M is W, Mo, V, or P. Bright luminescence from blue to green was observed.

Phosphors composed of a host lattice and an activator: In most cases, emission occurs from impure ions that generate the desired emission. When the absorption capacity of the activator ions is too weak, a second type of impurity can be added (sensitizer) to absorb energy and subsequently transfer energy to the activator.

(1) Phosphors are composed of an insulating host and an activator. The activator becomes excited and luminous. The absorption and emission bands of the activator depends on the crystal field (CF), site symmetry, and coordination number of activators in the host lattice. For example, the crystal structure of Sr_2SiO_4 provides two coordination environments for Sr^{2+}: Sr^{2+} with tenfold (SI) and Sr^{2+} with ninefold (SII) coordination by O^{2-}. The $Sr_2SiO_4{:}Eu^{2+}$ phosphor exhibits a broad absorption band in the wavelength range of 300–460 nm, which is mainly attributed to the *4f*–*5d* transition of Eu^{2+}. The ratio of 560-nm emission to 480 nm depends on the S(II)/S(I) ratio in the host lattice. Changing the lattice parameters by incorporating ions of different sizes into the lattices also alters the luminescence properties of the Eu^{2+} dopant ions [20].

(2) Phosphors are composed of a semiconductor host and an activator. During luminescence, the host crystal absorbs and transfers excitation energy to the excited activator as indicated in Fig. 12.2. The excited state returns to the ground state through radiation emission. The crystalline YVO_4 adopts a tetragonal structure with a space group of I_1/amd_1, which is composed of YO_8 dodecahedra (Y^{3+}ions in a non-centrosymmetric site) and VO_4 tetrahedra (T_d) [21]. This compound exhibits no luminescence at room temperature but produces a blue emission at the center at 420 nm under 330-nm excitation at temperatures <200 K. Luminescence cannot be observed at room temperature because of thermal quenching. However, an intense red luminescence can observed when trivalent rare-earth ions, such as Eu^{3+} and Dy^{3+}, are doped into the YVO_4 lattice. The YVO_4 host absorbs and transfers the energy to the dopant ions.

(3) Phosphors are composed of a semiconductor host and an activator, and both the host and the activator absorb and transfer excitation energy to the excited state of the activator. Double-perovskite $A_2B(Mo/W)O_6$ with a monoclinic or pseudo-cubic structure, are substantially absorbed in the NUV region [22]. The excitation spectra of the double-perovskite $Sr_2CaWO_6{:}Eu^{3+}$exhibit a broad absorption band to blue light in the NUV region. This phenomenon occurs because of the charge-transfer (CT) transitions of O to W in the WO_6 octahedral and the *f*–*f* electronic transitions of the Eu^{3+}activators.

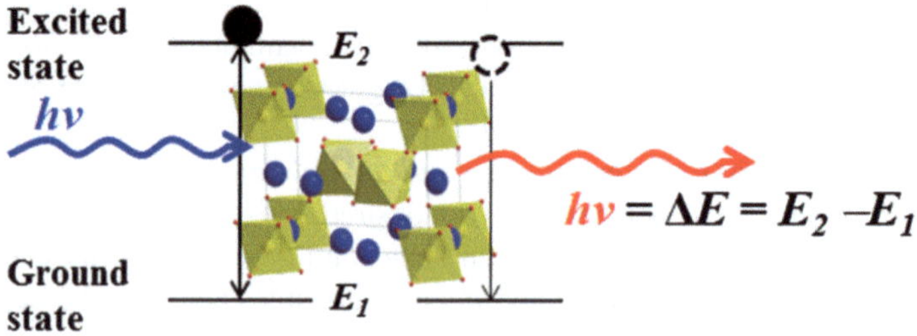

Fig. 12.2 Scheme of the excitation and emission of an activator doped into the crystal host

12.3 Luminescence Centers

During luminescence, only a few atoms are called "emission centers" or "activators." The wavelength emitted by the emission center is dependent on the atom itself, its electronic configuration, and the surrounding crystal structure.

Luminescence centers of ns²-type ions: The electronic configuration ns^2 for the ground state and nsnp for the first excited state (n = 4, 5, 6) are called ns^2-type ions, which include Sn^{2+}, Sb^{3+}, Tl^+, Pb^{2+}, and Bi^{3+}. The luminescence features of those five ions in the oxygen-dominated host lattice are as follows [23]:

- The luminescence is caused by the $^3P_1 \rightarrow {}^1S_0$ transition.
- The luminescence is usually associated with a large Stokes shift, and the spectra are considerably broad, particularly in cases of Sn^{2+} and Sb^{3+}.
- Luminescence decay can prolong in microseconds because of spin-forbidden emission transition.

Luminescence centers of rare-earth ions: Rare-earth elements usually comprise 17 elements: lanthanides (La to Lu), Sc, and Y. The lanthanides, from Ce^{3+} to Lu^{3+}, have 1–14 4*f* electrons added to their inner shell configuration, which is equivalent to Xe. Ions without single 4*f* electrons, such as Sc^{3+}, Y^{3+}, $La^{3+}$and Lu^{3+}, do not have electronic energy levels that can induce excitation and luminescence in or near the visible region. In contrast, Ce^{3+} to Yb^{3+}ions, which possess partially filled 4*f* orbitals with their own energy levels, show various luminescence properties in the UV and visible region. Many of these ions can be used as luminescent ions in phosphors, mostly by replacing Y^{3+}, Gd^{3+}, La^{3+}, and Lu^{3+} in various compound crystals. The 4*f* electronic energy levels of each lanthanide ion is not significantly affected by the environment because the 4f electrons are shielded from external electric fields by the outer $5s^2$ and $5p^6$ electrons. Most emitting levels are separated from the next lower level by at least 2×10^3 cm^{-1} or more. This outcome occurs because the excited states relax by way of two competitive paths: light emission and phonon emission [23].

Luminescence centers of transition-metal ions: The general outer electronic configuration of transition-metal ions (*d*-block elements) is $(n - 1)d^{1-10}$ ns^{1-2}. Transition-metal complexes often have spectacular colors caused by electronic transitions (*d–d* transitions or charge-transfer bands) by light absorption. In a *d–d* transition, an electron in a *d* orbital of the metal may jump to another d orbital of higher energy by way of photon excitation. During charge transfer, an electron from a metal-based orbital enters an empty ligand-based orbital resulting in metal-to-ligand charge-transfer transition. The converse situation also occurs. An electron from a ligand-based orbital jumps into an empty metal-based orbital (ligand-to-metal charge transfer). However, the energy levels of transition-metal complexes differ. The d orbitals can be split by CF strength, which depends on the oxidation state of a metal and the nature of the ligands. Thus, the optical properties of transition metal ions are strongly influenced by the host. The energy levels of various transition metals are often presented by Tanabe–Sugano diagrams [24].

12.4 Energy-Level Diagrams of $3d^3$ Transition-Metal Ions

Tanabe and Sugano theoretically calculated the energy levels of $3d^n$ ions according to the CF strength. Figure 12.3a depicts the Tanabe-Sugano energy diagram of Mn^{4+} ions in an octahedral CF, which is the most common depiction for transition-metal ions. The ligand-field splitting (the strength of the octahedral field) parameter (D_q) divided by the Racah parameter (B) is displayed through the x-axis of a Tanabe–Sugano diagram. The y-axis is in terms of energy (E/B). Three Racah parameters exist—A, B, and C—which describe various aspects of the Coulomb interaction between electrons. A is constant among the d-electron configurations. B and C correspond to individual *d*-electron repulsions. The ratio of C/B is expected to be almost the same as 4.5 for all the transition-metal ions. The octahedral field splits the five-fold degenerate *d* orbital into three- and two-fold degenerate t_{2g} and e_g states, respectively. The energy separation between e_g and t_{2g} is labeled $10D_q$. Each state is given a symmetry label (A_{1g}, T_{2g}, etc.). The symmetry label clearly shows that the energies of most multiplets are strongly dependent on CF strength, except for the $^2T_1$ and ^2E_g levels. The optical characteristics of Mn^{4+} with an octahedral symmetry can be inferred on the basis of their configurational-coordinate diagrams (Fig. 12.3b). The electronic t_2^3 orbital splits into the ^2E_g, $^2T_1$, $^2T_2$, and $^4A_2$ levels, whereas the $^4T_1$ and $^4T_2$ levels originate from another t_2^2e orbital. Thus, the large difference in the parabolas of the ground states $^4A_2$ and $^4T_1$ (or $^4T_2$) is observed, whereas a small displacement between the parabolas of $^4A_2$ and ^2E_g (or $^2T_1$, $^2T_2$) exists. Larger displacement means stronger electron–phonon interaction (namely, a larger Huang–Rhys factor) and thus larger spectral bandwidth of the transition. Moreover, the transitions between $^4T_1$ (or $^4T_2$) and $^4A_2$ levels are spin-allowed according to the spin-selection rule of $\Delta S = 0$. Thus, broad excitation or absorption spectra with a

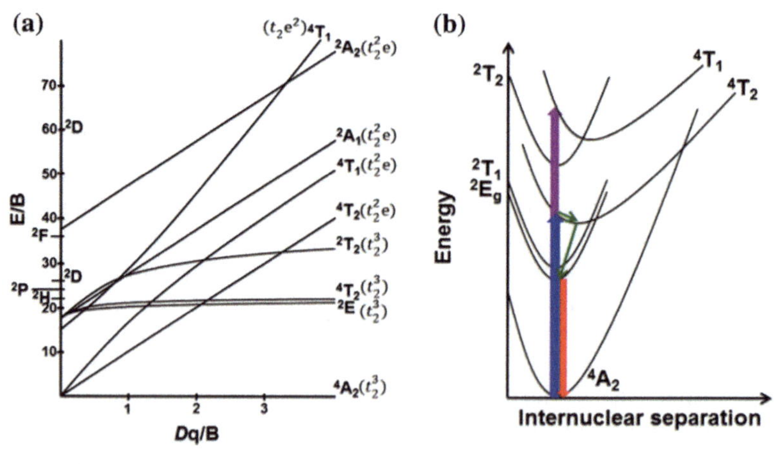

Fig. 12.3 **a** Tanabe-Sugano energy diagram of Mn^{4+} ions ($3d^3$ system) in an octahedral crystal field. **b** Configurational-coordinate diagram for Mn^{4+} in fluoride hosts

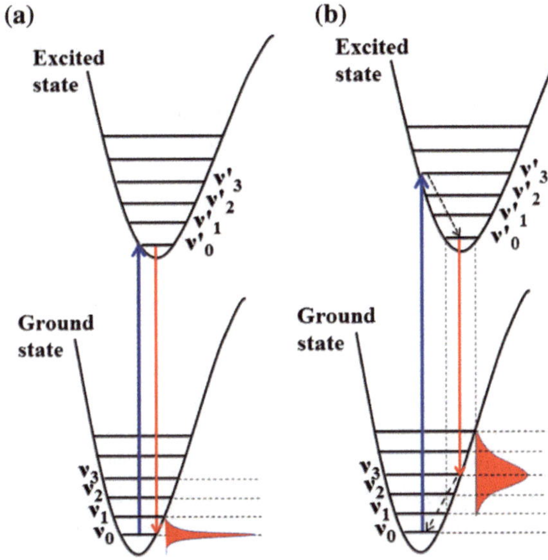

Fig. 12.4 Scheme of electronic transition with **a** no vibronic coupling and **b** vibronic coupling

high intensity are expected between these levels [25]. On excitation to $^4T_1$ or $^4T_2$ levels, the excited ions usually relax nonradiatively to ^2E_g.

In general, emission transitions between $3d^n$ states ($^2E_g \rightarrow {}^4A_2$) are forbidden because parity does not change. However, the parity-selection rule can be relaxed when the active ions are situated in the site without inversion symmetry by mixing a small amount of opposite-parity wave functions into the *d*-wave functions. Meanwhile, energy (a photon) can sometimes excite a molecule to an excited electronic and vibrational state because the energy required for the transition of one electronic state to another electronic state (i.e., electronic transition) is larger than that for the transition of one vibrational state to another vibrational state (i.e., vibrational transition). Figure 12.4 shows the electronic transition (no vibronic coupling) and the electronic transition coupled with the vibrational transition (vibronic coupling). Broadband emission can be observed for a strong vibronic coupling, whereas narrow-band emission can be observed for a weak vibronic coupling, such as that for the $^2E_g \rightarrow {}^4A_2$ emission transition of Mn^{4+} (Fig. 12.3b).

12.5 Synthesis Methods and Optoelectronic Properties of Mn^+-Doped Fluoride Phosphor A(I,II)M(IV)F_6

The valence states of Mn (2+, 3+, 4+, 6+ and 7+) are sensitive to synthesis temperature. Hence, the main difficulty depends on controlling the Mn valence state for synthesizing Mn^{4+}-activated fluoride complexes. A series of red fluoride phosphors —A_2MF_6:Mn^{4+} (A = K, Na, Cs, or NH_4; M = Si, Ge, Zr, Sn, or Ti) and $BSiF_6$:

Mn^{4+} (B = Ba or Zn)—were prepared by way of wet-chemical etching of silicon wafers in an aqueous HF solution with the addition of oxidizing agent $KMnO_4$ at room temperature [26–29] as presented in Fig. 12.5. However, the technique has significant drawbacks, namely, expensive pure Si wafers and metal shots used during synthesis, prolonged reaction time, tedious post-treatment, and low yield.

The optical properties of Mn^{4+}significantly depend on the CF exerted by the host. Doping Mn^{4+} ions into the A(I,II)M(IV)F_6 lattice containing the most electronegative fluorine elements should produce an ionic ligand. Hence, the crystal structures of metal fluorides are mainly governed by geometric and electrostatic principles. Mn^{4+} ions experience a strong CF because of their high effective positive charge. These ions exhibit sharp emission lines from 600 nm to 650 nm dominated by the spin-forbidden $^2E_g \rightarrow {}^4A_{2g}$ transitions under 460-nm excitation. The excitation spectrum indicates the two intense broad bands located at approximately 370–390 nm and 460–475 nm, which corresponds to the $^4A_{2g}$–$^4T_{2g}$ and $^4A_{2g}$–$^4T_{1g}$ spin-allowed transitions. The excitation spectra (PLE) of K_2SiF_6:Mn^{4+} is presented in Fig. 12.6A. A number of components with a spacing of a few hundred kaysers are observed in the 460-nm excitation band, which can be explained as a vibronic progression of the fundamental frequency combined with an unsymmetrical vibration of the SiF_6^{2-} (MnF_6^{2-}) octahedron superimposed on the electronic transition [26]. The relation between the intensities of the *n*th-order vibration sideband I_n^{ex} and the zero-photon line (ZPL)I_0^{ex} can be expressed by the following equation:

$$I_n^{ex} = I_0^{ex}\exp(-\bar{n})\frac{(\bar{n})^n}{n!}, \tag{1}$$

where $\bar{n}$ is the mean phonon local vibration number. The calculated results denote the ZPL ($\bar{n}$ = 0) energy at 20,300 cm^{-1} for the $^4A_2$–$^4T_2$ transitions. *D*q is approximately 2030 cm^{-1}, and the energy separation between each phonon replica is approximately 524 cm^{-1}. For the second broad excitation band observed at 350 nm for the $^4A_2$–$^4T_1$ transitions, the ZPL line can be estimated at 24,900 cm^{-1}. The Racah parameters can be determined from the ZPL energies of 16,100 cm^{-1} (^2E_g–$^4A_{2g}$), 20,300 cm^{-1}

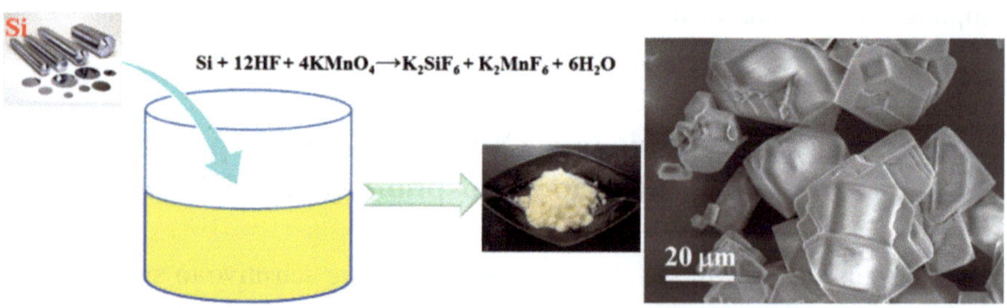

Fig. 12.5 Scheme of K_2SiF_6:Mn^{4+} preparation by wet-chemical etching and scanning electron micrograph of K_2SiF_6:Mn^{4+}. Reproduced with permission from *J. Electrochem. Soc.*, **155**, E183 (2008). Copyright 2008, The Electrochemical Society

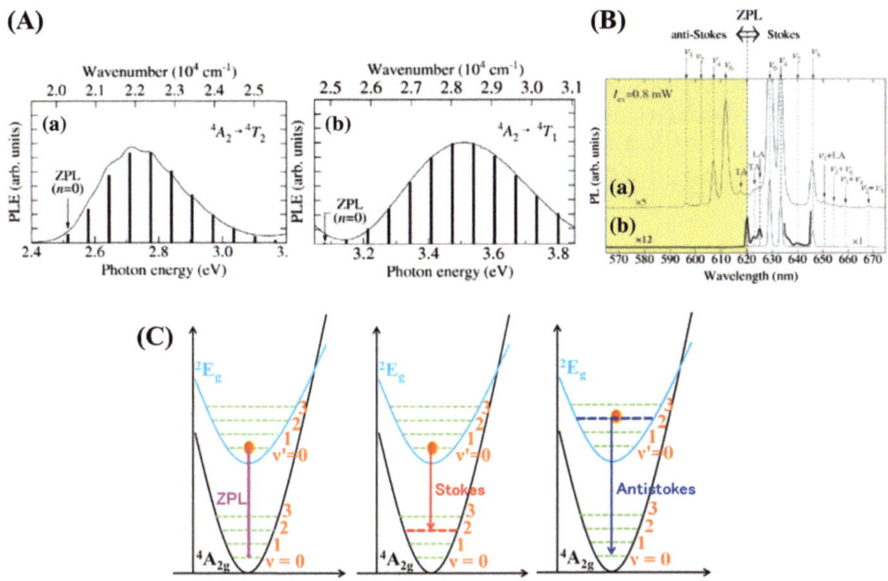

Fig. 12.6 **A** PLE spectra in the (**a**) $^4A_2$–$^4T_2$ and (**b**)$^4A_2$–$^4T_1$ transition regions of K_2SiF_6:Mn^{4+} phosphor at 300 K. The *vertical bars* in (**a**) and (**b**) present the calculated results of Eq. 1 with $\bar{n}$ = 4 and 7, respectively. The ZPL ($\bar{n}$ = 0) positions are indicated by the *vertical arrows*. **B** PL spectra for K_2SiF_6:Mn^{4+} phosphor measured at (**a**) 300 K and (**b**) 20 K with an He–Cd laser as an exciting light source. The exciting laser power is at I_{ex} = 0.8 mW. Reproduced with permission from *J. Electrochem. Soc.*, **155**, E183 (2008). Copyright 2008, The Electrochemical Society. **C** Scheme of emission transitions of ZPL, Stokes, and anti-Stokes

($^4A_{2g}$–$^4T_{2g}$), and 24,900 cm^{-1} ($^4A_{2g}$–$^4T_{1g}$) to be B approximately 770 cm^{-1}, C approximately 3470 cm^{-1}, and C/B = 4.5, respectively.

Figure 12.6B depicts the PL spectra of K_2SiF_6:Mn^{4+} at 300 and 20 K with ZPL emission at 620 nm. At T = 20 K, three long-wavelength lines peaking at 630, 634, and 646 nm were observed. At T = 300 K, short-wavelength lines can be observed at 595, 608, and 612 nm. The long- and short-wavelength emission lines are known as the Stokes and anti-Stokes lines, respectively (Fig. 12.6c). K_2SiF_6 possesses a cubic structure with a space group of $O_h^5 - Fm3m$, where an Si atom is on the 1a (O_h) position. Two K^- ions are located on 2c (Td), and six F^- ions are located on 6e (C_{4v}). Fundamental vibronic modes $\nu_2(e_g)$, $\nu_3(t_{1u})$, $\nu_4(t_{1u})$, and $\nu_6(t_{2u})$ for the SiF_6^{2-} octahedron are determined [26]. The intensity of the ZPL depends on the local symmetry of the Mn^{4+} surroundings. A Mn^{4+}-activated material with low crystal symmetry is expected to exhibit a high ZPL emission intensity [27]. The PLE and PL of cubic K_2SiF_6 ($O_h^5 - Fm3m$), orthorhombic $KNaSiF_6$ ($D162h - Pnma$), and trigonal Na_2SiF_6 ($D23 - P321$) are presented in Fig. 12.7A. ZPL emission intensity increased in the order K_2SiF_6:Mn^{4+}, $KNaSiF_6$:Mn^{4+}, and Na_2SiF_6:Mn^{4+}. No obvious differences in luminescent properties were observed among the A_2MF_6:Mn^{4+} (A = K, Na, Cs; M = Si, Ge, Ti) samples, as depicted in Fig. 12.7A through C.

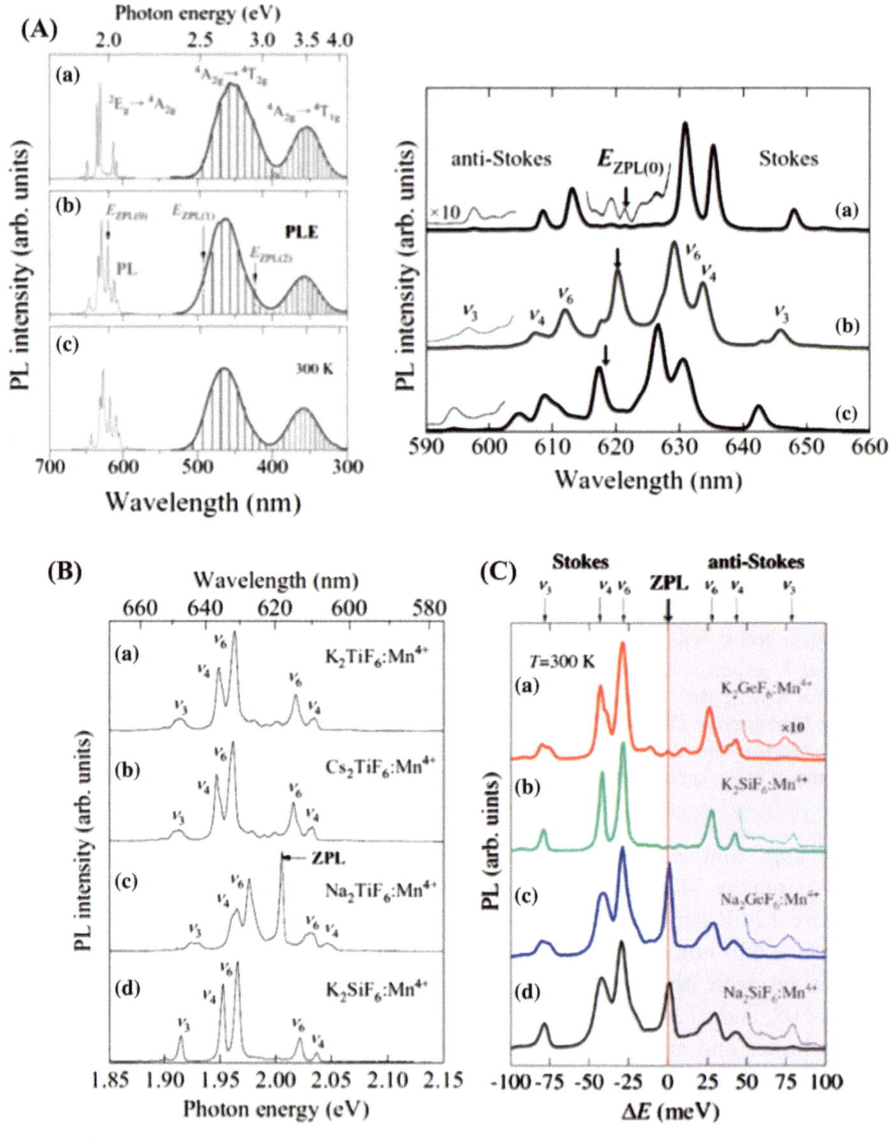

Fig. 12.7 **A** Room-temperature PLE and PL spectra of (**a**) $K_2SiF_6:Mn^{4+}$, (**b**) $KNaSiF_6:Mn^{4+}$, and (**c**) $Na_2SiF_6:Mn^{4+}$. Reproduced with permission from *J. Electrochem. Soc.*, **159**, J34 (2012). Copyright 2012, The Electrochemical Society. **B** Room-temperature PL spectra of (**a**) $K_2TiF_6:Mn^{4+}$, (**b**) $Cs_2TiF_6:Mn^{4+}$, (**c**) $Na_2TiF_6:Mn^{4+}$, and (**d**) $K_2SiF_6:Mn^{4+}$ red phosphors. Reproduced with permission from *J. Electrochem. Soc.*, **158**, J58 (2011). Copyright 2011, The Electrochemical Society. **C** PL spectra of (**a**) $K_2GeF_6:Mn^{4+}$, (**b**) $K_2SiF_6:Mn^{4+}$, (**c**) $Na_2GeF_6:Mn^{4+}$, and (**d**) $Na_2SiF_6:Mn^{4+}$ phosphors measured at 300 K with a He–Cd laser as an excitation light source. Reprinted with the permission from Ref. [29]. Copyright 2009, American Institute of Physics

However, a red shift in the $^2E_g \rightarrow {}^4A_{2g}$ transitions of $K_2SiF_6{:}Mn^{4+}$, $KNaSiF_6{:}Mn^{4+}$, and $Na_2SiF_6{:}Mn^{4+}$is evident in Fig. 12.7A. In addition, the large variation in the energy of the $^2E_g \rightarrow {}^4A_{2g}$ emission transitions was determined from 16,109 cm^{-1} (620.7 nm) in K_2SiF_6 to 13,826 cm^{-1} (723.3 nm) in $SrTiO_3$. Similarly, the CF-strength parameter (Dq) exhibited a significant variation from 1820 cm^{-1} in $SrTiO_3$ to 2400 cm^{-1} in the MgGeF phosphor [30]. The fluctuation in the position of the $^2E_g \rightarrow {}^4A_{2g}$ emission transition should not be attributed to the variation in the CF strength because the energy of the ^2E_g state in the d^3 electronic configuration is independent of CF (as shown in Fig. 12.3a). The relationship of bond distance between Mn^{4+} and the host lattice ligands was considered for this trend. The distance is the shortest for fluoride materials with highly ionic crystals, and this condition reduces the nephelauxetic effect. Thus, the fluorides possessed higher Racah parameter B values compared with oxides and garnets with more covalent bonds. Thus, the upward shift of the ^2E_g level on the energy scale increased with B values as illustrated in Fig. 12.8a, b.

Du [31] calculated the Mn^{4+} emission energies for several oxide and fluoride hosts using density functional theory with the standard Perdew–Burke–Ernzerh (DFT–PBE) method. The results indicated that the emission energy trend is reproduced effectively by the PBE calculations except for the transition from oxides to fluorides as shown in Fig. 12.8c. The DFT–PBE method can calculate the Mn^{4+}-emission energy trend effectively when the ligands in all host materials are the same, and the variation of Mn–ligand hybridization strength is caused by different bonding configurations. Mn–F hybridization is generally weak, and its variation in different fluorides is not significant. Hence, the variation in Mn^{4+}-emission energy is smaller in fluorides than in oxides. In particular, the emission wavelength of Mn^{4+}-doped fluorides in the range of 617–630 nm is more sensitive to human eyes than that of Mn^{4+}-doped oxides. Obviously, WLEDs using blue LED chips, commercial yellow phosphor, and Mn^{4+}-doped fluoride red phosphor exhibit more desirable optical properties than WLEDs using oxide compounds.

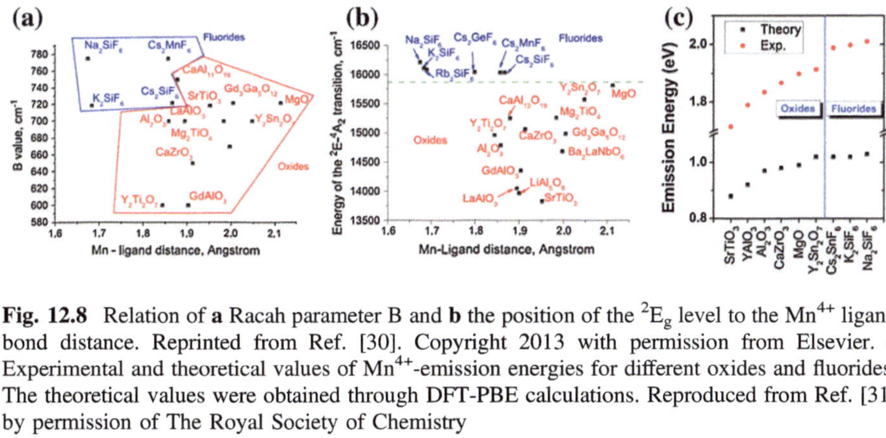

Fig. 12.8 Relation of **a** Racah parameter B and **b** the position of the ^2E_g level to the Mn^{4+} ligand bond distance. Reprinted from Ref. [30]. Copyright 2013 with permission from Elsevier. **c** Experimental and theoretical values of Mn^{4+}-emission energies for different oxides and fluorides. The theoretical values were obtained through DFT-PBE calculations. Reproduced from Ref. [31] by permission of The Royal Society of Chemistry

A previous study reported a novel strategy to improve the etching method for $K_2SiF_6:Mn^{4+}$ preparation by adding H_2O_2 into $HF/H_2O/KMnO_4$ solutions [32]. $K_2SiF_6:Mn^{4+}$ with micro-size and irregular shape exhibited a sharp red-emission peak at 630 nm and the strongest excitation peak at approximately 460 nm. Monochromatic green ($G_{B,M}$, $(Ba, Ca, Sr)_nSiO_4:Eu^{2+}$), amber ($A_{B,M}$, $(Ba, Ca, Sr)_3SiO_5:Eu^{2+}$), and narrow red ($R_{N,B,M}$, $K_2SiF_6:Mn^{4+}$) or wide red ($R_{1\text{-}5W,B,M}$, $(Sr, Ca)AlSiN_3:Eu^{2+}$) packages were further fabricated on the basis of blue diode chips and a long wavelength–pass dichroic filter (LPDF) as shown in Fig. 12.9a. These three monochromatic LEDs, together with a blue diode chip, were integrated to form WLEDs. White-light emissions with various CCTs were obtained by tuning the fractional applied current of each monochromatic LED. In addition, a four-package WLED with a CCT of 2700 K, a luminous efficacy of 107 lm/W, and a high CRI of 94 was achieved as indicated in Fig. 12.9b through d.

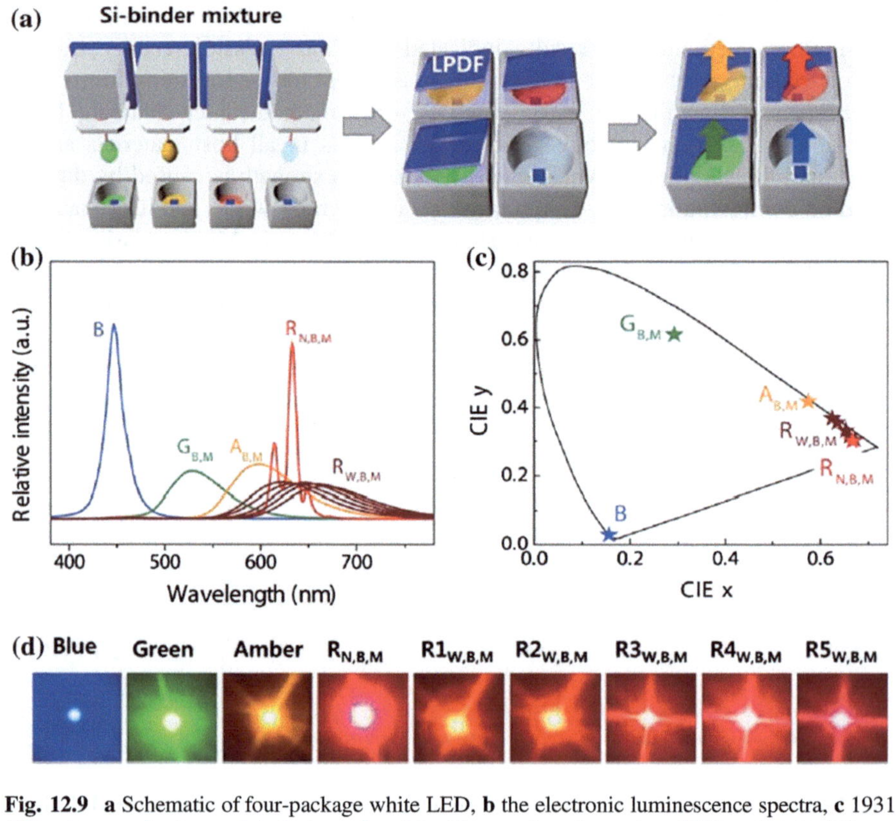

Fig. 12.9 **a** Schematic of four-package white LED, **b** the electronic luminescence spectra, **c** 1931 CIE color coordinates, and **d** photographs of the blue semiconductor-type LED and eight different colors of fully downconverted LPDF-capped pc-LEDs ($G_{B,M}$, $A_{B,M}$, $R_{N,B,M}$, $R_{1W,B,M}$, $R_{2W,B,M}$, $R_{3W,B,M}$, $R_{4W,B,M}$, and $R_{5W,B,M}$). Reproduced from Ref. [32] by permission of The Royal Society of Chemistry

Currently, convenient approaches for fluoride phosphor preparation are gaining considerable attention. $K_2TiF_6:Mn^{4+}$ and $K_2SiF_6:Mn^{4+}$ were prepared through cation exchange reaction by mixing fluoride hosts with an HF solution dissolved with K_2MnF_6 powders at room temperature in 20 min [25]. The prepared $K_2TiF_6:Mn^{4+}$ possesses a hexagonal structure, showing an excitation broadband peaking at 468 nm and a sharp red emission at 630 nm, with a photoluminescence quantum yield as high as 98 %. The optoelectronic structure of $K_2TiF_6:Mn^{4+}$ was consistent with those of $K_2SiF_6:Mn^{4+}$ and other Mn^{4+}-doped fluorides. Warm WLEDs with $K_2TiF_6:Mn^{4+}$ red phosphor and commercially available $Y_3Al_5O_{12}:Ce^{3+}$ (YAG) yellow phosphor on a blue-InGaN chip exhibited excellent CRI (Ra of approximately 81), a low CCT of approximately 3556 K, and a luminous efficacy of 116 lm/W^{-1}. Qiu et al. [33] prepared $K_2SiF_6:Mn^{4+}$ by way of redox reaction in HF/$KMnO_4$ solution at room temperature for 10 min (Fig. 12.10). SiO_2 powders were used as Si source for synthesis, and H_2O_2 was utilized to effectively reduce Mn^{7+} to Mn^{4+}. In addition, a hydrothermal (HT) method was employed for $K_2SiF_6:Mn^{4+}$ preparation using KF, $KMnO_4$, H_2O_2, and silicon fluoride solutions at 120 °C for 12 h [34]. The combination of yellow–green emission YAG phosphor and narrow-line red emission $K_2SiF_6:Mn^{4+}$ phosphor

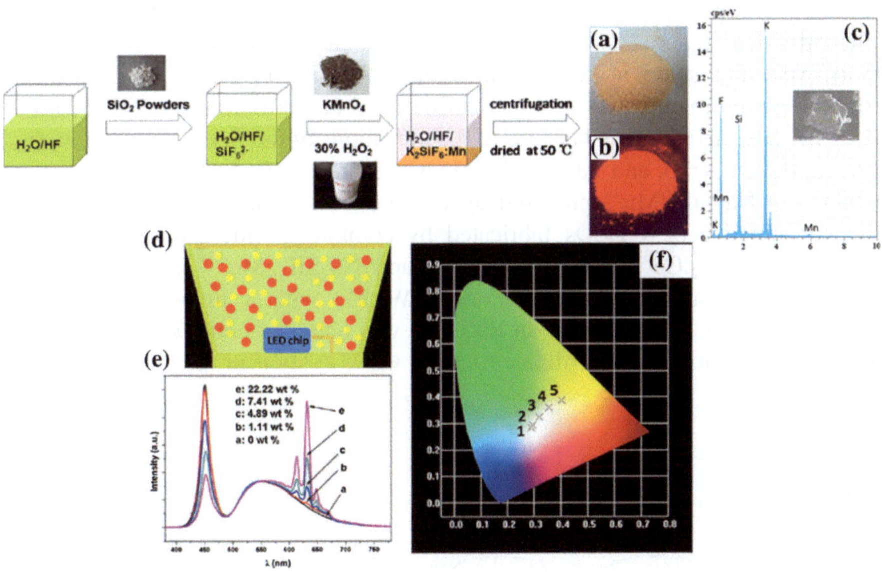

Fig. 12.10 Schematic showing syhthesis of $K_2SiF_6:Mn^{4+}$ red phosphor from silicon dioxide by redox reaction in HF/$KMnO_4$ solution and digital images of $K_2SiF_6:Mn^{4+}$ red phosphor **a** without and **b** with UV (λ_{ex} = 365 nm) excitation. **c** EDS spectrum of as-synthesized $K_2SiF_6:Mn^{4+}$ phosphors. **d** Schematic of a YAG:Ce and $K_2SiF_6:Mn^{4+}$-based WLED. **e** Emission spectra of a WLED prototype normalized to 544 with various amount of red phosphors under a working current of 20 mA. **f** CIE diagram of the corresponding $K_2SiF_6:Mn^{4+}$ phosphor concentration (1 through 5 corresponding to red $K_2SiF_6:Mn^{4+}$ phosphor amounts of 0, 1.11, 4.89, 7.41, and 22.22 wt%, respectively). Reproduced from Ref. [33] by permission of John Wiley & Sons Ltd.

generated white luminescence with a high CRI of 89.8–90.9, a CCT of approximately 3510–3900 K, and a high efficacy of 80 to 116 lm/W.

A simple coprecipitation method to synthesize $K_2GeF_6:Mn^{4+}$, $Rb_2SiF_6:Mn^{4+}$, $Na_2SiF_6:Mn^{4+}$, and $K_2SiF_6:Mn^{4+}$ was presented [35–37]. Chemical coprecipitation of $Rb^+/K^+/Na^+$ and GeF_6^{2-}/SiF_6^{2-} was performed by mixing HF/(Na,K,Rb)MnO_4 and silicon fluoride solutions using H_2O_2 to reduce Mn^{7+} to Mn^{4+} efficiently at room temperature in a few minutes, which is suitable for quantifiable production because of its high yield, repeatability, and low cost. $Na_2SiF_6:Mn^{4+}$ was also synthesized through a two-step coprecipitation method where K_2MnF_6, which was prepared from a silicon–fluoride solution in the presence of $KMnO_4$ and H_2O_2, was used as a starting material. The preparation schemes of $Na_2SiF_6:Mn^{4+}$ through one- and two-step coprecipitation are presented in Figs. 12.11 and 12.12. The coprecipitation reaction for $Na_2SiF_6:Mn^{4+}$ preparation can be expressed as follows:

$$NaOH + HF \rightarrow NaF + H_2O$$

$$(2-2x)SiO_2 + 2xNaMnO_4 + (4-2x)NaF + (8+2x)HF + (2-x)H_2O_2$$
$$\rightarrow 2Na_2Si_{1-x}Mn_xF_6 + 6H_2O + (1+x)O_2 \uparrow$$

$K_2GeF_6:Mn^{4+}$, $Rb_2SiF_6:Mn^{4+}$, $Na_2SiF_6:Mn^{4+}$, and $K_2SiF_6:Mn^{4+}$ phosphors exhibited sharp red emission lines from 600 to 650 nm originating from spin-forbidden $^2E_g \rightarrow {}^4A_{2g}$ transitions. The luminescent properties of these phosphors were the same at room temperature. In $Na_2SiF_6:Mn^{4+}$, the PL spectra exhibited seven main peaks in the emission band at approximately 594, 605, 609, 618, 627, 631, and 643 nm because of the transitions of the ν_3 (t_{1u}), ν_4 (t_{1u}), ν_6 (t_{2u}), ZPL, ν_6 (t_{2u}), ν_4 (t_{1u}), and ν_3(t_{1u}) vibronic modes, respectively. Similar results were obtained in $Na_2SiF_6:Mn^{4+}$ prepared by wet-chemical etching of an Si wafer [37]. The performance of WLEDs fabricated by combining blue InGaN chips, commercial $Y_3Al_5O_{12}:Ce^{3+}$ yellow phosphor, and $Na_2SiF_6:Mn^{4+}$ red phosphor was evaluated for commercial applications. AWLED with a luminous efficacy of 77.6 lm W^{-1}, a CRI of 86, and an R9 of 61 was obtained under a drive current of 200 mA. A chromaticity coordinate (0.3126, 0.2951) lay near the black-body locus

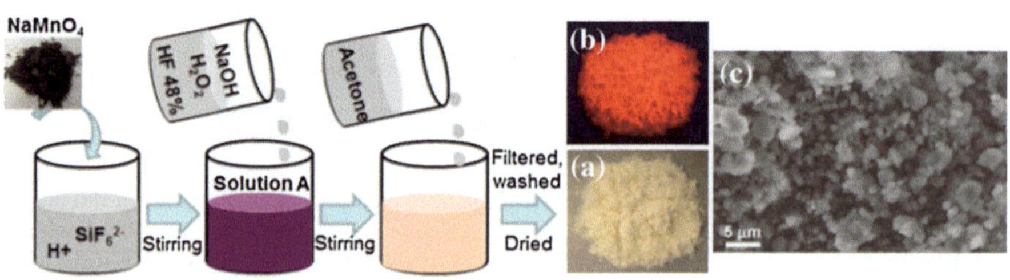

Fig. 12.11 Synthesis of $Na_2SiF_6:Mn^{4+}$ red phosphor by way of a one-step method and digital images of $Na_2SiF_6:Mn^{4+}$ under **a** warm room light and **b** UV light (λ_{ex} = 365 nm) excitation; **c** scanning electron micrographs of $Na_2SiF_6:Mn^{4+}$. Reproduced from Ref. [37] by permission of The Royal Society of Chemistry

Fig. 12.12 Synthesis of $Na_2SiF_6:Mn^{4+}$ red phosphor by way of a two-step method and digital images of $Na_2SiF_6:Mn^{4+}$ under **a** warm light and **b** UV light (λ_{ex} = 365 nm) excitation; **c** scanning electron micrographs of $Na_2SiF_6:Mn^{4+}$. Reproduced from Ref. [37] by permission of The Royal Society of Chemistry

in Commission Internationale de I'Éclairage (CIE) 1931 color spaces, and a color temperature of 6875 K was observed for the fabricated WLEDs. With a bright "warm" white light and a CRI of 89 under a drive current of 15 mA, the chromaticity coordinates of (0.4016, 0.4495) lay near the black body locus, and a CCT of 3974 K was observed using $K_2GeF_6:Mn^{4+}$ red phosphor.

However, the structure and luminescence of $K_2GeF_6:Mn^{4+}$ from other fluorides differed at high temperatures. A structural evolution from hexagonal $K_2GeF_6:Mn^{4+}$ with space group $P\bar{3}m1$ to $P6_3mc$ and $P6_3mc$ to cubic Fm3m was observed when the as-prepared $K_2GeF_6:Mn^{4+}$ phosphors were heated at various temperatures of approximately 400 and 500 °C, respectively (Fig. 12.13a). This result can be verified by the fact that $P6_3mc$ $K_2GeF_6:Mn^{4+}$ heated at 400 °C for 20 min in air exhibits an evident split behavior in the ν_5 vibration mode at approximately 330 cm^{-1} (Fig. 12.13b). Low-symmetry hexagonal $P6_3mc$ phase showed an obvious ZPL peaking at 621 nm. High-symmetry hexagonal $P\bar{3}m1$ exhibited an intense red emission without ZPL, whereas cubic Fm3 m phase showed no red emission

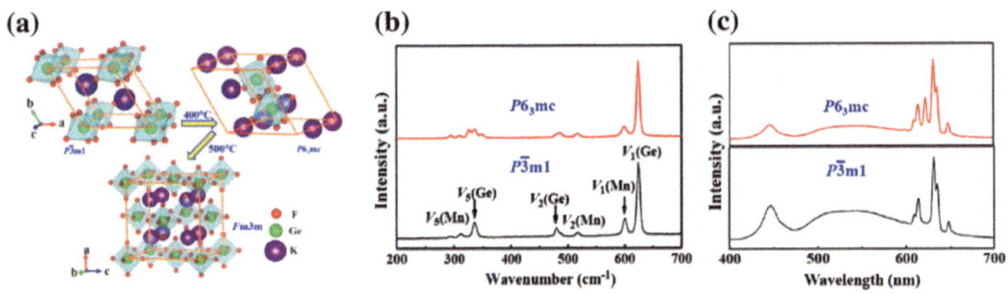

Fig. 12.13 **a** Schematic of structural evolution during heat treatment of K_2GeF_6. **b** Raman-scattering spectra of $K_2GeF_6:Mn^{4+}$ before and after heat treatment. **c** Luminescence spectra of WLEDs using $K_2GeF_6:Mn^{4+}$red phosphors. Reprinted with permission from Ref. [38]. Copyright 2015 American Chemical Society

because of Mn^{4+}oxidation at a high temperature. The ZPL emission properties of $K_2GeF_6:Mn^{4+}$ with low- and high-symmetry hexagonal were similar to those of hexagonal *P*321 $Na_2SiF_6:Mn^{4+}$ and cubic Fm3m $K_2SiF_6:Mn^{4+}$. Figure 12.13c presents the luminescent spectra of the WLEDs including blue InGaN chips, YAG: Ce^{3+} yellow phosphor, and $K_2GeF_6:Mn^{4+}$ red phosphors with different space groups $P\bar{3}m1$ and $P6_3mc$. The ZPL peak was evident in the $P6_3mc$ symmetry [38].

Other fluoride red phosphors, such as $BaSiF_6:Mn^{4+}$(trigonal structure) and $BaTiF_6:Mn^{4+}$(rhombohedral structure) were also synthesized by way of a one-step HT method [39, 40]. The starting materials $Ba(NO_3)_2$, TiO_2 or $Ti(OC_4H_9)_4$, NH_4SiF_6, $KMnO_4$, HF (wt 40 %), and distilled water were mixed thoroughly and heated at 120 °C for 12–20 h. The different morphologies of $BaSiF_6:Mn^{4+}$ and $BaTiF_6:Mn^{4+}$ prepared by way of various methods are described in Fig. 12.14a through d. The prepared phosphors exhibited a similar absorption of the blue region as well as intense red luminescence at approximately 632 nm (Fig. 12.14e). The

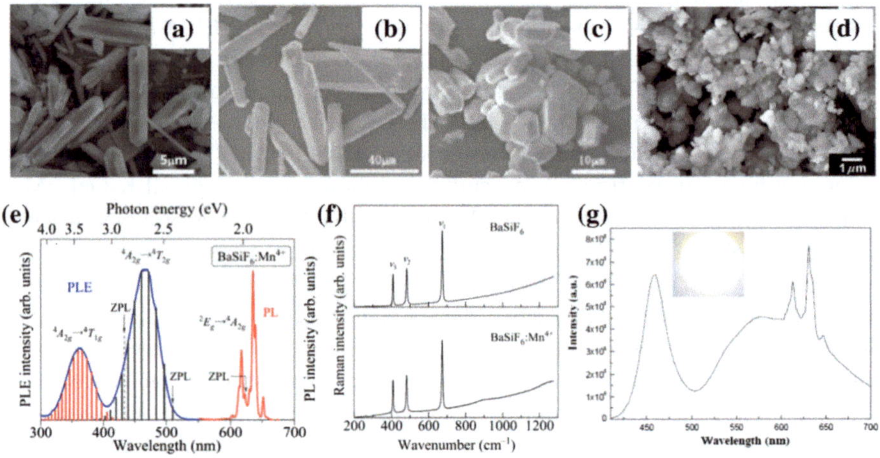

Fig. 12.14 SEM images of **a** $BaSiF_6:Mn^{4+}$prepared by way of an HT method. Reproduced from Ref. [39] by permission of The Royal Society of Chemistry. **b**, **c** $BaTiF_6:Mn^{4+}$prepared by way of an HT method using TiO_2 and $Ti(OC_4H_9)_4$, Reproduced from Ref. [40] by permission of The Royal Society of Chemistry. **d** $BaTiF_6:Mn^{4+}$prepared by way of wet-chemical etching. Reprinted with the permission from Ref. [41]. Copyright 2013, American Institute of Physics. **e** PLE and PL spectra of $BaSiF_6:Mn^{4+}$ at 300 K. The *vertical bars* denote the calculated results using Eq. (1). **f** Stokes Raman lines in pure $BaSiF_6$ and $BaSiF_6:Mn^{4+}$ at 300 K. Reprinted with permission from Ref. [41]. Copyright 2013, American Institute of Physics. **g** White-luminescence spectra of WLED fabricated with blue GaN chip, yellow phosphor YAG:Ce, and red phosphor $BaTiF_6:Mn^{4+}$. The *inset* shows image of WLED under current of 20 mA. Reproduced from Ref. [40] by permission of The Royal Society of Chemistry

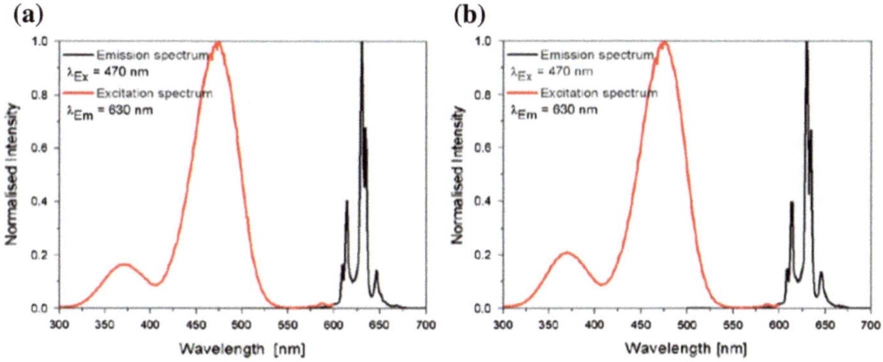

Fig. 12.15 PLE and PL spectra of $ZnSiF_6 \cdot 6H_2O:Mn^{4+}$ and $ZnGeF_6 \cdot 6H_2O:Mn^{4+}$ at 300 K. [42]. Reprinted from Ref. [42]. Copyright 2013 with permission from Elsevier

room-temperature Raman spectra of $BaSiF_6$ and $BaSiF_6:Mn^{4+}$ were measured, and the internal vibronic modes v_5 (t_{2g} [bending]), v_2 (e_g [stretching]), and v_1 (a_{1g} [stretching]) of the SiF_6^{2-} octahedron in $BaSiF_6$ were identified (Fig. 12.14f) [41]. The results indicate that the parameters above have highly similar optical features to that of $A_2MF_6:Mn^{4+}$ phosphors. Warm WLEDs with CCTs of 3931 and 4073 K, Ra values of 82.1 and 93.13, and luminous efficacies of 112 and 121 lm $W^{-}1$ were obtained under a 20-mA drive current using $BaSiF_6:Mn^{4+}$ and $BaTiF_6:Mn^{4+}$, respectively. The electroluminescent spectra of the warm WLEDs using $BaTiF_6$ are presented in Fig. 12.14g.

$ZnSiF_6 \cdot 6H_2O:Mn^{4+}$ and $ZnGeF_6 \cdot 6H_2O:Mn^{4+}$ phosphors were synthesized through co-crystallization [42, 43]. The solution of H_2SiF_6 or GeO_2 and K_2MnF_6 dissolved in HF was added into the solution of HF and ZnF_2. The mixed solution was cooled down until $ZnSiF_6 \cdot 6H_2O:Mn^{4+}$ precipitated. The trigonal structure with space group $R\bar{3}$ was observed for both $ZnSiF_6 \cdot 6H_2O$ and $ZnGeF_6 \cdot 6H_2O$. The excitation and emission peaks of the $ZnSiF_6 \cdot 6H_2O:Mn^{4+}$ and $ZnGeF_6 \cdot 6H_2O:Mn^{4+}$ phosphors are 470 and 630 nm, respectively (Fig. 12.15).

Table 12.1 presents a comparison of the luminescent properties of current Mn^{4+}-doped fluoride phosphors. The maximum absorption band in the blue region and strong red emission with a maximum wavelength at 627–630 nm of $A_2MF_6:Mn^{4+}$ (A = K, Na, Cs, Rbor NH_4; M = Si, Ge, Zr, Sn, or Ti), and (Ba/Zn)(Si/Ge)$F_6:Mn^{4+}$ qualify Mn^{4+}-doped fluoride phosphor as a potential candidate to improve the color reproducibility of WLED.

Table 12.1 Luminescent properties of typical Mn^{4+}-doped fluoride phosphors

Host lattice	Reference	Structure	ZPL emission	EX (nm)	EM (nm)	External QY (%)	PL lifetime (ms)
$(NH_4)_2SiF_6$	[44]	Cubic, *Fm*3*m*	–	472	633		
$(NH_4)_2TiF_6$	[44]	Trigonal, $P\bar{3}m1$	–	472	633		
$(NH_4)_2GeF_6$	[44]	Trigonal, $P\bar{3}m1$	–	472	633		
$(NH_4)_2SnF_6$	[44]	Trigonal, $P\bar{3}m1$	–	472	633		
Na_2SiF_6	[27, 37, 45]	Trigonal, *P*321	+	460	627		4.8–5.8
Na_2TiF_6	[27, 45]	Trigonal, *P*321 Trigonal, $P\bar{3}m1$	+ +	465 465	627 627		
Na_2GeF_6	[46]	Trigonal, *P*321	+	465	627		
Na_2SnF_6	[47]	Tetragonal, $P4_2/$ mnm	–	470	626		
K_2SiF_6	[27, 48–50]	Cubic, *Fm*3*m*	–	460	632	74	7.1
K_2TiF_6	[25, 51]	Trigonal, $P\bar{3}m1$	–	468	632		5.2–6.0
K_2GeF_6	[35, 38, 52]	Trigonal, $P\bar{3}m1$ Trigonal,$P6_3mc$	– +	475 475	632 632	54	
$K_2SnF_6 \cdot 6H_2O$	[53]	Orthorhombic, *Fddd*	+	470	630		
Rb_2SiF_6	[36]	Cubic, *Fm*3*m*	–	450	630		8.26
Cs_2SiF_6	[54]	Cubic, *Fm*3*m*	–	460	633		
Cs_2TiF_6	[55]	Trigonal, $P\bar{3}m1$	–	474	632		3.78
Cs_2GeF_6	[54]	Cubic, *Fm*3*m*	–	460	633		
Cs_2SnF_6	[47]	Trigonal, $P\bar{3}m1$	–	470	632		
$KNaSiF_6$	[27]	Orthorhombic, *Pmna*	+	460	629		5.2
$BaSiF_6$	[39, 41, 56]	Trigonal, $R\bar{3}m$	–	466	632		6.5
$BaTiF_6$	[42]	Trigonal, $R\bar{3}m$	–	466	630		
$BaGeF_6$	[57]	Trigonal, $R\bar{3}m$	–	460	634		5.89
$ZnSiF_6 \cdot 6H_2O$	[42, 43]	Trigonal, $R\bar{3}$	–	470	630		7.0
$ZnGeF_6 \cdot 6H_2O$	[42]	Trigonal, $R\bar{3}$	–	470	630		

– = *Low-intensity ZPL emission;* + = *high-intensity ZPL emission*

12.6 Morphology Control

Na_2SiF_6 micro-crystals with various morphologies were prepared using Na_2CO_3, SiO_2, and HF under microwave irradiation [58]. The thermodynamics of crystal growth along the prism faces and the kinetics of crystal growth along the prism were controlled by varying the microwave irradiation time and Na_2CO_3 concentration. As indicated in Fig. 12.16A, the Na_2SiF_6 crystals changed from long, thin hexagonal prisms to dodecagonal plates with pyramidal faces as Na_2CO_3 concentration was increased. The width of the hexagonal prisms increased with microwave irradiation

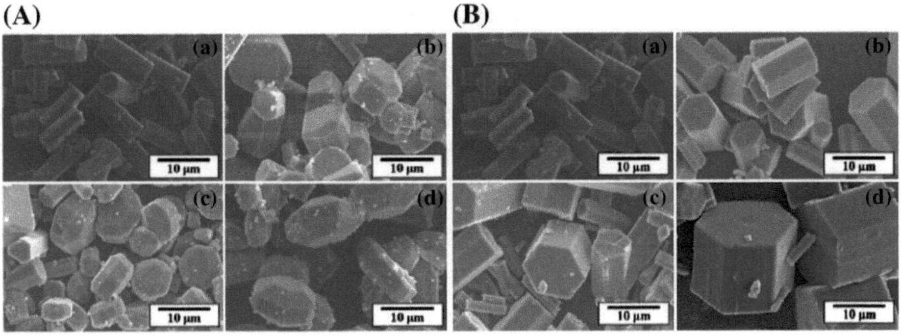

Fig. 12.16 SEM images of the Na_2SiF_6 crystals obtained at **A** various concentrations of Na_2CO_3: (**a**) 0.225 M, (**b**) 0.300 M, (**c**) 0.375 M, and (**d**) 0.450 M; at **B** several microwave irradiation times: (**a**) 4, (**b**) 6, (**c**) 8, and (**d**) 10 min. Reprinted from Ref. [58]. Copyright 2010 with permission from Elsevier

time, whereas the height remained almost constant (Fig. 12.16B). The change of the long, thin hexagonal prisms of Na_2SiF_6 crystals into thick hexagonal prisms under microwave irradiation thermal heating indicated that the activation energy of Na_2SiF_6 crystal growth in the prism morphology was greater than the activation energy of crystal growth in the basal or pyramidal morphologies.

Zhang et al. [59] proposed a facile polyol approach for the large-scale synthesis of water-soluble Na_2SiF_6 nano-tubes and nanos-pindles with polyacrylic acid (PAA) as a surfactant. The shape of the micelles formed in an aqueous solution, which plays a significant role in controlling the morphology of Na_2SiF_6 nano-structures, can be tuned by controlling the concentration of PAA (Fig. 12.17).

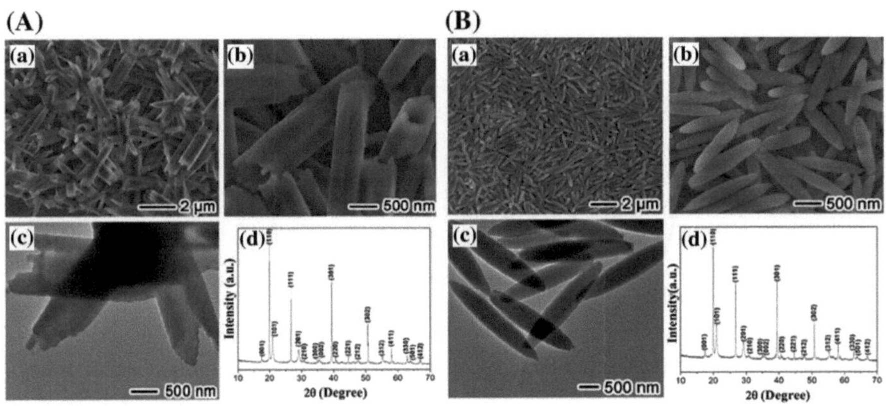

Fig. 12.17 **A** Na_2SiF_6 nano-tubes prepared at a PAA concentration of 12 g/L: (**a**) SEM image; (**b**) magnified SEM image; (**c**) TEM image. **B** Na_2SiF_6 nanospindles prepared at a PAA concentration of 2 g/L: (**a**) SEM image; (**b**) magnified SEM image; (**c**) TEM image. Reprinted from Ref. [59]. Copyright 2012 with permission from Elsevier

Hexagonal-shaped inverse micelles formed at a high PAA concentration of 12 g/L. Na_2SiF_6 nucleated and grew with hexagonal-shaped inverse micelles as the soft template. The hexagonal-shaped Na_2SiF_6 nano-tubes were finally prepared. When the concentration of PAA was decreased to 2 g/L, micelles with a spindle-like shape were obtained. As a result, the Na_2SiF_6 crystals possess the morphology of the spindle-shaped micelles.

12.7 Thermal Stability

Most Mn^{4+} fluoride phosphors exhibit high thermal stability [25, 33–57]. For example, the relative PL intensity of $K_2GeF_6:Mn^{4+}$ phosphor at 420 K remained approximately 96 % of that at 303 K, clearly indicating better thermal stability than the widely used YAG:Ce phosphor (88 % at 423 K) [35]. The temperature-dependent emission intensity of $K_2GeF_6:Mn^{4+}$ phosphor at 460-nm light excitation is presented in Fig. 12.18a. After increasing the temperature from 88 to 278 K, all of the emission lines became broader and red-shifted because of increased absorbed photons and enhanced vibration transition–coupling associated with the vibration modes of MnF_6^{2-}octahedron (Fig. 12.18b). The probability of nonradiative transition increased with temperature, and integrated PL intensity indicated thermal quenching, which can be fitted by $I_T/I_0 = [1 + D \exp(-E_a/kT)]^{-1}$, where I_0 is the intensity at $T = 0$ K, and D and activation energy E_a are the refined variables. Thermal activation energies of the nonradiative recombination obtained for Mn^{4+}fluoride red phosphors ranged from 0.62 to 0.93 eV. This range is three or four times higher than that of nitride compounds (approximately 0.25 eV) [37], thus confirming the excellent thermal stability of the fluoride phosphors. However, several fluoride compounds, $ZnXF_6 \cdot 6H_2O$(X = Si or Ge), were not stable at high

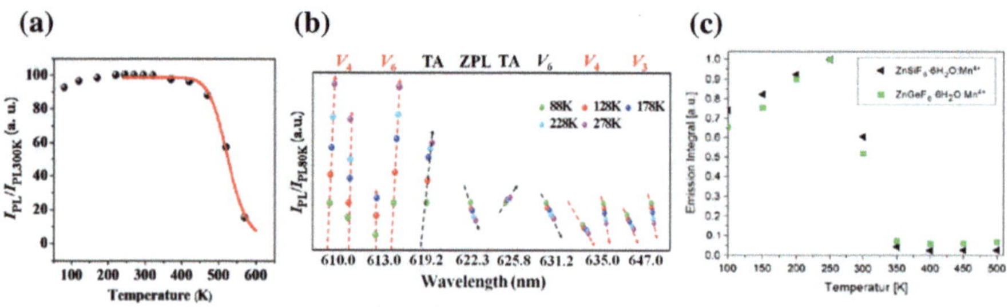

Fig. 12.18 **a** The wavelength position and relative intensity of each emission line obtained at different temperatures for $K_2GeF_6:Mn^{4+}$ red phosphor. **b** Integrated red PL intensity as a function of temperature (88–573 K) for $K_2GeF_6:Mn^4$. The *solid line* represents the fitting result with the expression of $I_T/I_0 = [1 + D \exp(-E_a/kT)]^{-1}$. Reproduced from Ref. [35] by permission of The Royal Society of Chemistry. **b** Integral emission intensity of $ZnSiF_6 \cdot 6H_2O:Mn^{4+}$ and $ZnGeF_6 \cdot 6H_2O:Mn^{4+}$ at temperatures of 100–500 K. Reprinted from Ref. [42], copyright 2013, with permission from Elsevier

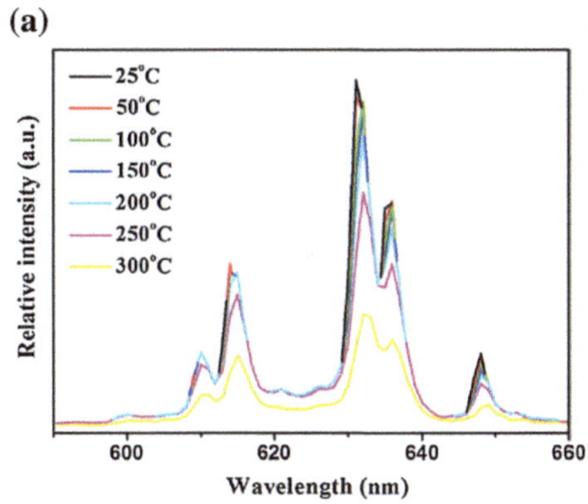

Fig. 12.19 **a** Temperature-dependent thermal luminescent spectrum of $Rb_2SiF_6:Mn^{4+}$ phosphor and **b** relative intensity of emission spectrum by integrating the spectral area. Reproduced from Ref. [36] by permission of The Royal Society of Chemistry

temperatures because of dehydration and decomposition of water molecules in the crystal structure as presented in Fig. 12.18c. Adachi et al. [43] observed that the PL intensity of $ZnSiF_6 \cdot 6H_2O:Mn^{4+}$ decreases with near-UV light illumination. Thus, these phosphors are not ideal for practical applications.

The excellent thermal stability of $Rb_2SiF_6:Mn^{4+}$ within 25–300 °C (298–573 K) is presented in Fig. 12.19. No emission peak position shift occurred when the temperature was increased. At 250 °C (523 K), the relative PL intensity of $Rb_2SiF_6:Mn^{4+}$ remained 100 % compared with that at 25 °C. The decrease in red-emission intensity of $Rb_2SiF_6:Mn^{4+}$ was observed at 300 °C with an integrated PL intensity of 70 %, but it was still higher than that of $K_2SiF_6:Mn^{4+}$ (50 %), $K_2TiF_6:Mn^{4+}$ (20 %), or $K_2GeF_6:Mn^{4+}$(15 %) [36].

12.8 High-Pressure Photoluminescence Properties

The optical properties of Mn^{4+} activated by fluoride phosphors strongly depend on the doping site symmetry and the CF strength because Mn^{4+} possesses a $3d^3$ electron configuration. Aside from chemical doping, pressure is also one of the state parameters that can effectively tune the CF strength and electronic structures by reducing the inter-atomic distances. A pressure effect is exerted on the structure and photoluminescence properties of $BaTiF_6:Mn^{4+}$ [60]. The $BaTiF_6:Mn^{4+}$crystal symmetry changed from trigonal at ambient pressure to monoclinic from 0.5 GPa and triclinic above 14 GPa, which corresponded to the distortion of $(Ti/Mn)F_6$ octahedral (Fig. 12.20a though c). The emission peaks monotonically red-shifted because of the reinforced CF strength within MnF_6 octahedral as pressure was increased (Fig. 12.20d) and blue-shifted back as pressure was decompressed

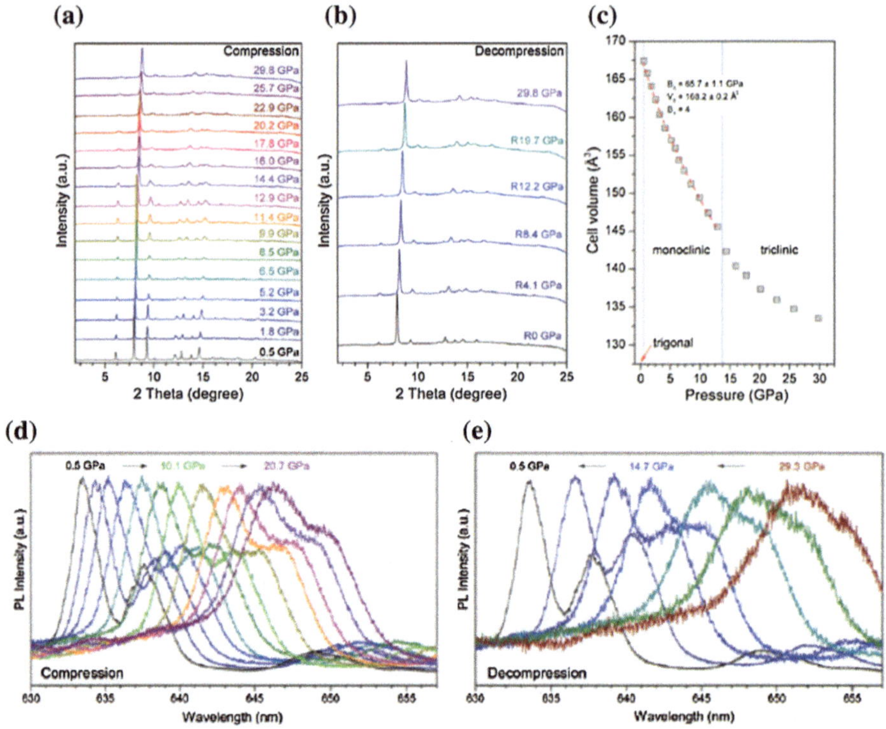

Fig. 12.20 In situ synchrotron powder XRD patterns of $BaTiF_6:Mn^{4+}$ at room temperature during **a** compression up to 29.8 GPa and **b** decompression. **c** Unit cell volume evolution of $BaTiF_6:Mn^{4+}$ as a function of pressure at room temperature. **d**, **e** Red-emission spectra of $BaTiF_6:Mn^{4+}$ as a function of pressure with excitation laser at a wavelength of 450 nm at room temperature. Reproduced from Ref. [60] by permission of The Royal Society of Chemistry

(Fig. 12.20e). The PL intensity and full width at half maximum of $BaTiF_6:Mn^{4+}$ and ruby $Al_2O_3:Cr^{3+}$ were compared using a neon pressure-transmission medium. $BaTiF_6:Mn^{4+}$ exhibited stable emission behavior under a high pressure up of to 40 GPa. This demonstration indicates that Mn^{4+}-doped fluorides are alternative candidates as a potential pressure gauge for high-pressure techniques.

12.9 Waterproof Properties

Although fluoride red phosphors exhibit excellent luminescent properties, their moisture instability limits their applications in the WLED industry. The degradation of fluoride phosphor can be attributed to the change-charged value of Mn ions or desorption of Mn from the crystal lattice because of the influence of environmental factors such as high temperature (HT) and high humidity (HH) as depicted in Fig. 12.21. A facile approach has been recently performed to coat fluoride red

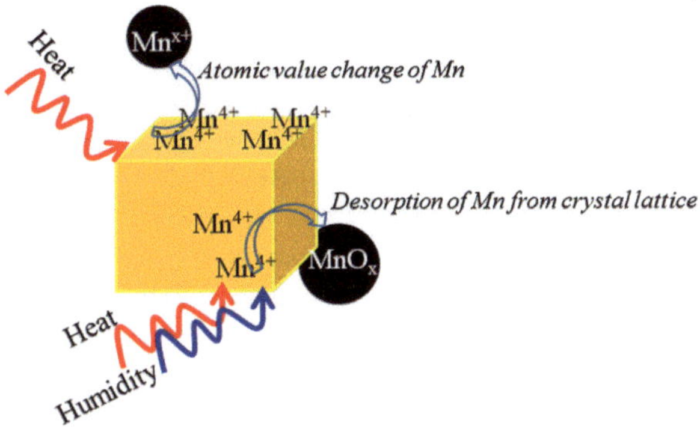

Fig. 12.21 Scheme of degradation of Mn^{4+}-activated fluoride phosphor

phosphors with a moisture-resistant organophosphate (OP) layer [61]. $K_2SiF_6:Mn^{4+}$ (KSFM) was prepared through the coprecipitation and esterification of P_2O_5 with alcohols. This two-step method was adopted to encapsulate the prepared phosphors using transition-metal ions as cross-linkers between alkylphosphates. The coated layers were abbreviated as (M/E/I)OP(Zn/Al/Ti) with M, E, and I, respectively, representing methanol, ethanol, and isopropanol; and Zn, Al, and Ti are the metal ions acting as cross-linkers between the alkylphosphates.

Decomposition of the KSFM surface under the electron beam of a transmission electron microscopy (TEM) system at high resolution indicated the unstable structure of KSFM. The 50- to 100-nm–thick coating layer on the phosphor surface was observed by way of TEM images [61]. The significant durability of KSFM-MOPAl in a hazardous environment with 85 % humidity and at 85 °C for 30 days was presented. The relative internal quantum efficiency ($RIQE = IQE_t/IQE_{t=0}$; approximately 85 %) and external quantum efficiency ($REQE = EQE_t/EQE_{t=0}$; approximately 87 %) of KSFM-MOPAl (0.05 M OP) were significantly higher than those of KSFM (*RIQE* approximately 57 % and *REQE* approximately 64 %) after 30 days under HH and HT conditions. The performances of WLEDs that were fabricated by combining blue InGaN chips, commercially available $YAG:Ce^{3+}$ yellow phosphor or β-$SiAlON:Eu^{2+}$ green phosphor, and KSFM-MOPAl red phosphor were evaluated for commercial application. Color-rendering indices ($R_a = 86$, $R_9 = 87$) and ($R_a = 86$, $R_9 = 93$) were observed for WLEDs using KSFM and KSFM-MOPAl, respectively. The chromaticity coordinates (0.4169, 0.4244) and (0.3995, 0.4091) near the black-body locus according to the CIE 1931 color spaces and the color temperatures at 3519 and 3766 K were also measured for WLEDs using KSFM and KSFM-MOPAl, respectively. The moisture-resistance values of warm WLEDs with commercially available KSFM (WLED/cKSFM), prepared-KSFM (WLED/pKSFM), KSFM-MOPAl (WLED/coated KSFM), and β-SiALON were also obtained. After 2016 h under HH and HT conditions and continuous application of 120 mA, the relative quantum efficiency of WLED/cKSFM significantly decreased (approximately 40 % of the initial intensity) compared with that of WLED/pKSFM (approximately 60 % of initial intensity). WLED-coated KSFM demonstrated excellent waterproof property by retaining 86 %

Table 12.2 Important photoelectric parameters for WLEDs using Mn^{4+}-doped fluoride phosphors

Blue LED (mA)	Phosphor	Chromaticity coordinates	CCTs (K)	Efficacy (lm/W)	R_a	R_9	Reference
350	$Na_2SiF_6:Mn^{4+}$	0.3126, 0.2951	6875	78	86	61	[37]
350	$K_2SiF_6:Mn^{4+}$	0.4169, 0.4244	3519	82	86	87	[61]
20	$K_2SiF_6:Mn^{4+}$	0.3828, 0.3331	3900	116	90	–	[34]
20	$K_2SiF_6:Mn^{4+}$	0.4034, 0.3869	3510	82	91	–	[33]
350	$K_2SiF_6:Mn^{4+}$ coated MOPAl	0.3995, 0.4091	3766	79	86	93	[61]
20	$K_2TiF_6:Mn^{4+}$	0.4575, 0.4124	2748	124	85	53	[25]
100	$K_2TiF_6:Mn^{4+}$	0.4550, 0.4172	2820	93	83	44	[25]
15	$K_2GeF_6:Mn^{4+}$	0.4016, 0.4495	3974	–	89	–	[35]
20	$BaTiF_6:Mn^{4+}$	0.3660, 0.3331	4073	121	93	–	[40]

of its initial quantum efficiency. The thermal stability and package results under HH and HT conditions indicate that the alkylphosphate-coated fluoride materials are promising red phosphors for warm WLED applications. Table 12.2 presents a comparison of the optoelectronic properties of WLEDs using blue LED chip (455 nm), yellow phosphor (YAG), and current red phosphor $A(I,II)M(IV)F_6:Mn^{4+}$.

12.10 Summary and Perspectives

The LED-lamp market is growing rapidly with the production of new devices that emit white light. In addition to backlighting color displays in cellular phones, other promising applications of WLEDs include household lighting. Consequently, the demand for warm WLEDs with high color reproducibility has increased. Commercially available red phosphors, such as rare-earth ion (Eu^{3+}, Ce^{3+})-doped oxy/nitride compounds, were employed to fabricate warm WLEDs. However, the broadband emission, low quantum efficiency, and high production cost limit the application of these red phosphors. Therefore, the narrow-band red-emission phosphors are a significant potential material to further improve the color reproducibility and luminous efficacy of warm WLEDs. Narrow-band red-emitting fluoride phosphors doped by Mn^{4+} exhibited potential applications in solid-state light sources because of their favorable thermal stability and low-cost preparation. Fluoride red phosphors $A_2MF_6:Mn^{4+}$ (A = K, Na; M = Si, Ge, Ti) and $BaMF_6:Mn^{4+}$ (M = Si, Ti) were prepared through conventional methods such as cation-exchange reaction, coprecipitation, and HT techniques. The morphology of fluoride compounds can be controlled by using a surfactant in the preparation process. Mn^{4+}-activated fluorides exhibited strong blue-light absorption and intense narrow-band red emission at approximately 630–650 nm, which is not sensitive to the human eyes. The structure and optical properties of fluoride phosphors can be tuned by employing high pressure or high temperature. $K_2SiF_6:Mn^{4+}$ red phosphors coated with hydrophobic alkylphosphate layers exhibited low thermal quenching and excellent moisture resistance with a relative external quantum efficiency of approximately 87 % under high temperature (HT, 85 °C) and high humidity (HH, 85 %) conditions. Approximately 86 % of the luminous efficacy of the WLED using coated $K_2SiF_6:Mn^{4+}$ was also retained after 2016 h of continuous current (120 mA) application at HT and HH conditions. The simple synthesis process, efficient red luminescence under blue-light excitation, low thermal quenching, and stable emission behavior under high pressure qualify Mn^{4+}-doped fluorides as potential candidates for optoelectronic materials to improve the color reproducibility of WLEDs.

References

1. Nakamura S, Fasol G (1997) The blue diode laser: GaN-based light emitters and lasers. Springer, Berlin, Germany
2. Nakamura S, Snoh M, Mukai T (1993) High-power InGaN/GaN double-heterostructure violet light emitting diodes. Appl Phys Lett 62:2390
3. Nakamura S, Mukai T, Senoh M (1994) Candela-class high-brightness InGaN/AlGaN double-heterostructure blue-light-emitting diodes. Appl Phys Lett 64:1687
4. Xie RJ, Hirosaki N, Kimura N, Sakuma K, Mitomo M (2007) 2-phosphor-converted white light-emitting diodes using oxynitride/nitride phosphors. Appl Phys Lett 90:191101
5. Sato Y, Takahachi N, Sato S (1996) Full-color fluorescent display devices using a near-UV light-emitting diode. Jpn J Appl Phys 35:L838
6. Mueller-March R, Mueller GO, Krames MR, Trottier T (2002) High-power phosphor-converted light-emitting diodes based on III-Nitrides. IEEE J Select Topics Quantum Electron 8:339
7. Taniyasu Y, Kasu M, Makimoto T (2006) An aluminium nitride light-emitting diode with a wavelength of 210 nanometres. Nature 441:325
8. Nishida T, Ban T, Kobayashi N (2003) High-color-rendering light sources consisting of a 350-nm ultraviolet light-emitting diode and three-basal-color phosphors. Appl Phys Lett 82:3817
9. Solid-State Lighting Research and Development Multi-Year Program Plan, (2013) US Department of Energy, Energy Efficiency and Renewable Energy
10. Haranath D, Chander H, Sharma P, Singh S (2006) Enhanced luminescence of $Y_3Al_5O_{12}:Ce^{3+}$ nanophosphor for white light-emitting diodes. Appl Phys Lett 89:173118
11. Isobe T (2006) Low-temperature wet chemical syntheses of nanocrystal phosphors with surface modification and their characterization. Phys Stat Sol 203:2686
12. Jang HS, Won YH, Jeon DY (2009) Improvement of electroluminescent property of blue LED coated with highly luminescent yellow-emitting phosphors. Appl Phys B 95:715
13. Park JK, Choi KJ, Yeon JH, Lee SJ, Kim CH (2005) Application of Ba^{2+} • Mg^{2+} Co-doped Sr2SiO4: Eu Yellow Phosphor for White-Light-Emitting Diodes. J Electrochem Soc 152: H121
14. Schlotter P, Schmidt R, Schneider J (1997) Luminescence conversion of blue light emitting diodes. Appl Phys A 64:417
15. Pust P, Weiler V, Hecht C, Tücks A, Wochnik AS, Hen AK, Wiechert D, Scheu C, Schmidt PJ, Schnick W (2014) Narrow-band red-emitting $Sr[LiAl_3N_4]:Eu^{2+}$ as a next-generation LED-phosphor material. Nature Mater 13:891

16. Krames MR, Shchekin OB, Mueller-Mach R, Mueller GO, Zhou L, Harbers G, Crafor MG (2007) Status and future of high-power light-emitting diodes for solid-state lighting. J Disp Tech 3:160
17. Smet PF, Parmentier AB, Poelman D (2011) Selecting conversion phosphors for white light-emitting diodes. J ElectrochemSoc 158:R37
18. George NC, Denault KA, Seshadri R (2013) Phosphors for solid-state white lighting. Annu Rev Mater Res 43:481
19. Spassky D, Mikhailin V, Nazarov M, Ahmad-Fauzi MN, Zhbanov A (2012) Luminescence and energy transfer mechanisms in $CaWO_4$ single crystals. J Lumin 132:2753
20. Catti M, Gazzaoni G, Ivaldi G (1983) Structures of twinned β-Sr_2SiO_4 and of α-$Sr_{1.9}Ba_{0.1}SiO_4$. Acta CrystC 39:29
21. Park WJ, Jung MK, Yoon DH (2007) Influence of Eu^{3+}, Bi^{3+} co-doping content on photoluminescence of YVO_4 red phosphors induced by ultraviolet excitation. Sensors and Actuators B 126:324
22. Blasse G, Bril A (1969) Luminescence of phosphors based on host lattices ABO_4 (A is Sc, In; B is P, V, Nb). J Chem Phys 50:2974
23. Shionoya S, Yen WM, Yamamoto H (1998) Phosphors handbook. CRC Press, New York, NY, USA
24. Tanabe Y, Sugano S (1954) On the absorption spectra of complex ions I. J Phys Soc Jpn 9:753
25. Zhu HM, Lin CC, Luo W, Shu ST, Liu ZG, Wang M, Kong JT, Ma E, Cao Y, Liu RS, Chen XY (2014) Highly efficient non-rare-earth red emitting phosphor for warm white light-emitting diodes. Nat Commun 5:4312
26. Takahashi T, Adachi S (2008) Mn^{4+}-Activated Red Photoluminescence in K_2SiF_6 Phosphor. J Electrochem Soc 155:E183
27. Adachi S, Abe H, Kasa R, Arai T (2012) Synthesis and Properties of Hetero-Dialkaline Heafluorosilicate Red Phosphor $KNaSiF_6$:Mn^{4+}. J Electrochem Soc 159:J34
28. Xu YK, Adachi S (2011) Properties of Mn^{4+}-Activated Hexafluorotitanate Phosphors. J Electrochem Soc 158:J58
29. Adachi S, Takahashi T (2009) Photoluminescent properties of K_2GeF_6:Mn^{4+} red phosphor synthesized from aqueous HF/$KMnO_4$ solution. J Appl Phys 106:013516
30. Brik MG, Srivastava AM (2013) On the optical properties of the Mn^{4+} ion in solids. JLumin 133:69
31. Du MH (2014) Chemical trends of Mn^{4+} emission in solids. J Mater Chem C 2:2475
32. Oh JH, Kang H, Eo YJ, Park HK, Do YR (2015) Synthesis of narrow-band red-emitting K_2SiF_6:Mn^{4+} phosphors for a deep red monochromatic LED and ultrahigh color quality warm-white LEDs. J Mater Chem C 3:607
33. Liao C, Cao R, Ma Z, Li Y, Dong G, Sharafudeen KN, Qiu J (2013) Synthesis of K_2SiF_6:Mn^{4+} phosphor from SiO_2 powders via redox reaction in HF/$KMnO_4$ solution and their application in warm-white LED. J Am Ceram Soc 96:3552
34. Lv L, Jiang X, Huang S, Chen X, Pan Y (2014) The formation mechanism, improved photoluminescence and LED applications of red phosphor K_2SiF_6:Mn^{4+}. J Mater Chem C 2:3879
35. Wei LL, Lin CC, Fang M-H, Brik MG, Hu SF, Jiao H, Liu RS (2015) A low-temperature co-precipitation approach to synthesize fluoride phosphors K_2MF_6:Mn^{4+} (M = Ge, Si) for white LED applications. J Mater Chem C 3:1655
36. Fang M-H, Nguyen H-D, Lin CC, Liu RS (2015) Preparation of a novel red Rb_2SiF_6:Mn^{4+} phosphor with high thermal stability through a simple one-step approach. J Mater Chem C 3:7277
37. Nguyen H-D, Lin CC, Fang MH, Liu RS (2014) Synthesis of Na_2SiF_6:Mn^{4+} red phosphors for white LED applications by co-precipitation. J Mater Chem C 2:10268

38. Wei LL, Lin CC, Wang YY, Fang MH, Jiao H, Liu RS (2015) Photoluminescent evolution induced by structural transformation through thermal treating in the red narrow-band phosphor $K_2GeF_6:Mn^{4+}$. ACS Appl Mater Interfaces 7:10656
39. Jiang X, Pan Y, Huang S, Chen X, Wang J, Liu G (2014) Hydrothermal synthesis and photoluminescence properties of red phosphor $BaSiF_6:Mn^{4+}$ for LED applications. J Mater Chem C 2:2301
40. Jiang X, Chen Z, Huang S, Wang J, Pan Y (2014) A red phosphor $BaTiF_6:Mn^{4+}$: reaction mechanism, microstructures, optical properties, and applications for white LEDs. Dalton Trans 43:9414
41. Sekiguchi D, Nara J-I, Adachi S (2013) Photoluminescence and Raman scattering spectroscopies of $BaSiF_6:Mn^{4+}$ red phosphor. J ApplPhys 113:183516
42. Kubus M, Enseling D, Justel T, Meyer H-J (2013) Synthesis and luminescent properties of red-emitting phosphors: $ZnSiF_6{\cdot}6H_2O$ and $ZnGeF_6{\cdot}6H_2O$ doped with Mn^{4+}. J Lumin 137:88
43. Hoshino R, Adachi S (2013) Optical spectroscopy of $ZnSiF_6{\cdot}6H_2O:Mn^{4+}$ red phosphor. J ApplPhys 114:213502
44. Xu YK, Adachi S (2012) Photoluminescence and Raman scattering spectra in $(NH_4)_2XF_6$: Mn^{4+}(X = Si, Ge, Sn, and Ti) red phosphors. J Electrochem Soc 159:E11
45. Wang Z, Liu Y, Zhou Y, Zhou Q, Tan H, Zhang Q, Peng J (2015) Red-emitting phosphors $Na_2XF_6:Mn^{4+}$ (X = Si, Ge, Ti) with high colour-purity for warm white-light emitting Diodes. RSC Adv 5:58136
46. Adachi S, Xu YK (2009) Properties of $Na_2SiF_6:Mn^{4+}$ and $Na_2GeF_6:Mn^{4+}$ red phosphors synthesized by wet chemical etching. J Appl Phys 105:013525
47. Adachi S, Arai Y (2011) Optical properties of Mn^{4+}-activated Na_2SnF_6 and Cs_2SnF_6 red phosphors. J Lumin 131:2652
48. Kasa R, Adachi S (2012) Red and deep red emissions from cubic $K_2SiF_6:Mn^{4+}$ and hexagonal K_2MnF_6 synthesized in HF/$KMnO_4$/KHF_2/Si Solutions. J Electrochem Soc 159:J89
49. Kim M, Park WB, Chang BB, Kim H, Sohn K-S (2015) Radiative and non-radiative decay rate of $K_2SiF_6:Mn^{4+}$ phosphors. J Mater Chem C 3:5484
50. Sijbom HF, Joos JJ, Martin LIDJ, Eeckhout KV, Poelman D, Smet PF (2016) Luminescent Behavior of the $K_2SiF_6:Mn^{4+}$ Red Phosphor at high fluxes and at the microscopic level. ECS J Sol Stat Sci Technol 5:R3040
51. Lv L, Chen Z, Liu G, Huanga S, Pan Y (2015) Optimized photoluminescence of red phosphor $K_2TiF_6:Mn^{4+}$ synthesized at room temperature and its formation mechanism. J Mater Chem C 3:1935
52. Ogasawara K, Novita M (2012) Comparative study of multiplet structures of Mn^{4+} in K_2SiF_6, K_2GeF_6, and K_2TiF_6 based on first-principles configuration–interaction calculations. Jpn J ApplPhys 51:022604
53. Adachi S, Arai Y (2011) Photoluminescent properties of $K_2SnF_6{\cdot}H_2O:Mn^{4+}$ hydrate phosphor. J Electrochem Soc 158:J81
54. Adachi S, Arai Y (2011) Optical transitions and internal vibronic frequencies of MnF_6^{2-} ions in Cs_2SiF_6 and Cs_2GeF_6 red phosphors. J Electrochem Soc 158:J179
55. Zhou Q, Zhou Y, Liu Y, Wang Z, Chen G, Peng J, Yan J, Wu M (2015) A new and efficient red phosphor for solid-state lighting: $Cs_2TiF_6:Mn^{4+}$. J Mater Chem C 2015:9615
56. Zhou Y, Zhou Q, Liu Y, Wang Z, Yang H, Wang Q (2016) Hydrothermal synthesis and luminescent properties of $BaTiF_6:Mn^{4+}$ red phosphor for LED backlighting. Mater Res Bull 73:14
57. Zhou Q, Zhou Y, Liu Y, Luo L, Wang Z, Peng J, Yan J, Wu M (2015) A new red phosphor $BaGeF_6:Mn^{4+}$: hydrothermal synthesis, photo-luminescence properties, and its application in warm white LED devices. J Mater Chem C 3:3055
58. Jeong H-L, Huh Y-D (2010) Synthesis of hexagonal prisms and hexagonal plates of Na_2SiF_6 microcrystals. Mater Lett 64:1816

59. Liu J, Du N, Wu P, Wang J, Zhang H, Yang D (2012) Large-scale synthesis of water-soluble Na_2SiF_6 nanotubes with polyacrylic acid as a surfactant. Mater Res Bull 47:3923
60. Wang Y, Wen T, Tang L, Yang L, Yang W, Zhao Y (2015) Impact of hydrostatic pressure on the crystal structure and photoluminescence properties of Mn^{4+}-doped $BaTiF_6$ red phosphor. Dalton Trans 44:7578
61. Nguyen H-D, Lin CC, Liu RS (2015) Waterproof alkyl phosphate coated fluoride phosphors for optoelectronic materials. Angew Chem Int Ed 54:10862

Chapter 13
Novel Phosphors for UVLEDs

Irish Valerie B. Maggay and Wei-Ren Liu

Abstract Phosphor is a key component in the recent advances in light-emitting diodes. It converts the light emitted by ultraviolet or blue LEDs into visible light. In the past decade, scientists and researchers have developed various phosphors for general illumination and optical displays, and they continue to develop new phosphors for w-LEDs in order to overcome conventional lighting systems. The phosphors should have strong absorption of the light emitted by the UVLEDs and be able to efficiently downconvert it into longer wavelengths (i.e., visible light). Among the number of developed phosphors, silicate-, borate-, phosphate-, and nitride-based phosphors for have been studied for UVLED applications. This chapter will discuss novel phosphors for UVLEDs according to their chemical compositions and their pertinent applications in white light–emitting diodes.

13.1 Introduction

Phosphor or, also known as a luminescent material, absorbs the light emitted by ultraviolet (UV) (360–410 nm) or blue LEDs (420–480 nm) and downconverts it into longer wavelengths; hence, it is an essential component in light-emitting diodes (LEDs). Currently, there exist a huge number of known phosphors; however, only few are suitable for LED applications. Thus, novel phosphors with high chemical stability, strong absorption in the UV region, low thermal quenching, and high conversion efficiency are being pursued.

Phosphors can be classified according to their chemical components, and some of the recent phosphors for UVLEDs are based *on silicates, borates, phosphates, and nitrides*. In addition, phosphors can be classified depending on their emission colors (i.e., blue-, green-, and red-emitting phosphors). In this chapter, we will

I.V.B. Maggay · W.-R. Liu (✉)
Energy and Opto-Electronics Materials Laboratory,
Department of Chemical Engineering, Chung Yuan Christian University,
Chung Li 32023, Taiwan
e-mail: WRLiu1203@gmail.com

R.S. Liu (ed.), *Phosphors, Up Conversion Nano Particles, Quantum Dots and Their Applications*, DOI 10.1007/978-3-662-52771-9_13

discuss the photoluminescence properties of phosphors excited by UVLEDs based on their chemical compositions.

13.2 Silicate-Based Phosphors

Owing to their high chemical and thermal stability, low cost, and good optical properties, scientists have taken a great interest in investigating and developing silicate-based phosphors. Silicate phosphors can be categorized into a binary system [e.g., $MO–SiO_2$ (M = Ca, Sr, Ba) and a ternary system [e.g., $MO–M'M–SiO_2$ (M = Ca, Sr, Ba; M′ = Li, Mg, Al, Zn, etc.) [1]. Among the number of silicate-based phosphors used in UVLEDs, some of the commercially available binary system include $M_2SiO_4:Eu^{2+}$ (M = Ba, Ca, Sr), $Sr_3SiO_5:Eu^{2+}$, and $(M_{1-x}M'_x)_2SiO_4:Eu^{2+}$ (M/M′ = Mg, Ca Sr, Ba) due to their strong crystal fields that result in red-shifted luminescence. In the ternary system, the well-studied and commercially known phosphors include $M_2MgSi_2O_7:Eu^{2+}$, $Ba_2(Mg, Zn)Si_2O_7:Eu^{2+}$, $M_3MgSi_2O_8:Eu^{2+}$, Mn^{2+} (M = Ba, Ca, Sr), and $Ca_8Mg(SiO_4)_4Cl_2:Eu^{2+}$, Mn^{2+} [2]. In this chapter, we will discuss recent developments on novel silicate-based phosphors that exhibit potential applications for UVLEDs.

The pursuit for new phosphors has recently focused on compounds that form complex solid solutions, which provide a mechanism to tune the luminescent properties. The emission color of the binary system—$M_{1-x}M'_x)_2SiO_4:Eu^{2+}$ (M/M′ = Mg, Ca Sr, Ba)—can be tuned from violet to red-orange, and the thermal stability is dependent on the components and ratio of M/M′. The bond distance of M/M′ = O and the volume of the unit cell greatly affects the optical properties of this system. A longer M/M′ = O bond distance and large unit cell provide a weaker crystal field, which results in a blue-shift emission. Moreover, a distortion of the crystal structure can also provide tunable optical properties of the phosphor [3].

Liu et al. [4] studied the structure and photoluminescence properties of green-emitting $NaBaScSi_2O_7:Eu^{2+}$ phosphors. Figure 13.1 shows the photoluminescence excitation (PLE), photoluminescence emission (PL), and ultraviolet-visible diffuse reflection (UV-Vis) spectra of the $NaBaScSi_2O_7:0.01Eu^{2+}$. As shown in the reflection spectrum, the phosphor exhibits a strong and broad absorption band in the UV region, which is ascribed to the $4f \rightarrow 5d$ transition of Eu^{2+}. When the sample is monitored at 501 nm, the excitation spectrum shows a broad band from 250 to 460 nm, which is in agreement with the UV-Vis spectrum. When $NaBaScSi_2O_7:0.01Eu^{2+}$ is excited at 365 nm, the sample exhibits a broad band from 450 to 600 nm with maximum peak at 501 nm displaying green emission. The results indicate that the phosphor is suitable for UVLEDs applications. The inset displays how the phosphor appears after synthesis.

The thermal luminescence of $NaBaScSi_2O_7:0.01Eu^{2+}$ was also studied in this work. Figure 13.2a shows the temperature-dependent emission spectra of the sample excited at 365 nm and (b) displays the comparison of the relative emission intensities and peak wavelengths as a function of temperature of

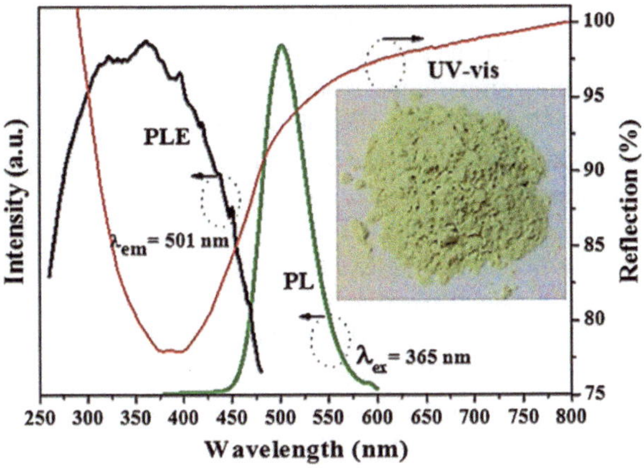

Fig. 13.1 Emission, excitation, and UV-Vis diffuse reflectance of $NaBaScSi_2O_7$:0.01Eu^{2+} phosphor. The *inset* shows the appearance of the as-obtained sample. Reproduced from Ref. [4] by permission of The Royal Society of Chemistry

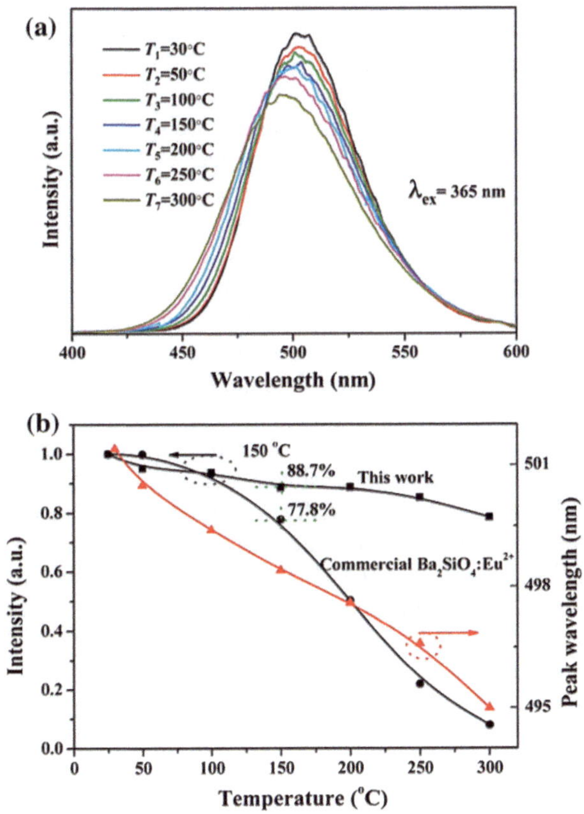

Fig. 13.2 **a** Temperature-dependent emission spectra of $NaBaScSi_2O_7$:0.01Eu^{2+} from 30 to 300 °C at 365 nm excitation and **b** comparison of the relative intensities of $NaBaScSi_2O_7$:0.01Eu^{2+} and commercial Ba_2SiO_4:Eu^{2+}, as well as the emission peaks of $NaBaScSi_2O_7$:0.01Eu^{2+}, all as a function of temperature. Reproduced from Ref. [4] by permission of The Royal Society of Chemistry

$NaBaScSi_2O_7$:0.01Eu^{2+} and commercial Ba_2SiO_4:Eu^{2+}. The relative emission intensity decreased gradually as the temperature is increased from room temperature to 300 °C. When the temperature reached 150 °C, the relative intensities of $NaBaScSi_2O_7$:0.01Eu^{2+} and commercial phosphor Ba_2SiO_4:Eu^{2+} decreased to 88.7 and 77.8 %, respectively. The emission wavelength shows blue shifting with increasing temperature, which is due to the thermally active phonon-assisted excitation from the excited states of the lower-energy emission band to the higher-energy emission band in the excited state of Eu^{2+}.

Liu et al. [4] also investigated the electroluminescence (EL) of $NaBaScSi_2O_7$:0.01Eu^{2+}. Figure 13.3 shows the EL spectrum of the combination of UV-chip emission at 370 nm, a blue-emission band corresponding to $BaMgAl_{10}O_{17}$:Eu^{2+} with maximum at approximately 450 nm, and green emission from $NaBaScSi_2O_7$:Eu^{2+} at approximately 500 nm and red emission corresponding to CaAlSiN3:Eu^{2+} with a maximum at 610 nm. The color-rendering index (R_a), color temperature was 2528 K (warm light), and the CIE coordinates of the packaged LED were 86.5, 2528 K (warm light), and (0.353, 0.324), respectively. The obtained results highly indicate that $NaBaScSi_2O_7$:Eu^{2+} is a potential phosphor for UVLEDs.

Ternary system $M_3MgSi_2O_8$ has received great attention due to its high efficacy in UVLED applications. Yang et al. [5] synthesized and studied the luminescence and energy-transfer properties of Eu^{2+}- and Mn^{2+}- activated $BaCa_2MgSi_2O_8$ novel phosphor. As seen in Fig. 13.3, when the samples of $BaCa_2MgSi_2O_8$:xEu^{2+} (x = 0.01–0.12) were monitored at 446 nm, the excitation spectra displayed a broad band ranging from 240 to 430 nm with a maximum peak at 360 nm, which matches well with the emission of the UVLED chip. The excitation spectra can be attributed to the parity-allowed $4f^7(^8S_{7/2}) \rightarrow 4f^65d$ transitions of Eu^{2+}. On excitation at 360 nm, the samples present a broad asymmetrical emission band from 390 to 600 nm with maximum peak at 446 nm. The asymmetrical emissions of Eu^{2+} suggest that it occupies both the Ba^{2+} and Ca^{2+} sites. Moreover, the emission spectrum is red-shifted, which indicates that the crystal field is strong.

Yang et al. [5] also introduced Mn^{2+} into the system, as observed in Fig. 13.4, at 360 nm excitation. When the concentration of Mn^{2+} is varied from 0 to 30 mol%, it is evident that the emission intensity of Eu^{2+} is decreased. This phenomenon is due to the energy transfer between the sensitizer (i.e., Eu^{2+}) and activator (i.e., Mn^{2+}). The emission intensity of Mn^{2+} initially increased until it reached its optimum concentration of 10 mol%. Further increase results in a decrease of emission intensity of Mn^{2+} which is due to the Mn^{2+}–Mn^{2+} internal concentration quenching. Moreover, the addition of Mn^{2+} caused a distortion of the crystal lattice, which resulted in a larger crystal-field splitting of Mn^{2+} 3d energy levels, which allows the phosphor to be tuned.

Figure 13.5 presents the CIE chromaticity diagram with its corresponding phosphors under the UV box and the fabricated w-LED using a 380-nm chip. The figure shows the

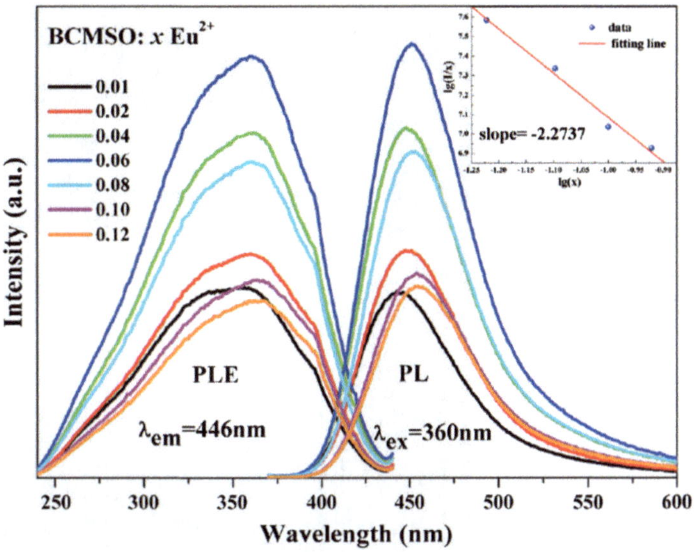

Fig. 13.3 Emission and excitation spectra of $BaCa_2MgSi_2O_8$:xEu^{2+} (x = 0.01–0.12) excited at 360 nm and monitored at 446 nm, respectively. The *inset* shows the log (1/x) versus log(x) fitting of $BaCa_2MgSi_2O_8$:xEu^{2+} samples. Reproduced from Ref. [5] by permission of The Royal Society of Chemistry

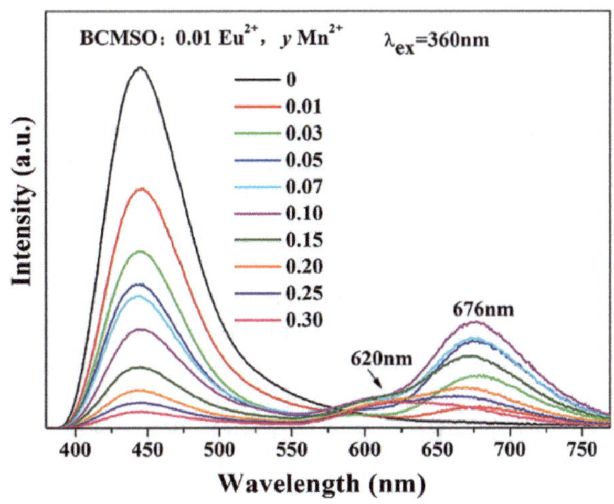

Fig. 13.4 Emission spectra of $BaCa_2MgSi_2O_8$:0.01Eu^{2+}, yMn^{2+} (y = 0–30 mol%) phosphors as a function of Mn^{2+} doping concentration. Reproduced from Ref. [5] by permission of The Royal Society of Chemistry

color tunability of the phosphors as the concentration of Mn^{2+} is varied. Moreover, the quantum efficiencies of $BaCa_2MgSi_2O_8$:0.01Eu^{2+} and of $BaCa_2MgSi_2O_8$:0.01Eu^{2+}, 0.20 Mn^{2+} were compared with commercial $BaMgAl_{10}O_{17}$:Eu^{2+} and were calculated to be 93.5, 53.6, and 16.5 %, respectively.

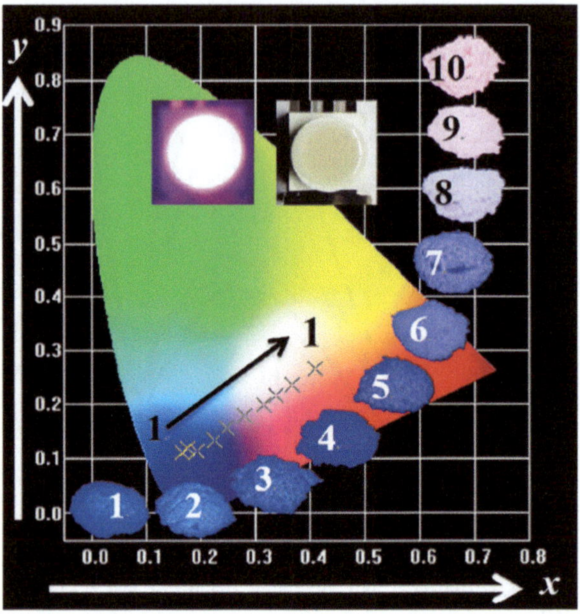

Fig. 13.5 Chromaticity diagram of $BaCa_2MgSi_2O_8$:0.01Eu^{2+}, yMn^{2+} (y = 0–30 mol%) phosphors with corresponding images under the UV box and the fabricated $BaCa_2MgSi_2O_8$:0.01Eu^{2+}, 0.20 Mn^{2+} phosphor. Reproduced from Ref. [5] by permission of The Royal Society of Chemistry

13.3 Borate-Based Phosphors

Borate compounds have long been studied due to their variety of structure type, transparency to a wide range of wavelengths, high laser-damage tolerance, and high optical quality [6]. Due to these qualities, they are generally used in nonlinear optics as well as piezo- and scintillation techniques and phosphors for w-LEDs [7]. Moreover, borate compounds are known for their low synthesizing temperature and high physical and chemical synthesis [8]. In the recent study by Maggay et al. [7] $Li_6Lu(BO_3)_3$:Ce^{3+}, Tb^{3+} is synthesized at a low temperature of 700 °C, and they were able to obtain high purity. Liu et al. [8] synthesized $NaCaBO_3$ at 850 °C; Reddy et al. [9] synthesized and obtained pure phased $KCa_4(BO_3)_3$ at 800 °C; and Sun et al. [10] $Sr_2B_2O_5$:Ce^{3+}, Tb^{3+} synthesized at 800 °C. Due to this, scientists have drawn their attention to borate-based phosphors. Some of the commercially available borate-based phosphors applied in UVLEDs are $LnBO_3$:Ce^{3+}, Tb^{3+} (Ln = Sc, Y, La, Gd, Lu), $Na_2(RE_{1-y-z}Ce_yTb_z)_2B_2O_7$ (RE = La, Y, Gd, Lu, Sc), and $(Ln_{1-x}Eu_x)M_3B_4O_{12}$ (Ln = Y/Gd) (M = Al, Sc, Ga) [2]. Notice that trivalent lanthanide (Ln) ions—such as Eu^{3+}, Ce^{3+}, Tb^{3+}, Dy^{3+}, Er^{3+}, and Gd^{3+} etc.—are generally used as activators for borate-based phosphors, and they usually replace rare-earth elements (RE) such as Y^{3+}, Lu^{3+}, La^{3+}, Gd^{3+}, and Sc^{3+}. Other studies have used divalent RE as activators such as $Ba_2Mg(BO_3)_2$:Eu^{2+}, Mn^{2+} [11], and $KSr_4(BO_3)_3$:Eu^{2+} [12].

Due to the forbidden 4f → 4f transitions, some lanthanide ions, such as Tb^{3+} and Eu^{3+}, display narrow excitation peaks in the UV region, which is not suitable for

UVLED application. Tb^{3+} is co-doped with Ce^{3+}, whereas Eu^{3+} is co-doped with Bi^{3+} or Sm^{3+} to enhance their luminescence properties and provide strong excitation peaks in the UV region. Some of the recent studies include the synthesis of $Li_6Lu(BO_3)_3$:Ce^{3+}, Tb^{3+} [7], $Sr_6YSc(BO_3)_6$:Ce^{3+}, Tb^{3+} [13], $Sr_3Gd(BO_3)_3$:Ce^{3+}, Tb^{3+} [14], $Ba_3Y_2(BO_3)_4$:Ce^{3+}, Tb^{3+} [15], $NaSr_4(BO_3)_3$:Ce^{3+}, Tb^{3+} [16], and ZnB_2O_4:Bi^{3+}, Eu^{3+} [17].

Figure 13.6 presents the PL and PLE of (a) $Li_6Lu(BO_3)_3$:Ce^{3+}, (b) $Li_6Lu(BO_3)_3$:Tb^{3+}, and (c) $Li_6Lu(BO_3)_3$:Ce^{3+}, Tb^{3+} [7]. We can see that with singly doped Tb^{3+}, the excitation spectrum is comprised of weak and narrow peaks in the UV region, which will result in inefficient absorption of UVLED. However, when Ce^{3+} is co-doped with Tb^{3+}, the excitation spectrum was broadened due to parity-allowed transition of Ce^{3+}, which makes the phosphor suitable for UVLED applications.

Huang and Chen [18] developed a novel single-composition trichromatic white light by co-doping Ce^{3+}, Mn^{2+}, and Tb^{3+} into a $Ca_3Y(GaO(BO_3)_4$ host phosphor. When $Ca_3Y(GaO(BO_3)_4$:$0.01Ce^{3+}$, $0.03Mn^{2+}$, yTb^{3+} (y = 0–0.10) phosphors were excited at 365 nm, they showed a series of emission spectra comprised of six broad bands in the visible region: one peak located at 409 nm due to the $5d^1 \rightarrow 4f^1$ transition of Ce^{3+}; four peaks at the green region with peaks located at 488, 544, 581, and 620 nm ascribed to the $^5D_4$ to $^7F_6$, $^7F_5$, $^7F_4$, and $^7F_3$ transitions of Tb^{3+}; and one broad orange red–emitting band situated at 589 nm attributed to the d-level spin-forbidden transition of Mn^{2+} as presented in Fig. 13.7. Moreover, it can be deduced from the figure that no energy transfer transpires between Tb^{3+} and Mn^{2+}

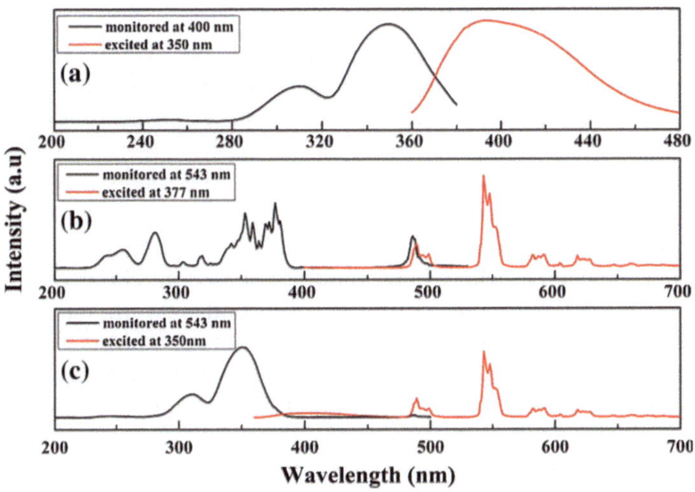

Fig. 13.6 Excitation and emission spectra of **a** $Li_6Lu(BO_3)_3$:Ce^{3+} (λ_{ex} = 350 nm, λ_{em} = 400 nm), **b** $Li_6Lu(BO_3)_3$:Tb^{3+} (λ_{ex} = 377 nm, λ_{em} = 543 nm), and **c** $Li_6Lu(BO_3)_3$:Ce^{3+}, Tb^{3+} (λ_{ex} = 350 nm, λ_{em} = 543 nm). Reproduced from Ref. [7] by permission of The Royal Society of Chemistry

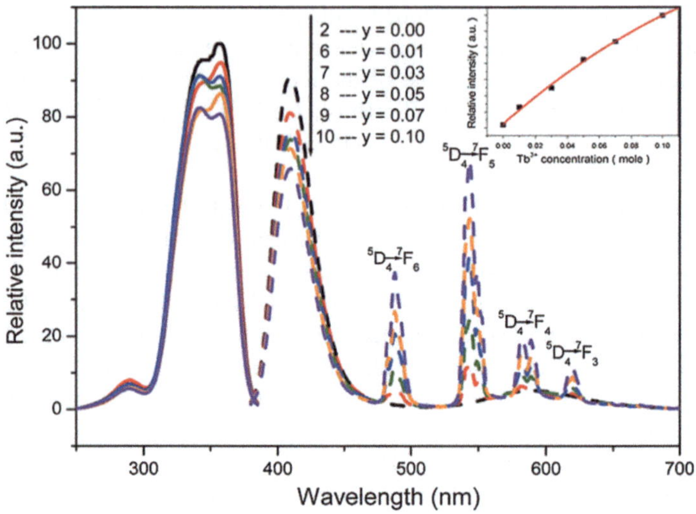

Fig. 13.7 Emission and excitation spectra of $Ca_3Y(GaO(BO_3)_4$:0.01Ce^{3+}, 0.03Mn^{2+}, yTb^{3+} (y = 0–0.10). The *inset* displays the relative emission intensity as a function of Tb^{3+} concentration when excited at 365 nm. Reprinted from Ref. [18] by permission of American Chemical Society

ions because the emission intensity of Mn^{2+} did not increase or decrease with the increasing Tb^{3+} dopant concentration. However, the emission intensity of Ce^{3+} decreased when Tb^{3+} concentration is increased; thus, the energy absorbed by Ce^{3+} ions were transferred to its neighboring Tb^{3+} ions.

$Ca_3Y(GaO(BO_3)_4$:0.01Ce^{3+}, 0.03Mn^{2+}, 0.07 Tb^{3+} phosphor was combined with a 365-nm UV chip to exhibit its potential application as a w-LED. Figure 13.8 shows the electroluminescence (EL) spectrum of the fabricated w-LED (driven by 350 mA) as well as the chromaticity coordinates and color temperature, which were found to be (0.31, 0.33) and 6524 K, respectively. The results indicate that Ca_3Y $(GaO(BO_3)_4$:0.01Ce^{3+}, 0.03Mn^{2+}, 0.07 Tb^{3+} phosphor demonstrates a great potential for UVLED applications.

A single-composition white light–emitting phosphor, $NaSrBO_3$:Ce^{3+}, Mn^{2+}, Tb^{3+}, was developed by Lin and colleagues [19]. In their work, they evaluated the applicability of their synthesized phosphor, $NaSrBO_3$:Ce^{3+}, Mn^{2+}, Tb^{3+}, by combining it with a UVLED chip (λ_{em} = 365 nm) to produce a w-LED. Figure 13.9 shows the fabricated w-LED (20 mA) of (a) $NaSr_{0.99}BO_3$:$Ce^{3+}_{0.01}$, (b) $NaSr_{0.92}BO_3$: $Ce^{3+}_{0.01}$, $Tb^{3+}_{0.07}$, and (c) $NaSr_{0.92}BO_3$:$Ce^{3+}_{0.01}$, $Tb^{3+}_{0.07}$, $Mn^{2+}_{0.03}$ phosphors with their corresponding CIE chromaticity coordinates. Based on these, it can be deduced that $NaSrBO_3$:Ce^{3+}, Mn^{2+}, Tb^{3+} phosphor is a potential candidate for UV-excited devices.

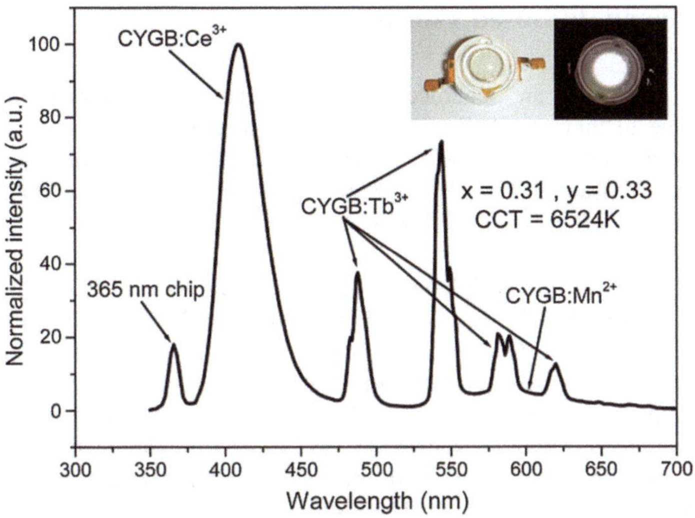

Fig. 13.8 EL spectrum of fabricated w-LED based on $Ca_3Y(GaO(BO_3)_4$:0.01Ce^{3+}, 0.03Mn^{2+}, 0.07Tb^{3+} phosphor combined with a 365-nm UVLED chip. Reprinted from Ref. [18] by permission of American Chemical Society

Fig. 13.9 CIE chromaticity coordinates of **a** $NaSr_{0.99}BO_3$:$Ce^{3+}_{0.01}$, **b** $NaSr_{0.92}BO_3$:$Ce^{3+}_{0.01}$, $Tb^{3+}_{0.07}$, and **c** $NaSr_{0.92}BO_3$:$Ce^{3+}_{0.01}$, $Tb^{3+}_{0.07}$, $Mn^{2+}_{0.03}$ phosphors. The *insets* display the fabricated w-LED under a 20-mA forward bias current. Reprinted from Ref. [19] by permission of American Chemical Society

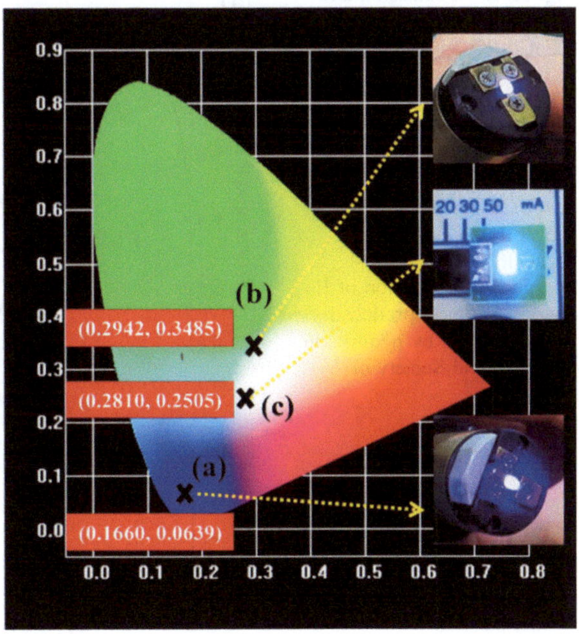

13.4 Phosphate-Based Phosphors

Phosphate phosphors are regarded as one of the most important luminescent materials due to their good thermal stability, high emission intensity, and lower temperature synthesis [20]. Some of the known phosphate-based phosphors include, $LnPO_4$ (Ln = Y, La, Gd, Lu) [21], orthophosphate phosphors, e.g., $MM'PO_4:Eu^{2+}$ (M = K, Na, Li; M′ = Ba, Ca, Sr) [20, 22–29] $LaPO_4:Ce^{3+}$, Tb^{3+} [2], and $Ba_2Ca(PO_4)_2:Eu^{2+}$ [30].

A single-phased white light–emitting phosphor, $KCaGd(PO_4)_2:Eu^{2+}$, Tb^{3+}, Mn^{2+}, was synthesized by Liu et al. [31]. When the $KCaGd(PO_4)_2:0.01Eu^{2+}$ is monitored at 463 nm, it displays a broad absorption band from 268 to 420 nm with peaks located at approximately 280 nm and 365 nm attributed to the $4f^6 \rightarrow 5d^1$ electronic dipole allowed transitions of Eu^{2+} ions. The broad excitation spectrum indicates that the phosphor is excitable by UVLED. Meanwhile, when the sample is excited at 365 nm, a blue broadband emission with maximum peak situated at approximately 463 nm is obtained. The inset reveals the intense blue appearance of $KCaGd(PO_4)_2:0.01Eu^{2+}$ under a UV box with 365-nm excitation as shown in Fig. 13.10a. The emission intensity of $KCaGd(PO_4)_2$ as a function of Eu^{2+} is presented

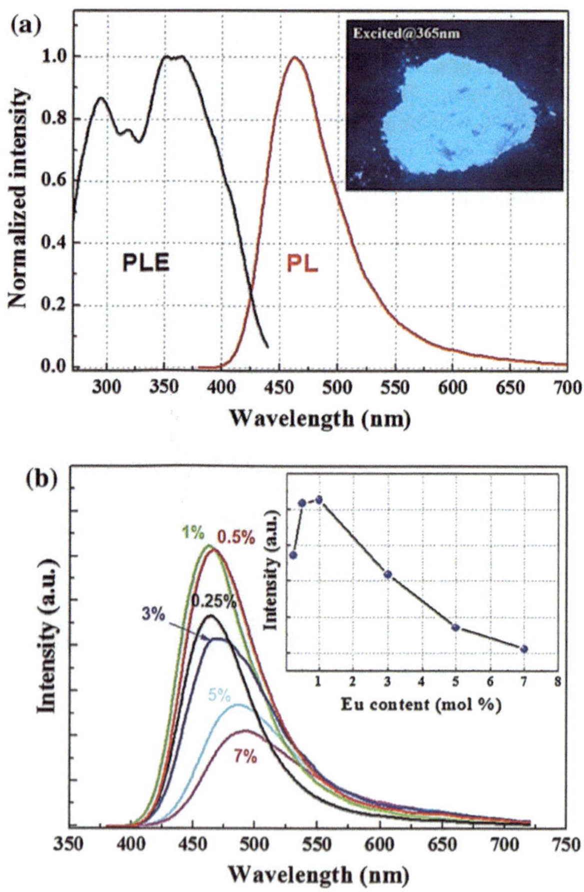

Fig. 13.10 **a** Emission and excitation spectrum of $KCaGd(PO_4)_2:0.01Eu^{2+}$ with its corresponding image under a UV box at 365-nm excitation and **b** emission spectra of $KCaGd(PO_4)_2:xEu^{2+}$ phosphors (x = 0.25–0.07). The *inset* shows the PL intensity of $KCaGd(PO_4)_2:xEu^{2+}$ phosphor as a function of Eu concentration. Reproduced from Ref. [31] by permission of The Royal Society of Chemistry

in Fig. 13.10b. It is evident that the optimum doping concentration of Eu^{2+} is 1 mol %, and further increase of the concentration causes a decrease in emission intensity due to concentration quenching.

The thermal quenching of $KCaGd(PO_4)_2$:$0.01Eu^{2+}$ was investigated and compared with that of the commercial phosphor, $BaMgAl_{10}O_{17}$:Eu^{2+} (BAM:Eu). Figure 13.11a, b shows that the of $KCaGd(PO_4)_2$:$0.01Eu^{2+}$ has a relatively high thermal stability with an activation energy (E_a) of 0.038 eV; although in comparison with the commercial BAM:Eu, it still has lower thermal stability. However, it could still be used for high-power LED applications.

The CIE coordinates of single-phased emission tunable $KCaGd(PO_4)_2$:Eu^{2+}, Tb^{3+}, Mn^{2+} phosphor under 365 nm are presented in Fig. 13.12 and Table 13.1 summarizes the CIE chromaticity diagram. Based on these results, we can infer that $KCaGd(PO_4)_2$:Eu^{2+}, Tb^{3+}, Mn^{2+} is a promising single-composition phosphor for w-LED applications.

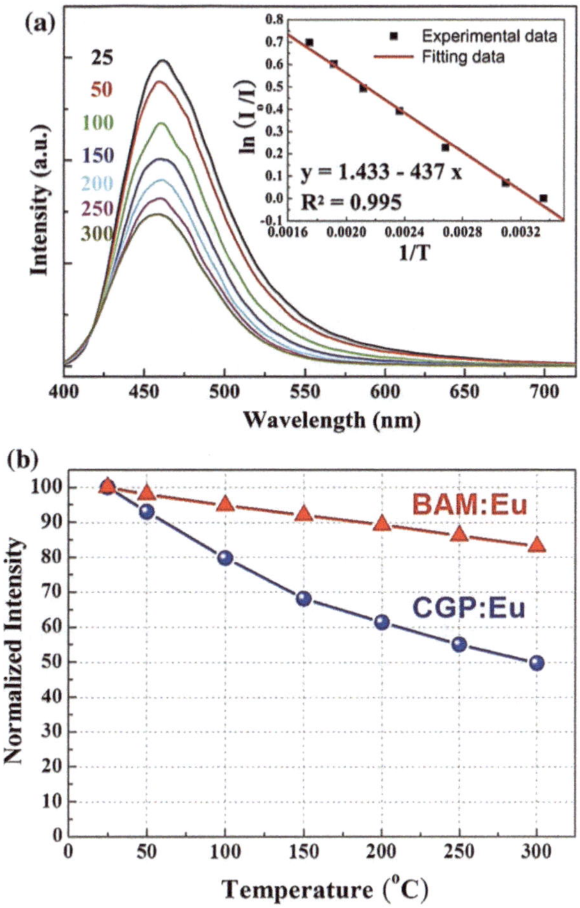

Fig. 13.11 **a** Thermal quenching of $KCaGd(PO_4)_2$:$0.01Eu^{2+}$ excited at 365 nm. The *inset* shows the plot of $\ln(I_o/I)$ versus 1/T. **b** Emission intensity of $KCaGd(PO_4)_2$:$0.01Eu^{2+}$ (CGP:Eu) versus commercial $BaMgAl_{10}O_{17}$:Eu^{2+} (BAM: Eu) as a function of temperature. Reproduced from Ref. [31] by permission of The Royal Society of Chemistry

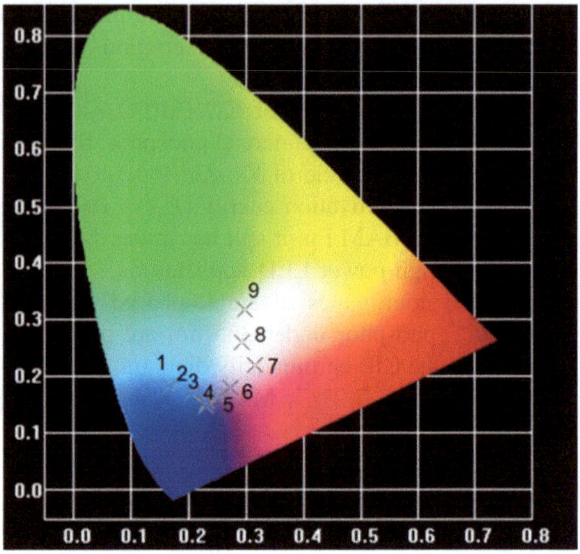

Fig. 13.12 CIE chromaticity-coordinates diagram of $KCaGd(PO_4)_2$: Eu^{2+}, Tb^{3+}, Mn^{2+} phosphors at 365 nm excitation. Reproduced from Ref. [31] by permission of The Royal Society of Chemistry

Table 13.1 CIE chromaticity coordinates $KCaGd(PO_4)_2{:}Eu^{2+}$, Tb^{3+}, Mn^{2+} phosphors excited at 36 nm

No.	Compositions	CIE (*x, y*)
1	$K(Ca_{0.99}Eu_{0.01})Gd(PO_4)_2$	(0.1669, 0.1823)
2	$K(Ca_{0.98}Eu_{0.01}Mn_{0.01})Gd(PO_4)_2$	(0.2042, 0.1701)
3	$K(Ca_{0.96}Eu_{0.01}Mn_{0.03})Gd(PO_4)_2$	(0.2314, 0.1542)
4	$K(Ca_{0.94}Eu_{0.01}Mn_{0.05})Gd(PO_4)_2$	(0.2291, 0.1466)
5	$K(Ca_{0.92}Eu_{0.01}Mn_{0.07})Gd(PO_4)_2$	(0.2518, 0.1644)
6	$K(Ca_{0.89}Eu_{0.01}Mn_{0.1})Gd(PO_4)_2$	(0.2726, 0.1802)
7	$K(Ca_{0.89}Eu_{0.01}Mn_{0.1})$ $(Gd_{0.99}Tb_{0.01})(PO_4)_2$	(0.3153, 0.2195)
8	$K(Ca_{0.89}Eu_{0.01}Mn_{0.1})$ $(Gd_{0.95}Tb_{0.05})(PO_4)_2$	(0.2934, 0.2590)
9	$K(Ca_{0.89}Eu_{0.01}Mn_{0.1})$ $(Gd_{0.9}Tb_{0.1})(PO_4)_2$	(0.2984, 0.3171)

Reproduced from Ref. [31] by permission of The Royal Society of Chemistry

Komuro et al. [32] synthesized a novel blue-green phosphor, $Ca_6BaP_4O_{17}{:}Ce^{3+}$, by way of solid-state synthesis. In this study, Si^{4+} was used to co-substitute with P, which resulted in enhancement of the luminescence. Komuro et al. also fabricated a w-LED by combining UVLED (λ_{em} = 400 nm), $Ca_6BaP_4O_{17}{:}Ce^{3+}$, Si^{4+} (blue-green), $Ca_6BaP_4O_{17}{:}Eu^{2+}$ (yellow), and CaAlSiN3:Eu^{2+} (red) phosphors as shown in Fig. 13.13. For this phosphor, the obtained R_a is 93, which is remarkably high, and when the operating currents were varied from 20 to 360 mA, the CIE coordinates were within the range $x = 0.370$–0.380 and $y = 0.403$–0.408; however, the R_a remained unchanged. These results highly suggest that $Ca_6BaP_4O_{17}{:}Ce^{3+}$, Si^{4+} is a promising phosphor for UVLEDs with expected good performance.

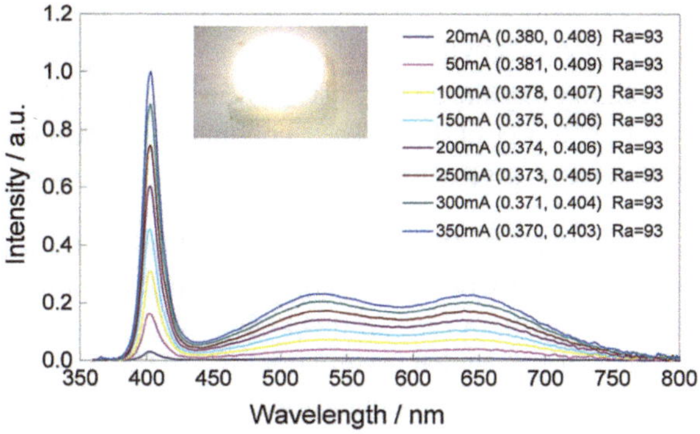

Fig. 13.13 EL spectrum of fabricated w-LED based on $Ca_6BaP_4O_{17}$:Ce^{3+}/Si^{4+}, $Ca_6BaP_4O_{17}$:Eu^{2+}, and CaAlSiN3:Eu^{2+} phosphors combined with UVLED. The *inset* shows the image of the w-LED in operation, CIE chromaticity coordinates, and R_a of each forward bias driving current. Reproduced from Ref. [32] by permission of The Royal Society of Chemistry

13.5 Nitride-Based Phosphors

Nitride compounds are formed by combining nitrogen with less electronegative elements. In general, nitrides can be grouped according to their chemical bond characteristics: metallic, ionic, and covalent. Metallic nitrides are produced by combining nitrogen with transition metals. Meanwhile, ionic nitrides take the form M–N (M = alkali, alkaline-earth metal, or rare-earth metals), whereas covalent nitrides are formed by combining nitrogen with IIIB/VB–group metals. As luminescent materials, metallic and ionic nitrides both have narrow band gaps and are either electrical or ionic conductors; meanwhile, covalent nitrides can be considered as host lattices for phosphors due to their wide band gaps. In addition, they have the characteristics of an insulator or semiconductor. The covalent chemical bonding in nitrides results in a strong nephelauxetic effect, thus reducing the excited-state energy of the 5*d* electrons of the activators yielding a long excitation/emission wavelength and low thermal quenching [33].

Nitride compounds can also be categorized as binary, ternary, quarternary, and multinary based on the number of elements included. Binary covalent nitrides, such as GaN and AlN, are not easily considered as host lattices for phosphors due to the fact that they lack suitable crystal sites for activators. Ternary, quarternary, and multinary covalent nitride compounds, typically silicon-based nitrides, have distinctive and rigid crystal structures. They have suitable crystal sites for activators and have a versatile structure, which allows the doped RE ions to exhibit useful photoluminescence [33].

Nitride/oxynitride phosphors have drawn much attention due to their promising optical properties for w-LED applications. Recent studies include $M_2Si_5N_8$:Eu^{2+} (M = Ba, Sr, Ca) [34], $CaAlSiN_3$:Eu^{2+} [35], β-SiAlON:Eu^{2+} [36], α-SiAlON:Eu^{2+}

[37], yellow-orange–emitting $CaAlSiN_3:Ce^{3+}$ [38], β-SiAlON:Yb^{2+} [39] $Ba_3Si_6O_{12}N_2:Eu^{2+}$ [40] Caα-SiAlON:Eu^{2+} [41], and $Sr_xCa_{1-x}AlSiN_3:Eu^{2+}$ [42].

Green-emitting silicon-oxynitride $Ca_3Si_2O_4N_2:Eu^{2+}$ was synthesized by way of solid-state reaction by Chiu et al. [43]. They conducted a series of experiments to determine its luminescence properties, thermal stability, and possible applications for w-LEDs.

A pure-phase $Ca_3Si_2O_4N_2:Eu^{2+}$ was successfully obtained by way of solid-state synthesis as shown in Fig. 13.14. The major peaks of the compound match well with the standard JCPDS pattern (no. 38-0944). Figure 13.15 illustrates the crystal structure of $Ca_3Si_2O_4N_2:Eu^{2+}$, which is similar to that of $Ca_3Al_2O_6$. It crystallizes in cubic space group Pa $\bar{3}$ with unit cell dimensions of $a = b = c = 15.07$ Å and has 24 formula units/unit cell. It is composed of isolated 12-membered rings comprising $Si_{12}(O, N)_{36}$. Ca^{2+} ions can occupy 7 different crystallographic sites, which are located in the voids of the packed 12-membered rings. The ionic radius of Eu^{2+} is close to that of Ca^{2+}; thus, it is expected that Eu^{2+} will occupy a Ca^{2+} lattice [43].

The PL and PLE spectra of $Ca_3Si_2O_4N_2$ along with varying concentrations of Eu^{2+} are shown in Fig. 13.16. The absorption spectra are comprised of a broad band from 250 to 430 nm with peaks located at approximately 289, 328, 368, and 405 nm attributed to the $4f^6 \rightarrow 5d^1$ transitions of Eu^{2+}. The phosphor exhibits an asymmetric green broadband emission on 330-nm excitation due to the $4f^65d^1 \rightarrow 4f^7$ transition of Eu^{2+}. Moreover, the broad band characteristic of the emission spectra corresponds to the high covalency of the Ca_{Eu}–N bond and large crystal field–splitting effect. The crystal-field splittings of Eu^{2+} are presented in Table 13.2, and they were estimated at 1873 to approximately 21,050 cm^{-1} with estimated Stokes shifts of 10,973–12,500 $cm^{-1.}$ It was observed that the emission band shifted to a longer wavelength as the concentration of Eu^{2+} increased, which can be ascribed to the change in the crystal-field splitting of Eu^{2+} and energy transfer between the Eu^{2+} ions at the higher 5*d* levels and the lower levels. This phenomenon causes a

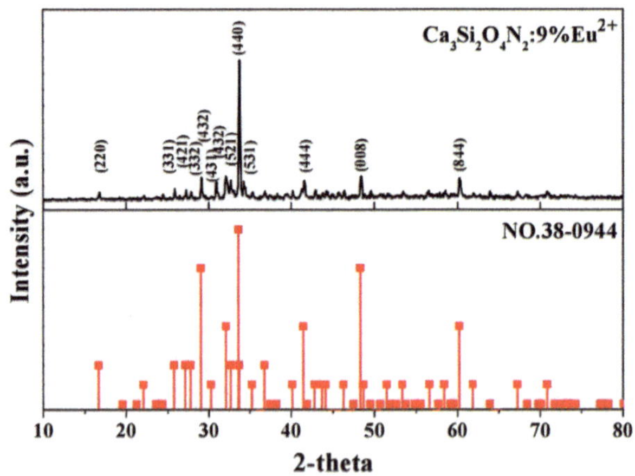

Fig. 13.14 XRD patterns of $Ca_3Si_2O_4N_2:0.09Eu^{2+}$ and standard $Ca_3Si_2O_4N_2$ pattern (JCPDS 38-0944). Reprinted from Ref. [43] by permission of OSA Publishing

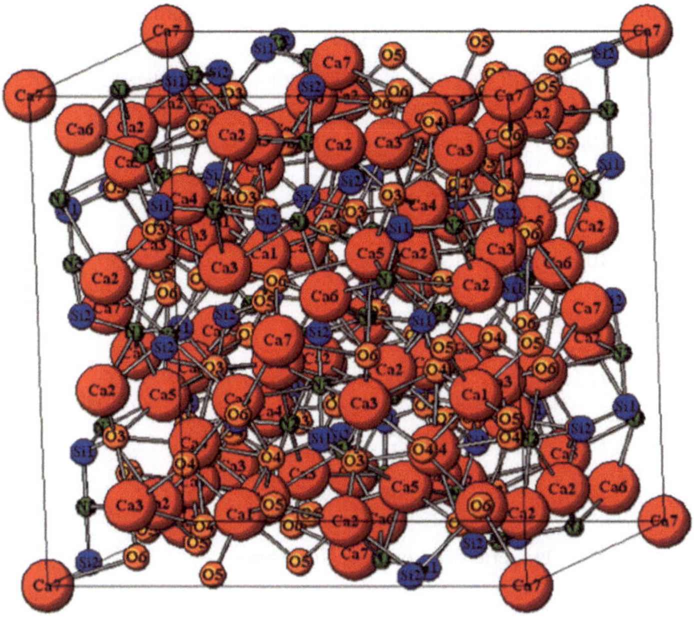

Fig. 13.15 Crystal structure of $Ca_3Si_2O_4N_2$. Reprinted from Ref. [43] by permission of OSA Publishing

decrease of the emission energy from the 5*d* excitation state to the 4*f* ground state resulting in longer-wavelength emissions [33].

Chiu et al. [43] developed a w-LED based on $Ca_3Si_2O_4N_2{:}Eu^{2+}$ combined with blue-emitting $BaMgAl_{10}O_{17}{:}Eu^{2+}$ phosphor, red-emitting $CaAlSiN_3{:}Eu^{2+}$, and a 380-nm UV LED chip. Figure 13.17 shows the EL spectrum of the fabricated LED device with its corresponding CCT (6029 K), CIE coordinates (0.322, 0.330), and

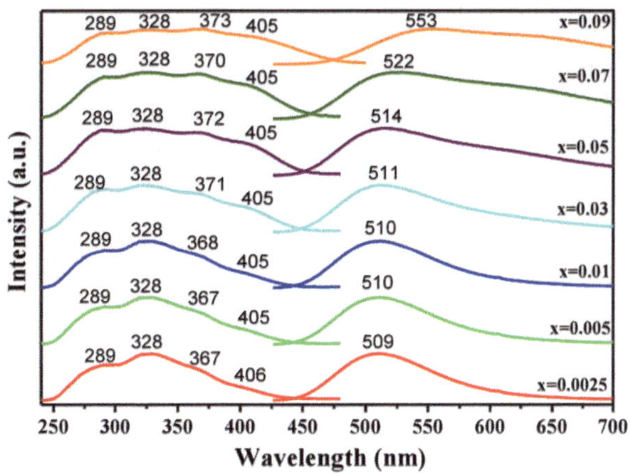

Fig. 13.16 PL/PLE spectra of $Ca_3Si_2O_4N_2{:}0.xEu^{2+}$ ($x = 0.0025–0.09$). Reprinted from Ref. [43] by permission of OSA Publishing

Table 13.2 Emission, Stokes shift, crystal-field splitting, normalized PL intensity, and CIE-chromaticity coordinates of $Ca_3Si_2O_4N_2$:xEu^{2+}

x	λ_{em} (nm)	Stokes shift (cm^{-1})	Crystal field splitting (cm^{-1})	Normalized PL intensity (%)	CIE (x, y)
0.0025	509	10,841	18,939	87	(0.25, 0.50)
0.0050	510	10,879	19,244	94	(0.26, 0.50)
0.0100	510	10,849	19,494	100	(0.26, 0.51)
0.0300	511	10,918	20,068	93	(0.29, 0.49)
0.0500	514	11,032	20,161	84	(0.33, 0.51)
0.0700	522	11,330	20,435	50	(0.35, 0.52)
0.0900	553	12,403	21,258	49	(0.42, 0.52)

Reprinted from Ref. [43] by permission of OSA Publishing

average R_a (88.25). The luminous efficiency of the w-LED is 20.1 lm/W at 350 mA. The results indicate that $Ca_3Si_2O_4N_2$:0.Eu^{2+} is a prospective green-emitting phosphor.

Red-emitting $CaAlSiN_3$:Eu^{2+} phosphor is frequently combined with blue- and green- emitting phosphors to generate a w-LED. Hu et al. [35] synthesized $CaAlSiN_3$:Eu^{2+} with the addition of fluxes such BaF_2, CaF_2, NH_4 and H_3BO_3. Figure 13.18 depicts the XRD patterns of standard $CaAlSiN_3$, $CaAlSiN_3$:Eu^{2+}, and $CaAlSiN_3$:Eu^{2+} + flux. It can be seen that even with the addition of fluxes,

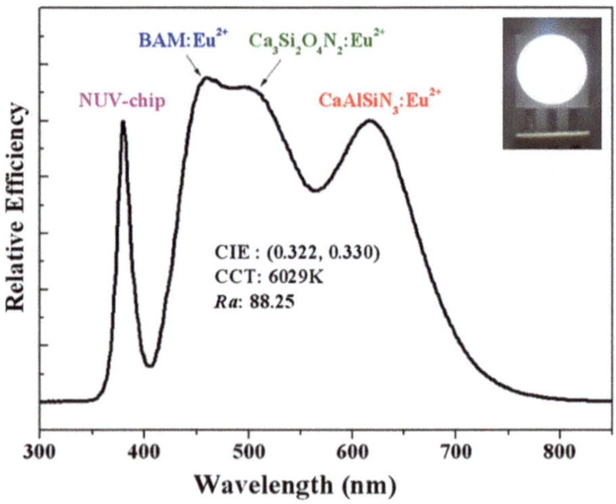

Fig. 13.17 EL spectra of fabricated w-LED driven by a 350-mA forward bias current. Reprinted from Ref. [43] by permission of OSA Publishing

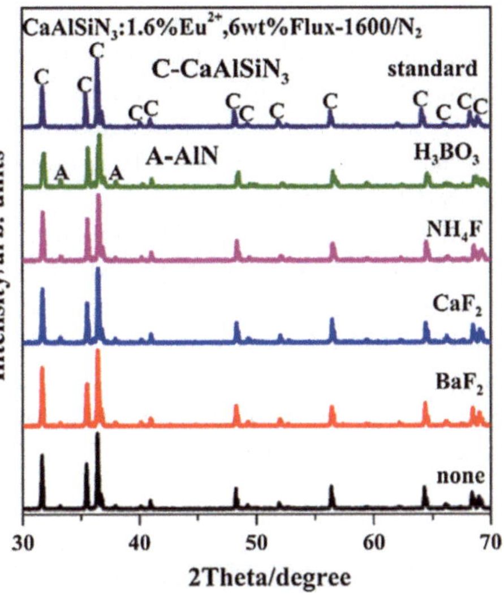

Fig. 13.18 XRD patterns of standard $CaAlSiN_3$ and $CaAlSiN_3$:0.016Eu^{2+} without and with 6 wt% of different fluxes (BaF_2, CaF_2, NH_4, and H_3BO_3). Reprinted from Ref. [35] by permission of Elsevier

pure-phased $CaAlSiN_3$ was obtained although weak peaks of AlN were evident; however, no additional peaks attributed to the fluxes were seen.

The emission and excitation spectra $CaAlSiN_3$:0.016Eu^{2+} with and without the presence of fluxes are presented in Fig. 13.19. The excitation spectra display a broad band from the UV region to the visible region. The peak at 240 nm is due to the electronic transition between the valence and the conduction band of the $CaAlSiN_3$ host, whereas the peaks at 320 and 460 nm are attributed to the allowed $4f^65d^1 \rightarrow 4f^7$ transition of Eu^{2+}. It is evident from the graph that the emission intensity of the phosphor was increased due to the addition of BaF_2, CaF_2, and NH_4. However, the addition of H_3BO_3 resulted in a decrease in the emission intensity and a shift in the shorter wavelength. The increased intensity of the

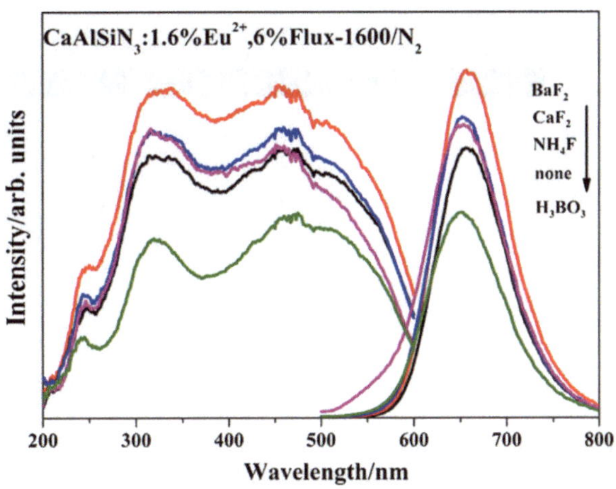

Fig. 13.19 PL and PLE spectra of $CaAlSiN_3$:0.016Eu^{2+} phosphors flux-free and with the addition of 6 wt% of BaF_2, CaF_2, NH_4, and H_3BO_3 fluxes. Reprinted from Ref. [35] by permission of Elsevier

emission due to the addition of flux can be assigned to the growth of the particle size, narrowing of the particle-size distribution, and increase of crystallinity. Large grains decrease light scattering due to the small particles, and high crystallinity decreases defects in the lattice and on the surface of the phosphor. The blue shift caused by the addition of H_3BO_3 flux is due to the weakening of the crystal field around Eu^{2+} [35].

SEM images of $CaAlSiN_3$:0.016Eu^{2+} both with and without BaF_2, CaF_2, NH_4, and H_3BO_3 fluxes are displayed in Fig. 13.20. Without the fluxes, the sample contains small particles, the round edges of which indicate poor crystallization. Adding (b) BaF_2, (c) CaF_2, and (d) NH_4 produced uniform particles with larger sizes, thus enhancing the luminescence properties. $CaAlSiN_3$:0.016Eu^{2+} with H_3BO_3 flux (e) also contains large particles along with a large number of small particles. The particle sizes of $CaAlSiN_3$:0.016Eu^{2+} with BaF_2, CaF_2, NH_4, and

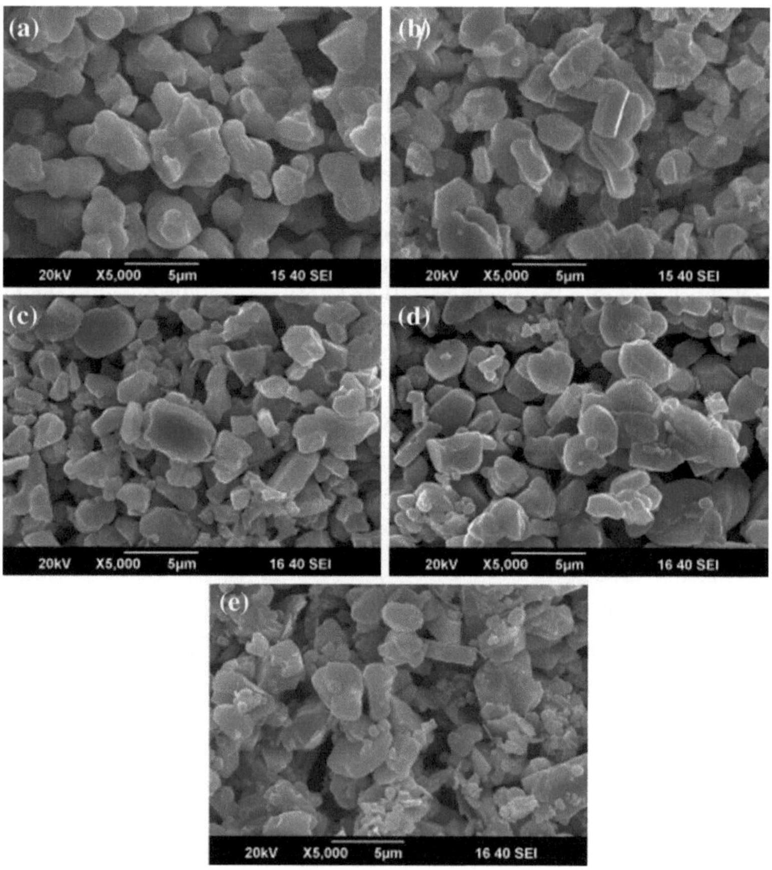

Fig. 13.20 SEM images of $CaAlSiN_3$:0.016Eu^{2+} **a** without flux and with 6 wt% of **b** BaF_2, **c** CaF_2, **d** NH_4, and **e** H_3BO_3 fluxes. Reprinted from Ref. [35] by permission of Elsevier

H_3BO_3 were 11.44, 11.20, 11.03, and 9.46 μm, respectively. With these results, it can be deduced that a good flux could promote increased growth and narrow grain-size distribution [35].

13.6 Summary

Phosphors absorb the energy emitted by ultraviolet LEDs (UVLEDs) and downconvert the energy into longer wavelengths. Since the emergence of white light–emitting diodes (w-LEDs), scientists and researchers have discovered and developed a number of phosphors to downconvert the energy emitted by UVLEDs to visible light. Phosphors can be classified based on their chemical composition and color emissions. Among these phosphors, silicate-, borate-, phosphate-, and nitride-based phosphors have been investigated and developed to efficiently convert UVLED emission to longer wavelengths.

To efficiently downconvert the emission wavelength of UVLED, the phosphor should have a strong absorption in the UV region (350–410 nm); therefore, the excitation spectrum of the phosphor should match the emission spectrum of the UVLED. In addition, the phosphor should have high physical and chemical stability. Because UVLEDs emit high-energy photons, the phosphor should be able to withstand degradation. Moreover, the phosphor should be chemically stable at ambient temperature and not easily react with CO, H_2O, CO_2, and air.

In this chapter, we saw examples of fabricated w-LEDs. Not only did the phosphors convert UV into visible light, the combination of UVLED and the phosphors generated white light with greater *Ra* ($Ra > 80$), greater CCT than the commercially known w-LEDs, and CIE coordinates closer to the ideal light source (0.33, 0.33).

References

1. Xie RJ, Li YQ, Hirosaki N, Yamamoto H (2011) Nitride phosphors and solid state lighting
2. Liu WR (2014) The fundamentals, characteristics and applications of LED phosphors
3. Brgoch J, Borg CKH, Denault KA, Mikhailovsky A, DenBaaras SP, Seshadri R (2013) An efficient, thermally stable cerium-based silicate phosphor for solid state white lighting. J Am Ceram Soc 52:8010
4. Liu C, Xia Z, Lian Z, Zhou J, Yan Q (2013) Structure and luminescence properties of green-emitting $NaBaScSi_2O_7:Eu^{2+}$ phosphors for near UV-pumped light emitting diodes. J Mater Chem 1:7139
5. Yang Z, Lin PC, Guo CF, Liu WR (2015) Color-tunable luminescence and energy transfer properties of Eu^{2+}- and Mn^{2+}- activated $BaCa_2MgSi_2O_8$ phosphor for ultraviolet light-emitting diodes. RSC Adv 5:13184
6. Wu L, Chen XL, Zhang Y, Kong YF, Xu JJ, Xu P (2009) Ab initio structure determination of novel borate $NaSrBO_3$. J Solid State Chem 179:1219

7. Maggay IVB, Lin PC, Liu WR (2015) Investigation of luminescence properties and energy transfer mechanism of Li6Lu(BO3)3: Ce3+, Tb3+ green-emitting phosphors. RSC Adv 5:5591
8. Liu WR, Huang CH, Wu CP, Chiu YC, Yeh YT, Chen TM (2011) High efficiency and high color purity blue-emitting $NaSrBO_3$:Ce^{3+} phosphor for near-UV light-emitting diodes. J Mater Chem 21:6869
9. Reddy AA, Das S. Goel A, Sen R, Siegel R, Mafra L, Prakash, GV, Ferreira JMF (2013) KCa4(BO3)3:Ln3 + (Ln = Dy, Eu, Tb) phosphors for near UV excited white–light–emitting diodes. AIP Adv 3:1
10. Su J, Lai J, Zhu J, Xia Z, Du H (2012) Luminescence properties and energy transfer investigation of $Sr_2B_2O_5$:Ce^{3+}, Tb^{3+} phophors. Ceram Int 38:5341
11. Yuan S, Yang Y, Zhang X, Tessier F, Chevire F, Adam J-L, Moine B, Chen G (2008) Eu^{2+} and Mn^{2+} codoped $Ba_2Mg(BO_3)_2$-new red phosphor for white LEDs. Opt Lett 33:2865
12. Zhang J, Chen G (2014) Luminescence properties of $KSr_4(BO_3)_3$ Eu^{2+} for light-emitting diodes. Chin J Lumin 12:1432
13. Chen X, Xia Z (2014) Synthesis and color-tunable luminescence of Ce^{3+}, Tb^{3+} Codoped $Sr_6YSc(BO_3)_6$ phosphor. J Solid State Lighting 1:1
14. Wu W, Xia Z, Zhuang J, Liao L (2012) Luminescence property and energy transfer in Sr3Gd1-x-y(BO3)3:xCe3+, yTb3+ phosphor. J Solid State Sci Technol 1:R46
15. Han B, Zhang J, Lu Y (2013) Novel emitting color tunable phosphors Ba3Y2(BO3)4:Ce3+, Tb3+ with efficient energy transfer for near UV light-emitting diodes. J Am Ceram Soc 96:179
16. Guo C, Ding X, Seo HJ, Ren Z, Bai J (2011) Double emitting phosphor NaSr4(BO3)3:Ce3+, Tb3+ for near UV light emitting diode. Opt Laser Technol 43:1351
17. Liu WR, Lin CC, Chiu YC, Yeh YT, Jang SM, Liu RS (2010) ZnB_2O_4:Bi^{3+}, Eu^{3+}: a highly efficient red-emitting phosphor. Opt Exp 18:2936
18. Huang CH, Chen TM (2011) A novel single-composition trichromatic white-light Ca3Y(GaO) 3(BO3)4:Ce3+, Mn2+, Tb3+ phosphor for UV-light emitting diodes. J Phys Chem 115:2349
19. Lin CC, Liu YP, Xiao ZR, Wang YK, Cheng BM, Liu RS (2014) All-in-one light-tunable borated phosphors with chemical and luminescence dynamical control resolution. ACS Appl Mater Interfaces 6:9160
20. Kim M, Kobayashi M, Kato H, Kakihana M (2013) A highly luminous $LiCaPO_4$:Eu^{2+} phosphor synthesized by a solution method employing a water-soluble phosphate ester. Opt Photon J 3:13
21. Lai H, Bao A, Yang Y, Tao Y, Yang H, Zhang Y, Han L (2008) UV luminescence property of YPO_4:RE (RE = Ce^{3+}, Tb^{3+}). J Phys Chem C 112:282
22. Yang ZP, Li XN (2009) A novel blue-emitting phosphor NaBaPO4:Eu2+ for white LEDs. Photon Optoelect 13. 10.1109/SOPO.2009.5230105
23. Jeong J, Lee HB, Jang K, Yi SS, Jeong JH (2009) Luminescence properties of $NaCaPO_4$:Eu^{2+} green phosphor for white-light UV-LEDs. J Korean Phys Soc 55:1572
24. Yim DK, Song HJ, Cho IS, Kim JS, Hong K (2011) A novel blue-emitting $NaSrPO_4$:Eu^{2+} phosphor for near UV based white light-emitting-diodes. Mater Lett 65:1666
25. Zhang S, Nakai Y, Tsuboi T, Huang Y, Seo HJ (2011) The thermal stabilities of luminescence and microstructures of Eu^{2+}-doped $KBaPO_4$ and $NaSrPO_4$ with ß-K2SO4 type structure. Inorg Chem 50:2987
26. Guan L, Liu C, Li X (2011) Synthesis and optical properties of $KCaPO_4$:Eu^{2+} phosphor. Mater Res Bull 46:1496
27. Zhang X, Mo F, Gong M (2013) Properties–structure relationship research on $LiCaPO_4$:Eu^{2+} as blue phosphor for NUV LED application. J Alloys Compd 575:314
28. Zhang S, Nakai Y, Tsuboi T, Huang Y, Seo HJ (2011) Luminescence and microstructural features of Eu-activated $LiBaPO_4$ phosphor. Chem Mater 23:1216
29. Wu ZC, Shi JX, Wang J, Gong ML, Su Q (2006) A novel blue-emitting phosphor $LiSrPO_4$: Eu^{2+} for white LEDs. J Solid State Chem 179:2356

30. Yu H, Deng D, Zhou D, Yuan W, Zhao Q, Hua Y, Zhao S, Huang L, Xu S (2013) $Ba_2Ca(PO_4)_2:Eu^{2+}$ emission-tunable phosphor for solid-state lighting: luminescent properties and application as white light emitting diodes. J Mater Chem C 1:5577
31. Liu WR, Huang CH, Yeh CW, Chiu YC, Yeh YT, Liu RS (2013) Single-phased white-light-emitting $KCaGd(PO_4)_2:Eu^{2+}$, Tb^{3+}, Mn^{2+} phosphors for LED applications. RSC Adv 3:9023
32. Komuro N, Mikami M, Shimomura Y, Bithell EG, Cheetham AK (2015) Synthesis, structure and optical properties of cerium-doped calcium barium phosphate—a novel blue-green phosphor for solid-state lighting. J Mater Chem C 3:204
33. Xie RJ, Hirosaki N (2007) Silicon-based oxynitride and nitride phosphors for white LEDs-a review. Sci Technol Adv Mater 8:588
34. Zeuner M, Hintze F, Schnick W (2009) Low temperature precursor route for highly efficient spherically shaped LED-phosphors $M_2Si_5N_8:Eu^{2+}$ (M = Eu, Sr, Ba). Chem Mater 21:336
35. Hu WW, Cai C, Zhu QQ, Xu X, Hao LY (2014) Preparation of high performance $CaAlSiN_3$: Eu^{2+} phosphors with the aid of BaF_2 flux. J Alloys Compd 613:226
36. Hirosaki N, Xie RJ, Kimoto K, Sekiguchi T, Yamamoto Y, Suehiro T, Mitomo M (2005) Characterization and properties of green-emitting ß-SiAlON:Eu^{2+} powder phosphors for white light-emitting diodes. Appl Phys Lett 86:1
37. Shioi K, Hirosaki N, Xie RJ, Takeda T, Li YQ, Matsushita Y (2010) Synthesis, crystal structure, and photoluminescence of Sr-α-SiAlON:Eu^{2+}. J Am Ceram Soc 93:456
38. Li YQ, Hirosaki N, Xie RJ, Takeda T, Mitomo M (2008) Yellow-orange-emitting $CaAlSiN_3$: Ce^{3+} phosphor: structure, photoluminescence, and application in white LEDs. Chem Mater 20:6704
39. Liu L, Xie RJ, Horosaki N, Takeda T, Zhang CN, Li J, Sun X (2011) Photoluminescence properties of ß-SiAlON: Yb^{2+}, a novel green-emitting phosphor for white light-emitting diodes. Sci Technol Adv Mater 12:1
40. Li W, Xie RJ, Zhou T, Liu L, Zhu Y (2014) Synthesis of the phase pure Ba3Si6O12N2:Eu2+ green phosphor and its application in high color rendition white LEDs. Dalton Trans 43:6132
41. Ryu JH, Park YG, Won HS, Kim SH, Suzuki H, Lee JM, Yoon C, Nazarov M, Noh DY, Tsukerblat B (2008) Luminescent properties of Ca-α-SiAlON: Eu^{2+} phosphors synthesized by gas-pressured sintering. J Electrochem Soc 155:J99
42. Watanabe H, Wada H, Seki K, Itou M, Kijima N (2008) Synthetic method and luminescence properties of $Sr_xCa_{1-x}AlSiN_3:Eu^{2+}$ mixed nitride phosphors. J Electrochem Soc 155:F31
43. Chiu YC, Huang CH, Lee TJ, Liu WR, Yeh YT, Jang SM, Liu RS (2011) Eu^{2+}-activated silicon-oxynitride $Ca_3Si_2O_4N_2$: a green-emitting phosphor for white LEDs. Opt Exp 9:A331

Chapter 14
Bismuth-Doped Photonic Materials: Are They Promising Phosphors for WLEDs?

Mingying Peng and Qinyuan Zhang

Abstract For phosphor-converted white light-emitting diodes (pc-WLED), phosphors are obviously indispensable for devices, and therefore they influence their eventual performance such as color-rendering index (CRI), color temperature, etc. For pc-WLEDs, which are based on near-ultraviolet chips, the phosphors should be excitable by lights with wavelengths from 380 to 420 nm and at the same time not by visible light. This can avoid the reabsorption between phosphors and therefore increase the overall efficiency of the device. Most efforts have been made, but they have mainly focused on rare-earth phosphors. Unfortunately, few green to red phosphors, e.g., Eu^{2+}-doped nitrides, can fully meet the requirements of the application. In this chapter, Bi^{3+}-doped phosphors will be reviewed, which could possibly be one of new solutions to this problem. At the end of the chapter, the future challenges of these phosphors will be discussed with respect to pc-WLEDs.

14.1 Introduction

Ever since the emergence of civilization, lighting has been a subject of perennial interest for mankind. In recent years, considerable attention has been paid to white lighting based on LEDs. This is primarily because of the overwhelming advantages compared with incandescent and fluorescent lighting, such as high luminous efficiency, low power consumption, long lifetime, and small size [1–5]. The high efficiency confers energy savings and environmental benefits to WLEDs. And these have promoted the use of LEDs in widespread applications such as in domestic and

M. Peng (✉) · Q. Zhang
The China-Germany Research Center for Photonic Materials and Device,
The State Key Laboratory of Luminescent Materials and Devices and Guangdong Provincial Key Laboratory of Fiber Laser Materials and Applied Techniques, The School of Materials Science and Engineering, South China University of Technology, Guangzhou 510641, People's Republic of China
e-mail: pengmingying@scut.edu.cn

R.S. Liu (ed.), *Phosphors, Up Conversion Nano Particles, Quantum Dots and Their Applications*, DOI 10.1007/978-3-662-52771-9_14

commercial lighting, automobiles, communications, imaging, agriculture, and medicine [1–3].

The mainstream technology for WLED is based on the combination of a single LED chip with one or more conversion phosphors. Currently the commercial and most popular approach is the conjunction of an InGaN blue LED (typically with a wavelength between 450 and 480 nm) with $Y_3Al_5O_{12}:Ce^{3+}$ (YAG:Ce^{3+}) [6–8]. However, YAG:Ce^{3+} exhibits weak emission in the red spectral region, and this results in poor color-rendering ability and a higher color temperature of the dichromatic system [6–11]. The addition of a red phosphor, especially with strong blue absorption, is one of best strategies for generating warm white light similar to that of incandescent light. Another alternative approach involves blending the emissions from blue, green, and red phosphors with a near-UV chip (typically with a wavelength in the range of 380–420 nm) [8, 12–17]. This scheme can improve the color rendering index and simultaneously keep the light color stable, especially when the driving current changes.

Thus, the creation of a warm WLED with either of the two above approaches necessitates the development of efficient phosphors. This challenge has stimulated researchers to revisit the commercial lamp and cathode-ray tube phosphors. The approach has proved to be unfruitful because of the absence of effective absorption in the spectral range of 380–480 nm or because of poor resistance of the phosphor against thermal and environmental attack. For example, Eu^{3+}-based phosphors absorb at 396 and 465 nm due to the transitions from the $^7F_0$ ground state to the $^5L_6$ and $^5D_2$ multiplets, respectively [18–20]. The first-order electric dipole-forbidden nature of the transitions limits the absorption strength, and the width of these absorption bands are <10 nm [21–23], which is much narrower than the emission band of the blue or UV LED chips. On the one hand, this lowers the overall efficiency of the device because only a fraction of the photons emitted from LED chips can be harvested; in contrast, the temperature or current-based shift of the emission spectrum of the chips is not tolerated [7, 8]. Furthermore, the weak oscillator strengths of the $4f^6 \rightarrow 4f^6$ transitions cannot prevent the leakage of excitation light efficiently, which is harmful to human health.

The demand for efficient WLED phosphors stimulated investigations of new Eu^{2+}-doped phosphors, e.g., oxynitrides, nitrides, aluminates, and aluminosilicates. Among these, nitrides and oxynitrides have been the subjects of intensive studies because of their outstanding performances such as quantum yield >80 %, and high thermal and chemical stability [6–8, 22, 24–27]. However, the synthesis of these hosts must be performed at high temperature and high pressure, e.g., 1900 °C and 10 atm N_2 atmosphere for Eu^{2+}-doped β-SiAlON [24]. The harsh preparation conditions and high price of raw materials destine these phosphors to be costly.

Figure 14.1a, b depict the typical excitation and emission spectrum of red phosphor $Sr_2Si_5N_8:Eu^{2+}$ and green phosphor β-Sialon:Eu^{2+} [24, 28]. They exhibit the absorption not only in near ultraviolet but also in visible range. For the newly found $Sr[LiAl_3N_4]:Eu^{2+}$ red phosphor [29], the absorption even extends to the red spectral range. Thus, as these phosphors are applied in WLEDs, the reabsorption

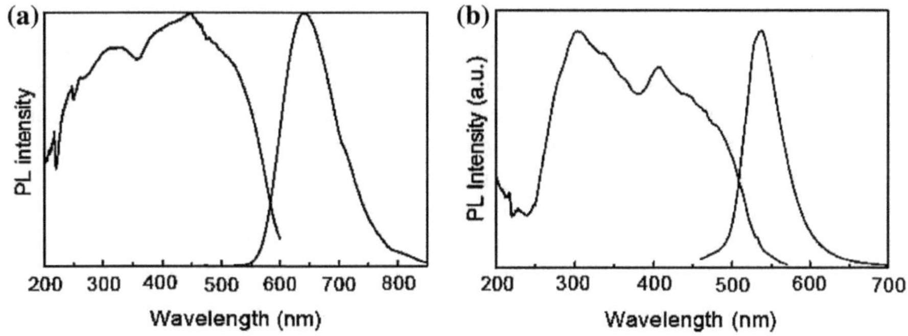

Fig. 14.1 **a** The excitation (λ_{em} = 640 nm) and emission (λ_{ex} = 450 nm) spectrum of red phosphor $Sr_2Si_5N_8{:}Eu^{2+}$ and **b** the excitation (λ_{em} = 535 nm) and emission (λ_{ex} = 450 nm) spectrum of green phosphor β-Sialon:Eu^{2+}. Reprinted from Ref. [28] by permission of Elsevier

will be unavoidable between phosphors and/or light just generated [6, 7]. This will lower the overall efficiency of the device.

In view of the sensitivity of Ce^{3+} luminescence to the crystal field, efforts have been made to modify the local coordination around Ce^{3+} by intentional codoping [11, 30]. Setlur et al. [30] implanted N^{3-} (with the lower electronegativity N, 3.04 compared with O, 3.44) into the surroundings of Ce^{3+}, and a red-shift of absorption and emission of YAG:Ce^{3+} was observed. The Ce^{3+} emission maintains the broadband characteristics, but the peak shifts from 560 to 600 nm when 20 % N^{3-} is codoped [30]. The drawback is that this modification leads to stronger luminescence quenching at high temperature. Jia et al. [11] recently presented a novel scheme incorporating Mn^{2+}–Si^{4+} into the matrix of YAG:Ce^{3+} to finely tune the Ce^{3+} emission spectrum. The CIE color coordinates change gradually between (0.425, 0.550) to (0.502, 0.482) due to the energy transfer from Ce^{3+} to Mn^{2+}. This is a promising route to modify YAG:Ce^{3+} because it does not significantly change the surroundings of Ce^{3+}, as reflected by the almost unshifted Ce^{3+} emission peak [8], and it may not result in increased thermal-luminescence quenching. This type of modification can also be applicable to the case for Eu^{2+} luminescence [31, 32].

The aforementioned schemes are based on rare-earth elements [6–9, 11, 16–18, 21, 22, 24–26, 31]. In recent years, non-rare earth-based ecofriendly phosphors, which can be prepared under milder conditions, have received burgeoning interest. This is driven by the soaring price of rare-earths elements and the strong desire to minimize the cost of LED devices, which is one of the main obstacles for widespread adoption. Lin et al. found that doping Ba^{2+} into BPO_4 can create oxygen-related defects, which emit a bluish white light with a quantum yield as high as 31 % [15], and can tune the color of emission by changing the concentration of Ba^{2+}. Although the excitation spectrum peaks at approximately 310 nm, the tail of the excitation extends to 450 nm [15]. This means it is nearly impossible to combine it with a near UV chip for an efficient WLED.

In the following, we are going to focus only on phosphors for WLEDs fabricated with a near ultraviolet (NUV) chip (NUV-WLED, hereafter). For efficient NUV-WLEDs, the general requirements of potential phosphors are (1) excitable by NUV rather than visible light, e.g., for red phosphor, it is better to have no absorption between 410 and approximately 600 nm; (2) strong excitation (absorption) peak intensity at NUV, which can guarantee full use and therefore avoid leakage of light emitted from the chip, which can reduce the chance to expose under ultraviolet lights; (3) emission peaking in the visible range, but the peak better lying at <650 nm because human eyes are not sensitive to longer-wavelength lights; (4) high quantum yield; (5) good resistance of the luminescence to thermal impact; (6) no luminescence degradation; (7) good chemical and thermal stability especially once exposed to strict ambient conditions such as high humidity or temperature; (8) raw materials and synthesis procedure that are benign to the environment and compatible with present and future environmental regulations, e.g., free of toxic species, "green" compositions; and (9) cost-effective, etc.

Checking over all of the rare earth-doped green to red phosphors with the requirements, we found that none of the rare-earth phosphors can fully meet them. For instance, very recent studies show that either Eu^{2+}-doped nitride or oxynitride exhibits permanent failure in luminescence and phase structure [33, 34]. When exposed to high-pressure water steam, degradation starts at 150 °C, and the luminescence is quenched very quickly due to the phase decomposition [33]. For long-term applications, it is definitely meaningful to find candidates that can perfectly match all of the requirements.

We the found Bi^{3+}-doped vanadate phosphors can be possible candidates [35–37]. They can be excited by NUV rather than visible light and they can emit yellow or red lights with high efficiency. For the yellow phosphor $LuVO_4$:Bi^{3+}, the efficiency is approximately 68 %, whereas it is 56 % for the red phosphor $ScVO_4$: Bi^{3+}. In the following text, we discuss the nature of Bi^{3+} luminescence first, which is followed by a short review on the luminescence in different types of hosts. The emphasis will be on bismuth-doped vanadates. At the end, the open questions will be identified.

14.2 The Nature of Bi^{3+} Luminescence

Bismuth, despite lying amongst toxic heavy metals in the periodic table, is generally considered as a green element because it is nontoxic and noncarcinogenic. Many bismuth compounds are even less toxic than sodium chloride [38, 39]. For this, it has been found applications in tremendous areas such as medicine (e.g., bismuth subsalicylate for antidiarrheal), paints (e.g., yellow $BiVO_4$), cosmetics (e.g., BiOCl), high-temperature superconductor [e.g., $Bi_{2.1}(Ca, Sr)_{n+1}Cu_nO_{2n+4+\delta}$ with $n = 1,2$ and 3], and low-melting special alloy (e.g., Bi–Cd–Pb–Sn), etc. [40–42]. Very recently, a bismuth-based electrocatalyst was found to be able to reduce CO_2 to CO with a Faradaic efficiency of approximately 95 % on an inert

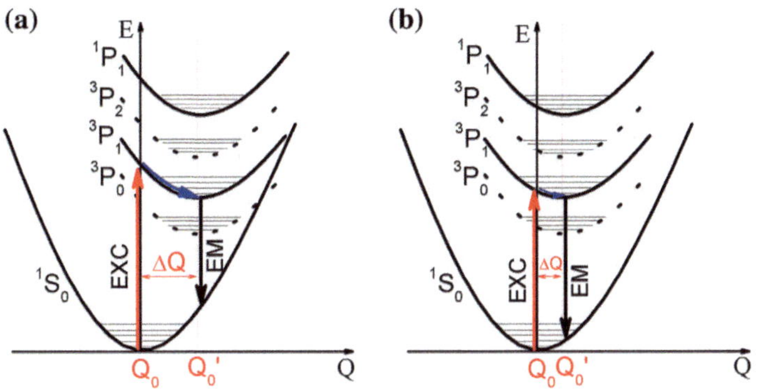

Fig. 14.2 Configurational coordinate diagrams of bismuth-doped alkali borate glasses: **a** for lower and **b** for higher content of alkali oxide. $^3P_0$ and $^3P_2$ states are marked as broken lines because the transitions from $^1S_0$ to the states are forbidden. Horizontal lines denote the vibrational levels. The blue solid lines refer to the nonradiation relaxations. Upward-directed red solid lines refer to the excitation processes, and downward-directed black solid lines refer to the emission process. ΔQ indicates the offset of parabolas. Reprinted from Ref. [44] by permission of OSA Publishing

glassy carbon electrode, comparable with those using expensive silver or gold cathodes [43].

Trivalent Bi^{3+} ion has an electronic configuration of $[Xe]4f^{14}5d^{10}6s^2$. The ground state $^1S_0$ is from the $6s^2$ configuration, whereas the excited states are $^3P_0$, $^3P_1$, $^3P_2$, and $^1P_1$ from $6s6p$ configuration (see Fig. 14.2). The energy of these states increases in a sequence. $^1S_0 \rightarrow {}^3P_0$ and $^1S_0 \rightarrow {}^3P_2$ are spin-forbidden transitions. And $^1S_0 \rightarrow {}^3P_1$ and $^1S_0 \rightarrow {}^1P_1$ can be lifted by spin-orbit coupling. Therefore, the absorption strength of these two transitions is much higher than the former. $^1P_1$ lies at a higher-energy state, and $^1S_0 \rightarrow {}^1P_1$ cannot be observed with an ordinary spectrometer. Therefore, the transition of $^1S_0 \rightarrow {}^3P_1$ is often reported (see Fig. 14.2). The Laporte-allowed radiative transition of $^3P_1 \rightarrow {}^1S_0$ has a typical decay time between 10^{-6} and 10^{-8} s. For instance, Peng et al. found the typical decay time of Bi^{3+} to be approximately 500 ns in different oxide glasses at room temperature [44].

14.3 Bi^{3+}-Doped Glasses

14.3.1 Bi^{3+}-Doped Borate and Silicate Glasses

Parke and Webb examined [45] the absorption, emission, and decay times of bismuth-doped $x Na_2O \cdot (100 - x)B_2O_3$ ($x = 10, 15, 20, 30, 35$), $20Li_2O \cdot 80B_2O_3$, $20K_2O \cdot 80B_2O_3$, $3Na_2O \cdot 7SiO_2$, and $M_xO \cdot P_2O_5$ (M = Li, Na, Mg, Ca, Sr, Zn) glasses. For bismuth-doped borate glasses $10Na_2O \cdot 90B_2O_3$ and $15Na_2O \cdot 85B_2O_3$,

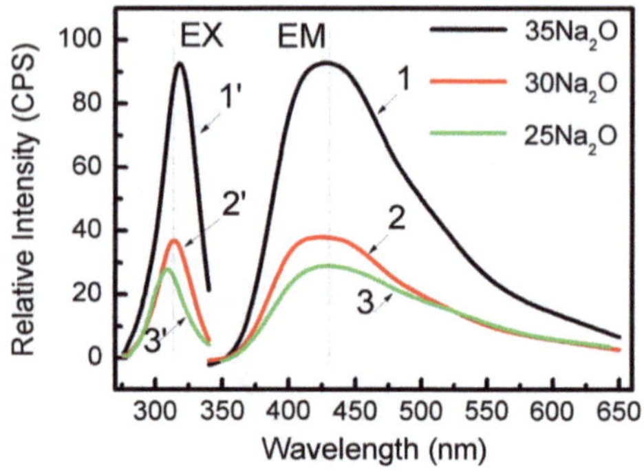

Fig. 14.3 Excitation (1′, 2′, 3′, λ_{em} = 429 nm) and emission (1, 2, 3, λ_{ex} = 305 nm) spectra of $x\text{Na}_2\text{O} \cdot (99.5 - x)\text{B}_2\text{O}_3 \cdot 0.5\text{Bi}_2\text{O}_3$ (x = 25, 30, 35) glasses. Reprinted from Ref. [44] by permission of OSA Publishing

they found brown coloration that was not uniformly distributed throughout the samples. We extended the study to a wider range $x\text{M}_z\text{O} \cdot (100 - x - y)\text{B}_2\text{O}_3 \cdot y\text{Bi}_2\text{O}_3$ (M = Li, Na, K, Ba; x = 5, 10, 15, 20, 25, 30, 33, 35; y = 0, 0.25, 0.5, 1.0, 1.5, 2.0, 2.5); however, we could only obtain clear colorless samples in the series. For bismuth-doped Na_2O–B_2O_3 binary glasses, only weak luminescence exists when soda content x increases from 5 to 20, and it is intensified markedly as x keeps increasing to 35 as can be seen in Fig. 14.3. This basically agrees with what Parke and Webb found in the glasses. They noticed that only when x is >15 can relatively stronger fluorescence be recorded. They also found that the absorption peak depends more on the concentration than the nature of the alkali ion. For instance, the absorption peaks at 230 nm (43,500 cm^{-1} due to the transition of ${}^1\text{S}_0 \rightarrow {}^1\text{P}_1$), which is the same as in $20\text{Li}_2\text{O} \cdot 80\text{B}_2\text{O}_3$, $20\text{Na}_2\text{O} \cdot 80\text{B}_2\text{O}_3$, and $20\text{K}_2\text{O} \cdot 80\text{B}_2\text{O}_3$ glasses. However, it moves toward the lower-energy side [225 nm (44,400 cm^{-1}) $\rightarrow$ 240 nm (41,700 cm^{-1})] as the concentration of alkali increases from 15 to 35 in sodium borate glasses [45]. Nevertheless, this comment does not hold all true for excitation and emission peaks as we will discuss below (see Figs. 14.3, 14.4 and 14.5). Because of instrumentation limitations, excitation in the spectral range of 200–250 nm cannot be recorded meaningfully.

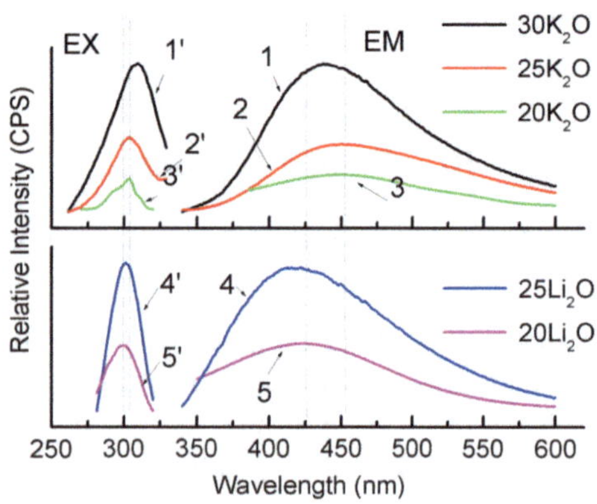

Fig. 14.4 Excitation (curves 1′, 2′, 3′, λ_{em} = 438 nm; 4′, 5′, λ_{em} = 416 nm) and emission (curves 1, 2, 3, 4, 5, λ_{ex} = 302 nm) spectra of $x\text{K}_2\text{O} \cdot (99.5 - x)\text{B}_2\text{O}_3 \cdot 0.5\text{Bi}_2\text{O}_3$ (x = 20, 25, 30) and $x\text{Li}_2\text{O} \cdot (99.5 - x)\text{B}_2\text{O}_3 \cdot 0.5\text{Bi}_2\text{O}_3$ (x = 20, 25) glasses. Reprinted from Ref. [44] by permission of OSA Publishing

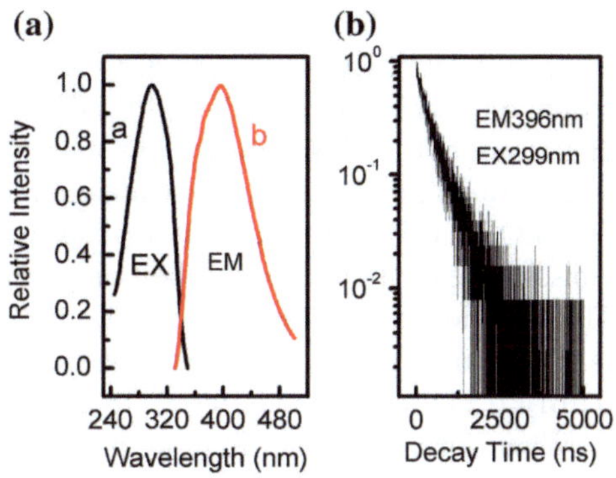

Fig. 14.5 **a** Excitation (a, $\lambda_{em} = 396$ nm) and emission (b, $\lambda_{ex} = 299$ nm) spectra and **b** fluorescence decay curve ($\lambda_{ex} = 299$ nm, $\lambda_{em} = 395$ nm) of $30Na_2O \cdot 69.5SiO_2 \cdot 0.5Bi_2O_3$ glass. Reprinted from Ref. [44] by permission of OSA Publishing

Figure 14.3 denotes the excitation and emission spectra of $xNa_2O \cdot (99.5 - x) B_2O_3 \cdot 0.5Bi_2O_3$ ($x = 25, 30, 35$) glasses. When x increases, emission and excitation peak exhibits different behaviors. The emission shifts blue albeit slightly (430 nm → 428 nm → 426 nm), whereas the excitation obviously shifts red [309 nm → 315 nm → 319 nm (see Fig. 14.1)]. At the same time the Stokes shift decreases (9106 cm^{-1} → 8382 cm^{-1} → 7874 cm^{-1}). The same scenario appears in bismuth-doped $Li_2O–B_2O_3$, $K_2O–B_2O_3$ and $BaO–B_2O_3$ glasses. The results are listed in Table 14.1. From Figs. 14.3, 14.4 and Table 14.1, we can notice that changes of alkali or alkali-earth content does not cause a significant shift of

Table 14.1 Excitation (λ_{ex} in nm), emission (λ_{em} in nm), Stokes shift (SS in cm^{-1}), Huang–Rhys parameter S, and fluorescence lifetime (τ_1 in ns by this work; τ_2 in μs by Parke and Webb) of Bi^{3+}-doped glasses. – means unobserved or unreported

Composition	λ_{ex}	λ_{em}	SS	S	τ_1	τ_2
$20Li_2O \cdot 79.5B_2O_3 \cdot 0.5Bi_2O_3$	298	427	10,138	4.25	–	–
$25Li_2O \cdot 74.5B_2O_3 \cdot 0.5Bi_2O_3$	300	419	9467	4.01	443	–
$20Na_2O \cdot 79.5B_2O_3 \cdot 0.5Bi_2O_3$	–	–	–		–	3.8
$25Na_2O \cdot 74.5B_2O_3 \cdot 0.5Bi_2O_3$	309	430	9106	3.87	519	–
$30Na_2O \cdot 69.5B_2O_3 \cdot 0.5Bi_2O_3$	315	428	8382	3.60	438	2.7
$35Na_2O \cdot 64.5B_2O_3 \cdot 0.5Bi_2O_3$	319	426	7874	3.42	475	2.7
$20K_2O \cdot 79.5B_2O_3 \cdot 0.5Bi_2O_3$	302	453	11,038	4.59	–	–
$25K_2O \cdot 74.5B_2O_3 \cdot 0.5Bi_2O_3$	304	450	10,673	4.45	526	–
$30K_2O \cdot 69.5B_2O_3 \cdot 0.5Bi_2O_3$	310	439	9478	4.01	452	–
$20BaO \cdot 79.5B_2O_3 \cdot 0.5Bi_2O_3$	302	432	9965	4.19	475	–
$25BaO \cdot 74.5B_2O_3 \cdot 0.5Bi_2O_3$	304	432	9747	4.11	444	–
$33BaO \cdot 66.5B_2O_3 \cdot 0.5Bi_2O_3$	308	432	9320	3.95	443	–
$30Na_2O \cdot 69.5SiO_2 \cdot 0.5Bi_2O_3$	299	395	8129		430	3.2
$49.5Na_2O \cdot 50P_2O_5 \cdot 0.5Bi_2O_3$	–	–	–		–	3.9

Reprinted from Ref. [44] by permission of OSA Publishing

emission peaks for the glass series of $BaO–B_2O_3$ and $Na_2O–B_2O_3$, particularly compared with $Li_2O–B_2O_3$ and $K_2O–B_2O_3$. For glasses with equal content of alkali oxide, e.g., $25M_2O \cdot 74.5B_2O_3 \cdot 0.5Bi_2O_3$ where M stands for Li, Na, and K, the emission of Bi^{3+} shifts toward low energy in the sequence Li → Na → K.

Figure 14.5a denotes the excitation and emission spectrum of $30Na_2O \cdot 69.5SiO_2 \cdot 0.5Bi_2O_3$ glass. It reveals one excitation peak at 299 nm and one emission peak at 395 nm for the transparent colorless sample (see inserted image). The emission peak differs from 385 nm by Parke and Webb [45]. The Stokes shift is 8129 cm^{-1}, which is comparable with the 8382 cm^{-1} of $30Na_2O \cdot 69.5B_2O_3 \cdot 0.5Bi_2O_3$ but smaller than the 9478 cm^{-1} of $30K_2O \cdot 69.5B_2O_3 \cdot 0.5Bi_2O_3$ glass.

When measuring the lifetime of Bi^{3+} in borate and silicate glasses, Parke and Webb selected pulsed exciting light with pulse width, i.e., FWHM of 200 ns, and they obtained a decay time ranging from 2.7 to 3.9 μs [45]. In this study based on the technology of time-correlated single photon counting, we used a hydrogen-filled gas nanosecond flash lamp with a typical pulse width of 1 ns. The typical decay curve of Bi^{3+} emission of doped silicate glass is illustrated in Fig. 14.5b, and it complies well with the single exponential decay equation, fit to which, respectively, yields 427 and 475 ns for the two glasses. The rest of the data are summarized as column "τ1" in Table 14.1. The typical lifetime of Bi^{3+}-doped borate and silicate glasses is approximately 500 ns [44].

In addition to types and concentrations of alkali or alkali earth, dopant concentration also affects the excitation and emission of Bi^{3+} as shown in Fig. 14.6. An increase in concentration leads to weakening intensities of emission and excitation peaks and simultaneous red shift of these peaks. The excitation peak varies from approximately 309 nm to approximately 330 nm, and the emission moves from approximately 425 to 440 nm, perhaps due to the self-absorption of Bi^{3+}. The lifetimes lie between 470 and 516 ns. This is very similar to the NIR emission of Bi-doped germanate glasses, but the dependence of optical properties on bismuth content is much stronger. For example, an increase of bismuth content from 0.01 to 2 % even leads to the shift of emission from 1100 to 1310 nm [46].

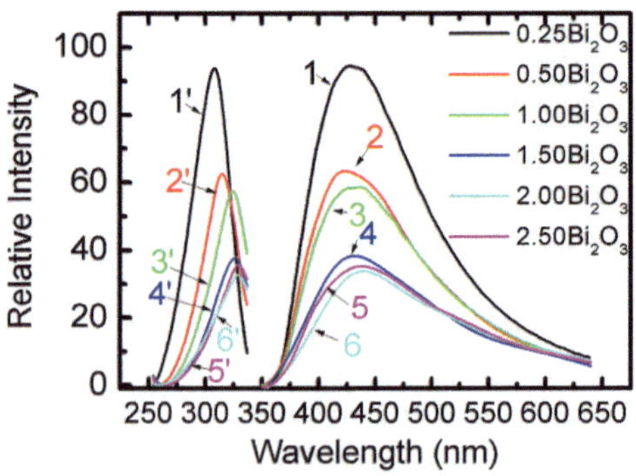

Fig. 14.6 Excitation (curves 1′, 2′, 3′, 4′, 5′, 6′, $\lambda_{em} = 430$ nm) and emission (curves 1, 2, 3, 4, 5, 6, $\lambda_{ex} = 305$ nm) spectra of $35Na_2O \cdot (65 - x)B_2O_3 \cdot xBi_2O_3$ ($x = 0.25, 0.5, 1.0, 1.5, 2.0, 2.5$) glasses. Reprinted from Ref. [44] by permission of OSA Publishing

14.3.2 *Mechanism for the Compositional Dependence of Bi^{3+} Emission in Borate and Silicate Glasses*

On the basis of spectroscopic data reported on Bi^{3+} doped crystals and glasses, it can be noticed that the first excitation band clearly corresponds to the dominant absorption transition from $^1S_0$ to $^3P_1$ at either room or lower temperature, e.g. 4.2 K, and the nature of emission transition strongly depends on temperature. After electrons have been raised to some vibrational level of the excited state of $^3P_1$, part of them will, at lower temperatures, relax to the lower-lying $^3P_0$ state by way of a nonradiative transition, and therefore the forbidden $^3P_0 \rightarrow {}^1S_0$ transition can be observed. However, at higher temperatures, electrons on the state of $^3P_0$, if there has ever been any electron populated on the state, will tend to be thermally depleted to $^3P_1$. As a consequence, the emission due to the transition of $^3P_1 \rightarrow {}^1S_0$ preponderates. Thus, at room temperature the excitation peak in borate and silicate glasses (see column "λ_{ex}" of Table 14.1) is attributed to the transition of $^1S_0 \rightarrow {}^3P_1$, and the emission to $^3P_1 \rightarrow {}^1S_0$ (see column "λ_{em}" of Table 14.1) [44].

Inspecting Figs. 14.3, 14.4 14.5 and Table 14.1 shows that, as stated earlier for Bi^{3+}-doped alkali or alkali-earth borate binary glasses, the excitation peak regularly shifts red and the emission peak shifts blue with increasing concentration of alkali or alkali-earth elements. As a result, the Stokes shift decreases. How does all this happen? To elucidate it, information on glass microstructural changes is helpful.

Figure 14.7 shows the FTIR spectra of $x\mathrm{BaO} \cdot (99.5 - x)\mathrm{B_2O_3} \cdot 0.5\mathrm{Bi_2O_3}$ ($x = 20, 25, 33$) and $x\mathrm{Na_2O} \cdot (99.5 - x)\mathrm{B_2O_3} \cdot 0.5\mathrm{Bi_2O_3}$ ($x = 25, 30, 35$) glasses. The spectra show the almost same trend: As the content of modifier ions increases, the amount of BO_4 anion groups grows at the expense of BO_3, possibly due to the introduction of additional oxygen along with the modifiers. This should be the reason why the Stokes shift decreases. It is known that the transitions of Bi^{3+} occur between the $6s^2$ and $6s6p$ configurations. Unlike rare-earth ions, they are easily influenced by ligand fields around the Bi^{3+} ion. The coupling strength of the electron and the phonon can be qualitatively measured by the Huang–Rhys factor S and it can be estimated from $S = 1/2(SS/\hbar\omega + 1)$ where $\hbar\omega$ is the maximum phonon energy, and SS means Stokes shift. Here, if the maximum phonon is taken as approximately 1350 cm^{-1} corresponding to the typical B–O stretching vibration of trigonal BO_3 unit in borate glasses, S can then be calculated and listed in Table 14.1. Based on the previous data, schematic configurational coordinate diagrams are constructed for two extreme cases—(a) lower content of alkali oxide with higher S and (b) higher content of alkali oxide with lower S—as denoted by Fig. 14.2a, b, respectively. For each series of doped samples, in M_2O–B_2O_3 (M = Li, Na, K), for case (a), the Bi^{3+} ion will experience stronger interaction with environment, and this leads to a larger offset ΔQ between $^1S_0$ and $^3P_1$, and shorter wavelength absorption, longer wavelength emission, and hence larger Stokes shift. Case (b) is the opposite of case (a) and results in a smaller Stokes shift. Figure 14.2 can also qualitatively apply to Bi^{3+}-doped BaO–B_2O_3 glass.

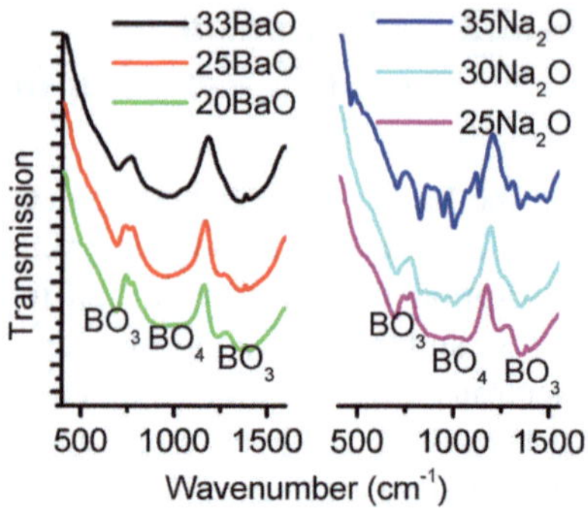

Fig. 14.7 The FTIR spectra of xBaO · (99.5 −x)B_2O_3 · 0.5Bi_2O_3 (x = 20, 25, 33) and $x$$Na_2O$ · (99.5 − x)B_2O_3 · 0.5Bi_2O_3 (x = 25, 30, 35) glasses. Reprinted from Ref. [44] by permission of OSA Publishing

14.4 Bi^{3+}-Doped Crystals

14.4.1 *Bi^{3+}-Doped Borate, Aluminate, Gallate, and Indate Crystals*

With an ordinary spectrometer, it is very difficult to study optical properties <200 nm, which lie in vacuum ultraviolet. Chen et al. tried a synchrotron radiation instrument to record the spectrum in the range. During their measurements, the electron energy was kept at 800 meV in storage ring, whereas the beam current was 100–250 mA. The pressure of the sample chamber was kept at approximately 1×10^{-3} Pa. They calibrated the excitation spectra with sodium salicylate. They found the absorptions due to $^1S_0 \rightarrow {}^1P_1$ and $^1S_0 \rightarrow {}^3P_1$ in addition to host absorption at approximately 170 nm (see Table 14.2) [47, 48]. As Tables 14.2 and 14.3 show, there are few reports on the absorption of $^1S_0 \rightarrow {}^1P_1$, perhaps due to the limits of instrumentation.

On excitation into host absorption, the Bi^{3+} emissions were observed at 294 and 330 nm, respectively. This indicates energy transfer from the host to the dopant. Because there are two crystallographic sites of Y^{3+} with symmetry of C_3 and S_6 in the compound, two sets of absorption and emission were found [47, 48].

We compiled the excitation and emission data of Bi^{3+}-doped borate, aluminate, gallate, and indate compounds in Table 14.2. According to these, we can find that few of these Bi^{3+}-doped compounds have absorption at wavelength longer than 340 nm. Most of the emissions lie in ultraviolet, and a few lie in blue or green.

14.4.2 *Other Bi^{3+}-Doped Compounds*

The excitation and emission data of Bi^{3+}-doped rare-earth oxides, zirconates, stannates, tungstate, niobate, sulfate, phosphate, silicate, germanate, and vanates, etc., are summarized in Table 14.3. Most emissions, some of which are yellow or

Table 14.2 Excitation (λ_{ex} in nm) and emission (λ_{em} in nm) of Bi^{3+}-doped crystals

Compounds	λ_{ex}		λ_{em}	References
	$^1S_0 \rightarrow {}^1P_1$	$^1S_0 \rightarrow {}^3P_1$		
$YBO_3:Bi^{3+}$	190	247	294	[47, 48]
	200	265	330	
$ScBO_3:Bi^{3+}$	–	284	299	[59]
$LaBO_3:Bi^{3+}$	–	248	365	[59, 60]
$LuBO_3:Bi^{3+}$	–	287	305	[59, 60]
$GaBO_3:Bi^{3+}$	–	275	290	[60]
		279	423	
$InBO_3:Bi^{3+}$	–	285	400	[59, 60]
$YAl_3B_4O_{12}:Bi^{3+}$		272	290	[59]
$Ca_2B_2O_5:Bi^{3+}$	216	293	357	[61]
$Ca_3B_2O_6:Bi^{3+}$	216	261	350	[61]
$Sr_3B_2O_6:Bi^{3+}$	217	271	364	[61]
$LiCaBO_3:Bi^{3+}$		304	378	[62]
$LaAlO_3:Bi^{3+}$		285	375	[63]
$MgAl_2O_4:Bi^{3+}$		335	400	[62]
$Ca_{12}Al_{14}O_{32}Cl_2:Bi^{3+}$		330	483	[64]
$Sr_3Al_2O_6:Bi^{3+}$		295	440	[65]
		330	400	
$Ca_3Al_2O_6:Bi^{3+}$		285	490	[66]
		305	375	
$Sr_2GdAlO_5:Bi^{3+}$		302	452	[67]
$Lu_3Al_5O_{12}:Bi^{3+}$		272	303	[67]
$Y_3Al_5O_{12}:Bi^{3+}$		274	307	[67]
$Y_3Ga_5O_{12}:Bi^{3+}$		285	320	[67]
$ZnGa_2O_4:Bi^{3+}$		315	520	[68]
$LaGaO_3:Bi^{3+}$		308	378	[62, 69]
$LaInO_3:Bi^{3+}$		340	420	[63]

even red, are in the visible range. For instance, $NaLaO_2:Bi^{3+}$ exhibits red emission at 620 nm, and $YVO_4:Bi^{3+}$ shows yellow emission at 566 nm.

Blasse et al. found in $CaSO_4:Bi^{3+}$ three structureless excitation bands at 354, 303 and 243 nm due to the transitions from $^1S_0$ to $^3P_1$, $^3P_2$, and $^1P_1$, respectively, and an intensive emission at 378 nm with a lifetime of 60 ± 10 ns due to the transition of $^3P_1 \rightarrow {}^1S_0$ at room temperature [49]. This is the only report we have found so far on the absorption of $^1S_0 \rightarrow {}^3P_2$. The Stokes shift is 2000 cm^{-1}, which is fairly low compared with the others (see Tables 14.1, 14.2 and 14.3). When the temperature was lowered to liquid helium, the structureless excitation and emission of $CaSO_4$: Bi^{3+} became structured. The emission red shifts from 378 to 400 nm, and the decay time becomes 330 ± 20 μs. This is the typical forbidden transition of $^3P_0 \rightarrow {}^1S_0$, which predominates over $^3P_1 \rightarrow {}^1S_0$ at liquid-helium temperature. The zero phonon

Table 14.3 Excitation (λ_{ex} in nm) and emission (λ_{em} in nm) of Bi^{3+}-doped crystals

Compounds	λ_{ex}		λ_{em}	References
	$^1S_0 \rightarrow {}^1P_1$	$^1S_0 \rightarrow {}^3P_1$		
$Y_2O_3:Bi^{3+}$		370	408	[57]
		340	520	
		368[1]	414[1]	[70]
$Gd_2O_3:Bi^{3+}$		372[1]/370	420[1]/458	[70, 71]
$La_2O_3:Bi^{3+}$		302[1]	450[1]	[70]
$YOCl:Bi^{3+}$		257	400	[59]
$LaOCl:Bi^{3+}$		272	345/444	[59]
$YOF:Bi^{3+}$		265	330	[59]
$La_2Zr_2O_7:Bi^{3+}$		290	385/515	[72]
$CaSb_2O_6:Bi^{3+}$		337	443	[73]
$SrSb_2O_6:Bi^{3+}$		352	487	[73]
$Gd_2GaSbO_7:Bi^{3+}$		290	370	[74]
$Y_2Sn_2O_7:Bi^{3+}$		280/285	330/510	[75]
		325	620	
$Lu_2WO_6:Bi^{3+}$		345[2]	510	[76]
$Y_2WO_6:Bi^{3+}$		341[2]	515	[59]
$CaWO_4:Bi^{3+}$		~300	464	[77]
$YNbO_4:Bi^{3+}$		310[2]	468/444[4]	[59, 78]
$GdNbO_4:Bi^{3+}$		307	445	[79]
$NaLaO_2:Bi^{3+}$		359	620	[80]
$NaGdO_2:Bi^{3+}$		326	400	[80]
$NaScO_2:Bi^{3+}$		314	413	[80]
$CaSO_4:Bi^{3+}$	243	303[3]/354	410	[49]
$CaS:Bi^{3+}$	316		450/550[5, 6]	[81, 82]
$SrS:Bi^{3+}$			490	[83]
$Sr_5(PO_4)_3Cl:Bi^{3+}$		370	440	[84]
$SrBiO_2Cl$			430	[85]
$BaBiO_2Cl$			490	[85]
$BaBiO_2Br$			500	[85]
$Bi_4Ge_3O_{12}$			459	[85]
$Ba_5SiO_4Cl_6:Bi^{3+}$		~300	480	[86]
$X_1–Y_2SiO_5:Bi^{3+}$		306	426	[87]
		372	414	
$X_2–Y_2SiO_5:Bi^{3+}$		293	343	[87]
			340/597	[88]
		295/326/370	355/408/504	[89]
$Lu_2SiO_5:Bi^{3+}$			348/566	[88]
$Gd_{4.67}Si_3O_{13}:Bi^{3+}$		370	455	[90]
$Zn_2SiO_4:Bi^{3\pm}$ Ga^{3+}			374–410[6]	[91]

(continued)

Table 14.3 (continued)

Compounds	λ_{ex}		λ_{em}	References
	$^1S_0 \rightarrow {}^1P_1$	$^1S_0 \rightarrow {}^3P_1$		
$YInGe_2O_7:Bi^{3+}$		302	457–496[5]	[92]
$KBaBP_2O_8:Bi^{3+}$		241	387	[62]
$YVO_4:Bi^{3+}$		~300[2, 7]	566	[37]
$LuVO_4:Bi^{3+}$		305[2, 7]	576	[36]
$ScVO_4:Bi^{3+}$		330[2, 7]	635	[35]

[1]On 1–2 × 10^6 V/cm
[2]Overlap with host absorption
[3]Due to $^1S_0 \rightarrow {}^3P_2$
[4]From Ref. [78]
[5]Change with bismuth content
[6]Long-lasting luminescence
[7]Extending to NUV

line (ZPL) of $^3P_0 \rightarrow {}^1S_0$ is very weak and lies at 384.0 nm, which is followed by two sets of emission lines. One of the series starts from ZPL, and they are separated by 490 cm^{-1}, which is due to the symmetric Bi–O stretching vibration. Another starts at a line at 320 cm^{-1} away from the ZPL, and it is stronger than the former progression. This frequency corresponds to the bending vibration, which has the symmetry to induce the intensity of $^3P_0 \rightarrow {}^1S_0$. For the transition of $^3P_1 \rightarrow {}^1S_0$, only one vibrational progression was found starting at the ZPL lying at 367.6 nm, and the energy interval is 490 cm^{-1} between neighboring lines, exactly corresponding to the symmetric Bi–O stretching mode. Because $^3P_1 \rightarrow {}^1S_0$ is electric dipole-allowed, it requires no vibronic coupling to gain intensity [49].

The excitation spectrum of $CaSO_4:Bi^{3+}$ shows structured features at liquid-helium temperature. The absorption progression of $^1S_0 \rightarrow {}^3P_1$ starts from ZPL at 366.6 nm, and the vibrational lines are separated by an energy interval of approximately 430 cm^{-1}, which is due to the stretching mode of Bi–O. The ZPL difference between 366.6 and 367.6 nm is perhaps due to self-absorption [49]. The Huang–Rhys factor is 1.5, and it shows a weak coupling between the electron and the phonons in the host. It therefore leads to a small offset of the excited state and then a small Stokes shift.

14.4.3 Bi^{3+}-Doped Vanadates

Different from other Bi^{3+}-doped other, we noticed that Bi^{3+}-doped rare-earth vanadates are promising phosphor candidates for NUV-WLED [35–37]. Here are the examples.

Yellow phosphor $LuVO_4:Bi^{3+}$ Members of the family of $LnBO_4$ compounds, in which Ln represents a lanthanide and B denotes P, V, or Nb, crystallize in the zircon crystal structure with the space group *I*4_1/*amd*, have long been known and

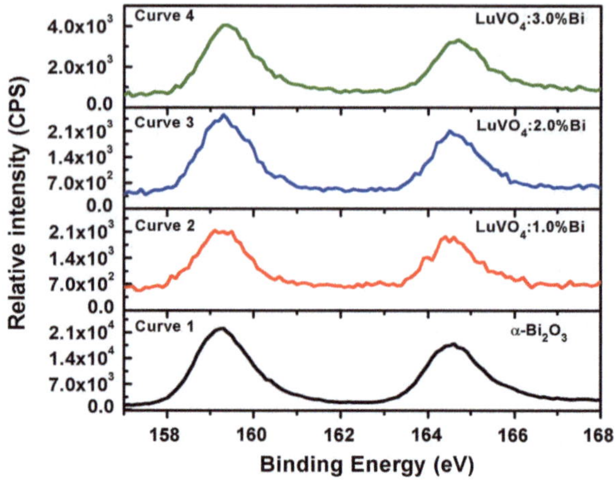

Fig. 14.8 XPS spectra of α-Bi_2O_3 (*curve 1*), $Lu_{0.99}VO_4$:1.0%Bi^{3+} (*curve 2*), $Lu_{0.98}VO_4$:2.0%Bi^{3+} (*curve 3*), and $Lu_{0.97}VO_4$:3.0%Bi^{3+} (*curve 4*). All of the curves shown in this figure have been shifted vertically for clarity. Reproduced from Ref. [36] by permission of John Wiley & Sons Ltd.

exploited as excellent laser hosts (e.g., for Er^{3+}, Nd^{3+}, or Tm^{3+}) by virtue of their outstanding thermal and chemical stability. If the energy levels lie at appropriate positions between the valence band (VB) and the conduction band (CB) of the $LuVO_4$ host, different colored phosphors can be generated by incorporating a dopant. Much effort has been made toward these phosphors, albeit mainly centered on rare-earth dopants [35–37].

Because the luminescence of bismuth strongly depends on its valence state, it is necessary to define its state in doped samples before further discussion. Therefore, we examined all of the samples by XPS. Selected spectra are shown in Fig. 14.8, which correspond to samples $Lu_{1\langle M\text{-}\rangle x}VO_4$:$x$Bi with $x = 1.0$, 2.0, and 3.0 %, respectively. For direct comparison, 99.999 % pure α-Bi_2O_3 was chosen as a reference, the XPS of which shows two characteristic Bi^{3+} peaks at approximately 159.3 and 164.6 eV due to $4f_{7/2}$ and $4f_{5/2}$, respectively. Clearly, the XPS positions of $Lu_{1-x}VO_4$:xBi match well with those of α-Bi_2O_3, thus implying the dominance of Bi^{3+} in the doped $LuVO_4$ samples. In addition, it is evident from Fig. 14.8 that, as expected, the intensity of the XPS peaks increases with increasing bismuth concentration [36].

As bismuth was doped into the compound, an expansion of unit cell occurred. For instance, the lattice parameters of $a = 7.0263$ Å, $c = 6.2352$ Å is for $Lu_{0.97}VO_4$:3.0%Bi^{3+}, and $a = 7.0268$ Å, $c = 6.2368$ Å is for $Lu_{0.95}VO_4$:5.0%Bi^{3+}. These parameters are larger than the $a = 7.0254$ Å, $c = 6.2347$ Å for the unmodified sample $LuVO_4$, which can be ascribed to the successful substitution of the smaller Lu^{3+} sites ($r = 0.977$ Å for coordination number 8) by larger Bi^{3+} ions ($r = 1.17$ Å for coordination number 8). This substitution is also promoted by the charge match between the two ions [36].

When exposed to UV light, a blank sample of $LuVO_4$ exhibits a blue emission peaking at 440 nm due to the VO_4^{3-} transition of $^1B(^1T_2) \rightarrow {}^1A_1$ (see Figs. 14.9 and 14.10) [36]. Monitoring at the emission at 440 nm produces the excitation spectrum of the blank sample, which comprises two broad bands with the maximum

Fig. 14.9 Room-temperature photographs of blank $LuVO_4$ (*top*) and $Lu_{0.98}VO_4$:2.0% Bi^{3+} (*bottom*) exposed to daylight, 254-nm lamp, 365-nm lamp, and 380–385 nm LED chip, respectively. Reproduced from Ref. [36] by permission of John Wiley & Sons Ltd.

at approximately 265 and 305 nm due to host absorptions from the ground state of $^1A_1$ to the excited states $^1E(^1T_1)$ and $^1B(^1T_2)$, respectively (see Fig. 14.10a). Once Bi^{3+} ions are incorporated into the $LuVO_4$ host, the excitation spectrum is broadened, and the excitation at longer wavelength side extends from 355 to 410 nm, whereas the two excitation positions of $VO_4{}^{3-}$ remain almost unchanged [36]. The broadening is due to the absorption of Bi^{3+} ions originating from the $^1S_0 \rightarrow {}^3P_1$ transition. When excited into the charge-transfer state of the $VO_4{}^{3-}$ group, for instance, at either 265 or 305 nm, there is always a Bi^{3+} emission at 576 nm (see

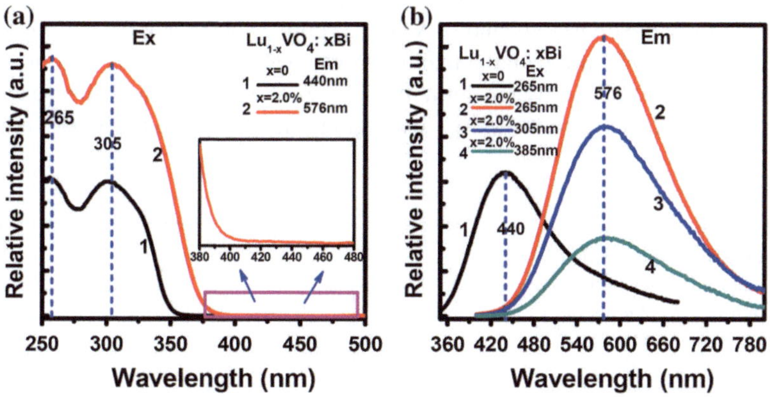

Fig. 14.10 **a** Excitation spectra of blank $LuVO_4$ (*curve 1*, $\lambda_{em} = 440$ nm) and $Lu_{0.98}VO_4$:2.0% Bi^{3+} (*curve 2*, $\lambda_{em} = 576$ nm) along with the enlarged excitation range of 380–480 nm for $Lu_{0.98}VO_4$:2.0%Bi^{3+} sample (*inset*); **b** emission spectra of blank $LuVO_4$ and $Lu_{0.98}VO_4$:2.0%Bi^{3+}. The excitation wavelengths are labeled beside each curve. Reproduced from Ref. [36] by permission of John Wiley & Sons Ltd.

Fig. 14.10). The emission maintains the same position as in he case when it is excited into the absorption of Bi^{3+} at 385 nm (see Fig. 14.10b). This is a clear sign of energy transfer from VO_4^{3-} to Bi^{3+}. The transfer can be further confirmed by the obvious shortening of the VO_4^{3-} lifetime from 9.89 to 7.21 μs as bismuth content increases from 0 to 5 % [36]. The independence of the emission on excitation wavelength and the constant position of the Bi^{3+} emission as the delay time is prolonged imply that there is only one type of Bi-emission center in $LuVO_4$. This agrees well with the Rietveld refining results. When exposed to 254- or 365-nm lamps, or even a UV LED chip with an emission in the spectral range of 380–385 nm, $Lu_{0.98}VO_4$:2.0%Bi^{3+} samples exhibit bright yellow emissions as depicted in Fig. 14.9 [36]. This means that $LuVO_4:Bi^{3+}$ may serve as a potential yellow phosphor for ultraviolet-converted WLEDs.

As known, quantum efficiency (QE) is crucial to assess the luminescence performance of one phosphor. We therefore measured the QE of the samples $LuVO_4$: xBi^{3+} (x = 0.5, 1.0, 2.0, 3.0 and 5.0 %). Figure 14.11 shows the dependences of QE on Bi^{3+} concentration on two different pump schemes: 380 nm (**curve 1**) and 330 nm (**curve 2**). It clearly reveals the optimal concentration of Bi^{3+} as 2.0 mol%. At this concentration, QE is 68 and 46 % for the excitations into 330 and 380 nm, respectively. It is higher than yellow phosphors of $Zn_3V_2O_8$ (QE = 52.0 %) [50] and $Ca_5Mg_4(VO_4)_6$ (QE = 41.6 %) [51] on excitation into 350 nm and even comparable with commercial YAG:Ce (QE = 70.0 %) [52]. In addition, one can see in the figure that the change tendency of QE shows similarity to the Bi^{3+}-emission intensity (**curve 3**). Namely, all values are initially enhanced with the Bi^{3+} concentration, and they reach the maximum at x = 2.0 %. Once the content of Bi^{3+} exceeds the optimal concentration, emission quenching appears and is accompanied by a rapid decrease in QE and intensity [36].

Red phosphor $ScVO_4:Bi^{3+}$ Fig. 14.12 shows the emission spectra (λ_{ex} = 265 nm) of $ScVO_4:xBi^{3+}$ (x = 0, 0.5, 1.0, 1.5, 2.0, and 3.0 %) samples. In a blank sample of $ScVO_4$, only the emission band of VO_4^{3-} groups is observed, presenting a broad band from 350 to 750 nm with the maximum value at approximately 465 nm. When small amounts of Bi^{3+} ions are built into the $ScVO_4$ matrix and then excited at 265 nm (which belongs to the VO_4^{3-} excitation), in addition to the VO_4^{3-} emission

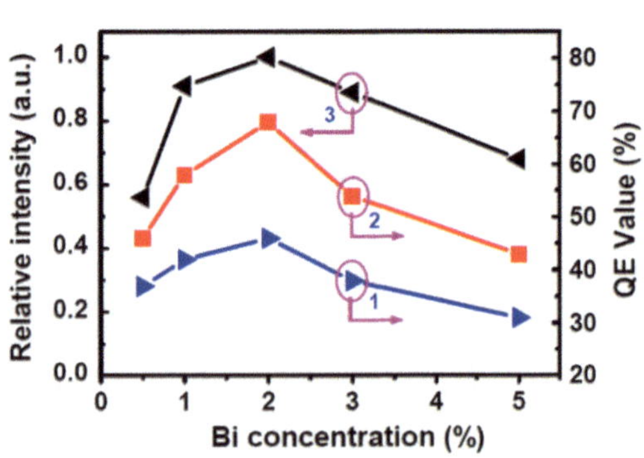

Fig. 14.11 QE values of $Lu_{1-x}VO_4:xBi^{3+}$ (x = 0.5, 1.0, 2.0, 3.0 and 5.0 %) samples on excitation wavelengths of 380 nm (*curve 1*) and 330 nm (*curve 2*) together with the corresponding emission intensity excited at 265 nm (*curve 3*). Reproduced from Ref. [36] by permission of John Wiley & Sons Ltd.

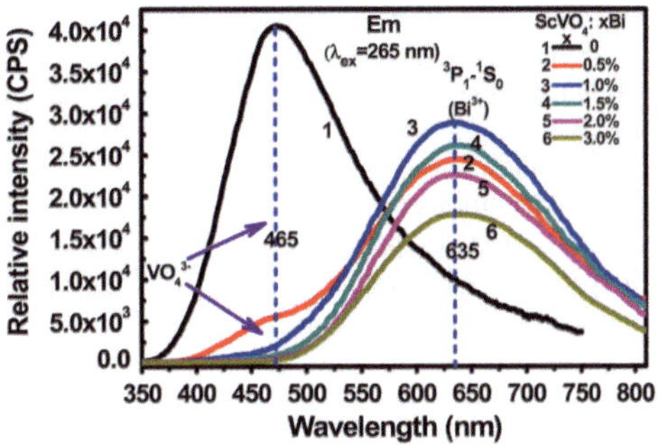

Fig. 14.12 Emission spectra of $ScVO_4$:$x$$Bi^{3+}$ samples (x = 0, 0.5, 1.0, 1.5, 2.0, and 3.0 %) on the excitation of 265 nm. Reproduced from Ref. [35] by permission of American Chemical Society

we can observe the characteristics emission of Bi^{3+} ions (450–810 nm) with the maximum value at approximately 635 nm. This broad band is ascribed to the $^3P_1 \rightarrow {}^1S_0$ transition of Bi^{3+} ions. Moreover, an increment of the Bi^{3+}-doping concentration can sharply decrease the VO_4^{3-}-emission intensity and quenches its luminescence completely when the Bi^{3+} concentration is up to 3.0 % [35]. Regarding the emission intensity of Bi^{3+}, however, it initially increases with increasing Bi^{3+} concentration until the Bi^{3+} content is up to 1.0 %, and then the intensity begins to decrease with further increment of Bi^{3+} concentration. This means that the Bi^{3+} emission can be excited by the VO_4^{3-} groups and that the corresponding emission positions do not depend on the Bi^{3+} concentration. The increase and decrease in the Bi^{3+} emission intensity can be attributed to the enhancement of the energy-transfer probability from VO_4^{3-} groups to Bi^{3+} and the Bi^{3+} concentration-quenching effect, respectively [35].

For blank sample of $ScVO_4$, two absorption bands within the range of 240–275 and 280–350 nm, which are ascribed to the absorption bands of the VO_4^{3-} groups, are observed. However, addition of 1.0 % molar Bi^{3+} into the $ScVO_4$ matrix induces another absorption band ranging from 360 to 430 nm [35]. The two absorption bands basically reconcile with the excitation bands of the corresponding samples as shown in Fig. 14.13. The two excitation bands at 265 and 330 nm are assigned to the matrix absorption from the $^1A_1$ ground state to ^1E ($^1T_1$) and ^1B ($^1T_2$) excited states within the VO_4^{3-} groups, respectively. Furthermore, the excitation profile of $ScVO_4$:1.0%Bi^{3+} monitoring at 635 nm is similar to that monitoring at 465 nm, which returns to prove the existence of energy-transfer progress from the VO_4^{3-} groups to Bi^{3+} [35]. However, because the existence of the Bi–O component referring to the $^1S_0 \rightarrow {}^3P_1$ transition of Bi^{3+} in the longer excitation-wavelength region, this excitation edge (approximately 430 nm), shown in Fig. 14.13 (curve 3), moves slightly toward longer wavelength, presenting much broader than that of the blank sample of $ScVO_4$. This indicates that doping of Bi^{3+} into the $ScVO_4$ matrix can tune the absorption band to meet the emission range of the NUV LEDs chip. Of course, monitoring at the emission wavelength of 465 nm slightly causes the appearance of Bi^{3+} excitation bands in $ScVO_4$:1.0%Bi^{3+} (see curve 1 and curve 2), which is due to the inevitable effect of the overlapped emission bands of the VO_4^{3-} groups and Bi^{3+}. Consequently, a broader excitation band (curve 1) emerges.

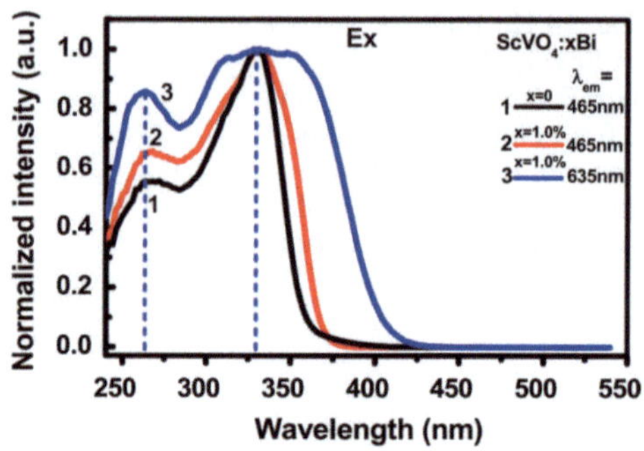

Fig. 14.13 Excitation spectra of $ScVO_4$:$x$$Bi^{3+}$ ($x = 0$, 1.0 %) samples: *curve 1* $\lambda_{em} = 465$ nm; *curve 2* $\lambda_{em} = 465$ nm; and *curve 3* $\lambda_{em} = 635$ nm. Reproduced from Ref. [35] by permission of American Chemical Society

Furthermore, the overlapped spectral range between the emission spectra of the VO_4^{3-} groups and the absorption bands of Bi^{3+} indicates that the energy transfer belongs to a radiative and subsequent reabsorption process [35].

Once bismuth is built into the matrix of $ScVO_4$, the sample color changes instantly to dark yellow as shown by Fig. 14.14. Accordingly, the emission changes from blue to brilliant red when exposed to either 254- or 365-nm UV lamps.

Quantum efficiency (QE) is obtained on 330-nm excitation at room temperature and is listed against bismuth content as curve 2 in Fig. 14.15. As can be seen in this figure, the QE value of blank $ScVO_4$ sample is 32 %, which is slightly larger than 30 % as reported in a previous article [35, 53]. When Bi^{3+} ions are incorporated into the $ScVO_4$ matrix, the QE value increases with increasing Bi^{3+} concentration, and a maximum QE value of approximately 56 % is observed at 1.0 % doping level. It is subsequently observed that the QE value decreases with further

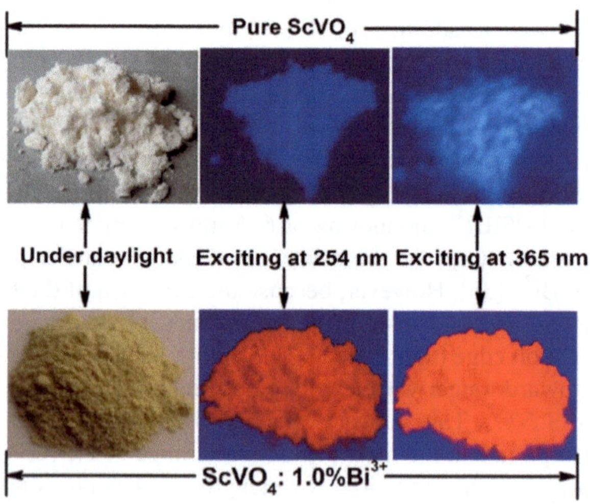

Fig. 14.14 Photoluminescence photographs of $ScVO_4$:$x$$Bi^{3+}$ ($x = 0$, 1.0 %) samples exposed to daylight, 254 and 365 nm, respectively. Reproduced from Ref. [35] by permission of American Chemical Society

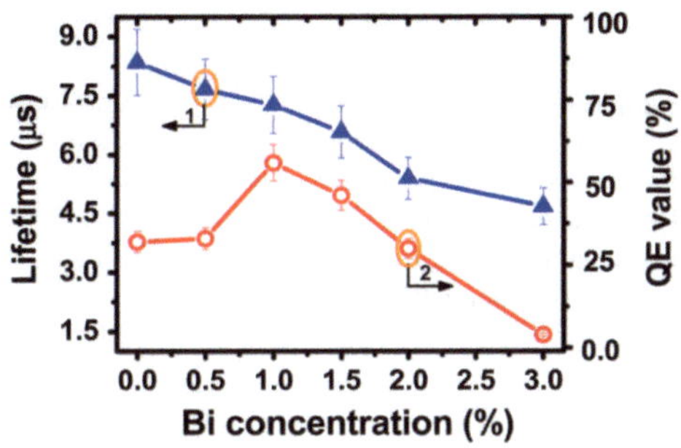

Fig. 14.15 The fitted VO_4^{3-} lifetimes (*curve 1*: $\lambda_{ex} = 265$ nm, $\lambda_{em} = 465$ nm) and quantum efficiency (QE) (*curve 2*: $\lambda_{ex} = 330$ nm) measured at room temperature plotted as Bi^{3+} concentration. Reproduced from Ref. [35] by permission of American Chemical Society

increasing Bi^{3+} concentration. This is consistent with the observed Bi^{3+}-emission intensity in Fig. 14.12, implying that the QE of Bi^{3+} in this work strongly depends on the efficiency of the energy transfer from the VO_4^{3-} groups to the Bi^{3+} ions. When the Bi^{3+} ions are at low concentration, the energy transfer from VO_4^{3-} groups to the Bi^{3+} ions is inefficient. Consequently, the QE at 0.5 % doping level is only 33 % ($\lambda_{ex} = 330$ nm). With the increase in concentration of Bi^{3+} ions, the energy transfer from the VO_4^{3-} groups increases gradually, thus leading to the higher quantum yield. However, if the Bi^{3+} concentration is too high, self-quenching of the Bi^{3+} ions begins. Consequently, the optimal doping concentration of Bi^{3+} ions is determined to 1.0 %. Noteworthy, excitation at a wavelength of 380 nm produces a QE value of approximately 47 % for $ScVO_4$:1.0% Bi^{3+} [35].

To obtain more information about the kinetics of luminescence, the time-resolved emission spectra of $ScVO_4$:1.0%Bi^{3+} were recorded on excitation into the absorption of VO_4^{3-}, e.g., at 265 nm at room temperature. The collected delay time ranges from 0 to 50 μs as plotted in Fig. 14.16. It shows there are at least two types of emission centers in $ScVO_4$:1.0%Bi^{3+}, viz., Bi^{3+} and VO_4^{3-} groups, and energy transfer does occur from the VO_4^{3-} groups to Bi^{3+}. The emission of the VO_4^{3-} groups appears first at 465 nm, and it increases initially along the delay time

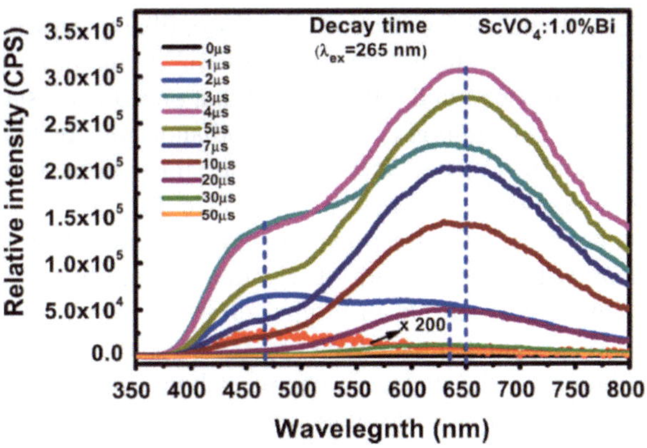

Fig. 14.16 Time-resolved spectra of the sample $ScVO_4$:1.0%Bi^{3+} on excitation into 265 nm. The collected time ranges from 0 to 50 μs. Reproduced from Ref. [35] by permission of American Chemical Society

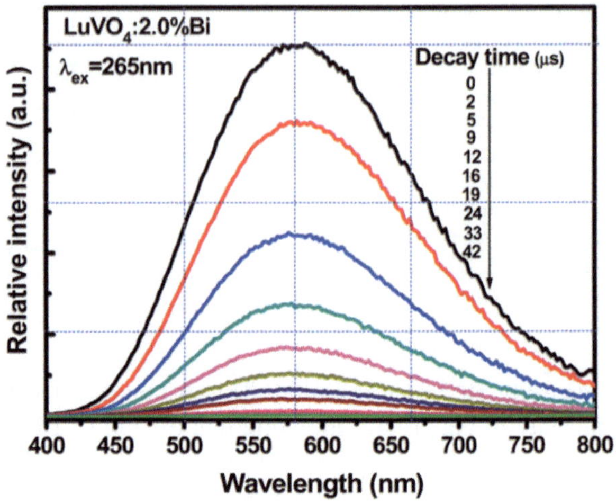

Fig. 14.17 Time-resolved spectra of $LuVO_4$:2.0%Bi sample on excitation into 265 nm. The collected time ranges from 0 to 42 μs. Reproduced from Ref. [37] by permission of The Royal Society of Chemistry

t and reaches the maximum at $t = 3$ μs. It disappears visually at $t = 20$ μs. At the expense of the VO_4^{3-} emission, the characteristic emission peak of Bi^{3+} emerges at 635 nm as $t = 2$ μs, and its intensity gradually increases with the increase of the delay time. When *t* is prolonged to 4 μs, the intensity reaches the maximum, and it is continuously enhanced more than five times. At the same time, the peak shifts to approximately 650 nm. After $t = 4$ μs, the intensity continuers to decrease. The peak gradually shifts back to 635 nm when $t = 20$ μs, and the peak position has since then remained constant [35]. The mechanism will be discussed later. Although $LuVO_4$ is isostructural to $ScVO_4$, the time-resolved spectra of Bi^{3+} emission are very different, and there is only one emission center in $LuVO_4$:Bi^{3+} as Fig. 14.17 shows.

14.4.4 Tuning Bi^{3+} PL

Bismuth naked 6 s electrons are susceptible to crystal field, particularly surrounding Bi^{3+}, and this enables the tuning of Bi^{3+} photoluminescence (PL) as it is doped into different compounds. We noticed that once the crystal field around Bi^{3+} is tuned by gradual substitution of smaller lanthanon ion for larger lanthonon ion in the solid solution of (Y, Lu, Sc)VO_4:Bi, the emission peak position can be modulated linearly between 566 and 635 nm (see Fig. 14.18) [37]. The inset shows digital photographs of corresponding samples exposed to natural light (upper) and a UV lamp with a wavelength of 254 nm. From left to right, the sample emission color changes gradually from bright yellow to orange and red.

We calculated the covalency (f_c) of lanthanon (Ln) to oxygen bond in the solid solution by the dielectric chemical bond theory of complex ionic crystals on the basis of the Rietveld refined crystallographic data and tried to show its underlying

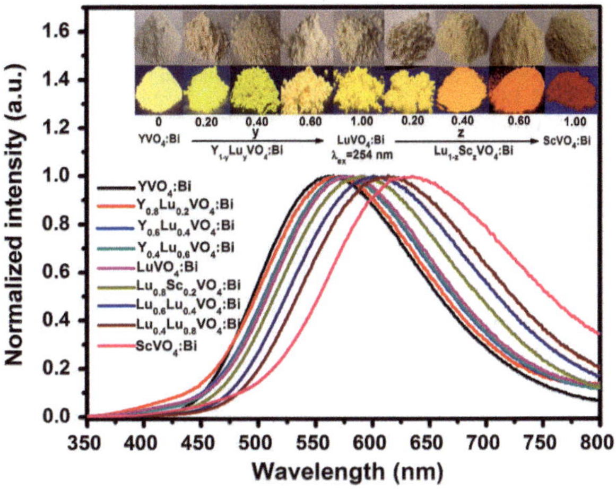

Fig. 14.18 Emission spectra of solid solution compounds $(Y_x,Lu_y,Sc_z)_{1-a}VO_4$:$a$Bi where x, y, or z changes between 0 and 1.0, respectively, whereas a equals 0.02 on excitation into 265 nm. The inset shows digital photographs of corresponding samples exposed to natural light (upper) and a UV lamp with a wavelength of 254 nm. Reproduced from Ref. [37] by permission of The Royal Society of Chemistry

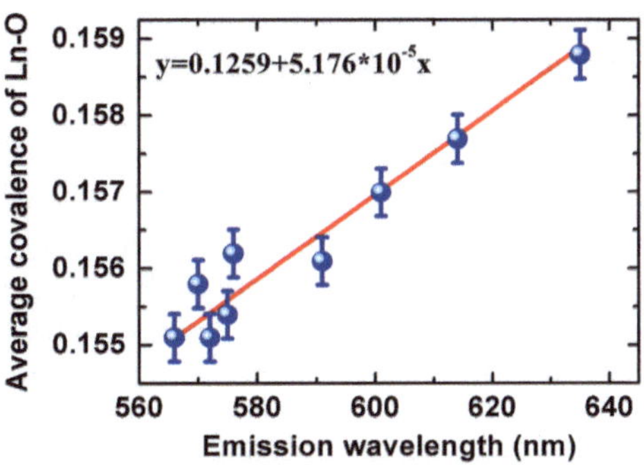

Fig. 14.19 Dependence of Bi^{3+} emission position on the average covalency of Ln–O in the solid solution. The *red line* is generated with the equation $y = 0.1259 + 5.176 \times 10^{-5}x$. Reproduced from Ref. [37] by permission of The Royal Society of Chemistry

relationship to the emission. Figure 14.19 illustrates the average covalence f_c of Ln–O against the emission peak, and it indicates that the increase of f_c basically leads to the red shift of the Bi^{3+} emission [37]. A linear equation can be established between them: $y = 0.1259 + 5.176 \times 10^{-5}x$ as depicted in Fig. 14.19. This implies again the sensitivity of Bi^{3+} to the surrounding environment, which governs the tunability of Bi^{3+} emission. Once bismuth was introduced into the samples, the samples become light brown (see inset of Fig. 14.18), and the excitation peaks are broadened particularly at in longer-wavelength region due to superimposition of

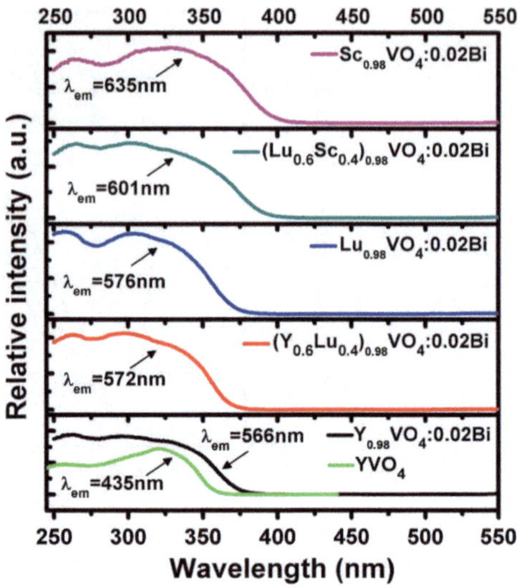

Fig. 14.20 Excitation spectra of solid solution (Y_x, Lu_y, Sc_z)$_{1-a}$$VO_4$:$a$Bi where x, y, or z changes between 0 and 1.0, respectively, whereas a varies between 0 and 0.02. Monitored emission wavelengths are labeled beside each curve, and they correspond to the emission peak of each sample. Reproduced from Ref. [37] by permission of The Royal Society of Chemistry

bismuth excitation peak ($^1S_0 \rightarrow {}^3P_1$) (see Fig. 14.20) [37]. For instance, the tail of excitation of YVO_4 stops at approximately 360 nm, whereas $Y_{0.98}VO_4$:2%Bi extends to approximately 380 nm. As shown Fig. 14.20 from bottom to top, i.e., along with the decrease of lattice parameters and therefore the increase of covalency, the tail of the bismuth excitation peak moves continuously to approximately 410 nm from approximately 380 nm [37].

As Table 14.4 depicts, there are similar reports of tuning Bi^{3+} PL by forming a solid solution in $In_{1-x}Lu_xBO_3:Bi^{3+}$, $Y(P_xV_{1-x})O_4:Bi^{3+}$, $Y_3(Al, Ga)_5O_{12}:Bi^{3+}$, and (Ca, M^+)$MoO_4:Bi^{3+}$ (M = Li, Na, K) [39, 54–56]. However, the tunable range is much narrower than in the case of (Y, Lu, Sc)VO_4:Bi, and the peaks mostly lie in the shorter-wavelength region. Because our concern focuses on the phosphors for NUV-WLED, we will not discuss these phenomena in details.

Table 14.4 Excitation (λ_{ex} in nm) and emission (λ_{em} in nm) of Bi^{3+}-doped solid solution compounds

Compounds	λ_{ex}	λ_{em}	References
	Tunable range	Tunable range	
$In_{1-x}Lu_xBO_3:Bi^{3+}$	284–287	370–400	[54]
$Y(P_xV_{1-x})O_4:Bi^{3+}$	315–355	558–573	[56]
$Y_3(Al, Ga)_5O_{12}:Bi^{3+}$	275–288	304–317	[55]
(Y, Lu, Sc)$VO_4:Bi^{3+}$	375–410[a]	566–635	[37]
(Ca, M)$MoO_4:Bi^{3+}$ (M = Li, Na, K)	266–287[1] 320[2]	554–586	[39]

[a]Absorption edge
[1]Host absorption
[2]Bi^{3+} absorption

14.4.5 The Resistance of Bi^{3+} PL to Thermal Impact

As well known, temperature can greatly influence the PL properties of one phosphor, and some peculiar performances perhaps appear at high or low temperatures. In view of this, we performed a study of the influence of temperature on PL properties. Here, for example, we present study results only on the yellow phosphor $LuVO_4$:Bi [36].

Figure 14.21 presents the excitation spectra of $Lu_{0.98}VO_4$:2.0%Bi^{3+} sample between 10 and 300 K. As the temperature increases from 10 to 300 K, the excitation intensity of VO_4^{3-} monitored at 440 nm (a) decreases, whereas the excitation intensity monitored at 576 nm increases (b) [36]. This hints at the enhancement of energy transfer from VO_4^{3-} to Bi^{3+} with increasing temperature. It can also be evidenced in Fig. 14.22, which illustrates the emission spectra in the same range of temperature. As temperature increases to 300 K from 10 K, the emission of VO_4^{3-} on the excitation into its charge-transfer state by 265 nm monotonically decreases and eventually emerges into the emission of Bi^{3+}. In contrast, the Bi^{3+} emission becomes dominant at the expense of the VO_4^{3-} emission. This vividly reveals the strong dependence of energy transfer from VO_4^{3-} groups to Bi^{3+} on temperature. At low temperature, the process seems frozen, but it will be enabled at higher temperature (see Fig. 14.22). An emission spectrum at medium temperature of 100 K, as illustrated in Fig. 14.22 (inset b), records the gradually growing emission of Bi^{3+} ion and the disappearing emission of the VO_4^{3-} group. It is obvious that the energy transfer is a thermal-activated process. Surprisingly, as exciting at 380 nm, temperature seems have less influence on the Bi^{3+} emission at 576 nm as can be seen in Fig. 14.22 (inset a). It remains almost unchanged in the range of 10–300 K, and the mechanism differs from the scheme of excitation into the charge-transfer state of the VO_4^{3-} group [36].

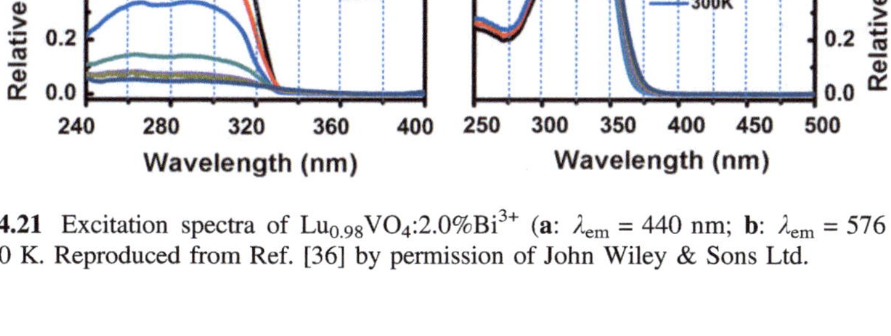

Fig. 14.21 Excitation spectra of $Lu_{0.98}VO_4$:2.0%Bi^{3+} (**a**: λ_{em} = 440 nm; **b**: λ_{em} = 576 nm) at 10–300 K. Reproduced from Ref. [36] by permission of John Wiley & Sons Ltd.

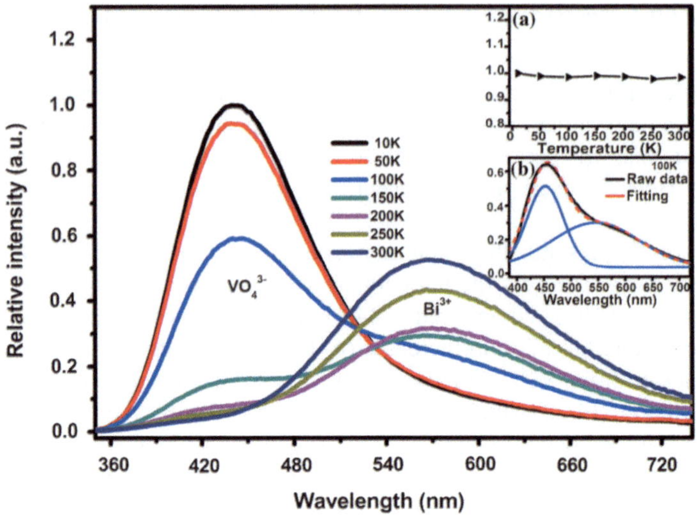

Fig. 14.22 Emission spectra of $Lu_{0.98}VO_4$:2.0%Bi^{3+} excited at 265 nm at 10 to 300 K. Inset **a**: Dependence of the relative emission intensity of $Lu_{0.98}VO_4$:2.0%Bi^{3+} on temperature on excitation of 380 nm; Inset (**b**). Gaussian-peak decomposition of the emission spectrum at 100 K. Reproduced from Ref. [36] by permission of John Wiley & Sons Ltd.

Figure 14.23 depicts the emission spectra of $Lu_{0.995}VO_4$:0.5%Bi^{3+} at temperatures ranging from 25 to 300 °C on excitation at 385 nm. As the sample was heated up, the emission intensity increases at first; then it reaches maximum at 100 °C, and decreases afterward [36]. This is accompanied by the simultaneous blue shift of emission from 576 to 558 nm (see Figs. 14.23 and 14.24). It happens to all of the samples activated by different content of bismuth. The temperature $T_{50\ \%}$ at which the emission intensity decreases to 50 % of that at room temperature, is approximately 250 °C for $Lu_{0.995}VO_4$:0.5%Bi^{3+}. For samples doped with higher content of Bi^{3+}, for instance,

Fig. 14.23 Emission spectra (λ_{ex} = 385 nm) of $Lu_{0.995}VO_4$:0.5%Bi^{3+} measured at 25–300 °C. Insets show the photographs of $Lu_{0.98}VO_4$:2.0%Bi^{3+} exposed to daylight at 25 °C (*left*) (RT: room temperature); (*middle*) 254 nm at 25 °C; and (*right*) 254 nm at 200 °C. Reproduced from Ref. [36] by permission of John Wiley & Sons Ltd.

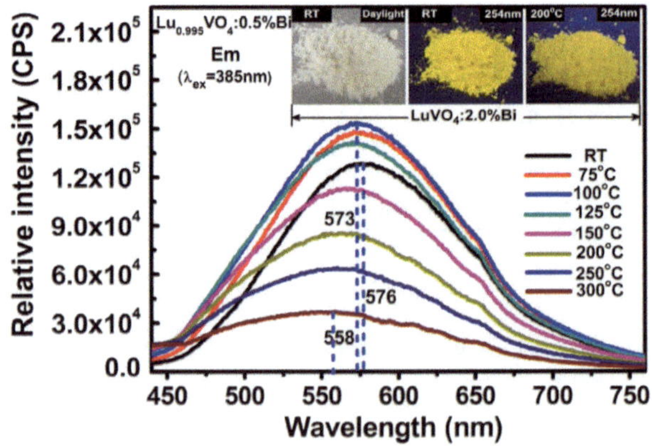

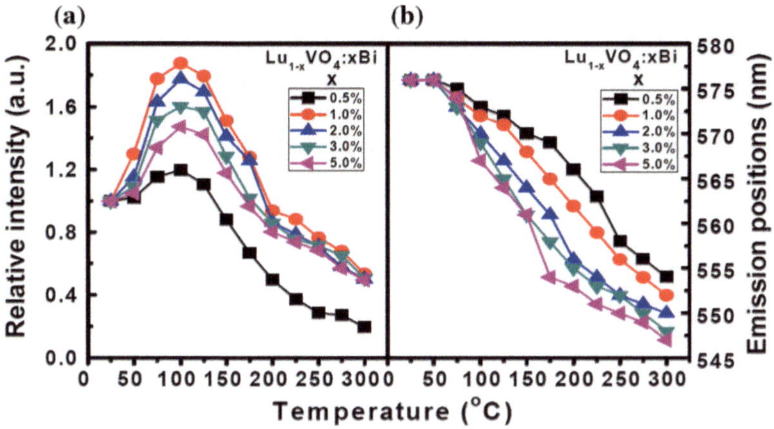

Fig. 14.24 The dependence of **a** emission intensity and **b** emission positions of $Lu_{1-x}VO_4$:xBi^{3+} (x = 0.5, 1.0, 2.0, 3.0 and 5.0 %) on temperature on excitation of 385 nm. The temperature is from 25 to 300 °C. Reproduced from Ref. [36] by permission of John Wiley & Sons Ltd.

1–5 %, $T_{50\ \%}$ even increases to approximately 300 °C (see Fig. 14.24). This can explain why the sample of $Lu_{0.98}VO_4$:2.0%Bi^{3+} remains brilliant at 250 °C [see Fig. 14.23 (inset)]. Carefully inspecting the initial stage from 25 to 100 °C reveals that the emission intensity of Bi^{3+} in the sample of $Lu_{0.995}VO_4$:0.5%Bi^{3+} at 100 °C is approximately 1.196 times stronger than that at 25 °C. Interestingly, this can be further improved to 1.878, 1.778, 1.604, and 1.473 times of that at 25 °C as the bismuth content increases to 1, 2, 3, and 5 %, respectively [36]. The anti-thermal-quenching phenomenon is extremely rare in phosphors.

Figure 14.25 depicts the heating–cooling cycle experiments on $Lu_{0.995}VO_4$:0.5% Bi^{3+} on excitation of 385 nm. The first heating reproduces Fig. 14.23, and slight

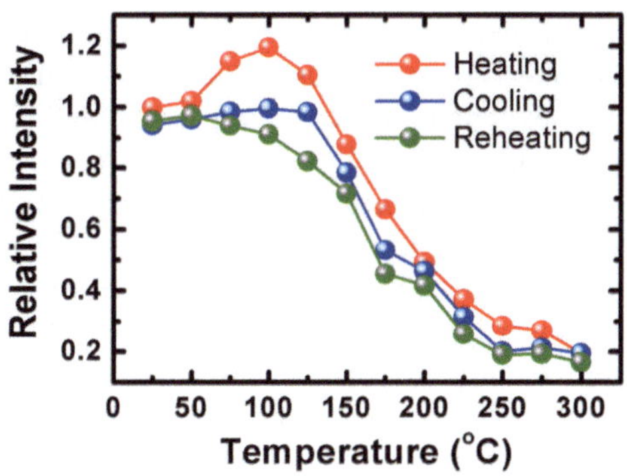

Fig. 14.25 Temperature-dependent relative emission intensity of $Lu_{0.995}VO_4$:0.5%Bi^{3+} in a repeated heating and cooling processes. Reproduced from Ref. [36] by permission of John Wiley & Sons Ltd.

increase in emission intensity was observed followed by a decrease. Consequent cooling does not restore the emission to the original state of first heating. At each temperature between 25 and 300 °C, the emission intensity is little lower in the cooling. The second heating also leads to an initial tiny increase and a subsequent decrease in emission intensity [36]. If the random error of measurement is concerned, the intensity of the emission almost recovers to the initial states at 25 or 300 °C after the cycle experiments, although the intensity is different at the intermediate state (see Fig. 14.25). To figure out the mechanism behind the phenomenon, we measured the TL spectra of these samples. Blank $LuVO_4$ sample does not exhibit TL peak, whereas there is a TL band ranging from 40 to 150 °C in the sample of $Lu_{0.995}VO_4$:0.5%Bi, which peaks at 86 °C. Reheating $Lu_{0.995}VO_4$:0.5%Bi can erase the TL peak. These results encouraged us to correlate the anti-thermal-quenching, at least partially, to the existence of the trap with a depth of 86 °C [36].

Carefully examining Figs. 14.23 and 14.24b, we can determine that the emission shifts blue at higher temperature or concentration of bismuth. The blue shift becomes more obvious particularly for the compound doped with higher content of Bi^{3+} at higher temperature. We can also see that there is still a slight increase in Bi^{3+} emission during either cooling or the second heating of $Lu_{0.995}VO_4$:0.5%Bi. How are all of these phenomena occurring? To determine the mechanism dominating the phenomena, we further measured the decay curves of the Bi^{3+} emission at 576 nm of $Lu_{0.98}VO_4$:2.0%Bi^{3+} at a temperature range of 10–500 K (see Fig. 14.26) [36]. The results show that as the temperature increases, the dominating decay behavior of Bi^{3+} changes from a double to single exponential decay equation. When the temperature is $\leq$300 K (see Fig. 14.26), the decay curves can well be fitted by the double exponential equation as follows:

$$I(t) = A_1 \exp(-t/\tau_1) + A_2 \exp(-t/\tau_2)$$

where τ_1 and τ_2 are the short and long decay components, respectively, and parameters A_1 and A_2 are the fitting constants, respectively. Clearly, as temperature increases, τ_2 decreases sharply from 243.78 to 10.70 μs, but τ_1 does not go through a dramatic decrease and instead fluctuates between 13.46 and 5.30 μs. However, it is unexpectedly found that further increasing the temperature from 350 to 500 K only produces the single exponential decay curve as shown in Fig. 14.26. The fitted lifetimes based on the equation $I(t) = A \exp(-t/\tau)$ show the shortest Bi^{3+} lifetime as approximately 542 ns (namely, approximately 0.542 μs) at 500 K. We proposed Fig. 14.27 to explain this as well as the changes in emission intensity and position along with temperature. It is based on the population redistribution between $^3P_0$ and $^3P_1$ at different temperatures [36].

In either scheme of excitation, for instance by 385 nm or an energy transfer from VO_4^{3-} to Bi^{3+}, electrons will be preferably lifted to the excited state of $^3P_1$. The electrons prefer relaxing to the bottom of the $^3P_1$ parabola along path 1, from where some of them first release energy in the form of yellow emission. At lower temperature, most electrons will flow across B along path 2 and relax to the bottom of the lower level of $^3P_0$. Some electrons on the $^3P_0$ state will emit yellow lights

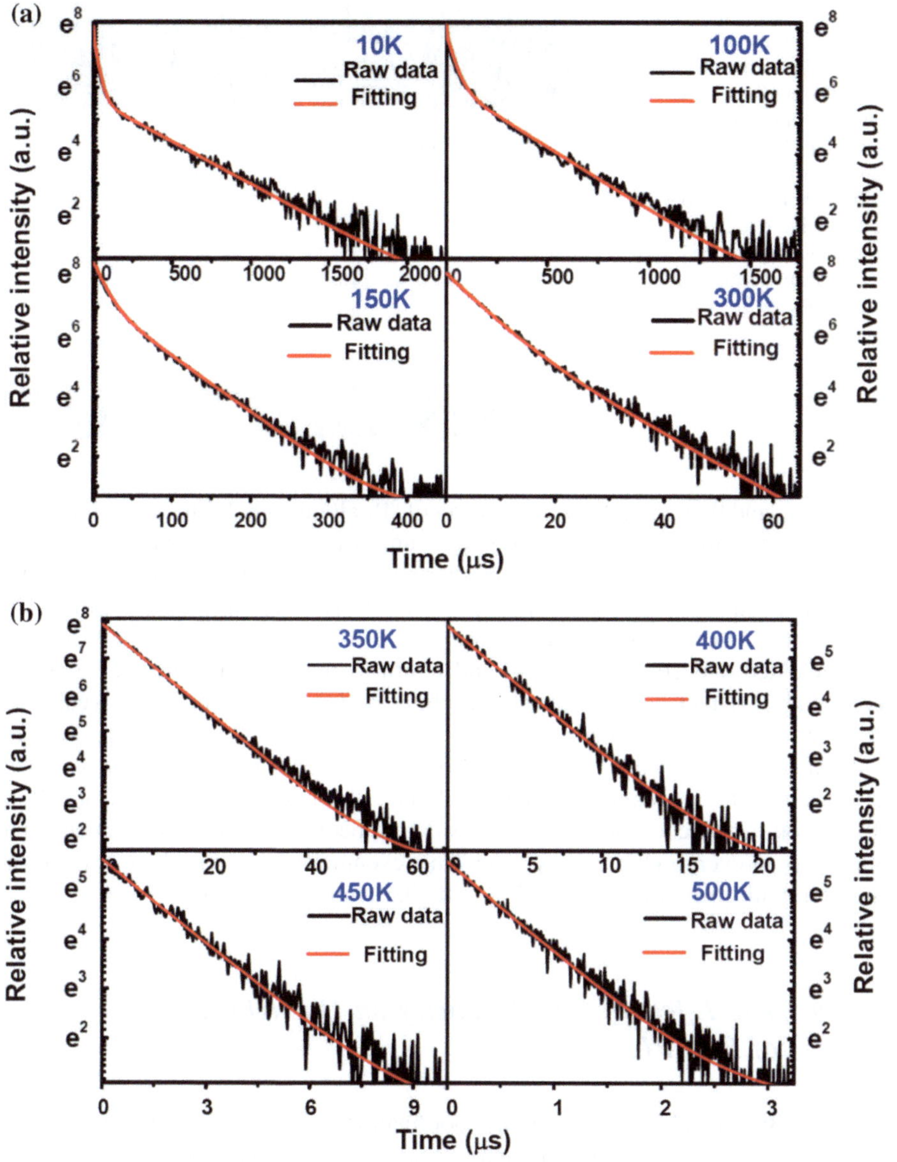

Fig. 14.26 PL decay curves (λ_{ex} = 265 nm, λ_{em} = 576 nm) of $Lu_{0.995}VO_4$:0.5%Bi^{3+} within the temperature range of **a** 10–300 K and **b** 350–500 K. Fittings were performed with double- and single-decay exponential equation for curves in **a** and **b**, respectively. Reproduced from Ref. [36] by permission of John Wiley & Sons Ltd.

peaking at longer wavelength with a lifetime up to more than 200 μs (see Fig. 14.26). Such long lifetime can lead to the excited state absorption. The electrons on the $^3P_0$ state prefer being thermally excited as the temperature continues

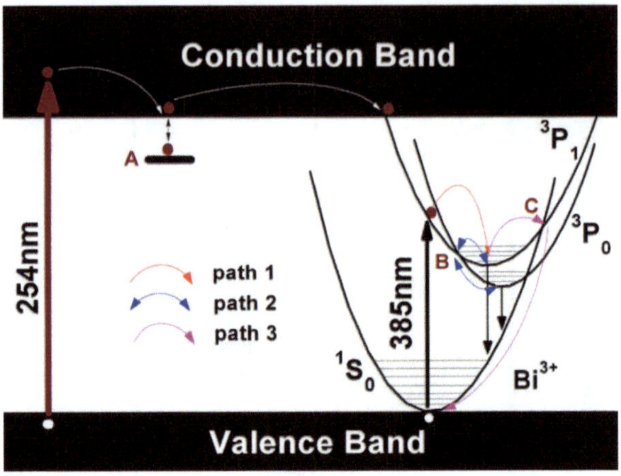

Fig. 14.27 Scheme for the anti-thermal-quenching of Bi^{3+} luminescence in $LuVO_4$: Bi^{3+}. The *white solid circle* stands for hole; the *wine solid circle* for electron; and the depth of trap A is 86 °C. Paths 1 though 3 denote the nonradiative relaxation processes. Reproduced from Ref. [36] by permission of John Wiley & Sons Ltd.

increasing, and they jump back to the level of $^3P_1$ along path 2. The transition of $^3P_1$ to $^1S_0$ again dominates the emission spectrum at higher temperature; at the same time, the emission lifetime is greatly shortened (see Fig. 14.13). Meanwhile, the emission is first intensified along with blue shifts (see Figs. 14.23, 14.24, and 14.25) [36]. At even higher temperature, e.g., 500 K, more electrons will be thermally excited and cross through C and return to the ground state nonradiatively along path 3. This leads to thermal quenching, and thus the emission is weakened. Its lifetime is correspondingly reduced to approximately 542 ns. Perhaps bismuth will experience a stronger interaction when its concentration is higher at higher temperature. On one hand, this could facilitate the population inversion from $^3P_0$ to $^3P_1$; in contrast, it could lead to further splitting between $^3P_1$ and $^3P_0$, which could lift $^3P_1$ to an even higher energy state. This eventually leads to a blue shift of the emission with bismuth concentration at higher temperature [36].

14.4.6 The Effect of Site Symmetry on Quenching of Bi^{3+} PL

Because Bi^{3+} resides in a similar coordination environment, the luminescent properties depend on the site symmetry because of the stereochemistry of the filled lone pair of 6*s* electrons. Here we show an example in bismuth-doped Y_2O_3. In this compound, there are two kinds of yttrium sites. Both of them are six-coordinated by oxygen atoms with symmetry of S_6 and C_2, respectively. Once bismuth ions are doped into the compound, they will preferentially occupy these sites, respectively. Bi^{3+} on S_6 site exhibits an emission at 408 nm and an excitation at 370 nm, and Bi^{3+} on C_2 site shows an emission at 520 nm and an excitation at 340 nm (see Table 14.3) [57]. The Stokes shift is much smaller for the former than the latter. However, the emission intensity of the latter is almost three times stronger than that

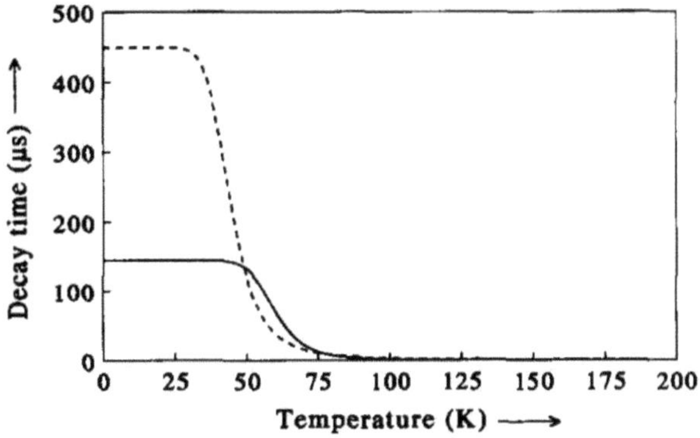

Fig. 14.28 Temperature dependence of the decay time of Bi^{3+} luminescence in $Y_2O_3:Bi^{3+}$: *curve* (—) indicates the S_6 site, and *curve* (- - -) indicates the C_2 site. Reprinted from Ref. [57] by permission of Elsevier

of the former. Both sites show a clear maximum of intensity at $T \approx 100$ K, which is followed by quenching as temperature increases to 275 K. At the same time, the decay time decreases rapidly in the temperature range, whereas the intensity increases (see Fig. 14.28).

Compared with the high symmetry of site, the asymmetrical fields have a drastic influence on the spectroscopic properties of Bi^{3+}. The lowest excitation level shifts toward higher energy, whereas the Stokes shift increases. The S_6 site has better resistance to thermal impact than the C_2 site at higher temperature [57]. A similar conclusion was reported in Bi^{3+}-doped yttrium or gadolinium niobates [58].

14.5 The Application of Bi^{3+}-Doped Phosphors in NUV-WLED

The above experimental results on Bi-doped rare-earth vanadates encouraged us to check the potential application of the phosphors in pc-WLED. Therefore, we fabricated a series of LED lamps by depositing for example the $Lu_{0.98}VO_4$:2.0%Bi^{3+} phosphor and the commercial (Ba, Eu)$MgAl_{10}O_{17}$:Mn phosphor on a UV LED chip (with an emission in 380–385 nm, 20 mA). The content of (Ba, Eu)$MgAl_{10}O_{17}$:Mn is fixed as 0.025 g, whereas that of $Lu_{0.98}VO_4$:2.0%Bi^{3+} increases in a step of 0.025 g from 0 to 2 g. It is found that gradually changing the $Lu_{0.98}VO_4$:2.0%Bi^{3+} concentration can tune the Commission International de l'Eclairage (CIE) chromaticity coordinates of the LEDs lamps [36].

Once the weight ratio between $Lu_{0.98}VO_4$:2.0%Bi^{3+} and (Ba, Eu)$MgAl_{10}O_{17}$: Mn phosphor reaches 5:1 [i.e., 0.125 g $Lu_{0.98}VO_4$:2.0%Bi^{3+} and 0.025 g (Ba, Eu) $MgAl_{10}O_{17}$:Mn], a near-white LED lamp can be obtained with satisfactory CIE chromaticity coordinates of (0.3241, 0.3403). The values of CRI and CT are 73.0 and 5580 K, respectively, and they are not as desirable as expected for illumination. This is due to the lack of the red light. We thus introduced a red phosphor—$Sr_4Al_{14}O_{25}$:Mn, which was recently reported by Peng et al. [5], into the device to

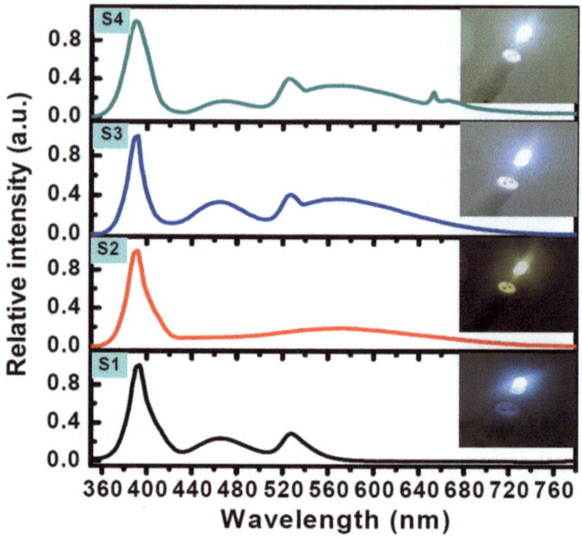

Fig. 14.29 EL spectra of LED lamps based on 380- to 385-nm LED chips. Lamp S1 is coated with (Ba, Eu) $MgAl_{10}O_{17}$:Mn, lamp S2 with $Lu_{0.98}VO_4$:2.0%Bi^{3+}, lamp S3 with (Ba, Eu)$MgAl_{10}O_{17}$:Mn and $Lu_{0.98}VO_4$:2.0%Bi^{3+}, and lamp S4 with (Ba, Eu) $MgAl_{10}O_{17}$:Mn, $Lu_{0.98}VO_4$:2.0%Bi^{3+}, $Sr_4Al_{14}O_{25}$:Mn, and $Lu_{0.98}VO_4$:2.0%Bi^{3+}. The image for each lamp is inserted beside the corresponding EL curve. Reproduced from Ref. [36] by permission of John Wiley & Sons Ltd.

improve the performance. The best result was achieved when the content of $Sr_4Al_{14}O_{25}$:Mn reached 0.3 g. The CRI could be improved to 85.2, and the color index shifted to (0.2823, 0.3174) whereas CT was lowered to 4536 from 5580 K. The electroluminescence (EL) spectra of the optimal WLED lamp is presented in Fig. 14.29 along with other three lamps fabricated by single coating the UV LED chip with or without $Lu_{0.98}VO_4$:2.0%Bi^{3+}, (Ba, Eu)$MgAl_{10}O_{17}$:Mn, and $Sr_4Al_{14}O_{25}$:Mn phosphors [36].

To study the impact of driving current on the EL of $Lu_{0.98}VO_4$:2.0%Bi^{3+}-coated LED, we measured the relative emission intensity as a function of the direct current, which changed from 10 to 80 mA, and illustrated it in Fig. 14.30 (curve 1). Obviously, the current increase initially enhances the emission intensity; as the current reaches 50 mA, the emission becomes the strongest. Afterward it decreases

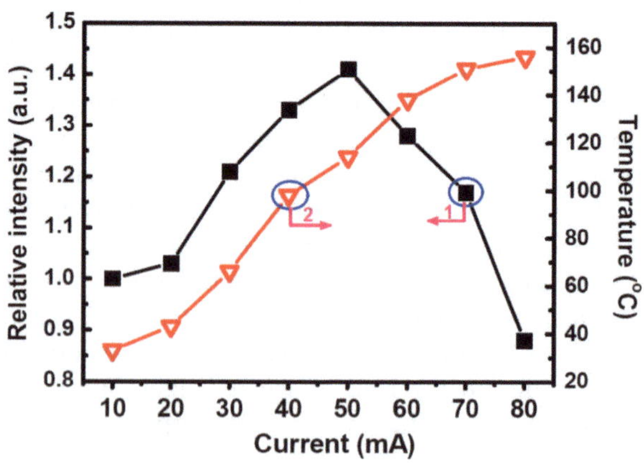

Fig. 14.30 Dependence of the relative emission intensity (*curve 1*) and temperature (*curve 2*) of lamp S2 on the direct current; the current changes from 10 to 80 mA; Reproduced from Ref. [36] by permission of John Wiley & Sons Ltd.

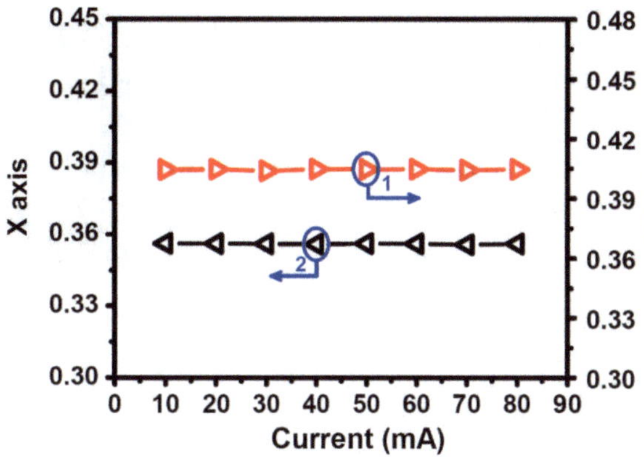

Fig. 14.31 The dependence of CIE chromaticity coordinates of the S2 lamp on current. Reproduced from Ref. [36] by permission of John Wiley & Sons Ltd.

as the current continues increasing. At the same time the chip temperature increases from 33 to 156 °C as the current increases from 10 to 80 mA (see curve 2 in Fig. 14.30). The enhancement of intensity at the initial state is perhaps due to (1) more energy of the LED chip can be converted into yellow light by $Lu_{0.98}VO_4$:2.0%Bi^{3+}; or (2) the electrons situated in traps as chip temperature increases can be thermally discharged, in which case the latter consequently excite the yellow phosphor and eventually is released as yellow light. As the chip temperature keeps increasing, electrons in traps will be depleted; in the meantime, the phosphor will experience thermal quenching of luminescence, and the emission intensity thereby decreases. Note that at current of 80 mA, as the chip temperature becomes 156 °C, the emission intensity still retains approximately 85 % of the initial value (10 mA, 33 °C). One hand, this reflects the good thermal stability of the $LuVO_4$:Bi^{3+} phosphor. On the other, in Fig. 14.31, which depicts the color coordinates of the lamp, it can be seen that they almost do not shift as the current increases [36].

The $LuVO_4$:Bi^{3+} phosphor does not show absorption in the blue or green spectral range, a quality that none of commercial rare earth-doped red or yellow phosphors (e.g., YAG:Ce, Eu^{2+}/Ce^{3+}-doped nitrides/oxynitrides phosphors or Eu-doped sulfide phosphors) can possess. In one way, this can be witnessed in the excitation spectrum; in another way, it can be supported by the EL spectra in Fig. 14.29. Lamp S1, which was made by (Ba, Eu)$MgAl_{10}O_{17}$:Mn, shows one blue and one green emission band, and the intensity ratio of the green band to the blue band is 1.19. The ratio keeps constant in lamp S3, which was fabricated with (Ba, Eu)$MgAl_{10}O_{17}$:Mn and $LuVO_4$:Bi^{3+}. It changes to 2.39 in lamp S4 made with (Ba, Eu)$MgAl_{10}O_{17}$:Mn, $LuVO_4$:Bi^{3+}, and $Sr_4Al_{14}O_{25}$:Mn. The change is due to the blue absorption of the red phosphor $Sr_4Al_{14}O_{25}$:Mn. Therefore, $LuVO_4$:Bi^{3+} phosphor can effectively avoid white-light distortion, thus impelling its potential application in our daily life [36].

As proof of concept, we also fabricated a series of WLED lamps by combing a UV LED chip peaking at approximately 390 nm with various contents of the red-emitting phosphor $ScVO_4$:1.0%Bi^{3+} and a commercial phosphor (Ba, Eu)

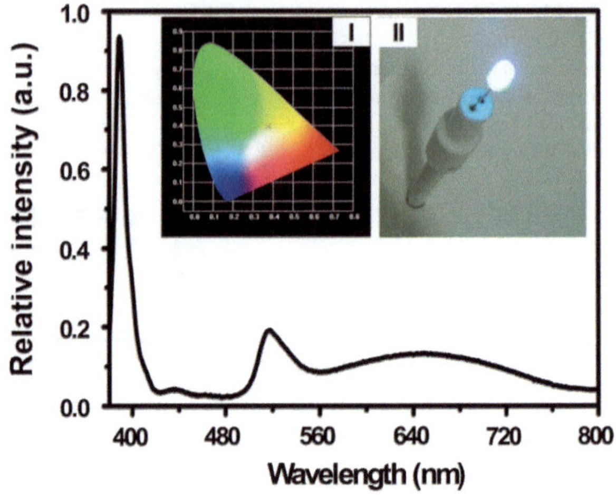

Fig. 14.32 Emission spectra of the WLEDs lamp combing an UV LED chip with the red-emitting phosphor $ScVO_4$:1.0%Bi^{3+} and the commercial phosphor (Ba, Eu)$MgAl_{10}O_{17}$:Mn. *Inset I* CIE chromaticity coordinates of the WLED lamp. *Inset II* Image of the WLED lamp; the input voltage and current are 5.0 V and 20 mA, respectively. Reproduced from Ref. [35] by permission of American Chemical Society

$MgAl_{10}O_{17}$:Mn (peaked at 436 and 516 nm). Consequently, satisfactory values of CIE chromaticity coordinates (0.3744, 0.3991), CRI (approximately 90), and CT (approximately 4306 K) are obtained [35]. Figure 14.32 shows the corresponding emission spectrum for such a lamp together with the CIE chromaticity coordinates profile (inset I) and the corresponding digital photograph for a current of 20 mA (inset II). For comparison, a reference lamp that comprises only the UV LED chip and the commercial (Ba, Eu)$MgAl_{10}O_{17}$:Mn was produced. Addition of $ScVO_4$:1.0%Bi^{3+} improves the CRI and CT values from 33.2 to 91.2 and >100,000 to 4036 K, respectively, thus testifying to the application in WLED [35].

14.6 Summary and Perspectives

This chapter first reviews briefly the progress on phosphors for NUV-WLED and then Bi^{3+} doped glasses and crystals. We found that Bi^{3+}-doped rare-earth vanadates are promising phosphors for the application because they can be excited by NUV simultaneously rather than visible lights and because they have tunable visible emissions with high QE and better resistance to thermal quenching. We made prototype of NUV-WLED with these phosphors. The color coordinates of the device do not shift obviously with driving currents. In future practical application of the new type of Bi^{3+}-doped phosphors, challenges exist, and they require thorough study. First is how to stabilize bismuth to the valence state of +3 in different compounds because the valence states and species of bismuth are diverse, and transformations easily happen from species to species and valence to valence. Second is how to enhance the absorption in NUV. Currently it is no problem that these phosphors can be excited by NUV lights, but the problem, as you can see in Figs. 14.10, 14.13 and 14.20, is that the absorption is not strong in the area. So once applied in the device, it cannot guarantee zero leakage of NUV lights, which, however, is essential to human health. Third is long-term degradation of the phosphors. It seems after several cooling and heating processes, Bi^{3+}-doped vanadates (see Fig. 14.25) can restore to their initial state. However, as we examine the figure more carefully, the emission intensity cannot be recovered in the intermediate states because after each process, the intensity keeps decreasing although the magnitude of degradation is not significant. Preliminary study has shown that this might correlate to the intrinsic defects inside the compounds. The fourth challenge is how to solve or ameliorate degradation of the phosphors. After these problems are solved, we believe the future of Bi^{3+}-doped phosphors is bright regarding their application of NUV-WLEDs.

Acknowledgments The authors would like to acknowledge the financial supports from National Natural Science Foundation of China (Grant No. 51322208), Guangdong Natural Science Foundation for Distinguished Young Scholars (Grant No. S20120011380), Department of Education of Guangdong Province (Grant No. 2013gjhz0001), Fundamental Research Funds for the Central Universities, Key Program of Guangzhou Scientific Research Special Project (Grant No. 201607020009), Hundred, Thousand and Ten Thousand Leading Talent Project in Guangdong Program for Special Support of Eminent Professionals.

References

1. Schubert EF, Kim JK (2005) Solid-state light sources getting smart. Science 308:1274
2. Pimputkar S, Speck JS, DenBaars SP, Nakamura S (2009) Prospects for LED lighting. Nat Photonics 3:179
3. Service RF (2009) The quest for white LEDs hits the home stretch. Science 325:809
4. Zhang F, Wang YH, Tao Y (2013) Investigation of the luminescence properties of Tb^{3+}-doped $Li_6Y(BO_3)_3$ phosphors in VUV-VIS range. J Lumin 136:51
5. Peng MY, Yin XW, Tanner PA, Liang CQ, Li PF, Zhang QY, Qiu JR (2013) Orderly-layered tetravalent manganese-doped strontium aluminate $Sr_4Al_{14}O_{25}:Mn^{4+}$: an efficient red phosphor for warm white light emitting diodes. J Am Ceram Soc 96:2870
6. Xie RJ, Hirosaki N, Sakuma K, Kimura N (2008) White light-emitting diodes (LEDs) using (oxy) nitride phosphors. J Phys D Appl Phys 41:144013
7. Smet PF, Parmentier AB, Poelman D (2011) Selecting conversion phosphors for white light-emitting diodes. J Electrochem Soc 158:R37
8. Lin CC, Liu RS (2011) Advances in phosphors for light-emitting diodes. J Phys Chem Lett 2:1268
9. Song YH, Jia G, Yang M, Huang YJ, You HP, Zhang HJ (2009) $Sr_3Al_2O_5Cl_2:Ce^{3+}$, Eu^{2+}: a potential tunable yellow-to-white-emitting phosphor for ultraviolet light emitting diodes. Appl Phys Lett 94:091902
10. Peng M, Wondraczek L (2010) Photoluminescence of $Sr_2P_2O_7:Bi^{2+}$ as a red phosphor for additive light generation. Opt Lett 35:2544
11. Jia YC, Huang YJ, Zheng YH, Guo N, Qiao H, Zhao Q, Lv WZ, You HP (2012) Color point tuning of $Y_3Al_5O_{12}:Ce^{3+}$ phosphor via Mn^{2+}–Si^{4+} incorporation for white light generation. J Mater Chem 22:15146
12. Zhang CM, Lin CK, Li CX, Quan ZW, Liu XM, Lin J (2008) Enhanced luminescence of BPO_4 by mixing with SiO_2 and Al_2O_3. J Phys Chem C 112:2183

13. Zhang CM, Lian HZ, Kong DY, Huang SS, Lin J (2009) Structural and bluish-white luminescent properties of Li^+-doped BPO_4 as a potential environmentally friendly phosphor material. J Phys Chem C 113:1580
14. Peng M, Wondraczek L (2009) Bi^{2+}-doped strontium borates for white-light-emitting diodes. Opt Lett 34:2885
15. Lin CK, Luo Y, You H, Quan Z, Zhang J, Fang J, Lin J (2006) Sol-gel-derived BPO_4/Ba^{2+} as a new efficient and environmentally-friendly bluish-white luminescent material. Chem Mater 18:458
16. Guo H, Wei RF, Liu XY (2012) Tunable white luminescence and energy transfer in Cu^{+2}, Eu^{3+} codoped sodium silicate glasses. Opt Lett 37:1670
17. Guo H, Wang XF, Chen JD, Li F (2010) Ultraviolet light induced white light emission in Ag and Eu^{3+} co-doped oxyfluoride glasses. Opt Express 18:18900
18. Zhang Q, Wang J, Zhang M, Ding W, Su Q (2007) Enhanced photoluminescence of $Ca_2Al_2SiO_7:Eu^{3+}$ by charge compensation method. Appl Phys a-Mater 88:805
19. You HP, Hong GY (1999) The change of Eu^{3+}-surroundings in the system Al_2O_3–B_2O_3 containing Eu^{3+} ions. J Phys Chem Solids 60:325
20. Peng M, Pei Z, Hong G, Su Q (2003) The reduction of Eu^{3+} to Eu^{2+} in $BaMgSiO_4$:Eu prepared in air and the luminescence of $BaMgSiO_4:Eu^{2+}$ phosphor. J Mater Chem 13:1202
21. Ivanovskikh K, Meijerink A, Ronda C, Piccinelli F, Speghini A, Bettinelli M (2011) Fast UV luminescence in Pr^{3+}-doped eulytite double phosphates. Opt Mater 34:419
22. Liu BT, Wang YH, Zhang F, Wen Y, Dong QZ, Wang ZF (2010) Thermal stability and photoluminescence of S-doped $BaMgAl_{10}O_{17}:Eu^{2+}$ phosphors for plasma display panels. Opt Lett 35:3072
23. Chen DQ, Yu YL, Huang P, Lin H, Shan ZF, Wang YS (2010) Color-tunable luminescence of Eu^{3+} in LaF_3 embedded nanocomposite for light emitting diode. Acta Mater 58:3035
24. Hirosaki N, Xie RJ, Kimoto K, Sekiguchi T, Yamamoto Y, Suehiro T, Mitomo M (2005) Characterization and properties of green-emitting beta-SiAlON:Eu^{2+} powder phosphors for white light-emitting diodes. Appl Phys Lett 86:211905
25. Peijzel PS, Wegh RT, Meijerink A, Holsa J, Lamminmaki RJ (2002) High energy levels and high-energetic emissions of the trivalent holmium ion in $LiYF_4$ and YF_3. Opt Commun 204:195
26. Hu CW, Li J, Zhu YZ, Bai CS, Zhang JH, Xia SL, Li YF (2013) Effects of Al on the splenic immune function and NE in rats. Food Chem Toxicol 62:194
27. Bachmann V, Justel T, Meijerink A, Ronda C, Schmidt PJ (2006) Luminescence properties of $SrSi_2O_2N_2$ doped with divalent rare earth ions. J Lumin 121:441
28. Xie RJ, Hirosaki N (2007) Silicon-based oxynitride and nitride phosphors for white LEDs—a review. Sci Technol Adv Mat 8:588
29. Pust P, Weiler V, Hecht C, Tucks A, Wochnik AS, Henss AK, Wiechert D, Scheu C, Schmidt PJ, Schnick W (2014) Narrow-band red-emitting $Sr[LiAl_3N_4]:Eu^{2+}$ as a next-generation LED-phosphor material. Nat Mater 13:891
30. Setlur AA, Heward WJ, Hannah ME, Happek U (2008) Incorporation of Si^{4+}–N^{3-} into Ce^{3+}-doped garnets for warm white LED phosphors. Chem Mater 20:6277
31. Chen WT, Sheu HS, Liu RS, Attfield JP (2012) Cation-size-mismatch tuning of photoluminescence in oxynitride phosphors. J Am Chem Soc 134:8022
32. Bachmann V, Ronda C, Oeckler O, Schnick W, Meijerink A (2009) Color point tuning for (Sr, Ca, Ba)$Si_2O_2N_2:Eu^{2+}$ for white light LEDs. Chem Mater 21:316
33. Zhu J, Wang L, Zhou T, Cho Y, Suehiro T, Takeda T, Lu M, Sekiguchi T, Hirosaki N, Xie R-J (2015) Moisture-induced degradation and its mechanism of (Sr, Ca)$AlSiN_3:Eu^{2+}$, a red-color-converter for solid state lighting. J Mater Chem C
34. Wang C-Y, Xie R-J, Li F, Xu X (2014) Thermal degradation of the green-emitting $SrSi_2O_2N_2$: Eu^{2+} phosphor for solid state lighting. J Mater Chem C 2:2735
35. Kang FW, Yang XB, Peng MY, Wondraczek L, Ma ZJ, Zhang QY, Qiu JR (2014) Red photoluminescence from Bi^{3+} and the influence of the oxygen-vacancy perturbation in $ScVO_4$: a combined experimental and theoretical study. J Phys Chem C 118:7515

36. Kang FW, Peng MY, Zhang QY, Qiu JR (2014) Abnormal anti-quenching and controllable multi-transitions of Bi^{3+} luminescence by temperature in a yellow-emitting $LuVO_4:Bi^{3+}$ phosphor for UV-converted white LEDs. Chem Eur J 20:11522
37. Kang FW, Peng MY, Yang XB, Dong GP, Nie GC, Liang WJ, Xu SH, Qiu JR (2014) Broadly tuning Bi^{3+} emission via crystal field modulation in solid solution compounds (Y, Lu, Sc) VO4:Bi for ultraviolet converted white LEDs. J Mater Chem C 2:6068
38. Mohan R (2010) Green bismuth. Nat Chem 2:336
39. Kang FW, Peng MY, Xu SH, Ma ZJ, Dong GP, Qiu JR (2014) Broadly tunable emission from $CaMoO_4$:Bi phosphor based on locally modifying the microenvironment around Bi^{3+} ions. Eur J Inorg Chem 2014:1373
40. Goldenberg M, Honkomp L, Castellion A (1975) The antidiarrheal action of bismuth subsalicylate in the mouse and the rat. Digest Dis Sci 20:955
41. Tao DP, Yang B, Li DF (2002) Prediction of the thermodynamic properties of quinary liquid alloys by modified coordination equation. Fluid Phase Equilib 193:167
42. Tallon JL, Buckley RG, Gilberd PW, Presland MR, Brown IWM, Bowden ME, Christian LA, Goguel R (1988) High-Tc superconducting phases in the series $Bi_{2.1}(Ca, Sr)_{n+1}Cu_nO_{2n+4+}$. Nature 333:153
43. DiMeglio JL, Rosenthal J (2013) Selective conversion of CO_2 to CO with high efficiency using an inexpensive bismuth-based electrocatalyst. J Am Chem Soc 135:8798
44. Xu W, Peng M, Ma Z, Dong G, Qiu J (2012) A new study on bismuth doped oxide glasses. Opt Express 20:15692
45. Parke S, Webb RS (1973) The optical properties of thallium, lead and bismuth in oxide glasses. J Phys Chem Solids 34:85
46. Peng MY, Wang C, Chen DP, Qiu JR, Jiang XW, Zhu CS (2005) Investigations on bismuth and aluminum co-doped germanium oxide glasses for ultra-broadband optical amplification. J Non-Cryst Solids 351:2388
47. Chen L, Jiang Y, Zhang GB, Wu C, Yang GT, Wang C, Li GH (2008) Concentration and temperature dependences of $YBO_3:Bi^{3+}$ luminescence under vacuum ultraviolet excitation. Chin Phys Lett 25:1884
48. Chen L, Zheng HW, Cheng JG, Song P, Yang GT, Zhang GB, Wu C (2008) Site-selective luminescence of Bi^{3+} in the YBO_3 host under vacuum ultraviolet excitation at low temperature. J Lumin 128:2027
49. van der Voort D, Blasse G (1992) Luminescence of $CaSO_4:Bi^{3+}$, a small-offset case. J Solid State Chem 99:404
50. Nakajima T, Isobe M, Tsuchiya T, Ueda Y, Manabe T (2010) Correlation between luminescence quantum efficiency and structural properties of vanadate phosphors with chained, dimerized, and isolated VO_4 tetrahedra. J Phys Chem C 114:5160
51. Huang Y, Yu YM, Tsuboi T, Seo HJ (2012) Novel yellow-emitting phosphors of $Ca_5M_4(VO_4)_6$ (M = Mg, Zn) with isolated VO_4 tetrahedra. Opt Express 20:4360
52. Haranath D, Chander H, Sharma P, Singh S (2006) Enhanced luminescence of $Y_3Al_5O_{12}:Ce^{3+}$ nanophosphor for white light-emitting diodes. Appl Phys Lett 89:173118
53. Blasse G, Bril A (1969) Luminescence of phosphors based on host lattices ABO_4 (A is Sc, In; B is P, V, Nb). J Chem Phys 50:2974
54. Dotsenko VP, Berezovskaya IV, Efryushina NP (1996) Photoionization and luminescence properties of Bi^{3+} in $In_{1-x}Lu_xBO_3$ solid solutions. J Phys Chem Solids 57:437
55. Yousif A, Kumar V, Ahmed HAAS, Som S, Noto LL, Ntwaeaborwa OM, Swart HC (2014) Effect of Ga^{3+} doping on the photoluminescence properties of $Y_3Al_{5-x}Ga_xO_{12}:Bi^{3+}$ phosphor. Ecs J Solid State Sc 3:R222
56. Cavalli E, Angiuli F, Mezzadri F, Trevisani M, Bettinelli M, Boutinaud P, Brik MG (2014) Tunable luminescence of Bi^{3+}-doped $YP_xV_{1-x}O_4$ ($0 \leq x \leq 1$). J Phys Condens Matter 26:385503
57. van de Craats AM, Blasse G (1995) The quenching of bismuth(III) luminescence in yttrium oxide (Y_2O_3). Chem Phys Lett 243:559

58. Jing XP, Gibbons C, Nicholas D, Silver J, Vecht A, Frampton CS (1999) Blue luminescence in yttrium and gadolinium niobates caused by bismuth. The importance of non-bonding ns^2 valence orbital electrons. J Mater Chem 9: 2913
59. Blasse G, Bril A (1968) Investigations on Bi^{3+}-activated phosphors. J Chem Phys 48:217
60. Dotsenko VP, Efryushina NP, Berezovskaya IV (1996) Luminescence properties of $GaBO_3$: Bi^{3+}. Mater Lett 28:517
61. Zhiwu P, Qiang S, Jiyu Z (1993) Luminescence of Bi^{3+} and the energy transfer from Bi^{3+} to R^{3+} (R = Eu, Dy, Sm, Tb) in alkaline-earth borates. Solid State Commun 86:377
62. Han B, Zhang J, Li PJ, Li JL, Bian Y, Shi HZ (2014) Photoluminescence properties of novel $KBaBP_2O_8$:M (M = Pb^{2+} and Bi^{3+}) phosphors. Opt Mater 37:241
63. van Steensel LI, Bokhove SG, van de Craats AM, de Blank J, Blasse G (1995) The luminescence of Bi^{3+} in $LalnO_3$ and some other perovskites. Mater Res Bull 30:1359
64. Yang ZG, Zhao ZY, Wen Y, Wang YH (2013) Structure and luminescence properties of Bi^{3+} activated $Ca_{12}Al_{14}O_{32}Cl_2$ phosphors. J Alloy Compd 559:142
65. Ju HD, Liu J, Wang BL, Tao XT, Ma YH, Xu SQ (2013) Bi^{3+}-doped $Sr_3Al_2O_6$: an unusual color-tunable phosphor for solid state lighting. Ceram Int 39:857
66. Ju HD, Deng WP, Wang BL, Liu J, Tao XT, Xu SQ (2012) The structure and luminescence properties of green $Ca_3Al_2O_6$:Bi^{3+} phosphors. J Alloy Compd 516:153
67. Setlur AA, Srivastava AM (2006) The nature of Bi^{3+} luminescence in garnet hosts. Opt Mater 29:410
68. Kim WN, Park HL, Kim GC (2005) Lithium solubility limit in $ZnGa_2O_4$:$Bi0.001^{3+}$, Li^+ phosphor. Mater Lett 59:2433
69. Srivastava AM (1999) Luminescence of Bi^{3+} in $LaGaO_3$. Mater Res Bull 34:1391
70. Fukada H, Ueda K, J-i I, Miyata T, Minami T (2010) Blue PL and EL emissions from Bi-activated binary oxide thin-film phosphors. Thin Solid Films 518:3067
71. Zou Y, Tang L, Cai J-L, Lin L-T, Cao L-W, Meng J-X (2014) Combustion synthesis and luminescence of monoclinic Gd_2O_3:Bi phosphors. J Lumin 153:210
72. Srivastava AM, Beers WW (1999) On the impurity trapped exciton luminescence in $La_2Zr_2O_7$; Bi^{3+}. J Lumin 81:293
73. Xia YJ, Huang FQ, Wang WD, Wang AB, Shi JL (2009) A reinvestigation of luminescence properties of Bi^{3+}-activated MSb_2O_6 (M = Ca, Sr) phosphors. J Alloys Compd 476:534
74. Srivastava AM, Szarowski A (1999) On the quenching of Bi^{3+} luminescence in the pyrochlore Gd_2GaSbO_7. J Solid State Chem 146:494
75. Srivastava AM (2002) On the luminescence of Bi^{3+} in the pyrochlore $Y_2Sn_2O_7$. Mater Res Bull 37:745
76. Madej A, Witkowski ME, Wojtowicz AJ, Zych E (2015) Photo- and radioluminescent properties of undoped and Bi-doped Lu_2WO_6 powders at 10–300K. J Lumin 160:50
77. Kang FW, Peng MY (2014) A new study on the energy transfer in the color-tunable phosphor $CaWO_4$:Bi. Dalton Trans 43:277
78. Shin SH, Jeon DY, Suh KS (2001) Charge-transfer nature in luminescence of $YNbO_4$:Bi blue phosphor. J Appl Phys 90:5986
79. Liu XM, Lin J (2007) Enhanced luminescence of gadolinium niobates by Bi^{3+} doping for field emission displays. J Lumin 122–123:700
80. van der Steen AC, van Hesteren JJA, Roos A, Blasse G (1979) The occurrence of vibrational structure in the luminescence spectra of the Bi^{3+} ion. J Lumin 18–19, Part 1: 235
81. Park HL, Kim HK, Chung CH (1988) Oxygen related luminescence center in CaS:Bi^{3+} phosphor. Solid State Commun 66:867
82. Jia DD, Zhu J, Wu BQ (2000) Correction of excitation spectra of long persistent phosphors. J Lumin 90:33
83. Kim TW, Park HL (1993) Solid solubility of bismuth as a single phase in the SrS: Bi^{3+} phosphor. J Mater Sci Lett 12:1243
84. Xu XL, Yu XB, Mao LH, Yang SP, Peng ZF (2004) Preparation and photoluminescence of Bi^{3+}-doped strontium chloroapatite nano-phosphor. Mater Lett 58:3665

85. Porter-Chapman Y, Bourret-Courchesne E, Derenzo SE (2008) Bi^{3+} luminescence in $ABiO_2Cl$ (A = Sr, Ba) and $BaBiO_2Br$. J Lumin 128:87
86. Lai SF, Yang ZW, Wang RF, Wu HJ, Liao JY, Qiu JB, Song ZG, Yang Y, Zhou DC (2013) Preparation and blue-white luminescence properties of Bi^{3+}-doped $Ba_5SiO_4Cl_6$. J Mater Sci 48:8566
87. Lin J, Su Q, Wang S, Zhang H (1996) Influence of crystal structure on the luminescence properties of bismuth(III), europium(III) and dysprosium(III) in Y_2SiO_5. J Mater Chem 6:265
88. Zorenko Y, Gorbenko V, Zorenko T, Malinowski P, Jary V, Kucerkova R, Beitlerova A, Mares JA, Nikl M, Fedorov A (2014) Luminescent and scintillation properties of Bi^{3+} doped Y_2SiO_5 and Lu_2SiO_5 single crystalline films. J Lumin 154:525
89. Kang F, Zhang Y, Peng M (2015) Controlling the energy transfer via multi luminescent centers to achieve white light/tunable emissions in a single-phased X_2-type $Y_2SiO_5:Eu^{3+}$, Bi^{3+} phosphor for ultraviolet converted LEDs. Inorg Chem 54:1462
90. Zhou H, Jiang M, Jin Y (2014) A novel blue-emitting phosphor $Gd_{4.67}Si_3O_{13}:Bi^{3+}$ for near-UV LEDs. Rsc Adv 4:45786
91. Mei Y, Xu H, Zhang J, Ci Z, Duan M, Peng S, Zhang Z, Tian W, Lu Y, Wang Y (2015) Design and spectral control of a novel ultraviolet emitting long lasting phosphor for assisting TiO_2 photocatalysis: $Zn_2SiO_4:Ga^{3+}$, Bi^{3+}. J Alloys Compd 622:908
92. Tsai YY, Chen HL, Chai YL, Chang YS (2013) Photoluminescence properties of Bi^{3+}-doped $YInGe_2O_7$ phosphors under an ultraviolet irradiation. Opt Mater 35:317

Chapter 15
Design of Single-Phased Multicolor-Emission Phosphor for LED

Chongfeng Guo and Hao Suo

Abstract Phosphor-converted white light-emitting diodes (w-LEDs) are generally fabricated by combining a blue LED chip with a yellow phosphor or an ultraviolet (UV) LED chip with a tricolor-emitting phosphor. To obtain w-LEDs with excellent performance, many single-color emission phosphors have been used in these systems, which result in high cost and low luminous efficiency due to the complicated manufacture, reabsorption of emission colors, and different aging rates for each phosphor. Searching for a novel single-phased multicolor-emitting phosphor with excellent chemical and thermal stability for UV-pumped white LEDs is an important consideration, which arouses more attention from researchers. According to previous publications and our results, five methods were summarized to design the multicolor-emitting phosphor: (1) enhancing the red emission of YAG:Ce^{3+} yellow phosphor by introducing red-emitting dopants; (2) use of a single activator with several metastable multiplets offers the possibility of simultaneous emission in the blue, green, orange, red, and infrared wavelengths; (3) use of a multiple-ion codoped system based on energy transfer; (4) use of an up-conversion luminescence system, and (5) use of a semiconductor quantum dot and defects emission. Moreover, the challenges and future of multicolor-emitting phosphors are also discussed.

15.1 Introduction

15.1.1 History of Light-Emitting Diodes

A light-emitting diode (LED) is a device in which single-color electroluminescence light results from the motion of a charge across a PN junction of a semiconductor and the subsequent recombination when operated in a forward biased direction;

C. Guo (✉) · H. Suo
National Key Laboratory of Photoelectric Technology and Functional Materials (Culture Base) in Shaanxi Province, National Photoelectric Technology and Functional Materials & Application of Science and Technology International Cooperation Base, Institute of Photonics & Photon-Technology, Northwest University, Xi'an 710069, China
e-mail: guocf@nwu.edu.cn

R.S. Liu (ed.), *Phosphors, Up Conversion Nano Particles, Quantum Dots and Their Applications*, DOI 10.1007/978-3-662-52771-9_15

modern LEDs are available across the visible, ultraviolet, and infrared wavelengths [1]. The first commercially usable LEDs were developed based on GaAsP with a 655-nm red-light emission in the 1960s; those were most used as indicators because of their low light output and high cost [2]. After that, LED technology progressed gradually through the 1970s, and additional colors (such as green and orange) became widely available and appeared in appliances and equipment. However, the failure rate of early devices was much greater than with current technology partially due to the inexact actual component assembly and there being no refined materials. High numbers of defects in the crystal, substrate, and epitaxial layers resulted in decreased efficiency and shorter device lifetimes. Rapid growth in the application of LEDs began to occur until a new GaAlAs semiconductor was developed in the 1980s, thus leading to high brightness and low operating voltage. However, the basic structure of the material remained relatively unchanged until laser diode technology was used to produce high-brightness and high-reliability LEDs, which led to the development of InGaAlP visible LEDs and tunable color emission by adjusting the size of the energy-band gap. Thus, green, yellow, orange, and red LEDs all could be produced, and the light-output degradation of InGaAlP material was significantly improved [3, 4].

Because one of the primary tricolors, blue-emitting LEDs have become a cornerstone to an entire generation of new applications. Blue LEDs are difficult to manufacture because of their high photon energies (>2.5 eV), relatively low eye sensitivity, and different technology compared with other LED materials. In 1994, the first high-brightness InGaN-based blue LEDs were successful fabricated by Shuji Nakamura from the Nichia Corporation of Japan, which built on critical developments in GaN nucleation on sapphire substrates and the demonstration of p-type doping of GaN [5]. Using the same basic GaN technology and growth processes, other colors, including high-brightness green (approximately 500–570 nm), violet (400–450 nm), purple, and ultraviolet (<400 nm) LEDs have also been developed. This makes it possible to produce white light and excite new developments in LED technology. More importantly, the discovery of blue LED opened the door to solid-state lighting.

In summary, LEDs have gone from infancy (red) to adolescence (blue) through approximately 30 years of development and are experiencing some of the most rapid market growth of their lifetime. By using InGaAlP and InGaN material and varying the relative In/Ga, Al/Ga fraction with metal-organic chemical vapor deposition (MOCVD) technology as the growth process, some bright, efficient, inexpensive, and reliable monochromic LEDs with even shorter wavelengths in the ultraviolet (UV) and deep-UV ranges are now available. This technology, together with other novel LED structures, will ensure wide application of LEDs. Further developments on white-light output will also guarantee the continued increase in applications of these economical light sources and eventually replace standard incandescent and fluorescent lighting [6].

15.1.2 Basic Concepts of Phosphor

A phosphor, also called "luminescent material," is a solid material that converts certain types of energy into electromagnetic radiation over and above thermal radiation. They are widely used in fluorescent lamps, television and computer screens, X-ray photography, and lasers; you can encounter these applications at any time and anywhere in daily life. Luminescent materials can be divided into up-conversion (UC) and down-conversion emission according to the energy difference between the excitation and emission radiant energy. For a down-conversion phosphor, all fluorescence light-emitters usually follow the well-known principle of the Stokes law, which simply states that excitation photons have greater energy with shorter wavelength than those of emitted ones with longer wavelength, i.e., output photon energy is weaker than input photon energy. In contrast to down-conversion, the upconversion process, or anti-Stokes emissions, usually converts pump light with long wavelength to light with short wavelength, e.g., infrared or near-infrared (NIR) light, and then to visible or ultraviolet light. This process involves two or more low-energy photons converting into one high-energy photon. This means that the efficiency of UC cannot be >50 %. It is worth noting that what we call "luminescence" usually refers to the type of down-conversion [7].

What is the role of phosphor? Phosphor is converts other kinds of radiation energy—such as light, electricity, heat, X-rays, cathode rays, mechanical energy, chemical energy, and so on—to light, which strongly depends on the composition of phosphor. A phosphor is composed of a host crystal, or matrix, and a small amount of activator(s). A sensitizer is also necessary in some cases, which may absorb the exciting radiation and subsequently transfer it to the activator. The common representation of a phosphor formula is expressed as $Ca_5(PO_4)_3F{:}Sb^{3+}$, Mn^{2+} (a well-known lamp phosphor), where the first part ($Ca_5(PO_4)_3F$) is the host lattice; the latter part (Mn^{2+}) is the luminescent center (or activator); and Sb^{3+} ion is the sensitizer [8]. The host materials are typically oxides, chalcogenides, nitrides, oxynitrides, and halides, whereas the activators are usually including rare-earth ions, transitional-metal ions, and some atomic groups. For the atomic-group luminescent center, it is usually the main component of the host and can produce emission by absorbing the exciting radiation. It is also called the "host luminophore," e.g., molybdates, tungstate, vandate, titanate, niobate, tantalite, and so on. A good phosphor should also offer excellent physical and chemical stability, high efficiency, low cost and environment-friendly synthesis [9].

15.1.3 Methods to Obtain White Light-Emitting Diodes

As the basis of solid-state lighting, white light-emitting diode (w-LEDs) are on the way to replace traditional incandescent and fluorescent illuminating sources due to the fact that they offer many attractive advantages including high efficiency, environmental friendliness, long lifetime, high reliability, fast response, and low power consumption [10]. For example, it takes only approximately 10 % as much electricity to produce the same amount of light compared with the energy requirements of traditional incandescent and fluorescent lamps. Worldwide electricity for lighting would decrease by >50 %, and total electricity consumption would decrease by >10 %/year if LEDs could displace general illumination light sources such as incandescent and fluorescent lamps [11]. In addition, typical lifetimes quoted are 25,000–100,000 h, but heat and current settings can extend or shorten this time significantly because LEDs are subject to limited wear and tear if operated at low currents and low temperatures [12]. Thus, as a typical energy-saving device, w-LEDs are considered to be the next generation of illuminating sources, and many developed or developing countries, as well as internationally known companies, have invested large amounts of manpower and material resources to develop them. With the progress in recent years, w-LEDs are becoming increasingly competitive in the lighting market. At present, the luminous efficiency of w-LEDs has already surpassed that of fluorescent lamps, and the luminous efficiency of today's commercial available w-LEDs reaches 200 lm/W with figures increasing at a staggering pace [13].

Generally, there are two primary strategies to produce white light-emitting diodes: color mixing and phosphor down-conversion as shown in Fig. 15.1. One involves combining three individual LEDs that emit three primary colors [RGB (red/green/blue)] in a single device with power ratios adjusted to obtain white light, in which no phosphor is involved, and these are known as multi-LED types as shown in Fig. 15.1a. In the mid-1970s, multi-LEDs possessed high efficiency because energy was lost in phosphor-converted devices due to the Stokes shift and poor color-rendering index (CRI, Ra). Their correlated color temperature (CCT) is tuned by varying the power of the individual LED components, and complex electronics must be used due to the different driving currents required by the

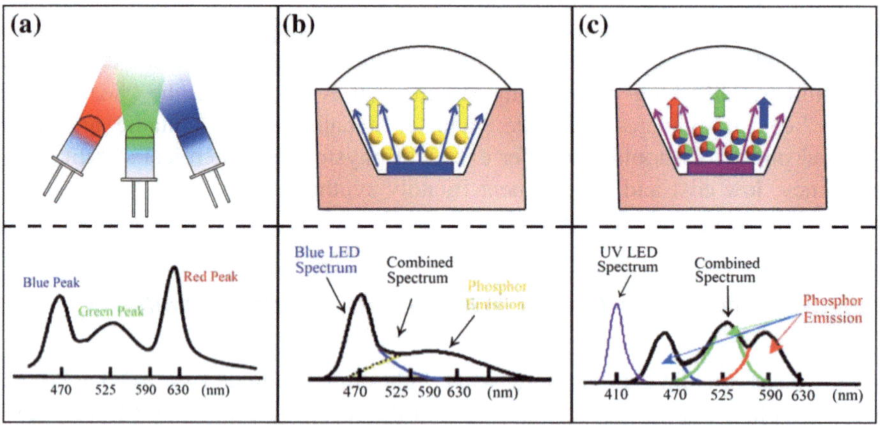

Fig. 15.1 Methods to obtain white-light LEDs

differently colored LEDs, which requires sophisticated software and hardware design to implement and enhance the cost and complexity. In addition, the overall light output of each RGB die being used degrades at a different rate eventually resulting in a color unbalance. Therefore, several technical problems must be solved before this type of LED can play a role in the market [14].

The other method is to use a blue or ultraviolet LED chip to excite phosphors, which emit wavelength–downconverted yellow or red/green/blue (RGB) light in a very inexpensive way, which offer more advantages such as simplicity, color stability, and the need for only one power source (known as "phosphor-converted LEDs" [pc-LEDs]) [15]. The first commercially available white LED was prepared by Nichia Corporation by combining a blue InGaN diode chip with Ce^{3+}-doped yttrium aluminum garnet yellow phosphor $Y_3Al_5O_{12}:Ce^{3+}$ (YAG:Ce^{3+}), in which the yellow light obtained by the excited yellow phosphor, using part of the blue light emitted from the LED chip, was combined with the rest of the blue light to generate white light as shown in Fig. 15.1b [16]. At present, the dominant white-LED products on the market are still fabricated by combining a blue LED with a yellow-emitting luminescent material such as YAG or a Eu^{2+}-doped ortho-silicates phosphor (Sr, Ba, Mg)$_2$SiO$_4$. The commercial use of orthosilicates is seriously limited due to their inability to overcome the stability issue. In addition to the method using blue LED + yellow phosphor, the ultraviolet (UV) or near ultraviolet (n-UV) LED +red/green/blue (RGB) method has been suggested for general illumination because UV-pumped phosphor white LEDs provide great color stability and high CRI and do not suffer from the strong change in chromaticity as shown in Fig. 15.1c [17–19]. Therefore, pc-converted w-LEDs could be realized by blue LED chip + yellow phosphor and an n-UV or UV LED chip + RGB phosphor.

15.2 Phosphor-Converted W-LEDs

15.2.1 Requirements for W-LED Phosphor

For phosphors used in LEDs, some basic requirements must be met.

- First, their excitation spectra should match well the LED chip and should emit efficient obligatory visible light with a high quantum yield; this means that the phosphor has bright emission and strong absorption in the emission region of the LED. For example, the YAG:Ce^{3+} excitation spectrum at 440–460 nm is well matched to the emission of blue LEDs and provides a dual ground-state broad-band emission in the yellow spectral region (500–700 nm). The excitation and emission spectra can be tuned by substitution of the lattice constituents, e.g. in $Y_3Al_5O_{12}:Ce^{3+}$ compound, Y may be replaced by Lu, Gd, or Tb, and Al may be replaced by Ga. Figure 15.2 shows the PL spectra of $(Y_{1-a}Gd_a)_{2.9}Al_5O_{12}$: $0.1Ce^{3+}$ and $Y_{2.9}(Al_{1-b}Ga_b)_5O_{12}:0.1Ce^{3+}$ as a function of the contents of Gd^{3+}

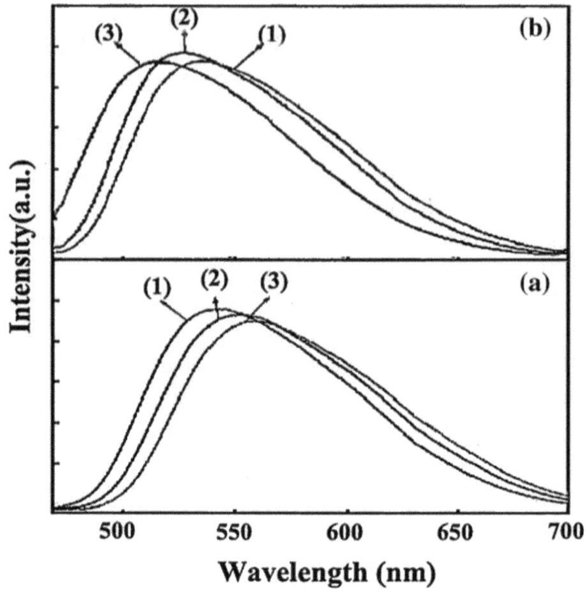

Fig. 15.2 PL spectra of sample as a function of substiution of **a** Gd^{3+} in the $(Y_{1-a}Gd_a)_{2.9}Al_5O_{12}:0.1Ce^{3+}$ phosphors: (1) $a = 0$, $\lambda_{ex} = 460$ nm; (2) $a = 0.25$, $\lambda_{ex} = 465$ nm; (3) $a = 0.45$, $\lambda_{ex} = 470$ nm; **b** Ga^{3+} in $Y_{2.9}(Al_{1-b}Ga_b)_5O_{12}:0.1Ce^{3+}$ phosphors (1) $b = 0$, $\lambda_{ex} = 460$ nm; (2) $b = 0.3$, $\lambda_{ex} = 455$ nm; (3) $b = 0.55$, $\lambda_{ex} = 450$ nm. Adapted with permission from Ref. [20]

and Ga^{3+}, respectively [20]. It has been found that the PL spectra shift to the shorter-wavelength region with increasing of Gd^{3+} contents; on the contrary, a significant red shift occurred with increasing Ga^{3+} concentrations.

- Second, phosphors should have excellent chemical and physical stability, especially thermal stability. For the application of phosphor in LEDs, especially high-power w-LEDs, the temperature of the LED chip increases up to approximately 150 °C in the working process of LEDs [21]. The thermal stability of the phosphor is one of the important issues to be considered because it can severely affect the efficiency of the device. A lower thermal-quenching effect is in favor of keeping the chromaticity and brightness of the white-light output. To evaluate the influence of temperature on the luminescence, the PL spectra at different temperature are usually measured. With the increase of temperature, there is little or no obvious red or blue shift in the peak of the emission bands (stable color point), thus indicating that high-quenching temperature $T_{0.5}$ (defined as the temperature at which the emission intensity is half of its original value) and the activation energies E_a are the typical characteristics for a perfect phosphor [22]. Generally, the luminescence thermal-quenching effect is attributed to the nonradiative relaxation through the crossing point between the excited state and the ground state in the configurational-coordinate diagram [23]. The nonradiative-transition probability is strongly dependent on temperature, thus resulting in decreased emission intensity. The decrease of phosphor emission intensity, depending on the temperature, can be described according to the following Eq. (1.1):

$$I(T) = \frac{I_0}{1 + c\exp(-E_a/kT)} \quad (1.1)$$

where I_0 is the initial PL intensity of the phosphor at room temperature; $I(T)$ is the PL intensity of the phosphor at different temperatures T; c is a constant; E_a is the activation energy for thermal quenching; and k is the Boltzmann constant (8.62×10^{-5} eV). According to Eq. (1.1), the plot of $\ln[(I_0/I) - 1]$ versus 10,000/T yields a straight line, and the activation energy E_a is obtained from the slope of the plot.

Thermal quenching of luminescence occurs in phosphors by way of a number of photophysical paths. In most cases, results from a crossover from at a high vibrational level in the excited state to the ground state. Other possible paths include thermally assisted photoionization and thermally induced charge transfer. These paths are located at relatively high energy level above a sequence of vibrational levels in excited state, and the thermal quenching thereby takes place at relatively high temperatures. In the case that vibrational transitions dominate the thermal transitions in both the ground state and excited state, the thermal stability of luminescence is mainly related to the bonding energy of the bonds between the activator atom and the host crystal of the phosphors. Furthermore, the mechanism of electron-transition events can be analyzed using a simple energy diagram in configurational coordinates that includes the activator and the surrounding bonds [24]. Consider a configurational-coordinate diagram that includes an activator and its bonds to the host crystal as shown in Fig. 15.3a. On the absorption of light, the electron transits from the ground state to the excited state. The system then relaxes to the lowest vibrational level followed by a radiation transition, Em, to the ground state. These two transition processes correspond to the fluorescence excitation (or absorption [Abs]) and emission (Em). The system can also return to the ground state through a crossover step at the crossing point between the excited state and the ground state. This process is more probable when the system is heated and therefore is populated at higher vibrational levels. After the crossover, the excited electron returns to the lowest vibrational level in the ground state through relaxation in a progression of vibrational levels. The process does not emit light but instead generates heat, which corresponds to the thermal quenching of luminescence. This would be realized by minimizing the displacement between the equilibrium position at the ground state and at the excited state, *i.e.*, $\Delta R = R'_0 - R_0$, as schematically shown in Fig. 15.3b [25].

- Third, suitable phosphor particle-size distribution (0.5–3 μm) and sphere-like shape would be in favor of the fabrication of LEDs with high luminous efficacy, which are related to the synthesis method. Almost all phosphors are prepared by solid-state reactions between raw materials at high temperatures, which require high temperatures, time-consuming heating processes, and subsequent grinding. The grinding process damages the phosphor surfaces resulting in the loss of emission intensity [26]. In addition, the aggregation and inhomogeneous shape are also unavoidable, both of which inhibit the absorption of excitation energy

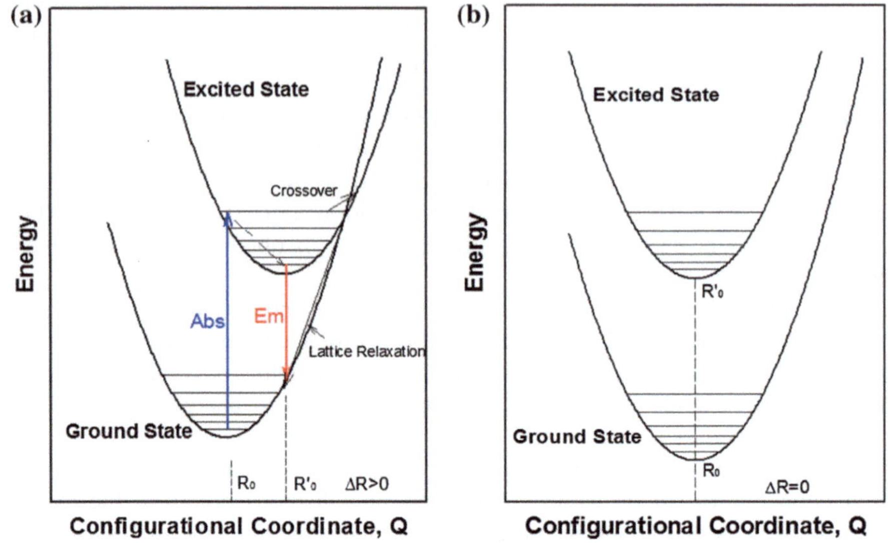

Fig. 15.3 Energy diagrams of fluorescence excitation and emission processes for displacement $\Delta R > 0$ (**a**) and $\Delta R = 0$ (**b**). Reprinted from Ref. [24] with kind permission from Springer Science + Business Media

and therefore decrease the emission intensity. Therefore, a simple and economical method for making high-quality phosphors is desirable. Some wet methods, such as sol-gel, co-precipitation, hydrothermal, spray pyrolysis, soft or hard template, and so on, offer many advantages compared with conventional solid-state method in the synthesis of fine powders, particularly phosphor materials, such as greater uniformity in particle-size distribution, nonagglomeration, and greater photo-luminescent intensity. The surface area of powders is very high, and the phosphors with greater surface area have better thermal conductivity when excited with high-energy photons [27], which are favorable to the lifetime of a white-light LED. As an example, Fig. 15.4 shows the SEM micrographs of phosphors using different methods [28–31]. Compared with phosphors prepared by wet methods, the particle-size distribution is broader, the average size is >10 μm, the sample surface is coarser; and the sample shape more irregular for the samples prepared by solid-state reaction.

In addition to the above-mentioned main requirements, phosphors used for LEDs also should possess high purity and not be hazardous and toxic. The cost of production should also be reasonable, and the processes of synthesis should also be environment-friendly.

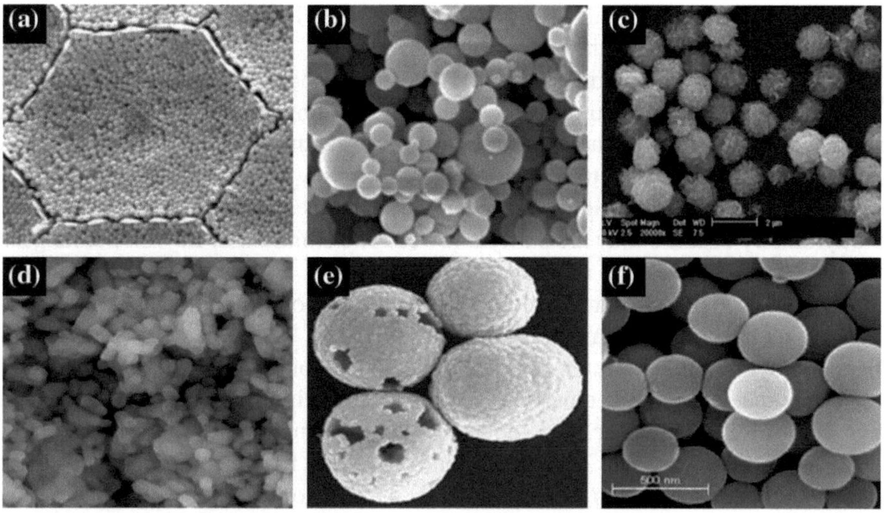

Fig. 15.4 SEM micrographs of phosphors with different methods **a** co-precipitation; **b** spray pyrolysis; **c** hydrothermal; **d** Sol-gel; **e** soft template; and **f** hard template

15.2.2 Two-Phosphor Method for Creating White LEDs

Currently, most commercially available white LEDs are generated by combining a blue LED with a yellow phosphor (double-band w-LEDs), but these w-LEDs give "cool" white light with unsatisfactory, high correlated color temperature (CCT) > 4000 K and poor color-rendering index (CRI, *Ra*) values of 70–80 because they lacking a red component [32]. For general indoor illumination, >90 % CRI is required, and the lower (i.e. warmer) CCT (2700–3200 K) are preferable for replacing traditional lamps. As expected, suitable w-LEDs could be produced using a two-phosphor approach by mixing a yellow or a green phosphor with a red one. The challenge in designing two-phosphor pc LEDs is to improve the CRI of the light source without a distinct decrease of luminous efficacy. The red-emitting luminescent materials suitable for a commercial blue chip are rather deficient owing to the requirements, which are difficult to meet. The common red-emitting centers are Eu^{2+}, Eu^{3+}, Sm^{3+}, Pr^{3+}, and some traditional metals such as Cr^{3+}, Mn^{4+}, which are beneficial to improve CRIs and low CCTs. Eu^{2+}-activated alkaline–earth binary sulfides MS (M = Ca, Sr, Ba, and Mg) and commercial nitridosilicates—such as (Ba, Sr, $Ca)_2Si_5N_8$ and (Ca, Sr)$AlSiN_3$ in the range of 570–690 nm—were generally selected as candidates for red phosphor [33, 34]. However, the low chemical stability and environmentally hazardous preparation methods limited the applications of sulfide-based phosphors, although improvement of their stability has been achieved by coating a layer of inert materials [35, 36]; the stability of nitrides or oxy-nitrides is sufficient due to the highly condensed, rigid framework structure. The rather broad emission of Eu^{2+}-activated red phosphors always leads to a trade-off between a CRI and the maximum achievable luminous efficacy because of longer emission wavelength >700 nm [37].

Eu^{3+} is the classical activator for red phosphor, and the red emission from the $^5D_0 \rightarrow {}^7F_2$ electric dipole transition is conducive to improve the color purity of the phosphor because it occupies the lattice sites without centro-symmetry. The intensity of electric dipole transitions depends strongly on the site symmetry in the host lattice, whereas the magnetic dipole *f–f* transitions are not affected much by the site symmetry because they are parity-allowed. The emission of Eu^{3+} in the range of 590–600 nm is due to the magnetic dipole transition $^5D_0 \rightarrow {}^7F_1$; the emission of approximately 610–630 nm is due to the electric dipole transition of $^5D_0 \rightarrow {}^7F_2$, induced by the lack of inversion symmetry. If Eu^{3+} occupies the site without inversion symmetry, the intensity of the electric dipole emission is stronger than that of the magnetic dipole emission, and the electric dipole emission dominates the emission of Eu^{3+}. However, Eu^{3+}-doped phosphors usually have intense absorption in the n-UV or UV region. Molybdates and tungstates have been suggested as host lattices showing strong *f–f* absorption ($^7F_0 \rightarrow {}^5D_2$) on the activator in the blue spectral region (approximately 465 nm), but the absorption bands are too narrow to allow application [38, 39]. Similar limitations apply to other rare-earth ions that display red emission lines, viz., Sm^{3+} and Pr^{3+}, which exhibit additional intrinsic limitations such as cross-relaxation.

Cr^{3+} and Mn^{4+} with $3d^3$ ions show line emission based on d-transitions. Due to the larger crystal-field splitting induced by its higher charge, the emission bands of Mn^{4+}-activated materials are usually favorably blue-shifted versus those of Cr^{3+}-activated materials. Both metals tend to emit in the deep red (>620 nm) region. Mn^{4+} ions are the activator of choice for lighting applications, and its emission in the octahedral environment can be explained by assuming electronic one ($^2E \rightarrow {}^4A_2$) or two transitions ($^2E \rightarrow {}^4A_2$ and $^2T_1 \rightarrow {}^4A_2$). In contrast to f-ions such as the lanthanides, d-ions may show relatively strong absorptions on the metal center, which is usually broad. Thus, Mn^{4+}-doped red-emitting phosphors—such as such as $3.5MgO \cdot 0.5MgF_2 \cdot GeO_2$, $CaAl_{12}O_{19}$, $Gd_3Ga_5O_{12}$, $YAlO_3$, $YAl_3(BO_3)_4$, $ZnGa_2O_4$, $SrAl_{12}O_{19}$, and Gd_2MgTiO_6—has been investigated greatly. Especially the Mn^{4+}-doped fluorides A_2XF_6 (A = K, Na, and Cs, X = Si, Ge, Ti, and Zr), in which Mn^{4+} substitutes the sites of Mg^{2+}, Al^{3+}, Ga^{3+}, or X^{4+}, are being given more attention for their potential applications in LEDs [40, 41]. When doped in a solid, Mn^{4+} ion experiences strong crystal field (CF) due to its high effective positive charge, and its emission spectrum is always dominated by the spin-forbidden $^2E_g \rightarrow {}^4A_{2g}$ transition (sharp line), whereas its PLE (absorption) spectra includes two broad bands corresponding to the $^4A_{2g} \rightarrow {}^4T_{2g}$ and $^4A_{2g} \rightarrow {}^4T_{1g}$ (^4F) spin-allowed transitions. The spin-allowed transition ($^4A_{2g}$- $\rightarrow$ T_{1g} (^4P)) is often hidden by the charge-transfer transitions or the host-lattice absorption. Despite that the energy-level structure of Mn^{4+} seems simple, there is a very large variation in the energy of the $^2E_g \rightarrow {}^4A_{2g}$ emission transition from 16,109 cm^{-1} (620 nm) in K_2SiF_6 to 13,826 cm^{-1} (723 nm) in $SrTiO_3$. Similarly, a large variation in the CF strength parameter Dq is also noticed: It ranges from approximately 1820 cm^{-1} in $SrTiO_3$ to 2400 cm^{-1} in the commercial $3.5MgO \cdot 0.5MgF_2 \cdot GeO_2$ phosphor. The

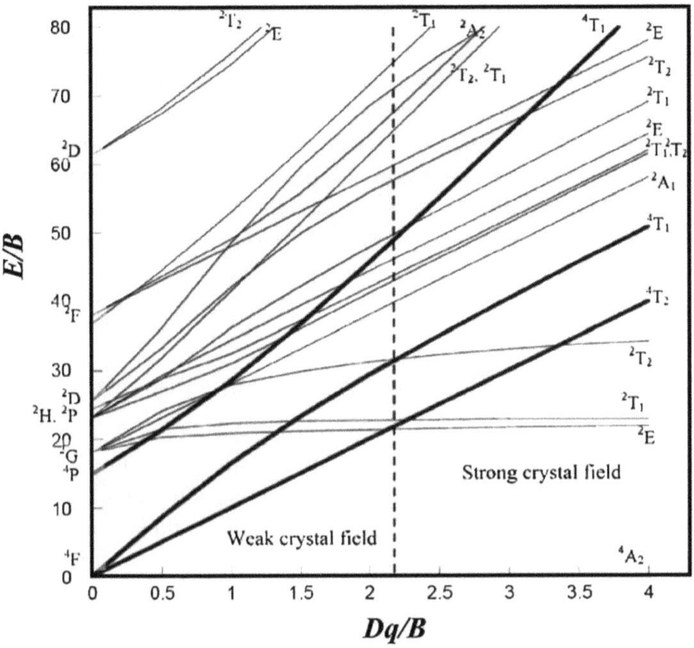

Fig. 15.5 Tanabe–Sugano diagram for the d^3 electron configuration in an octahedral crystal field. Reprinted from Ref. [42]. Copyright 2013 with permission from Elsevier

wide variation seen in the position of the $^2E_g \rightarrow {}^4A_{2g}$ emission transition is not due to the variation in CF strength because the energy of the ^2E_g state in the d^3 electronic configuration is independent of CF (Fig. 15.5) [42].

15.2.3 Tricolor Phosphor–Converted White LEDs

For the above-mentioned two-phosphor–converted w-LED, few phosphors match well with the blue LED chips. If the emission wavelength of the LED chip could be shifted to the UV region, more phosphors could be selected. To our delight, UV-, even deep UV-, wavelength LED chips (320 to 400 nm for UV-A, 280–315 nm for UV-B, and 180–280 nm for UV-C) are commercially available for $\lambda > 240$ nm or obtained in a laboratory for $\lambda < 240$ nm with the development of semiconductor fabrication technology [43, 44], which provides a solid foundation for tri-band phosphor–converted (pc) w-LEDs. In UV LED chip + red/green/blue (RGB) white LEDs, the visible components of the white light are generated only from the emission of tricolor phosphors, and they have been suggested for general illumination because of their great color stability, excellent color-rendering index, and tunable color temperature [45]. Presently the most frequently used commercial tricolor phosphors for UV LED are $BaMgAl_{10}O_{17}:Eu^{2+}$ ($BAM:Eu^{2+}$) for blue, ZnS: Cu^{+}, Al^{3+} for green, and $Y_2O_2S:Eu^{3+}$ for red. It is estimated that tri-band w-LEDs will dominate the market in the near future, but several single color-emitting phosphors that can be excited efficiently by UV light must be added in this scheme, thus leading to high cost and low luminous efficiency due to the complicated manufacture, reabsorption of emission colors and the different aging rates for each phosphor. Thus, it is possible to improve the efficiency of the white LEDs through the application of a novel single-phased multicolor-emitting phosphor with high chemical and thermal stability for UV-pumped white LEDs [46].

15.3 Approaches to Realize a Multicolor Phosphor

To avoid above-mentioned problems from mixing multiple phosphors, a single-phased multicolor-emitting phosphor that can simultaneously emit at least two colors with the excitation of UV light, blue light, and near infrared (NIR) light has proposed and investigated. The present review gives a simple summary of methods to design a multicolor-emitting phosphor for LED application even though the progress of the phosphor used for LEDs have been given by many scientists [47–51]. Until now, many methods have been developed to obtain a single-phased multicolor-emitting phosphor. Generally, single-phased multicolor-emitting phosphors are used in UV-LED, but the introduction of red spectral component in YAG yellow phosphor by co-doping with a red-emitting activator has received more attention. Therefore, two kinds of multicolor-emitting phosphors matched with blue or UV LED chip will be divided into five main methods to discuss in this review:

(1) Enriching the red emission of $YAG:Ce^{3+}$ yellow phosphor
(2) Single-activator doped system
(3) Multiple-ion co-doped system based on energy transfer
(4) Upconversion of a luminescence system
(5) Semiconductor quantum dot and defects emission

15.3.1 *Enriching the Red Emission of YAG:Ce^3+ Yellow Phosphor*

As mentioned in part 15.2.2, the white LED fabricated by a combination of yellow phosphor and blue LED chip could not meet the requirement of illuminating source due to the lack of a red spectral component. Generally, the methods to improve CRI and decrease color temperature is to introduce a red component in yellow phosphor by codoping Pr^{3+}, Cr^{3+}, Eu^{3+}, and Sm^{3+} ions with sharp-line emission in the red spectral region, in which rare-earth ions Eu^{3+}, Sm^{3+}, and Pr^{3+} enter into the sites of

Y^{3+}, whereas Cr^{3+} substitutes for Al^{3+} due to having a similar radius. As a typical red-emitting activator, Eu^{3+} was first introduced to supply a red component in YAG:Ce^{3+}. A sharp-line emission peaking at approximately 610 nm from the $^5D_0 \rightarrow {}^7F_2$ transition of Eu^{3+} appeared, in which the emission intensity of Eu^{3+} increases with increasing amount of Eu^{3+} concentration as shown in Fig. 15.6a [47]. In contrast, rare-earth Sm^{3+} was also used as a red-emitting activator due to its transitions from the $^4G_{5/2}$ level to the ground state $^6H_{5/2}$ and higher levels ^6H_J ($J > 5/2$). However, the high doping concentration of Sm^{3+} will lead to concentration quenching as shown in Fig. 15.6b [48]. Unfortunately, part of Eu^{3+} or Sm^{3+} could inevitably be reduced ($Eu^{3+} \rightarrow Eu^{2+}$ or $Sm^{3+} \rightarrow Sm^{2+}$) in the synthesis process of YAG:Ce due to reaction in the reducing atomsphere, resulting in a low red emission. In comparision, Pr^{3+} could show a red-emission peak in YAG, and there is no report for $Pr^{3+} \rightarrow Pr^{2+}$. Figure 15.6c shows the PL spectra of $Y_3Al_5O_{12}$: Ce^{3+}, Pr^{3+} phosphors; with varying the amount of Pr^{3+}, it is found that a additional sharp red line appears at approximately 608 nm originating from the $^1D_2 \rightarrow {}^3H_4$ transition of Pr^{3+} in addition to the broadband emission from Ce^{3+} under the excitation of 470-nm blue light [49]. When the amount of Pr^{3+} was 0.006 mol, a strong red-emission peak was observed without a decrease of Ce^{3+} emission. Regarding luminescence of Cr^{3+}, its emission color is red for high crystal-field strength. The evolution of the emission spectra of YAG co-doped with Cr^{3+}–Ce^{3+} as a function of Cr^{3+} concentration is shown in Fig. 15.6d. It is clearly observed that the typical emissions of Cr^{3+} contain an $^2E \rightarrow {}^4A_2$ zero-phonon line peaking at 685 nm with vibronic sidebands [50].

As shown in Fig. 15.6, both a broad emission band from Ce^{3+} and a sharp emission peak from R^{3+} (R = Pr^{3+}, Cr^{3+}, Eu^{3+}, and Sm^{3+}) were observed. The addition of Pr^{3+}, Cr^{3+}, Eu^{3+}, and Sm^{3+} activator in YAG:Ce^{3+} greatly enhance emission in the red spectral region, and the CRI of w-LED devices fabricated with YAG:Ce^{3+}, R^{3+} are also >80 [51]. The above-mentioned red emissions are all based on the efficient energy transfer from Ce^{3+} to R^{3+}. For example, Fig. 15.7 illumines the energy-level diagrams of Ce^{3+}, Pr^{3+}, or Eu^{3+} as well as the energy transfer routes from Ce^{3+} to Pr^{3+} or Eu^{3+}. When the YAG:Ce^{3+}, R^{3+} phosphor is illuminated by blue light, it strongly absorbs the blue light, and yellow light is emitted by the transition from the lowest $^2D(5d)$ band to the $^2F_{7/2}$ and $^2F_{5/2}$ states of the Ce^{3+} ion. Furthermore, a radiative energy transfer occurs from the relaxed lowest 5*d* energy band of Ce^{3+} to the $^5D_1$ energy level of Eu^{3+} and then relaxes to the $^5D_0$ energy level by a rapid nonradiative (NR) relaxation, thus resulting in a series of sharp-line emissions from the transitions of $^5D_0 \longrightarrow {}^7F_J$ (J = 1–6). Only the sharply peaked red emissions ascribed to the $^5D_0 \longrightarrow {}^7F_2$ transition of Eu^{3+} were observed because the other emission lines are weak and overlapped with the broadband emission from Ce^{3+}. For the process of energy transfer from $Ce^{3+} \longrightarrow Pr^{3+}$, radiative-energy transfer from the relaxed lowest 5*d* energy band of $Ce^{3} \longrightarrow {}^1D_2$ energy level of Pr^{3+} and nonradiative-energy transfer from the lowest 5*d* band of Ce^{3+} to the $^3P_0$ energy level of Pr^{3+} might occur, which will result a strong green emission ($^3P_0$–$^3H_4$) as well as a red emission ($^1D_2$–$^3H_4$). However, no green emission ($^3P_0$–$^3H_4$) from Pr^{3+} was present; the green emission may be reabsorbed by Ce^{3+},

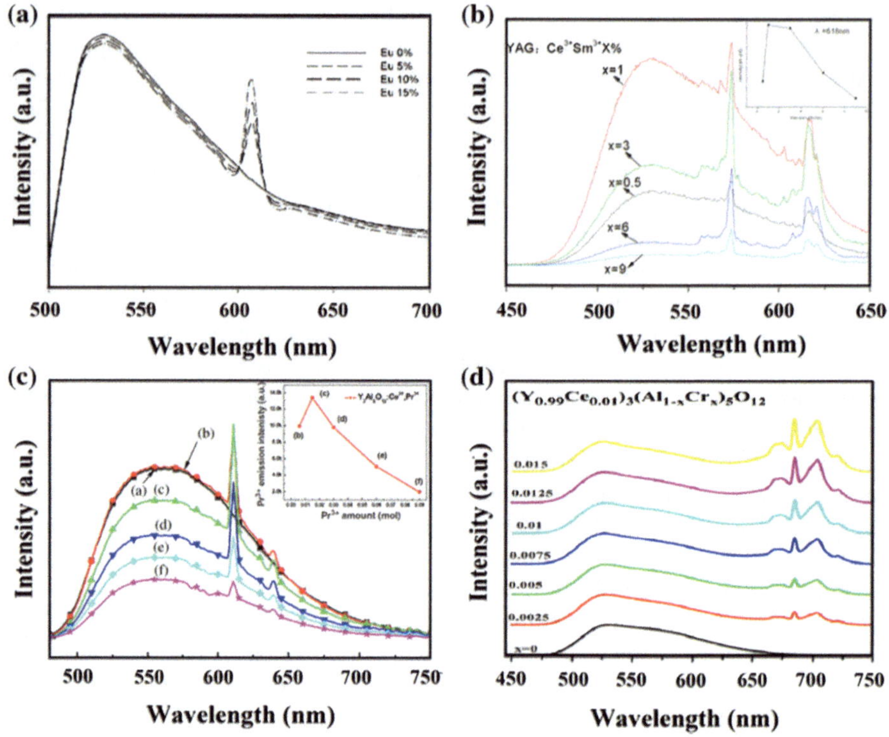

Fig. 15.6 PL spectra of YAG:Ce^{3+}, R^{3+} as a function of R^{3+} concentration with the excitation of blue light: **a** R = Eu, **b** R = Sm [48], **c** R = Pr, and **d** R = Cr [50]. **a** Adapted with permission from Ref. [47]. Copyright 2008 American Chemical Society. **c** Adapted from Ref. [49], copyright 2007, with permission from Elsevier

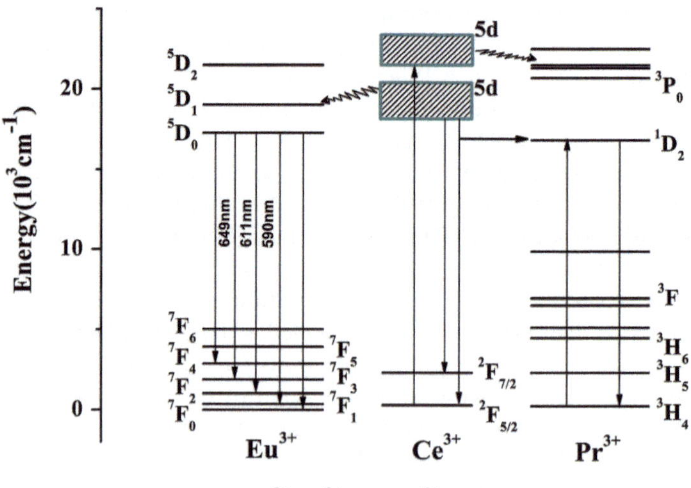

Fig. 15.7 Energy-level diagrams for Ce^{3+}, Pr^{3+}, and Eu^{3+} in $Y_3Al_5O_{12}$ as well as the proposed energy-transfer routes

and it can be negligible. Only a very weak red emission from $^3P_0$–$^3H_5$ was additionally observed at approximately 640 nm. In consequence, radiative-energy transfer may be the main factor for the decrease of the emission from Ce^{3+}.

15.3.2 *Single Activator–Doped Systems*

In this part, single activator–doped systems with multicolor emission are briefly introduced. The luminescent properties of a phosphor are basically determined by the host structure and the characteristics of the activator. Rare-earth ions and some transitional-metal ions are usually used as activators of phosphor. The energy-level spectra of some rare-earth ions, such as Eu^{3+}, Pr^{3+}, Tb^{3+}, and Dy^{3+} contain several metastable multiplets that can offer the possibility of simultaneous emission in the blue, green, orange, red, and infrared wavelengths in different crystalline hosts. In turn, these are strongly affected by the host lattice (phonon frequency as well as the crystal structure), and the concentration of the activators decide the different color-emission intensities from $4f$–$4f$ transitions. The emissions of other rare-earth ions with $d \rightarrow f$ transition (Eu^{2+}, Ce^{3+}), including transitional metal ions with d–d transitions, strongly depend on the host composition and structure such as radius.

15.3.2.1 Trivalent Rare-Earth Ions Doped with Single-Phased Sample (Pr^{3+}, Eu^{3+}, Tb^{3+}, Dy^{3+})

Pr^{3+}: Pr^{3+} ion has rich emission spectral lines in the UV, visible, and near-infrared regions because of its intricate energy level scheme with energy gaps of different magnitudes. It is well known that exoteric electric, magnetic, and ligand fields have almost no influence on the $4f$ shell of Pr^{3+} due to its well-shielding effect by the $5s$ and $5p$ outside shells, also leading to little effect of the crystal field on the energy levels. Pr^{3+} ion has unique properties because it offers the possibility of simultaneous blue ($^3P_0 \rightarrow {}^3H_4$), green ($^3P_1 \rightarrow {}^3H_5$), red ($^3P_0 \rightarrow {}^3F_2$), and deep red ($^3P_0 \rightarrow {}^3F_4$) emission. When the Pr^{3+} ions are pumped to the closely grouped ^3P_J ($J = 0, 1, 2$) multiplets, several emissions occur in the red, orange, green, and blue spectral regions. It is noteworthy that there are strong absorption peaks at approximately 450 nm for the $^3H_4 \rightarrow {}^3P_2$ transition, which benefit its application in LED due to good match with the blue LED chip [52]. Generally, the emission color of Pr^{3+} depends strongly on the host lattice, whereas the emission property of a phosphor is usually influenced by the crystal-field strength, which is a function of the structure of the host material, the coordination number of the activator, and the electro-negativity of the constituents. If the emission originates from the $^3P_0$ level, it may be green ($^3P_0$–$^3H_4$) as seen with $Gd_2O_2S{:}Pr^{3+}$, but red lines may also be strong ($^1D_2 \rightarrow {}^3H_4$, $^3P_0 \rightarrow {}^3H_6$, $^3F_2$) as seen with $LiYF_4$:Pr. If the emission originates from the $^1D_2$ level, it is in the red and near-infrared region [53, 54].

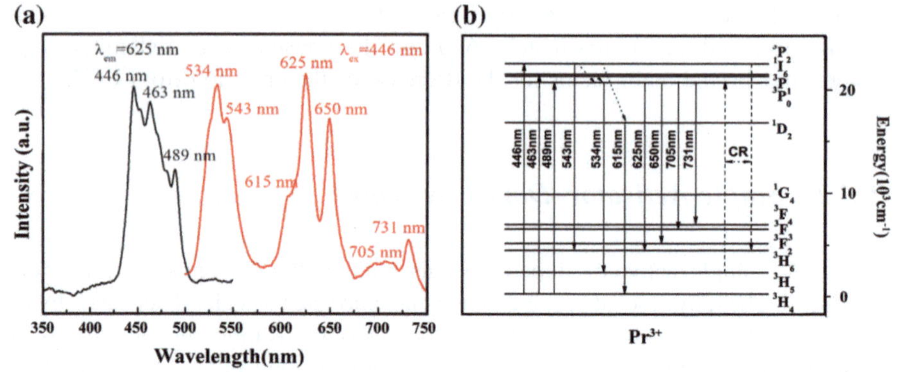

Fig. 15.8 **a** PLE and PL spectra of $LaMgAl_{11}O_{19}$:0.1 Pr^{3+} phosphors and **b** energy-level diagram of Pr^{3+}. Adapted from Ref. [55], copyright 2015, with permission from Elsevier

Figure 15.8 shows the schematic energy-level diagram of Pr^{3+} and its typical simultaneous blue, green, and red emissions in $LaMgAl_{11}O_{19}$:0.1 Pr^{3+} under the excitation of 446-nm blue light [55]. The PLE and PL spectra of $LaMgAl_{11}O_{19}$:0.1Pr^{3+} phosphors are presented in Fig. 15.8a. The excitation spectrum monitored at 625 nm is mainly composed of three peaks at 446, 463, and 48 9 nm, which originate from the transitions of $^3H_4 \rightarrow {}^3P_2$, $^3H_4 \rightarrow {}^1I_6$, and $^3H_4 \rightarrow {}^3P_0$, respectively. When the Pr^{3+} ion was excited to $^3P_2$ by the 446-nm light, it relaxed to the $^1D_2$ or $^3P_0$ level by way of nonradiative relaxation and then emitted the characteristic green and red emission (yellow for the naked eye) as shown in Fig. 15.8b. The emission spectrum includes a green emission band with two peaks at 534 and 543 nm due to the $^3P_0 \rightarrow {}^3H_5$ and $^3P_2 \rightarrow {}^3H_6$ transitions; a red emission band with three peaks at 615, 625, and 650 nm corresponding to the $^1D_2 \rightarrow {}^3H_4$, $^3P_0 \rightarrow {}^3H_6$, and $^3P_0 \rightarrow {}^3F_2$ transitions; and a weak near-infrared emission band at 705 and 730 nm corresponding to the $^3P_0 \rightarrow {}^3F_3$ and $^3P_0 \rightarrow {}^3F_4$ transitions. The changes in the characteristic emission peaks for Pr^{3+} ions mainly originate from the changing transition probability of the different energy levels for the rare-earth ions, which is directly related to their coordination environment.

Eu^{3+}: Eu^{3+} is a typical and efficient activator for red-emitting phosphor due to its transitions from the excited $^5D_0$ level to the ^7F_J ($J = 0$–4) levels of the $4f^6$ configuration. Its photoluminescence emission strongly depends on the symmetry of the crystal structure of Eu^{3+} occupied site in the host. The optical transitions of Eu^{3+} ions originating from the electronic dipole and magnetic dipole interaction of the internal $4f$ electrons are deeply affected by the crystal environment. If the Eu^{3+} ions occupy the sites with inversion symmetry, the emission will peak at 590–600 nm from the $^5D_0 \rightarrow {}^7F_1$ magnetic-dipole transition. This will dominate the emission, which is not affected much by the site symmetry. In contrast, the emission peaks at approximately 610–630 nm, due to the $^5D_0 \rightarrow {}^7F_2$ electronic dipole transition, will dominate the emission if the Eu^{3+} ion substitutes the site with no inversion symmetry [56]. Moreover, Eu^{3+}-doped phosphors usually have intense intrinsic absorption at approximately 395 ($^7F_0 \rightarrow {}^5L_6$) and 465 nm ($^7F_0 \rightarrow {}^5D_2$), which makes it a suitable activator for the phosphor used in UV-LEDs.

In addition to the above-mentioned $^5D_0 \rightarrow {}^7F_J$ red emission lines, its emission occurs from the higher excited states $^5D_1$ (green emission) $\rightarrow {}^7F_J$ (J = 1–3), $^5D_2$(blue–green) $\rightarrow {}^7F_J$ (J = 0–3), $^5D_3$(blue) $\rightarrow {}^7F_J$ (J = 1–3) depends critically on the Eu^{3+} concentration and the dominant vibration frequencies (υ_{max}) available in the host lattice, which offer the opportunity to generate multicolor emission by solely Eu^{3+} doping. The higher $^5D_{1,2,3}$ emissions of Eu^{3+} could be quenched by cross-relaxation occurring between two neighboring Eu^{3+} ions—Eu^{3+} ($^5D_1$) + Eu^{3+} ($^7F_0$) $\rightarrow$ Eu^{3+} ($^5D_0$) + Eu^{3+} ($^7F_3$)—if its concentration is high. In contrast, it might also be quenched by multi-phonon relaxation if the phonon energy (the highest vibration energy) of the host is high enough. In Y_2O_3:Eu^{3+}, all emissions from $^5D_0$ (red), $^5D_1$ (green), and $^5D_2$ (blue) of Eu^{3+} could be observed, and its$\upsilon_{max} \approx 600$ cm^{-1}, which means that Eu^{3+}-doped multicolor-emitting phosphors are obtained through low concentration (<1 mol%) in a host with low phonon energyυ_{max} < 600 cm^{-1} such as fluoride β-$NaYF_4$ (360 cm^{-1}), oxyfluoride LaOF (550 cm^{-1}), and some oxides such as $CaIn_2O_4$ (475 cm^{-1}) and $BaLn_2ZnO_5$ (370 cm^{-1}); however, the efficiency of these phosphors is low [57–61].

For instance, our group reported the color-tunable emitting phosphor $BaLn_2ZnO_5$:Eu^{3+} (Ln = La, Gd, and Y) with the excitation of n-UV light at 395 nm [60]. Figure 15.9 shows the PL spectra of BaY_2ZnO_5:Eu^{3+} and the CIE color-coordinate diagram as a function of Eu^{3+} contents as well as the energy-level diagram of Eu^{3+}. As mentioned previously, it is found that the emission spectra show a completely different ratio between the $^5D_{3,2,1}$ blue–green emission and the $^5D_0$ red emission for lower and greater Eu^{3+} concentrations, respectively. The dominant transition of $BaY_{2-x}Eu_xZnO_5$ phosphors for lower Eu^{3+} ion concentrations are $^5D_3 \rightarrow {}^7F_J$ blue emission and $^5D_2 \rightarrow {}^7F_J$ blue-green emission. The intensities of the emission peaks from the $^5D_{3,2}$ transition decrease with increasing Eu^{3+} concentration, and the intensities of the emission peaks from the $^5D_{1,0} \rightarrow {}^7F_J$ transition increase gradually due to concentration quenching by the cross-relaxation process [61], which also results in various color tones from blue through white to red (as shown in Fig. 15.9b). Figure 5.9c displays a partial energy-level diagram of Eu^{3+} consisting of the possible emissions in the red, green, and blue regions under the excitation of 395-nm light.

Tb^{3+}: Tb^{3+} is usually used as the activator of green-emitting phosphors due to transitions of $^5D_4$–^7F_J, but it also has blue emission from the higher-level emission $^5D_3 \rightarrow {}^7F_J$. The energy difference ΔE between the $^5D_3$ and $^5D_4$ excited states matches approximately the difference between the $^7F_6$ ground state and higher ^7F_J states. Similar to Eu^{3+}, the emissions from high excited states $^5D_3$–^7F_J are related to the concentration of dopant Tb^{3+} and the cross-relaxation. In diluted Tb^{3+} systems with large Tb–Tb distances, a process of cross-relaxation with a low rate and emission from both the $^5D_3$ and $^5D_4$ excited states is observed (unless the gap between these two states is bridged by phonon emission, for which υ_{max} is very high). The resulting emission spectrum has emission from the near-UV region into the red region. At higher Tb concentrations (on the order of 5 %), cross-relaxation

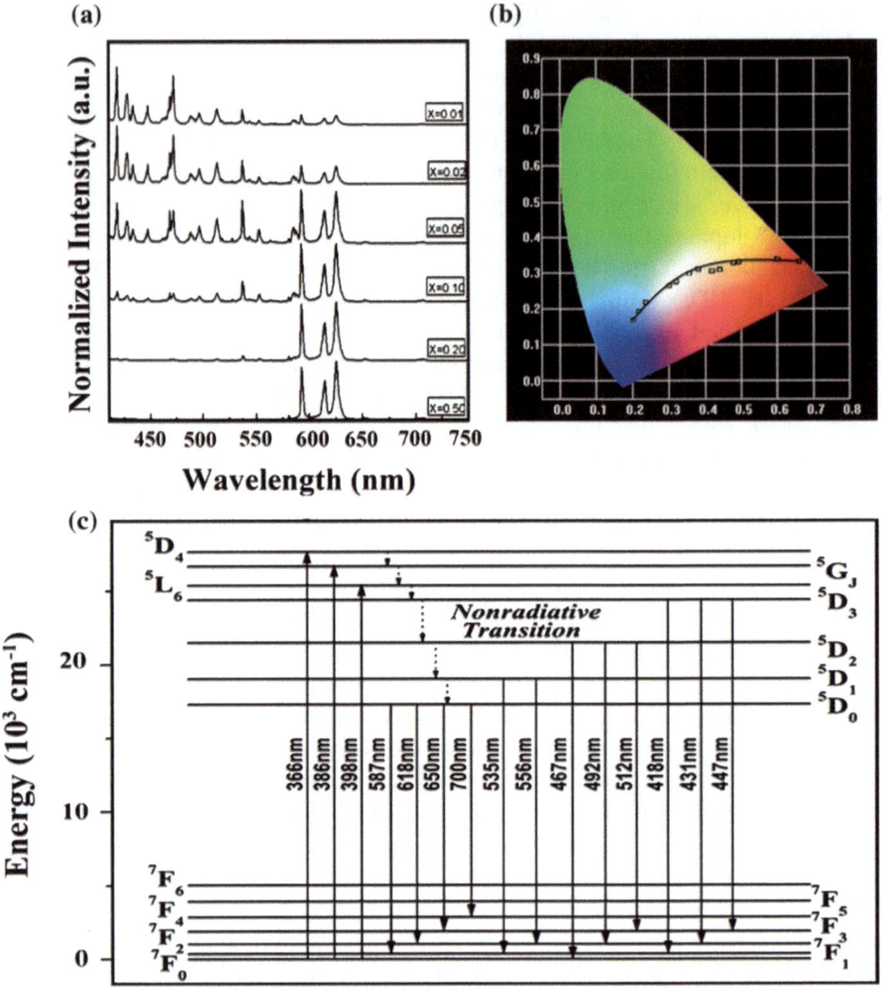

Fig. 15.9 PL spectra of $BaY_2ZnO_5{:}Eu^{3+}$ and CIE color-coordinate diagram as a function of Eu^{3+} contents (adapted with permission from Ref. [61]. Copyright 2008, American Institute of Physics) as well as the energy-level diagram of Eu^{3+}

quenches the emission from the $^5D_3$ level in favor of emission originating from the $^5D_4$ level: Tb^{3+} ($^5D_3$) + Tb^{3+} ($^7F_6$) → Tb^{3+} ($^5D_4$) + Tb^{3+} ($^7F_0$). The higher energy–level emission is quenched in favor of the low energy–level emission, implying that it is not possible to obtain blue Tb^{3+} emission in a phosphor with higher Tb^{3+} concentrations [62].

Figure 15.10 gives the PLE (λ_{em} = 544 nm) and PL (λ_{ex} = 375 nm) spectra of sample Tb^{3+} singly doped $Sr_2B_5O_9Cl$ and a partial energy-level diagram of Tb^{3+}. Under the excitation of 375 nm, the emission spectrum exhibits four line emissions peaking at 488, 543, 594, and 618 nm, which are assigned to the $^5D_4 \rightarrow {}^7F_J$ (J = 6,

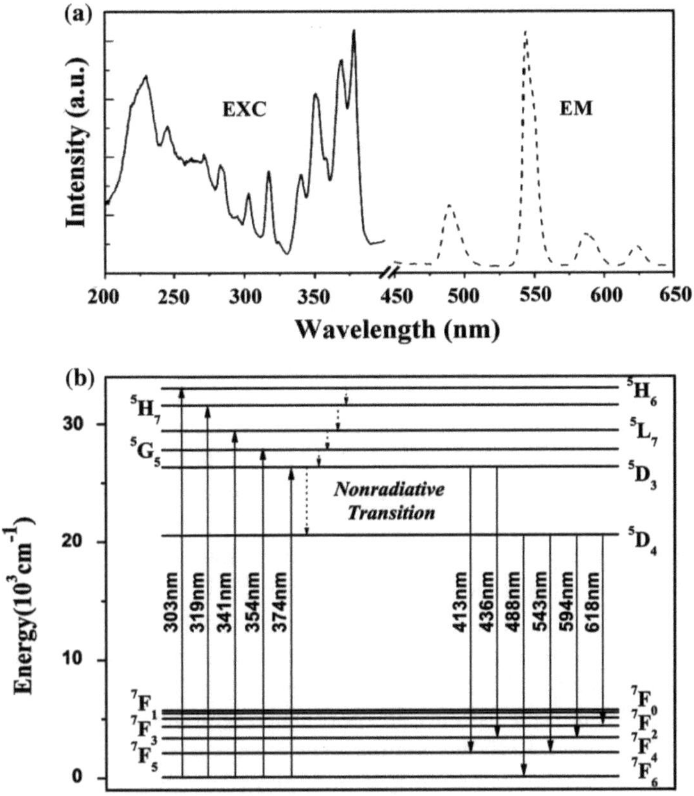

Fig. 15.10 PL and PLE spectra of $Sr_2Tb_{0.1}Li_{0.1}B_5O_9Cl$ and corresponding energy-level diagram. Adapted from Ref. [63], Copyright 2012, with permission from Elsevier

5, 4, and 3) characteristic transitions of Tb^{3+} ions, respectively. Monitored at 544 nm, the PLE spectrum consists of a group of absorption bands between 200 and 390 nm, among which the excitation bands observed in the range of 200–275 nm can be attributed to the spin-allowed $4f^8 \rightarrow 4f^7 5d^1$ transition of Tb^{3+} and those of sharp lines in the range of 300–400 nm owing to the forbidden $4f \rightarrow 4f$ transitions of Tb^{3+} ions from the $^7F_6$ to ^5I_J, ^5H_J, $^5D_{0,1}$, $^5G_{2,3,4}$,. $^5D_2$, $^5L_{10}$, and $^5D_3$ levels, respectively. Strong absorption in the region of 350–380 nm has been found, which matches well with the n-UV LED chip, but no obvious blue emissions are observed because of the high concentration and phonon energy. The nonradiative transition rate between levels of a $4f^n$ configuration of rare-earth ions is given by Van Dijk and Schuurmans as $W = \beta \exp[-(\Delta E - 2\,h\upsilon_{max})\alpha]$. Here W is the nonradiative rate; α and β are constants; ΔE is the energy difference between the levels involved; υ_{max} is the highest available vibrational frequencies of the surroundings of the rare-earth ions; and h is the Planck constant. The ions, such as Eu^{3+} and Tb^{3+}, can emit from higher excited states or not, and this depends critically on the host lattice. If υ_{max} is higher, the higher-level emission can be quenched by multiphonon emission; in contrast, if υ_{max} is low enough, these higher-level emissions can be observed [63–65].

Dy^{3+}: As a typical white light–emission ion, trivalent dysprosium (Dy^{3+}) ion generally has two dominant emission bands in the 460–500-nm blue regions and the 560–600-nm yellow regions. The emission has a whitish color that turns to yellow in host lattices where hypersensitivity is pronounced. The former is assigned to the $^4F_{9/2} \rightarrow {}^6H_{15/2}$ magnetic dipole transition, which is barely varies with the local symmetry in the crystal field; whereas the latter is attributed to the hypersensitive electric dipole transition $^4F_{9/2} \rightarrow {}^6H_{13/2}$ ($\Delta L = 2$, $\Delta J = 2$) where its intensity is strongly influenced by the surrounding chemical environment around Dy^{3+}. The intensity ratio of yellow-to-blue (Y/B) emissions in Dy^{3+}-doped phosphors is mainly influenced by the nephelauxetic effect between Dy^{3+} and the ligand. The greater the nephelauxetic effect (viz. covalency), the stronger the yellow emission of the Dy^{3+} ion. Therefore, Y/B could be tuned by adjusting the composition and local structure—e.g., ion radius, inequivalent substitution, local symmetry, etc.—of the sample, which leads to the varied emitting color and chromaticity coordinates [66].

Dy^{3+}-doped $Li_8Bi_2(MoO_4)_7$ (LBM) phosphor shows a near-white light generation under the excitation of n-UV 387-nm light [67]. The compound LBM crystallizes in a tetragonal structure with the space group $I\text{-}\overline{4}$ and is composed of tetrahedral units of MoO_4 and LiO_4, LiO_6 octahedra, and Bi atom surrounded by eight oxygen atoms [68]. Li atoms occupy four sites; Li(1) and Li(2) is coordinated by 6O; and Li(3) and Li(4) is fourfold coordinated with 4O. Because the radius and valence of Dy^{3+} and Bi^{3+} are similar, Dy^{3+} can easily enter into Bi^{3+} sites. Figure 15.11a shows the excitation spectra of phosphor LBM:0.1Dy^{3+} monitored at 573 nm, in which the excitation spectrum consists of a group of sharp lines absorbed from the *f–f* transitions of Dy^{3+} in 350–500-nm longer-wavelength region peaking at 353 ($^6H_{15/2} \rightarrow {}^6P_{7/2}$), 366 ($^6H_{15/2} \rightarrow {}^6P_{5/2}$), 387 ($^6H_{15/2} \rightarrow {}^4I_{13/2}$), 427 ($^6H_{15/2} \rightarrow {}^4G_{11/2}$), 453 ($^6H_{15/2} \rightarrow {}^4I_{15/2}$) and, 475 nm ($^6H_{15/2} \rightarrow {}^4F_{9/2}$) [69]. The two strongest peaks are centered at 387 and 453 nm, which match well with the commercially available near-UV and blue LED chips, respectively. Figure 15.11b displays the PL spectra of samples LBM:Dy^{3+} with different Dy^{3+} concentration: All of the PL profiles are similar, and no significant changes were presented except their intensities. It was found that the intensities of samples gradually increased with the increase of Dy^{3+} ion concentration at the beginning stage and then decreased after reaching the maximum at $\times = 0.20$. Moreover, the variation of Y/B ratio (the ratio of integrated area) as a function of Dy^{3+} ions concentration is also presented in Fig. 15.11b(inset) and increases from 1.5 to 1.8 when the Dy^{3+} concentration increases from 0.05 to 0.25 because the Y/B ratio could be affected by the site symmetry, and the distortion of the lattice increases with the growth of Dy^{3+} concentration [70]. This means that the emission colors and the microenvironment around the Dy^{3+} could be tuned through varying the Dy^{3+} concentration to enhance intensity of the yellow emission.

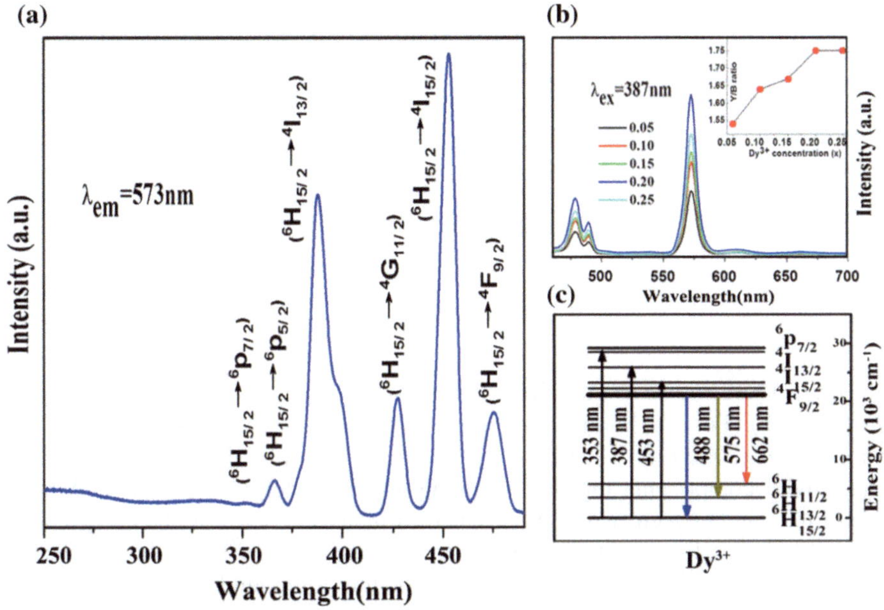

Fig. 15.11 **a** PLE (λ_{em} = 573 nm) spectra of LBM:0.1Dy^{3+} phosphors; **b** PL (λ_{ex} = 387 nm) spectra of LBM:$m$$Dy^{3+}$ (m = 0.05, 0.10, 0.15, 0.20, 0.25) phosphor annealed at 550 °C for 5 h (the *inset* shows the dependence of Y/B ratio on Dy^{3+} concentration); and **c** energy-level diagram of Dy^{3+}. Adapted from Ref. [67] by permission of The American Ceramic Society

It is worth noting that Pr^{3+}, Eu^{3+}, Tb^{3+}, and Dy^{3+} solely doped single-phased multicolor-emitting phosphors usually have low efficiency due to the forbidden *f*–*f* transitions in the n-UV region, thus resulting in narrow and weak absorption. In addition, low dopant concentration is another reason because high concentration will quench emission from the high excited level.

15.3.2.2 Multi-cation Sites Ce^{3+}, Eu^{2+}

Many luminescent ions (generally metal ions) show emission at different wavelengths in different host lattices, which opens up the possibility to change the emission color. The emission spectra of metal ions are strongly related to the crystal-field strength, which is affected by the metal ions' oxidation state, the arrangement of the ligands (coordination number), and the nature of the ligands. Therefore, any changes from the host composition could modify the emission color of the phosphor. This situation is encountered for many metal ions for the s^2 ions and the rare-earth ions showing $d \rightarrow f$ broadband emission Ce^{3+}, Eu^{2+}, Pr^{3+}, Nd^{3+}, and Er^{3+}. Here, we focus on the ions Ce^{3+} and Eu^{2+} with $d \rightarrow f$ emission because they are the last three ions only in the UV region [71].

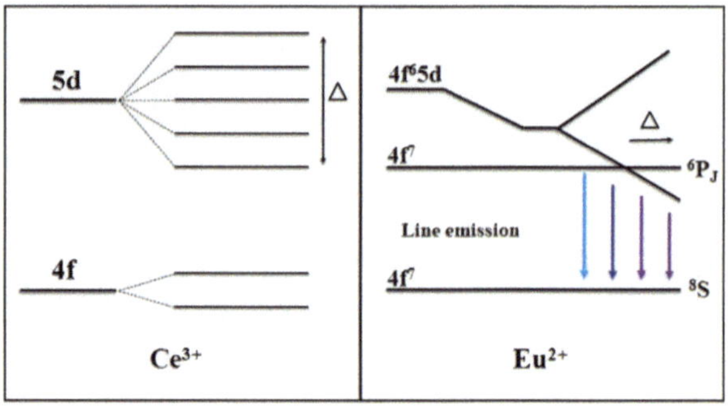

Fig. 15.12 Simplified energy-level scheme of the Ce^{3+} and Eu^{2+} ions

Figure 15.12 shows a simplified energy-level scheme of the Ce^{3+} ion (a) and the Eu^{2+} ion (b). For Ce^{3+} ion, spin-orbit coupling has split the 4*f* level into two components (2000 cm^{-1}), and the crystal field has split the 5*d* level into five crystal-field components together spanning approximately 15,000 cm^{-1}. Δ is the size of the gap between two or more sets of orbitals. For u^{2+} ion, the energy of the excited $4f^6 5d^1$ is lower than that of the lowest excited state ^6P in the $4f^7$ electron configuration. The Ce^{3+}/Eu^{2+} in the most of the compounds present broadband emission due to the $d \rightarrow f$ transitions. The energy difference between the *d*- and *f*-electrons is modified by the covalence of the $Ce^{3/}Eu^{2+}$–ligand bond and the crystal-field strength, which determine the spectral position of the emission. What determine the crystal-field strength? Many factors affect the crystal field, e.g., covalency (the nephelauxetic effect) will decrease the energy difference between the $4f^1$ and $5d^1$ configurations; crystal-field splitting of the 5*d* configuration (a large low-symmetry crystal field) will lower the lowest crystal-field component from which the emission originates), etc. Generally, the crystal-field strength is proportional to Δ. Ligands are ordered by the size of the splitting Δ such that they produce (small Δ to large Δ) $I^- < Br^- < S^{2-} < Cl^- < N^{3-} < O^{2-} < F^-$; an increase of the covalence of the Ce^{3+}/Eu^{2+}–ligand bond results in a lower energy difference of the 4*f*–5*d* energy separation (due to the nephelauxetic effect). Moreover, as the oxidation state increases for a given metal, the magnitude of Δ increases. Therefore, according to above-mentioned text, the crystal-field strength is influenced by site symmetry, covalency, ligand charge, and bond length. Crystal-field splitting (D_q) can be determined by the following equation:

$$D_q = \frac{1}{6} Z e^2 \frac{r^4}{R^5}$$

where D_q is a measure of the energy level separation; *Z* is the anion charge; *e* is the electron charge; *r* is the radius of the *d* wavefunction; and *R* is the bond length [72, 73].

In this way, an emission can be obtained extending from the UV part of the optical spectrum to the red part. Both are easily accessible by choosing appropriate host lattices, and for this reason broadband emitters can in general be tuned within a large spectral range as well as adapted to the application needs. It is possible to obtain a multicolor-emitting phosphor in a host with different cation sites, which could be substituted by Ce^{3}/Eu^{2+}.

Ce^{3+}: The Ce ion ($4f^1$) is the simplest example because it is a one-electron case. The excited configuration is $5d^1$. The $4f^1$ ground-state configuration yields two levels, viz. $^2F_{5/2}$ and $^2F_{7/2}$, separted by some 2000 cm^{-1} due to spin-orbit coupling. The $5d^1$ configuration is split by the crystal field in two to five components. The total splitting (Δ) amounts to approximately 15,000 cm^{-1}. The emission occurs from the lowest crystal-field component of the $5d^1$ configuration to the two levels of the ground state. This gives the Ce^{3+} emission its typical double-band shape. Because the $5d \rightarrow 4f$ transition is parity allowed and spin selection is not appropriate, the emission transition is a fully allowed one. Usually the Ce^{3+} emission is in the ultraviolet or blue spectral region, but in some hosts it could emit green, yellow, or red light [74, 75]. The decay time is longer if the emission is at longer wavelengths, and the decay time t is proportional to the square of the emission wavelength λ.

The Stokes shift of the Ce^{3+} emission is never very large and varies from a thousand to a few thousand wave numbers. The spectral position of the emission of the emission band depends on the crystal-filed strength. Any changes (mentioned previously) of the surroundings around the Ce^{3+} will result in changes of emission color. In one host, if Ce^{3+} enter two or more different sites, it will generate multicolor emission. Like in the host $Sr_5(PO_4)_2SiO_4$ (SPS), there are two different Sr^{2+} crystalline sites: one locates at the *6h* (C_s) site with seven-coordination (Sr2) and the other at the *4f* (C_3) site with nine-coordination (Sr1), providing a possibility of various site emission as doping Ce^{3+} ions in $Sr_5(PO_4)_2SiO_4$ compound [76]. Figure 15.13 shows the excitation and emission spectra of $Sr_5(PO_4)_2SiO_4{:}0.05Ce^{3+}$ and the emission spectra of $Sr_5(PO_4)_2SiO_4{:}xCe^{3+}$ as function of Ce^{3+} contention x. The emission spectrum (λ_{ex} = 356 nm) of SPS:$0.05Ce^{3+}$ covers the whole visible light area and consists of two broad bands located approximately 432 and 556 nm ($5d^1 \longrightarrow 4f^1$ transitions of Ce^{3+}). The excitation spectrum monitoring at 432 nm includes a broad band centered at approximately 356 nm together with a shoulder at 286 nm ($4f^1 \longrightarrow 5d^1$ transitions of Ce^{3+}) as shown in Fig. 15.13a, whereas the excitation spectrum has a noticeable red shift monitoring on longer wavelength (λ_{em} = 556 nm), which indicates that the emission of 432 and 556 nm come from different luminescent centers and it could be used in n-UV LED. In general cases, when Ce^{3+} locates at one specific lattice, a doublet-band emission with energy separation (ΔS) of approximately 2000 cm^{-1} is expected due to the transitions from the relaxed lowest *5d* excited state to the $^2F_{5/2}$ and $^2F_{7/2}$ spin orbit–split *4f* ground states. However, the calculated ΔS of the band at 432 and 556 nm is 5162 cm^{-1}, which is far higher compared with 2000 cm^{-1}, meaning that Ce^{3+} may occupy two sites in the SPS host. The emission spectrum was decomposed into four Gaussian sub-bands at 422, 455, 525, and 600 nm in Fig. 15.13a, respectively. The energy separation of the former two Gaussian bands was calculated to be 1719 cm^{-1} and the latter was 2381 cm^{-1}, thus showing that Ce^{3+} incorporates into two Sr^{2+} sites. According to crystal-field theory, a high-energy band at 432 nm is caused by Ce^{3+}

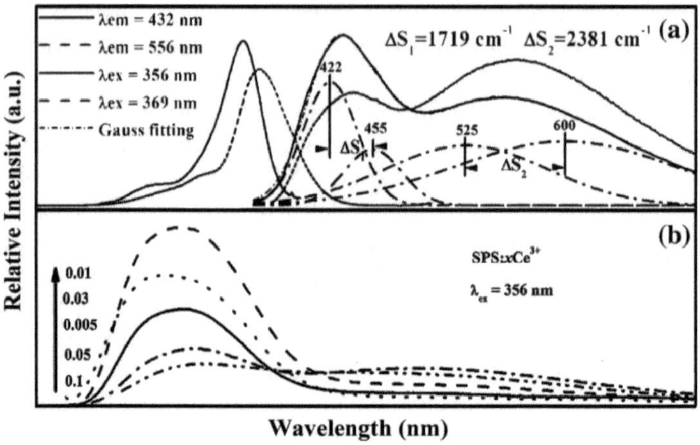

Fig. 15.13 Excitation and emission spectra of $Sr_5(PO_4)_2SiO_4$:0.05Ce^{3+} and emission spectra of $Sr_5(PO_4)_2SiO_4$:$x$$Ce^{3+}$ as a function of Ce^{3+} contention x. Adapted from Ref. [76], copyright 2013, with permission from Elsevier

occupying the nine-coordinated Sr1 site, whereas band 556 nm is for the Sr2 site. Under the excitation of a 356-nm n-UV light, Fig. 15.13b depicts the emission spectra of SPS:$x$$Ce^{3+}$ with different doping content x. It is found that the intensity of the Sr1 emission first increases then decreases with an increase of Ce^{3+} content, whereas the emission intensity of Sr2 reaches a maximum at × = 0.05 and then begins to decrease as a result of concentration quenching. According to the PL spectra, the CIE coordinates vary from near purple to greenish yellow for Ce^{3+} solely doped SPS.

Moreover, the blue and green double-color emission of Ce^{3+} in compound $Ba_2Ca(BO_3)_2$ was also investigated by Liang's group [77]. It was reported to crystallize in the monoclinic system with space group C2/m, and there are two cationic sits: Only one Ba^{2+} site with the C_5 symmetry is coordinated by nine oxygen atoms, and the Ca^{2+} atom occupies a slightly distorted octahedral site. Under near UV–light excitation, the emission of Ce^{3+} in $Ba_2Ca(BO_3)_2$ compound shows blue and green dichromatic emissions as Ce^{3+} ion occupy the sites of Ba^{2+} and Ca^{2+}, respectively.

Eu^{2+}: Eu^{2+} is the most well-known and widely applied example that shows a $5d \rightarrow 4f$ emission that can vary from long-wavelength ultraviolet to red. Its decay time is approximately 1 μs due to the fact that the emitting level contains (spin) octets and sextets, whereas the ground-state level (^8S from $4f^7$) is an octet. Thus, the spin-selection rule slows down the optional transition rate. Dependence of the host lattice on the emission color of the Eu^{2+} ion is determined by the same factors as in the case of the Ce^{3+} ion. If the crystal field is weak and the amount of covalency low, the lowest component of the $4f^65d$ configuration lies below it. Similar to the above-mentioned Ce^{3+}, Eu^{2+}, a solely doped single-phased multicolor-emitting phosphor could obtained by cooperated in a compound with more than one cationic

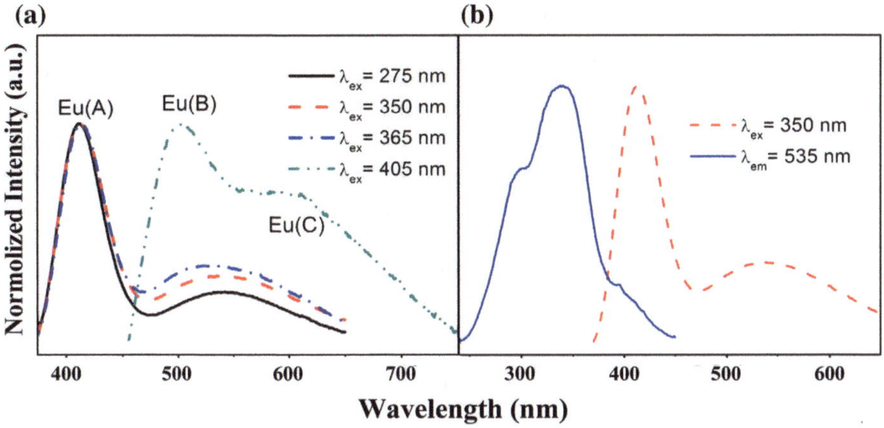

Fig. 15.14 Normalized PL spectra under different excitation wavelength (**a**) and normalized PLE (λ_{em} = 535 nm) and PL (λ_{ex} = 350 nm) spectra (**b**) of sample $Ca_{10}Li(PO_4)_7$:0.01Eu^{2+}. Adapted from Ref. [80], copyright 2015, with permission from Elsevier

site such as $BaSrMg(PO_4)_2$:Eu^{2+}, which could provide two cation sites (Ba^{2+} and Sr^{2+}) for Eu^{2+} and presents two emission bands at approximately 447 and 536 nm under the excitation of 350-nm n-UV light [78]. For some compounds with β-$Ca_3(PO_4)_2$ structure, there are five Ca^{2+} that yield various crystallographic environments; some alkaline–earth silicate systems—such as $M_3MgSi_2O_8$, $MMgSi_2O_7$ (M = Ca, Sr, Ba), and $Li_4SrCa(SiO_4)_2$—also have two different cationic sites and are suitable for Eu^{2+} ion doping.

When Eu^{2+} ions simultaneously enter into two cationic sites in one host, broadband emissions with different wavelength will be produced, and the positions of the emission could be tuned by changing the composition of the host. As a compound derived from β-$Ca_3(PO_4)_2$ structure, Eu^{2+}-doped $Ca_{10}Na(PO_4)_7$ was shown to display a broadband yellow emission peaking at 550 nm [79, 80], and the PL spectra of $Ca_{10}Li(PO_4)_7$:Eu^{2+} with different Eu^{2+} contents were composed of two broadband emissions peaking at 414 and 548 to approximately 504 nm (blue shift with increasing Eu^{2+} concentration, which implies that more than one Eu^{2+} emission center exists in the sample). As shown in Fig. 15.14a, it is found that at least three kinds of Eu^{2+} emission bands [marked as Eu(A), Eu(B), and Eu(C)], peaking at 414, 504, and 614 nm, exist in $Ca_{10}Li(PO_4)_7$:Eu^{2+} under the different excitation wavelength, and the long-wavelength excitation mainly corresponds to long-wavelength emission. Energy transfer among different Eu^{2+} centers may also exist. Figure 15.14b shows the PL and PLE spectra of $Ca_{10}Li(PO_4)_7$:0.01Eu^{2+}, and it is observed that a significant spectra overlap appears from 375 to 450 nm, i.e., the excitation band related to Eu(B) and Eu(C) is superimposed on the emission band of Eu(A). Thus, an ET could occur from Eu(A) to Eu(B) and Eu(C). The results indicate that the present sample could be excited efficiently and emits multicolor light, which could be used as a candidate phosphor for n-UV based LEDs.

15.3.2.3 Eu^{2+} and Eu^{3+} Coexistence

As is well known, the characteristic emission of Eu^{3+} ion gives a series of sharp lines in the spectral region of orange to red (575–750 nm) corresponding to the $^5D_0$–^7F_J ($J = 0$–4) transitions, whereas that of Eu^{2+} generally shows a broadband character from the blue to red region with a $4f^65d \rightarrow 4f^7$ transition in different host lattices. It is possible to generate multicolor emission through a strategy of combining intrinsic transitions of Eu ion in a different valence state in a single host lattice, which makes use of coexisting emissions of Eu^{3+} and Eu^{2+}. Eu^{3+} could be reduced to Eu^{2+} in some solid-state compound in the reducing atmosphere. However, in some special compounds, this reduction process could be thermally achieved in a nonreducing atmosphere, such as air, which has been investigated extensively by Su's group. Results indicate that four conditions seem to be necessary for the reduction of $Eu^{3+} \rightarrow Eu^{2+}$ in solid-state compounds when prepared at high temperature: (1) no oxidizing ions are present in the host compounds; (2) the dopant trivalent Eu^{3+} ion replaces a divalent cation in the host; (3) the substituted cation has a similar radius to the divalent Eu^{2+} ion; and (4) the host compound has an appropriate structure based on tetrahedral anion groups (BO_4, SO_4, PO_4, or AlO_4). In the following decades, this phenomenon was observed in BaB_4O_7, SrB_6O_{10}, $Sr_2B_5O_9Cl$, $BaMgSiO_4$, $Sr_4Al_{14}O_{25}$, $Sr_3P_4O_{13}$, and $Sr_2Y_8(SiO_4)_6O_2$, and a model based on the nature of defects caused by the substitution of cations was proposed. When trivalent Eu^{3+} ions were built into $Sr_2B_5O_9Cl$, they would replace the Sr^{2+} ions of $Sr_2B_5O_9Cl$ [81, 82]. To keep the electro-neutrality of the compound, two Eu^{3+} ions would substitute for three Sr^{2+} ions. Therefore, two positive defects of $[Eu_{Sr}]$* and one negative Sr^{2+} vacancy of $[V_{Sr}]$" would be created by each substitution for every two Eu^{3+} ions in the compound. By thermal stimulation, electrons of the $[V_{Sr}]$" vacancies would be then transferred to doped Eu^{3+} ions and reduce them to their Eu^{2+} from as shown in the following equation:

$$2[Eu_{Sr}]^{\cdot} + [V_{Sr}]'' = 2[Eu_{Sr}]^x + [V_{Sr}]^x.$$

If this model worked, we may imagine that the more electrons carried by negative defects were created, the more Eu^{3+} ions would be reduced to Eu^{2+} ions. As a result, the ratio of $Eu^{2+}$and Eu^{3+} emission intensities, I_{Eu2+}/I_{Eu3+}, would increase.

Recently, a tunable single-doped full color-emitting phosphor was obtained by valence-varied Eu-doped perovskite-type $LaAlO_3$, in which the valence distribution of Eu was also explored to modulate it by the addition of charge compensation agent Li^{+}., The PL spectra of phosphors can be tuned efficiently by varying the amount of Li^+ [83]. Figure 15.15 shows the PL and PLE spectra of $LaAlO_3$:Eu phosphor for full-color emission as well as the energy-transfer process from Eu^{2+} to Eu^{3+}. The PL spectrum exhibits full-color emissions peaking at 440 (blue), 515 (green), 592 (orange), and 618 nm (red) on excitation with ultraviolet (UV) light, and the PLE spectra broadly ranging from 250 to 400 nm. Both PLE and PL indicate that this phosphor is potentially suitable for application to UV-excited white light–emitting diodes or fluorescence lamps. The PL spectrum includes mixed

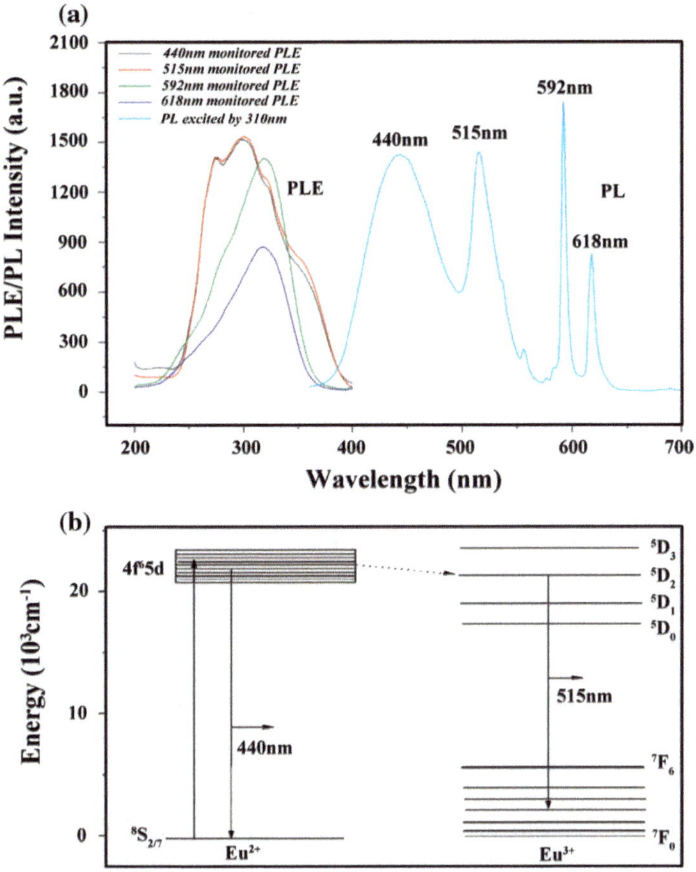

Fig. 15.15 PL and PLE spectra of $LaAlO_3$:Eu phosphor and the energy transfer from Eu^{2+} to Eu^{3+}. Adapted from Ref. [83] by permission of The Royal Society of Chemistry

emissions consisting of both band emission of Eu^{2+} and line emission of Eu^{3+}. It is reasonable to assign the 440-nm peaked band emission to the $4f^65d \rightarrow 4f^7(^8S_{7/2})$ transition of Eu^{2+}, whereas the 515-, 592-, and 618-nm line emissions were attributed to the $^5D_2 \rightarrow {}^7F_3$ (for 515 nm) and $^5D_0 \rightarrow {}^7F_{1,2}$ transitions of Eu^{3+}. Based on the PL and PLE spectra, energy transfer from Eu^{2+} to Eu^{3+} is proposed and shown in Fig. 15.15b. Eu^{2+} is excited from a ground state $^8S_{7/2}$ to an excited state $4f^65d$. Part of the energy is released in the inherent transition of Eu^{2+} to the ground state, giving a 440-nm blue-light emission, whereas another part of the energy is transferred to the closest level $^5D_2$ of Eu^{3+}; then 515-nm emission occurs through a transition to the $^7F_3$ ground state. It is worthy to illuminate that the PL intensities of red, green, and blue emissions could be tuned by adjusting the contents of Eu, and the relative ratio of Eu^{3+} to Eu^{2+} could be reduced by increasing the addition of Li^+.

15.3.3 Energy-Transfer Type of Downshift Phosphor

Among the methods to obtain the multicolor- or full color-emitting phosphor, most investigations are focused on the energy transfer between different activators and are thus designed according to this rule, in which the donor and the acceptor are considered sensitizer and activator, respectively. Energy transfer can only occur because the spectral overlap between the emission spectrum of the sensitizer and the excitation spectrum of the activator are significant, and a suitable interaction between both systems exists. The interaction may be either an exchange interaction or an electric or magnetic multipolar interaction. For this method, at least two luminescent emitting centers and energy transfer between them are necessary. According to the difference of the characteristic emission of the sensitizer, it was divided into a sharp-line emission from typical *f–f* transitions, a broadband emission from *f–d* or *d–d*, and a host emission three groups as discussed below.

15.3.3.1 Broadband Sensitizer: Ce^{3+}, Eu^{2+}

Ce^{3+} and Eu^{2+} usually shows a typical broadband emission accompanied by intense broad absorption from the *5d–4f* transitions. Their emission positions can vary from ultraviolet to red, and their corresponding excitation positions can also be tuned from ultraviolet to blue light, which depends on the crystal field strength of the host matrix. Therefore, Ce^{3+} and Eu^{2+} ion are efficient sensitizers and can transfer part of their energy to activator ions. In recent years, numerous single-phased multicolor-emitting phosphors have been developed based on the energy transfer from Ce^{3+} or Eu^{2+} to activators such as Mn^{2+}, Eu^{2+}, Tb^{3+}, Dy^{3+}, Pr^{3+}, and so on.

Mn^{2+} usually shows a broadband emission varying from green to red depending on the crystal field. Tetrahedrally coordinated (weak field) Mn^{2+} usually gives a green emission; however, octahedrally coordinated (strong field) Mn^{2+} gives an orange to red emission (500–700 nm). In addition, there is no remarkable change for the excitation spectrum of Mn^{2+} in different hosts, which usually includes several narrow absorption peaks in the UV and blue regions due to the transition from the from the ground state $^6A_1(^6S)$ to the excited states of $^4E(^4D)$, i.e., the $^4T_2(^4D)$, $[^4A_1(^4G)$, $^4E(^4G)]$, $^4T_2(^4G)$, and $^4T_1(^4D)$ levels of Mn^{2+}. The emission of sensitizer Ce^{3+}, Eu^{2+} in the n-UV or blue region could overlap with the excitation of Mn^{2+}; thus, co-doping $Ce^{3+} - Eu^{2+} - Mn^{2+}$ could be a multicolor-emitting phosphor. In these processes, Mn^{2+} ions are usually used as red luminescent–emitting centers by cooperating into the host with a strong field such as silicates, borates, phosphates, and nitrides.

As a typical example for single-phased multicolor-emitting phosphor based on energy transfer from Eu^{2+} to Mn^{2+}, the PL and PLE spectra of Eu^{2+} and Mn^{2+} solely and doubly doped $Sr_2Mg_3P_4O_{15}$ are shown in Fig. 15.16. Several narrow absorption bands locating at approximately 340, 355, 400, 416, and 470 nm are observed, which can be attributed to the forbidden transitions from the ground state $^6A_1(^6S)$ to

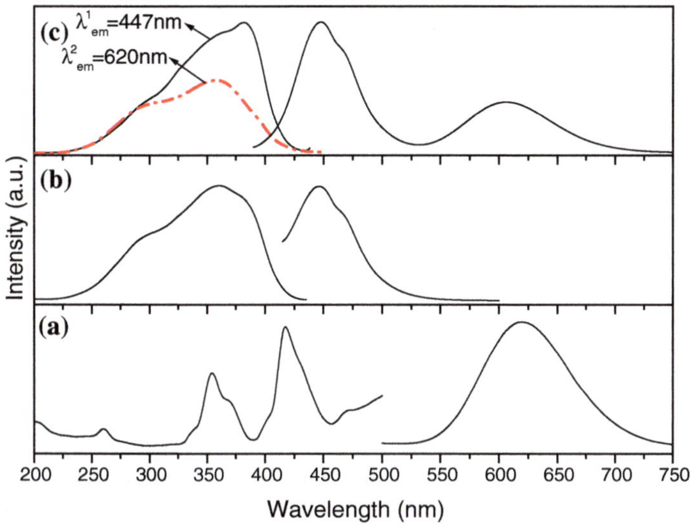

Fig. 15.16 PL and PLE spectra of $Sr_2Mg_3P_4O_{15}$:Mn^{2+} (**a**), $Sr_2Mg_3P_4O_{15}$:Eu^{2+} (**b**), and $Sr_2Mg_3P_4O_{15}$:Eu^{2+}, Mn^{2+} (**c**) phosphors. Reprinted from Ref. [88], copyright 2010, with permission from Elsevier

the excited states $^4E(^4D)$ (340 nm), $^4T_2(^4D)$ (355 nm), $[^4A_1(^4G), ^4E(^4D)]$ (400 nm), $^4T_2(^4G)$ (416 nm), and $^4T_1(^4G)$ (470 nm) levels of Mn^{2+}, respectively. A broad red-emission band centered at 620 nm from the $^4T_1$ (^4G) → $^6A_1$ (^6S) transition of Mn^{2+} are observed. The PL spectrum presents a broad blue-emission band centering at approximately 447 nm under the excitation of UV light for the Eu^{2+}-doped $Sr_2Mg_3P_4O_{15}$, which is attributed to the well-known $4f^65d^1(t_{2g}) \rightarrow 4f^7(^8S_{7/2})$ transition of Eu^{2+} as shown in Fig. 15.16b. Comparing the PL spectrum of $Sr_2Mg_3P_4O_{15}$:Eu^{2+} and the PLE spectrum of $Sr_2Mg_3P_4O_{15}$:Mn^{2+}, a significant spectral overlap is found, which indicates existence of the effective resonance-type energy transfer from Eu^{2+} to Mn^{2+} in $Sr_2Mg_3P_4O_{15}$. As expected, there are double-color blue- and red-emission bands in the PL spectrum of $Sr_2Mg_3P_4O_{15}$: Eu^{2+}, Mn^{2+} phosphor under the excitation of 380 nm, which are shown in Fig. 15.16c. The PLE spectra monitored at 447 or 620 nm are similar except for relative intensity, which also proves the existence of energy transfer from Eu^{2+} to Mn^{2+}. The PLE spectrum shows a strong absorption band in the near-UV region (325 to 400 nm), thus indicating that $Sr_2Mg_3P_4O_{15}$:Eu^{2+}, Mn^{2+} phosphor can be excited by the n-UV LED chip; thus, these phosphors could be a potential candidate for n-UV white LEDs as blue and red spectral components [84–88].

According to the investigation by Dorenbos [89], the energy of the Ce^{3+} or Eu^{2+} 5*d*–4*f* emission can be predicted according to the 5*d*–4*f* emission energy correlation between Eu^{2+} and Ce^{3+} in the same lattice site of a certain host:

$$E(Eu^{2+}) = (0.64 \pm 0.02)E(Ce^{3+}) + (0.53 \mp 0.06)eV.$$

where $E(Eu^{2+})$ and $E(Ce^{3+})$ are the energy of the $5d$–$4f$ emission of Eu^{2+} and Ce^{3+}, respectively. Therefore, the emission wavelength of Ce^{3+} or Eu^{2+} can be predicted if one of its emissions is known in the same host, which is useful to design a Ce^{3+} and Eu^{2+} codoped single-phased multicolor phosphor. Our group has designed a double-color blue- and yellow-emitting $Ca_2BO_3Cl:Ce^{3+}$, Eu^{2+} phosphor [90] as shown in Fig. 15.17.

As expected, the emission spectrum of Ce^{3+} and Eu^{2+} codoped Ca_2BO_3Cl shows not only a strong blue band of Ce^{3+} but also a strong yellow band of Eu^{2+}. Figure 15.17a shows the PL spectra for Ca_2BO_3Cl:6%Ce^{3+}, nEu^{2+} phosphor with different Eu^{2+} concentrations n, which were excited at 349 nm corresponding to the optimal excitation wavelength of Ce^{3+} (the energy donor). With increasing Eu^{2+} content (n), the PL intensity of Eu^{2+} (or the energy acceptor) was observed to increase, whereas that of Ce^{3+} was simultaneously found to decrease gradually, which confirms the existence of energy transfer between Ce^{3+} and Eu^{2+} in the host Ca_2BO_3Cl. The energy-transfer efficiency (η_T) from Ce^{3+} to Eu^{2+} can be expressed by

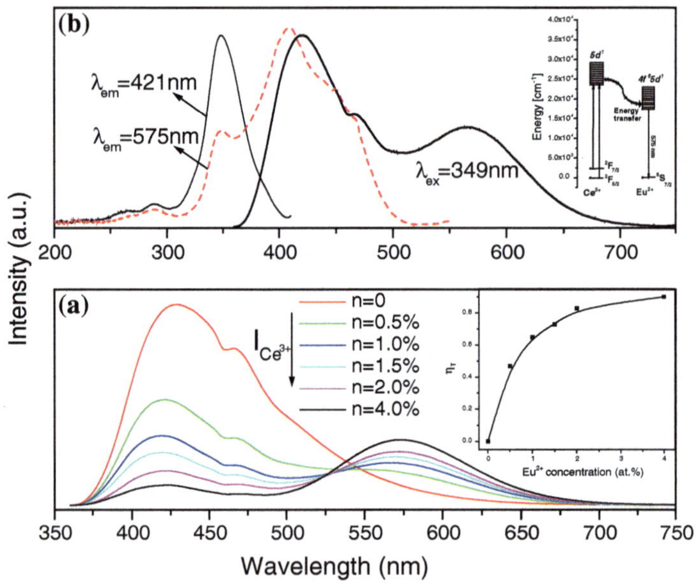

Fig. 15.17 **a** PL spectra for Ca_2BO_3Cl:6%Ce^{3+}, nEu^{2+} phosphors excited at 349 nm. The *inset* shows dependence of energy-transfer efficiency η_T on Eu^{2+} concentration n. **b** PL and PLE spectra for Ca_2BO_3Cl:6%Ce^{3+}, 1%Eu^{2+} phosphor. The *inset* shows the schematic of the energy-level system describing energy transfer in the $Ca_2BO_3Cl:Ce^{3+}$, Eu^{2+} phosphor. Adapted with permission from Ref. [90]. Copyright 2009, The Electrochemical Society

$$\eta_T = 1 - \frac{I_S}{I_{S_0}}.$$

where I_{S0} and I_S are the luminescence intensity of the sensitizer (Ce^{3+}) without and with the presence of an activator (Eu^{2+}). The η_T from Ce^{3+} to Eu^{2+} in Ca_2BO_3Cl is calculated as a function of Eu^{2+} concentrations n and presented in the inset of Fig. 15.17a, in which η_T was found to increase gradually with increasing Eu^{2+} doping content. Figure 15.17b illuminates the PLE and PL spectra for Ca_2BO_3Cl:6%Ce^{3+}, 1%Eu^{2+}. Because the emission wavelength was monitored at 421 nm, the PLE spectrum of $Ca_2BO_3Cl:Ce^{3+}$, Eu^{2+} is identical to that of solely Ce^{3+}-doped Ca_2BO_3Cl. In contrast, when the emission wavelength was monitored at 575 nm, the observed PLE spectrum resembled that of $Ca_2BO_3Cl:Eu^{2+}$; however, two broad emission bands centering at 421 and 575 nm in the PL spectrum are attributed to the emission of Ce^{3+} and Eu^{2+}, respectively. The inset of Fig. 15.17b shows the schematic of the energy-level system, which describes the energy transfer from Ce^{3+} to Eu^{2+}. The results indicate that the resonance-type energy transfer from Ce^{3+} to Eu^{2+} exists in the phosphor $Ca_2BO_3Cl:Ce^{3+}$, Eu^{2+}, and their emitting color can be easily turned from blue through white and eventually to yellow by varying the Eu^{2+} content. The white light–emitting phosphor was also realized by the adjustment of Ce^{3+} and Eu^{2+} concentration in the host Ca_2BO_3Cl under UV excitation.

In addition to the above-mentioned ions, other ions such as Tb^{3+}, Pr^{3+}, Dy^{3+}, and Sm^{3+} codoped with $Eu^{2\pm}$ Ce^{3+} can also generate multicolor emission, e.g., $Ba_2Ln(BO_3)_2Cl:Ce^{3+}$, Tb^{3+} (Ln = Gd and Y) [91], $Sr_3Y_2(Si_3O_9)_2:Ce^{3+}$, $Tb^{3\pm}$ $Mn^{2\pm}$ Eu^{2+} [92], $CaYAl_3O_7:Ce^{3+}$, Pr^{3+} [93], $Sr_{0.5}Ca_{0.5}Al_2O_4:Eu^{2+}$, Dy^{3+} [94], and $Sr_3Y(PO_4)_3:Eu^{2+}$, Tb^{3+}, Sm^{3+} [95]. Among these ions, Tb^{3+} is a typical green-emitting activator due to its predominant $^5D_4 \rightarrow {}^7F_5$ transition peaking at approximately 545 nm with narrow full width at half maximum (FWHM), thus leading to excellent color purity and the reproduction quality of optical properties of phosphor. Tb^{3+} ion can be activated either by directly exciting the $4f^n$ energy levels or by the energy-transfer process. However, the intensities of the Tb^{3+} absorption peaks in the *n*-UV region are very weak, and their widths are very narrow due to the strictly forbidden $4f$–$4f$ transitions, which makes it an imperfect activator from the perspective of application in LEDs. Ce^{3+} ion and Eu^{2+} are promising sensitizers to enhance the green emission, which can tune the emission color and widen the absorption band of Tb^{3+} due to its strong excitation band and efficient emission band originating from the allowed $4f$–$5d$ transition. Figure 15.18 displays the PL and PLE spectra of Tb^{3+} solely and Eu^{2+}/Tb^{3+}–codoped $Ba_3LaNa(PO_4)_3F$ (BLNPF) [96]. For the Tb^{3+} singly doped sample, the PLE spectrum monitored at 543 nm exhibits a broad band from f–d transitions of the Tb^{3+} ions and several weak peaks ranging for 280 to approximately 500 nm owing to the intra-$4f^8$ transitions. The emission peaks at 490, 543, 581, and 619 nm are assigned to the $^5D_4 \rightarrow {}^7F_J$ (J = 6, 5, 4, 3) transitions, whereas the emission peaks at 380, 410, and 430 nm are due to the $^5D_3 \rightarrow {}^7F_J$ transitions. Eu^{2+} can effectively sensitize Tb^{3+} due to the overlap

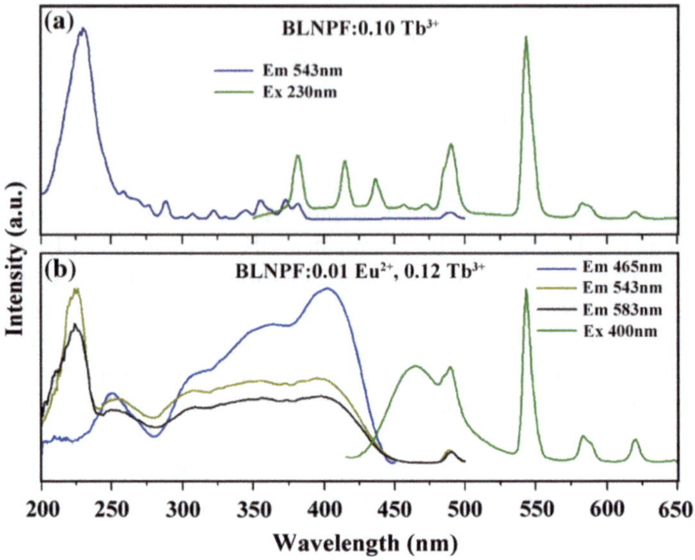

Fig. 15.18 PL and PLE spectra of Tb^{3+} solely (**a**) and Eu^{2+}/Tb^{3+}–codoped $Ba_3LaNa(PO_4)_3F$. Reprinted with the permission from Ref. [96]. Copyright 2013 American Chemical Society

between the PL spectra of BLNPF:Eu^{2+} and the PLE spectra of the BLNPF:Tb^{3+} phosphors. For the Eu^{2+}/Tb^{3+}-codoped sample, it can be efficiently excited in the range from 320 to 430 nm, which matches perfectly with the commercial n-UV LED chips. In addition, the emission spectrum is obviously composed of broadband emission from Eu^{2+} in the blue region and the characteristic line emissions of Tb^{3+}. The emission color can be tuned from blue to green with varying relative ratios of Eu^{2+} and Tb^{3+} ions. Compared with the Tb^{3+} singly doped phosphor, the codoped phosphors have more intense absorption in the n-UV range and stronger emission of the Tb^{3+} ions, both of which are attributed to the effective energy transfer from Eu^{2+} to Tb^{3+} ions.

15.3.3.2 Host Sensitizer Type

Some phosphor hosts containing polyhedral groups consisting of transitional-metal ions without *d* electron—such as Zr/Sn^{4+}, Ce^{4+}, Ti^{4+}, V^{5+}, W/Mo^{6+}, and Nb/Ta^{5+}—as well as oxygen ions, can emit broad bands in the blue-green region from the $M^{n+}\rightarrow O^{2-}$ ($n = 4$, 5, or 6) charge-transfer transitions. A multicolor-emitting phosphor could be achieved by combining these intrinsic host emissions with other rare-earth ions having typical red emission (such as Eu^{3+}, Pr^{3+} Sm^{3+}, Dy^{3+}, etc.), in which hosts transfer their energy to the rare-earth ion activator. The emission color can vary from the pure blue-green emission of the host to white light depending on the concentrations of rare-earth ions [97–104].

Vanadate-based compounds not only have excellent physical and chemical stabilities but also serve as a typical luminescent host. Usually, vanadates offer broad and intense excitation spectra ranging from the n-UV to visible region as well as emission ranging from blue to red due to their charge-transfer (CT) transitions in the VO_4^{3-} [105]. This typical broadband luminescence property makes vanadates suitable for phosphor-converted LEDs (pc-LEDs), which matches well with the n-UV chip and transfers the high-energy photons from sunlight (<400 nm) into blue to red light. Among the vanadates, compounds with formula $\{A_3\}[B_2]V_3O_{12}$ and garnet structure are known for their ability to tolerate extensive ionic substitution, and a large number of cations can be substituted on A and B sites [106]. Therefore, more investigations have been performed on its crystal structure, and a series of phase pure garnet $[Ca_2Na][M]_2V_3O_{12}$ (M = Mg^{2+},Co^{2+}, Ni^{2+}, Cu^{2+}, and Zn^{2+}) has been reported [107]. Previous results proved that these compounds are an excellent phosphor host and $[VO_4]^{3-}$ can emit an intense photoluminescence from green to yellow emission [108], thus, it is possible to obtain a single-phased multicolor-emitting phosphor by energy transfer from the host to rare-earth ion activators.

Here, Eu^{3+}/Sm^{3+} solely doped $Ca_2NaZn_2V_3O_{12}$ phosphor was used as a representative [109]. As shown in Fig. 15.19, the excitation spectrum of $Ca_2NaZn_2V_3O_{12}$ host gives broadband absorption from 250 to 400 nm, which is ascribed to CT transition of the VO_4^{3-} group. Similar PLE spectra were also observed for the $Ca_2NaZn_2V_3O_{12}$:A (A = Eu^{3+}, Sm^{3+}) phosphor, and the weak characteristic excitation peaks ascribed to Eu^{3+} (393 nm) or Sm^{3+} (411 nm) were also found. It is also observed that the excitation spectra yield a similar spectral profile corresponding to monitoring wavelengths at 497 nm for the emission from the VO_4^{3-} group, i.e., 613 nm for Eu^{3+} and 617 nm for Sm^{3+}, respectively, which verified that the energy transfer process should exist between VO_4^{3-} and Eu^{3+} or Sm^{3+} in the present system. The possible energy-transfer processes are also presented in Fig. 15.19d where new energy levels are created between the ground and

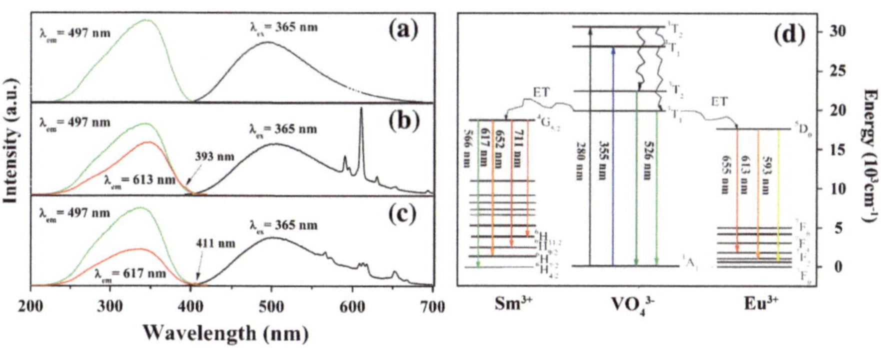

Fig. 15.19 PLE and PL spectra of $Ca_2NaZn_2V_3O_{12}$ (**a**), $Ca_2NaZn_2V_3O_{12}$:Eu^{3+} (**b**), and $Ca_2NaZn_2V_3O_{12}$:Sm^{3+} phosphors (**c**) and energy-transfer process (**d**). Adapted from Ref. [109], copyright 2013, with permission from Elsevier

^3T states in the RE-doped system. Part of the energy available with the ^3T states is used to pump the upper levels of $^5D_0$ and $^4G_{5/2}$ in Eu^{3+} and Sm^{3+}, the other part plus the excess energy available with the $^1T_2$ state causes the $^3T_{1,2}$ levels to decay in a nonradiative manner. Therefore, we can find broadband emission originating from the CT transition of VO_4^{3-} group and the characteristic *f–f* transition of Eu^{3+} or Sm^{3+} in a system.

$YTiTaO_6$ is a special group of materials with $A^{3+}B^{4+}C^{5+}O_6$ combination in which Y^{3+} ions occupy $C_S^{xy}(4)$ sites coordinated by six-fold oxygen to form an irregular YO_6^{9-} polyhedron, and Ta^{5+} and Ti^{4+} ions occupy $C_1(8)$ sites coordinated by six-fold oxygen randomly distributed on the same crystallographic positions in the framework, respectively. Actually, because of the promising chemical characteristics and optical properties, $YTiTaO_6$ can be used as host for a self-activated phosphor [110]. Eu^{3+}/Er^{3+}–doped $YTiTaO_6$ is a special single-phased multicolor-emitting phosphor, in which a broad emission band centered at 505 nm in the visible range of 400–600 nm due to the $Ta(Ti)O_6$ polyhedron and a strong broad band centered at 277 nm in the PLE spectra should attribute to host absorption. In addition to the host emission, the characteristic emission or absorption peaks from the *f–f* transitions of Eu^{3+} or Er^{3+} were also observed with the excitation of 277 nm, such as emission lines from the $^5D_1 \rightarrow {}^7F_1$ (538 nm), $^5D_0 \rightarrow {}^7F_1$ (597 nm), and $^5D_0 \rightarrow {}^7F_2$ (615 nm) transition of Eu^{3+} as well as the $^2H_{11/2} \rightarrow {}^4I_{15/2}$ (528 nm) and $^4S_{3/2} \rightarrow {}^4I_{15/2}$ (547 and 559 nm) transitions of Er^{3+} as shown in Fig. 15.20. The resonant-type energy transfer from the $Ta(Ti)O_6$ group of the host to Eu^{3+} and Er^{3+} in $YTiTaO_6$ phosphors by way of a dipole–dipole mechanism is shown in Fig. 15.20d. In this process, electronic transitions from the valence band (VB) to the conduction band (CB) of the $Ta(Ti)O_6$ polyhedron occur under the excitation of 277 nm. Then the recombination process leads to the broad emission band. Meanwhile, a part of the energy transfers to the $^5D_{0,1,2}$ levels of Eu^{3+} and the $^2P_{3/2}$ level of Er^{3+} by nonradiative transition and then produces characteristic transitions of Eu^{3+} ($^5D_1 \rightarrow {}^7F_1$ 538 nm, $^5D_0 \rightarrow {}^7F_1$ 597 nm,

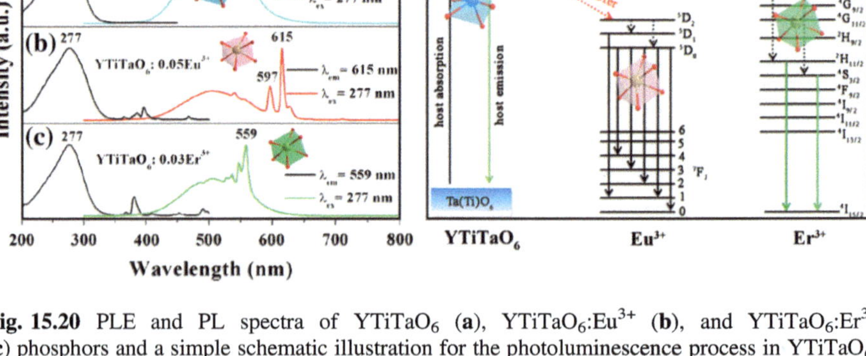

Fig. 15.20 PLE and PL spectra of $YTiTaO_6$ (**a**), $YTiTaO_6{:}Eu^{3+}$ (**b**), and $YTiTaO_6{:}Er^{3+}$ (**c**) phosphors and a simple schematic illustration for the photoluminescence process in $YTiTaO_6{:}Eu^{3+}/Er^{3+}$ (**d**). Reprinted with the permission from Ref. [110]. Copyright 2014 American Chemical Society

and $^5D_0 \rightarrow {}^7F_2$ 615 nm) and Er^{3+} ($^2H_{11/2} \rightarrow {}^4I_{15/2}$ 528 nm and $^4S_{3/2} \rightarrow {}^4I_{15/2}$ 547 and 559 nm), respectively.

As a famous and typical host emission or self-activated phosphor, Sr_2CeO_4 is novel blue luminescent material that has attracted much attention because it possesses some desirable qualities such as (i) ligands-to-metal charge transfer (CT) transition of Ce^{4+} at approximately 340 nm, (ii) efficient energy transfer that can occur from the Ce^{4+}–O^{2-} CT state to the trivalent rare-earth in $Sr_2CeO_4{:}RE^{3+}$, (iii) high efficiency for absorbing UV radiation, and (iv) large band gap and covalent bond energy prove its capability for using as a phosphor material. The Sr_2CeO_4 blue luminescence is believed to originate from a ligand-to-metal (O^{2-}/Ce^{4+}) charge-transfer mechanism. This process occurs between the lower coordination–number terminal oxygen atoms associated with the low dimensional structure in combination with an adjacent Ce^{4+} centre in the Sr_2CeO_4 structure. The emission is fairly broad (FWHM approximately 100 nm) at the blue region (450–500 nm), which makes this material a potential candidate for phosphor applications [111].

Figure 15.21 displays the PLE and PL spectra of $Sr_{2-x}Dy_xCeO_4$(x = 0.005–0.09) and a schematic energy-level diagram with possible energy-transfer pathways. The excitation spectrum is recorded by monitoring the emission of Dy^{3+} at 575 nm [112]. The excitation spectra consist of broad bands and some sharp narrow peaks. The broad bands consist of two peaks centered at 290 and 340 nm; this is attributed to the charge-transfer band (CTB) caused by an electron transfer from the O^{2-} (2*p*) orbital to the empty states of the $4f^6$ configuration of Ce^{4+} ($Ce^{4+} \rightarrow O^{2-}$ transition). The peak at 290 nm originated due to the transition from terminal oxygen (O_t) to Ce^{4+}. The peak at 340 nm is ascribed to the transition from equatorial oxygen (O) to Ce^{4+}. Apart from this broad CTB, the excitation spectra contain

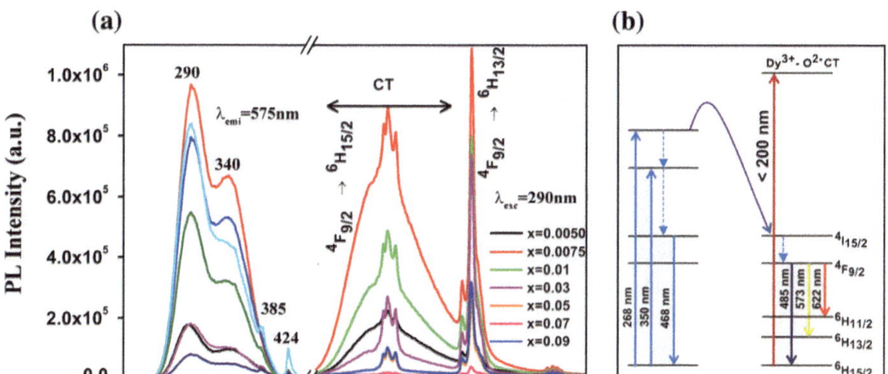

Fig. 15.21 PL and PLE spectra of Dy^{3+}-doped Sr_2CeO_4 (**a**) and a schematic diagram illustrating the energy and charge transfer pathways (**b**). Adapted from Ref. [112] by permission of The Royal Society of Chemistry

several other sharp peaks observable in the longer-wavelength region between 350 and 480 nm. These sharp features are due to intra-configuration *f–f* transitions of the Dy^{3+} dopant ions. In general, the direct excitation of the Dy^{3+} ground state to higher levels in the 4*f* configuration results in the electronic transitions ($^6H_{15/2} \rightarrow {}^6P_{3/2}$) at 326 nm, ($^6H_{15/2} \rightarrow {}^6P_{7/2}$) at 352 nm, ($^6H_{15/2} \rightarrow {}^6P_{5/2}$) at 366 nm, ($^6H_{15/2} \rightarrow {}^4I_{13/2}$) at 385 nm, ($^6H_{15/2}$ $^4G_{11/2}$) at 424 nm, ($^6H_{15/2} \rightarrow {}^4I_{15/2}$) at 453 nm, and ($^6H_{15/2} \rightarrow {}^4F_{9/2}$) at 475 nm. However we could observe sharp peaks only at 385 and 424 nm. This might be because the other expected peaks (expected at 326, 352 and 366 nm) are buried within the broad CTB of the host lattice. Regarding the PL spectra excited at 290 nm, their emission spectrum is similar to that excited at 424 nm except for relative intensity concluding with a broad band ranging from 400 to 500 nm corresponding to host emission. Along with this host emission there is overlap of Dy^{3+} emission at 488 nm. In addition, there is a prominent emission peak at 575 nm, and a weak feature is observed at 662 nm. The emission peaks of Dy^{3+} at 488, 575, and 662 nm can be assigned to the transitions from $^4F_{9/2}$ to the $^6H_{15/2}$, $^6H_{13/2}$, and $^6H_{11/2}$ states, respectively. Among these emissions, the yellow band (575 nm) and the blue band (488 nm) are the predominant transitions. It is worthy noting that the emission color can be tuned by adjusting the concentration of Dy^{3+}. The schematic energy-level diagram in Fig. 15.21b shows the excitation and emission mechanism of Sr_2CeO_4:Dy^{3+} phosphors. On host excitation at 290 nm, the photon energy is absorbed by the host leading to the charge transfer of O^{2-}/Ce^{4+}; this energy is transferred to the 4f shell of Dy^{3+}. Thus, the electrons of Dy^{3+} at the $^4F_{9/2}$ excited state can populate both from nonradiative charge-transfer feeding and nonradiative transitions from higher excited states. Then the characteristic emissions at 488, 575, and 662 nm correspond to the $^4F_{9}/6Hj$ (j = 15/2, 13/2, and 11/2) transitions, respectively. However, the direct excitation of Dy^{3+} ions induces transitions from the ground state to the metastable excited states. Part of the excited electrons in these metastable excited levels are repopulated into the $^4F_{9/2}$ level by multiphonon-assisted nonradative transitions. The transitions from $^4F_{9/2} \rightarrow 6Hj$ (j = 15/2, 13/2, and 11/2) are radiative and produce emission lines of 488, 575, and 662 nm, respectively

On the whole, for multicolor-emitting phosphors generated by the combination of RE^{3+} with red- or yellow-emitting in the visible region, with the typical blue-green emission from the special host emission in some compounds, their emitting color strongly depend on the contents of the rare-earth ion activators. If the contents are very low, the broadband blue-green emission from the intrinsic host emission will dominate the PL spectra; on the contrary, if the contents are very high, the emission from the host will be too weak to neglect and dominate by the emission of activators. Only if the content of the rare-earth ion is not too high or low can multicolor-emitting phosphor be obtained using the present methods.

15.3.3.3 Narrow Line Sensitizer: Tb^{3+} – Eu^{3+}

Generally, there are usually more luminescent centers in multicolor-emitting phosphors. Different rare-earth ions have abundant energy levels leading to different emissions colors such as the typical red emission from Eu^{3+}, the green emission form Tb^{3+}, and the yellow emission from Dy^{3+}. It is expected that multicolor emission could be generated by multiple ions co-doping in a single host, e.g., Eu^{3+}/ Dy^{3+} doped $Ca_2La_8(GeO_4)_6O_2$ [113], $NaGd(WO_4)_2$:Tm^{3+}, Dy^{3+}, Eu^{3+} [114], $YAl_3(BO_3)_4$:Eu^3/Tb^{3+} [115], GYAG:Bi^{3+}, Tb^{3+}, Eu^{3+} [116], and Eu^3/Sm^{3+}-codoped SrY_2O_4:Tb^{3+} [117], in which the emission color can be controlled by tuning the relative concentrations of dopant ions. Among these phosphors, most research has been focused on Tb^{3+}/Eu^{3+}-codoped systems to enhance the red emission in recent years due to the efficacy, and the CRI of Eu^{3+}-activated red-emitting phosphors are superior to that of Eu^{2+}-emitting nitride-phosphors [118]. However, Eu^{3+} suffers from low absorption strength in the blue spectral range, whereas Tb^{3+} exhibits comparatively strong absorption in this range and can be used as a sensitizer for Eu^{3+}. In these systems, several characteristic emission lines originating from the *f*–*f* transitions of Tb^{3+} and Eu^{3+} can be observed from the PL spectra, and the emission intensities' variation of the green emission peaking at approximately 544 nm from Tb^{3+} and red emission peaking at 617 nm induced the multicolor emission by adjusting the Tb^3/Eu^{3+} content ratio.

As a representative and for the sake of comparison, the PL and PLE spectra of Tb^3/Eu^{3+} solely or doubly doped $La_3GaGe_5O_{16}$, as well as the possible energy transfer mechanism, are shown in Fig. 15.22 [119, 120]. Figure 15.22a and b shows the Tb^{3+}/Eu^{3+} solely doped $La_3GaGe_5O_{16}$, and it is found that all of the excitation and emission peaks are from the characteristic *f*–*f* transitions of Tb^{3+} and Eu^{3+}. A significant spectral overlap is observed in comparing the excitation band of $La_{2.95}GaGe_5O_{16}$:$0.05Eu^{3+}$ with the emission band of $La_{2.85}GaGe_5O_{16}$:0.15 Tb^{3+},

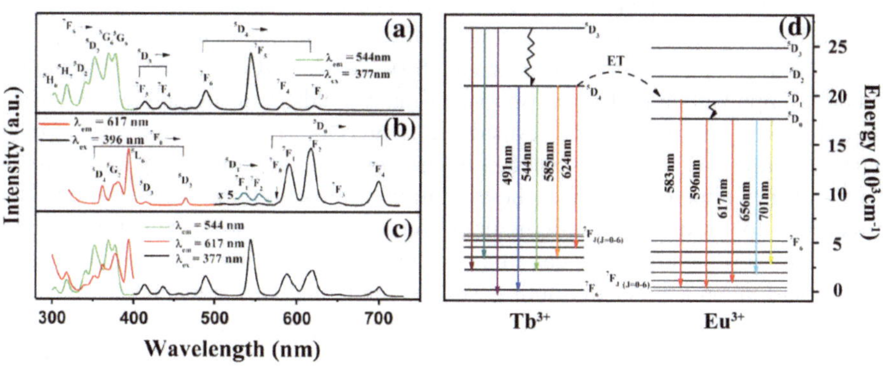

Fig. 15.22 PLE and PL spectra of Tb^3/Eu^{3+} solely or doubly doped $La_3GaGe_5O_{16}$ and the energy-transfer mechanism from Tb^{3+} to Eu^{3+}. **a**, **b**, **c** Adapted from Ref. [119] by permission of The Royal Society of Chemistry. **d** Adapted from Ref. [120] by permission of The Royal Society of Chemistry

which is in favor of the resonance-type energy transfer from Tb^{3+} to Eu^{3+}. Figure 15.22c illustrates the PLE and PL spectra of the typical $La_{2.8}GaGe_5O_{16}$:0.15 Tb^{3+}, 0.05Eu^{3+}. The excitation spectra that irradiated the wavelengths of 617 and 544 nm were identical to those of Tb^{3+} singly doped phosphor, which shows that doubly doped $La_3GaGe_5O_{16}$ can be used as a double-color green- and red-emitting phosphor in near UV-pumped white LEDs. In contrast, emission peaks of the Eu^{3+}/Tb^{3+}–codoped phosphor under 37-nm excitation were observed at 617 and 544 nm, which were attributed to the Eu^{3+} and Tb^{3+} ions, respectively. Therefore, the relative intensities of these two emissions can be varied by adjusting the concentrations of the two activators through the principle of energy transfer. Figure 15.22d shows energy diagrams of Tb^{3+} and Eu^{3+} as well as the sensitized Eu^{3+} luminescence mechanism. It can be observed that the energy level of Tb^{3+} ($^5D_4$) is a little higher than that of Eu^{3+} ($^5D_1$ and $^5D_0$), which makes energy transfer through nonradiative processes possible. In addition, the $^5D_4 \rightarrow {}^7F_{6,5,4,3}$ emission of Tb^{3+} effectively overlaps with the $^7F_{0,1} \rightarrow {}^5D_{0,1,2}$ absorption of Eu^{3+}; thus, the energy transfer from Tb^{3+} to Eu^{3+} is very efficient in general. For the emission of phonons in the $^5D_4$ energy level of Tb^{3+}, part of the energy can transfer to the $^5D_1$ or $^5D_0$ levels of Eu^{3+} by phonon-assisted electric dipole–dipole interaction, then relax to $^5D_0$ energy level, and finally transfer to the $^7F_1$ or $^7F_2$ level of Eu^{3+} by radiative transition.

15.3.4 Upconversion Multicolor-Emitting Phosphor

Upconversion (UC) is the spectroscopic process that converts two or more lower-energy photons with longer wavelength into one higher-energy photon with shorter wavelength through a multi-photon process [121]. Rare-earth ions (REI), such as Er^{3+}, Tm^{3+}, Ho^{3+}, Pr^{3+}, and Nd^{3+}, are suitable activator candidates for UC processes owing to their long-lived excited states and ladder-like energy levels. Driven by the popularization of low-cost efficient continuous-wave (CW) infrared diode lasers (e g., 980-nm InGaAs and 808-nm GaAs/GaAs) and the potential versatile applications of UC phosphors in many fields, considerable attention has been focused on trivalent REI-doped UC materials [122]. Tricolor (red, green, blue)–emitting UC phosphors can be obtained in rare-earth ions many Er^{3+}-, Ho^{3+}-, Tm^{3+}-, or Pr^{3+}doped hosts under the excitation of infrared light, which provides the possibility to produce white-light emission with an infrared excitation source. White-emission phosphors can be achieved by mixing red-, green-, and blue-emission phosphors or mixing yellow- and blue-emission phosphors—at a proper blending ratio. In addition, white-light generation can also be achieved by different codoped/tri-doped/multi-doped rare-earth ions at suitable concentrations in a single host matrix [123, 124]. Compared with downconversion phosphors, UC-luminescence phosphors benefitted in the 1980s from the development of inexpensive and efficient InGaAs infrared laser diodes (LD) emitting at approximately 980 nm. Furthermore, infrared excitation can effectively decrease the light-degradation effect from high-energy photons and eliminate damage from the leakage of UV light. For solid-state lighting and displaying technology based on infrared to visible UC luminescence, novel UC phosphors with high efficiency are desired [125, 126]. Here, multicolor-emitting UC phosphors are briefly reviewed.

Generally, the UC efficiency of RE^{3+} ions are poor due to the nature of the multiphoton process and low-absorption cross-section, and Yb^{3+} or Nd^{3+} ion is often chosen as a sensitizer to increase the UC emission efficiency and match well with the inexpensive commercial infrared LD emitting at 800 and 980 nm [127]. Almost all emission spectra of the UC phosphors are composed of more emitting peaks at different visible regions except for their different relative intensity. For example, Yb^{3+}–Er^{3+} and Yb^{3+}–Ho^{3+} yield green and red emissions, whereas there is full color emission in Yb^{3+}–Nd^{3+} and Yb^{3+}–Pr^{3+}, and so on, which strongly depends on the photon energy and dopant concentrations. To continue the example, our group has reported the multicolor-tunable emission properties of Er^{3}/Yb^{3+}–codoped $CaIn_2O_4$ [128]. Double-intense UC emission bands in the green and red region are observed resulting in a tunable ratio of red to green and the chromaticity coordinates by adjusting the concentration of Er^{3+} and Yb^{3+}. Figure 15.23 displays the UC spectra of $CaIn_2O_4$:Er^{3+}, Yb^{3+} and the variation of the intensity ratios of red to green as function of Yb^{3+} contents. Compared with the green emission, the UC intensity of the red emission was enhanced greatly with the increase of Yb^{3+} concentration. As shown in Fig. 15.23b, the intensity ratios of red to green emission increase monotonously from 0.073 to 2.376 with increasing the Yb^{3+} contents from 0 to 0.20. The emission color appears green for Er^{3+} solely doped $CaIn_2O_4$, but the emission color gradually changes to a yellowish shade with the addition of Yb^{3+} due to a combination of green and red emissions. The chromaticity coordinates of $CaIn_{2\text{-}0.01\text{-}y}Er_{0.01}Yb_yO_4$ (y = 0 approximately 0.20) are shown in Fig. 15.23c where it is observed that the CIE coordinates gradually shift from (0.284, 0.693) to (0.412, 0.565) with the increase of Yb^{3+} concentration due to the remarkable enhancement of the red-emission intensity. Accordingly, the UC emission intensity (I) depends on the laser pumping power (P) following the relationship $I \propto P^n$, where the n value is the number of pumping photons absorbed by REI at ground state to transit to the upper emitting state. Results indicated that the red and green emissions may come from both two- and three-photon processes.

To better comprehend the populating processes of the emitting states and the radiative transitions of green and red UC luminescence, the energy-level diagram and UC mechanisms of the Er^{3+}/Yb^{3+}-codoped $CaIn_2O_4$ system are schematically given in Fig. 15.23d. Under 980-nm excitation, Yb^{3+} absorbs an infrared photon and transits from the $^2F_{7/2}$ ground state to the excited state $^2F_{5/2}$. Er^{3+} ions are initially excited from ground state to the excited state $^4I_{11/2}$ through ground-state absorption, e.g., (GSA):Er^{3+} ($^4I_{15/2}$) + a photon (980 nm) $\rightarrow$ Er^{3+} ($^4I_{11/2}$) or the energy transfer (ET1), e.g., Er^{3+} ($^4I_{15/2}$) + Yb^{3+} ($^2F_{5/2}$) $\rightarrow$ Er^{3+} ($^4I_{11/2}$) + Yb^{3+} ($^2F_{7/2}$). For the above-mentioned two processes, the ET1 process plays a dominant role because the absorption cross-section of the $^2F_{7/2} \rightarrow {}^2F_{5/2}$ transition of Yb^{3+} (approximately 34.07×10^{-19} cm^2) is approximately eight times higher than that of the $^4I_{15/2} \rightarrow {}^4I_{11/2}$ transition of Er^{3+} (approximately 4.43×10^{-19} cm^2). Then

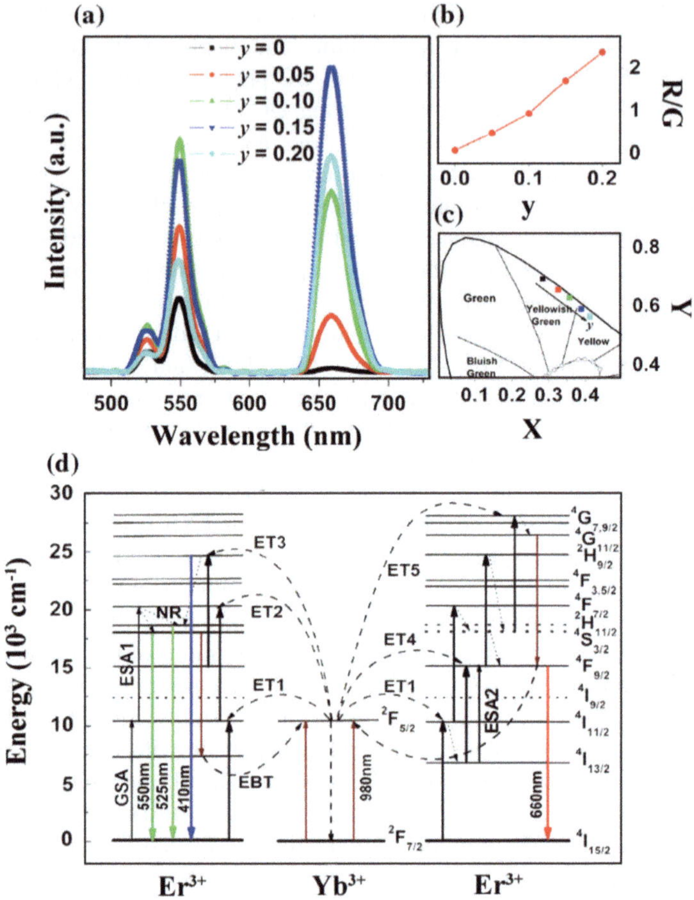

Fig. 15.23 UC spectra of $CaIn_2O_4:Er^{3+}$, Yb^{3+} with **a** the variation of the intensity ratios of red to green (**b**) and the CIE coordinates (**c**) as a function of Yb^{3+} contents as well as the possible UC processes. Adapted from Ref. [128]

the $^4F_{7/2}$ level of Er^{3+} is populated by way of ET2:Yb^{3+} ($^2F_{5/2}$) + Er^{3+} ($^4I_{11/2}$) → Yb^{3+} ($^2F_{7/2}$) + Er^{3+} ($^4F_{7/2}$) and the excited state absorption (ESA1), e.g., Er^{3+} ($^4I_{11/2}$) + a photon (980 nm) → Er^{3+} ($^4F_{7/2}$). The green-emitting states $^2H_{11/2}$ and $^4S_{3/2}$ levels are populated by rapid nonradiative relaxation (NR) from the upper $^4F_{7/2}$ level due to the small energy gap between these levels. Finally, the $^2H_{11/2} \rightarrow {}^4I_{15/2}$ and $^4S_{3/2} \rightarrow {}^4I_{15/2}$ transitions of Er^{3+} ions produce 525- and 550-nm green emissions, respectively. For a three-photon process related to the red-emitting level, $^4F_{9/2}$ is also involved in the green UCL, which is described as Er^{3+} ($^4F_{9/2}$) + Yb^{3+} ($^2F_{5/2}$) → Er^{3+} ($^2H_{9/2}$) + Yb^{3+} ($^2F_{7/2}$)(ET3) followed by Er^{3+} ($^2H_{9/2}$) → Er^{3+} ($^2H_{11/2}$, $^4S_{3/2}$)(NR). The integrated intensity of the red emission is greater than that of the green emission, thus indicating a larger population of the $^4F_{9/2}$ level than the $^2H_{11/2}$/$^4S_{3/2}$ levels, which is favorable to enhance the third energy-transfer process (ET3). The existence of a weak blue emission at 410 nm (it is too weak to be shown in the spectra) arising from the $^2H_{9/2} \rightarrow {}^4I_{15/2}$ transition can also confirm the ET3 process. For the red emission ($^4F_{9/2} \rightarrow {}^4I_{15/2}$), the long-lived intermediary $^4I_{13/2}$ level is populated by NR from the $^4I_{11/2}$ level. Population of the emitting state $^4F_{9/2}$ involves following some possible processes: Er^{3+} ($^4I_{13/2}$) + a photon (980 nm) → Er^{3+} ($^4F_{9/2}$) (ESA2), Yb^{3+} ($^2F_{5/2}$) + Er^{3+} ($^4I_{13/2}$) → Yb^{3+} ($^2F_{7/2}$) + Er^{3+} ($^4F_{9/2}$) (ET4), and NR from $^4S_{3/2}$ to $^4F_{9/2}$. A third energy transfer—Yb^{3+} ($^2F_{5/2}$) + Er^{3+} ($^4S_{3/2}$) → Yb^{3+} ($^2F_{7/2}$) + Er^{3+} ($^4G_{7/2}$) (ET5) —pumping to the high-lying $^4G_{7/2}$ level also contributes to population of the $^4F_{9/2}$ level. After reaching the $^4G_{7/2}$ state, the Er^{3+} decays to the $^4G_{11/2}$ state and then undergoes a cooperative decay to the red-emitting $^4F_{9/2}$ state by passing the green-emitting ones [129].

In addition, multiple ion–doped multicolor-emission UC phosphors have also been investigated by our group. It is well known that among the activators for UC phosphors, Tm^{3+} ions often show intense blue emission, and Ho^{3+} ions usually exhibit green and red emissions in UC processes. Therefore, tricolor emissions or white-light emission can be obtained by the combination of the blue emission from Tm^{3+} and the green and red emissions from Ho^{3+} through adjusting their relative concentrations. However, the red-emission intensity of Ho^{3+} is determined by the population of the $^5F_5$ level, which is affected strongly by the nonradiative relaxation of $^5I_6 \rightarrow {}^5I_7$ and $^5F_4/^5S_2 \rightarrow {}^5F_5$. Generally, the nonradiative relaxation rate closely depends on the phonon energy of the hosts. Low phonon energy not only results in the increase of the UC efficiency by hindering the nonradiative losses, it also decreases the red-emission intensity and causes a dominant green emission [130].

Our group has prepared $LiLa(MoO_4)_2$:$Tm^{3+}/Ho^{3+}/Yb^{3+}$ phosphors with tailorable upconversion (UC) emission colors. They were prepared according to a modified sol-gel method [131]. Under the excitation of a 980-nm laser diode, up-conversion red and green emissions in Ho^{3+}/Yb^{3+}–codoped and blue emission in Tm^{3+}/Yb^{3+}–codoped $LiLa(MoO_4)_2$ were observed, respectively. The intensities of the RGB (red/green/blue) emissions could be controlled by varying concentrations of Tm^{3+} or Ho^{3+} as shown in Fig. 15.24a, b. Under the excitation of 980-nm NIR light, it was clearly observed that there are three intense emission bands in the blue, green, and red regions peaking at 450 nm from the transition $^1G_4 \rightarrow {}^3H_6$ of Tm^{3+}, at 542 nm from the $^5F_4/^5S_2 \rightarrow {}^5I_8$ transition of Ho^{3+}, and 650 nm from the $^5F_5 \rightarrow {}^5I_8$ transition of Ho^{3+}; there was also a small contribution from the $^1G_4 \rightarrow {}^3F_4$ transition of Tm^{3+}, respectively in the $Yb^{3+}/Tm^{3+/}Ho^{3+}$ triply doped $LiLa(MoO_4)_2$. The relative intensity is strongly varied with the dopant concentrations, and the corresponding CIE coordinates of LiLa(MoO4)2 doped with *m* mol% Tm^{3+}/0.2 mol% Ho^{3+}/10 mol% Yb^{3+} (m = 0, 0.3, 0.5, 1.0, 1.5 and 2.0) and *n* mol % Ho^{3+}/1.0 mol% Tm^{3+}/10 mol% Yb^{3+} (n = 0, 0.1, 0.2, 0.3, 0.5, 0.7 and 0.9) are also calculated according to their UC spectra as shown in Fig. 15.24c. With the increase of Ho^{3+} concentration (*n* increases from 0 to 0.9), the emission color of $Tm^3/Ho^{3/}Yb^{3+}$–codoped samples changes from blue through white to the yellow region. When the Tm^{3+} content increases (*m* increases from 0 to 2.0) with fixed Ho^{3+} (0.2 mol%) and Yb^{3+} (10 mol%), the emission color of $Tm^{3+}/Ho^{3+}/Yb^{3+}$–

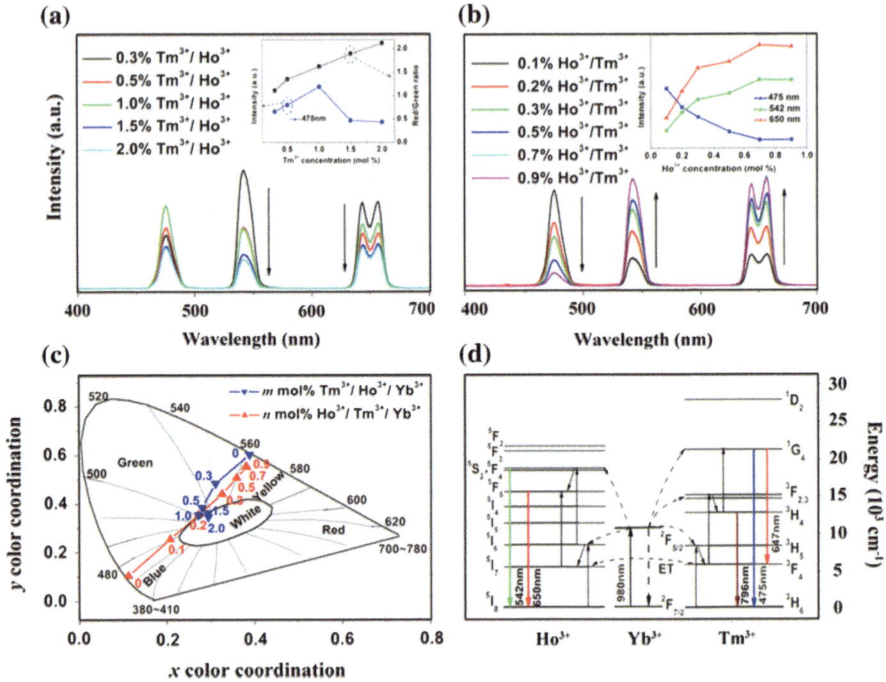

Fig. 15.24 UC spectra, CIE and possible UC processes as function of Tm^{3+} and Ho^{3+} contents in $LiLa(MoO_4)_2$. Adapted from Ref. [131] by permission of John Wiley & Sons Ltd

codoped samples changes from yellow (x = 0.388, y = 0.603) to white (x = 0.295, y = 0.348).

Figure 15.24d illuminates the energy-level diagrams of Yb^{3+}, Tm^{3+}, and Ho^{3+} ions as well as the UC mechanisms of visible and infrared emissions in $Tm^{3+}/Ho^{3+}/Yb^{3+}$-codoped $LiLa(MoO_4)_2$ phosphors. Generally, Yb^{3+} ions often act as the sensitizer in a codoped UC system due to its large-absorption cross-section at 980 nm and the large spectra overlap between Yb^{3+} and rare-earth ions (Er^{3+}, Ho^{3+}, and Tm^{3+}). In addition, Tm^{3+}/Ho^{3+} ions cannot absorb the 980-nm excitation energy due to the lack of a matched energy level. Therefore, the ET processes from Yb^{3+} to Tm^{3+}/Ho^{3+} play the dominant role in the UC emission in $Tm^{3+}/Ho^{3+}/Yb^{3+}$-codoped $LiLa(MoO_4)_2$. With the excitation of 980-nm infrared light, Yb^{3+} absorbs an infrared photon and transits from the ground state $^2F_{7/2}$ to the excited state $^2F_{5/2}$. For the green and red emissions from Ho^{3+}, the $^5I_6$ intermediary level of Ho^{3+} is populated by way of the energy transfer (ET) process of Ho^{3+} ($^5I_8$) + Yb^{3+} ($^2F_{5/2}$) → Ho^{3+} ($^5I_6$) + Yb^{3+} ($^2F_{7/2}$). Then the Ho^{3+} ions at the $^5I_6$ level can transit to the $^5F_4/^5S_2$ level by way of a second ET process: Yb^{3+} ($^2F_{5/2}$) + Ho^{3+} ($^5I_6$) → b^{3+} ($^2F_{7/2}$) + Ho^{3+} ($^5F_4/^5S_2$). Finally, the $^5F_4/^5S_2$ → $^5I_8$ transitions of Ho^{3+} ions produce 540-nm green emissions. For the red emission, the excited-state $^5F_5$ level is populated by way of two possible processes: One is a nonradiative transition from $^5S_2$ to $^5F_5$, and the other is a nonradiative transition from $^5I \rightarrow {}^5I_7$. Then $^5F_5$ is populated through the ET process Yb^{3+} ($^2F_{5/2}$) + Ho^{3+} ($^5I_7 \rightarrow b^{3+}$ ($^2F_{7/2}$) + Ho^{3+} ($^5F_5$), and the red emission of Ho^{3+} is produced from the $^5F_5 \rightarrow {}^5I_8$ radiative transition. For the visible and infrared emissions of Tm^{3+}, the $^3H_5$ level is populated by way of the ET process: Tm^{3+} ($^3H_6$) + Yb^{3+} ($^2F_{5/2}$) $\rightarrow$ m^{3+} ($^3H_5$) + Yb^{3+} ($^2F_{7/2}$). Then Tm^{3+} in $^3H_5$ level may relax nonradiatively to the $^3F_4$ level and transfers to the $^3F_{2,3}$ level by the second ET process: Tm^{3+} ($^3F_4$) + Yb^{3+} ($^2F_{5/2}$) $\rightarrow$ Tm^{3+} ($^3F_{2,3}$) + Yb^{3+} ($^2F_{7/2}$). The Tm^{3+} in $^3F_{2,3}$ level relaxes to the $^3H_4$ level by a multiphonon-assisted process. The infrared emission at 796 nm is generated from the transition of $^3H_4 \rightarrow {}^3H_6$. For the red and blue emissions of Tm^{3+}, the $^1G_4$ level is populated by way of a third ET process: Tm^{3+} ($^3H_4$) + Yb^{3+} ($^2F_{5/2}$) $\rightarrow$ Tm^{3+} ($^1G_4$) + Yb^{3+} ($^2F_{7/2}$). Finally, the $^1G_4 \rightarrow {}^3H_6$ and $^1G_4 \rightarrow {}^3F_4$ transitions generate the blue emission at approximately 475 nm and red emission at approximately 647 nm, respectively.

15.3.5 Quantum Dots and Defects

In addition to above-mentioned methods to obtain multicolor-emitting phosphors for LEDs tunable emissions based on semiconductor quantum dots with tunable size have also been investigated with an aim toward application in LEDs. A quantum dot is a nanoparticle made of any semiconductor material such as silicon, cadmium selenide, cadmium sulfide, or indium arsenide. Quantum dots are semiconductor nanoparticles that glow a particular color after being illuminated by light. The color they glow depends on the size of the nanoparticle. When the quantum dots are illuminated by UV light, some of the electrons receive enough energy to break free from the atoms. This capability allows them to move around the nanoparticle, creating a conduction band in which electrons are free to move through a material and conduct electricity. When these electrons drop back into the outer orbit around the atom (the valence band), they emit light. The color of that light depends on the energy difference between the conduction band and the valence band. The smaller the nanoparticle, the higher the energy difference between the valence band and conduction band, which results in a deeper blue color. The larger the dot, the lower energy its fluorescence spectrum; conversely, smaller dots emit bluer (higher-energy) light. The band-gap energy that determines the energy (and hence color) of the fluorescent light is inversely proportional to the size of the quantum dot. Many semiconductor substances can be used as quantum dots, and nanoparticles of any other semiconductor substance have the properties of a quantum dot. The gap between the valence band and the conduction band, which is present for all semiconductor materials, causes quantum dots to fluoresce. Thus, quantum dots of the same material, but with different sizes, can emit light of different colors because of the quantum-confinement effect. This shows that multicolor single-phased quantum dots can be obtained by adjusting their size distribution [132, 133].

Defect-induced luminescent materials are also an important kind phosphor as reviewed by Lin [134]. Some a defect emission, combined with the intrinsic emission, could result in both multicolor and full-color emission. For example, under UV excitation a strong broadband emission centered at 400 nm has been found in obtained fluoride nanocrystals prepared by hydrothermal method, which has also been reported by Lin's group [135, 136]. Because no any other activated ions were introduced in the experiments, the emission must be caused by some kind of defect or electronic centre.

15.4 Summary and Outlook

Multicolor-emitting single-phased phosphor for LEDs can be obtained by enriching the red emission of YAG:Ce^{3+} yellow phosphor, a single-activator doped system, a multiple-ion codoped system based on energy transfer, up-conversion of a luminescence system, semiconductor quantum dot and defect emission, and so on several methods. Many researchers have paid more attention and pursued hard work to generate multicolor a full color–emitting single-phased phosphor, which can be realized by the above-mentioned methods. However, these phosphors still face serious problem and have long way to go before they can be used as commercialized LED phosphors due to their low efficiency. Although strong intensity of LEDs is not needed in some applications, enhancing the efficiency of them is urgent for their application in daily life.

References

1. Shinde KN, Dhoble SJ, Swart HC, Park K (2012) Phosphate phosphors for solid state lighting. Springer Ser Mater Sci 174:25
2. Holonyak N, Bevacqua SF (1962) Coherent (visible) light emission from $Ga(As_{1-x}P_x)$ junctions. Appl Phys Lett 1:82
3. Reisfeld R (1987) Spectroscopy of solid state laser materials. Plenum Press, New York, p 343
4. Osbourn GC (1983) $In_xGa_{1-x}As$-$In_yGa_{1-y}As$ strained-layer superlattices: a proposal for useful, new electronic materials. Phys Rev B 27:5126
5. Nakamura S, Mukai T, Senoh M (1994) Candela-class high-brightness InGaN/AlGaN double-heterostructure blue-light-emitting-diodes. Appl Phys Lett 64:1687
6. Cooke M (2010) Going deep for UV sterilization LEDs. Semicond Today 5:82
7. Auzel F (2004) Upconversion and Anti-Stokes processes with f and d ions in solids. Chem Soc 104:139
8. Yen MW, Shionoya S, Yamamoto H (2006) Practical applications of phosphors. CRC press, Boca Raton
9. Shang MM, Li CX, Lin J (2014) How to produce white light in a single-phase host? Chem Soc Rev 43:1372
10. You CH (2005) Visual equivalence of light-emitting diode white light. Opt Eng 44:111307

11. Bergh A, Craford G, Duggal A, Haitz R (2001) The promise and challenge of solid state lighting. Phys Today 54:42
12. (2009) Lifetime of white LEDs at the wayback machine. US Department of Energy
13. Wu XT, Wen J, Li S, Huang S, Cheng J, Chen YH, Duan CK, Yin M (2014) Red-shift of vanadate band-gap by cation substitution fro application in phosphor-converted white light-emitting diodes. App Phys Lett 104:181904
14. Wold JH, Valberg A (2000) The derivation of XYZ tristimulus spaces: a comparison of two alternative methods. Color Res Appl 26:S222
15. Tanabe S, Fujita S, Yoshihara S, Sakamoto A, Yamamoto S (2005) YAG glass-ceramic phosphor for white LED (II): luminescence characteristics. In: Proceedings SPIE fifth international conference on solid state lighting 5941:594112
16. Nakamura S, Fasol G (1997) The blue laser diode: GaN based light emitters and lasers. Springer, Berlin, p 216
17. George NC, Denault KA, Seshadri R (2013) Phosphors for solid-state white lighting. Annu Rev Mater Res 43:481
18. Meyer J, Tappe F (2014) Photoluminescent materials for solid-state lighting: state of art and future challenges. Adv Opt Mater
19. Justel T, Nikol N, Ronda C (1998) New development in the fields of luminescent materials for lighting and Displays. Angew Chem Int Ed 37:3084
20. Zhang SS, Zhuang WD, Zhao CL, Hu YS, He HQ, Huang XW (2004) Study on $(Y, Gd)_3(Al, Ga)_5O_{12}$: Ce^{3+} phosphor. J Rare Earth 22:118
21. Kim JS, Park YH, Kim SM, Choi JC, Park HL (2005) Temperature-dependent emission spectra of M_2SiO_4: Eu^{2+} (M = Ca, Sr, Ba) phosphors for green and greenish white LEDs. Solid State Commun 133:445
22. Jing H, Guo CF, Zhang GG, Su XY, Yang Z, Jeong JH (2012) Photoluminescent properties of Ce^{3+} in compounds $Ba_2Ln(BO_3)_2Cl$ (Ln = Gd and Y). J Mater Chem 22:13612
23. Baginskiy I, Liu RS (2009) Significant improved luminescence intensity of Eu^{2+}-doped $Ca_3SiO_4Cl_2$ green phosphor fro white LEDs synthesized through two-stage method. J Electrochem Soc 156:G29
24. Blasse G, Grabmaier BC (1994) Luminescent materials. Springer, p 72
25. Tian YC (2014) Development of phosphors with high thermal stability and efficiency for phosphor-converted LEDs. J Solid State Lighting 1:11
26. Jung MK, Park WJ, Yoon DH (2007) Photoluminescence characteristics of red phosphor Eu^{3+}, Sm^{3+} co-doped Y_2O_3 for white light emitting diodes. Sensor Actuators B 126:328
27. Pang ML, Liu XM, Lin J (2005) Luminescence properties of R2MoO6: Eu (R = Gd, Y, La) phosphors prepared by Pechini sol-gel process. J Mater Res 20:2676
28. Rao RP (1996) Preparation and characterization of fine-grain yttrium-based phosphors by sol-gel method. J Electrochem Soc 143:189
29. Suo H, Guo CF, Li L (2015) Host Sensitized Spherical Up-conversion Phosphor Yb_2O_3: Er^{3+}. Ceram Int 41:7017
30. Koo HY, Hong SK, Han JM, Kang YC (2008) Eu-doped $Ca_8Mg(SiO_4)_4Cl_2$ phosphor particles prepared by spray pyrolysis from the colloidal spray solution containing ammonium chloride. J Alloy Compd 457:429
31. Xu Y, Guo CF, Luan L, Ding X (2010) Synthesis and characterization of spherical core-shell particles SiO_2 @ $AgEu(MoO_4)_2$. Appl Surf Sci 256:1798
32. Guo CF, Xu Y, Ding X, Li M, Yu J, Ren ZY, Bai JT (2011) Blue-emitting phosphor $M_2B_5O_9Cl$: Eu^{2+} (M = Sr, Ca) for white LEDs. J Alloy Compd 509:38
33. Guo CF, Huang DX, Su Q (2006) Methods to improve the fluorescence intensity of CaS: Eu^{2+} red-emitting phosphor for white LED. Mater Sci Eng B 130:189
34. Van Stun PHJ, Peeters MPJ, Bechtel HH, Heidemann M, Schoenmaekers JBM, Schmidt PJ (2014) Light source using remote phosphor and pink LED. US Patent Appl: 20140355242
35. Guo CF, Chu BL, Wu MM, Su Q (2003) Oxide coating for alkaline earth sulfide based phosphor. J Lumin 105:121

36. Guo CF, Chu BL, Su Q (2004) Improving the Stability of alkaline earth sulfide based phosphors. Appl Surf Sci 225:198
37. Meyer J, Tappe F (2015) Photoluminescent materials for solid state lighting: State of the art and future challenges. Adv Opt Mater 3:424
38. Guo CF, Gao F, Xu Y, Liang LF, Shi FG, Yan BH (2009) Efficient red phosphors $Na_5Ln(MoO_4)_4$: Eu^{3+}(Ln = La, Gd and Y) for white LED. J Phys D Appl Phys 42:095407
39. Dutta PS, Khanna A (2013) Eu^{3+} Activated Molybdate and Tungstate Based Red Phosphors with Charge Transfer Band in Blue Region Luminescence and Display Materials, Devices, and Processing. ECS J Solid State Sci Technol 2:R3153
40. Xu YK, Adachi S (2009) Properties of Na_2SiF_6: Mn^{4+} and Na_2GeF_6:Mn^{4+} red phosphors synthesized by wet chemical etching. J Appl Phys 105:013525
41. Kasa R, Adachi S (2012) Mn-activated K_2ZrF_6 and Na_2ZrF_6 phosphors: sharp red and oscillatory blue-green emissions. J Appl Phys 112:013506
42. Brik MG, Srivastava AM (2013) On the optical properties of the Mn^{4+} ion in solids. J Lumin 133:69
43. TaniyasuY KM, Makimoto T (2006) An aluminium nitride light-emitting diode with a wavelength of 210 nanometres. Nature 441:325
44. Kubota Y, Watanabe K, Tsuda O, Taniguchi T (2007) Deep ultraviolet light-emitting hexagonal boron nitride synthesized at atmospheric pressure. Science 317:932
45. Radkov E, Bompiedi R, Srivastava AM, Setlur AA, Becker C (2004) White light with UV LEDs. Proc SPIE 5187:171
46. Kim JS, Jeon PE, Choi JC, Park HL, Mho SI, Kim GC (2004) Warm-white-light emitting diode utilizing a single-phase full-color $Ba_3MgSi_2O_8$: Eu^{2+}, Mn^{2+} phosphor. Appl Phys Lett 84:2931
47. Lee JW, Lee JH, Woo EJ, Ahn H, Kim JS, Lee CH (2008) Synthesis of nanosized Ce^{3+}, Eu^{3+}-codoped YAG phosphor in a continuous supercritical water system. Ind Eng Chem Res 47:5994
48. Zhang L (2013) Luminescence and energy transfer of YAG: Ce^{3+}, M (M = Er^{3+}, Sm^{3+}, Cr^{3+}, Mn^{2+}). Dissertation, Chongqing University of Posts and Telecommunications
49. Jang HS, Im WB, Lee DC, Jeon DY, Kim SS (2007) Enhancement of red spectral emission intensity of $Y_3Al_5O_{12}$:Ce^{3+} phosphor via Pr co-doping and Tb substitution for the application to white LEDs. J Lumin 126:371
50. Wang L (2013) Study on energy transfer modified YAG: Ce^{3+}/R(R = Pr^{3+}, Cr^{3+}) phosphors for use in high color rendering white LEDs. Dissertation, Chinese Academy of science
51. Wang L, Zhang X, Hao ZD, Luo YS Wang XJ, Zhang JH (2010) Enriching red emission of $Y_3Al_5O_{12}$: Ce^{3+} by codoping Pr^{3+} and Cr^{3+} for improving color rendering of white LEDs. Opt Express 18:25177
52. Guo WJ, Lin YF, Gong XH, Chen YJ, Luo ZD, Huang YD (2009) Polarized spectral properties of Pr^{3+} ions in $NaGd(MoO_4)_2$ crystal. Appl Phys B 94:155
53. Lian JB, Sun XD, Gao T (2009) Preparation of Gd_2O_2S:Pr^{3+} scintillation ceramics by pressureless reaction sintering method. J Mater Sci Technol 25:254
54. Wang Q, Zhu G, Li YY, Wang YH (2015) Photoluminescent properties of Pr^{3+} activated Y_2WO_6 for light emitting diodes. Opt Mater 42:385
55. Min X, Fang MB, Huang ZH, Liu YG, Tang C, Wu XW (2015) Synthesis and optical properties of Pr^{3+}-doped $LaMgAl_{11}O_{19}$—a novel blue converting yellow phosphor for white light emitting diodes. Ceram Int 41:4238
56. Blasse G (1988) Luminescence of inorganic solids: From isolated centres to concentrated systems. Prog Solid State Chem 18:79
57. Liu XM, Li CX, Quan ZW, Cheng ZY Lin J (2007) Tunable luminescence properties of $CaIn_2O_4$: Eu^{3+} phosphors. J Phys Chem C 111:16601
58. Li CX, Zhang CM, Hou ZY, Wang LL, Quang ZW, Lian HZ Lin J (2009) β-$NaYF_4$ and β-$NaYF_4$: Eu^{3+} microstructures: Morphology control and tunable luminescence properties. J Phys Chem C 113: 2332

59. Shang MM, Li GG, Kang XJ, Yang DM, Geng DL, Peng C, Cheng ZY, Lian HZ, Lin J (2012) LaOF: Eu^{3+} nanocrystals: hydrothermal synthesis, white and color-tunging emission properties. Dalton Trans 41:5571
60. Guo CF, Ding X, Xu Y (2010) Luminescent properties of Eu^{3+}-doped $BaLn_2ZnO_5$ (Ln = La, Gd and Y) phosphors by sol-gel method. J Am Ceram Soc 93:1708
61. Liang HC, Chang YC, Chang YS (2008) Synthesis and photoluminescence characteristics of color-tunable BaY_2ZnO_5: Eu^{3+} phosphors. Appl Phys Lett 93:211902
62. Ronda C (2007) Luminescence from theory to applications. Wiley-VCH Verlag GmbH & Co, KGaA, p 19
63. Zheng JM, Guo CF, Ding X, Ren ZY, Bai JT (2012) Enhanced luminescence of Tb^{3+} by efficient energy transfer from Ce^{3+} in $Sr_2B_5O_9Cl$ host. Curr Appl Phys 12:643
64. Fu L, Xia HP, Dong YM, Li SS, Gu XM, Jiang HC, Chen BJ (2014) White light emission from Tb^{3+}/Sm^{3+} codoped $LiYF_4$ single crystal excited by UV light. IEEE Photonics Technol Lett 26:1485
65. Zheng YH, Lin JT, Wang QM (2012) Emissions and photocatalytic selectivity of $SrWO_4$: Ln^{3+} (Eu^{3+}, Tb^{3+}, Sm^{3+} and Dy^{3+}) prepared by a supersonic microwave co-assistance method. Photochem Photobiol Sci 11:1567
66. Su Q, Pei ZW, Chi LS, Zhang HJ, Zhang ZY, Zou F (1993) The yellow-to-blue intensity ratio (Y/B) of Dy^{3+} emission. J Alloys Compd 192:25
67. Zhao J, Guo CF, Su XY, Noh HM, Jeong JH (2014) Electronic structure and luminescence properties of phosphor $Li_8Bi_2(MoO_4)_7$: Dy^{3+}. J Am Ceram Soc 97:1878
68. Klevtsova RF, Solodovnikov SF, Glinskaya LA, Alekseev VI, Khalbaeva KM, Khaikina EG (1997) Syntheses and crystal sructure of the binary molybadate $Li_8Bi_2(MoO_4)_7$. J Struct Chem 38:1
69. Jia G, Song YH, Yang M, Huang YJ, Zhang LH, You HP (2011) Uniform YVO_4: Ln^{3+} (Ln = Eu, Dy, and Sm) nanocrystals: solvothermal synthesis and luminescence properties. Opt Mater 31:1032
70. Sun XY, Lin LW, Wang WF, Zhang JC (2011) White-light emission from $Li_2Sr_{1-3x/2}Dy_xSiO_4$ phosphors. Appl Phys A 104:83
71. Ronda C (2007) Luminescence from theory to applications. Wiley-VCH Verlag GmbH & Co, KGaA, p 27
72. Shriver DF, Atkins PW (2001) Inorganic Chemistry (4th ed.). Oxford University Press, p 227
73. Kim JS, Kang, JY Jeon PE, Choi JC, Park HL, Kim TW (2004) GaN-based white-light-emitting diodes fabricated with a mixture of $Ba_3MgSi_2O_8$: Eu^{2+} and Sr_2SiO_4: Eu^{2+} phosphors. Jpn J Appl Phys 43:989
74. Guo CF, Ding X, Seo HJ, Ren JY, Bai JT (2011) Luminescent properties of UV excitable blue emitting phosphors $MSr_4(BO_3)_3$: Ce^{3+} (M = Li and Na). J Alloys Compd 509:4871
75. Liu JQ, Wang XJ, Xuan TT, Wang CB, Li HL, Sun Z (2015) Lu3(Al, Si)5(O, N)12: Ce3+ phosphors with broad emission band and high thermal stability for white LEDs. J Lumin 158:322
76. Xin SY, Wang YH, Zhu G, Zhang F, Gong Y, Wen Y, Liu BT (2013) Tunable white light emitting from mono Ce^{3+} doped $Sr_5(PO_4)_2SiO_4$ phosphors for light emitting diodes. Mater Res Bull 48:1627
77. Lin HH, Liang HB, Han B, Zhong JP, Su Q, Dorenbos P, Birowosuto MD, Zhang GB, Fu YB, Wu WQ (2007) Luminescence and site occupancy of Ce^{3+} in $Ba_2Ca(BO_3)_2$. Phys Rev B 76:035117
78. Wu ZC, Liu J, Hou WG, Xu J, Gong ML (2010) A new single-host white-light-emitting $BaSrMg(PO_4)_2$: Eu^{2+} phosphor for white-light-emitting diodes. J Alloys Compd 498:139
79. Yu H, Deng DG, Li YQ, Xu SQ, Li YY, Yu CP, Ding YY, Lu HW, Yin HY, Nie QL (2013) Electronic structure and photoluminescence properties of yellow-emitting $Ca_{10}Na(PO_4)_7$: Eu^{2+} phosphor for white light-emitting diodes. J Lumin 143:132
80. Zhang J, Hua ZH, Wen SZ (2015) Luminescence of emission-tunable $Ca_{10}Li(PO_4)_7$: Eu^{2+}, Sr^{2+}, Mn^{2+} phosphors with various Eu^{2+} centers for LED applications. J Alloys Compd 637:70

81. Pei ZW, Su Q, Zhang JY (1993) The valence change from RE^{3+} to RE^{2+} (RE = Eu, Sm, Yb) in SrB_4O_7: RE prepared in air and the spectral properties of RE^{2+}. J Alloys Compd 198:51
82. Pei ZW, Zeng Q, Su Q (1999) A study on the mechanism of the abnormal reduction of $Eu^{3+} \rightarrow Eu^{2+}$ in $Sr_2B_5O_9Cl$ prepared in air at high temperature. J Solid State Chem 145:212
83. Mao ZY, Wang DJ, Lu QF, Yu WH, Yuan ZH (2009) Tunable single-doped single-host full-color-emitting $LaAlO_3$: Eu phosphor via valence state-controlled means. Chem Commun 3:346
84. Guo CF, Luan L, Ding X, Zhang FJ, Shi FG, Gao F, Liang LF (2009) Luminescent properties of $Sr_5(PO_4)_3Cl$: Eu^{2+}, Mn^{2+} as a potential phosphor for UV-LED based white LEDs. Appl Phys B 95:779
85. Guo CF, Luan L, Ding X, Huang DX (2008) Luminescent properties of $SrMg_2(PO_4)_2$: Eu^{2+}, Mn^{2+} as a potential phosphor for ultraviolet light-emitting diodes. Appl Phys A 91:327
86. Guo CF, Yu J, Ding X, Li M, Ren ZY, Bai JT (2011) A dual-emission phosphor $LiCaBO_3$: Ce^{3+}, Mn^{2+} with energy transfer for near-UV LEDs. J Electrochem Soc 158:J42
87. Ye S, Zhang JH, Zhang X, Lu SZ, Ren XG, Wang XJ (2007) Mn^{2+} concentration manipulated red emission in $BaMg_2Si_2O_7$: Eu^{2+}, Mn^{2+}. J Appl Phys 101:033513
88. Guo CF, Ding X, Luan L, Xu Y (2010) Two-color emitting of Eu^{2+} and Mn^{2+} co-doped $Sr_2Mg_3P_4O_{15}$ for UV LEDs. Sensors Actuat B Chem 43:712
89. Dorenbos P (2003) Relation between Eu^{2+} and Ce^{3+} f-d-transition energies in inorganic compounds. J Phys Condens Matter 15:4797
90. Guo CF, Luan L, Shi FG, Ding X (2009) White-emitting phosphor Ca_2BO_3Cl: Ce^{3+}, Eu^{2+} for UV light-emitting diodes. J Electrochem Soc 156:J125
91. Guo CF, Ding X, Seo HJ, Ren ZY, Bai JT (2011) Double emitting phosphor $NaSr_4(BO_3)_3$: Ce^{3+}, Tb^{3+} for near-UV light emitting diodes. Opt Laser Technol 43:1351
92. Zhang MF, Liang YJ, Tang R, Yu DY, Tong MH, Wang Q, Zhu YL, Wu XY, Li GG (2014) Highly efficient Sr3Y2(Si3O9)2: Ce3+, Tb3+/Mn2+/Eu2+ phosphors for white LEDs: structure refinement, color tuning and energy transfer. RSC Adv 4:40626
93. Unithrattil S, Lee KH, Chung WJ, Im WB (2014) Full-color-emitting $CaYAl_3O_7$: Pr^{3+}, Ce^{3+} phosphor for near-UV LED-based white light. J Lumin 152:176
94. Das S, Yang CY, Lin HC, Lu CH (2014) Structural and luminescence properties of tunable white-emitting $Sr_{0.5}Ca_{0.5}Al_2O_4$: Eu^{2+}, Dy^{3+} for UV-excited white-LEDs. RSC Adv 4:64956
95. Jia YC, Lu W, Guo N, Lu WZ, Zhao Q, You HP (2013) Utilizing Tb^{3+} as an energy transfer bridge to connect $Eu^{2+} \rightarrow Sm^{3+}$ luminescent centers: realization of efficient Sm^{3+} red emission under near-UV excitation. Chem Commun 49:2664
96. Jiao MM, Guo N, Lü W, Jia YC, Lv WZ, Zhao Q, Shao BQ, You HP (2013) Tunable blue-green-emitting Ba3LaNa(PO4)3F: Eu2+, Tb3+ phosphor with energy transfer for near-UV white LEDs. Inorg Chem 52:10340
97. Dutta PS, Khanna A (2013) Eu^{3+} activated molybdate and tungstate based red phosphors with charge transfer band in blue region. ECS J Solid State Sci 2:R3153
98. Takahashi Y, Kitamura K, Iyi N, Inoue S (2006) Visible orange photoluminescence in a barium titanosilicate $BaTiSi_2O_7$. App Phys Lett 88:151903
99. Nakajima T, Isobe M, Tsuchiya T, Ueda Y, Kumagai T (2009) A revisit of photoluminescence property for vanadate oxides AVO_3 (A: K, Rb and Cs) and $M_3V_2O_8$ (M: Mg and Zn). J Lumin 129:1568
100. Gupta SK, Ghosh PS, Pathak N, Aryab A, Natarajana V (2014) Understanding the local environment of Sm^{3+} in doped $SrZrO_3$ and energy transfer mechanism using time-resolved luminescence: a combined theoretical and experimental approach. RSC Adv 4:29202
101. Zhu HL, Jin DL, Zhu LM, Yang H, Yao KH, Xi ZQ (2008) A general hydrothermal route to synthesis of nanocrystalline lanthanide stannates: $Ln_2Sn_2O_7$ (Ln = Y, La–Yb). J Alloys Compd 464:508
102. Bharathy M, Rassolov VA, zur Loye HC (2008) Crystal growth of Sr_3NaNbO_6 and Sr_3NaTaO_6: new photoluminescent oxides. Chem Mater 20:2268

103. Danielson E, Devenney M, Giaquinta DM, Golden JH, Haushalter RC, McFarland EW, Poojary DM, Reaves CM, Weinberg WH, Wu XD (1998) A rare-earth phosphor containing one-dimensional chains identified through combinatorial methods. Science 279:837
104. Monika DL, Nagabhushana H, Hari Krishna R, Nagabhushana BM, Sharmac SC, Thomas T (2014) Synthesis and photoluminescence properties of a novel Sr_2CeO_4: Dy^{3+} nanophosphor with enhanced brightness by Li^+ co-doping. RSC Adv 4:38655
105. Ronde H, Blasse G (1976) The nature of the luminescence transition of the vanadate group. J Solid State Chem 17:339
106. Bayer G (1965) Vanadate $A_3B_2V_30_{12}$ with garnet structure. J Am Ceram Soc 48:600
107. Chen X, Xia ZG, Yi M, Wu XC, Xin H (2013) Rare-earth free self-activated and rare-earth activated $Ca_2NaZn_2V_3O_{12}$ vanadate phosphors and their color-tunable luminescence properties. J Phys Chem Solids 74:1439
108. Song D, Guo CF, Li T (2015) Luminescence of the self-activated vanadate phosphors $Na_2LnMg_2V_3O_{12}$ (Ln = Y, Gd). Ceram Int 41:6518
109. Chen X, Xia ZG, Yi M, Wu XC, Xin H (2013) Rare-earth free self-activated and rare-earth activated $Ca_2NaZn_2V_3O_{12}$ vanadate phosphors and their color-tunable luminescence properties. J Phys Chem Solids 74:1439
110. Zhang Y, Geng DL, Li XJ, Fan J, Li K, Lian HZ, Shang MM, Lin J (2014) Wide-band excited $YTiTaO_6$: Eu^{3+}/Er^{3+} phosphors: structure refinement, luminescence properties, and energy transfer mechanisms. J Phys Chem C 118:17983
111. Li HF, Zhao R, Jia YL, Sun WZ, Fu JP, Jiang LH, Zhang S, Pang Ran, Li CY (2014) $Sr_{1.7}Zn_{0.3}CeO_4$: Eu^{3+} novel red-emitting phosphors: synthesis and photoluminescence properties. *ACS Appl Mater Interfaces 6:*3163
112. Monika DL, Nagabhushana H, Krishna RH, NagabhushanaBM SS, Thomasd T (2014) Synthesis and photoluminescence properties of a novel Sr_2CeO_4: Dy^{3+} nanophosphor with enhanced brightness by Li^+ co-doping. RSC Adv 4:38655
113. Jeon YI, Bharat LK, Yu JS (2015) Synthesis and luminescence properties of Eu^{3+}/Dy^{3+} ions co-doped $Ca_2La_8(GeO_4)_6O_2$ phosphors for white-light applications. J Alloy Compd 620:263
114. Liu Y, Liu GX, Wang JX, Dong XT, Yu WS (2014) Single-component and warm-white-emitting phosphor $NaGd(WO_4)_2$: Tm^{3+}, Dy^{3+}, Eu^{3+}: synthesis, luminescence, energy transfer, and tunable color. Inorg Chem 53:11457
115. Reddy GVL, Moorthy LR, Chengaiah T, Jamalaiah BC (2014) Multi-color emission tunability and energy transfer studies of $YAl_3(BO_3)_4$: Eu^{3+}/Tb^{3+} phosphors. Ceram Int 40:3399
116. Park JY, Jung HC, Raju GSR, Moon BK, Jeong JH, Kim JH (2010) Tunable luminescence and energy transfer process between Tb^{3+} and Eu^{3+} in GYAG: Bi^{3+}, Tb^{3+}, Eu^{3+} phosphors. Solid State Sci 12:719
117. Pavitra E, Raju GSR, Ko YH, Yu YS (2012) A novel strategy for controllable emissions from Eu^{3+} or Sm^{3+} ions co-doped SrY_2O_4: Tb^{3+} phosphors. Phys Chem Chem Phys 14:11296
118. Baur F, Glocker F, Jüstel T (2015) Photoluminescence and energy transfer rates and efficiencies in Eu^{3+} activated $Tb_2Mo_3O_{12}$. J Mater Chem C 3:2054
119. Zhou J, Xia ZG (2014) Multi-color emission evolution and energy transfer behavior of $La_3GaGe_5O_{16}$: Tb^{3+}, Eu^{3+} phosphors. J Mater Chem C 2:6978
120. Hou L, Cui SB, Fu ZL, Wu ZJ, Fu XH, Jeong JH (2014) Facile template free synthesis of Kla $(MoO_4)_2$: Eu^{3+}, Tb^{3+} microspheres and their multicolor tunable luminescence. Dalton Trans 43:5382
121. Auzel F (2004) Up-conversion and anti-stokes processes with f and d ions in solids. Chem Rev 104:139
122. Downing E, Hesselink L, Ralston J, Macfarlane R (1996) A three-color, solid-state, three-dimensional display. Science 273:1185
123. Giri NK, Rai DK, Rai SB (2008) White light upconversion emissions from Tm^{3+} + Ho^{3+} + Yb^{3+} codoped tellurite and germanate glasses on excitation with 798 nm radiation. J Appl Phys 104:113107

124. Etchart I, Berard M, Laroche M, Huignard A, Hernandez I, Gillin WP, Curryd RJ, Cheetham AK (2011) Efficient white light emission by upconversion in Yb^{3+}- Er^{3+}- and Tm^{3+} -doped Y_2BaZnO_5. Chem Commun 47:6263
125. Chen DQ, Wang YS, Yu YL, Huang P, Weng FY (2008) Novel rare earth ions-doped oxyfluoride nano-composite with efficient upconversion white-light emission. J Solid State Chem 181:2763
126. Mahalingam V, Mangiarini F, Vetrone F, Venkatramu V, Bettinelli M, Speghini A, Capobianco JA (2008) Bright White Upconversion Emission from $Tm^{3+}/Yb^{3+}/Er^{3+}$-Doped $Lu_3Ga_5O_{12}$ Nanocrystals. J Phys Chem C 112:17745
127. Leonidova II, Zubkova VG, Tyutyunnika AP, Tarakinaa NV, Surata LL, Koryakovab OV, Vovkotrubc EG (2011) Upconversion luminescence in Er^{3+}/Yb^{3+} codoped $Y_2CaGe_4O_{12}$. J Alloy Compd 509:1339
128. Li T, Guo CF, Li L (2013) Up-conversion luminescence of Er^{3+}-Yb^{3+} co-doped $CaIn_2O_4$. Opt Express 21:18281
129. Page RH, Schaffes KI, Waide PA, Tassano JB, Payne SA, Krupke WF, Bischel WK (1998) Up-conversion-pumped luminescence efficiency of rare-earth-doped hosts sensitized with trivalent ytterbium. J Opt Soc Am B 15:996
130. Sangeetha NM, van Veggel FCJM (2009) Lanthanum silicate and lanthanum zirconate nanoparticles co-doped with Ho^{3+} and Yb^{3+}: matrix-dependent red and green upconversion emissions. J Phys Chem C 113:14702
131. Li T, Guo CF, Li L, Jeong JH (2013) Tailorable multicolor upconversion emissions in Tm^{3+}/ Ho^{3+}/Yb^{3+} Co-Doped $LiLa(MoO_4)_2$. J Am Ceram Soc 96:1193
132. Murray CB, Kagan CR, Bawendi MG (2000) Synthesis and characterization of monodisperse nanocrystals and close-packed nanocrystal assemblies. Annu Rev Mater Res 30:545
133. Ekimov AI, Efros AL, Onushchenko AA (1985) Quantum size effect in semiconductor microcrystals. Solid State Commun 56:921
134. Zhang CM, Lin J (2012) Defect-related luminescent materials: synthesis, emission properties and applications. Chem Soc Rev 41:7938
135. Zhang XM, Quan ZW, Yang J, Yang PP, Lian HZ, Lin J (2008) Solvothermal synthesis of well-dispersed MF_2 (M = Ca, Sr, Ba) nanocrystals and their optical properties. Nanotechnology 19:075603
136. Yang Z (2015) Rare earth doped alkaline earth fluoride micro/nano luminescent materials by hydrothermal controllable synthesis. Dissertation, Northwest University

Chapter 16
Crystal Structure and Luminescence Properties of Some Fluorides, (Oxy) nitrides and Oxides Phosphors

Yuhua Wang, Deyin Wang, Zhengyan Zhao, Chunfang Wu, Dan Wang and Zhipeng Ci

Fluoride materials provide some distinct advantages owing to their very low phonon frequencies, which makes them as good hosts for downconversion/upconversion phosphors. Oxides with wide band gap and good chemical stability are also easily prepared, which makes them the prior host matrix for phosphors used in LED, long-lasting phosphor, and plasma display panels (PDPs), in which vacuum ultraviolet (VUV) radiation from noble gas discharge is used as the excitation source. In contrast, because of the strong nephelauxetic effect and large crystal-field splitting caused by the coordination of nitrogen atoms, (oxy)nitrides exhibit broad excitation bands covering the emission of the InGaN near-UV and blue LED chips. Moreover, (oxy)nitrides materials feature high thermal and chemical stability due to the stiff frameworks of (Si, Al)–(N, O) tetrahedral in the host lattice. These features lead (oxy)nitrides to be excellent hosts for phosphors, especially those used in LED. In this section, we briefly summarize the crystal structures of some fluorides, (oxy)nitrides, and oxides and present their optical properties as well as give examples for their applications.

16.1 Downconverson/Upconversion Fluoride Phosphors

Owing to their features, such as high refractive index and low phonon energy (<400 cm^{-1}), rare-earth fluorides—including binary LnF_3 and complex $ALnF_4$ (Ln = lanthanide ions; A = alkali)—are considered excellent hosts for downconversion and upconversion phosphors [1]. Downconversion is the process by which one high-energy photon is converted into two or more lower-energy photons, whereas upconversion is the generation of higher-energy photons from lower-energy radiation on the basis of sequential adsorption and energy-transfer steps. Downconversion phosphors are widely applied in modern lighting and

Y. Wang (✉) · D. Wang · Z. Zhao · C. Wu · D. Wang · Z. Ci
Key Laboratory for Special Function Materials and Structural Design of the Ministry of Education, School of Physical Science and Technology, Lanzhou University, Lanzhou 730000, People's Republic of China
e-mail: wyh@lzu.edu.cn

R.S. Liu (ed.), *Phosphors, Up Conversion Nano Particles, Quantum Dots and Their Applications*, DOI 10.1007/978-3-662-52771-9_16

display fields such as white light-emitting diodes (WLEDs), plasma display panels (PDPs), solar cells, etc. Upconversion phosphors are important for applications in areas such as photonics, biomedicine, and biological imaging as well as solar cells. As a consequence, rare-earth fluorides are of significant importance in modern chemistry and materials science [1].

16.1.1 Crystal Structure of Some Selected Fluorides

CaF_2

CaF_2 occurs naturally as the mineral fluorite and in the cubic system with the $Fm\bar{3}m$ group. In this structure, the calcium ions are surrounded by eight neighboring fluoride ions, whereas the fluoride ions are surrounded by a tetrahedron of four calcium ions [2] as shown in Fig. 16.1. Although perfectly packed crystalline samples are colorless, the mineral is often deeply colored due to the presence of F-centers [2].

$NaGdF_4$

Depending on the synthesis condition, $NaGdF_4$ can crystallize either in a cubic α-phase or a hexagonal β-phase. The cubic structure α-$NaGdF_4$ has the space group $Fm\bar{3}m$ and is isomorphic with CaF_2 where the calcium ions have been substituted randomly by sodium and gadolinium ions [3]. The hexagonal structure β-$NaGdF_4$ has the space group $P\bar{6}$, and there are two kinds of crystallographic cation sites that exhibit the same symmetry in β-$NaGdF_4$. One site (1a) is fully occupied by the Gd^{3+} ion, and the other site (1f) is randomly occupied by the Gd^{3+} and Na^+ ions (each occupying half of the sites). The Gd^{3+} ion on site (1a) has three fluorine first

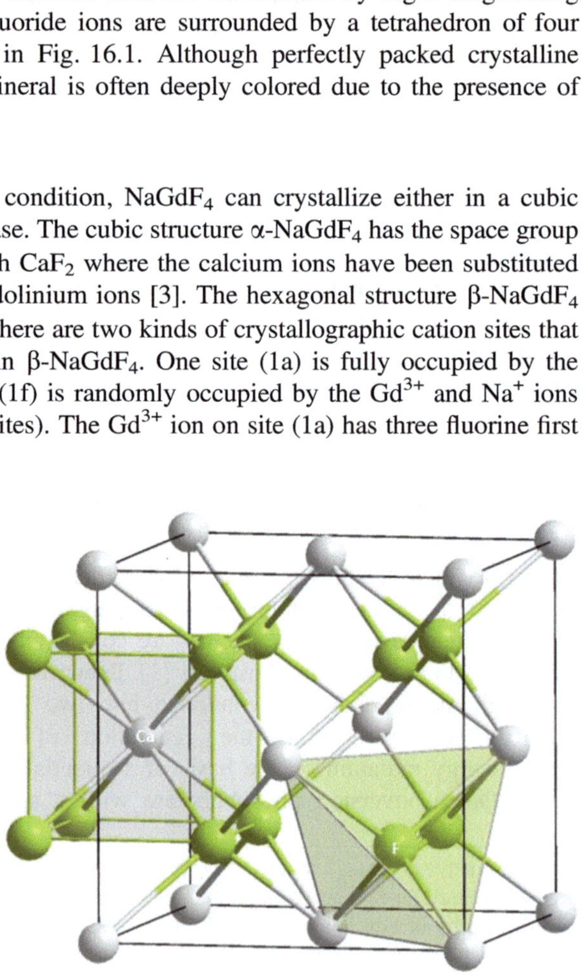

Fig. 16.1 Crystal structure of CaF_2 [2]

neighbors at 2.345 Å and six fluorine second neighbors at 2.371 Å, whereas these distances become 2.362 and 2.448 Å for the Gd^{3+} ion on the site (1f), which is shown in Fig. 16.2 [3].

YF_3

YF_3 crystallizes in the orthorhombic system with the Pnma group. In this structure, each yttrium atom is coordinated by nine fluorine atoms with eight fluorine neighbors at approximately 2.3 Å and another at 2.60 Å [5]. The arrangement of these nine neighboring fluorine atoms around the yttrium atom is shown in Fig. 16.3. The trifluorides of all the rare-earth elements, from samarium to lutetium, are isostructural with YF_3.

K_3LnF_5(Ln = Gd, Y)

The room-temperature structure of K_3LnF_6 (Ln = Gd, Y) is monoclinic with the space group $P2_1/n$ [6].The crystal structure of K_3LnF_6 (Ln = Gd, Y) is composed of distorted LnF_6 and $K(1)F_6$ octahedra and $K(2)F_4$ tetrahedra as shown in Fig. 16.4. The Ln and K(1) atoms occupy positions in the center of octahedra with C_i local symmetry. The average Gd–F distance in K_3GdF_6 is 2.21 Å and that for Y–F distance in K_3YF_6 is 2.16 Å. The first neighboring Gd atoms in K_3GdF_6 are located at an average distance of 6.475 Å, whereas the average distance between neighboring Y atoms is 6.429 Å in K_3YF_6 [6].

K_2GdF_5

The compound K_2GdF_5 crystallizes in the orthorhombic system with space group Pnma [7]. The structure contains linear chains of trans-linked GdF_7 polyhedra running along the b axis that are separated by eight-coordinate K atoms

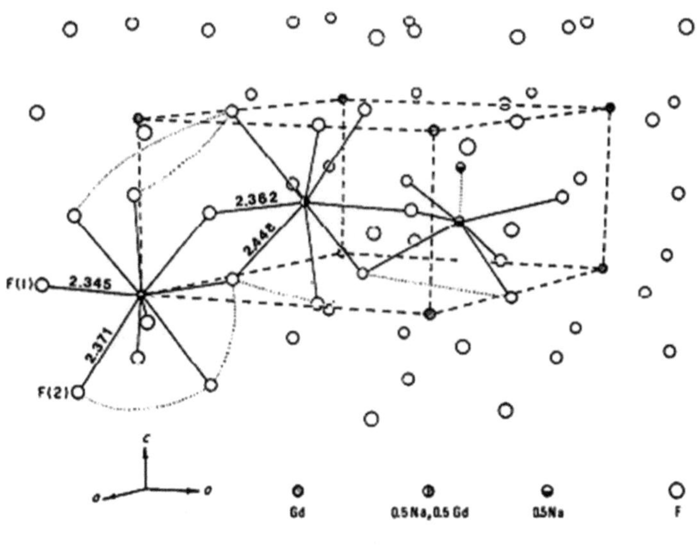

Fig. 16.2 Crystal structure of $NaGdF_4$. Reprinted from Ref. [3], Copyright 1987, with permission from Elsevier

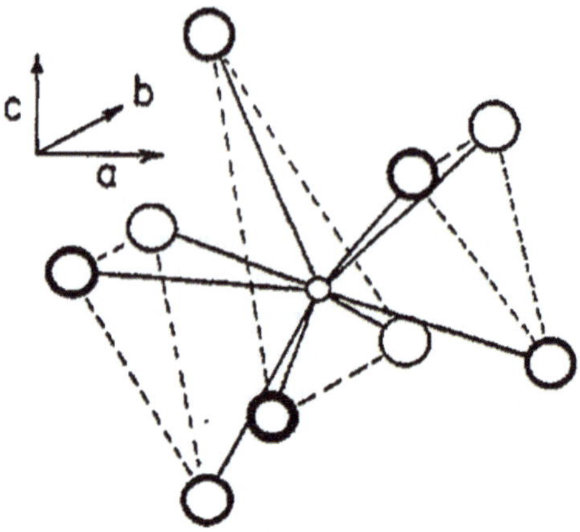

Fig. 16.3 Coordination of Y^{3+} in YF_3. Reprinted with the permission from Ref. [5]. Copyright 1953 American Chemical Society

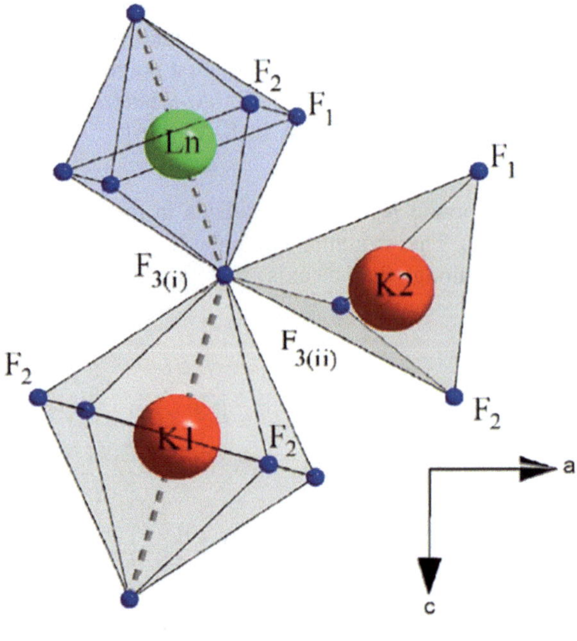

Fig. 16.4 Coordination polyhedra of Ln, K(1) and K (2) atoms along the *b* direction. Reprinted from Ref. [6], Copyright 2006, with permission from Elsevier

(see Fig. 16.5). Each Gd atom is coordinated to four crystallographically equivalent F_1 atoms and three nonequivalent atoms F_2, F_3, and F_4. Pairs of F_1 atoms bridge adjacent Gd atoms, and all of the F atoms serve as interconnections between the Gd and K atoms. All the K atoms are eight-coordinate with the resulting polyhedra forming by sharing the faces of atoms F_2, F_3, and F_4 [7].

KY_3F_{10}

At ambient conditions, KY_3F_{10} crystallizes in a cubic structure with space group Fm3m. Its crystal structure consists of two ionic groups, $[KY_3F_8]^{2+}$ and $[KY_3F_{12}]^{2-}$, alternating along the three crystallographic directions. The basic building unit for KY_3F_{10} consists of six square anti-prisms that share common corners in such a way that an empty cuboctahedron is formed at the centre of the unit [8] as shown in Fig. 16.6. The Y cations are incorporated into distorted square anti-prisms, and the K is coordinated tetrahedrally by 4 inner fluorine atoms at distances of approximately 2.7 Å; 1q2 additional anions are positioned at longer distances (≈3.2 Å) [8].

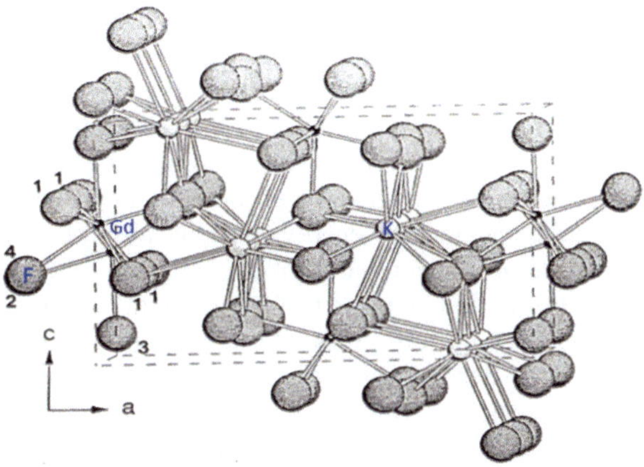

Fig. 16.5 Unit cell of K_2GdF_5. The *small black circles* represent Gd atoms, the *large dark circles* F atoms, and the *mid-sized light circles* K atoms [7]

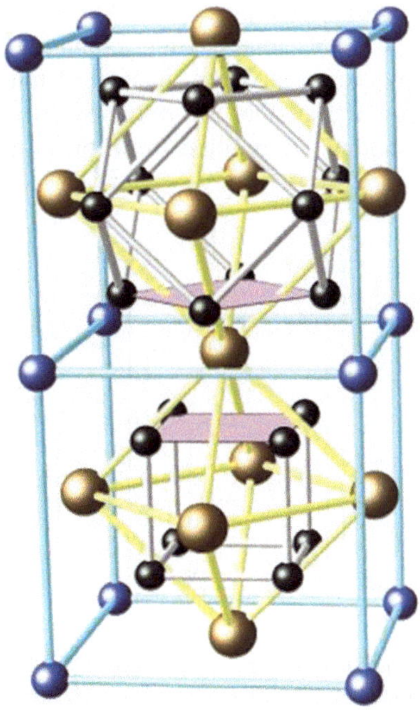

Fig. 16.6 Unit cell of the Fm3m structure of KY_3F_{10}. The large bronze (*light gray*) spheres substitute for yttrium atoms, the *medium blue* (*gray*) ones for potassium atoms, and the *small black* ones for fluorine atoms. Reprinted with permission from Ref. [8]. Copyright 2012 by American Physical Society

16.1.2 Downconversion Fluoride Phosphors

In this section, the downconversion fluoride phosphors mainly refer to phosphor that can convert one vacuum ultraviolet (VUV) photon into two UV-visible photons. The research for a downconversion phosphor is mainly promoted by enhancing the conversion efficiency of phosphors used in Hg-free fluorescent lamps and color plasma display panels, in which VUV radiation from noble gas discharge is used as the excitation source [9, 10]. Apart from this motivation, the creation of phosphors with a quantum efficiency >1 also forms an interesting scientific challenge. In theory, quantum cutting can be realized by a single luminescent center or a combination of luminescent centers. The concept of downconversion is illustrated in Fig. 16.7 using three types of ions, I through III, with imaginary energy-level schemes. Four different concepts are illustrated in Fig. 16.7. The first concept is based on using a single type of ion with three energy levels. After transition of the center to the highest energy level by absorption of one high-energy photon, a sequence of two emissions takes place. The second concept involves two types of ions. In the first step, part of the excitation energy is transferred from ion I to a neighboring ion II by cross-relaxation, thus bringing ion II to an excited state. After energy transfer, ion I still is in an excited state, and it transfers the remaining energy to another neighboring ion II. The two excited ion II return to the ground state by emitting two photons. The third concept also involves two types of ions and is similar to the second concept. The difference is that after ion I transfers part of the absorbed energy to a neighboring ion II, ions I and II return to the ground state by emitting one photon each. The fourth concept is based on a combination of three types of ions. After obtaining part of the excitation energy from ion I by-cross-relaxation, instead of returning to the ground state by emitting a photon, ion II transfers the obtained energy to a neighboring ion III. Subsequently, two photons are emitted from ion I and ion III.

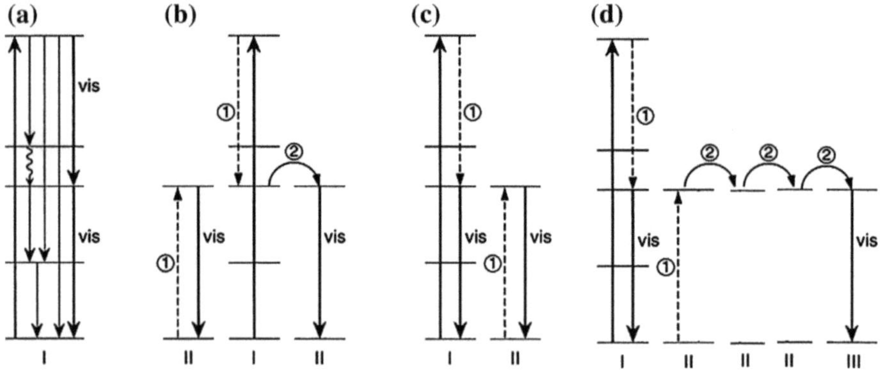

Fig. 16.7 Energy level diagrams for three (hypothetical) types of RE ions (*I* through *III*) showing the concept of downconversion. ① and ② denote energy transfer steps. Reprinted from Ref. [21], Copyright 1999, with permission from Elsevier

A. **Downconversion in single Pr^{3+}-doped fluorides**

Until now, downconversion based on the four concepts mentioned previously has been demonstrated. The first example was demonstrated in single Pr^{3+}-doped phosphor. In this case, it requires the $^1S_0$ level of Pr^{3+} to be located below its lowest 4*f*5*d* state. When Pr^{3+} is pumped to its $^1S_0$ level, two-photon emission is completed by two subsequent emissions of 405 nm for the $^1S_0 \rightarrow {}^1I_6$ transition followed by radiation of 480 nm for the $^3P_0 \rightarrow {}^3H_4$ transition or of 610 nm for the $^3P_0 \rightarrow {}^3H_6$ transition as shown Fig. 16.8. Until now, cascade emission from Pr^{3+} has been found in a number of fluoride hosts [11–20] such as $SrAlF_5$ [11, 12], LnF(Ln = Y, La, Lu) [13–16], $BaMgF_4$ [15], AYF_4(A = Na, K) [13, 17], $AMgF_3$(A = Na, K) [18, 19], and $LiSrAlF_6$ [20]. Figure 16.9 shows the emission spectrum of single

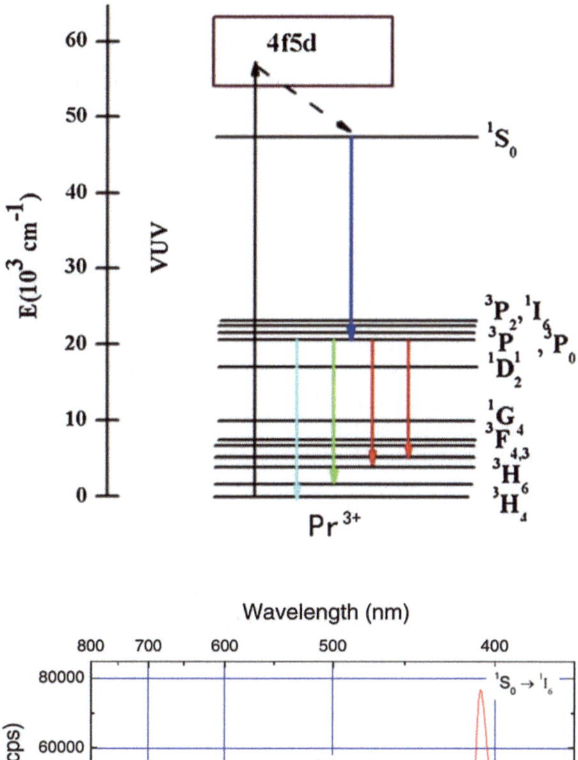

Fig. 16.8 Schematic energy-level diagram of Pr^{3+} ion, showing the concept of the two photon emission process under excitaion of one VUV photon. Reprinted from Ref. [10], Copyright 2010, with permission from Elsevier

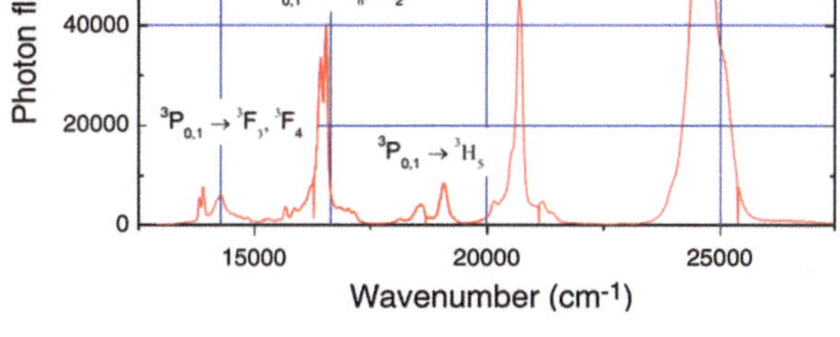

Fig. 16.9 Emission spectrum of YF_3:0.01 %Pr^{3+} at room temperature under excitation at 191 nm. Reprinted from Ref. [16], Copyright 2005, with permission from Elsevier

Pr^{3+}-doped YF_3. The presence of the emission from $^1S_0$ indicated that the $^1S_0$ level in YF_3:Pr^{3+} is lower than its lowest 5*d* state, and the observation of emission from $^1S_0 \rightarrow {}^1I6,{}^3P_{0,1} \rightarrow {}^3H_4$, and $^3P_{0,1} \rightarrow {}^3H_6$ of Pr^{3+} indicates that a two-step down-conversion emission occurs in YF_3.

B. Downconversion in Eu^{3+}/Gd^{3+}-doped fluorides

Quantum cutting based on the second concept is demonstrated in a combination of Gd^{3+} and Eu^{3+} in $LiGdF_4$ [9, 21–28]. Figure 16.10 shows the energy-level diagram and quantum-cutting model in a Gd^{3+}–Eu^{3+} system proposed by Weghet et al. [9]. It was noticed that the energy gap between the $Gd^{3+6}G_J \rightarrow {}^6P_J$ (approximately 590 nm) transition matches well with that of the $Eu^{3+7}F_J \rightarrow {}^5D_0$ (approximately 590 nm) transition. On excitation in the ^6G_J levels of Gd^{3+}, part of the excitation energy is transferred from Gd^{3+} to a neighboring Eu^{3+} ion first by cross-relaxation resulting in Eu^{3+} in the $^5D_0$ excited state and Gd^{3+} in the ^6P_J state. In the second step, the Gd^{3+} ion in the ^6P_J state transfers the remaining excitation energy to a second neighboring Eu^{3+} ion, which is followed by fast relaxation to the ^5D_J state.

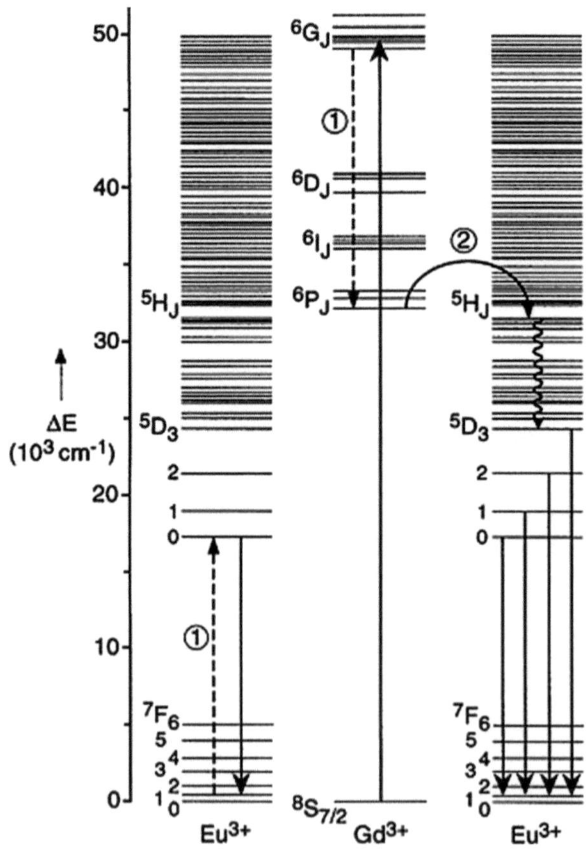

Fig. 16.10 Energy level diagram for a Gd^{3+}–Eu^{3+} system, showing the visible quantum cutting process by two-step energy transfer upon excitation in the ^6G_J levels of Gd^{3+}. The two energy-transfer steps are indicated by ① (*CR* cross relaxation step) and ② (*DT* direct energy transfer) Reprinted from Ref. [21], Copyright 1999, with permission from Elsevier

Both steps result in the emission of a visible photon due to the $^5D_J \rightarrow {}^7F_J$ transitions on Eu^{3+}. Figure 16.11 shows the emission spectra of $LiGdF_4$:0.5%Eu^{3+} at room temperature on excitation in the ^6G_J and ^6I_J levels, respectively [9, 21]. The emission spectra on excitation in ^6G_J and ^6I_J were recorded because in the case the two-step energy transfer occurs, a difference in intensity ratio between $^5D_0$ and $^5D_1$ is expected. From the emission spectra in Fig. 16.11, it is clear that the $^5D_0/^5D_1$ emission intensity ratio is indeed much higher on ^6G_J excitation compared with ^6I_J excitation, thus showing the occurrence of visible quantum cutting by downconversion. Later on, quantum cutting using Gd^{3+}–Eu^{3+} system was observed in other fluoride hosts [22–28]. Table 16.1 lists the quantum efficiency of these phosphors on excitation in the ^6G_J level of Gd^{3+}. The quantum efficiency is calculated by evaluating the $^5D_0$ and $^5D_{1,2,3}$ integrated emission intensities using the following equation [9, 21]:

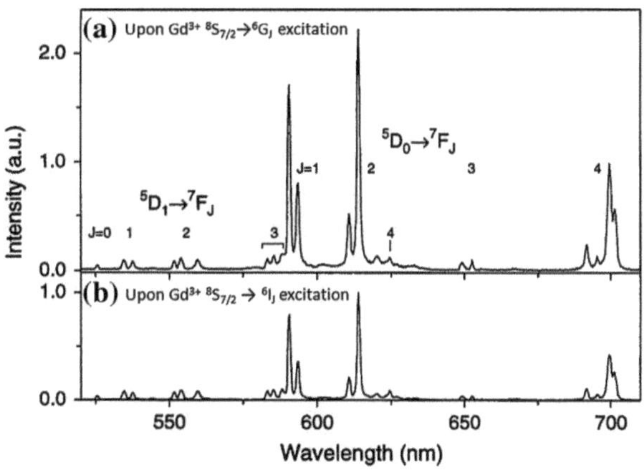

Fig. 16.11 Emission spectra of $LiGdF_4$:0.5%Eu^{3+} on **a** $^8S_{7/2}$–^6G_J excitation on Gd^{3+} (202 nm) and **b** $^8S_{7/2}$–^6I_J excitation on Gd^{3+} (273 nm) measured at 300 K. The spectra are scaled on the $^5D_1$–^7F_J emission intensity. Reprinted from Ref. [21], Copyright 1999, with permission from Elsevier

Table 16.1 Quantum cutting efficiency of some Eu^{3+}-doped fluoride phosphors

Host lattice	Synthesis methods	Eu^{3+} doping concentration (%)	Quantum efficiency (%)	References
$LiGdF_4$	Solid-state reaction	0.5	190	[9]
$NaGdF_4$	Hydrothermal method	0.5	160	[22]
BaF_2	Solid-state reaction	1	194	[23]
KGd_3F_{10}	Solid-state reaction	2	165	[24]
$RbGd_3F_{10}$	Hydrothermal method	0.5	150	[25]
$KLiGdF_5$	Solid-state reaction	2	140	[26]
$CsGd_2F_7$	Solid-state reaction	0.5	150	[27]
GdF_3	Hydrothermal method	0.5	170	[28]

$$\frac{P_{CR}}{P_{CR}+P_{DT}}=\frac{R(^5D_0/^5D_{1,2,3})_{^6G_J}-R(^5D_0/^5D_{1,2,3})_{^6I_J}}{R(^5D_0/^5D_{1,2,3})_{^6I_J}+1}$$

where P_{CR} is the probability for cross-relaxation (step 1 in Fig. 16.10); P_{DT} is the probability for the direct energy from Gd^{3+} to Eu^{3+}(step 2 in Fig. 16.10); $R(^5D_0/^5D_{1,2,3})$ is the ratio of the $^5D_0$ and the $^5D_{1,2,3}$ emission intensities; and the subscript (^6G_J or ^6I_J) indicates the excitation level for which the ratio is obtained. This formula is applicable for all the Gd^{3+}–Eu^{3+} systems.

C. Downconversion in Tb^{3+}–Gd^{3+}-doped fluorides

Quantum cutting based on the third concept is demonstrated in Tb^{3+}-doped K_2GdF_5 and $BaGdF_5$ [29, 30]. Figure 16.12 shows the energy-level diagram and quantum-cutting model in a Gd^{3+}–Tb^{3+} pair system. In this model, on $Gd^{3+6}I_J$ excitation, no quantum cutting occurs, whereas on Tb^{3+} 4*f*5*d* excitation, the quantum cutting can be explained by a two-step process: cross-relaxation and direct energy transfer [29, 30]. On excitation to the Tb^{3+} 5*d* level, one Tb^{3+} ion in is excited from the ground state to its 4*f*5*d* state, and it subsequently relaxes from the 4*f*5*d* state to an intermediate state $^5D_3$. The energy released during the relaxation process is transferred to a neighboring Tb^{3+} ion by cross-relaxation, thus exciting the neighboring Tb^{3+} ion to its $^5D_4$ level resulting in the first green emission due to the $^5D_4 \rightarrow {}^7F_J$ transition (step 1). The transition from $^5D_3$ state of the original Tb^{3+} generates a second photon (step 2). It is also possible that while Tb^{3+} relaxes from

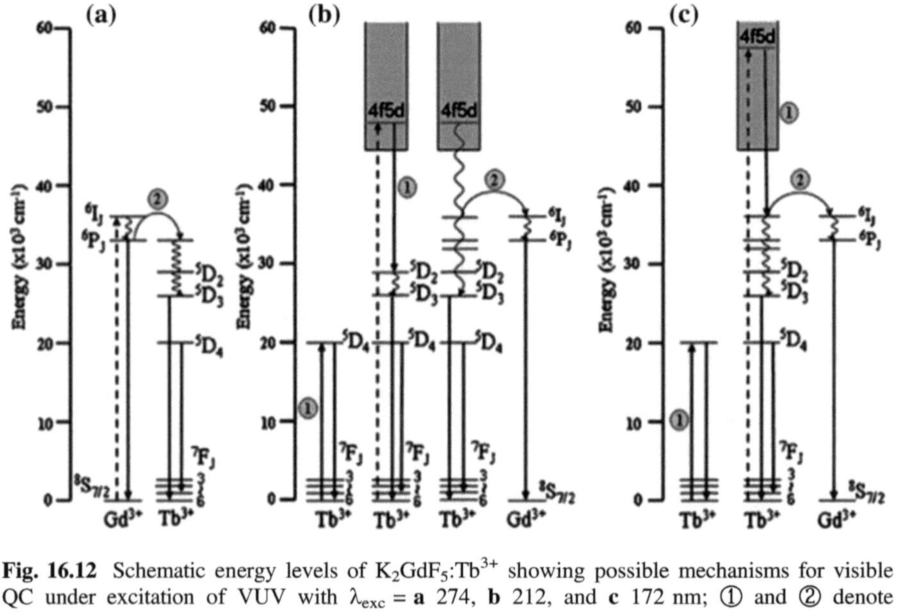

Fig. 16.12 Schematic energy levels of K_2GdF_5:Tb^{3+} showing possible mechanisms for visible QC under excitation of VUV with λ_{exc} = **a** 274, **b** 212, and **c** 172 nm; ① and ② denote cross-relaxation and direct energy transfer, respectively. Reprinted with permission from Ref. [29]. Copyright 2006, American Institute of Physics

the $4f5d$ state to the $^5D_3$ and $^5D_4$ levels, the released energy is transferred directly to a neighboring Gd^{3+} ion, resulting in $Gd^{3+6}P_J \rightarrow {}^8S_{7/2}$ emission. According to the quantum-cutting model shown in Fig. 16.12, in the case quantum cutting occurs, a relative increase in the $Tb^{3+5}D_4$ emission with respect to the $Tb^{3+5}D_3$ emission and $Gd^{3+6}P_J$ emissions is expected. Figure 16.13 shows the emission spectra of K_2GdF_5:5%Tb^{3+} excited at the $Gd^{3+6}I_J$ level (274 nm) and the Tb^{3+} $4f5d$ states (212 and 172 nm) [29]. The spectra are scaled to the $^5D_3 \rightarrow {}^7F_5$ excitation intensity. The relative intensity of the emission from the $^5D_4$ level on excitation at 172 and 224 nm is stronger than that on the 273-nm excitation providing clear evidence that quantum cutting takes place in K_2GdF_5:5%Tb^{3+}. The quantum efficiency in this system is calculated by the following equation [29, 30]:

$$\frac{P_{CR}}{P_{CR} + P_{DT}} = \frac{R(^5D_4/\text{Rest})_{Tb^{3+}} - R(^5D_4/\text{Rest})_{Gd^{3+}}}{R(^5D_4/\text{Rest})_{Tb^{3+}} + 1}$$

where P_{CR} represents the probability for cross-relaxation; and P_{DT} is the probability for direct energy transfer. R($^5D_4$/rest) is the ratio of the emission intensity of $^5D_4$ to that attributed to $^5D_3$ of Tb^{3+} and ^6P_J of Gd^{3+}. The subscript indicates excitation from Tb^{3+} or Gd^{3+}. For K_2GdF_5:11 %Tb^{3+}, the calculated visible QE was found to be 189 and 187 % for VUV excitations at 212 and 172 nm, respectively. In the case of $BaGdF_5$:15%Tb^{3+}, the calculated QE was found to be 168 and 180 % on excitation with 215 and 187 nm, respectively [29, 30].

D. **Downconversion in $Er^{3+}/Gd^{3+}/Tb^{3+}$-doped fluorides**

Quantum cutting based on the fourth concept is demonstrated in a combination of $Er^{3+}/Gd^{3+}/Tb^{3+}$ in $LiGdF_4$ [31]. Figure 16.14 shows the simplified energy-level diagram of the Er^{3+}–Gd^{3+}–Tb^{3+} system. In this model, on VUV excitation in the $4f^{10}5d$ levels of Er^{3+}, part of the excitation energy is transferred to Gd^{3+} by

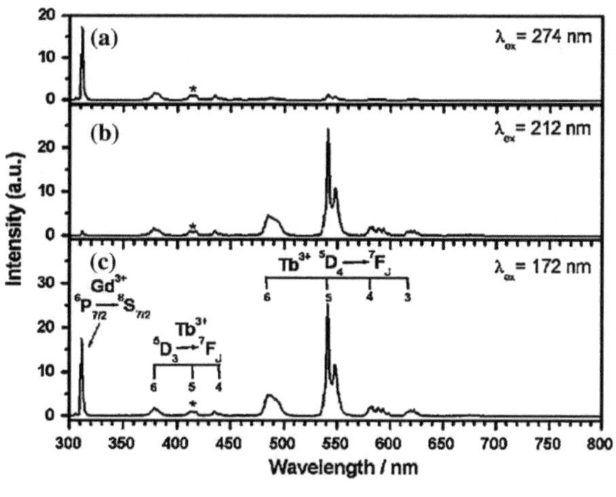

Fig. 16.13 Emission spectra of K_2GdF_5:5%Tb^{3+} excited in the $Gd^{3+6}I_J$ level at 274 nm (**a**) and Tb^{3+} $4f5d$ states at 212 nm (**b**) and 172 nm (**c**). The spectra are scaled to the $^5D_3$–$^7F_5$ excitation intensity. Reprinted with permission from Ref. [29]. Copyright 2006, American Institute of Physics

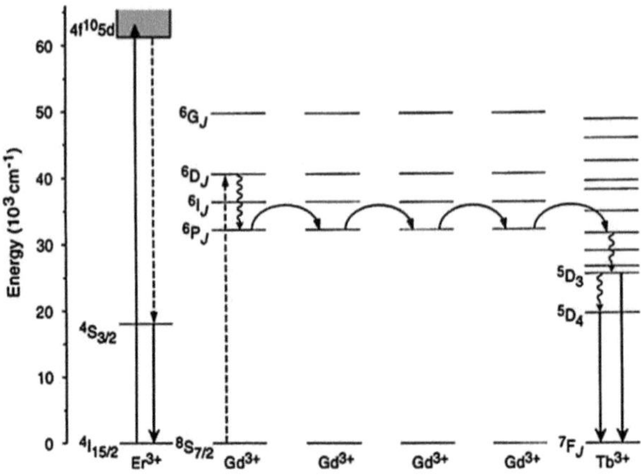

Fig. 16.14 Simplified energy-level diagram of the Er^{3+}–Gd^{3+}–Tb^{3+} system showing the visible quantum cutting process upon excitation in the $4f^{10}5d$ state of Er^{3+}. Reprinted from Ref. [31], Copyright 2000, with permission from Elsevier

cross-relaxation resulting in both Er^{3+} and Gd^{3+} being in an excited state. The excited Gd^{3+} transfers the excited energy among the Gd^{3+} sublattice and finally reaches a Tb^{3+} ion. The excited Er^{3+} and Tb^{3+} returned to the ground state by emitting two photons: One is from Er^{3+} due to the $4S_{3/2} \rightarrow {}^4I_{15/2}$ transition, and the other is from Tb^{3+} due to the ${}^5D_J \rightarrow {}^7F_J$ transition. Figure 16.15 shows the emission spectra of $LiGdF_4$:1.5% Er^{3+},0.3 %Tb^{3+} on $Er^{3+} \rightarrow 4f5d$ excitation (145 nm) and $Gd^{3+8}S_{7/2} \rightarrow {}^6I_J$ excitation (273 nm). The emission spectra are scaled on the $Tb^{3+5}D_3 \rightarrow {}^7F_J$ emission intensity. A significant increase of the $Er^{3+4}S_{3/2}$ emission is observed relative to the other emissions, thus indicating that downconversion occurs in Er^{3+}–Gd^{3+}–Tb^{3+}. The quantum efficiency for $LiGdF_4$:1.5%Er^{3+}, 0.3 %Tb^{3+} has been determined to be 130 %, which is calculated by the following equation [31]:

$$\frac{p_{\mathrm{ET}+}}{p_{\mathrm{ET}+} + p_{\mathrm{ET}-}} = \frac{R({}^4S_{3/2}/\mathrm{rest})_{\mathrm{Er}^{3+}} - R({}^4S_{3/2}/\mathrm{rest})_{\mathrm{Gd}^{3+}}}{R({}^4S_{3/2}/\mathrm{rest})_{\mathrm{Gd}^{3+}} + 1}$$

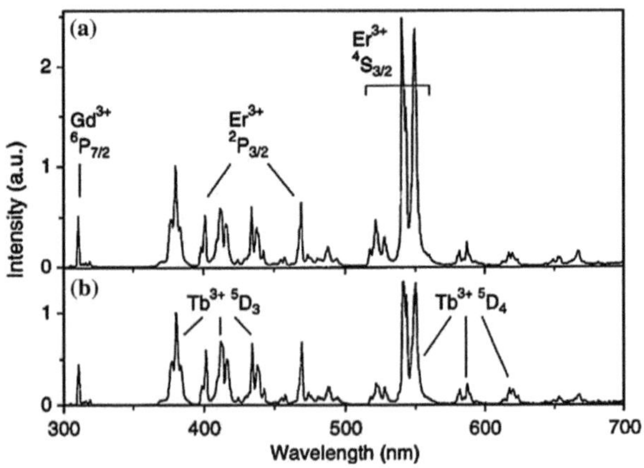

Fig. 16.15 Emission spectra of $LiGdF_4$:1.5%Er^{3+}, 0.3 % Tb^{3+} on Er^{3+} $4f5d$ excitation (145 nm) and $Gd^{3+8}S_{7/2} \rightarrow {}^6I_J$ excitation (273 nm). The emission spectra are scaled on the $Tb^{3+5}D_3 \rightarrow {}^7F_J$ emission intensity. Reprinted from Ref. [31], Copyright 2000, with permission from Elsevier

where p_{ET+} and p_{ET-} are the probabilities of the desired and the undesired energy-transfer possibilities from the $4f^{10}5d$ states of Er^{3+} to Gd^{3+}, respectively. $R(^4S_{3/2}/\mathrm{rest})_{Er^{3+}}$ and $R(^4S_{3/2}/\mathrm{rest})_{Gd^{3+}}$ are the intensity ratios of the $Er^{3+4}S_{3/2}$ emission to all remaining emissions on $Er^{3+}4f$-$5d$ excitation and on $Gd^{3+6}I_J$ excitation, respectively.

16.1.3 Upconversion Phosphors

A. **Principles of upconversion luminescence (UCL)**

Upconversion (UC) is a nonlinear, sequential absorption phenomenon of two or more photons by way of long-lived excited states followed by the emission of light with a shorter wavelength than that of the pump light source. This general concept was first recognized and formulated independently by Auzel, Ovsyankin, and Feofilov in the mid-1960s [32]. Since then, conversion of infrared radiation into the visible region has generated much of the interest in UC research, thus progressively generating and incorporating novel areas of investigation (such as infrared quantum counter detectors, temperature sensors, compact solid-state laser, bio-analysis, medical therapy [33, 34], display technologies, and light harvesting in photovoltaic cells).

The UC processes are mainly divided into three broad classes: excited-state absorption (ESA), energy-transfer upconversion (ETU), and photon avalanche (PA). All of these processes involve the sequential absorption of two or more photos (Fig. 16.16). Thus, UC processes are different from the multi-photon process where the absorption of photons occurs simultaneously.

In the case of ESA, excitation takes the form of successive absorption of pump photons by a single ion. The general energy diagram of the ESA process is shown

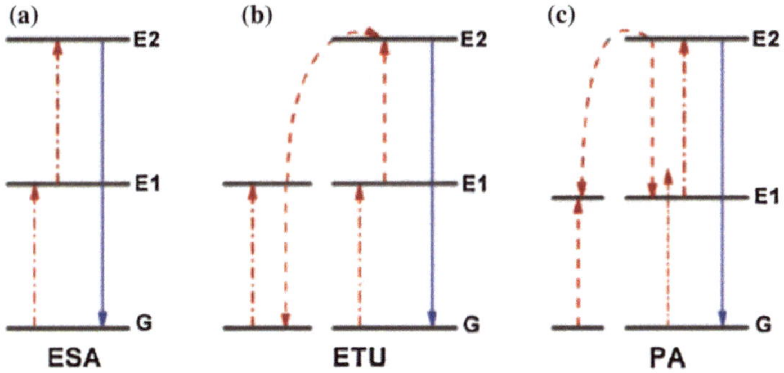

Fig. 16.16 Principal UC processes for lanthanide-doped crystals: **a** excited state absorption, **b** energy-transfer up-conversion, and **c** photo avalanche. The *dashed/dotted*, and *full arrows* represent photon excitation, energy transfer, and the emission process, respectively. Reproduced from Ref. [34] by permission of The Royal Society of Chemistry

in Fig. 16.16a for a simple three-level system. If the excitation energy is resonant with the transition from ground level G to excited metastable level E1, phonon absorption occurs and populates E1 from G in a process known as "ground-state absorption" (GSA). A second pump photo, which promotes the ion from E1 to the higher-lying state E2, results in UC emission corresponding to the E2 → G optical transition. ETU is similar to ESA in that both processes utilize sequential absorption of two photos to populate the metastable level. The essential difference between ETU and ESA is that the excitation in ETU is realized through energy transfer between two neighboring ions, In an ETU process, each of two neighboring ions can absorb a pump phonon of the same energy, thereby populating the metastable level E1 (Fig. 16.16b). A nonradiative energy-transfer process promotes one of the ions to the upper-emitting state E2, whereas the other ion relaxes back to the ground state G. The dopant concentration that determines the average distance between the neighboring dopant ions has a strong influence on the UC efficiency of an ETU process. The phenomenon of PA was first discovered by Chivian et al. in Pr^{3+}-based infrared quantum counters [35]. The PA-induced UC features an unusual pump mechanism that requires a pump intensity above a certain threshold value. The PA process starts with population of the level E1 by nonresonant weak GSA followed by resonant ESA to populate the upper visible-emitting level E2 (Fig. 16.16c). After the metastable-level population is established, cross-relaxation energy transfer (or ion-pair relaxation) occurs between the excited ion and a neighboring ground-state ion, thus resulting in both ions occupying the intermediate level E1. The two ions readily populate level E2 to further initiate cross-relaxation and exponentially increase level E2 population by ESA producing strong UC emission as an avalanche process.

B. **Composition characteristics of UC phosphors**

A UC phosphor consists of a host and dopant (activator). The dopant acts as luminescent centers, and the host provides a matrix to bring these centers into optimal position. A large number of suitable hosts doped with actinide [36, 37] and transition-metal ions—such as Cm^{3+}, U^{3+}, Mo^{3+}, Os^{3+}, Ni^{2+}, Ti^{2+}, and Re^{4+}—have been reported to show upconversion luminescence [38]. However, this occurs mainly in the RE elements due to its special $4f^n5d^{0-1}$ inner shell configurations, which are well-shielded by outer shells and have abundant and unique energy-level structures. Er^{3+}, Tm^{3+}, and Ho^{3+} are currently the most common activators in UC phosphors. Yb^{3+} is usually co-doped in the host as a sensitizer due to the larger absorption cross-section of Yb^{3+} in the NIR spectral region to improve upconversion luminescence efficiency. In the case of a Yb^{3+}, Er^{3+}-co-doped system, the energy interval of the ground $^2F_{7/2}$ and the $^2F_{5/2}$ excited state of Yb^{3+} matches well with the transition energy between the $^4I_{11/2}$ and $^4I_{15/2}$ states as well as the $^4F_{7/2}$ and $^4I_{11/2}$ states of Er^{3+}, thus allowing for efficient resonant energy transfer from Yb^{3+} to Er^{3+} ions. Yb^{3+} is not only a common sensitizer for Er^{3+} systems, it is also common for Tm^{3+}, Ho^{3+}, and Pr^{3+} ions. It is also very critical to choose an appropriate host material, which plays vital roles in the upconversion luminescent process. The choice of host lattice determines the distance between the dopant ions, the coordination numbers, the relative spatial position, and the type of anions around the dopant, thus resulting in different optical properties of the UC phosphors [39, 40]. Desirable host materials should have adequate transparency within a certain wavelength range, low phonon energy, and high optical-damage threshold. Until now, various host materials, including fluoride, oxide, chloride, bromide, iodide, oxysulfide, phosphate, and vanadate, etc., have been studied. Hexagonal-phase $NaYF_4$ crystal in bulk state has been reported to be the most efficient matrix for upconversion phosphors [41, 42]. Other kinds of fluorides and alkali fluorides, such as $NaGdF_4$, $NaLaF_4$, LaF_3, GdF_3, CeO_2, $LiNaF_4$, $Ca_3(PO_4)_2$, ZrO_2, and GdOF, etc., have also been considered as excellent host materials in recent years due to their high refractive index and high transparency arising from low-energy phonons. To enhance the UC luminescence efficiency, a sensitizer with a sufficient absorption cross-section in the NIR region is usually co-doped along with the activator to take advantage of the efficient ETU process between the sensitizer and activator. Trivalent Yb possesses an extremely simple energy-level scheme with only one excited 4*f* level of $^2F_{5/2}$ (Fig. 16.17). The absorption band of Yb^{3+}, which is located at approximately 980 nm due to the $^2F_{7/2} \rightarrow {}^2F_{5/2}$ transition, has a larger absorption cross-section than that of other lanthanide ions. Additionally, the $^2F_{7/2} \rightarrow {}^2F_{5/2}$ transition of Yb^{3+} is well resonant with many f–f transitions of typical upconverting lanthanide ions (Er^{3+}, Tm^{3+}, and Ho^{3+}), thus facilitating efficient energy transfer from Yb^{3+} to other ions. These optical characteristics make Yb^{3+} particularly suitable for use as a UC sensitizer. The sensitizer content is normally kept high (approximately 20 mol%) in doubly or triply doped

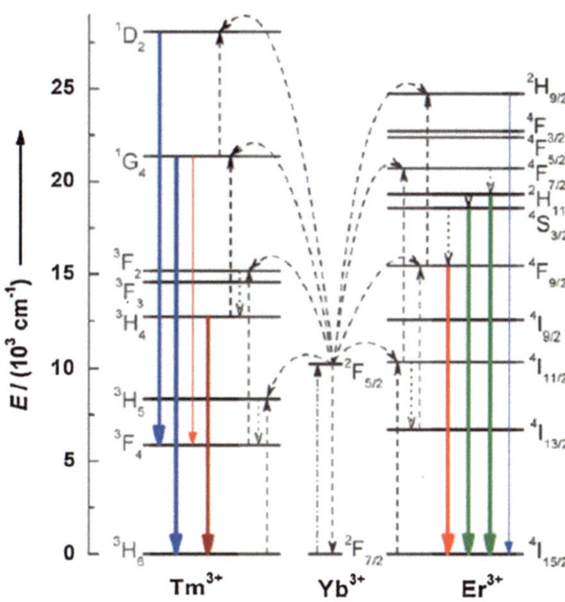

Fig. 16.17 Proposed energy-transfer mechanisms showing the UC processes in Er^{3+}-, Tm^{3+}-, and Yb^{3+}-doped crystals under 980-nm diode laser excitation. The *dashed-dotted*, *dashed*, *dotted*, and *full arrows* represent the photoexcitation, energy transfer, multiphonon relaxation, and emission processes, respectively. Reproduced from Ref. [34] by permission of The Royal Society of Chemistry

nanocrystals, whereas the activator content is relatively low (<2 mol%), thus minimizing cross-relaxation energy loss.

C. Application of upconversion phosphors

(a) Photovoltaic Applications

Sustainable energy production based on the capture and conversion of the solar radiation into useable forms, such as heat or electricity, and storage of the acquired energy, is expected to gain importance because it may be the only renewable source capable of generating sufficient energy to meet the long-term worldwide energy demand. PV (photovoltaic) cells are such attractive devices that can capture sunlight and convert it into electricity in a clean and sustainable way. A major problem limiting the conversion efficiency of PV cells is their insensitivity to a full solar spectrum. The spectral distribution of sunlight at air mass 1.5 global (AM 1.5G) consists of photons with wide wavelengths ranging from ultraviolet to infrared (280–2500 nm), but current PV cells only utilize a relatively small fraction of solar photons as shown in Fig. 16.18 [43]. This is attributed to the fact that each PV material responds to a narrow range of solar photons with energy matching the characteristic band gap of the materials. In principle, only the absorption of photons with energy higher than the band gap can generate electron-hole pairs contributing to electric current. Indeed, the transmission of sub-band-gap photons is one of the major energy-loss mechanisms in conventional solar cells. To this end, the use of upconversion materials may provide a solution to the transmission loss by converting two sub-band-gap photons into one above-band gap photon. A response of the cell at an energy of 1.391 eV under excitation of 1 W 0.039 cm^{-2} has been observed on a GaAs cell (the band gap is 1.43 eV) by applying a vitroceramic co-doped with Er^{3+} and Yb^{3+}. Theoretical analysis of energy conversion using the

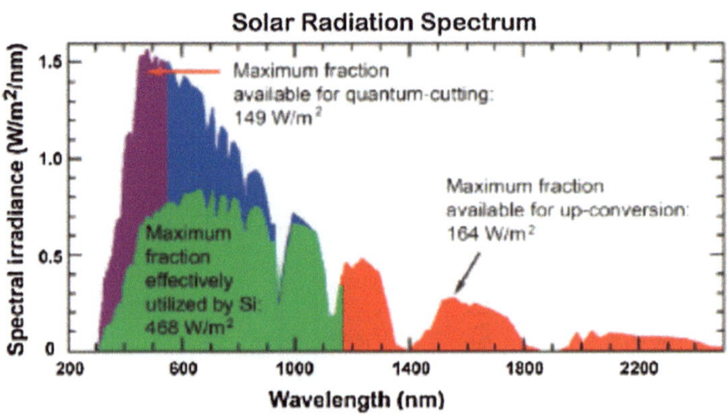

Fig. 16.18 AM 1.5G spectrum showing the fraction (highlighted in *green*) absorbed by a typical silicon-based PV cell and the spectral regions that can be utilized through quantum-cutting and upconversion processes (highlighted in *purple* and *red*, respectively). Reproduced from Ref. [43] by permission of The Royal Society of Chemistry

UC of sub-band gap photons for a conventional single-junction bifacial solar cell was presented by Trupke et al. in 2002 [44]. The upconverter, consisting of Er^{3+}-doped $NaYF_4$, was located on the rear side of a bifacial cell, thus leading to a response of the cell when it was excited at 1500 nm. From then on, significant improvement, including both theoretical analysis and experimental achievements, has been reported in the field of solar UC. According to the literatures, UC is predicted to enhance the efficiency of solar cells when mounted on the rear of the solar cell [45, 46].

Strümpel et al. summarized some of the key requirements for UC materials that must be met for the effective application of UC in solar cells [47]. First, the upconverter should be excited effectively at low excitation intensity by photons with wavelength longer than that corresponding to the band gap of the solar cell, i.e., 1100 nm for c–Si systems, and the upconverted emission should be shorter in wavelength than the band gap; also the UC materials should have highly transmittivity for unconverted light. Er^{3+}-doped upconversion materials are the most promising upconverters for c-Si solar cells due to the GSA of Er^{3+} in the range of 1480–1580 nm ($^4I_{15/2} \rightarrow {}^4I_{13/2}$ transition). The GSA centered at approximately 1540 nm results in upconversion by way of an ETU mechanism giving rise to four emission bands: $^4I_{11/2} \rightarrow {}^4I_{15/2} = 980$ nm, $^4I_{9/2} \rightarrow {}^4I_{15/2} = 810$ nm, $^4F_{9/2} \rightarrow {}^4I_{15/2} = 660$ nm, and $^4S_{3/2} \rightarrow {}^4I_{15/2} = 550$ nm (Fig. 16.19). These emission bands match well with the absorption of c-Si. Note that the exact emission wavelength can vary by ± 10 nm depending on the host material. Shalav et al. in 2005 reported the application of $NaYF_4$:20 %Er^{3+} phosphors as the upconverters in a bifacial c-Si solar cell [48]. These phosphors were mixed into an optically transparent acrylic adhesive medium at a concentration of 40 wt% and then placed on the rear of a bifacial c-Si solar cell. Reflective white paint was used as a reflector on the rear of the system. An external quantum efficiency of 2.5 % was obtained for the solar cell under excitation at 1523 nm with a 5.1-mW laser. More recently, Fischer et al. also investigated the

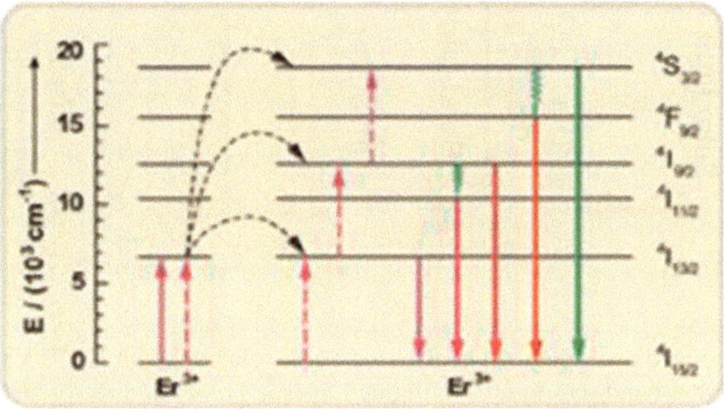

Fig. 16.19 Upconversion processes by way of the ETU mechanism between two Er^{3+} ions, resulting in c–Si above band gap emissions under c–Si sun-band gap pump excitation (1520 nm). *Solid*, *dotted*, and *wavy arrows* represent photon absorption/emission, energy transfer, and multiphonon relaxation, respectively. Reproduced from Ref. [43] by permission of The Royal Society of Chemistry

potential of $NaYF_4:Er^{3+}$ to significantly decrease the sub-band gap losses of c-Si solar cells [49]. The optical upconversion efficiency of $NaYF_4$:20 %Er^{3+} was determined to be approximately 5.1 % on excitation at 1523 nm with a power density of 1880 W/m^2. The c-Si solar cell device combined with an upconverter showed an external quantum efficiency of 0.34 % on irradiation at 1522 nm with a power density of 1090 W/m^2.

The Ho^{3+} ion has a relatively wide absorption band in the 1150–1225-nm spectral range due to $^5I_8 \rightarrow {}^5I_6$ transition. The irradiation power density of sunlight in this spectral range is approximately 40 W/m^2, which is approximately twice more intense than that in the 1480–1580-nm range. Lahoz in 2008 reported the use of Ho^{3+} singly doped oxyfluoride glass ceramics as promising upconverters for efficiency enhancement in c-Si solar cells [50]. Under sub-band gap excitation at 1170 nm, upconversion emissions in the visible (approximately 650 nm, $^5F_5 \rightarrow {}^5I_8$ transition) and NIR (approximately 910 nm, $^5I_5 \rightarrow {}^5I_8$ transition) spectral ranges have been obtained by way of the ETU mechanism. In a following paper, Lahoz et al. [51] co-doped Ho^{3+} with Yb^{3+} and found enhanced emission intensity of NIR-to-NIR upconversion as a result of the ET from Ho^{3+} to Yb^{3+} ions. Importantly, owing to the high transparency to the excitation wavelength of approximately 1540 nm, these Ho^{3+}-based glass ceramics could be used in combination with Er^{3+}-doped upconverters. The authors proposed a double-layer design with the Ho^{3+}-doped upconverting layer placed directly at the rear surface of a bifacial c-Si solar cell followed by the Er^{3+}-doped upconverter and a mirror (Fig. 16.20) [43]. The integration of both upconverters would enlarge the absorption of the solar sub-band-gap spectral range exploited for enhancing the c-Si solar cell response. Significant progress, including theoretical analysis and experimental

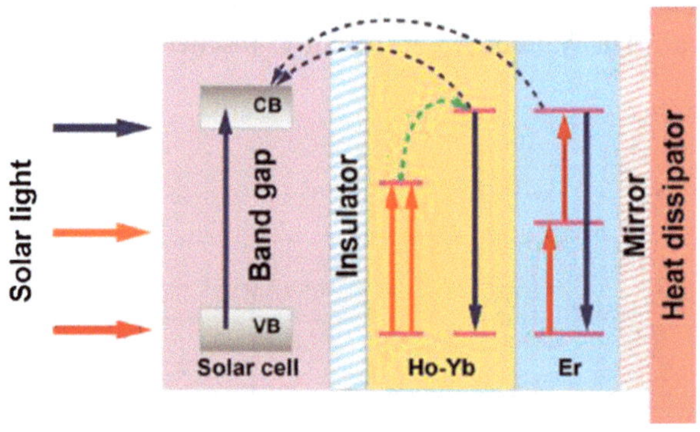

Fig. 16.20 A proposed operating mechanism for a c-Si solar cell with Ho^{3+}–Yb^{3+} and Er^{3+}-doped up-converters. Above band-gap light (highlighted in *blue color*) is directly absorbed by the solar cell, which is electronically isolated from the upconverter. The sub-band-gap light (highlighted in *orange* and *red colors*) transmitted by the solar cell is stepwise upconverted into high-energy photons, which are subsequently absorbed in the solar cell. A mirror reflector is located behind the upconverter. CB and VB are the conduction and valence bands of c–Si, respectively. The energy gap between VB and CB is approximately 1.12 eV for c–Si. Reproduced from Ref. [43] by permission of The Royal Society of Chemistry

achievement, has occurred in the field of UC for PV applications. However, the performance of the currently investigated UC materials is not efficient enough to improve the practical efficiency of solar cells under normal solar irradiation. More efforts are required to design and develop higher- efficiency UC materials. In contrast, more sophisticated engineering concepts with respect to the architecture of the solar cells must be developed to enable the exploitation of UC. Based on the recent improvements in UC efficiency of RE-doped phosphors, a promising future for this concept can be foreseen.

(b) **Bioimaging**

The development of upconversion nanophosphors (UCNPs) during the past decade has facilitated the translation of fluorescence imaging from the microscopic to macroscopic imaging. Although the uses of other conventional fluorescent materials —including organic dyes, fluorescent proteins, metal complexes, or semiconductor quantum dot as based biomarker—have achieved significant progress in real-time detection and bioimaging, they still have some drawbacks. These fluorescent materials are generally excited by ultraviolet (UV) or visible light, which may induce autofluorescence from the biological tissues, DNA damage, and cell death to biological samples, thus resulting in low signal-to-noise ratio and limited sensitivity. In addition, the broad emission spectra of these fluorescent materials make them unsuitable for multiplex bio-labeling and often suffer from low photostability when exposed to external illumination. Quantum dots (QDs) that feature a large molar extinction coefficient, high quantum yield, narrow emission bandwidth, large Stokes shift, size-dependent tunable emission, and high photostability, which makes them attractive as alternative luminescent labels for imaging. However, the potential toxicity of QDs limit their biological applications.

In contrast, UCNPs have many good characteristics. Firstly, compared with conventional luminescent materials, UCNPs have relatively longer lifetime. This feature makes time-resolved luminescence detection feasible to minimize the interference of the undesired short-lived background fluorescence, which originates from biological tissues or other dopants under multiple-photon excitation. This property also greatly improved the signal-to-noise ratio and the detection sensitivity, which makes UCNPs more feasible for bioimaging applications. Second, lanthanide-doped UCNPs show a distinct set of sharp emission bands with 10–20-nm full width-at-half maximum value under a certain wavelength of 980-nm light excitation. This characteristic provides distinguishable spectroscopic fingerprints for accurate interpretation of the emission spectra in the environment of overlapping emission spectra. The emission colors are decided by the kind of co-doped lanthanide ions. The relative emission intensities are controlled by changing the kinds of the host, the dopant concentration, or the morphology of the nanoparticles. Third, the near-infrared light excitation provides deep penetration and low tissue damage for in vivo imaging. Recently increasingly more NIR-to-visible and NIR-to-NIR UCNPs have been developed because the wavelength of 650–1000 nm light provides higher biological-tissues penetration depth.

Finally, in contrast to the conventional anti-Stokes probes, such as two-phonon absorption or second harmonic generation, upconversion nanophosphors exhibit higher emission efficiency and can be excited by continuous-wave laser rather than the costly femtosecond pulsed laser.

All of these benefits mean that UCNPs can be used as probes for intra-vital imaging, which enables the investigation of physiological processes within the context of a living organism and provides a more complete picture of disease pathology and development. On excitation at 980 nm, $NaYF_4$:Yb, Er nanoparticles exhibit yellow to red emission by increasing the concentration of Yb^{3+} (25–60 %) (Fig. 16.21). As a new imaging technique, UCNPs imaging offers a unique

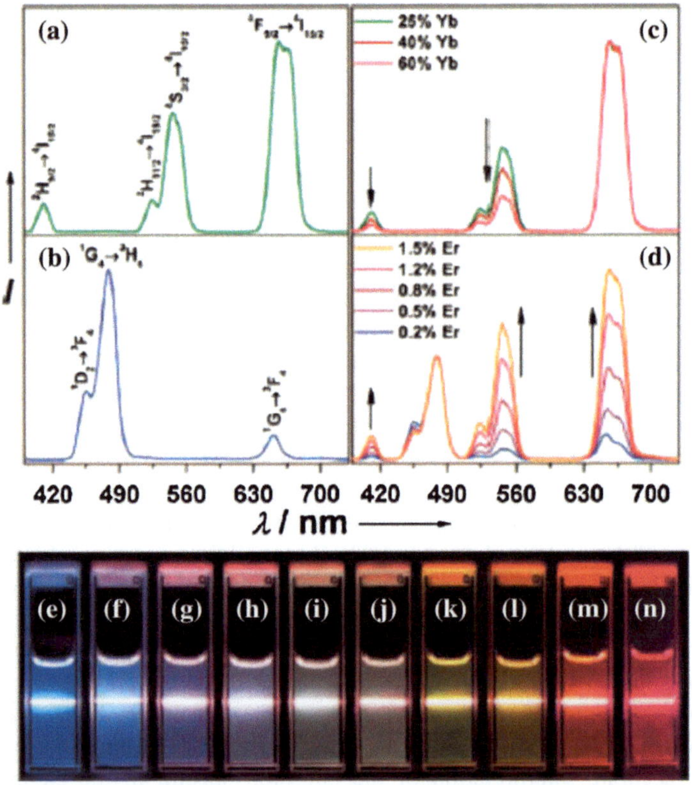

Fig. 16.21 Room temperature upconversion emission spectra of **a** $NaYF_4$:Yb/Er (18/2 mol%), **b** $NaYF_4$:Yb/Tm (20/0.2 mol%), **c** $NaYF_4$:Yb/Er (25–60/2 mol%), and **d** $NaYF_4$:Yb/Tm/Er (20/0.2/0.2–1.5 mol%) particles in ethanol solutions (10 mM). The spectra in (**c**) and (**d**) were normalized to Er^{3+} 650-nm and Tm^{3+} 480-nm emissions, respectively. Compiled luminescent photos showing corresponding colloidal solutions of **e** $NaYF_4$:Yb/Tm (20/0.2 mol%), **f–j** $NaYF_4$: Yb/Tm/Er (20/0.2/0.2–1.5 mol%), and **k–n** $NaYF_4$:Yb/Er (18–60/2 mol%). The samples were excited at 980 nm with a 600 mW diode laser. The photographs were taken with exposure times of 3.2 s for **e–l** and 10 s for *m* and *n* Reprinted from Ref. [52], with permission from Springer Science + Business Media

approach for visualizing morphological details in tissue with sub-cellular resolution and has become a powerful noninvasive tool for visualizing the full range of bio-species from living cells to animals. For example, most recently Li's group developed a cynine-modified nanocomposite (hCy7-UCNPs) that is capable of monitoring MeHg+ ex vivo and in vivo by UCL bioimaging with a detection limit of as low as 0.18 ppb (Fig. 16.22) [53].

(c) **Photodynamic therapy**

Photodynamic therapy (PDT) is a new method that uses light to excite a photosensitizer to produce photochemical effects that will kill tumor cells and pathogenic microorganisms. The method offers the advantages of simple operation, minimal damage (noninvasive), decreased side effects, and repeatability. It has been successfully applied for the treatment of tumors and other benign diseases (such as skin and eye diseases) and is regarded as one of the most promising modern medical technologies after chemotherapy, radiotherapy, and operative therapy. The photodynamic effect requires a photosensitizer, excitation light, and oxygen. After the photosensitizer is excited with light of a certain wavelength, its molecules absorb photon energy such that it reaches the excited state (tri-state), which has a longer lifespan than the ground state (singlet state). Such molecules in the excited state work on a substrate to produce free ions through electron transfer and then react with surrounding oxygen to produce cytotoxic substances (such as peroxide, hydroxyl radicals, and lipid-derived radicals among others). They may convert molecular oxygen in the ground state to produce singlet oxygen through energy transfer, which then reacts with various biological molecules (such as proteins,

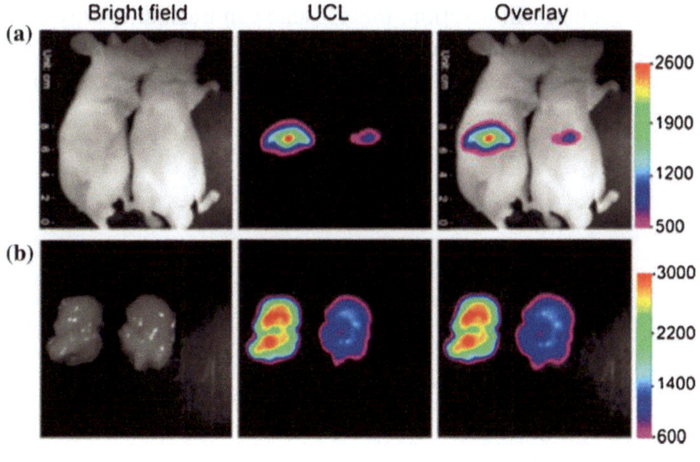

Fig. 16.22 **a** In vivo UCL images of 40 μg hCy7-UCNPs-pretreated living mice injected intravenously with 0.2 mL normal saline (*left mouse*) or 0.1 mM MeHg + solution (*right mouse*). **b** The corresponding UCL images of the livers which were isolated from the above dissected mice. The UCL emission was collected at 800 ± 12 nm on irradiation at 980 nm. Reprinted with the permission from Ref. [53]. Copyright 2013 American Chemical Society

lipids, and nucleic acid) to damage the cell structure or affect cell functions, thereby improving the therapeutic effect (Fig. 16.23) [55].

To improve the effect of photodynamic therapy, researchers have not only optimized the water solubility of the photosensitizer to make it more easily dispersed in the aqueous environment of the body, they have also attempted to use different carrier systems to improve the tissue-transport depth of photosensitizers and to prevent the photosensitizer from aggregating easily. However, these measures have failed to lessen or overcome the disadvantage of PDT because as its application is still limited to shallow lesions or small tumors. This limitation is mainly caused by the excitation wavelength of photosensitizers, which ranges from 600 to 900 nm. Because light in this wave band is strongly absorbed by biological tissues, the excitation wavelength cannot penetrate deeper tissues and organs, thus decreasing the therapeutic effectiveness of PDT in vivo.

To overcome this hurdle, researchers have made an attempt to exploit the NIR excitation feature of UCPs and modified UCPs by different methods in the past decades. In 2007, Zhanget et al. was the first to dope the photosensitizer erocyanine 540 into the silicon shell of Y_2O_3:Yb^{3+}, Er^{3+} particles [54]. On excitation with light at 980 nm, the emitted upconverted light at 550 nm was able to sensitize the photosensitizer to produce singlet oxygen. The photodynamic effect of the singlet oxygen produced was also demonstrated to be capable of killing cancer cells in vitro. When exposed to light of a specific wavelength, photosensitizer (PS) in its singlet ground state (S0) becomes activated to an excited singlet state (S1) followed by intersystem crossing to an excited triplet state (T1). The excited triplet state can also transfer a hydrogen atom or an electron to biomolecules (lipids, proteins, nucleic acids, etc.) or free radicals (R), which interact with O_2 to form ROS such as the superoxide anion (O^{2-}), the hydroxyl radical (OH), and hydrogen peroxide (H_2O_2) (type I reaction). In addition, the T1 state can transfer energy directly to the ground-state molecular oxygen in its natural triplet state ($^3O_2$), which then yields the nonradical but highly reactive singlet oxygen, $^1O_2$ (type II reaction) [55].

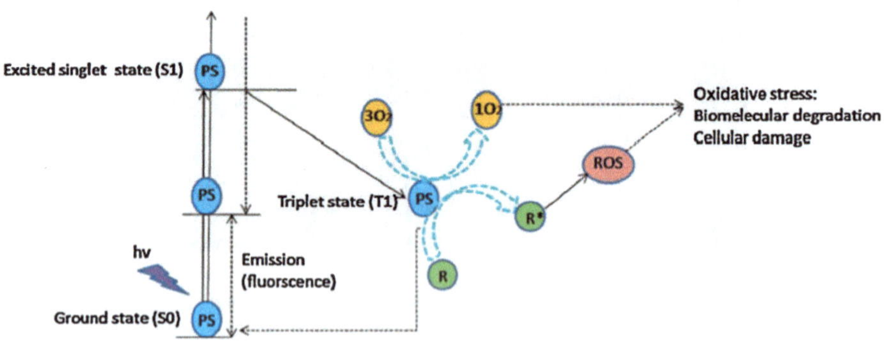

Fig. 16.23 Simplified scheme of photophysical processes and photosensitized reactions. Reproduced from Ref. [55] by permission of The Royal Society of Chemistry

16.1.4 Summary

In addition to the industrial motivation, the creation of phosphors with a quantum efficiency (QE) > 1 also forms an interesting scientific challenge. Depending on the excitation energy, phosphors with QE higher than unity have been realized by either a downconversion or upconversion process. Owing to their low phonon-energy character, fluorides are considered as the most appropriate hosts for downconversion and upconversion phosphors. The crystal structure of some selected fluorides were presented. Meanwhile, the downconverion/upconversion concepts realized by a single luminescent center or a combination of luminescent centers have been presented. They were all described by a number of consecutive steps, the most important ones of which are as follows: absorption of radiation, nonradiative energy transfer, and emission of radiation. Although it is still a long way to go for their commercialization, substantial progress has been made on downconverson/upconversion phosphors from their fundamental physics to optical spectroscopy and diverse proof-of-concept applications.

16.2 Nitridosilicate Phosphors

16.2.1 Introduction

Recently, nitride phosphors, typically nitridosilicates (M–Si–N, M = alkaline-earth metals and lanthanide metals), nitridoaluminosilicates (M–Si–Al–N), oxynitridosilicates (M–Si–O–N), and oxynitridoaluminosilicates (M–Si–Al–O–N) have become very attractive for white LEDs [56, 57]. Unlike the structural situation of oxygen in classical (oxo)silicates, nitrogen exhibits more varied crosslinking patterns in nitridosilicates. Although oxygen can only be terminally bound to Si denoted as $O^{[1]}$ or simply bridges $O^{[2]}$, nitrogen occurs as $N^{[1]}$, $N^{[2]}$, $N^{[3]}$, and even ammonium-type $N^{[4]}$, connecting up to four neighboring Si tetrahedral centers. Furthermore, in nitridosilicates, [SiN_4] tetrahedra can share both common corners and common edges, whereas oxosilicates nearly exclusively contain corner-sharing [SiO_4] tetrahedra [58]. Another important consequence of O/N substitution in silicates is the significantly extended range for the degree of condensation κ, i.e., the molar ratio Si:(O, N). The lowest value $\kappa = 1/4$ corresponds to noncondensed tetrahedral anions $[SiO_4]^{4-}$ and $[SiN_4]^{8-}$, whereas the maximum value is reached in case of the binary species SiO_2 ($\kappa = 1/2$) and Si_3N_4 ($\kappa = 3/4$), respectively. In silicate chemistry, no higher κ values seem possible because they would lead to cationic silicate substructures that have never been observed [58–63]. Substitution of O by N in silicates can be accompanied additionally by substitution of Si by Al, thus leading to so-called SiAlONs (oxonitridoalumosilicates), which by analogy to silicates are typically made up of [TX_4] tetrahedra (T = Al, Si, and X = O, N). From a systematic point of view, SiAlONs represent the superordinate class of compounds, and oxosilicates, nitridosilicates, oxonitridosilicates, and nitridoalumosilicates are subgroups thereof. Consequently, the manifold group of oxosilicates could be interpreted to form a subset of the even more varied class of SiAlONs.

Compared with oxosilicates, structures of nitridosilicates, oxonitridosilicates, and related SiAlONs can be even more complex and varied. Therefore, in this contribution the degree of condensation and the dimensionality of the (oxo)nitridosilicate substructures are employed as means to classify these compounds. For rare-earth ions (i.e., Eu^{2+} and Ce^{3+}) with 5*d* electrons unshielded from the crystal field by 5s and 5p electrons and in the excited state, the spectral properties (e.g., symmetry, covalence, coordination, bond length, site size, crystal-field strength, etc.) are strongly affected by the surrounding environment. Because of the higher formal charge of N^{3-} compared with O^{2-} and the nephelauxetic effect (covalence), the crystal-field splitting of 5*d* levels of rare-earth elements is larger, and the center of gravity of the 5*d* states is shifted to lower energies (i.e., longer wavelength) than in an analogous oxygen environment. Consequently, silicon-based oxynitride and nitride phosphors are anticipated to show longer excitation and emission wavelengths than their oxide counterparts. Furthermore, the Stokes shift becomes smaller in a rigid lattice with a more extended network of SiN_4 tetrahedra. A small Stokes shift leads to high conversion efficiency and small thermal quenching of phosphors [64]. With these promising photoluminescent properties, nitride phosphors are very suitable for use as downconversion luminescent materials in white LEDs, thus forming a fairly new and important family of phosphor materials for lighting. A number of nitride phosphors, e.g., Eu^{2+}-activated α-sialon, β-sialon, and $CaAlSiN_3$, have been commercialized and play key roles in producing highly efficient and reliable white-LED products.

16.2.2 Synthetic Approaches

The use of different starting powders leads to a variety of synthetic approaches or routines for nitrides. Table 16.2 summarizes the synthetic methods for silicon-based oxynitride and nitride phosphors. Among these methods, the solid-state reaction is now accepted by industries to synthesize commercial phosphor powders. All the methods have their own advantages and disadvantages in terms of powder natures (impurity, particle size, particle shape, and particle-size distribution), production cost, and production scale. Cost-effective and simple synthetic methods for the mass production of high-efficiency nitride phosphors are therefore continuously pursued by phosphor researchers and engineers. Until now, the major methods reported for the synthesis of nitridosilicates are high-temperature solid-state reaction, gas reduction and nitridation, carbothermal reduction and nitridation, and direct nitridation.

Table 16.2 Synthesis methods of nitrides and oxynitride phosphors [65]

Synthesis method	Phosphors	Raw materials	Temperature (K)	Pressure (MPa, N_2–H_2)	References
Solid-State reaction	$CaAlSiN_3:Ce^{3+}$	Ca_3N_2, α-Si_3N_4, AlN, Li_3N, CeN	1973	0.50	[66]
	$Ba_2AlSi_5N_9$: Eu^{2+}	Ba, Eu, Si_3N_4, AlN	1873	Ambient pressure	[67]
Direct nitridation	Ca-α-SiAlON: Ce^{3+}, Sm^{3+}, Dy^{3+}	α-Si_3N_4, AlN, $CaCO_3$, rare-earth oxides	2023	1.02	[68]
Gas reduction and nitridation	$LaSi_3N_5:Ce^{3+}$	La_2O_3, CeO_2, SiO_2	1723	Ambient pressure (NH_3–CH_4)	[69]
	Ca-α-SiAlON: Eu^{2+}	Si_3N_4, AlN, SiO_2, CaO	2023		[70]
Spray pyrolysis	$SrSi_2O_{2-\delta}N_{2+2/3\delta}$: Eu^{2+}	Si_3N_4, $Sr(NO_3)_2$, Eu_2O_3	1573	Ambient pressure	[71]
Calcium cyanamide	$CaSrSi_5N_8:Eu^{2+}$	$SrCO_3$, Si_3N_4, Eu_2O_3, $CaCN_2$	1823		[72]
Self-propagating high-temperature synthesis	$CaAlSiN_3:Eu^{2+}$	Si, Ca, Al, Eu	1823	Ambient pressure	[73]
Ammonothermal synthesis	$M_2Si_5N_8:Eu^{2+}$	Nanocrystalline silicon, $M(NH_2)_2$ M = Sr, Ba, Eu, EuN	1673	30	[74]
Carbothermal reduction and nitridation	$Sr_2Si_5N_8:Eu^{2+}$	$SrCO_3$, Si_3N_4, Eu_2O_3, C	1773	Ambient pressure	[75]
Alloy method + nitride mixture + hot isostatic pressing	$Sr_xCa_{1-x}AlSiN_3$: Eu^{2+}	Sr, Ca, Al, Si, Eu, Ca_3N_2, α-Si_3N_4, AlN	2173	190	[76]

16.2.2.1 High-Temperature Solid-State Reactions

The solid-state reaction is a common and simple method of synthesizing oxynitride and nitride phosphor powders. It usually involves reactions at high temperatures among powder precursors containing the corresponding chemical constituents. Frequently, these classical reactions have been performed at increased temperatures in the range of 1550–1750 °C; however, silicon nitride (Si_3N_4) and moreover rare-earth nitrides exhibit only low chemical reactivity [77] owing to low inter-diffusion coefficients in solid-state reactions. Therefore, synthetic approaches leading to ternary and multinary (oxo)nitridosilicates had to be developed, thus abandoning the use of binary nitrides and treating the pure metals with silicon diimide ($Si(NH)_2$) instead. This approach proved to be successful for the synthesis of a variety of nitridosilicates [78]. Therefore, for nitridosilicates, they are usually synthesized through chemical reactions (1) between silicon nitride and metal nitrides or (2) between silicon diimide and metals [57]. Using all-nitride starting powders, the red-emitting $CaAlSiN_3{:}Eu^{2+}$ was prepared by Uehda et al. [79]. The powders were mixed in a nitrogen-filled glove box with a concentration of oxygen and moisture <1 ppm. The powder mixtures were then launched into a gas-pressure sintering furnace and fired at 1600 °C under 1.0 MPa N_2. The chemical reaction can be expressed as

$$(1-x)Ca_3N_2 + Si_3N_4 + xEuN + 3AlN \rightarrow 3(Ca_{1-x}Eu_x)AlSiN_3 \quad (16.2.1)$$

Li et al. [80] synthesized the $(Sr, Ca)_2Si_5N_8{:}5\%Eu^{2+}$ red phosphors by firing the powder mixtures of alkaline earth-metal nitrides, europium nitride, and silicon nitride in an alumina horizontal tube furnace at 1300–1400 °C for 12–16 h under a flowing $90\%N_2$–$10\%H_2$ atmosphere. The reaction is shown as

$$\begin{aligned}&(1.90-1.90x)Sr_3N_2 + 1.90xCa_3N_2 + 3yEuN + 5Si_3N_4\\&\quad\rightarrow 3(Sr_{1-x}Ca_x)1.90Eu_{0.1}Si_5N_8\end{aligned} \quad (16.2.2)$$

As seen above, silicon nitride is commonly used as the starting powder for the preparation of nitridosilicates, but the reaction temperature is usually high (1500 °C to approximately 2000 °C) due to its chemical inertness (e.g., low diffusion coefficient). Moreover, to suppress the decomposition of silicon nitride at high firing temperatures (>1820 °C), high nitrogen gas pressure (>0.1 Mpa) is generally required. Alternatively, more reactive silicon diimide ($Si(NH)_2$) was used as a silicon source to prepare nitridosilicates.

Schnick et al. [77] developed a novel synthetic route to ternary and multinary nitridosilicates by high-temperature reactions of silicon diimide with pure metals instead of binary metal nitrides. This method has been proved to succeed in synthesizing many nitridosilicates such as $M_2Si_5N_8$ (M = Ca, Sr, Ba) and $La_3Si_6N_{11}$. The nitridosilicates were prepared by firing the powder mixture of silicon diimide and alkaline-earth/lanthanide metals in a radiofrequency furnace (type IG 10/200 Hy, frequency = 200 kHz, electrical output = 0 ± 12 kW; Huttinger, Freiburg) at 1500–1650 °C. These reactions are shown below:

$$2M + 5Si(NH)_2 \rightarrow M_2Si_5N_8 + N_2 + 5H_2 \quad (16.2.3)$$

$$3La + 6Si(NH)_2 \rightarrow La_3Si_6N_{11} + 1/2N_2 + 6H_2 \quad (16.2.4)$$

The chemical reaction between a metal and $Si(NH)_2$ is possibly interpreted as a dissolution of an electropositive metal in the nitride analogus, polymeric acid Si $(NH)_2$ accompanied by the evolution of hydrogen. The methods for nitridosilicates can be extended to synthesize oxonitridosilicates or oxoaluminonitridosilicates (M–Si–Al–O–N). For these oxygen-containing materials, metal oxides or metal carbonates, together with metal nitrides, are used as starting powders. Xie and et al. applied solid-state reaction to prepare a variety of oxynitride (oxonitridosilicate and oxoaluminonitridosilicate) phosphors including α-sialon:Eu^{2+}, β-sialon:Eu^{2+}, JEM:Ce^{3+}, La–Si–O–N:Ce^{3+}, $Sr_2Si_5N_8$:Eu^{2+}, and $CaAlSiN_3$:Eu^{2+} [57, 64].

Ca-α-sialon, with the composition of $Ca_{m/2}Si_{12\text{-}m\text{-}n}Al_{m+n}O_nN_{16\text{-}n}$, can be synthesized by firing the powder mixture of Si_3N_4, Al_2O_3, $CaCO_3$, AlN, and Eu_2O_3 in a gas-pressure sintering furnace at 1600–1800 °C under 0.5–1.0 MPa N_2. In addition, Eu_2O_3 is reduced to EuO in a nitriding atmosphere. The chemical reaction is described as

$$(12 - m - n)/3Si_3N_4 + m/2CaCO_3 + (2n - m)/6Al_2O_3 + (4m + n)/3AlN \rightarrow Ca_{m/2}Si_{12-m-n}Al_{m+n}O_nN_{16-n} + m/2CO_2 \quad (16.2.5)$$

and

$$6Eu^{3+} + 2N^{3-} \rightarrow 6Eu^{2+} + N_2 \quad (16.2.6)$$

For β-sialon ($Si_{6\text{-}z}Al_zO_zN_{8\text{-}z}$), it was synthesized by firing a powder mixture of Si_3N_4, Al_2O_3, and AlN at 1800–2000 °C under 1.0 MPa N_2 atmosphere.

$$(2 - z/3)Si_3N_4 + z/3AlN + z/3Al_2O_3 \rightarrow Si_{6-z}Al_zO_zN_{8-z} \quad (16.2.7)$$

A specially developed radiofrequency furnace, used for the inductive heating of the crucibles containing the reaction mixture, is advantageous for obtaining high yields and purities in these reactions [56]. Preparative amounts of nitridosilicates as coarsely crystalline single-phase products are accessible by this procedure in short reaction times.

16.2.2.2 Gas Reduction and Nitridation

The process of gas reduction and nitridation (GRN) is an effective and inexpensive way to synthesize oxynitride and nitride phosphors by using cheap and commercially available oxide starting powders. In this method, the reaction is generally performed in an alumina or quartz tube furnace through which NH_3 or NH_3–CH_4 gas flows. The NH_3 or NH_3–CH_4 gas acts as both the reducing and nitriding agents. The GRN method was used to prepare multinary oxynitride and nitride phosphors by Suehiro et al. and Li et al. These phosphors include Ca-α-sialon:Eu^{2+} [81] (Eq. 16.2.8), $Sr_2Si_5N_8$:Eu^{2+} [82] (Eq. 16.2.9), $LaSi_3N_5$:Ce^{3+} [83], Y-α-sialon:Eu^{2+} [84], and $BaSi_7N_{10}$:Eu^{2+} [85]. They used multi-component oxides as the starting powder and heated them in a flowing NH_3–CH_4 gas atmosphere mixture.

$$mCaO + (24 - 2m - 2n)SiO_2 + (m + n)Al_2O_3 + (32 - 2n)NH_3 + (48 - n)CH_4 \rightarrow 2Ca_{m/2}Si_{12-m-n}Al_{m+n}O_nN_{16-n} + (144 - 5n)H_2 + (48 - n)CO \quad (16.2.8)$$

$$2SrO + 5SiO_2 + 4NH_3 + 12CH_4 \rightarrow Sr_2Si_5N_8 + 25H_2 + 12CO \quad (16.2.9)$$

The phase purity of nitride phosphors after the GRN reaction is strongly related to processing parameters such as the heating rate, firing temperature, holding time, gas-flowing rate, and postannealing. By controlling these parameters carefully, one can achieve highly efficient phosphors with a small particle size (1 to approximately 2 μm) and narrow particle-size distribution.

16.2.2.3 Carbothermal Reduction and Nitridation

The carbothermal reduction and nitridation (CRN) is also an inexpensive method of synthesizing silicon-based oxynitride and nitride compounds. It differs from GRN in that (i) carbon powder is applied as the reducing agent and (ii) N_2 instead of NH_3 or NH_3–CH_4 is used as the nitriding agent. The precursor for CRN is a mixture of oxide, nitride, and carbon powders [64]. Zhang and Piao et al. used the CRN method to synthesize Ca-α-sialon:Eu^{2+} [86] and $M_2Si_5N_8$:Eu^{2+} [87]. The reactants for this method include $CaCO_3$, Si_3N_4, Eu_2O_3, Al_2O_3, and carbon (graphite) for Ca-α-sialon:Eu^{2+}, and MCO_3 (M = Ca, Sr), Si_3N_4, Eu_2O_3, and carbon oxides, silicon nitride/silica, alumina, and carbon for $M_2Si_5N_8$:Eu^{2+} (Eq. 16.2.10):

$$6SrCO_3(s) + 5Si_3N_4(s) + 18C(s) + 12N_2(g) \rightarrow 3Sr_2Si_5N_8(s) + 18CO(g) \quad (16.2.10)$$

The nitride phosphors prepared by the CRN method always contain some amount of residual carbon, which significantly decreases the absorption and luminescence of the phosphor powder itself. Therefore, it is necessary to remove the excess carbon after the carbothermal-reduction process. The common way for the removal of carbon is to fire the powder in an oxidizing atmosphere at temperatures >600 °C. However, this process may also lead to the oxidation of the phosphor powders, which decreases the luminescence of phosphors.

16.2.2.4 Ammonothermal Synthesis

The ammonothermal technique was first investigated to synthesize nitride phosphors by Li et al. [88–90]. In their synthesis of red-emitting $CaAlSiN_3$:Eu^{2+}, the CaAlSi alloy powder precursor was first prepared by arc-melting Si, Ca, Al, and Eu metal shots in an Ar atmosphere. Then the alloy was converted to sodium ammonometallates (referring to metal amides, imidoamides, and their partially polymerized compounds) in 100 Mpa supercritical ammonia at 400 °C followed by decomposing into the nitride by releasing NH_3 at 800 °C (Eq. 16.2.11).

$$
\begin{aligned}
&2Na + 2NH_3 \rightarrow 2NaNH_2 + H_2 \\
&CaxEu_{1-x}AlSi + NaNH_2 + NH_3 \rightarrow \text{amides and imidoamides} \\
&\quad + H_2 \rightarrow \text{more polymerized compounds} + NH_3 \rightarrow CaxEu_{1-x}AlSiN_3 + NH_3
\end{aligned} \tag{16.2.11}
$$

Zeuner et al. [74] applied the ammonothermal technique to synthesize $M_2Si_5N_8{:}Eu^{2+}$ (M = Ca, Sr, Ba) phosphors. In their synthesis, metal amides $M(NH_2)_2$ (M = Eu, Ca, Sr, Ba) were first prepared by dissolution of the respective metals in supercritical ammonia at 150 °C and 300 bar. Then the target $M_2Si_5N_8{:}Eu^{2+}$ phosphors were synthesized by reacting metal amides with silicon diimide at 1150–1400 °C (Eq. 16.2.12).

$$
\begin{aligned}
&M + 2NH_3 \rightarrow M(NH_2)_2 + H_2 \\
&2M(NH_2)_2 + 5Si(NH)_2 \rightarrow M_2Si_5N_8 + 6NH_3
\end{aligned} \tag{16.2.12}
$$

Nitridosilicate phosphors prepared by the ammonothermal synthesis method exhibit high phase purity and fine particle size.

16.2.2.5 Direct Nitridation

Piao et al. [91, 92] investigated the synthesis of red-emitting $CaAlSiN_3{:}Eu^{2+}$ (Eq. 16.2.13) and $Ba_2Si_5N_8{:}Eu^{2+}$ phosphors by using the direct nitridation process. The (Ca, Eu)AlSi or $(Ba, Eu)_2Si_5$ alloy powder was used as the starting powder, and reacted with flowing nitrogen gas at approximately 1050 °C. The resultant powder was then calcinated at 1350–1550 °C to form final nitride phosphors:

$$
2Ca_{1-x}Eu_xAlSi + 3N_2 \rightarrow 2Ca_{1-x}Eu_xAlSiN_3 (x = 0\text{–}0.2) \tag{16.2.13}
$$

The synthesized $CaAlSiN_3{:}Eu^{2+}$ powder consisted of irregular particles with a particle size of 6–9 μm, which is larger than that prepared by the solid-state reaction. The luminescence intensity of the phosphor prepared by the direct nitridation was comparable with that prepared by the solid-state reaction.

16.2.3 Nitride Phosphors in White LEDs

The abundance of (oxo)nitrido(alumino)silicates thus makes it possible to develop phosphors with different emission colors. Eu^{2+} and Ce^{3+} are obviously the two most used dopants in luminescent (oxy)nitrides because their $4f \rightarrow 5d$ emissions are broad, highly efficient, and structure-dependent. The optical properties of Eu^{2+}- and Ce^{3+}-doped (oxy)nitrides reported are summarized in Tables 16.3 and 16.4. In this section, we will provide an overview of nitride luminescent materials suitable for white-LED applications.

Table 16.3 Photoluminescence properties of Eu^{2+}-doped (Oxy)nitride phosphors

Phosphor	λ_{ex}/nm	λ_{em}/nm	FWHM/nm	η/%	References
AlN	290	465	52	63	[94]
$BaSi_6N_8$	370	450	44	38	[95]
$BaSi_7N_{10}$	300	482–500	80–110	52	[85, 96]
$Ca_2Si_5N_8$	450	605–615	104	50–55	[80, 97]
$Sr_2Si_5N_8$	450	650–690	92	75–88	[87, 98]
$Ba_2Si_5N_8$	450	570–680	125	75–80	[56, 97]
$CaSiN_2$	410	630		28.5	[99]
$Ba_3Ga_3N_5$		638	84.7		[100]
$SrScSi_4N_7$	390	520–526			[101]
$SrYSi_4N_7$	390	548–570			[101, 102]
$BaYSi_4N_7$	385	503–527			[103]
$CaSi_2O_2N_2$	446	563	98	72	[104]
$SrSi_2O_2N_2$	456	544	83	69	[104]
$BaSi_2O_2N_2$	458	492	36	41	[94]
$Si_{6-z}Al_zO_zN_{8-z}$	450	535	55	49.8	[105, 107]
$Ba_3Si_6O_{12}N_2$	400	527	65–68		[108]
$Ca_3Si_2O_4N_2$	330	510–550			[109]
$Ba_2AlSi_5N_9$	450	584	100	22.3	[110]
$Sr_3Si_2O_4N_2$	460	600	80		[111]
$Ca_{15}Si_{20}O_{10}N_{30}$	460	618–642			[112]
$CaAlSiN_3$	450	650	90	70	[57, 113]
$SrAlSi_4N_7$	450	632		53	[114, 115]
$La_{4-x}Ca_xSi_{12}O_{3+x}N_{18-x}$	460	565	82		[116]
$BaSi_3Al_3O_4N_5$	300	470		79	[117]
$Sr(Al_{0.3}Si_{0.7})_4(N_{0.8}O_{0.2})_6$	344	390	97		[118]
$Sr_5Al_{5+x}Si_{21-x}N_{35-x}O_{2+x}$	365	510	39		[119]
$Sr_{14}Si_{68-s}Al_{6+s}O_sN_{106-s}$	370	508			[120]
$Sr_3Si_{13}Al_3O_2N_{21}$	365	515	66	67	[121]
$Ca_{m/2}Si_{12-m-n}Al_{m+n}O_nN_{16-n}$	450	583–603	94	55.9	[122, 123]
$Li_mSi_{12-m-n}Al_{m+n}O_nN_{16-n}$	450	563–586	92	50	[124]
$Sr_{m/2}Si_{12-m-n}Al_{m+n}O_nN_{16-n}$	400	575			[125]

16.2.3.1 α-Sialon

Crystal Chemistry of α-Sialon

Silicon nitride, Si_3N_4, occurs in two different crystallographic modifications denoted α and β. It is generally accepted that α-Si_3N_4 is the low-temperature modification and that β–Si_3N_4 is the stable high-temperature modification at normal pressure. The space group is P31c of trigonal symmetry, with each unit cell

Table 16.4 Photoluminescence properties of Ce^{3+}-doped (Oxy)nitride phosphors

Phosphor	λ_{ex}/nm	λ_{em}/nm	FWHM/nm	η/%	References
$LaSi_3N_5$	360	440	~100	67	[83]
$La_3Si_6N_{11}$	460	577–581	~120	42.4	[71, 126]
$CaSiN_2$	535	625	80	40	[72, 127]
$Y_5Si_3O_{12}N$	325	475			[73, 128]
$Y_4Si_2O_7N_2$	390	500			[73, 128]
$YSiO_2N$	350	440			[73, 128]
$Y_2Si_3O_3N_4$	390	500			[73, 128]
$La_5Si_3O_{12}N$	358	478	108		[74, 129]
$La_4Si_2O_7N_2$	317	488	117		[75, 130]
$LaSiO_2N$	356	416			[75, 130]
$La_3Si_8O_4N_{11}$	365	425			[75, 130]
$CaSiAlN_3$	483	570	148	50	[76, 131]
$LaAl(Si_{6-z}Al_z)N_{10-z}O_z$	405	470–500	110	50	[77, 132]

containing four Si_3N_4 units. Both structures are built up of SiN_4 tetrahedra, which are joined by sharing corners in such a way that each N is common to three tetrahedral; thus, each Si atom has four N atoms as nearest neighbors. The idealized silicon nitride structure can be described as a stack of Si–N layers in either an ABAB… (β) or an ABCD… (α) sequence as shown in Fig. 16.24. In α, the c-glide plane relating the layer CD with AB replace the continuous channels with of the β structure by large closed interstices at 1/3, 2/3, and 2/3, 1/3, and 7/8. In the α unit cell ideally containing $Si_{12}N_{16}$ there are two sites large enough to accommodate other atoms or ions [133].

The general chemical formula of α-sialon can be given as $Me_xSi_{12-m-n}Al_{m+n}O_nN_{16-n}$, where x is the solubility of metal Me in the lattice of α-sialon, and x = m/v (v is the valence of Me); m and n denote the number of the Si–N bonds replaced by the Al–N and Al–O bonds, respectively. Because there are only two cages in one unit cell, the solubility of Me is not more than 2 ($x \leq 2$). The α-sialon is derived from α-Si_3N_4 by partial replacement of Si^{4+} by Al^{3+} and of

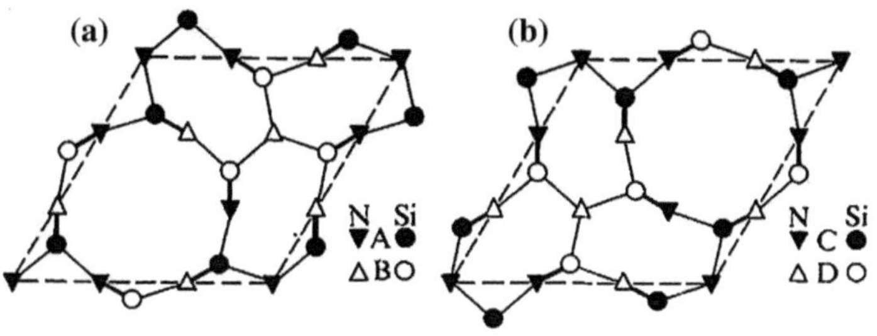

Fig. 16.24 Idealized Si–N layers in α and β silicon nitrides: **a** AB layers; **b** CD layers. β-structure, ABAB; α-structure, ABCD. Reprinted by permission from Macmillan Publishers Ltd: Ref. [133], copyright 1978

N^{4-} by O^{2-} and stabilized by trapping “modifying” cations, such as Li, Mg, Ca, Y, and some lanthanide elements, except La and Ce, in the voids of the (Si, Al)(O, N)$_4$ tetrahedral network. With bond lengths Si–N approximately 1.74 Å, Al–O approximately 1.75 Å, and Al–N approximately 1.78 Å, the lattice expansion occurs in α-sialon with the substitution. There are m (Si–N) replaced by m (Al–N) and n (Al–O). If the unit-cell dimensional changes corresponding to these respective replacements are in the ratio 5:1, the observed increment in unit cell dimensions of α-sialon phases fit the following relationships:

$$\Delta_a = 0.045\text{m} + 0.009\text{n} \qquad (16.2.2.1)$$

$$\Delta_c = 0.0040\text{m} + 0.008\text{n} \qquad (16.2.2.2)$$

Both x-ray and neutron diffraction studies of α-sialon have shown that the Me cations occupy the large interstitial sites (2b) at positions (1/3, 2/3, z) and (2/3, 1/3, z + 1/2) [134, 135]. Si and Al are distributed over 6c sites, whereas O and N occupy 2a, 2b, and 6c sites. The extended x-ray absorption fine structure (EXAFS) study of α-sialon has shown that the general preference for the modifying Me cations is to locate along the c-axis [136]. α-sialon is a nitrogen-rich compound with some oxygen dissolved in the structure. Izumi et al. [134] reported that the N/O ratio in the seven-fold coordination was 6:1, and later Cole et al. [136] suggested that the value was 5:2. These results indicate that most of the chemical bonds (71 or 86 %) between the modifying metal Me and the anions are covalent. The bond length of Me-(O, N) is in the range of 2.360–2.694 Å for Ca-α-sialon [134], with the bond parallel to the c-axis being much shorter than the other six ones. The crystal structure and seven-fold coordination of Ca atoms are illustrated in Fig. 16.25.

Photoluminescence of α-Sialon:Ce^{3+}

The photoluminescence of Ce^{3+} in Y- or Ca-α-sialon was reported by Xie et al. [137, 138]. As shown in Fig. 16.26, the emission spectrum centered at 495 nm

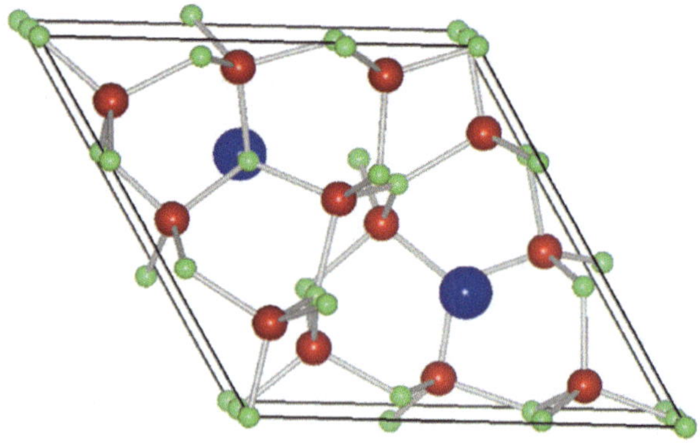

Fig. 16.25 Crystal structure of Ca-a-sialon viewed along the [0 0 1] direction. The *blue*, *red*, and *green spheres* represent Ca, Si/Al, and O/N atoms, respectively. Reprinted from Ref. [64], Copyright 2007, with permission from Elsevier

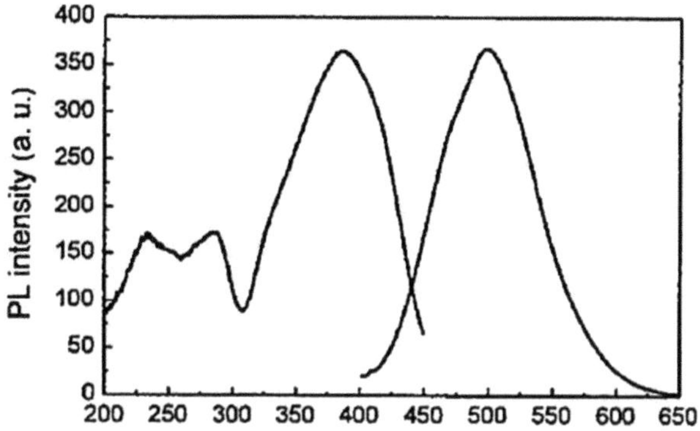

Fig. 16.26 Excitation and emission spectra of Ca-a-sialon:Ce^{3+}. Reproduced from Ref. [137] by permission of John Wiley & Sons Ltd

extends from 400 to 650 nm on 385 nm excitation. The peak emission wavelength is red-shifted from 485 to 503 nm when the Ce concentration increases from 5 to 25 mol% [68, 139]. Moreover, the emission of α-sialon:Ce^{3+} can also be tuned by varying the values of m and n. The excitation spectrum shows a broad band with a peak located at 389 nm, which closely matches the emission wavelengths of UV or NUV LEDs.

Photoluminescence of Ca-α-Sialon:Eu^{2+}

Eu^{2+}-doped Ca-α-sialon was reported to be a promising yellow-emitting phosphor [122, 128, 140, 141]. Eu^{2+}-doped Ca-α-sialon shows strong absorptions in the UV-to-visible light spectral region, which is due to the absorption of Eu^{2+}. The photoluminescence spectra of Ca-α-sialon:Eu^{2+} are given in Fig. 16.27. The excitation spectrum consists of two broad bands peaking at 302 and 412 nm, respectively. The emission spectrum shows a broad and symmetric band peaking at

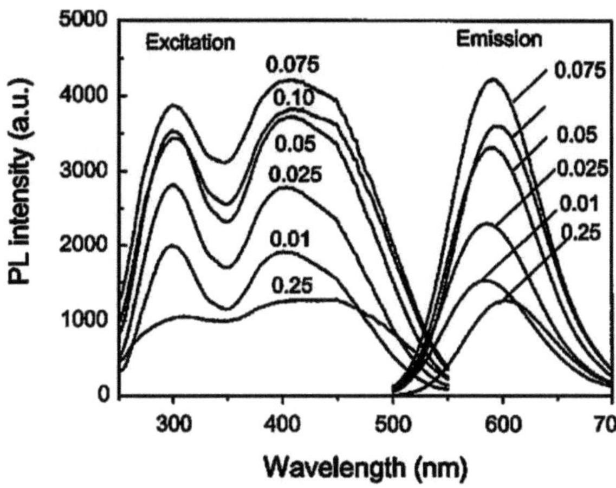

Fig. 16.27 Excitation and emission spectra of Ca-α-sialon. The excitation and monitoring wavelengths are 450 and 590 nm, respectively. Reprinted with permission from Ref. [122]. Copyright 2004, American Institute of Physics

580 nm with an FWHM of 94 nm. A Stokes shift of approximately 3300 cm^{-1} is then estimated for Ca-α-sialon:Eu^{2+}. In comparison with yellow YAG:Ce^{3+}, Ca-α-sialon:Eu^{2+} exhibits smaller thermal quenching. At 150 °C, the luminescence is quenched by 23 and 14 % of the initial intensity for YAG:Ce^{3+} and Ca-α-sialon: Eu^{2+}, respectively. The absorption and external quantum efficiency of Ca-α-sialon: Eu^{2+} (m = 1.4, n = 0.7, 7.5 mol% Eu) are 75 and 56 %, respectively. This indicates that Ca-α-sialon:Eu^{2+} is a very attractive yellow phosphor for white LEDs when combined with blue LEDs.

Photoluminescence of Ca-α-sialon:Yb^{2+}

The photoluminescence of Yb^{2+} in Ca-α-sialon was reported by Xie et al. [141]. As seen in Fig. 16.28, the excitation spectrum of Ca-α-sialon:Yb^{2+} consists of three bands centered at 299, 344, and 444 nm and two shoulders at 258 and 285 nm, respectively. The emission spectrum consists of a single and symmetric broad band with the maximum at 549 nm, which is assigned to the transition between the allowed $4f^{13}5d$ and $4f^{14}$ configurations of the Yb^{2+} ion. Xie et al. stated that the compositional dependence of luminescence properties is dominantly attributed to the changes in crystallinity, phase purity, and particle morphology of α-sialon with varying m [141]. With lower m values, the crystallization of the powders is poor, and there are some defects in each particle that can trap or scatter the emitted light.

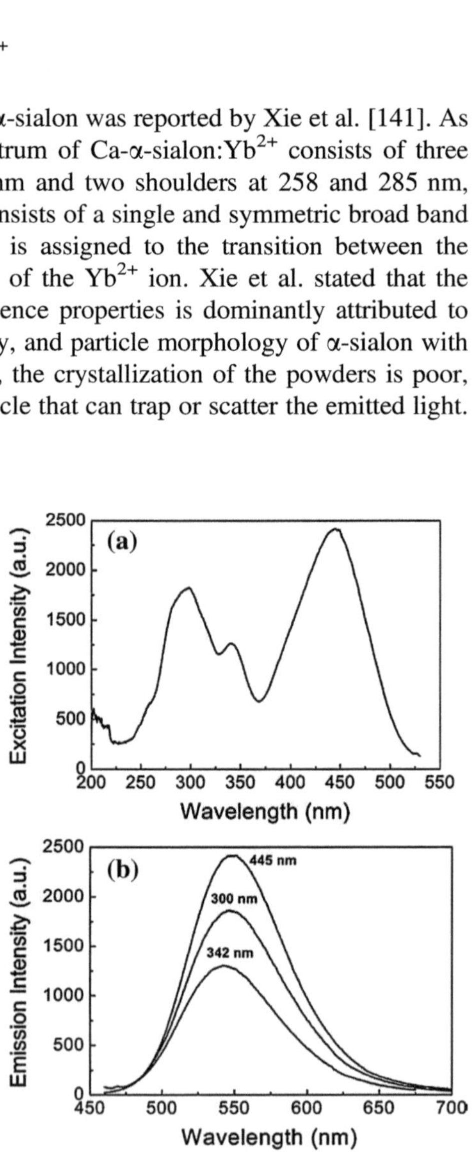

Fig. 16.28 **a** Excitation and **b** emission spectra of the Yb^{2+}-doped Ca-α-SiAlON (*m*) 2, (*x*) 0.005). The excitation spectrum was monitored at 549 nm and the emission spectrum was recorded at 342, 300, and 445 nm. Reprinted with the permission from Ref. [141]. Copyright 2005 American Chemical Society

With high m values, large hard agglomerates form, and this results in low packing density of the powder, which causes strong light scattering.

16.2.3.2 β-*sialon*

Crystal Chemistry of β-sialon

In the early 1970s, Oyama [142, 143] and Jack [143] observed that Al^{3+} can be dissolved in the silicon nitride lattice without changing its crystal structure by substituting Si^{4+} if N^{3-} is replaced by O^{2-} simultaneously. Such a solid solution is then named "β-sialon," which is derived from β–Si_3N_4 by simultaneous equivalent substitution of Al–O for Si–N. β-sialon is commonly described by the formula $Si_{6-z}Al_zO_zN_{8-z}$. In this formula, the z value denotes the solubility of Al^{3+} in the β–Si_3N_4 lattice, which varies in the range of 0–4.2 [143, 144].

In β-sialon, there are continuous channels that are parallel to the c direction as seen in Fig. 16.29. It is accepted that the several types of modifying cations can be accommodated in the α-sialon structure for achieving charge neutralization, whereas no cations, except Al^{3+}, can enter into the β-sialon lattice because the charge balance is not required. However, Hirosaki et al. have reported very intense luminescence of Eu^{2+} in β-sialon, which obviously indicates that Eu^{2+} is indeed dissolved in the lattice of β-sialon [105, 107].

Photoluminescence of β-sialon:Eu^{2+}

Hirosaki et al. [105] first report the green photoluminescence of Eu^{2+} in β-sialon; later, Xie et al. made a comprehensive investigation of the Eu^{2+}-doped β-sialon phosphor [107]. The β-sialon:Eu^{2+} phosphor produces intense green emission with a peak located at 538 nm as can be seen in Fig. 16.30. The broad emission spectrum has a full width-at-half maximum value of 58 nm. Two well-resolved broad

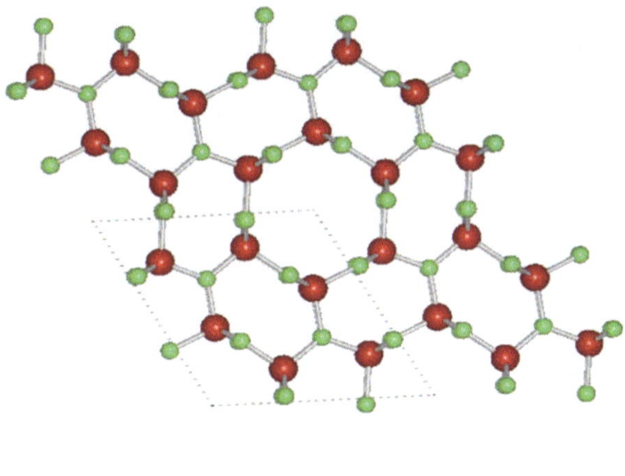

Fig. 16.29 Crystal structure of β-sialon viewed along the [0 0 1] direction. The *red* and *green spheres* represent Si/Al and O/N atoms, respectively. Reprinted from Ref. [64], Copyright 2007, with permission from Elsevier

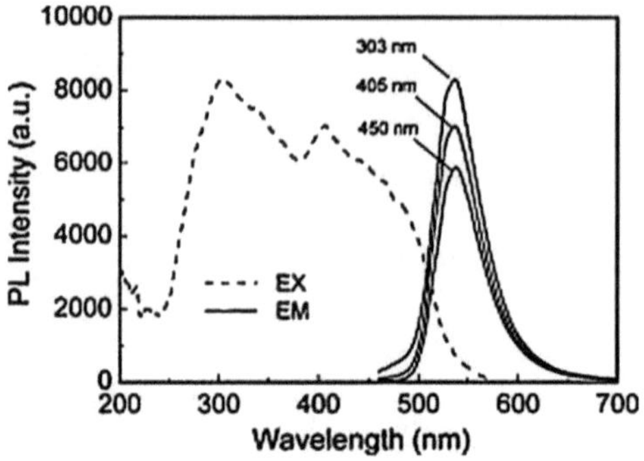

Fig. 16.30 Excitation and emission spectra of β-sialon: Eu^{2+}. The excitation spectrum was monitored at 535 nm and the emission spectrum was recorded at 303, 405, and 450 nm. Reprinted with permission from Ref. [105]. Copyright 2005, American Institute of Physics

bands, centered at 303 and 400 nm, are observed in the excitation spectrum. The broad excitation range enables the β-sialon:Eu^{2+} phosphor to emit strongly under NUV (400–420 nm) or blue (420–470 nm) light excitation. Xie et al. [107] investigated the effects of the z-value and Eu^{2+} concentration on the phase formation and luminescent properties of β-sialon:Eu^{2+} phosphors. The results showed that the concentration quenching occurs at 0.5–0.7 mol% for $z = 0.1$ and 0.5 and at 0.3 mol% for $z = 1.0$–2.0. The concentration quenching occurring at higher Eu^{2+} concentration in samples with low z-values could be due to the high solubility of Eu^{2+} in β-sialon with smaller z-values [107]. It is observed that β-sialon:Eu^{2+} shows higher luminescence for compositions with smaller z-values. This is related to the changes in particle size, particle size distribution, and phase purity with varying z-values. The compositions with lower z-values usually exhibit fine particle size and narrow particle-size distribution as well as higher phase purity.

The temperature-dependent luminescence intensity of β-sialon:Eu^{2+} is shown in Fig. 16.31. At 150 °C the luminescence of β-sialon:Eu^{2+} remains 83 % of the initial

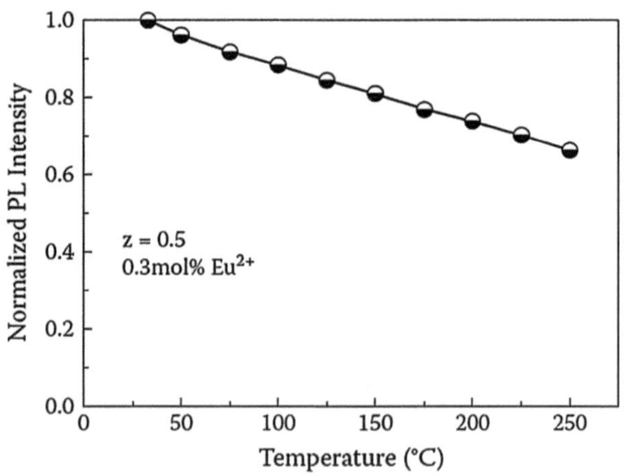

Fig. 16.31 Thermal quenching of β-sialon:Eu^{2+}. Reproduced with permission from Ref. [107]. Copyright 2007, The Electrochemical Society

intensity measured at room temperature, which is indicative of small thermal quenching. In addition, the optimized β-sialon:Eu^{2+} has the absorption and external quantum efficiency of 68.2 and 49.8 % when it is excited at 455 nm. Furthermore, there are no changes in chromaticity coordinates when β-sialon:Eu^{2+} experiences the high-temperature and high-humidity test (85 °C, 86 % RH) for 6000 h; is irradiated by 365-nm light with a powder of 120 mW/cm^2 for 10000 h; or is heated up to 150 ° C for 2500 h [106]. These excellent photoluminescent properties and high reliability enable β-sialon:Eu^{2+} one of the most important green phosphors for white LEDs.

16.2.3.3 $MSi_2O_2N_2$:Eu^{2+} (M = Ca, Sr, Ba)

Crystal Chemistry of $MSi_2O_2N_2$ (M = Ca, Sr, Ba)

The crystal structure of alkaline-earth oxonitridosilicates, $MSi_2O_2N_2$ (M = Ca, Sr, Ba), was reported by Schnick et al. (Fig. 16.32) [145–147]. The structures of these oxonitridosilicates are closely related and consist of highly dense and corner-sharing $SiON_3$ tetrahedra (Q3 type) layers that are separated by alkaline-earth metal layers. $CaSi_2O_2N_2$ crystallizes in the monoclinic structure with the space group $P2_1$ (No. 4), $SrSi_2O_2N_2$ in the triclinic system with the space group P1 (no. 1), and $BaSi_2O_2N_2$ in the orthorhombic system with the Pbcn (no. 60) space group. In the anion $[Si_2O_2N_2]^{2-}$layer of $MSi_2O_2N_2$ (M = Ca, Sr, Ba), every N atom is connected to three neighboring Si tetrahedron centers ($N^{[3]}$), and the O atoms are exclusively bonded terminally to Si atoms. The configuration of the Q3 type $SiON_3$ tetrahedra

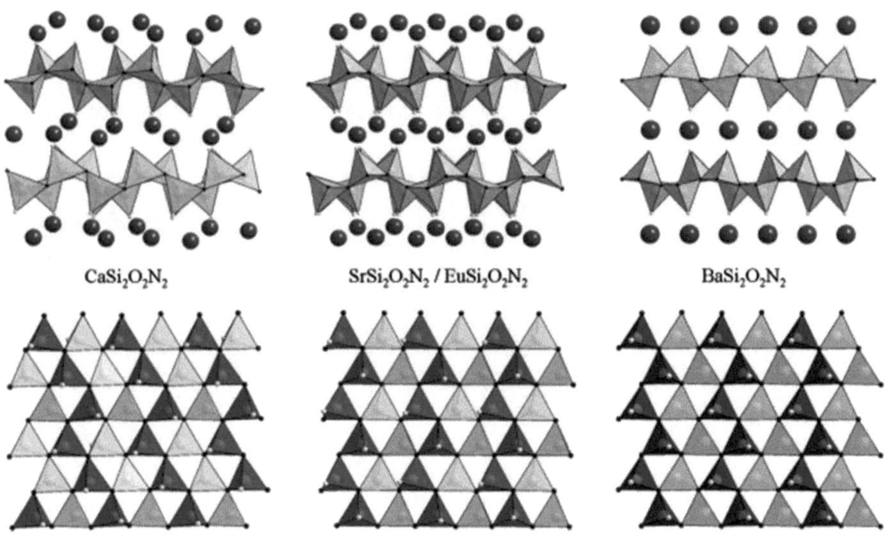

Fig. 16.32 Structure and silicate layers of $MSi_2O_2N_2$ (M = Ca, Sr, Ba). For clarification the *zigzag lines*, tetrahedra and vertices up are depicted in *dark gray*, with vertices down in *light gray* (N^{3-}: *black*, O^{2-}: *light gray*). Reprinted from Ref. [145], Copyright 2008, with permission from Elsevier

in $SrSi_2O_2N_2$ (up–down sequence) is quite analogous to $BaSi_2O_2N_2$, but the silicate layers are shifted against each other in the Sr compound. In $CaSi_2O_2N_2$, however, the $SiON_3$ tetrahedra are connected with a different up–down sequence [145]. In $CaSi_2O_2N_2$ and $SrSi_2O_2N_2$, Ca/Sr atoms are coordinated to six O atoms forming a distorted trigonal prism that is capped by a single N atom (coordination number 6 + 1), whereas in $BaSi_2O_2N_2$, Ba atoms are surrounded by six atoms forming a cuboid that is additionally capped by two N atoms (coordination number 6 + 2).

Photoluminescence of $MSi_2O_2N_2:Eu^{2+}$ (M = Ca, Sr, Ba)

Generally, $SrSi_2O_2N_2:Eu^{2+}$ is a green-emitting phosphor, $CaSi_2O_2N_2:Eu^{2+}$ a yellow-green phosphor, and $BaSi_2O_2N_2:Eu^{2+}$ a blue-green phosphor. Figure 16.33 presents the photoluminescence spectra of $MSi_2O_2N_2:Eu^{2+}$ (M = Ca, Sr, Ba). The

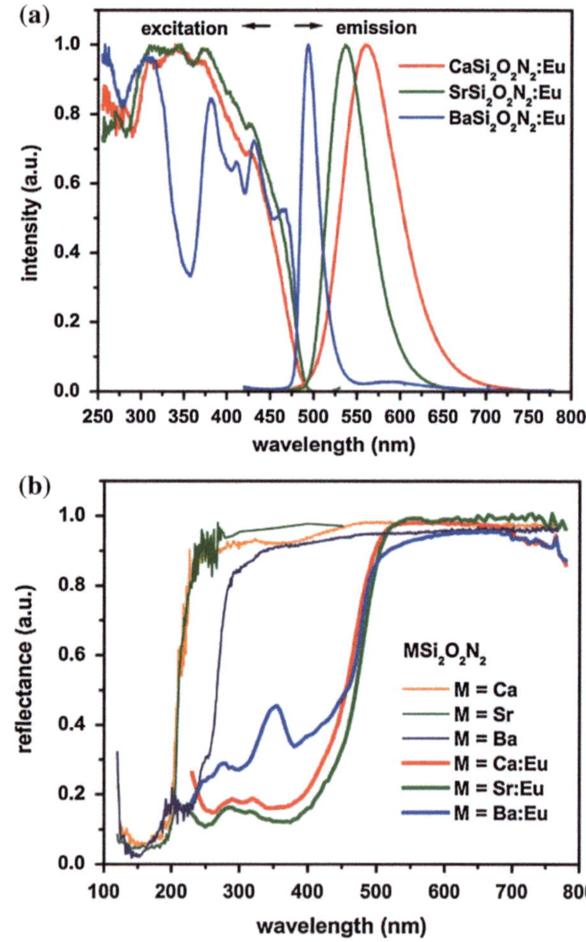

Fig. 16.33 **a** Excitation and emission spectra of $CaSi_2O_2N_2$, $SrSi_2O_2N_2$, and $BaSi_2O_2N_2$, each doped with 2 % Eu^{2+} taken at 4 K; emission spectra were taken at 450 nm excitation wavelength; excitation spectra were taken at the emission maxima; **b** diffuse reflection spectra of undoped and 2 % Eu^{2+}-doped $CaSi_2O_2N_2$, $SrSi_2O_2N_2$, and $BaSi_2O_2N_2$. Reprinted with the permission from Ref. [104]. Copyright 2009 American Chemical Society

emission maxima are at 494, 537, and 560 nm for the Ba, Sr, and Ca compounds, respectively. The onset of the $4f^6 5d^1$ band in the excitation spectra shifts to higher energies going from Ba to Ca. Under excitation of 450 nm, the absorptions of these alkaline earth oxonitridosilicates are 83, 80, and 78 % for $CaSi_2O_2N_2$:Eu^{2+}, $SrSi_2O_2N_2$:Eu^{2+}, and $BaSi_2O_2N_2$:Eu^{2+}, respectively. The external quantum efficiencies for the corresponding phosphors are 72, 79, and 41 %, respectively. The excellent photoluminescence properties of these alkaline-earth oxonitridosilicates make them very suitable for use as yellow-green ($CaSi_2O_2N_2$:Eu^{2+}), green ($SrSi_2O_2N_2$:Eu^{2+}), and blue-green ($BaSi_2O_2N_2$:Eu^{2+}) phosphors in white LEDs.

16.2.3.4 $CaAlSiN_3$

Crystal Structure of $CaAlSiN_3$

$CaAlSiN_3$ is built up of a three-dimensional network structure comprising [(Si, Al) N_4] tetrahedra where one third of the nitrogen atoms connect two Si ($N^{[2]}$), and the remaining two thirds ($N^{[3]}$) connect three tetrahedral centers (Fig. 16.34). $CaAlSiN_3$ crystallizes in the orthorhombic structure with the space group of $Cmc2_1$ (no. 36), which is isotypic with $LiSi_2N_3$ and $NaSi_2N_3$. The Al and Si atoms are randomly distributed on the same tetrahedral sites forming $(Si, Al)_6N_6$ rings. The structure of $CaAlSiN_3$ can be regarded as a superstructure variant of the wurtzite-type binaries AlN or GaN. The Ca atoms are located in the large channels along the c-axis, which are coordinated by five nitrogen atoms with an average distance of approximately 2.50 Å [56, 57].

Photoluminescence Properties of $CaAlSiN_3$:Ce^{3+}

The photoluminescence of Ce^{3+} in $CaAlSiN_3$ was reported by Li et al. [131]. Figure 16.35 presents the luminescence properties of $CaAlSiN_3$:Ce^{3+}. There are

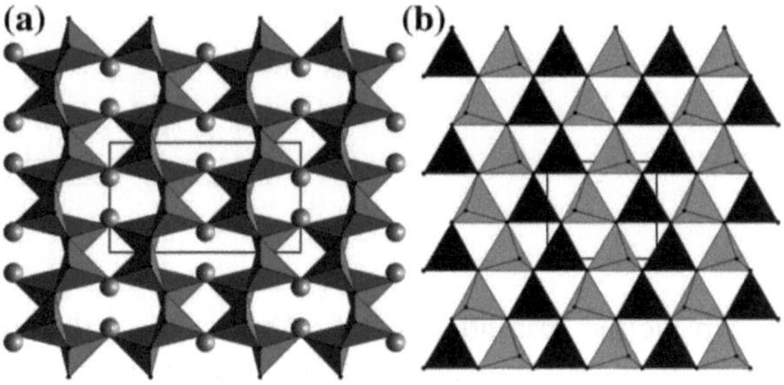

Fig. 16.34 Structure of $CaAlSiN_3$. **a** Viewed along [001], Ca *light gray spheres*, [SiN_4] and tetrahedra *gray*. **b** View showing layers of condensed "dreier" rings; [SiN_4] tetrahedra up (*gray*) and down (*black*). Reproduced from Ref. [56] by permission of John Wiley & Sons Ltd

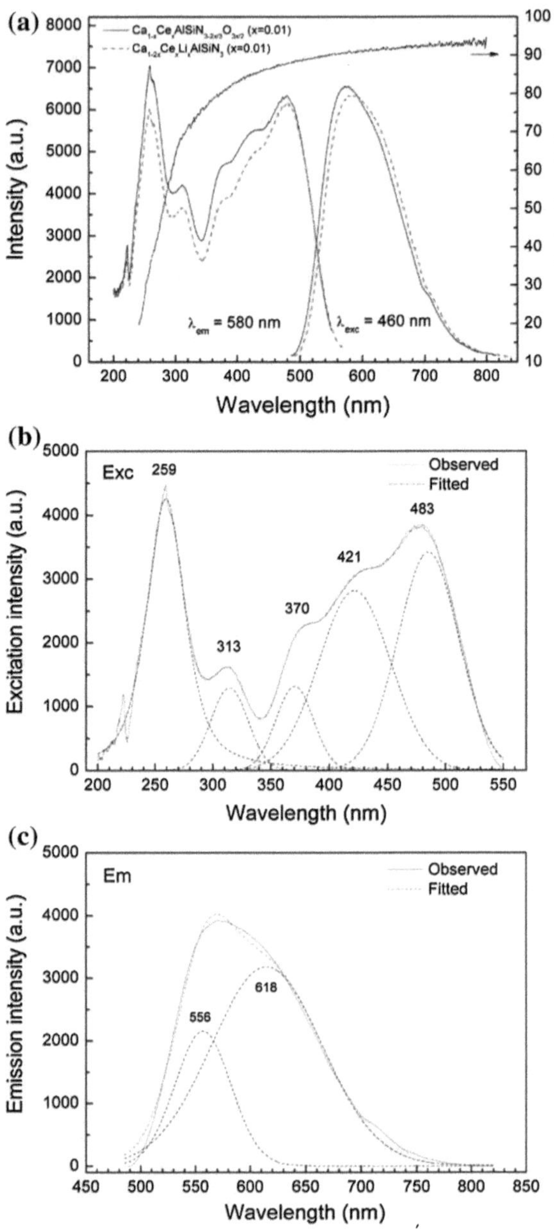

Fig. 16.35 **a** Diffuse reflection spectrum of $CaAlSiN_3$ and excitation and emission spectra of $Ca_{1-2x}Ce_xLi_xAlSiN_3$ (x = 0.01) and $Ca_{1-x}Ce_xAlSiN_{3-2x/3}O_{3x/2}$ (x = 0.01); luminescence spectra of $Ca_{1-2x}Ce_xLi_xAlSiN_3$ (x = 0.01) with deconvolved Gaussian subbands for the excitation **b** and emission **c** spectra. Reprinted with the permission from Ref. [66]. Copyright 2008 American Chemical Society

two principle excitation bands: One is located under UV (200–330 nm) with a shoulder at approximately 310 nm, and another one is in the visible spectral range (350–560 nm) with several sub-bands (or shoulders). The fundamental absorption edge of $CaAlSiN_3$ is estimated to be approximately 235 nm (5.3 eV) from its diffuse reflection spectrum, which partially overlap with the top of the excitation band in UV. At excitation under UV and/or the blue spectral range, $CaAlSiN_3:Ce^{3+}$ emits efficient yellow-orange light with a maximum at approximately 580 nm. The excitation band of Ce^{3+} consists of up to five Gaussian components at approximately 259, 313, 370, 421, and 483 nm with larger crystal-field splitting (CFS) of the $4f^0 5d^1$ excited configure of Ce^{3+} of approximately 14,000 cm^{-1}. The emission band of $CaAlSiN_3:Ce^{3+}$ shows a broad nonsymmetrical band as expected for Ce^{3+}. Similarly, this emission band of Ce^{3+} can be fitted into two sub-bands peaking at approximately 556 nm (greenish-yellow) and 618 nm (red) approximately corresponding to the transitions of the 5*d* excited states to its two ground-state configurations of $^2F_{7/2}$ and $^2F_{5/2}$ of Ce^{3+}, respectively, with a difference of approximately 1800 cm^{-1}. Under 460-nm excitation, the absorption and external quantum efficiency of $CaAlSiN_3:Ce^{3+}$ are 70 and 56 %, respectively. The yellow $CaAlSiN_3:Ce^{3+}$ phosphor also exhibits a small thermal quenching, and its luminescence at 150 °C is quenched by 10 % of the initial intensity. The obtained warm white-LEDs has been given a luminous efficacy of approximately 50 lm/W, a color temperature of 3722 K, and a color-rendering index of 70.

Photoluminescence Properties of $CaAlSiN_3:Eu^{2+}$

Figure 16.36 presents the photoluminescence spectra of $CaAlSiN_3:Eu^{2+}$ [92]. The excitation spectrum shows a broad band covering the region from the near-UV to visible part. The first band, located at approximately 240 nm, is caused by the electronic transition between the valence and the conduction band of the $CaAlSiN_3$

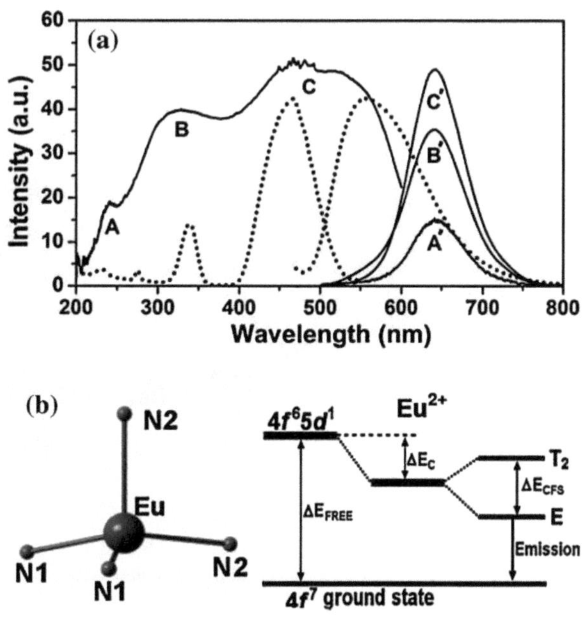

Fig. 16.36 Typical photoluminescence excitation (λ_{mon} = 649 nm) and emission (λ_{ex} = 460 nm) spectra of $Ca_{0.98}Eu_{0.02}AlSiN_3$ (**a**) and the schematic picture of the influence of the environment of a Eu^{2+} on the positions of electronic states (**b**). Reprinted with the permission from Ref. [92]. Copyright 2007 American Chemical Society

host. Another two bands, peaking at 320 and 470 nm, are due to the excitation of the Eu^{2+} ions ($4f^7 \rightarrow 4f^6 5d$). The emission spectrum shows a broad band centered at 650 nm [92]. When the phosphor is excited by blue light, the absorption is 86 %, and the external quantum efficiency is ≤ 70 %. In addition, it has a small thermal quenching. The luminescence is quenched by 10 % of the initial intensity when the phosphor is heated to 150 °C and excited at 450 nm. These excellent photoluminescence properties and small thermal quenching enable $CaAlSiN_3:Eu^{2+}$ to be a very attractive red phosphor for white-LED applications [57].

16.2.3.5 $M_2Si_5N_8$ (M = Ca, Sr, Ba)

Crystal Structure of $M_2Si_5N_8$ (M = Ca, Sr, Ba) [56]

$M_2Si_5N_8$ with M = Sr, Ba, and Eu crystallizes isotypically in the space group Pmn21. The crystal structure is based on a network of corner-sharing $[SiN_4]$ tetrahedral, in which half of the nitrogen atoms connect two ($N^{[2]}$) and the other half connect three neighboring Si atoms ($N^{[3]}$). The $[SiN_4]$ tetrahedra form corrugated layers of highly condensed dreier rings with $N^{[3]}$ atoms. These layers are three-dimensionally interconnected through further $[SiN_4]$ tetrahedra (Fig. 16.37). Carrying the higher formal charge, the $N^{[2]}$ atoms mainly coordinate the metal ions, which are situated in channels formed by sechser rings. $Ca_2Si_5N_8$ is composed of vertex-sharing $[SiN_4]$ tetrahedral exhibiting a similar network compared with $M_2Si_5N_8$ (M = Sr, Ba, Eu), in which half of the nitrogen atoms connect two and the other half connect three silicon atoms. However, it crystallizes in the monoclinic space group Cc and differs in the distribution of the Si/N ring sizes. Furthermore, significantly fewer corrugated layers have been observed compared with $M_2Si_5N_8$ (M = Sr, Ba, Eu) (Fig. 16.39).

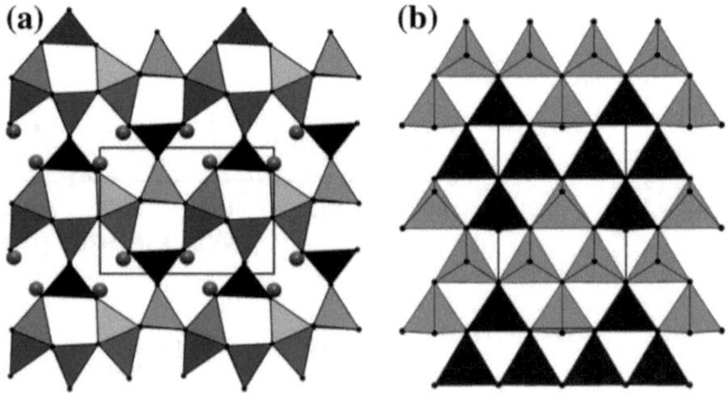

Fig. 16.37 Crystal structure of $M_2Si_5N_8$ (M = Sr, Ba, Eu). **a** Viewed along [100]; corrugated $[SiN_4]$ layers are *gray*, the interconnecting $[SiN_4]$ tetrahedra are *black*, the metal ions are *gray spheres*, and the nitrogen atoms are *black spheres*. **b** View showing layers of condensed dreier rings; $[SiN_4]$ tetrahedra up (*gray*) and down (*black*). Reproduced from Ref. [56] by permission of John Wiley & Sons Ltd

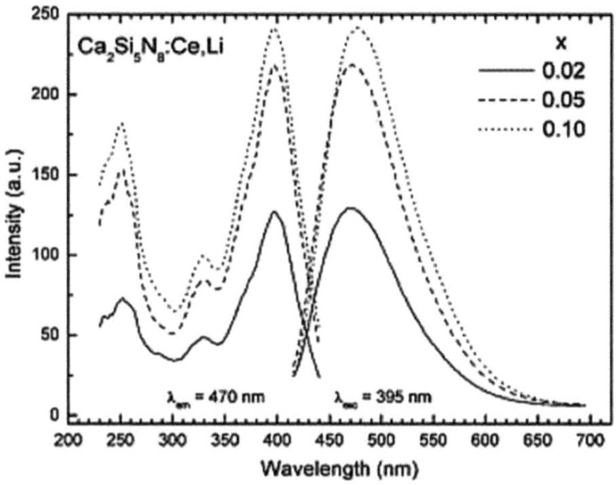

Fig. 16.38 Excitation and emission spectra of $Ca_{2-2x}Ce_xLi_xSi_5N_8$ (x = 0.02, 0.05, 0.1). Reprinted from Ref. [149], Copyright 2006, with permission from Elsevier

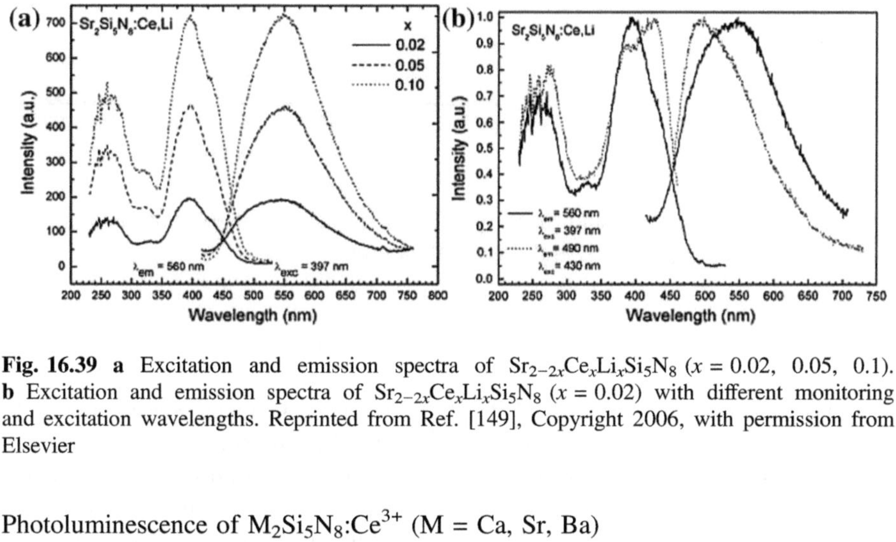

Fig. 16.39 a Excitation and emission spectra of $Sr_{2-2x}Ce_xLi_xSi_5N_8$ (x = 0.02, 0.05, 0.1). b Excitation and emission spectra of $Sr_{2-2x}Ce_xLi_xSi_5N_8$ (x = 0.02) with different monitoring and excitation wavelengths. Reprinted from Ref. [149], Copyright 2006, with permission from Elsevier

Photoluminescence of $M_2Si_5N_8$:Ce^{3+} (M = Ca, Sr, Ba)

The luminescence properties of Ce^{3+}-doped $M_2Si_5N_8$ (M = Ca, Sr, Ba) have been reported by Li [149] (Figs. 16.38 through 16.40). The Ce^{3+}-activated $M_2Si_5N_8$ phosphors exhibit broad emission bands with maxima at 470, 553, and 451 nm for M = Ca, Sr, and Ba, respectively, due to the 5*d*-4*f* transition of Ce^{3+}. In addition, it obviously shows two Ce^{3+} emission centers due to the fact that the Ce^{3+} ions occupy two M sites. With the ionic radius of M^{2+} increasing going from Ca to Ba, the ratio of the emission intensity to the absorption intensity (at 400 nm) of Ce^{3+} decreases. A strong absorption and excitation band in the UV-blue range of 370–

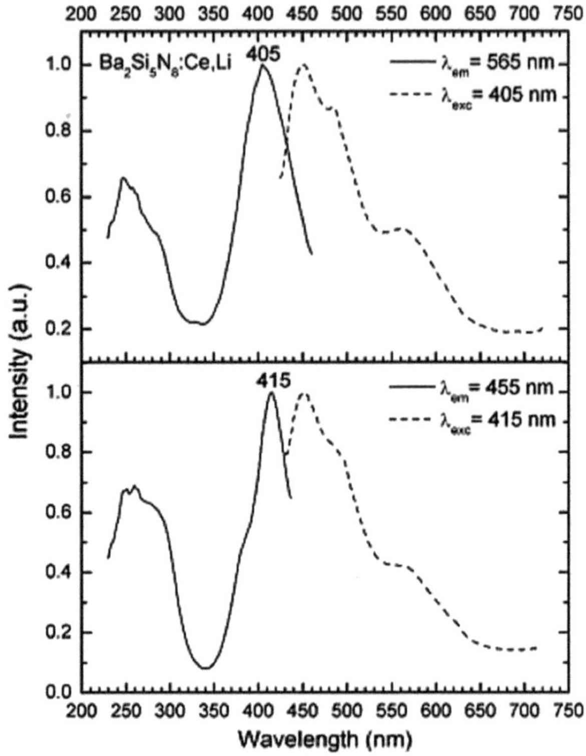

Fig. 16.40 Excitation (*solid line*) and emission (*dashed line*) spectra of $Ba_{2-2x}Ce_xLi_xSi_5N_8$ ($x = 0.02$). Reprinted from Ref. [149], Copyright 2006, with permission from Elsevier

450 nm in $Ca_2Si_5N_8:Ce^{3+}$ and $Sr_2Si_5N_8:Ce^{3+}$ demonstrates these materials to be of high potential for white-LED lighting applications.

Photoluminescence of $M_2Si_5N_8:Eu^{2+}$ (M = Ca, Sr, Ba)

The photoluminescence of $M_2Si_5N_8:Eu^{2+}$ (M = Ca, Sr, Ba) was investigated by Li et al. (Fig. 16.41) [97, 150]. $M_2Si_5N_8:Eu^{2+}$ (M = Ca, Sr) shows a typical broad-band emission in the orange to red spectral range (600–680 nm) depending on the type of M and the europium concentration. $Ba_2Si_5N_8:Eu^{2+}$ shows yellow to red emission with maxima from 580 to 680 nm with increasing Eu^{2+} content. The long-wavelength excitation and emission is attributed to the effect of high covalency and large crystal-field splitting on the 5*d* band of Eu^{2+} in the nitrogen environment. With increasing europium concentration, the emission band shows a red shift for all $M_2Si_5N_8:Eu^{2+}$ compounds due to changing Stokes shift and the reabsorption by Eu^{2+}. The conversion (i.e., quantum) efficiency increases going from Ca to Ba and Sr under excitation at 465 nm. In particular, $Sr_2Si_5N_8:Eu^{2+}$, with a quantum efficiency of 75–80 % and a thermal quenching of only a few percent at 150 °C, proves to be a highly promising red-emitting conversion phosphor for white-LED applications.

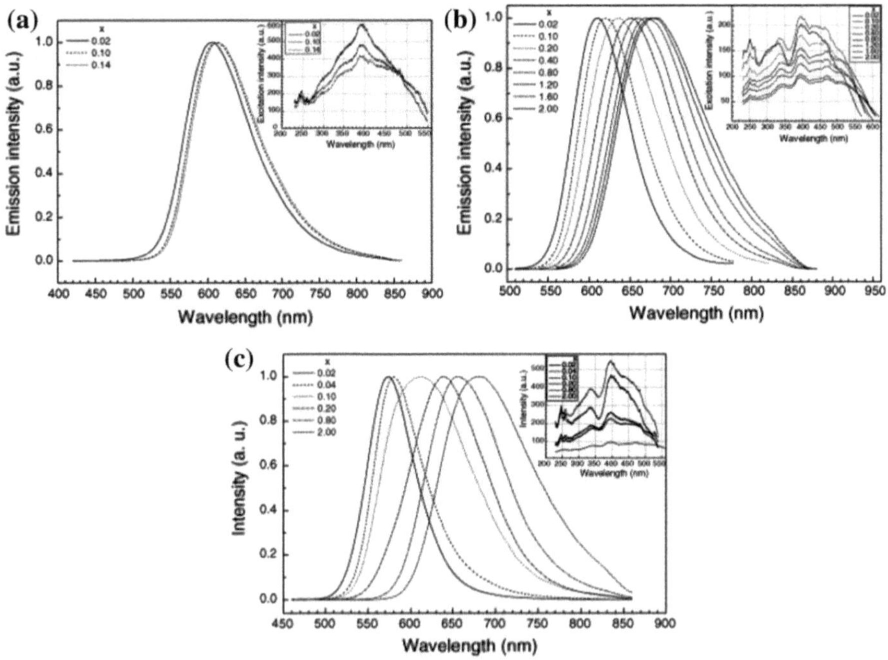

Fig. 16.41 Excitation (*inset*) and emission spectra of $M_{2-x}Eu_xSi_5N_8$: **a** M = Ca, **b** M = Sr and **c** M = Ba. Reprinted from Ref. [97], Copyright 2006, with permission from Elsevier

16.2.3.6 $LaAl(Si_{6-z}Al_z)(N_{10-z}O_z):Ce^{3+}$ (z = 1)

$LaAl(Si_{6-z}Al_z)N_{10-z}O_z$, which is also called the "JEM phase," crystallizes in the orthorhombic system with the space group of Pbcn, a = 9.4303 Å, b = 9.7689 Å, and c = 8.9386 Å. The Al atoms and (Si, Al) atoms are tetrahedrally coordinated by (N, O) atoms, thus yielding an $\mathrm{Al(Si, Al)_6(N, O)_{10}^{3-}}$ network (Fig. 16.42). The La atoms are accommodated in tunnels extending along the [0 0 1] direction and are irregularly coordinated by seven (N, O) atoms at an average distance of 2.70 Å. As shown in Fig. 16.43, the emission spectrum of JEM:Ce^{3+} displays a broad band extending from 420 to 700 nm under 405 nm excitation with a peak located at 475 nm. The broad excitation spectrum extending from 200 to 450 nm is due to the 4*f*-5*d* electronic transition of Ce^{3+}. Both spectra are red-shifted when the concentration of Ce^{3+} or the z-value increases, thus enabling this blue phosphor to be excited efficiently by UV (370–400 nm) or NUV (400–410 nm) LEDs [64, 132].

16.2.3.7 AlN:Eu^{2+}

Figure 16.44 shows the excitation and emission spectra of the sample with 2.9 mol % Si and 0.24 mol% Eu^{2+}. The emission spectrum of AlN:Eu^{2+} consists of a single

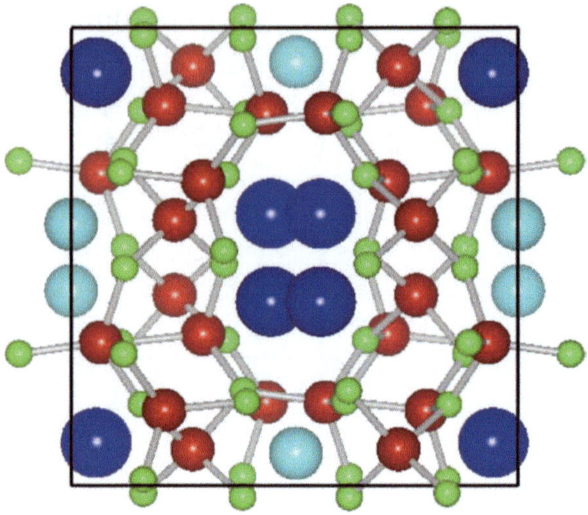

Fig. 16.42 Crystal structure of JEM viewed along the [0 0 1] direction. The *blue*, *pale blue*, *red*, and *green spheres* represented are La, Al, Si/Al, and O/N atoms, respectively. Reprinted from Ref. [64], Copyright 2007, with permission from Elsevier

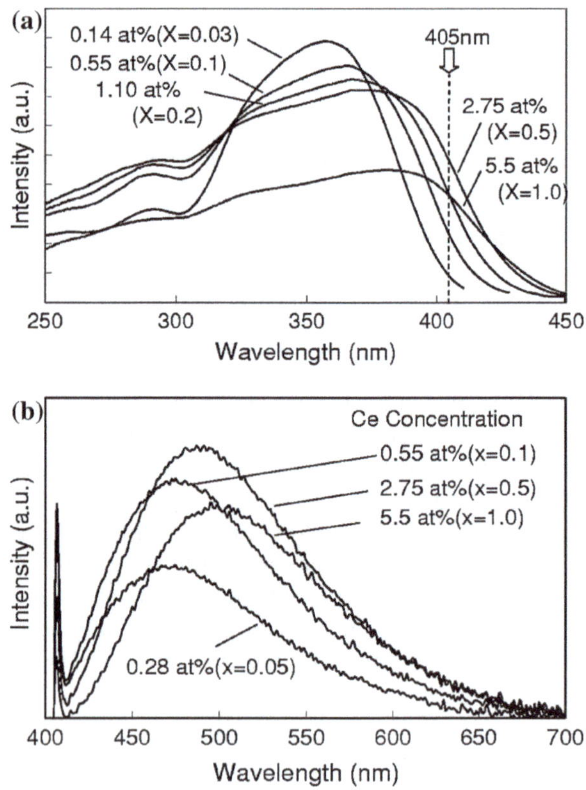

Fig. 16.43 Excitation and emission spectra of JEM phosphors with various Ce concentrations. The samples were excited at wavelength of 405 nm. Reprinted with permission from Ref. [132]. Copyright 2007, American Institute of Physics

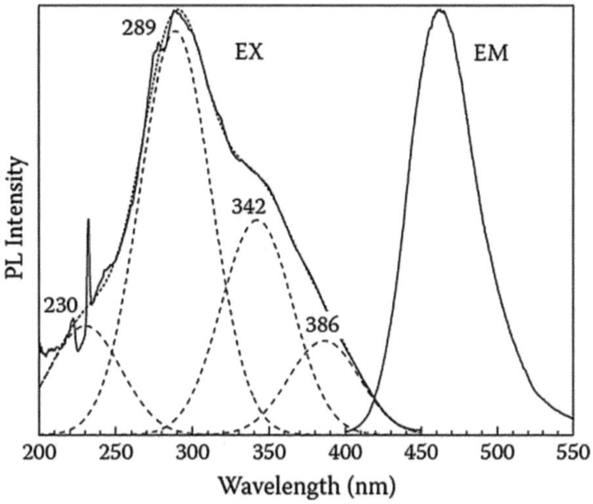

Fig. 16.44 Excitation and emission spectra of AlN:Eu^{2+}. Reprinted with the permission from Ref. [94]. Copyright 2009 American Chemical Society

and symmetric broad band centered at 465 nm when the sample was excited at 300 nm. The full width-at-half maximum (FWHM) value of the band is approximately 52 nm. The blue emission can be ascribed to the allowed $4f^65d \rightarrow 4f^7$ transitions of Eu^{2+}. The excitation spectrum, monitored at 465 nm, covers a broad range of 250–450 nm. Several peaks or shoulders at 280, 290, and 350 nm are observed. The structure in the excitation spectrum is due to the crystal-field splitting of the 5*d* level of Eu^{2+} ions. The excitation band can be fitted into four Gaussian sub-bands centered at 230, 289, 342, and 386 nm, respectively. The crystal-field splitting is estimated to be 17,600 cm^{-1}, and the Stokes shift is calculated to be 4400 cm^{-1}. Under 365-nm excitation, the internal quantum efficiency (η_i), external quantum efficiency (η_0), and absorption (α) of AlN: Eu^{2+} are 76, 46, and 63 %, respectively. The high absorption and quantum efficiency of AlN:Eu^{2+} at the excitation wavelength of 365 nm indicate that it matches well with the emission wavelength of InGaN UV LED chips and is a promising blue-emitting downconversion phosphor in white LEDs [57].

16.2.3.8 Ln–Si–O–N:Ce^{3+} (Ln = Y, La)

Van Krevel et al. [128] investigated the luminescence of Ce^{3+} in several Y-Si-O-N compounds such as $Y_5(SiO_4)_3N$, $Y_4Si_2O_7N_2$, $YSiO_2N$, and $Y_2Si_3O_3N_4$. Generally, these compounds emit a blue color with a peak emission wavelength of 400–500 nm and show maximum excitation bands at 325–400 nm. They demonstrated that the N/O ratio and the crystal structure had a strong effect on the emission, Stokes shift, and crystal-field splitting. The excitation and emission spectra shifted toward the long-wavelength side when Ce^{3+} was coordinated to more N^{3-} vs. O^{2-} in the sequence of $Y_2Si_3N_3N_4$ > $YSiO_2N$ > $Y_4Si_2O_7N_2$ > $Y_5(SiO_4)_3N$. The luminescence of $YSiO_2N:Ce^{3+}$ and $Y_2Si_3O_3N_4:Ce^{3+}$ is higher than that of the other two, which have large Stokes shifts, thus making them interesting blue phosphors for UV or near-UV LEDs.

The photoluminescence of Ce^{3+}-doped La-Si-O-N was investigated by Dierre et al. [130]. The emission spectrum of $La_5Si_3O_{12}N$ doped with 2 mol% Ce^{3+} extends from 400 to 650 nm on the excitation at λ_{ex} = 358 nm showing a broad structureless band. It has a maximum at approximately 472 nm with a FWHM of intensity of 108 nm. The excitation spectrum monitored at λ_{mon} = 472 nm consists of five bands in the range of 220–400 nm. On excitation at 317 nm, the emission spectrum of $La_4Si_2O_7N_2$ doped with 6 mol% Ce^{3+} shows a broad structureless band with a maximum at 488 nm and a FWHM of 117 nm. The excitation spectrum, monitored at 488 nm, starts at 400 nm and extends to 275 nm. A single-emission band centered at 416 nm is observed on excitation at 356 nm for $LaSiO_2N$ doped with 4 mol% Ce^{3+}. $La_3Si_8O_4N_{11}$ doped with 6 mol% Ce^{3+} shows an emission spectrum consisting of a broad band situated at 424 nm and a shoulder at 458 nm when it is excited at 365 nm. The FWHM of the emission band is approximately 80 nm. Among these four La-Si-O-N:Ce^{3+} phosphors, $La_3Si_8O_4N_{11}:Ce^{3+}$ shows the most interesting luminescent properties and small thermal quenching. This blue phosphor thus has the greatest potential to be used in white-LED applications.

16.2.3.9 $LaSi_3N_5:Ce^{3+}$

$LaSi_3N_5$ was obtained by the reaction of Si_3N_4 and La_2O_3 at 2000 °C and 725 psi (50 bar) N_2 pressure [151]. Nitridation of two-phase alloys $RESi_3$ led to $RESi_3N_5$ with RE=Ce, Pr, Nd [152]. The structure is built up exclusively of vertex-sharing [SiN_4] tetrahedra forming vierer rings, which are interconnected by chains of [SiN_4] tetrahedra forming a three-dimensional network (Fig. 16.45). According to the molar ratio of Si:N = 3:5, two fifths of the nitrogen atoms connect three silicon atoms ($N^{[3]}$), whereas three fifths are connected with two silicon atoms ($N^{[2]}$) [56].

Suehiro et al. investigated the photoluminescence of Ce^{3+}-doped $LaSi_3N_5$ phosphors [83]. As seen in Fig. 16.46, the excitation spectrum covers the spectral range of 240–400 nm with two distinct bands centered at 255 and 355 nm, respectively. The emission spectrum of $La_{0.9}Ce_{0.1}Si_3N_5$ showed a broad band centered at 440 nm and a full width-at-half maximum (FWHM) value of 95. The emission band of $La_{1-x}Ce_xSi_3N_5$ is red-shifted when the Ce^{3+} concentration increases. This red shift can be ascribed to the enhanced Stokes shift because the excitation spectra of these samples basically do not alter in shape with the Ce^{3+} concentration. The external quantum efficiency of the optimal composition (x = 0.10) is 67.0 and 34.3 % under excitation of 355 and 380 nm, respectively. This sample also shows small thermal quenching, and its emission intensity measured at 150 °C under 355-nm excitation decreases approximately 15 % of the initial intensity. The excellent photoluminescent properties of $LaSi_3N_5:Ce^{3+}$ make it a promising blue-emitting luminescent material for UV LEDs [57].

Fig. 16.45 Crystal structure of $LaSi_3N_5$ (viewed along [100], "vierer" *rings* are *dark gray*, interconnecting [SiN_4] tetrahedra are *light gray*, metal ions are *gray spheres*, nitrogen *black spheres*). Reproduced from Ref. [56] by permission of John Wiley & Sons Ltd

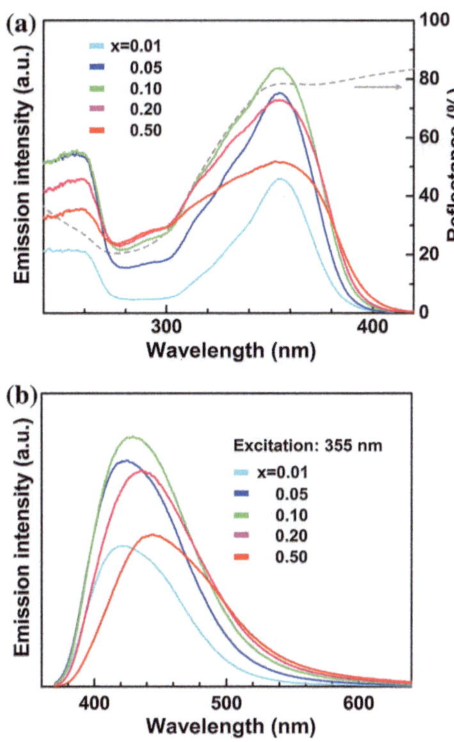

Fig. 16.46 PL **a** excitation and **b** emission spectra of $La_{1-x}Ce_xSi_3N_5$ samples. Reprinted with permission from Ref. [83]. Copyright 2009, American Institute of Physics

16.2.3.10 $Ba_3Si_6O_{12}N_2$:Eu^{2+}

$Ba_3Si_6O_{12}N_2$ crystallizes in the trigonal crystal system with the space group of P-3 (no. 147). The crystal structure of $Ba_3Si_6O_{12}N_2$ has fused rings-sheets,

$2_{\infty}[(Si_6\ ^{[4]}O_6^{[2]}N_2^{[3]})O_6^{[1]}]^{6-}$, which are composed of 8-membered Si-(O, N) and 12-membered Si-O rings. It is built up of corrugated corner-sharing SiO_3N tetrahedron layers that were re-separated by Ba atoms. The Ba^{2+} ions occupy two different crystallographic sites; one is trigonal anti-prism (distorted octahedron) with six oxygen atoms, and the other is trigonal anti-prism with six oxygen atoms, which is further capped with one nitrogen atom (see Fig. 16.47) [57].

Uheda et al. [153] and Braun et al. [108] reported the photoluminescence of Eu^{2+} in $Ba_3Si_6O_{12}N_2$. $Ba_3Si_6O_{12}N_2:Eu^{2+}$ has a broad excitation band that covers a spectral range of 250–500 nm showing the strong absorption of UV blue light (Fig. 16.48). The emission spectrum shows a narrow band centered at 525 nm with a FWHM of 68 nm. The narrow green emission observed in $Ba_3Si_6O_{12}N_2:Eu^{2+}$ may be expected from the Ba2 site (coordinated with six oxygen atoms and one nitrogen atom) rather than the Ba1 site (with six oxygen atoms only). This green phosphor exhibits high color purity with the International Commission on Illumination (CIE) chromaticity coordinates of $x = 0.28$ and $y = 0.64$ as well as small thermal quenching. At 150 °C, the luminescence of $Ba_3Si_6O_{12}N_2:Eu^{2+}$ is quenched by 10 % of the initial intensity at room temperature, whereas it is reduced by 25 % for $(Sr, Ba)_2SiO_4:Eu^{2+}$. The useful photoluminescence and small thermal quenching enable $Ba_3Si_6O_{12}N_2:Eu^{2+}$ to be a promising green-emitting luminescent material in white LEDs for general lighting or liquid crystal display (LCD) backlights [57].

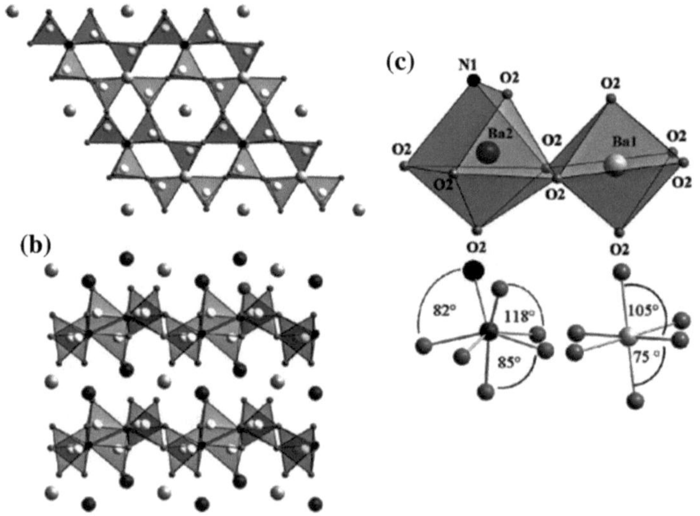

Fig. 16.47 Crystal structure of $Ba_3Si_6O_{12}N_2$ projected on **a** the (001) plane and **b** the (100) plane. **c** The coordination of Ba atoms by (O,N) atoms. Reproduced from Ref. [108] by permission of John Wiley & Sons Ltd

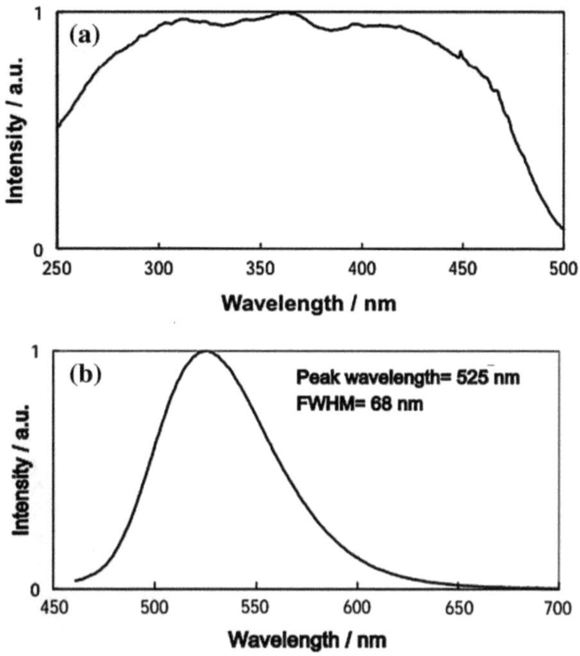

Fig. 16.48 Excitation **a** and emission **b** spectra of the new green phosphor, $(Ba,Eu)_3Si_6O_{12}N_2$. Reproduced with permission from Ref. [153]. Copyright 2008, The Electrochemical Society

16.2.3.11 $SrAlSi_4N_7:Eu^{2+}$ [115, 154]

The nitridoalumosilicate $SrAlSi_4N_7$ contains a highly condensed network structure built up of $[SiN_4]$ and $[AlN_4]$ tetrahedral [154]. $SrAlSi_4N_7$ is a nitridoalumosilicate featuring infinite one-dimensional chains of edge-sharing tetrahedral that are most probably centered solely by aluminum. Particularly interesting is the geometric situation inside these chains, which are almost linear. These trans-linked chains are connected exclusively through common corners with the highly corrugated $[SiN_4]$ layers forming a three-dimensional network (Fig. 16.49) pervaded with channels alongside the chains. The channels host two different Sr^{2+} positions that are coordinated by irregular polyhedra made up of six or eight nitrogen atoms [56]. Figure 16.50 presents the photoluminescence spectra of $SrAlSi_4N_7:Eu^{2+}$. The reflectance spectrum shows broad absorption bands in the UV to green spectral region. For blue excitation, a broadband spectrum peaking at 635 nm is obtained showing a spectral width of FWHM of 2955 (75) cm^{-1}. Values for Stokes shift and the lowest-lying absorption band position are thus $\Delta S = 4823(\pm 100)\ cm^{-1}$ and approximately 494 nm, respectively [154]. When the sample is excited at 450 nm, the absorption and external quantum efficiency are 76 and 53 %, respectively. In addition, this red-emitting phosphor shows high absorption and quantum efficiency under UV to near-UV irradiation, thus indicating that $(Sr, Eu)AlSi_4N_7$ is an attractive red phosphor candidate for white LEDs when coupled to blue, near-UV, or UV LEDs [57].

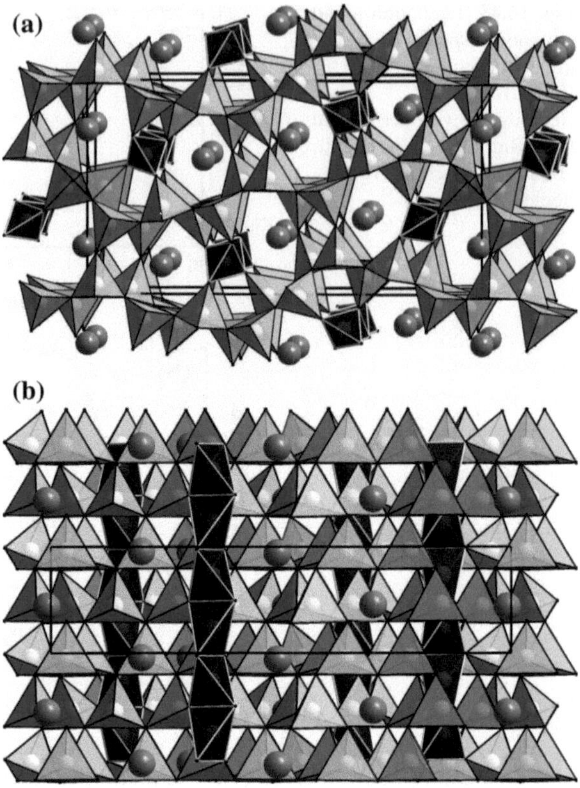

Fig. 16.49 **a** Structure of $SrAlSi_4N_7$ viewed along [001]; chains of edge sharing tetrahedra are marked in *black*, Sr are *gray spheres*, and $[SiN_4]$ tetrahedra are *light gray*. **b** viewed along [100]. The chains of edge-sharing tetrahedra are marked in *dark gray*. Reprinted with permission from Ref. [154]. Copyright 2009 American Chemical Society

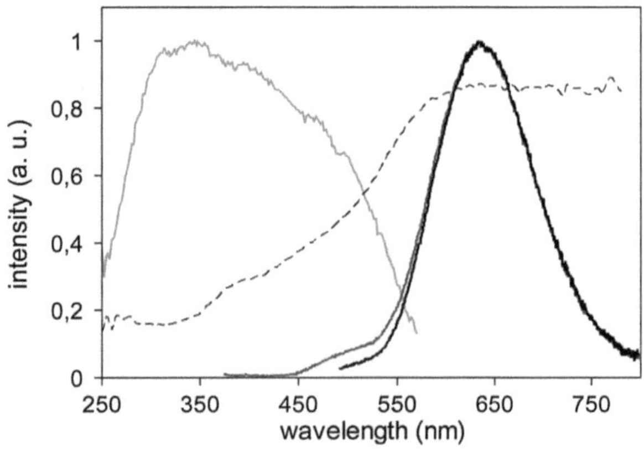

Fig. 16.50 The emission spectrum for 450 and 350 nm excitation (*black* and *dark gray curves*, respectively), excitation spectrum monitored at 635 nm (*light gray curve*), and reflection spectrum (*dashed line*) of $Sr_{0.98}AlSi_4N_7:Eu_{0.02}$. Reprinted with permission from Ref. [154]. Copyright 2009 American Chemical Society

16.2.3.12 $MSiN_2:Eu^{2+}/Ce^{3+}$

The crystal structures of $BaSiN_2$ and $SrSiN_2$ are illustrated in Fig. 16.51. The structure of $SrSiN_2$ is closely related to that of $BaSiN_2$ containing both edge-sharing

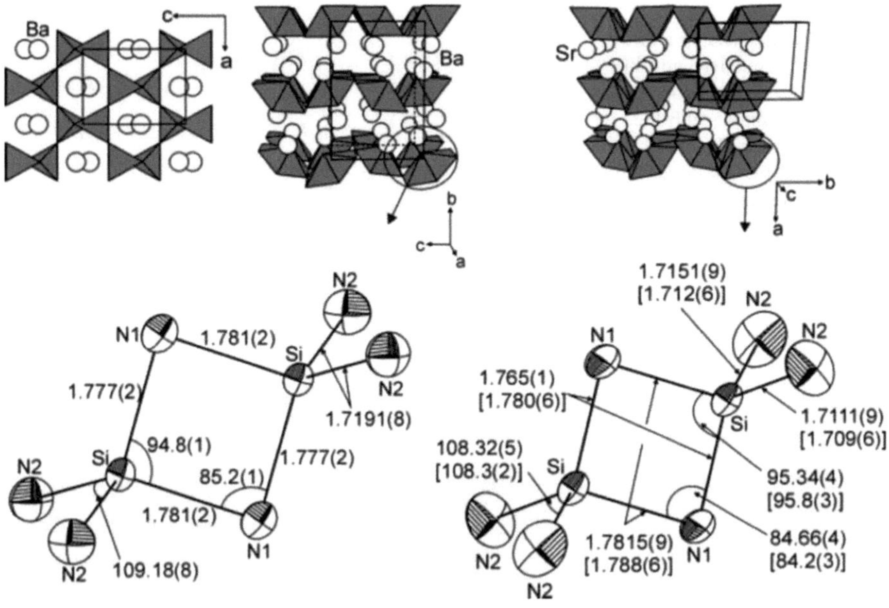

Fig. 16.51 Crystal structures of $BaSiN_2$ and $SrSiN_2$. Reprinted with the permission from Ref. [154]. Copyright 2009 American Chemical Society

and corner-sharing SiN_4 tetrahedra. Pairs of SiN_4 tetrahedra share edges to form "bow tie" units of Si_2N_6, and these units condense by vertex sharing of the four remaining unshared vertexes to form puckered two-dimensional sheets separated by Sr or Ba atoms. There is only one crystallographic site for Sr and Ba atoms, which are both connected to eight N atoms [57].

As seen in Fig. 16.52, $CaSiN_2$ contains highly condense SiN_4 tetrahedra that are joined only at their vertexes. The Ca atoms are hosted in the channels along [100], which are formed by the tetrahedral network. There are two different crystallographic positions for Ca atoms. The Ca1 atom is surrounded by four N atoms located between 2.40 and 2.49 Å and by two further N atoms with distances of 2.79 and 3.02 Å. The Ca2 atom is in a highly distorted octahedral environment, again connected to four N atoms with shorter distances (between 2.43 and 2.48 Å) and to two N atoms with longer distances (between 2.79 and 2.83 Å). Because all SiN_4 tetrahedra are corner-shared, all of the N atoms are coordinated to two Si atoms [57]. The photoluminescence of the Ce^{3+}-doped $CaSiN_2$ phosphor with cubic structure was reported by Le Toquin and Cheetham [127], and that of both Eu^{2+}- and Ce^{3+}-doped $MSiN_2$ (M = Sr, Ba) was further investigated by Duan et al. [156] and Li et al. [99]. Both $SrSiN_2$:Eu^{2+} and $BaSiN_2$:Eu^{2+} show red emission under blue-light irradiation. The 5*d* energy levels were split into three dominant band, resulting in a broad excitation spectrum for both samples. The emission spectrum shows a single broad band centered at 670–685 nm ($SrSiN_2$:Eu^{2+}) and 600–630 nm ($BaSiN_2$:Eu^{2+}), depending on the Eu concentration. The FWHM of $BaSiN_2$:Eu^{2+} is

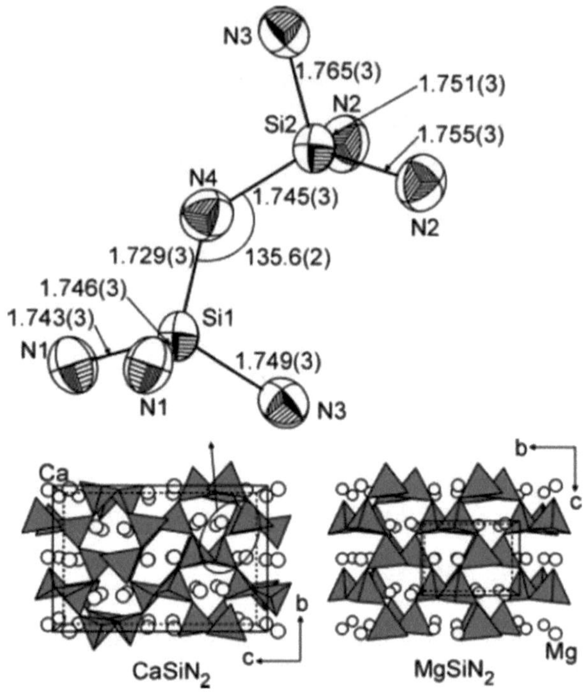

Fig. 16.52 The Cristobalite-derived structure of $CaSiN_2$ compared with the wurtzite-derived structure of $MgSiN_2$ Reprinted with the permission from Ref. [154]. Copyright 2009 American Chemical Society

smaller than that of $SrSiN_2$:Eu^{2+} due to the higher symmetry of Ba atoms. Duan et al. reported that quantum efficiency of $BaSiN_2$:Eu^{2+} and $SrSiN_2$:Eu^{2+} is 40 and 25 % with respect to that of YAG:Ce^{3+} [156]. The Ce^{3+} luminescence is quite different for $MSiN_2$ with different alkaline-earth metals. The cubic $CaSiN_2$:Ce^{3+} emits red color; $SrSiN_2$:Ce^{3+} emits green; and $BaSiN_2$:Ce^{3+} emits blue. This difference can be related to the strength of the crystal field acting on the alkaline-earth metal ions in the structure. Among these materials, $CaSiN_2$:Ce^{3+} exhibits the largest crystal-field splitting leading to a broad excitation band. Furthermore, $CaSiN_2$:Ce^{3+} can be excited either by UV to near-UV light or by blue light, whereas both $SrSiN_2$:Ce^{3+} and $BaSiN_2$:Ce^{3+} can only be excited by near-UV light [57].

16.2.4 Summary

Nitride phosphors are playing an increasingly important role in producing highly efficient and reliable white LEDs. Although searching for new host materials is continuously pursued, it is actually an urgent task to improve the performance of the developed phosphors. Compositional tailoring is a commonly used method to improve/modify photoluminescence properties, which has been addressed in the article. By applying the post-treatment, the particle size, morphology, and particle size distribution, and the surface defects can be controlled. In addition, the co-doping also provides an effective way to enhance the luminescence intensity in some cases by taking advantage of the energy-transfer principle. Research on the performance enhancement of phosphors cannot go far if one does not have deep insight into the structural analysis. Moreover, the origins of the property improvement or thermal degradation can only be understood by the comprehensive structure analysis. Therefore, it is extremely essential to clarify the surface and local structures, nanoscale structure, dopant valence and distribution, and band-gap structure by means of various analytic techniques and facilities.

16.3 Long-Persistent Phosphors

16.3.1 Introduction to Long-Persistent Phosphors

Long-persistent phosphors are also called "long lasting," "long duration," "long lived," or "long afterglow" phosphors. Long-persistent phosphors are one of the most important states luminescent materials, and they have very long afterglow emission after the excitation (or irradiation) source is switched off. The excitation resource used may be visible light or UV, X-ray, or gamma radiation. Long-persistent luminescence as a phenomenon dates from the beginning of the 17th century. In 1602, an Italian shoemaker, V. Casciarolo, observed strong luminescence from $BaSO_4$ mineral barite. It appears, however, that the material actually was not the barite mineral itself but rather the (BaS) barium sulfide [157]. At that time, during the following decades, the bright emission from the barium sulfide was discovered: The phenomenon aroused the interest of both scientists and laymen, and even several books were written on this miraculous phenomenon. However, this phenomenon clearly went beyond the knowledge of the scientists of that time, and no explanation was found for the persistent luminescence produced without any evident excitation source. Until today, the research for new persistent luminescent compounds has become increasingly popular (see Fig. 16.53). Quite surprisingly, the number of known compounds with a decent afterglow brightness and lifetime is still rather limited. Fortunately, scientists have confirmed that the long decay time of persistent luminescence is due to the storage of the excitation energy by traps, which is then released with thermal energy. Thus, the term "thermally stimulated luminescence" (TSL) is appropriate; however, for the sake of brevity, herein the phenomenon is called "persistent luminescence." Scientists still do not know clearly the luminescence mechanisms of long-persistent phosphors. Different luminescence mechanisms of persistent phosphors have been put forward. The investigation on the long persistent luminescence still continues.

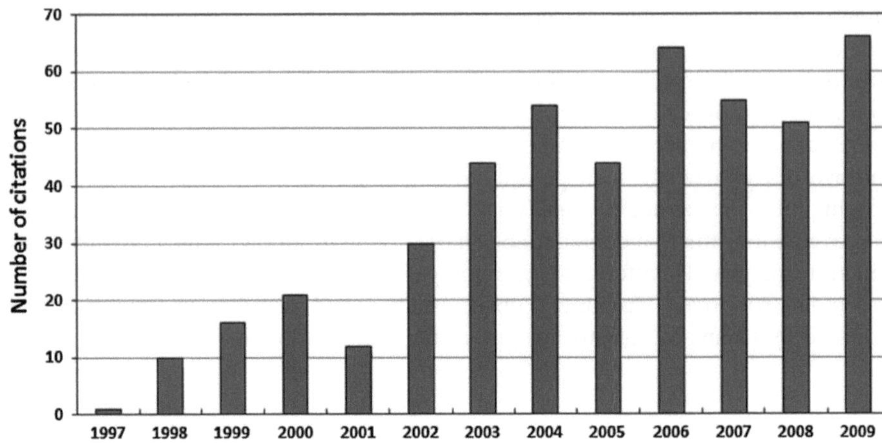

Fig. 16.53 Number of citations in the 1996 paper by Matsuzawa et al. [158]

16.3.2 Luminescent Properties of Long-Persistent Phosphors

According to different host materials, long-persistent phosphors are divided into sulfide, oxysulfide, aluminates, and silicate phosphors, etc. In the next paragraphs, we introduce the luminescent properties of these long-persistent phosphors.

a. **Long-persistent sulphide phosphors**

CaS:Bi, (Ca, Sr)S:Bi, (Zn, Co)S:Cu, and ZnS:Cu are the traditional long-persistent sulfide phosphors [159, 160]. The most famous long-persistent phosphor is ZnS doped with Cu. It has been used in many commercial products including watch dials, luminous paints, and glow-in-the-dark toys. The emission was modified with a partial substitution of Zn with Co. However, at its best, persistent luminescence from these materials was both weak and short, i.e., lasting for a few hours only. Some environmentally dubious "tricks," such as the doping of ZnS with Eu^{2+}, Dy^{3+}, and Er^{3+} elements, were used to prolong the duration of persistence with external excitation. These tricks were only retarding the final rejection of ZnS-based phosphors also due to their pronounced instability against humidity. The main luminescent properties of long-persistent sulfide phosphors are shown in Table 16.5.

b. **Long-persistent oxysulfide phosphor**

$Y_2O_2S:Eu^{3+}$ oxysulfide is a new kind of long-persistent phosphor. Figures 16.54 and 16.55 display the emission spectrum of $Y_2O_2S:Eu^{3+}$ and the afterglow decay curves of Mg^{2+}- and Ti^{4+}-sensitized $Y_2O_2S:Eu^{3+}_{0.06}$, respectively. Under 330-nm excitation, $Y_2O_2S:Eu^{3+}$ shows a strong red emission. After doping with Mg^{2+} and Ti^{4+} ions, the afterglow decay time of $Y_2O_2S:Eu^{3+}$ extends >2 h. Despite the high

Table 16.5 Luminescent properties of long-persistent sulfide phosphors

Composition	Emission color	Emission peak (nm)	After-glow brightness (after 10 min) ($mcd \cdot m^{-2}$)	After-glow persistence time (min)
CaS:Eu,Tm	Red	650	1.2	~45
(Ca,Sr)S:Bi	Blue	450	5	~90
ZnS:Cu	Yellow-green	530	45	~200
(Zn,Co)S:Cu	Yellow-green	530	40	~500

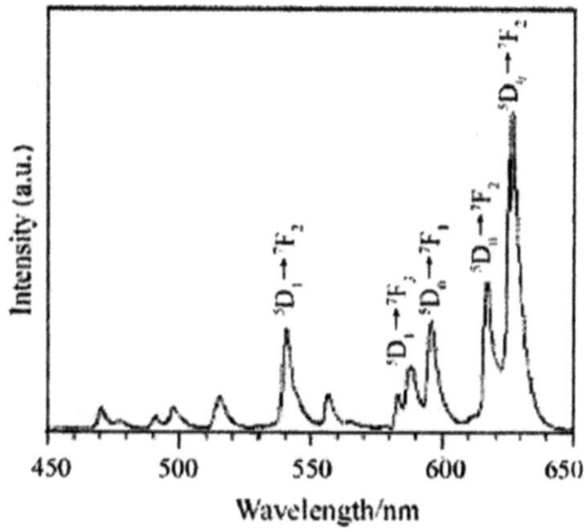

Fig. 16.54 Emission spectrum of $Y_2O_2S:Eu^{3+}$ (λ_{ex} = 330 nm). Reprinted from Ref. [161], Copyright 2006, with permission from Elsevier

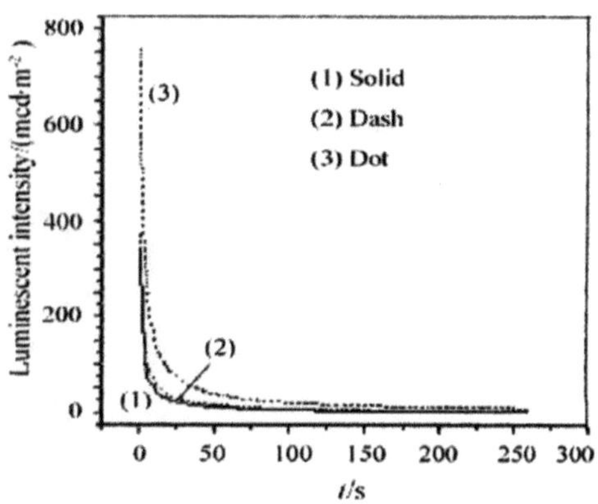

Fig. 16.55 Afterglow decay curves of *1* Y_2O_2S:6 %Eu^{3+}, 2 %Mg^{2+}; *2* Y_2O_2S:6 %Eu^{3+}, 2 %Ti^{4+}; and *3* Y_2O_2S:6 % Eu^{3+}, 2 %Mg^{2+}, 2 %Ti^{4+}. Reprinted from Ref. [161], Copyright 2006, with permission from Elsevier

initial persistent luminescence of $Y_2O_2S:Eu^{3+}$, Mg^{2+}, Ti red-emitting material, its long-term luminescence is much weaker than the duration of its blue-emitting ($Sr_2MgSi_2O_7:Eu^{2+}$, Dy^{3+}, approximately 25 h) and green-emitting ($SrAl_2O_4:Eu^{2+}$, Dy^{3+}; approximately 15 h) counterparts.

c. **Long-persistent aluminates phosphors**

Aluminates phosphors have been used as the important persistent phosphor from 1996. In August 1996, Matsuzawa et al. published an article reporting that by co-doping $SrAl_2O_4:Eu^{2+}$ with the rare-earth element dysprosium (Dy^{3+}), they were able to create a material, $SrAl_2O_4:Eu^{2+}$, that was able to emit bright light for hours after the excitation was ended. They found that the afterglow had both a far higher initial intensity and a much longer persistence compared with ZnS:Cu, Co (Fig. 16.56). Their discovery attracted researchers' focus on alkaline-earth aluminates. The good emission intensity and long-afterglow properties of a series of Eu^{2+} doped MAl_2O_4 (M = Ca, Sr, Ba) are reported. Under the radiation of sunlight, $SrAl_2O_4:Eu^{2+}$, Dy^{3+} can emit green light in the region of 400–520 nm due to the *4f-5d* transition of Eu^{2+}. Adding Dy^{3+} as the auxiliary activator dramatically increases the initial brightness. The brightness is approximately 60 m cd/m^2 after 1 h after the excitation resource is switched off. The afterglow persistence is >2000 min.

$CaAl_2O_4:Eu^{2+}$ and $CaAl_2O_4:Eu^{2+}$, Nd^{3+} are other important long-persistent phosphors. They can be efficiently excited by a photon with the wavelength at 225–400 nm and emit a broadband blue emission in the region of 400–500 nm due to the *4f-5d* transition of Eu^{2+}. Adding Nd^{3+} as the auxiliary activator increases the after-glow persistence time as shown in Fig. 16.57. The luminescent characteristic of the main long-persistent aluminates is shown in Table 16.6. It is obvious that long-persistent aluminates phosphors have the perfected emission and long after-glow properties, but the brightness color is monotonous, and the emission wavelength locates only in the region of 440–520 nm. In addition, the photoluminescent properties are unstable in a damp environment.

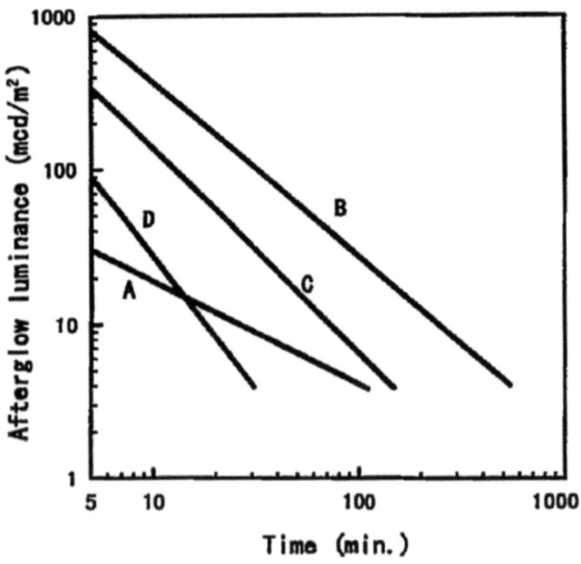

Fig. 16.56 Comparison of afterglow characteristics measured after 10 min of exposure to 200 lx of D_{65} light. *A* $SrAl_2O_4:Eu^{2+}$; *B* $SrAl_2O_4:Eu^{2+},Dy^{3+}$; *C* $SrAl_2O_4:Eu^{2+},Nd^{3+}$; *D* ZnS:Cu,Co [158]

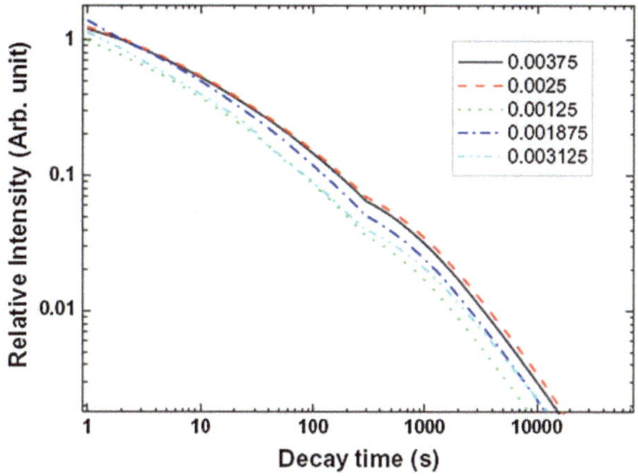

Fig. 16.57 Afterglow characteristics of $Ca_{(1-x-y)}Al_2O_4:Eu_x^{2+}$, Nd_y^{3+}(x = 0.00125, 0.00125 $\leq y \leq$ 0.00375). Reprinted with permission from Ref. [162]. Copyright 2007, American Institute of Physics

Table 16.6 Luminescent characteristic of long persistent aluminates phosphors

Composition	Emission maximum (nm)	After-glow time (min)	References
$CaAl_2O_4:Eu^{2+},Nd^{3+}$,	440	>1000	[162]
$Sr_4Al_{14}O_{25}:Eu^{2+},Dy^{3+}$	490	>2000	[163]
$SrAl_2O_4:Eu^{2+},Dy^{3+}$	520	>2000	[164]
$BaAl_2O_4:Eu^{2+},Dy^{3+}$	500	>120	[165]
$SrAl_4O_7:Eu^{2+},Dy^{3+}$	480	>180	[166]
$SrAl_{12}O_{19}:Eu^{2+},Dy^{3+}$	400	>180	[166]
$SrMgAl_{10}O_{17}:Eu^{2+},Dy^{3+}$	460	>3	[167]
$BaMgAl_{10}O_{17}:Eu^{2+},Dy^{3+}$	450	>5	[168]

d. **Long-persistent silicate phosphors**

Silicate phosphors have attracted more attention and have been widely used in the lighting and display field for their thermally and chemically, inexpensive raw material (SiO_2). In 2001, Lin et al. reported a bright and long-lasting afterglow in $Sr_2MgSi_2O_7:Eu^{2+}$, Dy^{3+}; shortly afterward other doped silicates were found to exhibit an equally long afterglow. After 2001, many new long-persistent silicate phosphors were reported and are shown in Table 16.7. $aMO{\cdot}bM'O{\cdot}cSiO_2{\cdot}dR$: Eu_x, Ln_y is a kind of new long-persistent silicate phosphor that was prepared by Chinese scientists. M and M′ is the alkaline-earth metal elements. R, such as B_2O_3 、P_2O_5 is the flux. Ln is the rare-earth ions or transition elements. a, b, c, d, x, and y is the molar coefficients, $0.6 \leq a \leq 6, 0 \leq b \leq 5, 1 \leq c \leq 9, 0 \leq d \leq 0.7, 0.00001 \leq x \leq 0.2$, $0 \leq y \leq 0.2$. These kinds of phosphors have stable, good afterglow properties and multiple brightness colors. SB and SBG are the grade of the product by the Luming Technology Group Co., Ltd, which includes two persistent phosphors, and their compositions are $aMO{\cdot}bM'O{\cdot}cSiO_2{\cdot}dR:Eu_x$, Ln_y. The main luminescent properties of SB and SBG are shown in Table 16.8. The

Table 16.7 Long persistent silicate phosphors

Composition	Emission maximum (nm)	Afterglow duration (h)	References
$Sr_2MgSi_2O_7:Eu^{2+},Dy^{3+}$	470 (blue)	>10	[166]
$Ca_2MgSi_2O_7:Eu^{2+}$ $(,Tb^{3+})$	515/535 (green)	>5	[169, 170]
$Ba_2MgSi_2O_7:Eu^{2+},Tm^{3+}$	505 (green)	>5	[171]
$Sr_3MgSi_2O_8:Eu^{2+},Dy^{3+}$	460 (blue)	>10	[172, 173]
$Ca_3MgSi_2O8:Eu^{2+},Dy^{3+}$	470 (blue)	>5	[172]
$Ba_3MgSi_2O_8:Eu^{2+},Dy^{3+}$	440 (blue)	>1	[172]
$CaMgSi_2O_6:Eu^{2+},Dy^{3+}$	445 (blue)	>4	[174]
$Sr_3Al_{10}SiO_{20}:Eu^{2+},Ho^{3+}$	465 (blue)	>6	[175]
$CaAl_2Si_2O_8:Eu^{2+},Dy^{3+}$	435 (blue)	>3	[176]
$Sr_2Al_2SiO_7:Eu^{2+},Dy^{3+}$	485 (blue/green)	>2	[177]
$Sr_2ZnSi_2O_7:Eu^{2+},Dy^{3+}$	460 (blue)	>5 min	[178]
$Sr_2SiO_4:Eu^{2+},Dy^{3+}$	480 (green)	>5 min	[179]

Table 16.8 Luminescent properties of SB and SBG

Composition	Brightness colour	Luminescence at peak (nm)	After-glow brightness (mcd/m^2)				After-glow persistence time (min)
			1 min	10 min	30 min	60 min	
SB	Blue	469	500	87	28	14	>2000
SBG	Blue-green	490	650	121	39	19	>1000

afterglow persistence time is >2000 min The afterglow brightness is approximately 20 mcd/m^2 at 60 minters after the excitation is switched off.

e. **Other long-persistent phosphors**

In addition to the previously discussed phosphors, only a few compounds are known to exhibit long-persistent phosphors as seen in Table 16.9. Many of these originate from LED research and are also commonly used as conversion phosphors.

Table 16.9 Other long persistent luminescent phosphors

Composition	Brightness color	Afterglow duration (h)	References
p-$Zn_3(PO_4)_2$:Mn	Red	>2	[180]
$MTiO_3$:Eu (M = Ca, Sr, Ba)	Red	>4	[181]
$Sr_2P_2O_7:Eu^{2+},Y^{3+}$	Blue	>8	[182]
$Ca_2P_2O_7:Eu^{2+},Y^{3+}$	Blue	>6	[183]
$SrMg_2P_2O_8:Eu^{2+},Ce^{3+}$	Blue	>2	[184]
$Ca_2Si_5N_8:Eu^{2+},Tm^{3+}$	Orange	>1	[185]
$CaAl_2B_2O_7:Eu^{2+},Nd^{3+}$	Blue	>1	[186]
$SrB_2O_4:Eu^{2+}$	Blue	unknown	[187]

16.3.3 Luminescent Mechanisms of Long-Persistent Phosphors

At present, it has been widely accepted that persistent phosphors luminesce though the following processes. First, the persistent phosphors are excited, and then charge carriers form and could get caught by so-called "traps," which are energy levels inside the forbidden band gap that have a very long lifetime. The charge carriers are only gradually released from these traps, after which they can return to the activators and produce luminescence. Two kinds of traps exist in persistent phosphors. One is the electronegative center, corresponding to hole trapping (HT). The other is the electropositivity center, corresponding to electron trapping (ET). The traps play an important role in the persistent luminescence. One of their main properties is their "depth," which the activation energy needs to release a captured charge carrier. To observe persistent luminescence at room temperature, a trap depth of approximately 0.65 eV is considered optimal [158].

i. **Transfer of positive-hole mode**

The transfer-of-positive-hole mod was first suggested by Matsuzawa et al. to explain the extraordinary persistent luminescence of $SrAl_2O_4:Eu^{2+}$, Dy^{3+}, so this model is also called the "Matsuzawa model." In this model, the investigators explained the persistent luminescence using the theory of luminescence in semiconductors, i.e., they mainly analyzed the captured electrons and holes, the local level caused by the impurity ions, and the energy transfer between the conduction bands and the valence bands. A schematic picture of the transfer-of-positive-hole mode for $SrAl_2O_4:Eu^{2+}$, Dy^{3+} is shown in Fig. 16.58. Holes are assumed to be the main charge carriers. This assumption is based on earlier measurements by Abbruscato on non-co-doped $SrAl_2O_4:Eu^{2+}$. When a Eu^{2+} ion is excited by an incident photon (ultraviolet or sunlight), the *4d-5f* energy level transition of Eu^{2+} occurs; there is a possibility that a hole escapes to the valence band, thereby leaving behind a Eu^{+} ion. Parts of holes recombine Eu^{+} to emit.

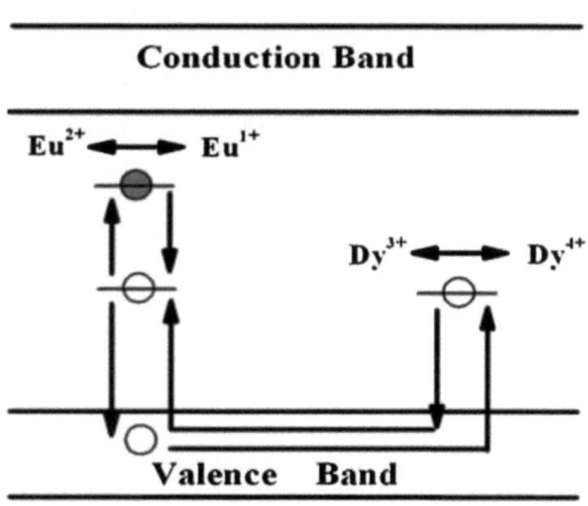

Fig. 16.58 Schematic picture of the transfer of positive hole mode for $SrAl_2O_4:Eu^{2+}$, Dy^{3+} [158]

Another hole is then captured by a Dy^{3+} ion, thus creating a Dy^{4+} ion. After a while, thermal energy causes the trapped hole to be released into the valence band again. From there it can move back to a Eu^{+} ion, allowing it to return to the Eu^{2+} ground state with the persistent emission of a photon [158].

The energy-level depth of the doping ions decides the persistence time and the emission intensity. When the depth is low, the ions rapidly release the holes and do not actively take part in persistent emission processes at room temperature. When the depth is deeper, more energy is required, and the ions slowly release the holes. Therefore, to observe persistent luminescence at room temperature, the depth should have appropriate activation energy somewhere between these two extremes. The hole in the valence bands has been confirmed by experiments, but the transfer-of-positive-hole mode has not been accepted completely. The main question is whether or not the Eu^{+} ion exists. It is well known that the trivalent ion of lanthanide is stable and that the tetravalence ion of lanthanide is hard to form under visible-light excitation. There is no evidence to indicate that Eu^{+}, Er^{4+}, Ho^{4+}, Dy^{4+} ions exist in the host lattice.

ii. **Configuration coordinate model**

To solve the defect of the transfer-of-positive-hole mode, Qiu and Qiang et al. suggested the configuration-coordinate model. In this model, persistent phosphors are excited first, and then the excitation electron transits from the high-energy to the low-energy level followed by emission of a photon. The longer afterglow time is due to a decrease of the electron-transition efficiency or the extension of electron-transition time. Figure 16.59 shows a simple configuration coordinate model for Eu^{2+}: A is the ground state of Eu^{2+}; B is the excitation state of Eu^{2+}; and C is the defect level, which locates between A and B. The defect level could capture the excitation electron formed by the lattice distortion because of the doping ions, e.g., Sr^{2+} ion replacing M^{2+} ion, or the defect of the host lattice, e.g., oxygen vacancies. Qiu et al. believe that this defect level captures not only the electron but also the hole. The energy released on recombination of the electron and the hole is delivered directly to the europium ions by means of energy transfer; thus persistent luminescence occurs. The numbered lines in Fig. 16.59 respectively display the electron transits from the ground state to the excitation state (line 1), part of the excitation electrons returning to the ground state (line 2), and the excitation electron relaxing and storing in the defect energy level (line 3). The afterglow time is decided by both the electron quantities and the absorbed energy. The quantities of

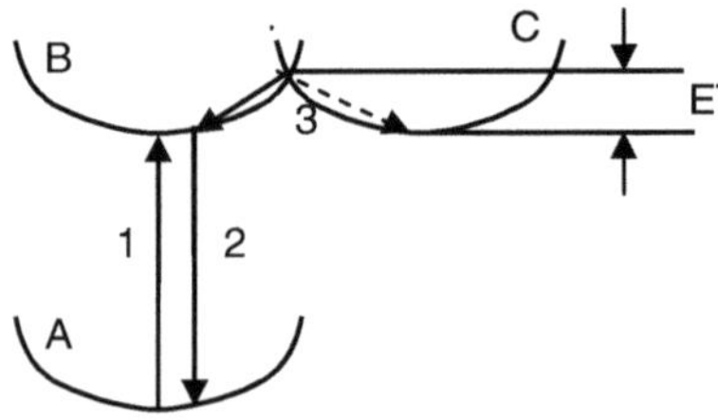

Fig. 16.59 Configuration coordinate model. Reprinted from Ref. [189], Copyright 1998, with permission from Elsevier

electrons in the defect energy level become greater with longer afterglow time. The absorbed energy is used to facilitate the electron transition from C to B (ET) and thus realize the persistent luminescence. However, this does not mean that persistence time is longer with greater absorbed energy. The configuration-coordinate model is unable to explain the light- and thermal-conductance phenomena. In addition, the explanation of how the hole is produced and the role of the hole in afterglow luminescence is not sufficiently clear.

iii. **Energy-transfer model**

The energy-transfer mode was first put forward by Tuomas Aitasalo et al.; thus, this model is also called the "Aitasalo model" [190]. In this model, the electron defect, hole defect, and Eu^{2+} combine a interrelated system because of the electrostatic forces. Electrons are excited directly from the valence band into trap levels of unspecified origin. The hole created in this way migrates toward a vacancy where it is caught. The electron is removed from the trap level by thermal energy and ends up at an oxygen-vacancy level. Because the conduction band is located too high above the energy level of the oxygen-vacancy trap to enable a thermally assisted transition to the conduction band, the investigators assumed that the energy released on recombination of the electron and the hole was delivered directly to the europium ions by means of energy transfer (see Fig. 16.60) [190]. The energy-transfer model lacks experimental data and is unable to explain the light- and thermal-conductance phenomena. In addition, this model does not consider the electron transfer in the conduction band and the role of Dy^{3+} in the afterglow luminescence.

iv. **Thermoluminescence model**

Jorma Holsa et al. investigated the luminescence properties of different trivalent rare-earth ions on $SrAl_2O_4{:}Eu^{2+}$, Dy^{3+} and suggested the thermoluminescence

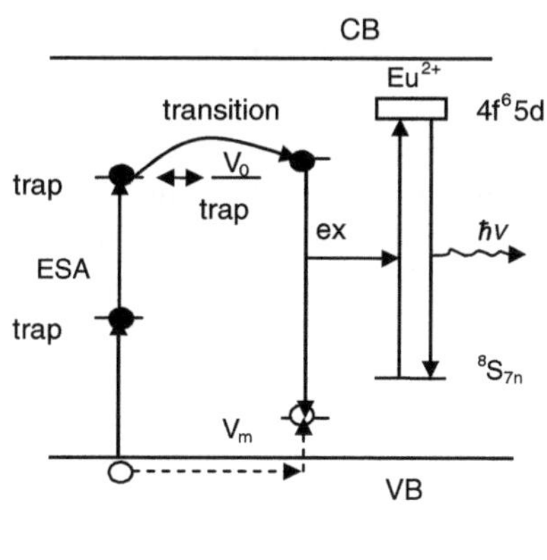

Fig. 16.60 Energy-transfer model. Reprinted from Ref. [190], Copyright 2001, with permission from Elsevier

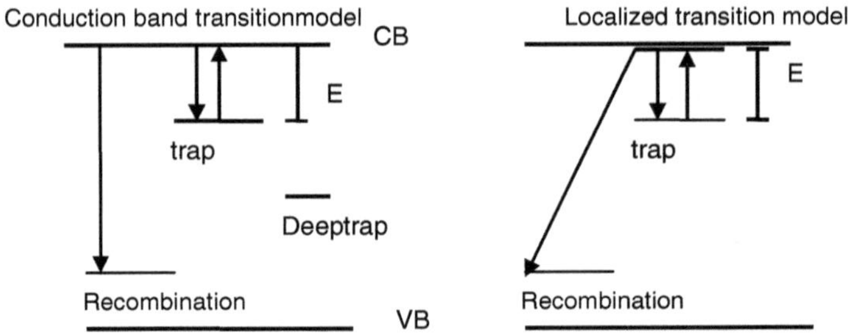

Fig. 16.61 Thermoluminescence model. Reprinted from Ref. [191], Copyright 2001, with permission from Elsevier

model. In this model, afterglow luminescence is the result of the energy released on recombination of the electron and the hole by thermal energy and delivered directly to the europium ions. Two thermoluminescence models were put forward as seen in Fig. 16.61. One is the conduction band-transition model, and the other is the localized-transition model. In the conduction band-transition model, the trapped electron and hole ore of uniform distribution in the crystal. However, the distance between them is father. The energy released from the recombination of the trapped electron by way of the conduction band and the hole by way of the valence band by thermal energy transfers the energy to the luminescence centre Eu^{2+}. In the localized-transition model, the trapped electron and hole are paired close to each other. The energy released from the recombination of the trapped electron and the hole by thermal energy is directly transferred to the luminescence centre Eu^{2+}. In the thermoluminescence model, the reason why the electron-hole pair appears is not explained perfectly. The effect of Dy^{3+} and Nd^{3+} for the persistent phosphors is not clear enough, and the provided evidence is insufficient.

16.3.4 Summary

Currently, the investigation on long-persistent phosphors has attracted more attention. In the future study, the main work is to choose good chemical stability, strength afterglow, and persistent phosphors with long afterglow time. The activator and sensitizer ions should have the appropriate trap and the optimal energy transfer and energy storage. The excitation spectrum extends to the visible area as much as possible. In addition, combining a large number of experimental results can put forward a reasonable model of long-afterglow luminescence, which will play a guiding role to synthesize or design new long-persistence luminescent materials. Meanwhile, how to save energy and protect the environment must be considered.

16.4 Oxide Phosphors for White LEDs and Plasma-Display Panels (PDPs)

16.4.1 Oxide Phosphor for White LEDs

$Y_3Al_5O_{12}$:Ce^{3+}(YAG:Ce^{3+})

This phosphor is referred to in many patents for use as a color converter using blue-LED excitation. It emits yellow light, which, when combined with more blue light from the LED, gives a passable white light. The excitation spectrum of YAG: Ce^{3+} has its peak at approximately 470 nm (Fig. 16.62). In the same figure, the emission spectrum is shown to be asymmetrical because it is made up of two transitions $^5D \rightarrow {}^2F_{5/2}$ and $^5D \rightarrow {}^2F_{7/2}$. Both the asymmetry and the position of this band varies depending on the Ce^{3+} concentration, although the center of the peak is usually at approximately 550 nm. Also, Fig. 16.63 shows the optimum position of an exciting blue emission (as from a blue LED emitting at 470 nm) [192–194]. In Fig. 16.63, the position of the color coordinates of YAG:Ce^{3+} are shown at point 7, whereas those of the exciting blue LED light are shown at point 1. Points 2–6 are generated by putting layers of YAG:Ce^{3+} phosphor particles on glass and exciting them with blue light so some blue light also comes through the phosphor layer along with the yellow emission. Point 2 is created by one layer of YAG:Ce^{3+} particles; point 3 is created by two layers and so on to point 6, which contains five layers [192–195]. Clearly, points 3 and 4 in Fig. 16.63 are close to the white color point (x = 0.33, y = 0.33), and a line between these points represents the best "whites" that can be generated from a blue LED emitting at 470 nm and YAG:Ce^{3+} phosphor particles. To improve these whites, it is necessary to chemically modify the YAG:Ce^{3+} phosphor particles. There are several ways this can be performed as shown in Fig. 16.64 [193]. It can be seen that by adding Pr^{3+} cations to YAG:Ce^{3+}, an emission peak at approximately 620 nm (in the red) is added to the emission spectrum of YAG: Ce^{3+}. In addition, by adding some Gd^{3+} to replace some Y^{3+}, the main YAG:Ce^{3+} emission moves toward the red spectrum. As can be

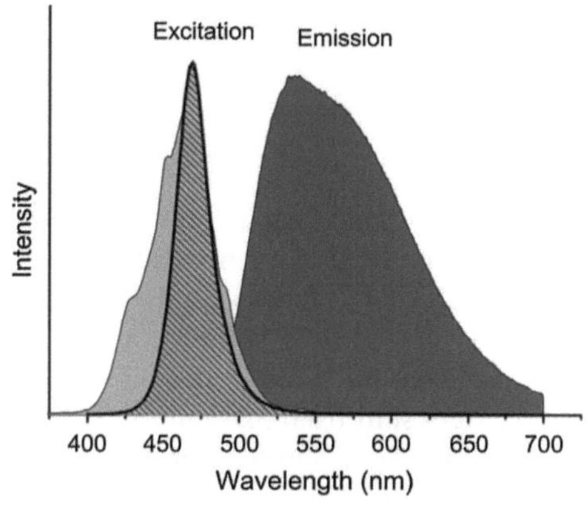

Fig. 16.62 Excitation and emission spectra of YAG: Ce^{3+}. The hatched peak depicts an exciting emission from a blue LED emitting at 470 nm. Reprinted with permission from Ref. [193]. Copyright 2005 Society of Photo Optical Instrumentation Engineers

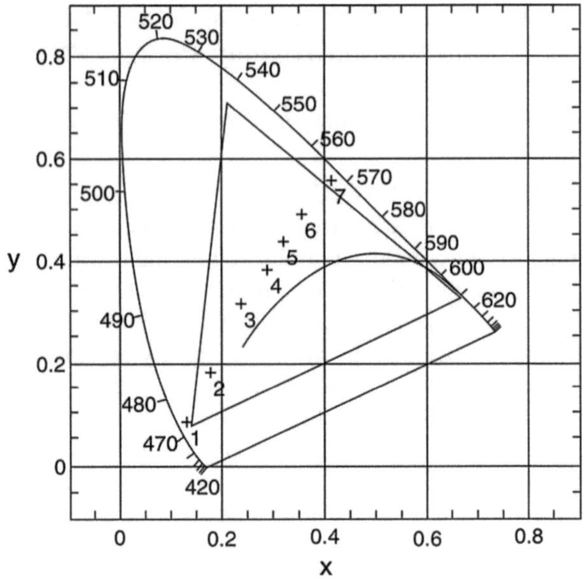

Fig. 16.63 CIE chromaticity diagram showing the position of the emission of YAG:Ce^{3+} phosphor (point 7) and the emission of a blue 470 nm LED (point 1). The other points, i.e., 2 through 6 are explained in the text. The continuous *curved line* in the center of the diagram is known as the Planckian locus, which is the locus of blackbody emitters. Reprinted with permission from Ref. [193]. Copyright 2005 Society of Photo Optical Instrumentation Engineers

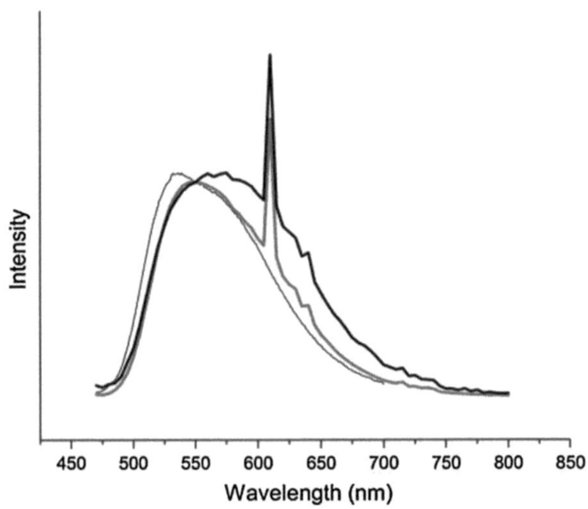

Fig. 16.64 Emission spectra of YAG:Ce^{3+} (*thin black line*), $Y_3Al_5O_{12}$:Ce^{3+}, Pr^{3+} (*gray line*) and (Y, Gd)$_3Al_5O_{12}$:Ce^{3+}, Pr^{3+} (*thick black line*). Reprinted with permission from Ref. [193]. Copyright 2005 Society of Photo Optical Instrumentation Engineers

seen in the same figure, both of these effects can be engineered in the same phosphor particles. An empirical rule has been put forward for Ce^{3+} luminescence in $Y_3Al_5O_{12}$:Ce^{3+}: Increasing the diameter of the ion on the dodecahedral (Y^{3+}) site increases crystal-field splitting, while increasing the diameter on the octahedral (Al^{3+}) site has the reverse effect [194, 195]. Thus, on the dodecahedral site there is a wavelength shift from 550 nm for Y^{3+} to 585 nm for the larger Gd^{3+} ion as well as and a shift from 550 nm for Y^{3+} to 510 nm for the smaller Lu^{3+} ion. This is apparent for (Y, Gd)$_3$Al$_5$O$_{12}$:Ce^{3+}, Pr^{3+} in Fig. 16.64. In contrast, when the cation is located on the octahedral site, there is a wavelength shift from 550 nm for Al^{3+} to 505 nm for the larger Ga^{3+} ion. However, substituting Ga^{3+} for Al^{3+} decreases the emission intensity. The basic YAG:Ce^{3+} lattice can accommodate >60 different elements, so many different combinations are possible.

Silicate garnets and related phosphors

Ca$_3$Sc$_2$Si$_3$O$_{12}$:Ce is excited at 446 nm and emits a broad band similar to that of YAG:Ce at approximately 570 nm in Fig. 16.65 [196]. A related phosphor system is BaY$_2$SiAl$_4$O$_{12}$:Ce. In a patent [197], a wide range of substitution is claimed based on the host lattice general formula MLn$_2$QR$_4$O$_{12}$, where M is at least one element selected from the group consisting of Mg, Ca, Sr and Ba; Ln is at least one rare-earth element selected from the group consisting of Sc, Y, La, Ce, Pr, Nd, Sm, Eu, Gd, Tb, Dy, Ho, Er, Tm, Yb, and Lu; Q is at least one element selected from the group consisting of Si, Ge, Sn, and Pb; and R is at least one element selected from the group consisting of B, Al, Ga, In, and Tl [197]. The luminescence of the vanadate garnet Ca$_2$NaMg$_2$V$_3$O$_{12}$:Eu^{3+} has been evaluated and discussed as a potential ultraviolet light-emitting diode (UV-LED) phosphor [198]. This single phosphor has been claimed to be capable of converting the ultraviolet emission of a UV-LED into white light with good luminosity and color-rendering index. The luminescence of this material at increased temperatures is of interest because the junction temperatures of typical LEDs can be >100 °C. Indeed, it has been reported [198] that there is significant thermal quenching of this phosphor and an emission color shift at temperatures >100 °C. This has been explained as energy migration and transfer to nonradiative traps and Eu^{3+} within the host lattice.

4. YCa$_3$M$_3$B$_4$O$_{15}$:Eu^{3+}

The photoluminescence properties of red-emitting YCa$_3$M$_3$B$_4$O$_{15}$:Eu^{3+} (M = Al or Ga) phosphors under UV (395 nm) irradiation have been reported. A strong

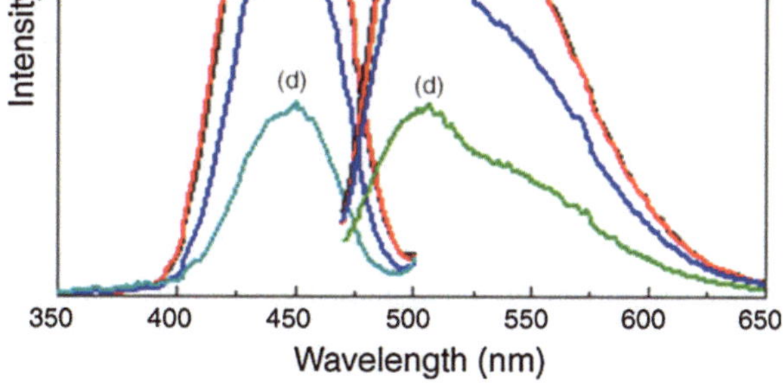

Fig. 16.65 PL excitation and emission spectra of Ca$_3$Sc$_2$Si$_3$O$_{12}$:Ce^{3+} samples at different firing temperatures. Reprinted from Ref. [196], Copyright 2010, with permission from Elsevier

emission at 622 nm and several weaker peaks near this emission were observed. The emission intensity increased with increasing Eu^{3+} content up to $x = 0.75$ in $Y_{1-x}Eu_xCa_3M_3B_4O_{15}$. The investigators claim that these phosphors are better than $CaS:Eu^{3+}$ for LED applications [199].

5. Alkaline-earth metal silicates $(Ba_{1-x-y}Sr_xCa_y)SiO_4:Eu^{2+}$
Barry [200] reported that the incorporation of Eu^{2+} in the compounds $Ca_3MgSi_2O_8$, $Sr_3MgSi_2O_8$, and $Ba_3MgSi_2O_8$ produces phosphors with high luminescence yield. The peak of the emission band occurs at progressively shorter wavelength and narrows as the radius of the major alkaline-earth ion (Ca, Sr or Ba) is increased. A 1200 °C, isotherm on compositions intermediate to $Sr_3MgSi_2O_8$ and $Ba_3MgSi_2O_8$ shows a complete series of crystalline solutions to exist at this temperature. Using an orthorhombic cell, the d-spacings of the (222) reflections vary continuously from 1.956 Å for $Sr_3MgSi_2O_8:Eu^{2+}$ to 2.012 Å for $Ba_3MgSi_2O_8:Eu^{2+}$. However, the emission spectra of these samples vary in a discontinuous manner. In general, the broadening of spectral energy distributions is observed as compositions move into the ternary section. All of the compositions prepared resulted in phosphors of relatively high efficiency [200]. Following from this work, a complete series of solid solutions, e.g., the Ba_2SiO_4–Sr_2SiO_4 system with a continuous shift of luminescent color from green (λ_{max} = 505 nm) to nearly orange (λ_{max} = 575 nm) for europium-doped materials, Starick et al. [201] reported interesting phosphors in the ternary $(Ba_{1-x-y}Sr_xCa_y)SiO_4:Eu^{2+}$ system. In contrast to Barry's work [200], regarding the isothermal run for all prepared phosphors, they preferred an individual optimisation of the preparation conditions for every promising member of the series [200].

It was found that the luminescent properties of these phosphors strongly depend on the conditions of their synthesis. Critical factors for getting a high-product quality include:

- careful selection of suitable raw materials (purity, reactivity);
- selection of good flux materials;
- control of the fi ring atmosphere (N_2/H_2);
- optimization of the heating and cooling regimes; and
- the posttreatment.

Phosphors with high crystal perfection, favorable particle-size distribution. and improved luminescent efficiencies have been reported. Their excitation spectra are broad and cover from <300–500 nm, whereas their broadband emission spectra range from 525 nm green to 595 nm orange. These alkaline-earth metal silicates of the general formula $M_2SiO_4:Eu^{2+}$ (where M = Ba, Sr or Ca) are color tunable (by suitable combinations of Ba, Sr, and Ca) green to yellow phosphors. Quantum efficiencies >90 % are said to be feasible. They have narrow emission bands for green phosphors, but they suffer from low quenching temperatures and are not stable at ambient temperatures [202–206]. Figure 16.66 shows the photoluminescence properties of $Sr_{1.99-x}Ca_xSiO_4:0.01Eu^{2+}$ samples.

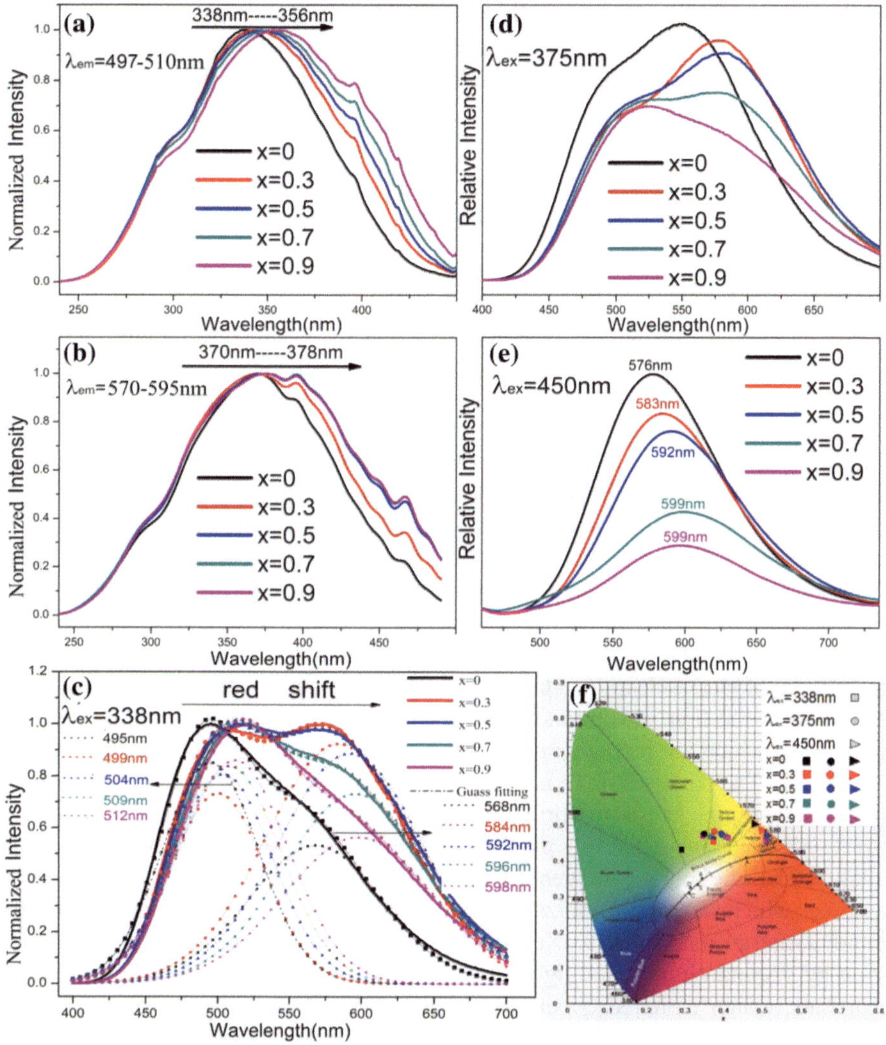

Fig. 16.66 **a** and **b** PLE spectra monitored at 497–510 and 570–595 nm of $Sr_{1.99\text{-}x}Ca_xSiO_4$: $0.01Eu^{2+}$ samples (x = 0, 0.3, 0.5, 0.7, 0.9); **c**, **d** and **e** PL spectra excited at 338, 375 and 450 nm of $Sr_{1.99\text{-}x}Ca_xSiO_4$: $0.01Eu^{2+}$ samples (x = 0, 0.3, 0.5, 0.7, 0.9); **f** CIE chromaticity coordinates of $Sr_{1.99\text{-}x}Ca_xSiO_4$: $0.01Eu^{2+}$ samples (x = 0, 0.3, 0.5, 0.7, 0.9). Reprinted from Ref. [202], Copyright 2015, with permission from Elsevier

White light-emitting phosphors with a nominal composition of $(Sr_{2.55}Ba_{0.4})MgSi_{1.7}O_8$:$Eu^{2+}$ that can be excited in the near-UV region have recently been reported. Their emission spectra consist of two bands in the visible region (one a blue band at 460 nm and the other a yellow band at 575 nm); these are thought to originate from Eu^{2+} in both $Sr_3MgSi_2O_8$ and Sr_2SiO_4, respectively. A white LED was fabricated from combining this phosphor with a near-UV InGaN chip (λ_{em} = 405 nm). The LED is claimed to manifest high CRI (>80), high color stability, very good color coordinates (CIE, $x = 0.34$, $y = 0.33$), and high luminous efficiency (14.31 lm/W) [207].

16.4.2 Oxide Phosphor for PDPs

From the point of view of luminescence applications under VUV excitation, large band-gap materials are suitable matrices into which a small concentration of foreign ions called "activators" can be incorporated [208–210]. New technologies are now emerging where rare-earth activators are applied in the VUV and XUV excitation energy range and concern lighting with the development of new Hg-free fluorescent lamps and colour plasma display panels where VUV light from a Penning mixture of xenon and neon excites phosphors deposited on special glass screens [210–212]. As far as medical imaging is concerned, there is still a big claim in the development of new inorganic scintillators that should meet the basic requirements of important light emission yield and fast response. The improvement of X-ray storage phosphors, and by extension thermoluminescence dosimeters for ionizing radiation monitoring, is also a permanent challenge [213–216]. The basic structure of PDPs is composed of two pieces of glass substrate, a transparent electrode, and an electrode, in which the electrode in the first substrate support system. The substrate is provided with the electrodes, which are perpendicular to each other; then the substrate is coated with phosphors. The front substrate is the anode substrate as a PDP cathode. The rib for isolation and the substrate form a discharge space, and the space constitutes a pixel unit to prevent pixel-to pixel cross-talk causing deterioration of the image quality. The working gas is inert gas PDP Ne + Xe, the collision of metastable Ne and Xe, which makes Xe ionization avalanche ionization Xe+, again through the internal energy relaxation, to Xe the ^1S_n excited state. It transitions to the ground state by vacuum ultraviolet light (147 and 172 nm) to stimulate light-emitting materials, which emit red, green, blue, and other colors; the principle of PDP is shown in Fig. 16.67. Using metastable atoms and ions, high-energy photons bombard the secondary electrons from the cathode surface and the gas ionization can be self-maintained when the secondary electron satisfies, Regarding the emission coefficient of the secondary electron, formula U is the applied voltage, and ETA is ionization coefficient. The volt ampere curve of PDP has strong nonlinearity; location for a short time; and its location in the matrix number can reach 2 K, pixel spacing of 0.33 mm, a barrier width of 30 m, physical resolution $\leq$ 1280*1024 cd/m^2, brightness of 50 to approximately 500, contrast of 200 to approximately 400, and angle $\leq 160°$. Thus, the PDP achieves an image comparable the digital TV standard.

In PDP, a rare gas radiation of UV wavelengths is mainly concentrated in the 100- to approximately 200-nm vacuum ultraviolet (VUV) band, whereas Ar, Kr, Xe, and other discharges can produce VUV light; however, the Ar, Kr spectral

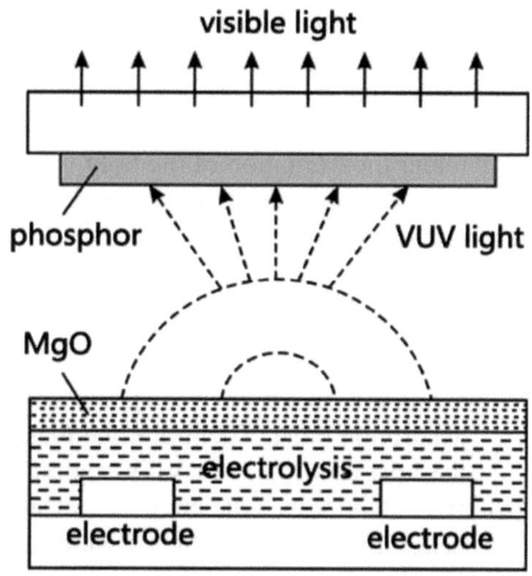

Fig. 16.67 Working principles of plasma display panel (PDP) devices

linewave length is shorter, and the light effect is not that of Xe, so the color PDP used is the Xe 147-nm spectral line. The spectral lines of 147 nm are composed of $^1S_4$ level to the ground-state transition radiation outward. When the pressure increases at 172 nm, the continuous radiation spectrum is obvious, and the wavelength of light radiation pressure is related with gas. The search for such new promising phosphors for excitation by VUV and XUV radiations will certainly progress with a better quantitative understanding of luminescence-excitation mechanisms. With the gain of full knowledge of the processes occurring in rare earth-activated inorganic lattices within the special energy excitation range considered here, the behavior of new luminescent materials can be predicted and optimized for different applications.

Oxide-based insulators doped with lanthanide ions have received the major focus of attention toward improved VUV phosphors essentially because of their attractive chemical properties as well as their easy preparation. Fluorides are still being investigated, especially for quantum-cutting materials, but they are much more sensitive to oxygen-surface contamination, which may influence their emission properties. Both materials are wide band-gap lattices, and absorption may occur in the host or in the luminescent center itself depending on the photon energy and the relative positioning of the energy levels of the luminescent impurity within the forbidden band gap or optical window of the host. High-efficiency phosphors require intense and wide absorption bands characterized by important oscillator strengths in primary photon inelastic interaction and an efficient energy transfer from the excited levels to the emitting center in the energy-dissipation process. This last point is certainly the main factor determining to a large extent the efficiency of the host-sensitized luminescence, which occurs through nonradiative or radiative exciton-energy transfer to the luminescent center or the sequential capture of free charge carriers, which in general are thermalized by impact mechanisms and phonon relaxation. Electron- or hole-capture by the rare-earth center is related to well-known charge-transfer mechanisms and is promoted by the ability of the lanthanide ion to change its oxidation state. Because the vacuum ultraviolet wavelength is 100 to approximately 200 nm, the 254-nm UV fluorescent power for PDP under the excitation of the light is quite different with fluorescent powder. The excitation wavelength of high energy requires the PDP phosphor to meet the following requirements:

(a) Broad band gap: The PDP phosphor excitation wavelength in vacuum ultraviolet fluorescent powder requires a strong absorption in the vacuum ultraviolet region. In PDP vacuum ultraviolet fluorescent powder, for external excitation spectrum we usually see a strong excitation band from the host absorption band. Therefore, the matrix of PDP fluorescent powder must select the host absorption in VUV region-wide material.
(b) High luminous efficiency: To clearly show the image, the plasma display panel must have a light emission luminance.
(c) Good color purity: To achieve a PDP full-color display with good color purity, three primary colors of fluorescent powder are required: PDP green phosphor with good color purity, the PDP display color gamut than the CRT, but the PDP powder and PDP powder blue color purity is still better than CRT phosphor corresponding slightly worse. Only by improving the blue PDP phosphor, especially the color purity of red, can the PDP full-color display be improved.
(d) Good stability: PDP phosphors must have good stability. There is a temperature >500 °C in the production of the baking screen in the PDP-production process, but the thermal stability of PDP fluorescent powder is greater than this.
(e) Short decay time: When used for display, fluorescent powder for the 1/e value must be <5 ms or the 1/10 value must be <10 ms; otherwise, it easily lead to image overlap and horizontal cross-talk (the so-called "cross-talk effect").

According to the National Television Standards Committee (NTSC), in order to ensure the greatest degree of image quality, the image should comprise red, blue, green; three colors; and NTSC standard red, blue, and green fluorescence powder. The standard color coordinates are listed in Table 16.10. In the past 20 years, foreign researchers have screened out the following phosphors from 600 kinds of borate red fluorescent material according to their luminous efficiency and color coordinates: (Y, Gd) $BO_3:Eu^{3+}$ (YGB), aluminate blue phosphor $BaMgAl_{10}O_{17}$: Eu^{2+}(BAM), and silicate green phosphor $Zn_2SiO_4:Mn^{2+}$ (ZSM) phosphor for PDP. The three-color phosphor emission spectra are shown in Fig. 16.68. The three primary colors used in the mixing of fluorescent powder, the color coordinates of the mixed powder, and the relative luminous efficiency of the mixed powder are presented in Table 16.11. It can be seen from Table 16.11 that the PDP phosphor combination (one mix) and the color coordinates of the NTSC standard are the most close and the relative luminous efficiency the highest.

Table 16.10 Main VUV excited commercial phosphors

Phosphor	Color coordinates	Efficiency	Decay time (ms)
NTSC blue	0.14, 0.08		
$BaMgAl_{10}O_{17}:Eu^{2+}$	0.147, 0.067	1.6	1
$BaMgAl_{14}O_{23}:Eu^{2+}$	0.142, 0.087	1.1	1
$Y_2SiO_5:Ce^{3+}$	0.16, 0.09	0.74	
$CaWO_4:Pb^{2+}$	0.17, 0.17		
NTSC green	0.21, 0.71		
$Zn_2SiO_4:Mn^{2+}$	0.21, 0.72	1.0	11.9
$BaAl_{12}O_{19}:Mn^{2+}$	0.182, 0.732	1.1	7.1
$SrAl_{12}O_{19}:Mn^{2+}$	0.16, 0.75	0.62	
$CaAl_{12}O_{19}:Mn^{2+}$	0.15, 0.75	0.34	
$ZnAl_{12}O_{19}:Mn^{2+}$	0.17, 0.74	0.54	
$YBO_3:Tb^{3+}$	0.33, 0.61	0.61	
$GdBO_3:Tb^{3+}$	0.33, 0.61	0.61	
$ScBO_3:Tb^{3+}$	0.35, 0.60	0.60	
$LuBO_3:Tb^{3+}$	0.33, 0.61	0.61	
$Sr_4Si_8O_6Cl_4:Eu^{2+}$	0.14, 0.33	1.3	
NTSC red	0.67, 0.33		
$Y_2O_3:Eu^{3+}$	0.65, 0.34	0.67	1.3
$(Gd,Y)BO_3:Eu^{3+}$	0.65, 0.35	1.2	4.3
$YBO_3:Eu^{3+}$	0.65, 0.35	1.0	
$GdBO_3:Eu^{3+}$	0.66, 0.34	0.94	
$ScBO_3:Eu^{3+}$	0.61, 0.39	0.94	
$LuBO_3:Eu^{3+}$	0.63, 0.37	0.74	
$Y_2SiO_5:Eu^{3+}$	0.66, 0.34	0.67	
$Y_3Al_5O_{12}:Eu^{3+}$	0.63, 0.37	0.47	
$Zn_3(PO_4)_2:Mn^{2+}$	0.67, 0.33	0.34	

The commercial PDP primary colors of fluorescent powder have many shortcomings such as the low red fluorescent pink purity, the red-yellow-green fluorescence powder, and the long afterglow time and poor stability of blue phosphor. In addition, because the fluorescent powder is high energy and has strong radiation and VUV excitation, the fluorescent pink hearts surface defect caused by VUV radiation can lead to decreased luminescence properties. PDP phosphor luminescence efficiency is low and is mainly related to the quality of fluorescent powder. Fluorescent-powder quality depends on two aspects: On one hand, due to the PDP display-panel manufacturing process, the sealing temperature is 45 °C and the temperature of the fluorescent powder because of the heat and the brightness decay. For example, a bright green barium aluminate phosphor therefore has manganese decreased by 10 %, BAM blue phosphor, or even 20 % decline. In contrast, phosphor luminance will decay under working conditions. The causes are as follows: vacuum ultraviolet radiation, gas-discharge sputtering, ionization of gas

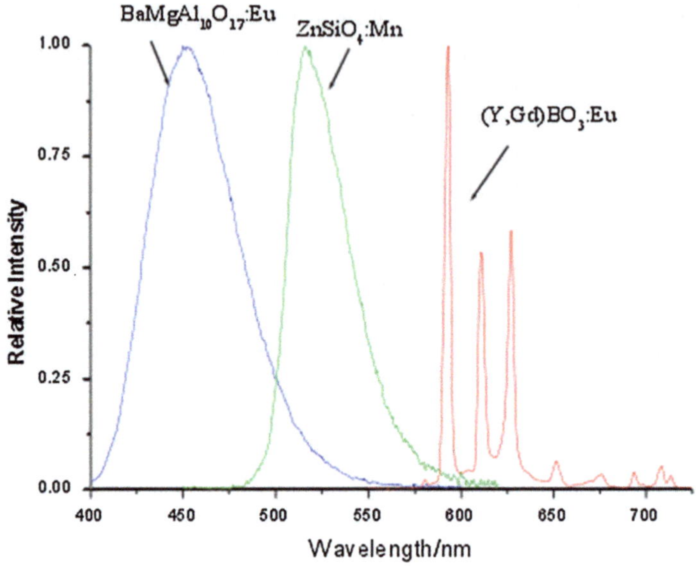

Fig. 16.68 Emission spectra of the main vacuum ultraviolet (VUV)—excited phosphors

Table 16.11 Emission efficiency and colorcoordinates of the main VUV excited phosphors

No	Red	Blue	Green	Coordinates	Efficiency
1	YGB	BAM	ZSM	0.31, 0.31	182
2	YGB	BAM	$BaAl_{12}O_{19}$:Mn^{2+}	0.30, 0.31	172
3	YBO_3:Eu^{3+}	BAM	ZSM	0.29, 0.31	150
4	Y_2O_3:Eu^{2+}	Y_2SiO_5:Ce^{3+}	ZSM	0.31, 0.38	100
5	YBO_3:Eu^{3+}	$YP_{0.85}V_{0.15}O_4$	ZSM	0.31, 0.31	83
6	Y_2O_3:Eu^{2+}	$CaWO_4$:Pb^{2+}	ZSM	0.34, 0.44	58

molecular collisions, fluorescent-powder molecular dissociation, and impurity adsorption of fluorescent-powder surface-state changes. In addition, ion bombardment and vacuum ultraviolet radiation may also have an effect on fluorescent powder. However, the factors influencing different powder are also different, e.g., the impact of green fluorescent powder by ion bombardment and blue fluorescent powder by vacuum ultraviolet radiation is greater: After 15,000-h blue phosphor luminance, attenuation can reach 30–40 %. Although some progress has been made in the research of vacuum ultraviolet light-emitting materials, there is still emitting material in the vacuum ultraviolet excitation from the high-energy photons produced by the visible photons; the energy conversion efficiency is low; and the varieties of vacuum ultraviolet light-emitting materials are old and stagnant, etc.

16.5 Concluding Remarks

Phosphors play a crucial role in lighting and display including PDP, Hg-free fluorescent lamps, and white LEDs, etc. In this chapter, the crystal structures of some fluorides, (oxy)nitrides, and oxides, along with their optical properties applications, are briefly presented. A great number of concepts and models have been proposed to increase the quantum efficiency of phosphors and explain the luminescence mechanism under different excitation by photons with different energy. Although new host materials are continuously being pursued, it is an urgent task to improve the performance of the developed phosphors so far. Emphasis should also be placed on the relationships among crystal structure, luminescence properties, and device performances. Research on the performance enhancement of phosphors cannot go far if one does not have deep insight into their structural analysis.

References

1. Li CX, Lin J (2010) Rare earth fluoride nano-/microcrystals: synthesis, surface modification and application. J Mater Chem 20:6831–6847
2. http://en.wikipedia.org/wiki/Calcium_fluoride
3. Mahiou R, Arbus R, Cousseins JC, Fournier MT (1987) Relationship between spectroscopic and crystallographic data in $NaGdF_4$. J Less-Common Metals 136:9–18
4. Ptacek P, Schäfer H, Kömpe K, Haase M (2007) Crystal phase control of luminescing $NaGdF_4:Eu^{3+}$ nanocrystals. Adv Funct Mater 17:3843–3848
5. Zalkin A, Templeton DH (1953) The crystal structures of YF_3 and related compounds. J Am Chem Soc 75(10):2453–2458
6. Gusowski MA, Gagor A, Trzebiatowska-Gusowska M, Ryba-Romanowski W (2006) Crystal structure and vibrational properties of new luminescent hosts K_3YF_6 and K_3GdF_6. J Solid State Chem 179:3145–3150
7. Christopher BO (1993) Master thesis: synthesis and study of new optical hosts p1–p52
8. Pytalev DS, Chukalina EP, Popova MN, Shakurov GS, Malkin BZ, Korableva SL (2012) Hyperfine interactions of Ho^{3+} ions in KY_3F_{10}: electron paramagnetic resonance and optical spectroscopy studies. Phys Rev B 86:115124–115132
9. Wegh RT, Donker H, Oskam KD, Meijerink A (1999) Visible quantum cutting in $LiGdF_4$: Eu^{3+} through downconversion. Science 283:663–666
10. Zhang QY, Huang XY (2010) Recent progress in quantum cutting phosphors. Progress in Mater Sci 55:353–427
11. Vink AP, Dorenbos P, van Eijk CWE (2003) Observation of the photon cascade emission process under 4f1-5d1and host excitation in several Pr^{3+}-doped materials. J Solid State Chem 171:308–312
12. Vink AP, Dorenbos P, de Haas JTM, Donker H, Rodnyi PA, Gavanesov A, van Eijk CWE (2002) Photon cascade emission in $SrAlF_5:Pr^{3+}$. JPhys Condens Matter 14:8889–8899
13. Sommerdijk JL, Bril A, de Jager AW (1974) Two photon luminescence with ultraviolet excitation of trivalent praseodymium. J Lumin 8:341–344
14. Sommerdijk JL, Bril A, de Jager AW (1974) Luminescence of Pr^{3+}-activated fluorides. J Lumin 9:288–296
15. Kück S, Sokólska I (2002) Observation of photon cascade emission in Pr^{3+}-doped LuF_3 and $BaMgF_4$. Chem Phys Lett 364:273–278

16. Kück S, Sokólska I, Henke M, Osiac E (2005) Quantum efficiency of $^1S_0$ and $^3P_0$,1 levels of Pr^{3+} doped YF_3. Chem Phys 310:139–144
17. Wang DW, Huang SH, You FT, Qi SQ, Fu YB, Zhang GB (2007) Vacuum ultraviolet spectroscopic properties of Pr^{3+} in MYF_4[M = Li, Na, and K] and $LiLuF_4$. J Lumin 122–123:450–452
18. Sokólska I, Kück S (2001) Observation of photon cascade emission in Pr^{3+}-doped pervoskite $KMgF_3$. Chem Phys 270:355–362
19. Masson NJML, Vink AP, Dorenbos P, Bos AJJ, Van Eijk CWE, Chaminade JP (2003) Ce^{3+} and Pr^{3+} 5d-energy levels in the [pseudo]perovskites $KMgF_3$ and $NaMgF_3$. J Lumin 101:175–183
20. Kück S, Sokólska I (2002) High energetic transitions in Pr^{3+}-doped polycrystalline $LiCaAlF_6$ and $LiSrAlF_6$. J Electrochem Soc 149:J27–J30
21. Wegh RT, Donker H, Oskam KD, Meijerink A (1999) Visible quantum cutting in Eu^{3+}-doped gadolinium Fuoridesvia downconversion. J Lumin 82:93–104
22. You FT, Wang Y, Lin JH, Tao Y (2002) Hydrothermal synthesis and luminescence properties of $NaGd_{F4}$:Eu. J Alloys Compd 343:151–155
23. Liu B, Chen Y, Shi C, Tang H, Tao Y (2003) Visible quantum cutting in BaF_2:Gd, Eu via downconversion. J Lumin 101:155–159
24. Kodama N, Watanabe Y (2004) Visible quantum cutting through downconversion in Eu^{3+}-doped KGd_3F_{10} and KGd_2F_7 crystals. Appl Phys Lett 84:4141–4144
25. You FT, Huang SH, Liu SM, Tao Y (2004) Synthesis, structure and VUV luminescent properties of rubidium rare-earth fluorides. J Solid State Chem 177:2777–2782
26. Kodama N, Oishi S (2005) Visible quantum cutting through downconversion in $KLiGdF_5$: Eu^{3+} crystals. J Appl Phys 98:103515–103520
27. Karbowiaka M, Mecha A, Romanowski WR (2005) Optical properties of Eu^{3+}:$CsGd_2F_7$ downconversion phosphor. J Lumin 114:65–70
28. Hua RN, Niu JH, Chen BJ, Li MT, Yu TZ, Li WL (2006) Visible quantum cutting in GdF_3: Eu^{3+} nanocrystals via downconversion. Nanotechnology 17:1642–1647
29. Lee TJ, Luo LY, Diau EWG, Chen TM, Cheng BM, Tung CY (2006) Visible quantum cutting through K_2GdF_5:Tb^{3+} phosphors. Appl Phys Lett 89:131121–131124
30. Tzeng HY, Cheng BM, Chen TM (2007) Visible quantum cutting in green-emitting $BaGdF_5$: Tb^{3+} phosphors via downconversion. J Lumin 122–123:917–920
31. Wegh RT, van Loef EVD, Meijerink A (2000) Visible quantum cutting via downconversion in $LiGdF_4$:Er^{3+}, Tb^{3+} upon Er^{3+} $4f^{11} \rightarrow 4f^{10}5d$ excitation. J Lumin 90:111–122
32. Auzel F (2004) Upconversion and anti-stokes processes with f and d ions in solids. Chem Rev 104:139–173
33. Eliseeva SV, Bünzli JG (2010) Lanthanide luminescence for functional materials and biosciences. Chem Soc Rev 39:189–227
34. Wang F, Liu X (2009) Recent advances in the chemistry of lanthanide-doped upconversion nanocrystals. Chem Soc Rev 38:976–989
35. Chivian JS, Case WE, Eden D (1979) Photo avalanche-new pnenomenon in Pr-3(+)-based infrared quantum counters. Appl Phys Lett 35:124–125
36. Deren PJ, Strek W, Zych E, Drozdzynski J (2000) Up-conversion in elpasolite crystals doped with U^{3+}. Chem Phys Lett 332(3):308–312
37. Stump NA, Murray GM, Delcul GD, Haire RG, Peterson JR (1993) Stokes and anti-stokes luminescence from the trihalides CM-248. Radiochim Acta 61(3–4):129–136
38. Auzel F (2003) Upconversion and anti-stokes process with f and d ions in solids. Chem Rev 104(1):139–174
39. Dou Q, Zhang Y (2011) Tuning of the structure and emission spectra of upconversion nanocrystals by alkali ion doping. Langmuir 27(21):13236–13241
40. Patra A, Friend CS, Kapoor R, Prasad PN (2002) Upconversion in Er^{3+}:ZrO_2 nanocrystals. J Phys Chem B 106(8):1909–1912
41. Gorris HH, Wolfbeis OS (2013) Photo-upconverting nanoparticles for optical encoding and multiplexing of cells, biomolecules, and microspheres. Angew Chem Int Ed 52:3584–3600

42. Krämer KW, Biner D, Frei G, Güdel HU, Hehlen MP, Lüthi SR (2004) Hexagonal sodium yttrium fluoride based green and blue emitting upconversion phosphors. Chem Mater 16:1244–1251
43. Huang X, Han S, Huang W, Liu X (2013) Enhancing solar cell efficiency: the search for luminescent materials as spectral converters. ChemSoc Rev 42:173–201
44. Trupke T, Green M, Würfel P (2002) Improving solar cell efficiencies by upconversion of sub-band-gap light. J Appl Phys 92(7):4117–4122
45. Trupke T, Shalav A, Richards BS, WürfelP Green MA (2006) Efficiency enhancement of solar cells by luminescent upconversion of sunlight. Sol Energy Mater Sol Cells 90(18–19):3327–3338
46. Badescu V, Badescu AM (2009) Improved model for solar cells with upconversion of low-erengy photons. Renewable Energy 34(3):1538–1544
47. Strümpel C, McCann M, Beaucarne GA, Slaoui ASV (2007) Modifying the solar spectrum to enhance silicon solar cell efficiency- an overview of available materials. Sol Energy Mater Sol Cells 91(4):238–249
48. Shalav A, Richards BS, Trupke T, Kramer KW, Gudel HU (2005) Application of $NaYF_4$: Er^{3+} up-converting phosphors for enhanced near-infrared silicon solar cell response. Appl Phys Lett 86:013505
49. Fischer S, Goldschmidt JC, Loper P, Bauer GH, Brugemann R, Kramer K, Biner D, Hermle M, Glunz SW (2010) Enhancement of silicon solar cell efficiency by upconversion: optical and electrical characterization. J Appl Phys 108(4):044912
50. Lahoz F (2008) Ho^{3+}-doped nanophase glass ceramics for efficiency enhancement in silicon solar cells. Opt Lett 33:2982–2984
51. Lahoz F, Perez-Rodriguez C, Hernandez SE, Martin IR, Lavin V, Rodriguez-Mendoza UR (2011) Upconversion mechanisms in rare-earth doped glasses to improve the efficiency of silicon solar cells. SolEnergy Mater Sol Cells 95:1671–1677
52. Yang Y (2014) Upconversion nanophosphors for use in bioemaging, therapy, drug delivery and bioassays. Microchim Acta 181:263–294
53. Liu Y, Chen M, Cao T, Sun Y, Li C, Liu Q (2013) A cyanine-modified nanosystem for in vivo upconversion luminescence bioimaging of methylmercury. J Am Chem Soc 135 (26):9869–9876
54. Zhang P, Steelant W, Kumar M, Scholfield M (2007) Veratile photosensitizers for photodynamic therapy at infrared excitation. J Am Chem Soc 129:4526–4527
55. Guo H, Sun S (2012) Lanthanide-doped upconverting phosphors for bioassay and therapy. Nanoscale 4:6692–6706
56. Zeuner M, Pagano S, Schnick W (2011) Nitridosilicates and oxonitridosilicates: from ceramic materials to structural and functional diversity. Angew Chem Int Ed 50(34):7754–7775
57. Xie R-J, Li Y Q, Hirosaki N, Yamamoto H (2011) Nitride phosphors and solid-state lighting. CRC Press
58. Schnick W, Huppertz H (1997) Nitridosilicates—a significant extension of silicate chemistry. Chem Eur J 3(5):679–683
59. Liebau F (1985) Structural chemistry of silicates. Springer Berlin etc.
60. Liebau F (1999) Silicate und Perowskite: zwei Themen mit Variationen. Angew Chem 111 (12):1845–1850
61. Liebau F (1999) Silicates and perovskites: two themes with variations. Angew Chem Int Ed 38(12):1733–1737
62. Schnick W (1993) Festkörperchemie mit Nichtmetallnitriden. Angew Chem 105(6):846–858
63. Schnick W (1993) Solid-state chemistry with nonmetal nitrides. Angew Chem Int Ed Engl 32(6):806–818
64. Xie R-J, Hirosaki N (2007) Silicon-based oxynitride and nitride phosphors for white LEDs—A review. Sci Technol Adv Mater 8(7–8):588

65. Ye S, Xiao F, Pan Y, Ma Y, Zhang Q (2010) Phosphors in phosphor-converted white light-emitting diodes: recent advances in materials, techniques and properties. Mat Sci Eng R: Rep 71(1):1–34
66. Li Y, Hirosaki N, Xie R, Takeda T, Mitomo M (2008) Yellow-orange-emitting $CaAlSiN_3$: Ce^{3+} phosphor: structure, photoluminescence, and application in white LEDs. Chem Mater 20(21):6704–6714
67. Kechele JA, Hecht C, Oeckler O, Schmedt auf der Günne J, Schmidt PJ, Schnick W (2009) $Ba_2AlSi_5N_9$—a new host lattice for Eu2+-Doped luminescent materials comprising a nitridoalumosilicate framework with corner-and edge-sharing tetrahedra. Chem Mater 21 (7):1288–1295
68. Xie RJ, Hirosaki N, Mitomo M, Suehiro T, Xu X, Tanaka H (2005) Photoluminescence of rare-earth-doped Ca-α-SiAlON phosphors: composition and concentration dependence. J Am Ceram Soc 88(10):2883–2888
69. Suehiro T, Hirosaki N, Xie RJ, Sato T (2009) Blue-emitting LaSiN: Ce fine powder phosphor for UV-converting white light-emitting diodes. Appl Phys Lett 95:051903
70. Suehiro T, Hirosaki N, Xie R J, Sakuma K, Mitomo M, Ibukiyama M, Yamada S (2008) One-step preparation of Ca-α-SiAlON: Eu 2+ fine powder phosphors for white light-emitting diodes. Appl Phys Lett 92(19):191904-191904-191903
71. Jung KY, Seo JH (2008) Preparation of fine-sized $SrSi_2O_{2\text{-}\delta}\,N_{2+\,2/3\delta}$: Eu^{2+} phosphor by spray pyrolysis and its luminescent characteristics. Electrochem Solid-State Lett 11(7):J64–J67
72. Piao X, Machida K, Horikawa T, Hanzawa H (2008) Synthesis of nitridosilicate $CaSr_{1-\,x}\,Eu_{\,x}\,Si_5N_8$ (x = 0–1) phosphor by calcium cyanamide reduction for white light-emitting diode applications. J Electrochem Soc 155(1):J17–J22
73. Piao X, Machida K, Horikawa T, Hanzawa H, Shimomura Y, Kijima N (2007) Preparation of $CaAlSiN_3$: Eu^{2+} phosphors by the self-propagating high-temperature synthesis and their luminescent properties. Chem Mater 19(18):4592–4599
74. Zeuner M, Schmidt PJ, Schnick W (2009) One-pot synthesis of single-source precursors for nanocrystalline LED phosphors $M_2Si_5N_8$: Eu2+(M = Sr, Ba). Chem Mater 21(12): 2467–2473
75. Piao X, Horikawa T, Hanzawa H, K-i Machida (2006) Characterization and luminescence properties of Sr[sub 2]Si[sub 5]N[sub 8]:Eu[sup 2+] phosphor for white light-emitting-diode illumination. Appl Phys Lett 88(16):161908
76. Watanabe H, Wada H, Seki K, Itou M, Kijima N (2008) Synthetic method and luminescence properties of Sr_xCa_{1-x} AlSiN3: Eu^{2+} mixed nitride phosphors. J Electrochem Soc 155(3): F31–F36
77. Schnick W, Huppertz H, Lauterbach R (1999) High temperature syntheses of novel nitrido-and oxonitrido-silicates and sialons using rf furnaces. J Mater Chem 9(1):289–296
78. Schnick W, Huppertz H (1997) Nitridosilicates—a significant extension of silicate chemistry. Chem -Eur J 3(5):679–683
79. Uheda K, Hirosaki N, Yamamoto H (2006) Host lattice materials in the system Ca_3N_2–AlN–Si_3N4 for white light emitting diode. Phys Status Solidi (a) 203(11):2712–2717
80. Li YQ, De With G, Hintzen H (2008) The effect of replacement of Sr by Ca on the structural and luminescence properties of the red-emitting $Sr_2Si_5N_8$: Eu^{2+} LED conversion phosphor. J Solid State Chem 181(3):515–524
81. Suehiro T, Hirosaki N, Xie R-J, Mitomo M (2005) Powder synthesis of Ca-α′-SiAlON as a host material for phosphors. Chem Mater 17(2):308–314
82. Li H-L, Xie R-J, Hirosaki N, Yajima Y (2008) Synthesis and photoluminescence properties of $Sr_2Si_5N_8$: Eu^{2+} Red phosphor by a gas-reduction and nitridation method. J Electrochem Soc 155(12):J378–J381
83. Suehiro T, Hirosaki N, Xie R-J, Sato T (2009) Blue-emitting $LaSi_3N_5$:Ce^{3+} fine powder phosphor for UV-converting white light-emitting diodes. Appl Phys Lett 95(5):051903
84. Suehiro T, Onuma H, Hirosaki N, Xie R-J, Sato T, Miyamoto A (2009) Powder synthesis of Y-α-SiAlON and its potential as a phosphor host. J Phys Chem C 114(2):1337–1342

85. Li H-L, Xie R-J, Zhou G-H, Hirosaki N, Sun Z (2010) A cyan-emitting $BaSi_7N_{10}$: Eu^{2+} phosphor prepared by gas reduction and nitridation for UV-pumping white LEDs. J Electrochem Soc 157(7):J251–J255
86. Zhang H, Horikawa T, Hanzawa H, Hamaguchi A, Machida K-i (2007) Photoluminescence properties of α-SiAlON: Eu^{2+} prepared by carbothermal reduction and nitridation method. J Electrochem Soc 154(2):J59–J61
87. Piao X, Horikawa T, Hanzawa H, Machida K-i (2006) Characterization and luminescence properties of $Sr_2Si_5N_8$:Eu^{2+} phosphor for white light-emitting-diode illumination. Appl Phys Lett 88(16):161908
88. Li J, Watanabe T, Sakamoto N, Wada H, Setoyama T, Yoshimura M (2008) Synthesis of a multinary nitride, Eu-doped $CaAlSiN_3$, from alloy at low temperatures. Chem Mater 20(6):2095–2105
89. Li J, Watanabe T, Wada H, Setoyama T, Yoshimura M (2007) Low-temperature crystallization of Eu-doped red-emitting $CaAlSiN_3$ from alloy-derived ammonometallates. Chem Mater 19(15):3592–3594
90. Li J, Watanabe T, Wada H, Setoyama T, Yoshimura M (2009) Synthesis of Eu-Doped $CaAlSiN_3$ from ammonometallates: effects of sodium content and pressure. J Am Ceram Soc 92(2):344–349
91. Piao X, Machida K-i, Horikawa T, Hanzawa H (2007) Self-propagating high temperature synthesis of yellow-emitting $Ba_2Si_5N_8$:Eu^{2+} phosphors for white light-emitting diodes. Appl Phys Lett 91(4):041908-041908-041903
92. Piao X, K-i Machida, Horikawa T, Hanzawa H, Shimomura Y, Kijima N (2007) Preparation of $CaAlSiN_3$:Eu^{2+} phosphors by the self-propagating high-temperature synthesis and their luminescent properties. Chem Mater 19(18):4592–4599
93. Xie RJ (2013) Optical properties of (oxy) nitride materials: a review. J Am Ceram Soc 96(3):665–687
94. Inoue K, Hirosaki N, Xie R-J, Takeda T (2009) Highly efficient and thermally stable blue-emitting AlN: Eu^{2+} phosphor for ultraviolet white light-emitting diodes. J Phys Chem C 113(21):9392–9397
95. Shioi K, Hirosaki N, Xie R-J, Takeda T, Li Y (2008) Luminescence properties of $SrSi_6N_8$: Eu^{2+}. J Mater Sci 43(16):5659–5661
96. Li YQ, Delsing A, Metslaar R, de With G, Hintzen H (2009) Photoluminescence properties of rare-earth activated BaSi 7 N 10. J Alloys Compd 487(1):28–33
97. Li YQ, Van Steen J, Van Krevel J, Botty G, Delsing A, DiSalvo F, De With G, Hintzen H (2006) Luminescence properties of red-emitting $M_2Si_5N_8$: Eu^{2+}(M = Ca, Sr, Ba) LED conversion phosphors. J Alloys Compd 417(1):273–279
98. Xie R-J, Hirosaki N, Suehiro T, Xu F-F, Mitomo M (2006) A simple, efficient synthetic route to $Sr_2Si_5N_8$: Eu^{2+}-based red phosphors for white light-emitting diodes. Chem Mater 18 (23):5578–5583
99. Li YQ, Hirosaki N, Xie R-J, Takada T, Yamamoto Y, Mitomo M, Shioi K (2010) Synthesis, crystal and local electronic structures, and photoluminescence properties of red-emitting $CaAl_zSiN_{2+z}$:Eu^{2+} with orthorhombic structure. Int J Appl Ceram Technol 7(6):787–802
100. Hintze F, Hummel F, Schmidt PJ, Wiechert D, Schnick W (2012) $Ba_3Ga_3N_5$A novel host lattice for Eu2+-doped luminescent materials with unexpected nitridogallate substructure. Chem Mater 24(2):402–407
101. Horikawa T, Fujitani M, Hanzawa H, Machida K-i (2012) Structure and photoluminescence properties of $M_{II}M_{III}Si_4N_7$: Eu^{2+} (M_{II} = Ca, Sr, Ba/M_{III} = Sc, Y, La) phosphors prepared by carbothermal reduction and nitridation. ECS J Solid State SC 1(4):R113–R118
102. Li YQ, Fang CM, de With G, Hintzen H (2004) Preparation, structure and photoluminescence properties of Eu^{2+} and Ce^{3+}-doped $SrYSi_4N_7$. J Solid State Chem 177 (12):4687–4694
103. Li YQ, de With G, Hintzen H (2004) Synthesis, structure, and luminescence properties of Eu^{2+} and Ce^{3+} activated $BaYSi_4N_7$. J Alloys Compd 385(1):1–11

104. Bachmann V, Ronda C, Oeckler O, Schnick W, Meijerink A (2008) Color point tuning for (Sr, Ca, Ba)$Si_2O_2N_2$:Eu^{2+} for white light LEDs. Chem Mater 21(2):316–325
105. Hirosaki N, Xie R-J, Kimoto K, Sekiguchi T, Yamamoto Y, Suehiro T, Mitomo M (2005) Characterization and properties of green-emitting β-SiAlON:Eu^{2+} powder phosphors for white light-emitting diodes. Appl Phys Lett 86(21):211905
106. Yamada S, Emoto H, Ibukiyama M, Hirosaki N (2012) Properties of SiAlON powder phosphors for white LEDs. J Eur Ceram Soc 32(7):1355–1358
107. Xie R-J, Hirosaki N, Li H-L, Li YQ, Mitomo M (2007) Synthesis and photoluminescence properties of β-sialon: Eu^{2+} ($Si_{6-z}Al_zO_zN_{8-z}$: Eu^{2+}): a promising green oxynitride phosphor for white light-emitting diodes. J Electrochem Soc 154(10):J314–J319
108. Braun C, Seibald M, Börger SL, Oeckler O, Boyko TD, Moewes A, Miehe G, Tücks A, Schnick W (2010) Material properties and structural characterization of $M_3Si_6O_{12}N_2$:Eu^{2+} (M = Ba, Sr)—a comprehensive study on a promising green phosphor for pc-LEDs. Chem EurJ 16(31):9646–9657
109. Chiu Y-C, Huang C-H, Lee T-J, Liu W-R, Yeh Y-T, Jang S-M, Liu R-S (2011) Eu $^{2+}$-activated silicon-oxynitride $Ca_3Si_2O_4N_2$: a green-emitting phosphor for white LEDs. Opt Express 19(103):A331–A339
110. Kechele JA, Hecht C, Oeckler O, Schmedt auf der Günne Jr, Schmidt PJ, Schnick W (2009) $Ba_2AlSi_5N_9$A New host lattice for Eu^{2+}-Doped luminescent materials comprising a nitridoalumosilicate framework with corner-and edge-sharing tetrahedra. Chem Mater 21 (7):1288–1295
111. Wang X-M, Wang C-H, Kuang X-J, Zou R-Q, Wang Y-X, Jing X-P (2012) Promising oxonitridosilicate phosphor host $Sr_3Si_2O_4N_2$: Synthesis, structure, and luminescence properties activated by Eu^{2+} and Ce^{3+}/Li^+ for pc-LEDs. Inorg Chem 51(6):3540–3547
112. Park WB, Singh SP, Yoon C, Sohn K-S (2012) Eu^{2+} luminescence from 5 different crystallographic sites in a novel red phosphor, $Ca_{15}Si_{20}O_{10}N_{30}$: Eu^{2+}. J Mater Chem 22 (28):14068–14075
113. Uheda K, Hirosaki N, Yamamoto Y, Naito A, Nakajima T, Yamamoto H (2006) Luminescence properties of a red phosphor, $CaAlSiN_3$: Eu^{2+}, for white light-emitting diodes. Electrochem Solid-State Lett 9(4):H22–H25
114. Hecht C, Stadler F, Schmidt P, Schnick W (2008) $SrAlSi_4N_7$: Eu^{2+}—a nitridoalumosilicate phosphor for warm white-light LEDs with edge-sharing tetrahedrons. Z Anorg Allg Chem 634(11):2044-2044
115. Ruan J, Xie RJ, Hirosaki N, Takeda T (2011) Nitrogen gas pressure synthesis and photoluminescent properties of orange-red $SrAlSi_4N_7$: Eu^{2+} phosphors for white light-emitting diodes. J Am Ceram Soc 94(2):536–542
116. Park WB, Shin N, Hong KP, Pyo M, Sohn KS (2012) A new paradigm for materials discovery: heuristics-assisted combinatorial chemistry involving parameterization of material novelty. Adv Funct Mater 22(11):2258–2266
117. Tang J-Y, Xie W-J, Huang K, Hao L-Y, Xu X, Xie R-J (2011) A high stable blue $BaSi_3Al_3O_4N_5$: Eu^{2+} phosphor for white LEDs and display applications. Electrochem Solid-State Lett 14(8):J45–J47
118. Yamane H, Shimooka S, Uheda K (2012) Synthesis, crystal structure and photoluminescence of a new Eu-doped Sr containing sialon $(Sr_{0.94}Eu_{0.06})(Al_{0.3}Si_{0.7})_4(N_{0.8}O_{0.2})_6$. J Solid State Chem 190:264–270
119. Oeckler O, Kechele JA, Koss H, Schmidt PJ, Schnick W (2009) $Sr_5Al_{5+x}Si_{21-x}N_{35-x}O_{2+x}$: Eu^{2+} ($x \approx 0$)—a novel green phosphor for white-light pcleds with disordered intergrowth structure. Chem -EurJ 15(21):5311–5319
120. Shioi K, Michiue Y, Hirosaki N, Xie R-J, Takeda T, Matsushita Y, Tanaka M, Li YQ (2011) Synthesis and photoluminescence of a novel Sr-SiAlON:Eu^{2+} blue-green phosphor ($Sr_{14}Si_{68-s}Al_{6+s}O_sN_{106-s}$:$Eu^{2+}$ ($s \approx 0.7$)). J Alloys Compd 509(2):332–337
121. Fukuda Y, Ishida K, Mitsuishi I, Nunoue S (2009) Luminescence properties of Eu^{2+}-doped green-emitting Sr-sialon phosphor and its application to white light-emitting diodes. Appl Phys Express 2(1):012401

122. Xie R-J, Hirosaki N, Sakuma K, Yamamoto Y, Mitomo M (2004) Eu^{2+}-doped Ca-α-SiAlON: A yellow phosphor for white light-emitting diodes. Appl Phys Lett 84(26):5404–5406
123. Xie R-J, Hirosaki N, Mitomo M, Yamamoto Y, Suehiro T, Sakuma K (2004) Optical Properties of Eu^{2+} in α-SiAlON. J Phys Chem B 108(32):12027–12031
124. Xie R-J, Hirosaki N, Mitomo M, Sakuma K, Kimura N (2006) Wavelength-tunable and thermally stable Li-α-sialon: Eu^{2+} oxynitride phosphors for white light-emitting diodes. Appl Phys Lett 89(24):241103-241103-241103
125. Shioi K, Hirosaki N, Xie RJ, Takeda T, Li YQ, Matsushita Y (2010) Synthesis, crystal structure, and photoluminescence of Sr–α–SiAlON: Eu^{2+}. J Am Ceram Soc 93(2):465–469
126. Suehiro T, Hirosaki N, Xie R-J (2011) Synthesis and photoluminescent properties of (La, Ca) $_3Si_6N_{11}$: Ce^{3+} fine powder phosphors for solid-state lighting. ACS Appl Mater Interf 3 (3):811–816
127. Le Toquin R, Cheetham AK (2006) Red-emitting cerium-based phosphor materials for solid-state lighting applications. Chem Phys Lett 423(4–6):352–356
128. van Krevel JWH, Hintzen HT, Metselaar R, Meijerink A (1998) Long wavelength Ce^{3+} emission in Y-Si–O–N materials. J Alloys Compd 268(1–2):272–277
129. Xie R-J, Hirosaki N, Takeda T (2009) Wide color gamut backlight for liquid crystal displays using three-band phosphor-converted white light-emitting diodes. Appl Phys Express 2 (2):022401
130. Dierre B, Xie R-J, Hirosaki N, Sekiguchi T (2007) Blue emission of Ce^{3+} in lanthanide silicon oxynitride phosphors. J Mater Res 22(07):1933–1941
131. Li YQ, Hirosaki N, Xie RJ, Takeda T, Mitomo M (2008) Yellow-orange-emitting $CaAlSiN_3$: Ce^{3+} phosphor: structure, photoluminescence, and application in white LEDs. Chem Mater 20(21):6704–6714
132. Takahashi K, Hirosaki N, Xie R-J, Harada M, Yoshimura K-i, Tomomura Y (2007) Luminescence properties of blue $La_{1-x}Ce_xAl(Si_{6-z}Al_z)(N_{10-z}O_z)(z \sim 1)$ oxynitride phosphors and their application in white light-emitting diode. Appl PhysLett 91(9):091923
133. Hampshire S, Park H, Thompson D, Jack K (1978) α′-Sialon ceramics. Nature 274 (5674):880–882
134. Izumi F, Mitomo M, Suzuki J (1982) Structure refinement of yttrium α-sialon from X-ray powder profile data. J Mater Sci Lett 1(12):533–535
135. Cao GZ, Metselaar R, Haije WG (1993) Neutron diffraction study of yttrium α′-sialon. J Mater Sci Lett 12(7):459–460
136. Cole M, O'Reilly KPJ, Redington M, Hampshire S (1991) EXAFS study of a hot-pressed α′-sialon ceramic containing erbium as the modifying cation. J Mater Sci 26(19):5143–5148
137. Xie R-J, Hirosaki N, Mitomo M, Yamamoto Y, Suehiro T, Ohashi N (2004) photoluminescence of cerium-doped α-SiAlON materials. J Am Ceram Soc 87(7):1368–1370
138. Xie R-J, Hirosaki N, Mitomo M, Suehiro T, Xu X, Tanaka H (2005) Photoluminescence of rare-earth-doped Ca-α-SiAlON phosphors: composition and concentration dependence. J Am Ceram Soc 88(10):2883–2888
139. Xie RJ, Hirosaki N, Mitomo M, Yamamoto Y, Suehiro T, Ohashi N (2004) photoluminescence of cerium-doped α-SiAlON materials. J Am Ceram Soc 87(7):1368–1370
140. Xie R-J, Mitomo M, Uheda K, Xu F-F, Akimune Y (2002) Preparation and luminescence spectra of calcium- and rare-earth (R = Eu, Tb, and Pr)-Codoped α-SiAlON ceramics. J Am Ceram Soc 85(5):1229–1234
141. Xie RJ, Hirosaki N, Mitomo M, Uheda K, Suehiro T, Xu X, Yamamoto Y, Sekiguchi T (2005) Strong green emission from α-SiAlON activated by divalent ytterbium under blue light irradiation. J Phys Chem B 109(19):9490–9494
142. Oyama Y, Kamigaito O (1971) Solid solubility of some oxides in Si_3N_4. Jpn J Appl Phys 10 (11):1637
143. Jack KH (1976) Sialons and related nitrogen ceramics. J Mater Sci 11(6):1135–1158
144. Ekström T, Nygren M (1992) SiAlON ceramics. J Am Ceram Soc 75(2):259–276

145. Kechele JA, Oeckler O, Stadler F, Schnick W (2009) Structure elucidation of $BaSi_2O_2N_2$–A host lattice for rare-earth doped luminescent materials in phosphor-converted (pc)-LEDs. Solid State Sci 11(2):537–543
146. Höppe HA, Stadler F, Oeckler O, Schnick W (2004) $Ca[Si_2O_2N_2]$—a novel layer silicate. Angew Chem Int Ed 43(41):5540–5542
147. Oeckler O, Stadler F, Rosenthal T, Schnick W (2007) Real structure of $SrSi_2O_2N_2$. Solid State Sci 9(2):205–212
148. Xie R-J, Hirosaki N, Kimura N, Sakuma K, Mitomo M (2007) 2-phosphor-converted white light-emitting diodes using oxynitride/nitride phosphors. Appl Phys Lett 90 (19):191101-191101-191103
149. Li YQ, de With G, Hintzen HT (2006) Luminescence properties of Ce^{3+}-activated alkaline earth silicon nitride M2Si5N8 (M = Ca, Sr, Ba) materials. J Lumin 116(1–2):107–116
150. Li YQ, van Steen JEJ, van Krevel JWH, Botty G, Delsing ACA, DiSalvo FJ, de With G, Hintzen HT (2006) Luminescence properties of red-emitting $M_2Si_5N_8:Eu^{2+}$ (M = Ca, Sr, Ba) LED conversion phosphors. J Alloys Compd 417(1–2):273–279
151. Inoue Z, Mitomo M, Ii N (1980) A crystallographic study of a new compound of lanthanum silicon nitride, $LaSi_3N_5$. J Mater Sci 15(11):2915–2920
152. Woike M, Jeitschko W (1995) Preparation and crystal structure of the nitridosilicates $Ln_3Si_6N_{11}$ (Ln = La, Ce, Pr, Nd, Sm) and $LnSi_3N_5$ (Ln = Ce, Pr, Nd). Inorg Chem 34 (21):5105–5108
153. Uheda K, Shimooka S, Mikami M, Imura H, Kijima N (2008) Synthesis and characterization of new green oxonitridosilicate phosphor, $(Ba, Eu)_3Si_6O_{12}N_2$, for White LED, in: meeting abstracts. J Electrochem Soc 3195–3195
154. Hecht C, Stadler F, Schmidt PJ, auf der Günne JS, Baumann V, Schnick W (2009) $SrAlSi_4N_7:Eu^{2+}$ – A nitridoalumosilicate phosphor for warm white light (pc)LEDs with edge-sharing tetrahedra. Chem Mater 21(8):1595–1601
155. Ruan J, Xie RJ, Hirosaki N, Takeda T (2011) Nitrogen gas pressure synthesis and photoluminescent properties of orange-red $SrAlSi4N7:Eu^{2+}$ phosphors for white light-emitting diodes. J Am Ceram Soc 94(2):536–542
156. Duan CJ, Wang XJ, Otten WM, Delsing ACA, Zhao JT, Hintzen HT (2008) Preparation, electronic structure, and photoluminescence properties of Eu^{2+}- and Ce^{3+}/Li^{+}-activated alkaline earth silicon nitride $MSiN_2$ (M = Sr, Ba). Chem Mater 20(4):1597–1605
157. (a) Gübelin E (1974) International World of Gemstones. ABC-Verlag, Zürich. (b) H J Müllen meister (1990), Faszination Edlelstein. H. J. Müllenmeister. Markt Schwaben
158. Eeckhout KV, Philippe F, Smet PF, Poelman D (2010) Persistent luminescencein Eu^{2+}-Doped compounds: a review. Materials 3:2536–2566
159. Jia DD, Jiang LH, Liu YL et al (1998) Investigation of $Ca_{1-x}Sr_xS$: Bi, m, Cu and Cu: Eu phosphor. Chin J Lumin 19(4):312–316
160. Newton HE (1957) A history of luminescence: from the earliest times until 1900, p. 305. Amer Phil Soc, Philadelphia
161. Wang YH, Wang ZL (2006) Characterization of $Y_2O_2S:Eu^{3+}_{0.06}$, $Ti^{4+}_{0.02}$ long-lasting phosphor synthesized by flux method. J Rare Earth 24:25–28
162. Wang YH, Wang L (2007) Defect states in Nd^{3+} doped $CaAl_2O_4:Eu^{2+}$. J Appl Phys 101:053108–053113
163. Lin Y, Tang Z, Zhang Z, Nan CW (2002) Anomalous luminescence in $Sr_4Al_{14}O_{25}$:Eu, Dy phosphors. Appl Phys Lett 81:996–998
164. Katsumata T, Nabae T, Sasajima K, Komuro S, Morikawa T (1997) Effects of composition on the long phosphorescent $SrAl_2O_4:Eu^{2+}$, Dy^{3+} phosphor crystals. J Electrochem Soc 144: L243–L245
165. Lin Y, Zhang Z, Tang Z, Zhang J, Zheng Z, Lu X (2001) The characterization and mechanism of long afterglow in alkaline earth aluminates phosphors co-doped by Eu_2O_3 and Dy_2O_3. Mater Chem Phys 70:156–159
166. Sakai R, Katsumata T, Komuro S, Morikawa T (1999) Effect of composition on the phosphorescence from $BaAl_2O_4$: Eu^{2+}, Dy^{3+} crystals. J Lumin 85:149–154

167. Preethi KRS, Lu CH, Thirumalai J, Jagannathan R, Natarajan TS, Nayak NU, Radhakrishna I, Jayachandran M, Trivedi DC (2004) $SrAl_4O_7:Eu^{2+}$ nanocrystals: synthesis and fluorescence properties. J Phys D-Appl Phys 37:2664–2669
168. Wanjun T, Donghua C, Ming W (2009) Luminescence studies on $SrMgAl_{10}O_{17}$:Eu, Dy phosphor crystals. Opt Laser Technol 41:81–84
169. Aitasalo T, Hölsä J, Kirm M, Laamanen T, Lastusaari M, Niittykoski J, Raud J, Valtonen R (2007) Persistent luminescence and synchrotron radiation study of the $Ca_2MgSi_2O_7:Eu^{2+}$, R3 + materials. Radiat Meas 42:644–647
170. Blasse G, Wanmaker WL, ter Vrugt JW, Bril A (1968) Fluorescence of Eu^{2+}-activated silicates. Philips Res Rep 23:189–200
171. Aitasalo T, Hreniak D, Hölsä J, Laamanen T, Lastusaari M, Niittykoski J, PelléF Strek W (2007) Persistent luminescence of $Ba_2MgSi_2O_7:Eu^{2+}$. J Lumin 122–123:110–112
172. Lin Y, Tang Z, Zhang Z, Nan CW (2003) Luminescence of Eu^{2+} and Dy^{3+} activated $R_3MgSi_2O_8$-based (R = Ca, Sr, Ba) phosphors. J Alloy Compd 348:76–79
173. Sabbagh Alvani AA, Moztarzadeh F, Sarabi AA (2005) Effects of dopant concentrations on phosphorescence properties of Eu/Dy-doped $Sr_3MgSi_2O_8$. J Lumin 114:131–136
174. Jiang L, Chang C, Mao D, Feng C (2004) Luminescent properties of $CaMgSi_2O_6$-based phosphors co-doped with different rare earth ions. J Alloy Compd 377:211–215
175. Kuang JY, Liu YL, Zhang JX (2006) Effects of RE3+ as a co-dopant in blue-emitting long-lasting phosphors, $Sr_3Al_{10}SiO_20$: Eu^{2+}. J Mater Sci 41:5500–5503
176. Wang Y, Wang Z, Zhang P, Hong Z, Fan X, Qian G (2004) Preparation of Eu^{2+} and Dy^{3+} co-activated $CaAl_2Si_2O_8$-based phosphor and its optical properties. Mater Lett 58:3308–3311
177. Ding Y, Zhang Y, Wang Z, Li W, Mao D, Han H, Chang C (2009) Photoluminescence of Eu single doped and Eu/Dy codoped $Sr_2Al_2SiO_7$ phosphors with long persistence. J Lumin 129:294–299
178. Jiang L, Chang C, Mao D, Zhang B (2004) A new long persistent blue-emitting Sr2ZnSi2O7: Eu^{2+}, Dy^{3+} prepared by sol-gel method. Mater Lett 58:1825–1829
179. Lakshminarasimhan N, Varadaraju UV (2008) Luminescence and afterglow in Sr2SiO4: Eu^{2+}, RE^{3+} [RE = Ce, Nd, Sm and Dy] phosphors-Role of co-dopants in search for afterglow. Mater Res Bull 43:2946–2953
180. Wang J, Su Q, Wang SB (2005) A novel red long lasting phosphorescent (LLP) material β-$Zn_3(PO_4)_2:Mn^{2+}$, Sm^{3+}. Mater Res Bull 40:590–598
181. Huang P, Wu YQ, Cui CE, Guo J (2011) Effect of different dopants on the long lasting phosphorescence behavior of $CaTiO_3:Pr^{3+}$. J Rare Earths 29:340–344
182. Pang R, C Li, Shi L, Su Q (2009) A novel blue-emitting long-lasting proyphosphate phosphor $Sr_2P_2O_7:Eu^{2+}$, Y^{3+}. J Phys Chem Solids 70:303–306
183. Pang R, Li C, Zhang S, Su Q (2009) Luminescent properties of a new blue long-lasting phosphor $Ca_2P_2O_7:Eu^{2+}$, Y^{3+}. Mater Chem Phys 113:215–218
184. Liu L, Li C, Wang S, Su Q (2006) Redshift phenomenon of the excitation light of long life emission phosphor. Appl Phys Lett 88:241107
185. Miyamoto Y, Kato H, Honna Y, Yamamoto H, Ohmi K (2009) An orange-emitting, long-persistent phosphor, $Ca_2Si_5N_8:Eu^{2+}$, Tm^{3+}. J Electrochem Soc 156:J235–J241
186. Li C, Su Q (2006) A new blue phosphorescent glass-ceramic: Rare-earth-doped calcium aluminoborate. J Alloy Compd 408–412:875–878
187. Zhang L, Li C, Su Q (2006) Long lasting phosphorescence in Eu^{2+} and Ce^{3+} co-doped strontium borate glasses. J Rare Earths 24:196–198
188. Matsuzawa T, Aoki Y, Takeuchi N, Murayama Y (1996) A new long phosphorescent phosphor with high brightness, $SrAl_2O_4:Eu^{2+}$, Dy^{3+}. J Electrochem Soc 143:2670–2673
189. Qiu J, Kawasaki M (1998) Phenomenon and mechanism of longlasting phosphorescence in Eu^{2+}-doped aluminosilicate glasses. J Phys Chem Solids 59:151–152
190. Aitasalo T, Hölsä J, Jungner H, Lastusaari M, Niittykoski J (2001) Mechanisms of persistent luminescence in Eu^{2+}, RE^{3+} doped alkaline earth aluminates. J Lumin 94–95:59–63
191. Hölsä J, Jungner H, Lastusaari M, Niittykoski J (2001) Persistent luminescence of Eu^{2+} doped alkaline earth aluminates, $MAl_2O_4:Eu^{2+}$. J Alloys Compd 323–324:326–330

192. Withnall R, Silver J, Rose J (2004) Phosphor screens excited by blue LEDs as a white light source for full colour head up displays in automobiles. In: Proceedings of EL 2004, Sept. 20–23, Toronto, Canada. pp. 325–328
193. Silver J, Withnall R, Rose JA (2005) Broad-band green phosphor screens as a light source for head up displays in moving platforms. technical abstracts, Opto Ireland, April 4–6, Dublin, Ireland. p. 87
194. Silver J, Withnall R, Rose JA (2005) Broad-band green phosphor screens as a light source for head up displays in moving platforms, proc SPIE. Opt Sens Spectro 5826:537–544
195. Silver J (2006) Phosphors for LEDs, Current state of the art, problems, and needs. invited lecture, conference entitled 'The Future of LEDs', June 26–27, London
196. Liu YH, Zhuang WD, Hua YH, Gao WG, Hao JH (2010) Synthesis and luminescence of sub-micron sized $Ca_3Sc_2Si_3O_{12}$: Ce green phosphors for white light-emitting diode and field-emission display applications. J Alloys Compd 504:488–492
197. Hirakata SO (2003) Inorganic oxide and phosphor. US Pat. 7, 029, 602 B2, Sept. 25
198. Setlur AA, Comanzo HA, Srivastava AM, Beers WW (2005) Evaluation of a White Light Phosphor for UV-LEDs–$Ca_2NaMg_2V_3O_{12}$: Eu^{3+}. J Electrochem Soc 152:H205–H208
199. Lee DW, Kwak CH, Jung HK (2006) Photoluminescence properties of red-emitting $YCa_3M_3B_4O_{15}$: Eu^{3+} (M = Al or Ga) phosphors under UV irradiation: a recent advance in EL, Displays and Lighting. JeJu, Korea, Tech. Digest pp. 363–364
200. Barry TL (1968) Equilibria and Eu^{2+} luminescence of subsolidus phases bounded by $Ba_3MgSi_2O_8$, $Sr_3MgSi_2O_8$, and $Ca_3MgSi_2O_8$. J Electrochem Soc 115:733–738
201. Starick D, Kempfert W, Limberg H-J (2003) Silicate phosphors and their performance in white LEDs, phosphor global summit, scottsdale, AR. USA. Conference Paper 19:1–3
202. Ci ZP, Sun MX, Li Y, Que MD, Wang Q, Zhu G, Wang YH (2015) Optimization and temperature-dependent photoluminescence properties of yellow-orange emitting $Sr_{1.99\text{-}x\text{-}y}Ca_xM_y(Si_{1\text{-}y}Z_y)O_4$: $0.01Eu^{2+}$ (M = Y^{3+}, K^+, Z = Al^{3+}, P^{5+}) phosphors for white light-emitting diodes. Opt Mater 39:188–194
203. Withnall R, Silver J, Rose J (2004) Phosphor Screens excited by blue LEDs as a white light source for full colour head up displays in automobiles. In: Proceedings of EL 2004, Sept 20–23, Toronto, Canada. pp 325–328
204. Levinson LM, Srivastava AM (2000) Light emitting device with $Ba_2MgSi_2O_7$: Eu^{2+}, Ba_2SiO_4: Eu^{2+}, or $(Sr_xCa_y\ Ba_{1\text{-}x\text{-}y})(A1_zGa_{1\text{-}z})_2Sr$: Eu^{2+} phospors. US Pat. 6,429,583, Nov. 30 (1998)
205. Ota K, Hirano A, Ota A (2000) Light emitting device having phosphor of alkaline earth metal silicate. US Pat. 6,943,380, Dec. 28
206. Tasch S, Pacher P, Roth G, Tews W, Kempfert W, Starick D (2003) Light source comprising a light-emitting element. US Pat. 6,809,347, Dec 22
207. Jae CK, Park JP, Kim KN, Kim CH, Kim HK (2006) White light-emitting diodes of InGaN-based $(Sr, Ba)_3MgSi_{1.7}O_8$: Eu and the luminescent properties: a recent advance in EL, displays and lighting. JeJu, Korea, Tech. Digest pp 334–335
208. Wegh TR, Meijerink A, Lammi nmäki RJ, Hölsä J (2000) Extending Dieke's diagram. J Lumin 87–89:1002–1004
209. Dorenbos P (2000) The 5d level positions of the trivalent lanthanides in inorganic compounds. J Lumin 91(3–4):155–176
210. Ge'rard I, Krupa JC, Simoni E, Martin P (1994) Investigation of charge transfer $O^{2-} \rightarrow Ln^{3+}$ and $F^- \rightarrow Ln^{3+}$ in LaF_3:(Ln^{3+}, O^{2-}) and YF_3:(Ln^{3+}, O^{2-}) systems. J Alloys Compd 207–208:120–127
211. Krupa JC, Queffelec M (1997) UV and VUV optical excitations in wide band gap materials doped with rare earth ions: 4f-5d transitions. J Alloys Compd 250:287–292
212. Mcclure DS, Pedrini C (1985) Excitons trapped at impurity centers in highly ionic crystals. Phys Rev B 32:8465–8468
213. Belsky AN, Krupa JC (1999) Luminescence excitation mechanisms of rare earth doped phosphors in the VUV range. Display 19:185–196

214. Matuszewski J, Kropiwnicka J, Znamierowska T (1988) The crystal structure of lanthanum metaphosphate LaP_3O_9. J Solid State Chem 74:39–46
215. Rao RP, Devine DJ (2000) RE-activated lanthanide phosphate phosphor for PDP applications. J Lumin 87–89:1260–1263
216. Zhang ZY, Zhang YH, Li XL, Xu JH, Huang Y (2008) VUV-UV luminescence of magnetoplumbite: $(Sr_{0.96-x}Ba_{0.04})Al_{12-y}Mg_yO_{19}:Tb_x$. J Lumin 128:476–480

MIX
Papier aus verantwortungsvollen Quellen
Paper from responsible sources
FSC® C105338

If you have any concerns about our products,
you can contact us on
ProductSafety@springernature.com

In case Publisher is established outside the EU,
the EU authorized representative is:
Springer Nature Customer Service Center GmbH
Europaplatz 3, 69115 Heidelberg, Germany

Printed by Libri Plureos GmbH
in Hamburg, Germany